# Pedestrian and Evacuation Dynamics

Richard D. Peacock · Erica D. Kuligowski ·
Jason D. Averill
Editors

# Pedestrian and Evacuation Dynamics

Springer

*Editors*
Richard D. Peacock
National Institute of Standards and Technology
Fire Research Division
Gaithersburg, MD, USA
richard.peacock@nist.gov

Erica D. Kuligowski
National Institute of Standards and Technology
Fire Research Division
Gaithersburg, MD, USA
erica.kuligowski@nist.gov

Jason D. Averill
National Institute of Standards and Technology
Fire Research Division
Gaithersburg, MD, USA
jason.averill@nist.gov

ISBN 978-1-4419-9724-1 e-ISBN 978-1-4419-9725-8
DOI 10.1007/978-1-4419-9725-8
Springer New York Dordrecht Heidelberg London

Library of Congress Control Number: 2011931166



Printed on acid-free paper

Springer is part of Springer Science+Business Media (www.springer.com)

## Preface

In many countries, an aging population, increasing obesity and more people with mobility impairments are bringing new challenges to the management of routine and emergency people movement. These population challenges, coupled with the innovative designs being suggested for both the built environment and other commonly used structures (e.g., transportation systems) and the increasingly complex incident scenarios of fire, terrorism, and large-scale community disasters provide even greater challenges to population management and safety. Key to effective management procedures is a better understanding of human performance in a variety of incident scenarios, tools that assess human performance in these scenarios, and the proper use of such tools. The Fifth International Conference on Pedestrian and Evacuation Dynamics (PED 2010), held at the National Institute of Standards and Technology in Gaithersburg Maryland On March 8-10 2010, addressed both pedestrian and evacuation dynamics and associated human behavior to provide answers for policy makers, designers, and emergency management to solve real world problems in this rapidly developing field.

PED 2010 was quite an amazing three days where a phenomenal amount of information was shared by participants from more than 20 countries around the world. On the first day, we had 20 presentations on data and data collection methods and noted the lifelong contributions of two giants in the field, Professor John Bryan and Dr. John Fruin. On the second day, we reflected on the recent loss of a preeminent researcher Dr. Guylène Proulx and saw nearly 30 papers on modeling, including some new models. On the last day, we looked at emergency events and the technology and human factors of elevators.

As we reflect on the conference, a few observations were apparent. We are deeply impressed with the passion and commitment of the work presented at the conference. Everyone is clearly committed to excellence and moving our field forward. We should be proud.

At the same time, as a multi-disciplinary community of professionals, we have a long way to go to get where we want to go. Our discipline requires a stronger technical foundation of data and theory to support the application and use of models. Such data will establish the bedrock upon which a strong foundation for our discipline can be based.

Our third observation is a rhetorical challenge provoked by a comment by Professor Bryan at the conference dinner. He recalled with great pride how he challenged his undergraduate students to work on more difficult, graduate level problems. So we ask our academics if we are asking enough of our students in the projects we design for them. Are our consultants demanding from the modelers solutions which answer the questions that need to be answered rather than accepting what we are given? Are our modelers expending enough effort serving the end-users or confining their problems to fit the available resources?

As conference organizers, we find ourselves energized by the presentations; walking away with a thousand new ideas and suspect that many of the attendees felt the same way. As you consider which of these many ideas you wish to work on nest, we should also remember to work on high-impact, challenging problems that will make real contributions. We have the opportunity to truly improve the world we live in and that is what makes this discipline truly satisfying.

Gaithersburg, Maryland
September 2010

*Jason D. Averill*
*Erica D. Kuligowski*
*Richard D. Peacock*

## Dedication

This proceedings is dedicated to the memory of our colleague, Guylène Proulx, PhD, who died in December, 2009, after a brief illness.

Guylène's mark on the field of evacuation research is particularly notable in the frequency with which her work was quoted or referenced in the papers and presentations contributed to this conference. Guylène's research covered a wide range of topics related to the evacuation of people in structure fires. Since joining the National Research Council of Canada in 1992, her work included investigation into various aspects of human behavior in fire, including human response to fire alarm signals, the influence of photoluminescent materials on evacuation, human movement during building evacuation drills, the effectiveness of voice communication messages on evacuation behavior, evacuation strategies for building evacuation (including the use of elevators), and the social and environmental factors that influence delay time during a fire. She studied the reactions and behaviors of occupants in real fires, and most notably participated in post-event investigations of the 1993 World Trade Center attacks, the 2001 World Trade Center disaster, and the 2003 Cook County Administration Building fire.

Over the course of her career, Guylène gave over 300 presentations and authored more than 90 publications. Her research findings on evacuation strategies, smoke alarms, and photoluminescent markings are used to ensure safer buildings in North America and around the world.

Guylène was committed to giving back to the research community, as well. She served as adjunct professor in the Department of Civil and Environmental Engineering at Carleton University and the Department of Fire Protection Engineering at Worcester Polytechnic Institute, and was an integral member of the team that developed the SFPE Short Course on Human Behavior in Fire Emergencies that she later co-instructed. She was an active member of SFPE and IAFSS, was a techical editor for NFPA's Fire Protection Handbook, served on the editorial board of Fire Technology, was a member of ASME's A17 task group on the use of elevators for occupant egress, and was a key member of the program committees for the first four International Symposia on Human Behaviour in Fire.

Guylène's passing is a huge loss to the evacuation research community; not just her research contributions, but her gifts as a communicator, teacher, mentor and friend.

*Rita Fahy, September 2010*

## Pedestrian and Evacuation Dynamics Awards

Beginning with the Fifth International Conference on Pedestrian and Evacuation Dynamics, the conference organizers have decided to create a Pedestrian and Evacuation Dynamics Award. The award is intended to recognize outstanding individuals or organizations that have advanced the empirical data, application, or modeling of pedestrian and evacuation dynamics. This year, we are honored to have selected Dr. John L. Bryan and Dr. John J. Fruin for the awards based on their lifelong dedication and passion to the field. Both awardees have passed that passion onto numerous current and emerging leaders in the study of building egress and pedestrian movement ensuring that their efforts will continue far into the future.

### *Dr. John L. Bryan*

Dr. John L. Bryan was presented the award for his lifelong study of human behavior in fires and dedication to the discipline of fire protection engineering. Beginning in 1956, Dr. Bryan served for 37 years as professor and then the first chairman of what was eventually named the Department of Fire Protection Engineering at the University of Maryland in 1976. Dr. Bryan retired from the university in 1993 and received his Professor Emeritus appointment. While teaching at Maryland, he also graduated from American University with a Doctorate in Educational Psychology.

Dr. Bryan has had a lifelong commitment to the practice of fire fighting, research, and education. Also, he is known as a founding father in the field of understanding the human component in fires. One of the initial projects involving an attempt to study the behavior of the occupants during a fire incident was performed by Dr. Bryan where he analyzed interview statements of the survivors of the Arundel Park Hall fire in MD published in 1957. Dr. Bryan has over 100 technical publications and has studied numerous fires in homes, health care facilities, apartments, and hotels, namely the MGM Grand fire and the Westchase Hilton Hotel fire. As part of Project People 1 and 2 from 1974 to 1980, Dr. Bryan analyzed data from 335 fire incidents in residential occupancies involving 584 participants (as part of Project People 1) and 59 health care occupancy fire incidents and 6 non-health care fire incidents (as part of Project People 2). In these research reports, Bryan identified the actions and behaviors in which his subjects engaged...with a focus on how individual, social, and environmental factors influenced these actions. Many of the concepts that are a part of the current fire safety narrative were introduced by Dr. Bryan's research, including convergence clusters and group behavior, altruistic behavior of occupants during emergencies, pre-evacuation actions (including fire fighting actions by the occupants), reentry actions, and walking distances through smoke. These concepts, his research, and other research in the field of human behavior are captured in chapters included in both the SFPE and NFPA handbooks, authored by Dr. Bryan.

*Erica Kuligowski, March 2010*

## *Dr. John J. Fruin*

Dr. John J. Fruin was selected for an award based on his lifelong accomplishments in the field of pedestrian movement. In addition to his seminal work quantifying the relationship between density and flow, his research has established much of the technical basis for building codes, computer models, and movement theory.

Dr. Fruin is a Traffic Engineer with nearly 50 years experience involving the research, analysis, design, application, operation, and safety of pedestrian facilities. He is author of a book and more than 50 publications on these subjects. The book, "Pedestrian Planning and Design", in its second edition, is a key resource in the field of pedestrian movement. He is the developer of the level of service and time-space methods of pedestrian traffic analysis and evaluation.

He has worked on pedestrian traffic and circulation analysis for a wide variety of buildings including air, bus, rail, and ferry terminals, high-rise buildings, museums, shopping centers, sports facilities, casinos, hotels, and an aquarium. Notable projects included pedestrian traffic analysis for the Louvre Museum entrance designed by architect I.M Pei, and many aspects of pedestrian movement in the New York World Trade Center. He was a consultant on crowd dynamics for the City of Cincinnati during the "Who" Concert crowd disaster investigation, and an expert witness on the San Juan Dupont Plaza Hotel fire and other litigations.

While at the Port Authority he was Project Director of the USDOT Accelerating Walkway Demonstration Project involving studies of the public acceptance and safety of this experimental technology, Project Manager and Principal Researcher of a USDOT study of falling accidents in transit terminals, and Project Manager of the USDOT study of the New York Staten Island Ferry System. He is a Fellow of the American Society of Civil Engineers, a Fellow of the Institute of Transportation Engineers, and a former member of the Human Factors and Ergonomics Society. In 1983 he was the recipient of the American Society of Civil Engineers annual Transportation Engineers Award based on his pedestrian traffic research.

*Jake Pauls, March 2010*

# Contents

# Five Grand Challenges in Pedestrian and Evacuation Dynamics

**J. D. Averill**

National Institute of Standards and Technology
Gaithersburg, MD 20899-8664 USA
Corresponding author: averill@nist.gov

**Abstract** This paper identifies five grand challenges in the multidisciplinary field of pedestrian and evacuation dynamics (PED). In order to maximize the effectiveness of limited resources, the PED community would benefit greatly from a prioritized, consensus-based research agenda. The five proposed research initiatives include (1) a general human behavior model with a theoretical foundation and numerical validity, (2) a database archiving actual building emergency evacuations, (3) methods to embrace the stochastic nature of inputs and outcomes in building evacuation, (4) a validated method to integrate distributions of egress calculations with fire hazard calculations, and (5) adoption of technology for people movement, data collection, and within modeling constructs. The list proffered in this paper may reflect the building evacuation perspective of the author and is intended merely as a spark for discussion amongst the greater PED community; a true consensus research agenda requires deliberation by leaders in the community.

## Problem

The emerging field of pedestrian and evacuation dynamics (PED) has the potential for tremendous impact on the built environment by achieving design elements which are safe, intuitive to use, and support the intended purpose of the structure or space. However, resources historically committed to PED research have limited progress in the science and simulation of PED. These resource limitations manifest in two primary ways: the geographic and disciplinary distribution and the magnitude of research infrastructure.

First, the research and implementation efforts are widely distributed. Note that these phenomena would be expected in an emerging discipline. With a few notable exceptions, progress is led by individuals at geographically diverse institutions and originates from diverse disciplinary backgrounds, both of which serve as obstacles to collaborative progress. Regular symposia such as the PED meetings, along with digital networking and productivity tools, minimize the geographic distribution problem. However, disciplinary diversity, while bringing novel approaches and perspectives to PED, can also lead to confused terminology, anchor-

R.D. Peacock et al. (eds.), *Pedestrian and Evacuation Dynamics*,
DOI 10.1007/978-1-4419-9725-8_1, 

ing around pre-existing physical or theoretical analogies, and competing goal structures.

Second, the PED community has limited resources relative to the magnitude of the problem. Careless or unknowing building and space design can create confusion, inefficiency, and safety concerns for users. On the other hand, the potential for positive impact is substantial. Well-designed buildings and spaces ensure safety during emergency situations, enable efficient and intuitive circulation, and promote the functions of the space (mercantile spaces promote commerce, museums promote art viewing, and mass transit stations enable a high throughput of objective-based movement, for example). Satisfying these disparate objectives, however, requires a fundamental understanding of human behavior, cultural and social norms, crowd dynamics, interaction of people with the built environment, human factors of technology interfaces, and human physiology. Prediction and analysis of people movement requires mathematical, statistical, and graphical sophistication. The current state of PED practice generally lacks a comprehensive grounded theory, employs a limited number of data sets, and often fails to capture the stochastic nature of both the inputs and outputs of models. Despite the numerous research opportunities in this field, there are perhaps a few hundred researchers and practitioners actively engaged in advancing the science and theory of PED. Only a handful of academic institutions support curricula oriented towards PED science and simulation. As a result of these resource limitations, progress is slower than would be expected given the potential for positive societal impact.

In the context of these structural barriers to progress, the PED community would benefit greatly from a prioritized, consensus-based research agenda. By marshalling limited resources towards collectively or systematically addressing significant issues, the field can mature more rapidly and maximize the impact of future efforts. A consensus research agenda approach has been successful in other disciplines at guiding both researchers during the proposal development stage, as well as agencies or organizations which fund research. If a research proposal has the magnitude of the problem validated by an objective, traceable publication linked to the consensus of disciplinary experts, confidence in successful outcomes is increased in both funding and receiving parties. One example of a successful consensus research agenda is the Firefighter Life Safety Initiatives [1]. A representative cross-section of fire service leaders gathered and achieved consensus on sixteen priority research needs. The document subsequently guided grant applications and awards from agencies of the U.S. Federal Government. A second example of a research agenda includes the six research priorities identified in "Grand Challenges for Disaster Reduction [2]," a document developed by the U.S. National Science and Technology Council's Subcommittee on Disaster Research.

The purpose of this paper is to suggest five grand challenges which could unify the PED field as a prompt for further discussion within our community. The scope we have used in this effort is that each of the challenges described herein must be addressable within the next ten years, given appropriate resources and, once solved, the solution must enable significant progress in the field. While this list is not comprehensive or complete, the purpose is to engender spirited discussion on necessary research priorities, along with what steps, metrics, and resources will be required for making progress. This paper may reflect the building evacuation perspective of the author; indeed, other significant challenges in the PED community warrant research and a true consensus research agenda requires deliberation by leaders in the community.

## Grand Challenge #1: Develop and validate a comprehensive theory which predicts human behavior during pedestrian or evacuation movement

While there are dozens of predictive computer models for people movement and building evacuation [3], physics-based and other movement models are inadequate to predict the full range of possibilities for pedestrian and evacuation scenarios. The pre-evacuation actions of occupants can account for a significant portion of the overall building evacuation time and evacuation decisions can impact the duration or success of the evacuation. Specific actions can be a function of many factors, including situation awareness, cue interpretation, prior experience, risk perception, and social context. There are three steps necessary to achieve a full accounting of human behavior in computational models: development of grounded theory, development of methods to collect behavioral data, and development on computational methods to implement the results of the first two steps.

The first step will be to develop and validate theoretical behavioral models, several variants of which already exist [4]. Kuligowski [5] reviewed the state-of-the-art in modeling of human behavior during building evacuation and proposed a general process model for occupant response to physical and social cues during emergency evacuation. Furthermore, the lack of a general theory of human behavior grounded in the science of psychology and sociology, results in two difficulties. First, experimental designs for data collection (observational or interrogatory) may lack the structure of being guided by theory. As a result, the design may lack completeness (specification error). Second, data analysis may be reduced to haphazard association or correlation of factors without understanding of causality or incorporation of statistical controls for other important factors.

Once a general theory is developed (or proposed), methods (beyond observational) should be developed which can validate the whole of or components of the theoretical behavioral models. However, in order to collect the data, research should extend the current methods to account for the myriad of complex factors which determine behaviors. Examples include expressed versus revealed preference, identification of unconscious motivations, and the variance in individual situation awareness, all of which affects the selection of data collection methods and validity.

The final step is to integrate the theoretical models into the pedestrian and evacuation computational models. Behavior, like individual walking speeds, is unique to an individual and must be accounted for with stochastic approaches. The subsequent grand challenge ("#3: Embrace variance") contains a more complete discussion of the difficulties of incorporating these phenomena into computational models.

## Grand Challenge #2: Create a comprehensive database of actual emergency data

The field of pedestrian and evacuation dynamics has developed largely on the foundation of a small number of data sets (primarily Predtechenskii and Milinskii [6], Fruin [7], Jin [8], and Pauls [9]). The foundational data are generally over 30 years old and are routinely extrapolated beyond the scope of the collection. A primary example of extrapolation is the application of pedestrian movement data or non-emergency evacuation data to emergency scenarios. Such extrapolation is necessary in the absence of a comprehensive set of emergency evacuation data. Several differences exist between non-emergency and emergency evacuations, possibly including:

- perception of imminent danger to self or others,
- visibility of evacuation path due to smoke, loss of lighting, or other event,
- blockage of escape routes due to heat, debris, or other event,
- full participation by all occupants, and
- knowledge that the building may not be reoccupied may encourage additional gathering of items (keys, purses, computers, phones, jackets, etc).

These differences may have the effect of lessening the building evacuation time or lengthening the building evacuation time, dependent upon the specific scenario. Currently, however, scant information exists to understand any differences between fire drill or experimental data and real emergency scenarios. A comprehensive database which catalogues the progress and outcomes for real emergency incidents (the crucible in which theory and drills are tested) is necessary condition for acceptance and validation of all knowledge in the field.

Presently, buildings with emergencies are black boxes, wrapped in a myriad of technical and bureaucratic obstacles. First, building evacuations involving injuries or deaths are often complicated by civil and criminal proceedings which may inhibit access to critical data. Data records may include security video, emergency responder radio communications, interviews with building staff and evacuees, or emergency planning records. Second, even if the researchers knew when and where an event would occur, the infrastructure to collect, analyze, and archive the data consistently across incidents has not yet been developed. Finally, reconstructions are resource intensive, requiring substantial labor and funding resources to properly document large events.

Gwynne [10], [11] has documented many of the challenges which impede progress in the collection of egress data. Gwynne has proposed a new framework which begins to address many of these challenges, including (1) standardized vocabulary within the egress community, (2) identification of the context and purpose of the evacuation, (3) characterization of the building and occupants prior to the building evacuation, (4) archival of raw data (across multiple formats such as spreadsheet and video), and (5) linkage of data to analyses or conclusions produced in published works. Galea has also produced databases for archiving aircraft evacuation data [12] and high-rise building evacuation data (starting with interviews of World Trade Center occupants) [13].

Finally, there are three potential opportunities for collecting emergency evacuation data: before, during, and after an incident. Before an incident, a research team could pre-install equipment to capture building evacuation. Motion sensors, connection to the fire alarm system, or remote manual activation could potentially initiate data collection. During an incident, the one agency certain to respond is the fire service. Therefore, deployment of data instrumentation on the fire apparatus or as part of the firefighter turnout gear would ensure deployment to the scene, though the incipient stages (if not the entire duration) could be completed prior to the arrival of the fire service and large segments of the evacuation process may be missed. Finally, the most common opportunity to collect emergency evacuation data is after the incident is complete. Among other problems, documenting an event afterwards suffers from well-documented difficulties of human memory recall, loss of building components during the incident, lack of situation awareness by key personnel, and incomplete cooperation. While the challenges are significant, it is critical to develop infrastructure to support capture of data from emergency building evacuations.

## Grand Challenge #3: Embrace variance

While the vast majority of current generation egress models are deterministic, pedestrian movement and building evacuations are highly stochastic processes. Evacuate the same building with the same people starting in the same places on consecutive days and the answers could vary significantly.

The PED community must move away from terms such as "average" for modeling inputs and "evacuation time" for modeling output. Tools and techniques which manage distributions of inputs and outputs should be developed and broadly implemented. Figure 1 shows the substantial variance inherent in stairwell evacuation speeds in four high-rise buildings [14]. The importance of variance is reflected in the dependence of one occupant's evacuation speed on the surrounding occupants. A slow-moving occupant will likely reduce the movement speed of surrounding occupants. Many simulation models currently employ singular values for movement speed (horizontal, descending, etc…). Model inputs which should be accounted for using distributions include number, location, and capabilities of occupants.

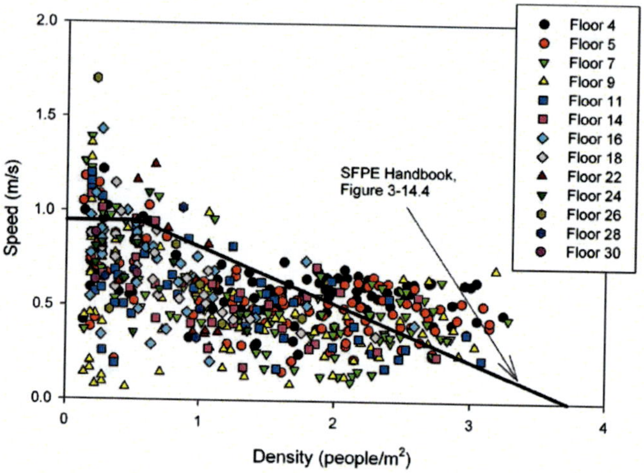


**Fig. 1. Relationship between speed and density in stairwells (from Peacock, et al).**

Lord, et al. discuss the role of distributions in modeling inputs and model results [15]. As shown in Figure 2, probabilities should be attached to the building evacuation time distributions and a discussion of acceptable risk should take place in every nation and community. For example, using the results of Figure 2, a building authority could specify a confidence level (99 percent) that a building would be evacuated by a certain time (in this case, roughly 400 seconds). Presumably, the evacuation time requirement is related to the onset of untenable conditions (which would, itself, have a distribution and interact with the evacuation terms), including a safety factor.

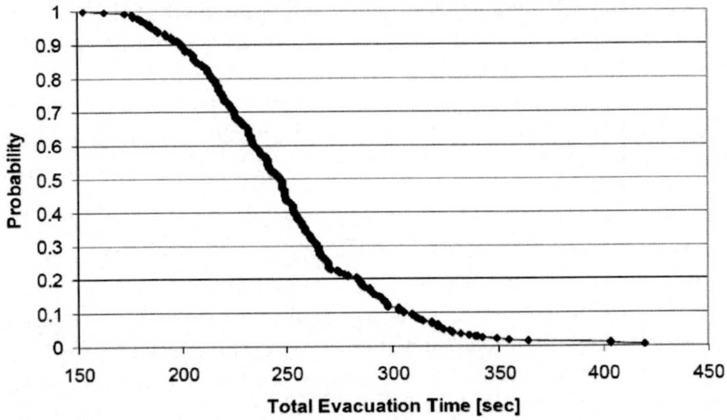


**Fig. 2. Probability that an evacuation of the building will be complete (from Lord, et al).**

Explicitly acknowledging that the design includes a finite probability that occupants will not successfully evacuate the building will encourage a more informed discussion about local level of acceptable risk.

In order for the PED community to fully embrace variance, there are several key steps: education, tools, and communication. First, the mathematics and logic of distributions must be taught to the stakeholders, including students, engineers, architects, regulators, and building owners and managers. A fluency in distributions and acceptable risk amongst all parties will inform decision-making in a more powerful and consistent manner. Second, tools for capturing, manipulating, and displaying distributions must be incorporated into simulation tools. When capturing and presenting building evacuation data, distributions (rather than means) should be utilized, when possible. Rather than singular movement values and a singular resulting building evacuation time, simulation models should require distribution parameters or links to public datasets (utilizing distributions). Monte Carlo techniques, fractional factorial, or other calculation methods could be used to produce probabilities of outcomes.

## Grand Challenge #4: Integrate results of evacuation models with fire models to enable accurate and reliable performance-based design

For the fire safety community (rather than the greater PED community), integration of egress modeling calculation with fire modeling calculations resides at the heart of performance based design. However, success requires several advances from the current state-of-the-art, including quantification of the ability of both the egress models and the modelers to produce reliable and repeatable results, development of standard egress modeling scenarios, and accounting for the interactions between the fire and the occupants.

The calculation of Available Safe Egress Time (ASET) through computational fire modeling is substantially more advanced than current-generation predictions of Required Safe Egress Time (RSET) using occupant egress models. While the increased investment in the scientific principles of fire science have exceeded the investment in building evacuation science, the fire modeling community has also engaged in activities which could also benefit the egress modeling community. The fire science community has initiated the first efforts to conduct round robin assessments of fire model performance to establish the strengths, weaknesses, and opportunities for both fire modeling methods and user capabilities [16]. Additionally, the fire science community has established a standard guide for evaluation of fire model predictive capability [17]. A round robin exercise and a validation guide would be informative to the egress modeling community in order to establish state-of-the-art and inform future scientific investments.

Evacuation scenarios, equivalent to the performance-based design required design fire scenarios [18], should be developed for building evacuation. Lower-probability, higher-consequence scenarios (such as a fully occupied building with 50 percent of the exits out-of-service) should be blended with higher-probability, lower-consequence scenarios ('typical' occupant load with all exits available). An example, based upon analysis of New Zealand fire incidents and casualties, was presented at the 8$^{th}$ Int. Conf. on Performance-based Codes and Fire Safety Design [19]. Methods for combining the outcome distributions from the fire scenarios and the evacuation scenarios in a meaningful way that can be understood by the design and regulatory communities for safe and cost-effective building design must be developed prior to realization of the full potential for performance based design.

Finally, a science-based bridge between the effects of the fire and the occupants must be established. The interaction of the occupants with the constraints imposed by the emergency (e.g., people evacuating through smoke) has implications for a host of disciplinary contributions, including toxicology, psychology, sociology, architecture, and engineering. Establishing the causal and distributional outcomes for interactions will enable more precise understanding of the (sub)lethal effects on individuals and the effects on overall evacuation outcomes.

## Grand Challenge #5: Embrace technology

Given the paucity of data on simple concepts (such as stairs), it is unsurprising that virtually no data exist for use of technology to improve people movement or building evacuation effectiveness. However, technology can be applied to improve movement speed (elevators, e.g.), improve situation awareness (integration of building sensor technologies, e.g.), or create new egress pathways (alternative escape devices, e.g.). The Rethinking Egress workshop [20] in 2008 identified over 400 alternatives for improving the state-of-the-art in building evacuation, often through novel application of existing technologies from other disciplines.

For the majority of these technologies, there are virtually no experimental data, incident data, theoretical models, or computational algorithms to encourage adoption of more effective strategies. A principal challenge associated with collecting data about the effectiveness of new technology, however, is a "chicken or egg" argument: technology is unlikely to be adopted until empirical evidence proves the utility and empirical evidence is difficult to gather until the technology is adopted. The PED community must lead the way in enabling the enhancements by proactively seeking and developing technologies through data and models.

A novel model for the widespread adoption of new technology was demonstrated by the American Society of Mechanical Engineers (ASME) A17.1 Task Groups on Use of Elevators by Occupants (and Firefighters) for Evacuation (or Access) During Fire Emergency. As a result of having little or no research data to base recommendations for changes to codes and standards, the open-participation task group used an ISO standard hazard analysis methodology [21] over a five year period to systematically address design and implementation hazards for use of elevators by occupants and firefighters during a fire in a high-rise office building. The rigorous process has involved diverse stakeholder participation, thousands of hours of analysis and discussion, and a hazard analysis stretching over 300 pages. The rigorous hazard analysis method was partially responsible for the ready adoption of this new technology in U.S. model codes (International Building Code [22] and Life Safety Code [23]). Individuals with expertise in PED –related disciplines should be significant leaders in efforts such as the ASME A17.1 task groups and future efforts to enable the adoption of PED technologies.

## Conclusions

The emerging discipline of PED has many challenges and limited resources. For the field to mature, the community must identify and adopt theoretical and computational approaches which have sound basis in experimental and emergency incident science. Designers and regulators must have standards of practice and computational tools which embrace the full distribution of scenarios and solutions. The preceding grand challenges are merely a starting point for a greater community-inclusive discussion. An interim goal is to develop a focused, consensus agenda on research priorities which will maximize the impact of limited research investments. Ultimately, when the outcomes of these identified goals are widely available and understood, occupants will be able to use and evacuate structures more reliably, efficiently, and with less total cost to society.

## References

1. Report of the National Fire Service Research Agenda Symposium, National Fire Service Research Agenda Symposium. Emmitsburg, MD (2005) (http://www.everyonegoeshome.com/initiatives.html)
2. Grand Challenges for Disaster Reduction. A Report of the Subcommittee on Disaster Reduction. National Science and Technology Council. Washington, D.C. (2005)
3. Kuligowski, E.; Peacock, R. "Review of Building Evacuation Models." NIST Technical Note 1471; Natl. Inst. Stand. Technol., Gaithersburg, MD (2005)
4. Proulx, G. "Occupant Behavior and Evacuation." NRCC-44983, Nat. Res. Council of Canada, Ottawa, Ontario (2001)
5. Kuligowski, E. "Modeling Human Behavior During Building Fires." NIST Technical Note 1619, Natl. Inst. Stand. Technol., Gaithersburg, MD (2008)
6. Predtechenskii, V.M. and Milinskii, A.I.: Planning for Foot Traffic in Buildings, Amerind Publishing Co. Pvt. Ltd., New Delhi (1978)
7. Fruin, J. J.: Pedestrain Planning and Design. (Revised Edition), Elevator World, Inc., Mobile, AL (1987)
8. Jin, T: Visibility Through Fire Smoke, Journal of Fire and Flammability, **9**, 135-155 (1978)
9. Pauls, J.: Movement of People. In: P.J.DiNenno, C. L. Beyler, R. L. P. Custer, W. D. Walton, J. M. Watts, D. Drysdale, and J. R. Hall (Eds.), The SFPE Handbook of Fire Protection Engineering (Second ed., pp. 3-263-3-285). Society of Fire Protection Engineers, Bethesda, MD (1995)
10. Gwynne, S.M.V. Conventions in the Collection and Use of Human Performance Data, NIST GCR 10-928, Natl. Inst. Stand. Technol., Gaithersburg, MD (2010)
11. Gwynne, S.M.V. The Standardization of Human Egress Data. In: 4th International Symposium on Human Behaviour in Fire 2009, Interscience Communications, Ltd, London (2009)
12. Galea E.R., Finney, K.M., Dixon, A.J.P., Siddiqui A., and Cooney D.P.: An analysis of human behaviour during aircraft evacuation situations using the AASK V3.0 database, The Aeronautical Journal, **107**, 219-231 (2003)
13. Galea, E. R., Hulse, L., Day, R., Siddiqui, A., and Sharp, G.: The UK WTC 9/11 Evacuation Study: An Overview of the Methodologies Employed and Some Analysis Relating to Fatigue, Stair Travel Speeds, and Occupant Response Times. In: Proceedings of the 4th International Symposium on Human Behaviour in Fire, Interscience Communications, Ltd, London (2009)
14. Peacock, R., Hoskins, B. and Kuligowski, E. Overall and Local Movement Speeds During Fire Drill Evacuations in Buildings Up to 31 Stories. In: Proceedings of the 5th Int. Conf. Pedestrian and Evac. Dynamics. Springer Publishing, Boston (2010)

15.Lord, J., Meacham, B., Moore., Fahy, R., and Proulx, G.: Guide for Evaluating the Predictive Capabilities of Computer Egress Models, NIST GCR 06-886, Natl. Inst. Stand. Technol., Gaithersburg, MD (2006)

16.Reina, G., Torero, J., Jahn, W., Stern-Gottfried, J., Ryder, N., Desanghered, S., Lázaro, M., Mowrer, F., Coles, A., Joyeux, D., Alveare, Capotee, J., Jowseya, A., Abecassis-Empisa, C., and Reszkaa, P.: Fire Safety Journal, **44**, Issue 4, 590-602 (2009)

17.ASTM E1355, Standard Guide for Evaluating Predictive Capability of Deterministic Fire Models, American Society for Testing and Materials, West Conshohoken, PA (2005).

18.NFPA 101 Life Safety Code, Ch. 5.5 Design Fire Scenarios, National Fire Protection Association, Quincy, MA (2009)

19.Robbins, A.P. and Wade, C. Characterizing Fire Scenarios Based on New Zealand Fire Incident Data. In: Proceedings of the 8th International Conference on Performance-Based Codes and Fire Safety Design Methods, 16-18 June 2010, Lund University, Sweden (2010)

20.Averill, J., Peacock, R., and Keeney, R.: Rethinking Egress: A Vision for the Future, NIST Technical Note 1647, Natl. Inst. Stand. Technol., Gaithersburg, MD (2009)

21.ISO 14798:2009: Lifts (elevators), escalators, and moving walks – Risk assessment and reduction methodology. International Organization for Standardization, Geneva (2009)

22.International Building Code. International Code Council, Country Club Hills, IL (2009)

23.NFPA 101: Life Safety Code. Natl. Fire Prot. Assoc., Quincy, MA (2009)

# Data Collection (Evacuation)

# Emergency Door Capacity: Influence of Population Composition and Stress Level

**W. Daamen and S.P. Hoogendoorn**

Delft University of Technology
Faculty of Civil Engineering and Geosciences
Department of Transport & Planning
PO Box 5048, 2600 GA Delft, the Netherlands
Corresponding author: w.daamen@tudelft.nl

**Abstract** For the new version of the Dutch design guidelines for buildings, a threshold value for the capacity of emergency doors needs to be set. Innovative large-scale laboratory experiments have been performed to investigate the capacity of emergency doors during evacuation conditions. This paper focuses in particular on the relation between capacity and the independent variables population composition and stress level.

It turns out that the average observed capacities are for all widths lowest for the lowest stress level and highest for the highest stress level. The population with a greater part of children has the highest capacity (on average 3.31 P/m/s). This is mainly due to the smaller physical size of children compared to adults and elderly, which makes it possible that more children can pass a door at the same time than adults. The lowest capacity (on average 2.02 P/m/s) is found for the experiment with 5% disabled participants.

## Introduction

Since 1992, the Dutch national building design guidelines ("Building decree") sets requirements to the width of emergency doors. Since 2003, these requirements depend on the number of persons that rely on an emergency door. According to the Building decree a door width of one meter is sufficient to let 135 persons pass during the period available for safe escape (1 minute). This value corresponds to research of Peschl [1], being similar to 2.25 P/m/s. The threshold of 135 persons per meter width during a safe escape time of one minute has been discussed for years between the Ministry for Housing, Regional Development and the Environment and the fire brigades that are used to allow a maximum of 90 persons per meter width during a safe escape time of one minute (1.5 P/m/s).

A literature research has been performed to find other research related to similar bottlenecks. This overview showed a wide variety of capacities ranging between 1.03 P/m/s and 3.23 P/m/s [2-8]. Most of the capacities found are lower than the capacities in the design guidelines (2.25 P/m/s), but the conditions differ

R.D. Peacock et al. (eds.), *Pedestrian and Evacuation Dynamics*,
DOI 10.1007/978-1-4419-9725-8_2, © Springer Science+Business Media, LLC 2011

widely. To collect further information on the capacity of doors during emergency conditions new, dedicated, experimental research has been performed.
In previous research, we developed a conceptual model of pedestrian behavior in relation to capacities of emergency doors. This model shows that doorway width, population composition, light intensity, the presence of an open door and stress level affect the capacity of emergency doors. This paper focuses on the relations between population composition and stress level and capacity.

The next section describes the set up of the experiments in more detail. In the third section the methodology is described to calculate the capacities, the results of which are shown in the subsequent section. We end with conclusions and recommendations for future research.

## Experimental Set Up

The capacity of an emergency door depends on several aspects, among which the composition of the population using the door, the conditions under which the door is used and the door width. Before describing these experimental variables in more detail, some boundary conditions are set.

In the experiments, an opening represents the emergency door: subjects pass a free passage of a certain width. In this opening, no doorstep is present, to reduce hindrance and prevent possibly dangerous situations for participants. In addition, the pedestrian flow is one-directional, implying that no counter flows are present caused by fire fighters and people from emergency services. In reality, these people will rarely enter a building when the evacuation process is still going on.

The experiments performed by Peschl [1] have been based on a student population. However, in practice, the population will not consist of persons being in good shape, but the persons will have different physical conditions. In this research, we will use age as an indication for a person's physical condition. Here, we distinguish four categories: children (under 18 years of age), adults (between 18 and 65 years of age), elderly (over 65 years of age) and disabled persons. With these categories, we are able to compose populations corresponding to a variety of situations, see Table 1.

The disabled people are represented by three persons in wheelchair and three blindfolded persons.

**Table 1. Overview of different populations in the experiments**

| | Population | Children | Adults | Elderly | Disabled |
|---|---|---|---|---|---|
| 1 | School | 90% | 10% | 0% | 0% |
| 2 | Station during peak hours | 0% | 100% | 0% | 0% |
| 3 | Home for the elderly | 5% | 20% | 75% | 0% |
| 4 | Work meeting | 5% | 90% | 5% | 0% |
| 5 | Shopping centre | 30% | 60% | 10% | 0% |
| 6 | Average | 25% | 55% | 20% | 0% |
| 7 | Disabled | 23% | 54% | 18% | 5% |

The conditions under which an emergency door is used may vary considerably. In the experiments, both the stress level of the participants and ambient conditions are varied. Not much is known on how to introduce stress in an experiment. In the past two methods have been considered favorable: enforcing participants to hurry e.g. by rewarding participants according to their performance and exposing participants to noise. Here, we have chosen to use for the latter option by sounding the slow-whoop signal. In addition, the stress level of the participants is raised by a combination of the slow-whoop signal and stroboscope light. In total, participants have been exposed to three stress levels: none, a slow-whoop signal and a combination of a slow-whoop signal and stroboscope light.

The sight is reduced by reducing illumination to a low level. Two alternative light situations are considered: full lighting (200 lux) and dimmed lighting (1 lux, corresponding to emergency lighting).

In the experiments, the opening width is varied between 50 cm (the minimal free passageway of an escape route in the Building decree for existing buildings) and 275 cm. In addition to an opening of 85 cm wide (minimal free passageway of an escape route in the Building decree for new estates) openings are a multiple of 55 cm. Furthermore, an opening of 100 cm is tested to see the correspondence with the normative capacity expressed as the number of persons passing an opening of one meter wide in one minute.

Ideally, all combinations of experimental variables should be investigated. Since this is not feasible due to time restrictions (the experiments should not last longer than a single day), for each experiment one variable is changed, while for the other variables the default value is maintained. By interpolation of the results of the various experiments, pronouncements can be made on the not performed experiments. The stress levels are varied for all experiments.

Each experiment will be performed multiple times to guarantee the reliability of the observations. To determine the number of repetitions, a total time of congestion of three minutes should be achieved. Since the time of congestion for wide doors is shorter than for narrow doors, more repetitions are performed for the wide doors. An overview of the experiments is shown in Table 2.

**Table 2. Overview of the performed experiments**

| Experiment | Opening width [cm] | Population | Light [lux] | Open door | Start time |
|---|---|---|---|---|---|
| 1 | 100 | Average | 200 | No | 9:58 |
| 2 | 220 | Average | 200 | No | 10:17 |
| 3 | 85 | Elderly home | 200 | No | 10:43 |
| 4 | 85 | Average | 200 | No | 10:58 |
| 5 | 165 | Average | 1 | No | 11:25 |
| 6 | 275 | Average | 200 | No | 11:52 |
| 7 | 85 | Work meeting | 200 | No | 12:49 |
| 8 | 85 | Disabled | 200 | No | 12:23 |
| 9 | 85 | School | 200 | No | 13:48 |
| 10 | 85 | Average | 1 | No | 14:08 |
| 11 | 50 | Average | 200 | No | 14:24 |
| 12 | 110 | Average | 200 | No | 14:39 |
| 13 | 85 | Shopping centre | 200 | No | 15:19 |
| 14 | 85 | Average | 200 | Yes | 15:40 |
| 15 | 165 | Average | 200 | No | 16:03 |
| 16 | 85 | Station | 200 | No | 16:24 |

A digital video camera and an infrared camera are used to observe the experiments. The infrared camera observes LED's, attached on top of the caps of the participants. This technique guarantees good observations for the dimmed conditions. For the other experiments a digital camera is used, which is attached to the ceiling next to the infrared camera.

In total 75 children of 11 years old (blue caps), 90 adults (red caps) and 50 elderly persons (yellow caps) have participated in the experiments. This leads to populations of between 90 and 150 persons, which are large enough to cause congestion upstream of the door to observe capacities.

To represent an emergency door, a wall has been built in the middle of a large hallway, perpendicular to the sidewall. In this wall, an opening is made, whose width is easy to vary. At the side of the wall, some space is left to walk from one side of the wall to the other without using the opening. Above the centre of the opening an emergency exit sign has been hung up. An overview of the experimental site is shown in

Fig. 1. To use the doorway more efficiently the participants use it in two directions: in the first experiment, they walk from one side of the wall to the other and in the next experiment they walk back again.

**Fig. 1. Overview of the experimental site**

## Methodology to Calculate Emergency Door Capacity

The images from the digital video camera form the basis to calculate the capacity of an emergency door. The movie of each repetition of an experiment is split into separate images with a frequency of 25 images per minute.

The flow through the door can be calculated by counting the number of pedestrians passing the door during a specific time period. When during this time period congestion occurs, the observed flow is equal to capacity. In our case, in all repetitions of the experiment this congestion occurs, so we will not mention this aspect further. To determine the capacity, we identify the passing moments of pedestrians at a cross-section directly downstream of the door using the similar idea behind a finish photo. From all images of a repetition, we take the part that corresponds to the indicated cross-section. By placing all these parts next to each other for increasing time moments, we see what happens at the cross-section over time (horizontal axis). On this new image, we then identify the persons (caps). The $x$-pixel of this cap indicates the time moment when the participant passes the cross-section, while the $y$-pixel indicates his lateral position.

Assuming that the capacity of the door does not change during a repetition of the experiment, a straight line is fit through the cumulative curve. The derivative of this line corresponds to the average capacity of this door during this repetition. The average capacity of the experiment is then the average of the capacities of all repetitions. If the average capacity of each experiment is known, the relations between the capacity and the various experimental variables (door width, population, stress level, etc.) can be determined.

## Relations between Capacities and Experimental Variables

Based on the methodology described in the previous section capacities have been calculated for all repetitions of all experiments. In this section, the influence of the experimental variables population composition and stress level are discussed.

### *Stress Level*

Fig. 2 shows that the average observed capacities over all doorway widths are lowest for the lowest stress level and highest for the experiments with slow-whoop and stroboscope considered as the highest stress level. The figure on the right shows some outliers for the experiments without stress. An explanation can be found in the time of the day this experiment has taken place (see also Fig. 4).

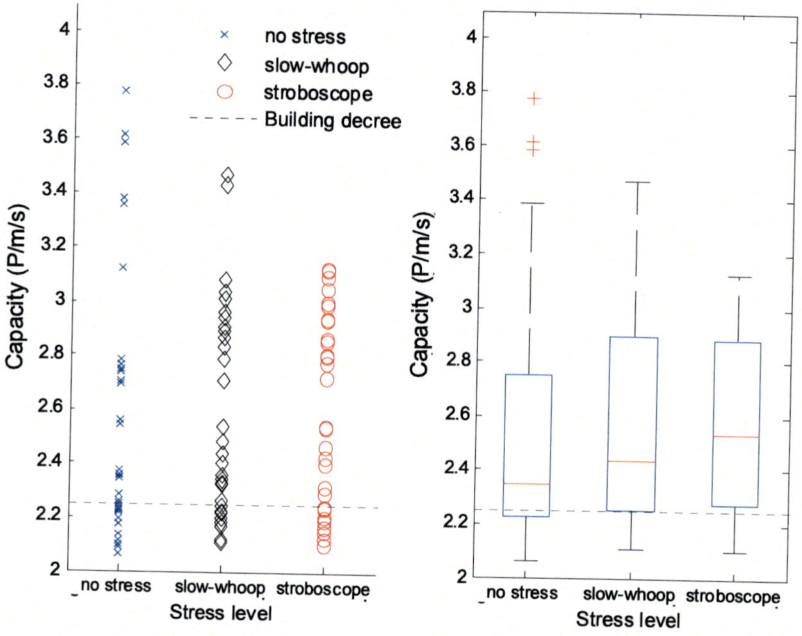


**Fig. 2. Capacity as function of stress level for various doorway widths. On the left, observations are shown per repetition, on the Right, a box plot.**

A closer look at the data shows that the experiments with the largest variance in capacity were the first two experiments performed. Moreover, the stress level appears to have an opposite effect for both experiments: for the first experiment the capacity is lowest for the lowest stress level, while in the second experiment the highest capacities occur at the lowest stress level. The other experiments do not show such a clear effect of the various stress level. This leads to the conclusion that the difference is not structural and can be attributed to the conditions (enthusiasm) during the first experiments. While the participants of the first experiment did not know what to deal with, the participants of the second experiment could wait and see what happened. Especially at the start of the experiment, these participants were very motivated and were in full focus to pass the door. In the first repetitions (without stress and with slow-whoop respectively) this lead to pushy behavior, which was clearly visible in the video images.

The capacity of most repetitions is higher than the capacity prescribed in the Building decree. Only most repetitions of the experiment with the widest opening are below the capacity threshold from the Building decree. Since the first experiments showed that the capacities appeared to be higher than the planned capacities all adults and elderly have joined the experiment. This lead to a slightly different population with more elderly participants than the average population, which has a negative effect on the capacity as will be shown in the following section.

## *Population Composition*

During the experiments also the population has been varied. These experiments have been performed with a doorway of 85 cm wide, a normal light intensity (200 lux) and without an open door. Fig. 3 shows the results of these experiments.

The figure shows that five out of six populations result in a capacity higher than the capacity threshold indicated in the Building decree. Only the population with 5% disabled persons results in a slightly lower capacity (2.0 P/m/s versus 2.25 P/m/s). The population with mainly children has the highest capacity. This is mainly caused by the physical fact that children are smaller than adults, which makes it possible for more children to pass a door opening at the same time. The populations representing a retirement home, a meeting and a shopping centre do not differ much. Conversely, the capacity of the population 'station' varies considerably from the population 'meeting'. The first population consists only of adults, while the second population consists of 90% adults, completed with 5% children and 5% elderly. However, the difference between both capacities is somewhat more than 8%. Also the population 'shopping centre' and 'average' have a substantially different capacity (15%), while the first population has only 5% more children, 5% more adults and 10% less adults. These differences might be explained by the moment of the day the experiment has been performed (see Fig. 4).

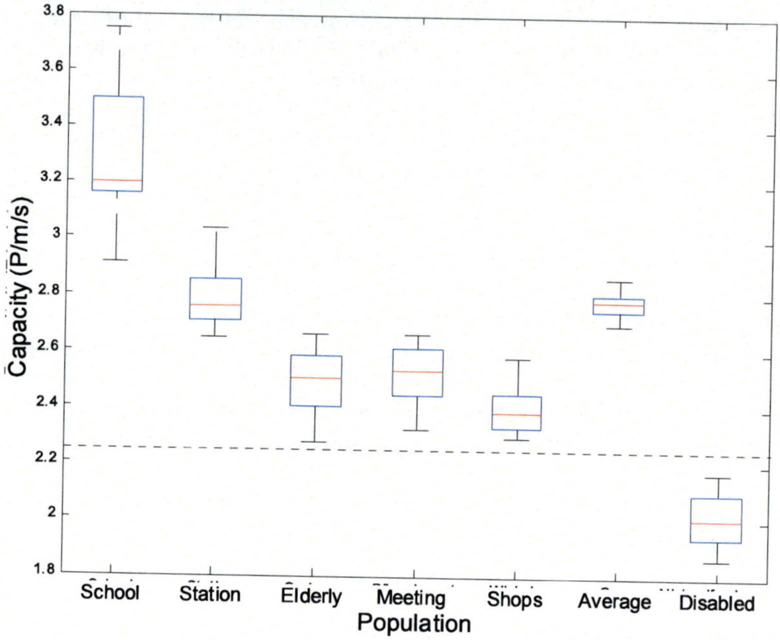


**Fig. 3. Capacity as a function of the population composition**

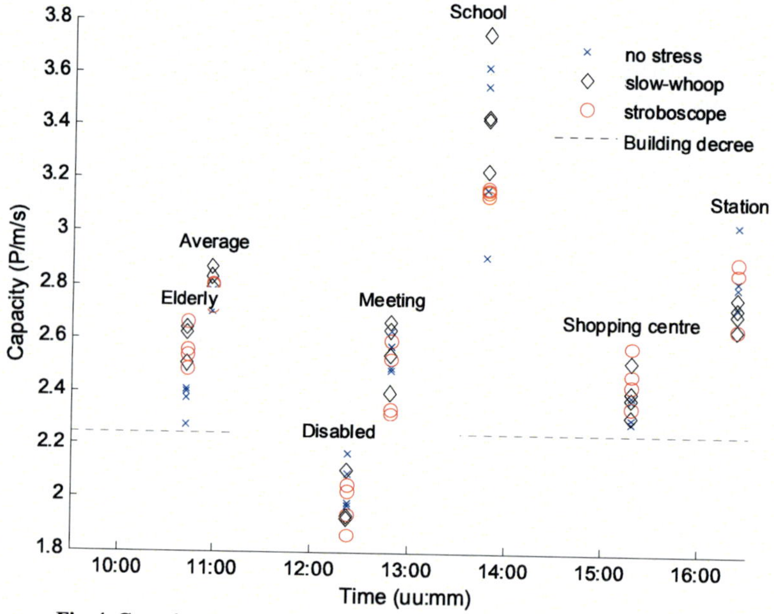


**Fig. 4. Capacity as a function of population composition and time of day**

For both situations mentioned above the performance moment of the experiment has a clear but opposite effect. The experiment with the average population was the fourth experiment of the day just before a short break, while the experiment with the shopping centre population occurred halfway the afternoon. At that moment the fatigue had increased considerably and the enthusiasm decreased, which lead to a lower capacity than the capacity of a comparable average population. Exactly the opposite causes the difference in capacity between the meeting population and the station population. The experiment with the meeting population occurred by one o'clock, when the participants were clearly in need of a lunch break, while the experiment with the station population occurred at the end of the day. To motivate the participants extra, the challenge was set to improve the highest capacity of the children. This led to a very strong motivation, resulting in a much higher capacity than the one of a similar population.

The variation in capacity is highest for the school population, which can be attributed to the fact that children strongly react to each other: if the first person passes the doorway very fast, the others will follow very fast as well, whereas if the first person passes the doorway very slow, the others will also take it easy. However, the variation between the experiments with the stroboscope was very small, probably because this unusual external condition makes the children focus more on the aim of the experiments (less distraction).

## Conclusions and Recommendations

The main conclusion to be drawn is that the capacity of most experiments is higher than 2.25 P/m/s. The experiments with the lowest capacity have a population with disabled persons and a very wide doorway (275 cm) with less children than the other experiments with different doorway widths.

Another conclusion is that in the performed experiments, more pushing does not lead to the 'faster-is-slower' effect. In the experiments a higher urgency (higher stress level) leads to higher speeds and to a higher capacity.

Many differences between the observed capacities can be explained by the different experimental variables. The images of the experiments indicate that an explanation can also be found in the individual behavior of the participants. When this microscopic behavior can be predicted, also the capacities can be predicted for a larger variety of conditions. This will be subject of future research.

The research described here has explicitly been focused on the capacity of emergency doors. This is only part of the total evacuation process. The previous process (pre-evacuation, route choice, walking towards the exit) has a direct influence on the arrival pattern of pedestrians at the emergency door, and thus whether or not capacity of the door will be reached. This is also subject of future research.

## References


1. Peschl, I.A.S.Z.: Evacuation Capacity of Door Openings in Panic Situations. Bouw. **26**, 62–67 (1971) (In Dutch)
2. Hoogendoorn, S.P., Daamen, W.: Pedestrian Behavior at Bottlenecks. Transportation Science. **39**(2), 147–159 (2005)
3. Kretz, T., Grünebohm, A., Schreckenberg, M.: Experimental Study of Pedestrian Flow through a Bottleneck. Journal of Statistical Mechanics. P10014, 1-23 (2006)
4. Müller, K.: The Design and Measurement of Routes for Evacuation of Persons from Buildings. Dissertation, Technische Hochschule Magdeburg (1981) (In German)
5. Nagai, R., Fukamachi, M., Nagatani, T.: Evacuation of crawlers and walkers from corridor through an exit. Physica A. **367**, 449–460 (2006)
6. Seyfried, A., Passon, O., Steffen, B., Boltes, M., Rupprecht, T., Klingsch, W.: New Insights into Pedestrian Flow Through Bottlenecks. Transportation Science Articles in Advance (2009). DOI: 10.1287/trsc.1090.0263
7. Daamen, W.: Modelling passenger flows at public transport facilities. PhD thesis, Trail Thesis Series, T2004/6, The Netherlands TRAIL Research School (2004)
8. Buchmüller, S., Weidmann, U.: Parameters of pedestrians, pedestrian traffic and walking facilities. IVT-Report Nr. 132, Institut for Transport Planning and Systems (IVT). Swiss Federal Institute of Technology Zurich (ETHZ) (2006)
9. Van den Berg, M.: Pedestrian Interaction Behaviour in Relation to the Capacity of Evacuation Doorways. Master Thesis, Delft University of Technology (2009)

# Overall and Local Movement Speeds During Fire Drill Evacuations in Buildings up to 31 Stories

**R. D. Peacock, B. L. Hoskins, and E. D. Kuligowski**

National Institute of Standards and Technology
Gaithersburg, MD 20899 USA
Corresponding author: richard.peacock@nist.gov

**Abstract** The time that it takes an occupant population to reach safety when descending a stairwell during building evacuations is typically described by measureable engineering variables such as stairwell geometry, speed, density, and pre-evacuation delay. In turn, engineering models of building evacuation use these variables to predict the performance of egress systems for building design, emergency planning, or event reconstruction. As part of a program to better understand occupant movement and behavior during building emergencies, the Building and Fire Research Laboratory at the National Institute of Standards and Technology (NIST) has been collecting stairwell movement data during fire drill evacuations of office buildings. These data collections are intended to provide a better understanding of this principal building egress feature and develop a technical foundation for future codes and standards requirements. To date, NIST has collected fire drill evacuation data in eight office building occupancies ranging from six to 62 stories in height that have included a range of stairwell widths and occupant densities.

While average movement speeds in the current study of 0.48 m/s ± 0.16 m/s are observed to be quite similar to the range of literature values, local movement speeds as occupants traverse down the stairwell are seen to vary widely within a given stairwell, ranging from 0.056 m/s to 1.7 m/s.

## Introduction

Timing of occupant descent down stairwells during building evacuations is typically described by measureable engineering variables such as stairwell geometry, speed, density, and pre-evacuation delay. In turn, engineering models of building evacuation use these variables to predict the performance of egress systems for building design, emergency planning, or event reconstruction. While there are dozens of models to simulate the evacuation of occupants from a given building geometry [1], there is limited contemporary data to support the model inputs or as-

R.D. Peacock et al. (eds.), *Pedestrian and Evacuation Dynamics*,
DOI 10.1007/978-1-4419-9725-8_3, 

sumptions and even less information available to validate the models for actual emergencies. While some models have had extensive validation efforts by the developers [2,3] and others have included uncertainty in the analysis for a few limited data sets [4], there is still a significant need for independent data on evacuation behavior both for further development of the models as well as independent validation efforts. Collection and analysis of basic evacuation data would also provide a basis for building code requirements, the practice of egress system design, and ensure robustness for analysis of emerging issues.

As part of a program to better understand occupant movement and behavior during building emergencies, the Building and Fire Research Laboratory at the National Institute of Standards and Technology (NIST) has been collecting stairwell movement data during fire drill evacuations of office buildings. These data collections are intended to provide a better understanding of this principal building egress feature and develop a technical foundation for codes and standards requirements. To date, NIST has collected fire drill evacuation data in eight office building occupancies ranging from six to 62 stories in height that have included a range of stairwell widths and occupant densities.

This paper builds on a paper from the previous conference [5] to examine evacuee movement in four additional buildings, local movement speeds in addition to overall movement speeds, and an initial examination of underlying factors that may influence occupant evacuation.

## Data Collection for Buildings Included in Current Study

While real emergency data is most desirable and might provide the most realistic predictor of behavior, it is not as readily available as fire drill data. For practical purposes, fire drill data is often used to represent emergency behavior. A key assumption, consistent with most of the data presented in the literature values discussed earlier, is that fire drill data can be used to approximate the response of individuals in an actual emergency [6]. This is, of course, dependent on whether the population is directly exposed to smoke and/or fire cues; meaning that fire drill data may best approximate the reaction and conditions experienced of those who are not close enough to the hazard to identify it as an emergency. In many high-rise evacuations, as is the case in this study, it is conceivable that a significant portion of the population has not been exposed to enough fire cues to be certain if it is an emergency. Information from real emergencies can inform fire drill data collections and provide a check of the validity of fire drill data.

## *Data Collection Procedures*

In this study, fire drill evacuation were collected by positioning video cameras out of the way of building occupants to record an overhead view of occupant movement in an exit stair during the evacuation. In most buildings, unless specified, the video cameras were placed on every other floor to capture a view of that floor's main landing, the door into the stair at that level, and 2-3 steps on each side of the main landing (leading to and from the main landing). This camera placement captured the times in which the occupant was seen moving past a particular floor landing as well as the time when he/she was seen moving into the stairs.

After video data was taken from each building evacuation drill, NIST transcribed specific data from the videos into a spreadsheet format for each stair monitored during the drill. For each stair recorded, data were collected 1) for each occupant evacuating in that stair and 2) for each time during the evacuation drill that the occupant was seen at a specific floor in the stair (a camera position), typically both entering and exiting the camera view. Additional information included gender, body size, location on the stairs, handrail usage, and whether anything was being carried by each occupant.

All of the buildings were typical office occupancies with up to 500 evacuees in a stairwell. These data are available on the NIST website at http://fire.nist.gov/egress/. A summary of the four buildings included in the current study is shown in Table 1, with additional details on each of the buildings available on the website.

**Table 1. Stairwell geometry and evacuation details for buildings included in the study**

| | 10-Story Building | 18-Story Building | 24-Story Building | 31-Story Building |
|---|---|---|---|---|
| Occupancy | Office | Office | Office | Office |
| Floors | 10 | 18 | 24 | 31 |
| Stair width[a] (m) | 1.27 | 1.12 | 1.12 | 1.38 |
| Stair riser (mm) | 178 | 191 | 178 | 178 |
| Stair tread (mm) | 279 | 254 | 279 | 273 |
| Exit width (m) | 0.91 | 0.83 | 0.91 | 0.91 |
| Evacuees | 436 / 368 | 255 / 292 / 340 / 197 | 249 / 356 | 704 / 538 |
| Evacuation time (s) | 1022 | 1192 | 1090 | 1002 |

[a] Full stair width including handrails

## Overall Movement Speed

A summary of pre-evacuation times and average stairwell descent speeds is shown in Table 2. The average evacuee speeds in all stairwells of the buildings are within experimental variability (as expressed by one standard deviation). Figure 1 shows average local movement speeds for the four fire drills, including data from the drills included in an earlier paper [5].

**Table 2. Pre-evacuation time and stairwell movement speeds in three fire drill evacuations**

| Building | Pre-evacuation delay time (s) | Average speed (m/s) |
|---|---|---|
| 10-Story | 171 ± 124 | 0.44 ± 0.19 |
| 18-Story | 224 ± 146 | 0.44 ± 0.15 |
| 24-Story | 137 ± 86 | 0.56 ± 0.12 |
| 31-Story | 149 ± 88 | 0.52 ± 0.10 |

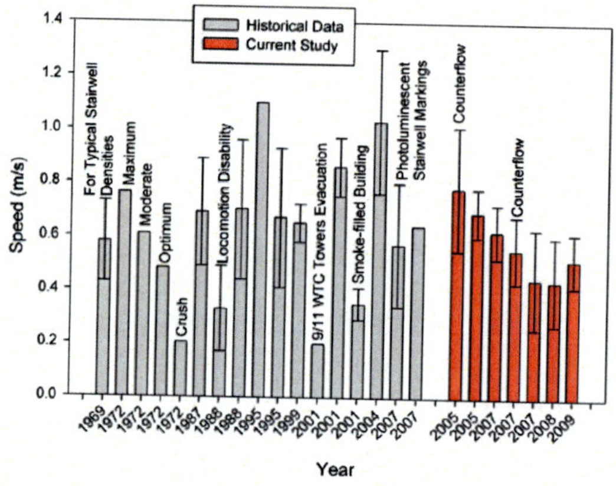


**Fig. 1.** Comparison of current study average stairwell descent speeds with literature values. Where available, data points include standard deviation of average movement speeds.

Figure 1 also compares the current study to historical data. It is important to recognize that all of these data were collected under differing conditions, with a range of building heights (ranging from a few stories to about 30 stories in height), occupant capabilities (one study looked specifically at occupants with locomotion disabilities), and evacuation conditions (many were fire drills, but actual events are also included). With the considerable variation in all the available data (as indicated by the standard deviation shown for many of the studies), the newer data are typically within the range of data in the literature and quite similar to the "optimum" or "moderate" movement speed of Fruin [7].

Values for very dense evacuations (Fruin's crush load [7] and the 9/11 World Trade Center evacuation [8,9]) are significantly lower than both the current study and average values from the literature. This may be indicative of the difference between fire drill evacuations and real emergency situations or due to higher occupant densities in the slower stairwells.

While the current study does not support recent concerns over slowing evacuation speeds resulting from increased obesity rates and lower fitness levels, additional study is needed, particularly to understand the impact of emergency conditions compared to fire drill evacuations.

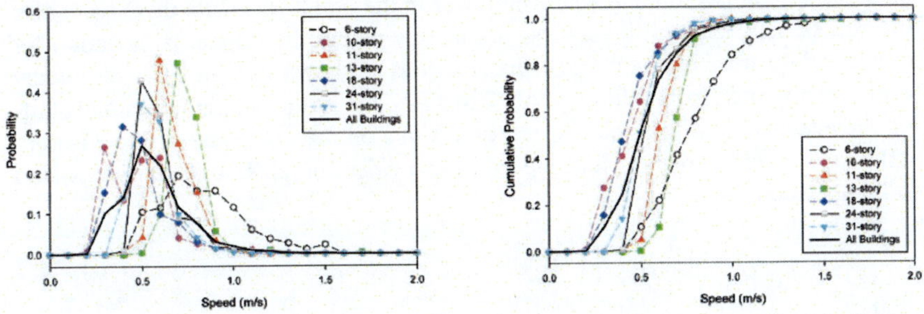


(a) Probability Distribution (b) Cumulative Probability Distribution

**Fig. 2.** Distribution of movement speeds down stairwells in several fire drill evacuations

The distribution of stairwell movement speeds in the buildings shown in Figure 2a and the cumulative distribution functions shown in Figure 2b provide additional details of the range of speeds in the evacuations. Overall, 19 % of the occupants move slower than 0.4 m/s (63 % of these are in the 18 story building; 99 % of these are in the 10-, 18-, and 31-story buildings) and just 2 % move faster than 1 m/s. With the exception of the 6-story building (data from the earlier paper [5]), the cumulative probability curves show similar shapes with the majority of speeds between 0.3 m/s and 0.7 m/s. The 6-story building tends towards faster movement speed, consistent with the higher overall average movement speed of 0.78 m/s ± 0.23 m/s compared to an average of 0.52 m/s ± 0.19 m/s for all of the buildings examined. However, with the overlapping standard deviations between the 6-story building and the other buildings, the difference is not likely to be significantly different.

## Local Movement Speeds

Overall movement speed, arguably the most commonly reported value for evacuee movement in stairwell evacuation, illustrates only a small part of the dynamics of movement during an evacuation. Though not surprising, there is considerable variation in movement speed not only among individuals involved in the evacuation, but also for each individual as they proceed down the stairwell during the evacuation.

Figure 3a shows the variation in local movement speeds (here an average speed for all evacuees passing each camera location). The average local movement speed varied by floor within a building. Fastest speeds are seen lower in the building, slower speeds on the middle floors, and typically somewhat faster speeds high in the building, but not to the levels seen lower in the building. While there were also differences between buildings, these are largely within the standards deviation for a floor and dwarfed by the range of local speeds as shown in Figure 3b. Individual local speed ranged from 0.06 m/s to 1.2 m/s (though with a single individual starting the evacuation with a local speed of 1.7 m/s on the top floors of the 31-story building). A wide range in local speeds was evident on all floors, stairwells, and buildings studied.

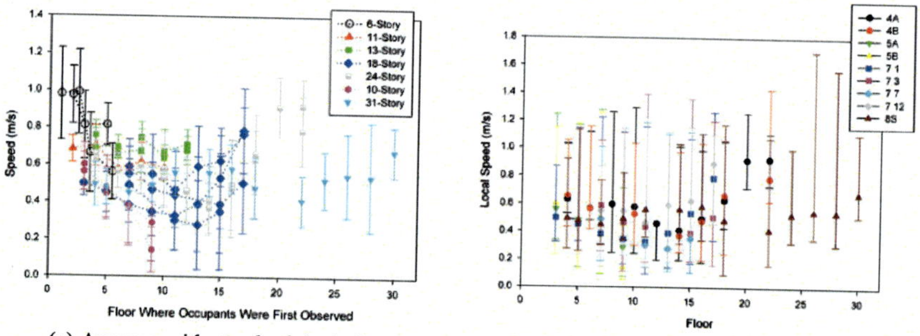


(a) Average with standard deviation (b) Range of local speeds

**Fig. 3.** Local occupant movement speed down stairwells in several fire drill evacuations.

Some of the variation in local speed is attributable to changing evacuee density in the stairwell as the evacuation proceeds. As each evacuee entered a camera view travelling down a stairwell, the local density is calculated from the number of other evacuees ahead in the path of evacuation and the total area of stairs and landing surfaces in the same camera view.

Figure 4 shows individual local speed as a function of density for all evacuees in the 24-, 10-, 18-, and 31-story buildings. Also shown in the figure is the correlation for evacuation speed as a function of density from the SFPE Handbook of Fire Protection Engineering [6] based on the data of Fruin [7], Pauls [10], and Predtechenskii and Milinskii [11]. While the correlation is contained within the data from the current study and the evident decrease in speed with increasing density is understandable, the fit of the correlation to the data (with an $R^2$ of about 0.2) again highlights the inherent variability in the data. A better understanding of the underlying theory and of evacuee behavior would be required to justify any particular correlation to the data.

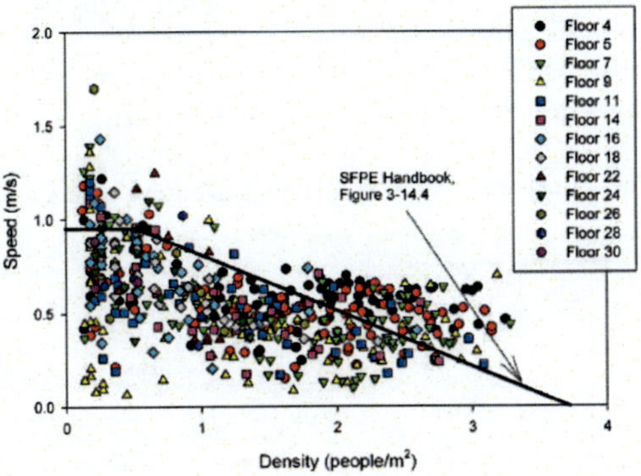


**Fig. 4.** Local speed as a function of local density for evacuees in all stairwells of 10-, 18-, 24-, and 31-story buildings during fire drill evacuations.

## Regression Modeling

A regression model was constructed to explore the components affecting occupant descent speeds in the stairwells for the 10-, 18-, 24-, and 31-story buildings. The dependent variable, local movement speed, was calculated based on the time difference between when the occupant was seen on adjacent cameras (typically 2 floors apart in a stairwell) and the known distance between the cameras. Eight independent variables, stairwell, gender, carrying objects, exit lane, handrail use, pre-evacuation time, density, and travel distance, were included in the model.. The first five variables were categorical variables and the final three variables were continuous variables.

SPSS Version 12.0.1[1] was used to estimate the linear regression model gauging the net effects of the independent variables on the local movement speed (Table 3). The correlation was significant at the less than 0.001 level. For the categorical variables, reference values were chosen simply to allow comparison (male, not carrying anything, middle exit lane, not using the handrail, and the 31-story

[1] Certain commercial entities, equipment, or materials may be identified in this document in order to describe an experimental procedure or concept adequately. Such identification is not intended to imply recommendation or endorsement by the National Institute of Standards and Technology, nor is it intended to imply that the entities, materials, or equipment are necessarily the best available for the purpose.

building stairwell being defined as 0). Table 4 includes the unstandardized and standardized coefficients for each variable as well as the standard error and significance for all the variables in the main effects model. For the categorical variables, the coefficients are interpreted as the increase in local movement speed for someone with that characteristic compared to someone with the reference characteristic independent of all other variables. For the continuous variables, the coefficients are interpreted as the increase in speed when the independent variable increases by one unit and all other variables are held constant.

**Table 3. Main effects regression model for local movement speed**

| | Unstandardized Coefficients | | Standardized Coefficients | |
|---|---|---|---|---|
| Model | B | Std. Error | Beta | Significance |
| (Constant) | .625 | .007 | | .000 |
| Density | -.082 | .002 | -.376 | .000 |
| Women | -.010 | .003 | -.029 | .003 |
| Carry | -.010 | .004 | -.027 | .006 |
| Travel Distance | .000 | .000 | -.012 | .219 |
| Pre-Evacuation Time | $8.3\times10^{-05}$ | .000 | .053 | .000 |
| Inner Exit Lane | -.007 | .005 | -.017 | .189 |
| Outer Exit Lane | -.009 | .004 | -.024 | .054 |
| Handrail | -.002 | .004 | -.007 | .514 |
| Stair 4A | .143 | .006 | .258 | .000 |
| Stair 4B | .131 | .006 | .223 | .000 |
| Stair 5A | .028 | .007 | .043 | .000 |
| Stair 5B | .014 | .007 | .021 | .039 |
| Stair 7-1 | -.025 | .006 | -.046 | .000 |
| Stair 7-3 | .011 | .007 | .016 | .120 |
| Stair 7-7 | -.107 | .008 | -.137 | .000 |
| Stair 7-12 | .044 | .007 | .059 | .000 |

The data was examined to ensure that the assumptions of regression modeling were met. Zero-order correlation matrices were examined and no multicollinearity was detected. Heteroscedasticity was tested by graphing the residuals versus the independent variables. Two of the variables, density and pre-evacuation time, failed the test. While the coefficients are still accurate, the standard deviations might not be. The coefficient for pre-evacuation time was found to be significant, but relatively small. Thus, treating it as if it was not significant leaves the findings unchanged. For density, the model was run again by changing the density variable to a categorical variable based on Fruin's Level of Service [7]. All levels were found to be significantly different from 0 at the <0.001 level, so the density variable is significant in this model.

Of the eight variables in the regression model, six were significant at the 0.10 level or lower. The final regression model explained 21 % of the variance in the local movement speeds included in this study.

As expected, the regression analysis shows that as density increases, speed decreases. Individuals carrying anything were slower than those who were not while using the handrail did not cause a significant change. Evacuees travelling near the edges of the stairwells travelled slower than those in the middle. Men travelled slightly faster than women. Movement speed from one stairwell to another was found to be different by up to 0.27 m/s. This implies that some variable(s) not included in this model causes occupants to move at different speeds.

The change in local movement speed based on travel distance was not significant and, for pre-evacuation time, the change was significant but relatively small for most occupants; the heteroscedasicity also could make this value insignificant, so no findings should be drawn from it.

Secondary interaction terms between the different variables were examined to see if combinations of variable levels were behaving differently. Two general trends in the interactions were noted. The first set involved interactions with density and the second set involved interactions with the different stairwells.

As density increased, all of the variation in other variables tended to approach zero. In essence, for a highly dense flow, the speed of all occupants was more uniform. In the main effects model, this difference will cause several of the terms to appear less significant than they are in reality. Based on these interactions with density, future research should look at how the interactions between people within the group alter the individual movement speeds.

While stairwells within a building were generally similar, differences between buildings (and in the 19-story building, between individual stairwells) were substantial. For example, women moved faster than men in one of the buildings. Individual characteristics varied in how significant they were in influencing local movement speeds. In one building, the speed based on the density was significantly different than the other buildings. As was the case with density, these differences across buildings lead to the coefficients in the main effect model to appear less significant than they might be in reality.

Several assumptions made for data analysis will limit the accuracy of the data. While the density could be changing throughout the time interval of the movement speed calculation, the density at the start of the interval is assumed to be the value throughout. Also, the measurements for travel distance and pre-evacuation time for individuals that did not enter on a floor with a camera are off by the time and distance travelled until the first camera. For the model itself, the regression model is based on the linear estimators that best fit the data; excluded variables that could be significant in determining movement speed will not be captured. In the main effects model, differences that were occurring based on different conditions (for example, at different densities) were not captured. Also, this model did not capture interactions with other occupants.

Overall, local movement speed could be predicted based on the eight variables used in this analysis. The speed depends on the characteristics of the occupants as well as the physical conditions within the stairs. There were also differences that were found to occur based on which building was being studied. Due to the similarities between these buildings, the exact cause of this difference is unknown.

## Conclusions

This paper has summarized the typical engineering variables used to describe stairwell movement during building evacuations, reviewed literature values for movement speeds, and presented data from several new fire drill evacuations. The following conclusions are evident from the study:

- Mean movement speed for the four buildings evacuations studied was 0.48 m/s ± 0.16 m/s.
- There is considerable variation in local movement speeds. Individual local movement speeds ranged from 0.056 m/s to 1.7 m/s. Using a distribution of movement speeds rather than a single value should provide more realistic representation of movement speed in stairwells.
- Data from the current study are reasonably consistent with historical data. Use of historical data may still be appropriate with the scope and limitations of the original collection.
- From the regression model, the two most significant variables were the stairwell that the occupant was in and the density. It is believed that the difference between stairwells comes from variables that were not included in this model. A clear relationship is evident in the data and regression analysis between density and speed. Algebraic formulas for prediction of speed as a function of density are a significant oversimplification of the process.
- This paper provides just a beginning in understanding the additional human behavior-related factors that impact movement beyond classic hydraulic calculation-based variables. Additional research is appropriate to better understand these factors.
- Data presented in this paper are available for review and/or further analysis at the NIST website, http://fire.nist.gov/egress.

## References

1 Kuligowski, E.D., R. D. Peacock: Review of building evacuation models. Technical Note 1471, National Institute of Standards and Technology, Gaithersburg, MD, 2005.
2 Gwynne, S., E. R. Galea, M. Owen, P. J. Lawrence: Further Validation of the buildingEXODUS Evacuation Model Using the Tsukuba Dataset. Rep. No. 98/IM/31, University of Greenwich, London, 1998.
3 Gwynne, S., E. R. Galea, M. Owen, P. J. Lawrence: Validation of the building-EXODUS Evacuation Model. Rep. No. 98/IM/29, University of Greenwich, London, 1998.
4 Lord, J., B. Meacham, B. Moore, R. Fahy, G. Proulx: Guide for evaluating the predictive capabilities of computer egress models. GCR 06-886, National Institute of Standards and Technology, Gaithersburg, MD, 2005.
5 Peacock, R.D., J. D. Averill, E. D. Kuligowski,: Stairwell evacuation from buildings: What we know we don't know. In *Pedestrian and Evacuation Dynamics 2008*, ed. W. W. F. Klingsch, C. Rogsch, A. Schadschneider, and M. Schreckenberg. NewYork: Springer, 2010.
6 G. Proulx. Movement of People: The Evacuation Timing, chapter 3-13, pages 3-341 – 3-366. In *The SFPE Handbook of Fire Protection Engineering*. Society of Fire Protection Engineers, Bethesda, MD, third edition, 2002.
7 Fruin, J.J. Pedestrian Planning and Design, Revised Edition). ElevatorWorld, Inc., Mobile, AL, 1987.
8 Averill, J. D., D. S. Mileti, R. D. Peacock, E. D. Kuligowski, N. Groner, G. Proulx, P. A. Reneke, and H. E. Nelson. Federal Building and Fire Safety Investigation of the World Trade Center Disaster: Occupant Behavior, Egress, and Emergency Communication. NIST NCSTAR 1-7. National Institute of Standards and Technology. Gaithersburg, MD, 2005.
9 Galea, E. D., L. Hylse, R. Day, A. Siddiqui, G. Sharp, K. Boyce, L. Summerfield, D. Canter, M. Marselle, and P. V. Greeall, "The UK WTC 9/11 Evacuation Study: An Overview of the Methodologies Employed and Some Preliminary Analysis." In *Pedestrian and Evacuation Dynamics 2008*, ed. W. W. F. Klingsch, C. Rogsch, A. Schadschneider, and M. Schreckenberg. NewYork: Springer, 2010.
10 Pauls, J. L., Effective-Width Model for Evacuation Flow in Buildings. In *Proceedings, Engineering Applications Workshop,* Society of Fire Protection Engineers, Boston, 1980.
11 Predtechenskii, V.M. and Milinskii, A.I., *Planning for Foot Traffic Flow in Buildings,* Amerind Publishing Company, Inc., New Delhi, 1978.

# Predicting the Probability of Evacuation Congestion Occurrence Relating to Elapsed Time and Vertical Section in a High-rise Building

**J. Choi, H. Hwang and W. Hong**

School of Architecture and Civil Engineering, Kyungpook National University
#1370 Sankyuk-dong, Buk-gu, Daegu, South Korea
Corresponding author: kissist@knu.ac.kr

**Abstract** It is very important to predict human behavior in fire and to take measures for evacuation time however it is difficult to predict individual responses and dangerous to perform a field experiment. According to the Final Report on the Collapse of the World Trade Center Towers from the NIST, 2005, it took an average of 48 seconds for evacuees to descend to the floor below. It was more than twice the time it would ordinary take. The evacuees needed to be rescued more than once during their descent due to physical fatigue.

The evacuation behavior of evacuees that have descended staircases repeatedly causes physical fatigue or dizziness which obstructs evacuation traffic leading to evacuation congestion and delay. Evacuation congestion is amplified in high-rise buildings due to a decline in physical strength of evacuees caused by the inordinate length of evacuation lines and the confluence of new evacuees at each floor.

In this paper, a field evacuation experiment with 351 participants was performed. Through the statistical analysis of the evacuation behaviors recorded by CCTVs, the probability of calculating the evacuation congestion from each floor was estimated and occurrence patterns relating to time and vertical space were analyzed.

## Introduction

Burj Dubai, the tallest building in the world was completed in January 2010. Many high-rise buildings have been built during the last decade with North America and Asia being central. However if a fire occurs in a high-rise building there will be a high degree of danger. The taller the building longer the evacuation route and the longer the evacuation time required because evacuation is the immediate and rapid movement of people away from the threat or actual occurrence of a hazard. In situations involving hazardous materials or possible contamination from a

R.D. Peacock et al. (eds.), *Pedestrian and Evacuation Dynamics*,
DOI 10.1007/978-1-4419-9725-8_4, © Springer Science+Business Media, LLC 2011

fire or smoke, evacuees may escape from the buildings and get to the ground level prior to being transported out of the contaminated area.

Most human egress behavior occurs in emergency staircases within a high-rise building. Evacuees move to the ground or refuge floor by way of a staircase. They descend or ascend numerous spiraling stairways. However, if there are many occupants who all evacuate together at the same time, they may experience and be faced with evacuation congestion in the staircases because it is natural for physical condition to vary between individuals along with body type, athletic ability and psychological response.

The phenomenon of evacuation congestion which may occur in a staircase under an emergency situation such as a fire, accident or terrorism is similar to that of a traffic jam which happens in rush hour. Evacuees can experience friction with the people surrounding them or become engaged in a physical confrontation because of evacuation congestion. Therefore it is necessary to shorten the delay time resulting from evacuation congestion because the longer the egress time the stronger the possibility for evacuees to be exposed to flames or smoke.

This study analyzed research done earlier to propose a solution for the phenomena known as evacuation congestion or delay occurrence. For this, a field evacuation experiment was conducted at a thirty-story high-rise apartment with 351 participants.

## Method of a trial evacuation experiment [1]

For this study a thirty-story high-rise apartment housing complex located in Daegu, South Korea was selected and a trial evacuation experiment with 351 participants was conducted in which a fire situation was assumed within the building. Firstly participants were dispersed in groups of 3 or 4 to 4 households on every floor.

Before starting the evacuation, all the participants were standing by in the household that they were previously assigned to. Then they stayed there for about an hour to experience living but they did not know how long they would have to wait until the fire alarm went off. Meanwhile an unexpected fire alarm was sounded and an announcement about safety was made and the participants started to escape from where they were to the ground level via the staircase.

This study only analyzed the characteristics of participants whose features were recorded by CCTVs. Throughout the experimental the staff recorded the time and behavior of individual or clusters of evacuees using CCTVs installed on the ceiling of every floor. They did not affect the evacuation process or disturb the movement of the participants.

However the effect that would have been caused by an actual fire and smoke could not be considered in this experiment due to the associated life safety hazard. Moreover, this trial experiment did not sufficiently consider the detection and pre-movement stage because the invited participants already knew that they would have to evacuate accidentally in a short time. The experiment was only focused on analyzing the vertical movements and behaviors of evacuees in the staircase.

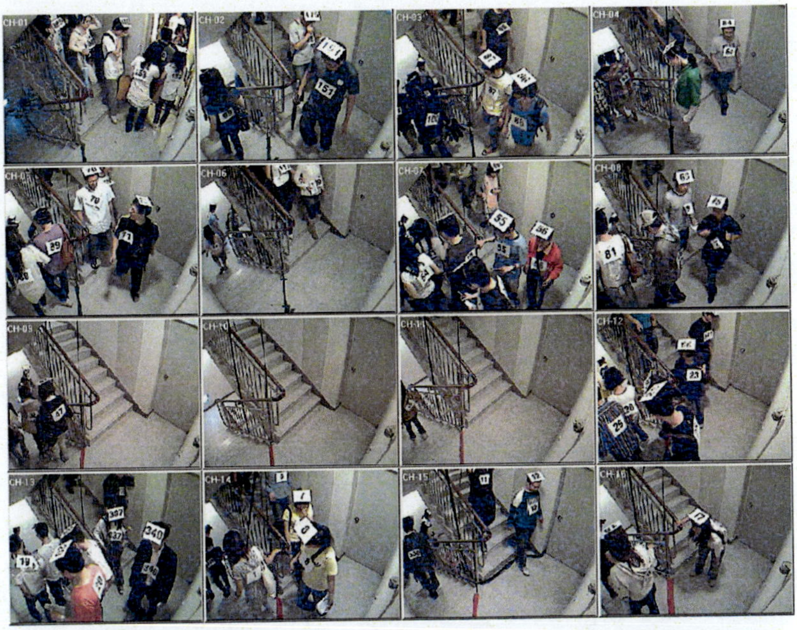

**Fig. 1. Scene of evacuees' descending egress behaviors and congestion occurrence**

The measurements of the staircase are shown in the following figures.

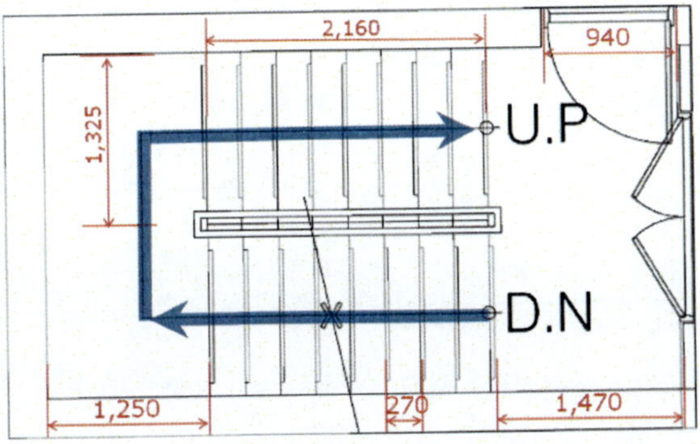


**Fig. 2. Floor plan of the staircase**

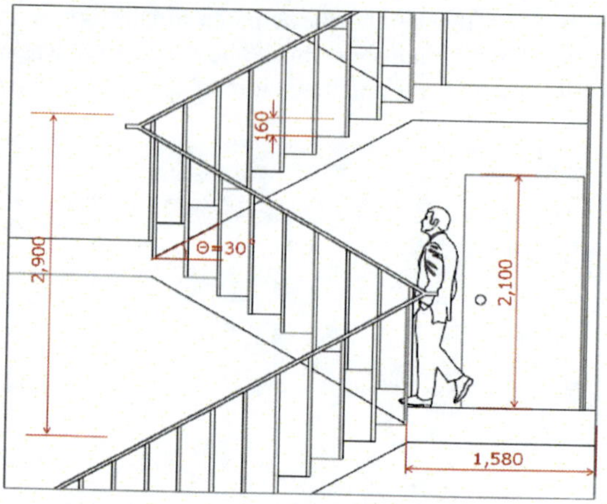


**Fig. 3. Sectional plan of the staircase**

## Analysis of Evacuation Time and Evacuee's Position

It took 422 seconds for all evacuees to finish escaping from their own households where they had been distributed before the evacuation experiment. When compared with their average ordinary walking speed (1.35 m/s), their average walking speed (0.73 m/s) decreased by 46 percent on the average while descending.

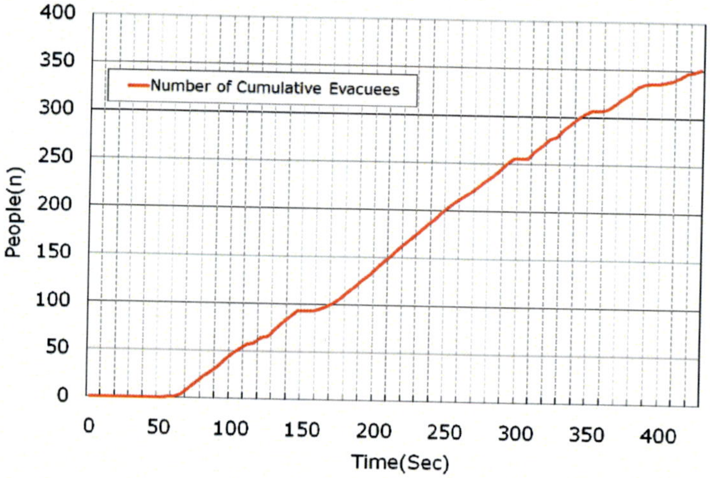


**Fig. 4. The number of cumulative evacuees (finish time: 422s)**

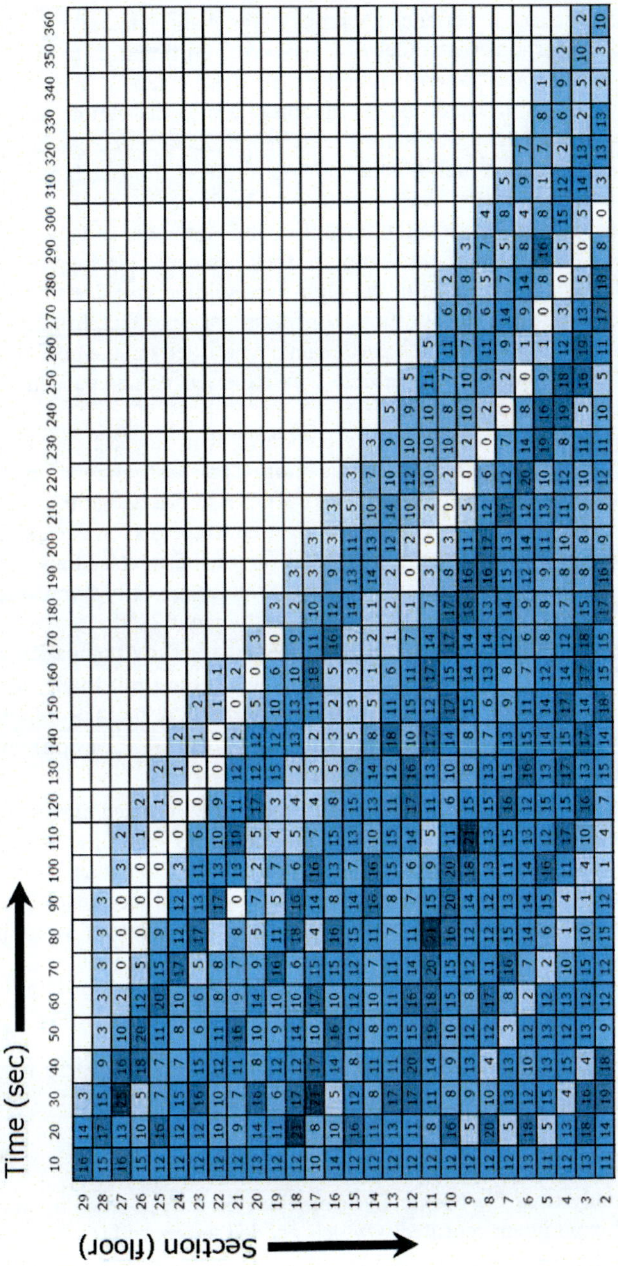


**Fig. 5. Frequency distribution of evacuees by elapsed time and vertical section: the darker color, the more crowded**

Most of the evacuees experienced a remarkable change in walking speed because they had interactions with other evacuees due to differences in their characters, physical abilities and patterns of action called 'fluctuation' and 'density' in physics. Fluctuation in the sense of evacuation refers to the various actions that disturb one's egress behavior; it means that an insignificant action can have a huge effect as time goes on. Although fluctuation was different from each person, it delayed the evacuation of all the evacuees. Density in this case refers to how many people were in staircase. It was determined that, the higher the density of the evacuees the greater the degree of fluctuation in this instance. This suggests that the evacuation delay that was caused in the high-rise building was due to downward and repeated evacuation in narrow staircases. As a matter of fact, from a result of the former study, 91.2 percent of the participants who filled out the post-survey after arriving at the ground floor stated that they experienced an evacuation delay. [1]

Congestion generally means excessive crowding. It is characterized by slower speeds, longer trip times, and increased queuing. The most common example is the physical use of roads by vehicles. When traffic is so heavy that the interaction between vehicles slows the speed of the traffic stream, congestion occurs. As traffic volume approaches the capacity of a road (or of the intersections along the road), extreme traffic congestion sets in. When vehicles are fully stopped for long periods of time, this is colloquially known as a traffic jam. [2]

In this study, evacuation congestion was characterized by population density per unit area. The unit area from the (n+1)th floor to the (n)th floor, was about 12㎡ except in the unapproachable areas. If there were 11 or more people in a unit of staircase, it was regarded as evacuation congestion because it was difficult for a person not to run into another person or to be affected by others in 1.1㎡ of area.

Fig. 5 shows the frequency distribution of evacuees by elapsed time and vertical section while they were escaping the building via stairs. The figure shows how many people were on every floor for every 10 seconds. The darker the shade of blue the more crowed a staircase. The pattern of congestion occurrence had a tendency to appear on the upper floor initially and to move downward from there as time progressed.

## Probability of Evacuation Congestion Occurrence: using Monte Carlo Method

To predict the probability of evacuation congestion occurrence in a high-rise building, this study used the Monte Carlo method because it was impractical to conduct a field experiment repeatedly which needed an entire high-rise building and hundreds of people.

The Monte Carlo method is a technique that involves using random numbers and probability to solve problems. The term Monte Carlo Method was coined by S. Ulam and Nicholas Metropolis in reference to games of chance, a popular attraction in Monte Carlo, Monaco (Hoffman, 1998; Metropolis and Ulam, 1949). Monte Carlo simulation is a method for iteratively evaluating a deterministic model using sets of random numbers as inputs. This method is often used when the model is complex, nonlinear, or involves more than just a couple uncertain parameters. A simulation can typically involve over 10,000 evaluations of the model, a task which in the past was only practical using super computers. The five simple steps of Monte Carlo simulation are listed below: [3]

Step 1: Create a parametric model, $y=f(x_1, x_2, ..., x_q)$.
Step 2: Generate a set of random inputs, $x_{i1}, x_{i2}, ..., x_{iq}$.
Step 3: Evaluate the model and store the results as $y_i$.
Step 4: Repeat steps 2 and 3 for $i=1$ to $n$.
Step 5: Analyze the results using summary statistics, confidence intervals, etc.

However, the data from Fig. 5 was based on the trial field evacuation experiment not on a set of random inputs. Instead of performing Step 2, this study reflects a new time and vertical section unit zone for interpreting and predicting the congestion probability. A new unit was created for only calculating the Monte Carlo simulation and was made up of sets of 4 stairs and a 30 second time window. The different numbers of people on the staircase (12㎡) were weighted: 90% (11 to 15 people), 100% (16 to 20 people) and 110% (21 to 25 people).

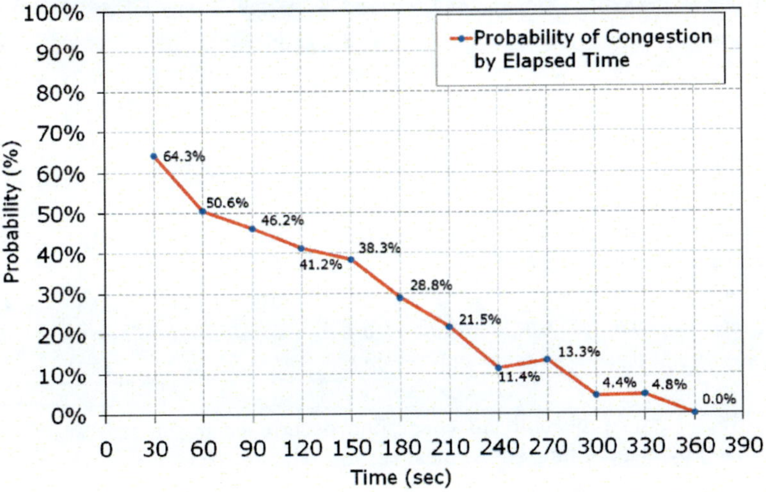


**Fig. 6. Probability of congestion by elapsed time**

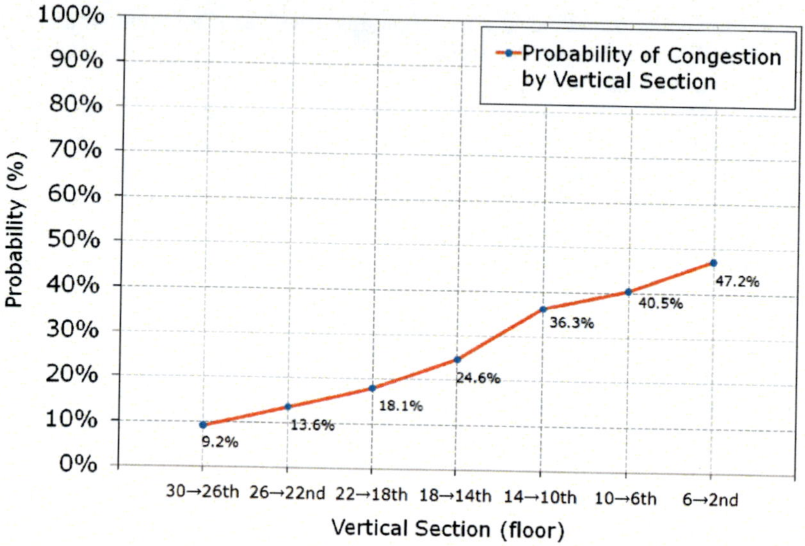


**Fig. 7. Probability of congestion occurrence by vertical Section**

**Table 1. Probability of congestion occurrence (%)**

| | 30-26th | 26-22nd | 22-18th | 18-14th | 14-10th | 10-6th | 6-2nd | Avg. |
|---|---|---|---|---|---|---|---|---|
| - 30s | 68.3 | 63.3 | 66.7 | 51.7 | 71.7 | 56.7 | 71.7 | 64.3 |
| 31- 60s | 42.2 | 35.0 | 41.7 | 51.7 | 73.3 | 41.7 | 68.3 | 50.6 |
| 61- 90s | 0.0 | 51.7 | 31.7 | 63.3 | 68.3 | 75.0 | 33.3 | 46.2 |
| 91-120s | 0.0 | 13.3 | 30.0 | 50.0 | 56.7 | 86.7 | 51.7 | 41.2 |
| 121-150s | 0.0 | 0.0 | 46.7 | 13.3 | 73.3 | 48.3 | 86.7 | 38.3 |
| 151-180s | 0.0 | 0.0 | 0.0 | 36.7 | 45.0 | 55.0 | 65.0 | 28.8 |
| 181-210s | 0.0 | 0.0 | 0.0 | 29.1 | 13.3 | 80.0 | 28.3 | 21.5 |
| 211-240s | 0.0 | 0.0 | 0.0 | 0.0 | 6.7 | 21.7 | 51.7 | 11.4 |
| 241-270s | 0.0 | 0.0 | 0.0 | 0.0 | 26.7 | 13.3 | 53.3 | 13.3 |
| 271-300s | 0.0 | 0.0 | 0.0 | 0.0 | 0.0 | 7.3 | 23.3 | 4.4 |
| 301-330s | 0.0 | 0.0 | 0.0 | 0.0 | 0.0 | 0.0 | 33.3 | 4.8 |
| 331-360s | 0.0 | 0.0 | 0.0 | 0.0 | 0.0 | 0.0 | 0.0 | 0.0 |
| Avg. | 9.2 | 13.6 | 18.1 | 24.6 | 36.3 | 40.5 | 47.2 | |

Table 1 shows probability of congestion occurrence in the staircase during evacuation by elapsed time and vertical section. The top 10 percent of the highest values of total sections were filled with cyan color. Probability of congestion occurrence decreased as time went by. [Fig.6.] However, it increased as evacuees descended from the upper floors to the ground floor. [Fig.7] According to Table.1, probability of evacuation congestion occurrence was generally relatively higher on the lower floors under the middle of the building at the early stage of evacuation. Fig. 8 and Fig. 9 are broken line graphs based on Table. 1.

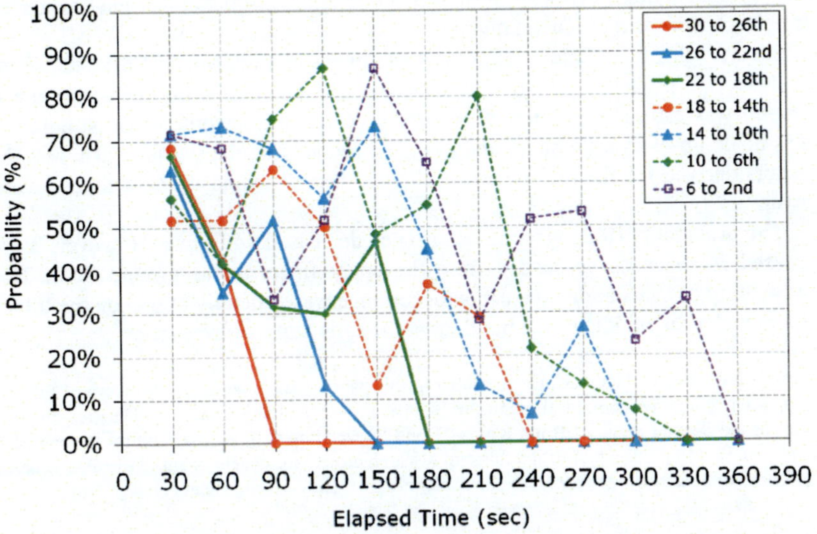


**Fig . 8. Probability of congestion occurrence by elapsed time and vertical section**

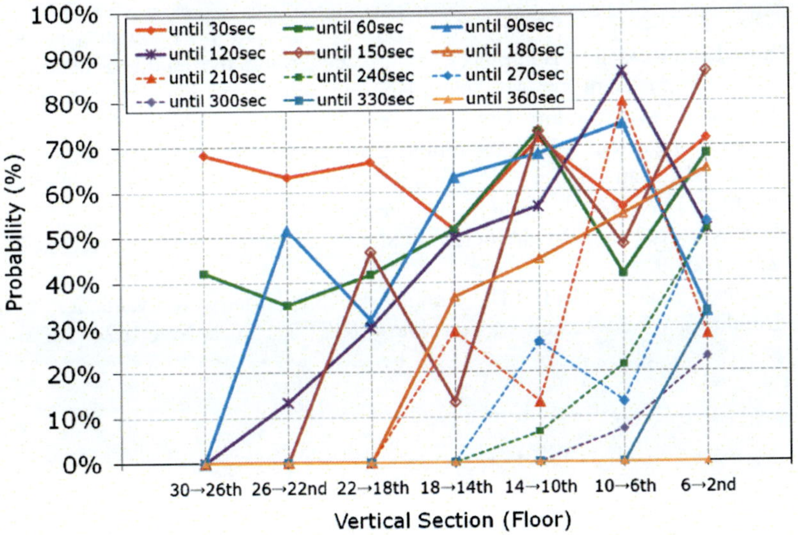


**Fig. 9. Probability of congestion occurrence by vertical section and elapsed time**

## Conclusion and Discussion

It was determined that evacuation congestion in the high-rise building was intensified by evacuation queues of an inordinate length, confluences of new evacuees from each floor and evacuee fatigue.

As a result of the field evacuation experiment and the Monte Carlo statistical analysis, the probability of having evacuation congestion from each time-space section was estimated. A new unit for only calculating Monte Carlo simulation was made up of every 4 stairs and every 30 seconds of elapsed time. The weight factors derived for the differences in the number of people in the staircase were given.

The probability of congestion occurrence decreased with time. However, it increased as evacuees descended from the upper floors to the ground floor. The probability of evacuation congestion occurrence was generally higher on the lower floors under the middle of the building at the early stage of evacuation.

**Acknowledgments** The research presented in this paper was supported by Ministry of Education, Science and Technology through the 'Global Expert Training Center for Disaster Prevention' of the Second Stage of Brain Korea 21 and 'Development of Smoke Control System of a High-rise Building', as a part of the flagship research project for development of the next generation fire safety technology funded by the National Emergency Management Agency, Korea.

## References

1. Choi, J., Hwang, H., Hong, W., Choi, Y.: Analysis on Occupants' Escape Speed and Reason of Bottle-neck Occurrence through the Trial Evacuation Experiment at a High-rise Apartment Housing. 4th International Symposium on Human Behaviour in Fire, Conference Proceedings, 123-134 (2009)
2. Traffic congestion. Wikipedia. http://en.wikipedia.org/wiki/Traffic_congestion. Accessed 22 Jan. 2010
3. Wittwer, J.W.: "Monte Carlo Simulation Basics" http://vertex42.com/ExcelArticles/mc/MonteCarloSimulation.html. Accessed 22 Jan. 2010
4. Peacock, R.D.; Averill, J.D., Kuligowski, E.D.: Stairwell Evacuation from Buildings: What we know we don't know. NIST Technical Note 1624 (2009)

# Employing Human Egress Data

**S.M.V. Gwynne**

Hughes Associates UK
Corresponding Author: sgwynne@haifire.com

**Abstract** The availability of credible, accurate and comprehensive human egress data is critical for responsible engineering. However, human egress data is particularly complex and susceptible to omissions and severe limitations. Currently, much of the data available is incompletely described, inconsistently represented or difficult to access, making the misuse of data more likely. This article outlines the issues that can arise from the compromised representation of data and associated background information. A Data Portal design is described that should help existing data-sets be described more comprehensively, and aid future data collection efforts. As such, the portal should aid in the development of more accurate theories and in more responsible engineering activities.

## Introduction

Empirical data provides the bridge between reality and understanding. Data-sets enable engineers and scientists alike to form theoretical models to represent our understanding of the real world, and then apply these models to engineering applications. As such, empirical data-sets form the foundation of the engineering process. Unfortunately, there are a number of issues with the data currently available to describe human egress.

Human egress data-sets are scarce. This scarcity is due to the relative immaturity of the field, the comparatively recent identification of the importance of human behavior in egress calculations, technical factors relating to data collection methods, matters of privacy and commercial sensitivity, and a range of other procedural and political factors. The data-sets currently available are relatively narrow in scope: they tend to focus on the physical component of egress performance, rather than the behavioral. The data-sets are derived from a range of sources and locations. This variety can add to the richness of the data-sets available; however, it includes data imported from adjacent fields (e.g., circulation movement) in order to fill the gaps in the data currently available. In addition, many of the egress data-sets collected (and frequently used), are several decades old. These problems lead to critical weaknesses in our understanding of real-world phenomena, in our attempts to develop theories and also in our attempts to model these phenomena. Data are often difficult to find, difficult to understand, and difficult to apply. This article outlines some of these issues, how these issues can

R.D. Peacock et al. (eds.), *Pedestrian and Evacuation Dynamics*,
DOI 10.1007/978-1-4419-9725-8_5, 

propagate through the engineering process, and the efforts made to combat this issue.

## Data Portal Development

As part of a NIST-funded project, guidance has been developed to inform the collection and representation of human egress data [1]. This work provides a *Data Acquisition Matrix* to inform the data collection process and a *Data Template* outlining the information required to adequately describe data collected. These two tools, along with several component tools, will then be provided in an online *Data Portal*. Once the data portal is fully implemented a central repository of data will be created that provides tools to facilitate the storage, representation and access to the data needed for researchers and engineers alike. As such, data should be represented in more clear, comprehensive and refined manner, and be more easily found. This can be used to inform the data acquisition and dissemination process: the use of data made available through the portal, and the interrogation of existing data using the template provided (see **Fig. 1**). All three uses (uploading data, downloading data, and interrogating data) should improve the understanding and use of the data available.

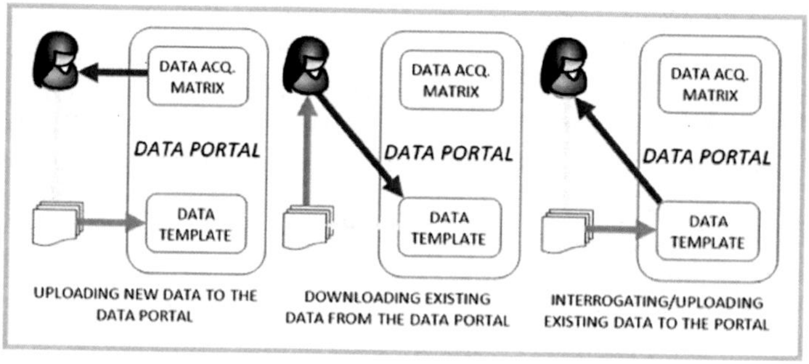


**Fig. 1. The various uses of the Data Portal [1]**

As a prelude to the implementation of the data portal design produced, this article describes how the representation of egress data can influence the engineering process in both beneficial and harmful ways. Initially, the complexities of human egress data are discussed: why this data, in particular, is susceptible to misrepresentation and misuse. Following this, several examples of incomplete data-sets are presented, along with a description of the potential consequences of their use. Finally, a brief description of the *Data Portal* is provided along with a description of how it might address the misuse of data.

## Difficulties with Human Egress Data

Three key elements make the collection and description of human egress data particularly difficult: the human decision-making process is complicated [2-5]; human egress does not occur in isolation [6-8], but in conjunction with other, equally complex human-related events; the data collection process is technically and organizationally difficult to conduct.

The individual decision-making process forms the basic component of human egress. It represents the underlying driver for the evacuation process. As such, it is critical that this is represented in a credible manner. The decision-making process is now reasonably well understood (see Fig. 2). It is now seen as a complex process involving several stages, rather than as an instantaneous stimulus-based response or an instinctive, potentially irrational, panic-based response. As such, to fully understand and represent this process accurately, data is needed to address each of the stages and interactions within this process. This is not currently the case. However, in the past, data was collected according to a range of different models of human behavior (e.g., panic models). These models influenced the design of the data collection methods used, the data collected and the associated description. It is important to understand that data collected in the past will have been collected according to a dated understanding of the decision-making process; and the data should be treated accordingly.

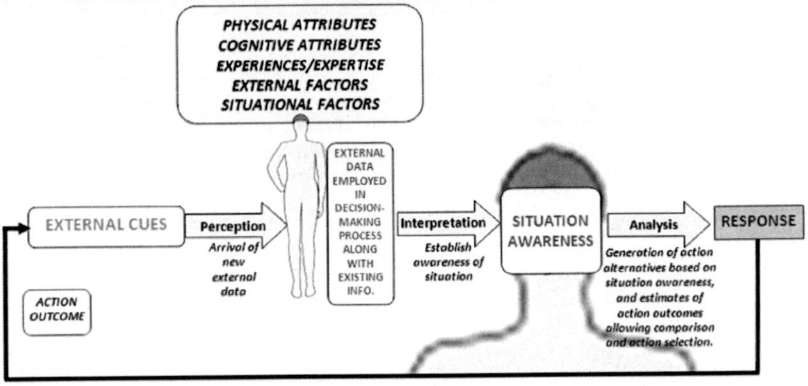


**Fig. 2. Simplified version of the individual model developed by Kuligowski and Gwynne [1,2].**

Engineering models are being applied to broader applications that go beyond emergency egress scenarios. This is because their value in understanding other people movement components (e.g., ingress, circulation, etc.) has been recognized. More importantly, it has been recognized that emergency egress does not occur in isolation: it can occur while other phases of movement and other procedures are in place (see Fig. 3). This 'coupled' movement increases the complexity of an event, and subsequently the data that is required to describe it.

Data acquisition in relation to people movement presents a number of challenges – primarily due to the subject matter involved; i.e., human participants. Data can be collected according to a range of different interests and disciplines. This focus adopted will influence the focus of the data acquisition, the underlying theories employed and the terminology used to describe the data. Data can be collected using a number of different techniques (e.g., video, survey, etc.) and from a range of different event scenarios (e.g., from a real incident, an experiment, a drill, etc.). Both the techniques employed to collect the data and the event scenario will have a direct impact on the format and content of the data collected. This will determine the credibility and applicability of the data itself. It is therefore critical that sufficient information is available to third parties interested in using the data regarding (1) the data acquisition techniques employed, (2) the event scenario, and (3) that the data are presented in as complete a form as possible to ensure that it is responsibly and reliably employed: that a third party can understand the limitations of the data and the event from which it was collected.

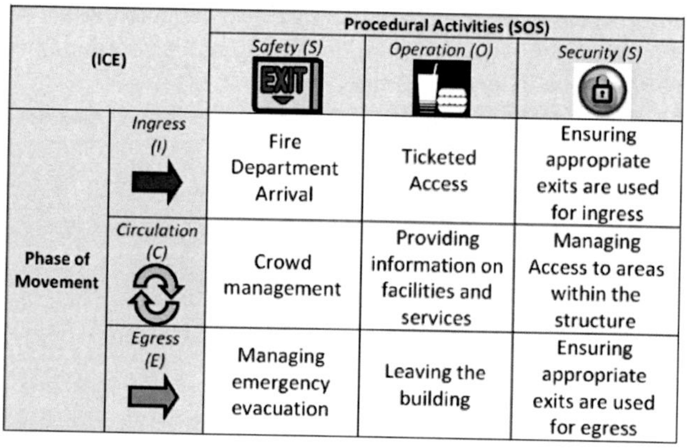


**Fig. 3. Interaction between various phases of movement and procedural activities [1,9].**

Data does not exist independently of the acquisition process; data are not collected in a vacuum. The data acquisition process requires decisions to be made at a number of stages, and these directly influence the scope, refinement and applicability of the data. Given this, it is important to understand the process by which data are produced. In doing so, we can attempt to remedy any deficiencies in this process. Initially, a decision has to be made to acquire or seek out data: there is a reason behind the acquisition of the data based on a research or engineering need. The data acquisition process is therefore *selective*. Data are then collected. The methods used during this process will influence the nature of the data and the availability of contextual (background) information. The methods selected may be based on their appropriateness, but also based on less rational reasons: available expertise, cost, convenience, etc. The data are then extracted and analyzed.

The data and the derived understanding are then described and presented; i.e., distilled into a representative state from a raw form and summarized along with the background information that is available (according to the information collected) and deemed worthy of inclusion. The data-sets are then shared with an audience of interested third parties. This may range from immediate colleagues to the general public. These parties attempt to understand the data according to the distilled format and the associated background information. On this basis, the data are then applied. Third parties may not necessarily have access to the most appropriate data for their application, but instead make judgments based on those sources with which they are familiar or to which they have access; i.e., not only is data acquisition selective, but data use is also selective and not necessarily based on the appropriateness of the data itself. This is compounded by the limited background information associated with the data and the data being provided in a summarized form. In such circumstances, the likelihood of data being inappropriately employed is increased.

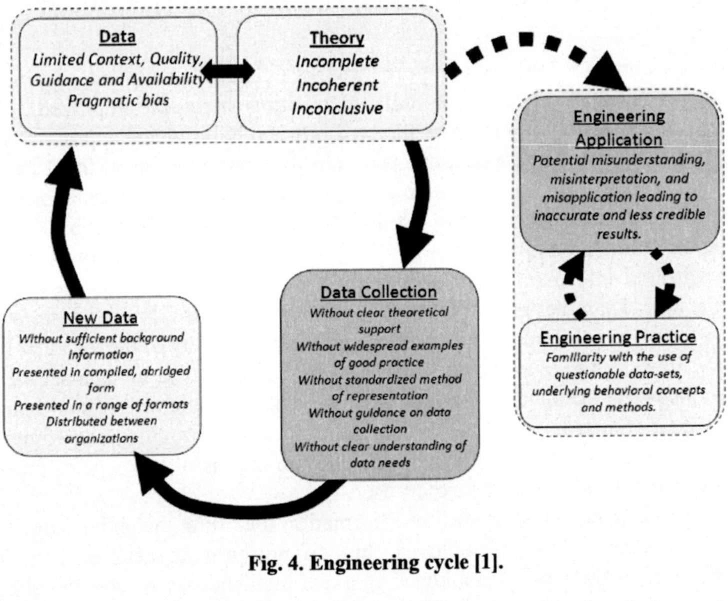


**Fig. 4. Engineering cycle [1].**

There are a number of opportunities within this process for the data to be misrepresented, misunderstood, and misapplied. In most instances, only a sub-set of the data collected is shared. It is shared in a reduced/distilled format, rather than in a complete format. Potentially more importantly, in the vast majority of cases only a limited amount of information is provided on the background conditions evident during the original event; i.e., the event described by the data. The reduced data-set and limited context requires a greater degree of interpretation by a third party. This increases the potential for the underlying causal factors being misunderstood, the results being misinterpreted, and the data-set being inappropriately applied. Not only does this type of error influence a particular application, but there is a high probability that it will propagate through a chain of different individuals. For instance, a researcher misusing third party data may then produce faulty theoretical understanding. This may then be embedded into a simulation tool. This tool may be validated against data inappropriately selected for comparison. The 'validated' model may then be used by an engineer who may select inappropriate data to configure the model for application. At this stage, a model that is based on a faulty theory, has been validated inappropriately, is then applied incorrectly. This can then lead to the engineering cycle shown in Fig. 4 that is based on faulty theory, which is derived from limited data, and then employed within engineering models that may then become standard practice.

## Consequences of Data Limitations

Data-sets consist of the data acquired and associated background information (i.e., the event scenario and the data acquisition methods employed). This combined information will influence the credibility, reliability and appropriateness of the data for a specific application. To help illustrate the impact that different data-sets might have, these two variables (data and background information) have each been categorized according to two levels: data (low/high detail); background information (limited/comprehensive). The four subsequent permutations of data and background information can potentially have different consequences on the use of the data. Data that is not represented in detail and has a limited description, provides the user with limited information on the applicability of the data and little flexibility in the manner in which it can be applied. This type of data-set can be found in historical accounts where, for instance, only a single measure is reported (e.g., evacuation time), and little if any information is provided regarding the event or the data collection methods used. The limitations of this type of data-set are usually fairly apparent to the expert user.

Some data-sets have a great deal of information regarding the underlying event and the collection methods employed, but do not provide detailed data. For instance, an investigative report addressing a real incident may be able to collect a great deal of information regarding the incident and describe in detail the interview (for instance) or survivors; however, it may not be able to provide detailed quantitative data on the outcome of the incident that could be used within an evacuation model application. Although, the low detail will restrict the application possible, the background information will at least allow the data to be applied appropriately by the expert user.

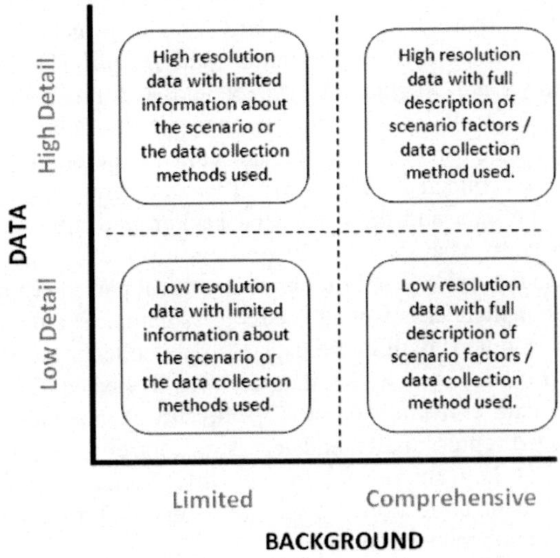


**Fig. 5. Categorization of data and background levels.**

Ideally data should be presented in a detailed and comprehensive manner and accompanied by a full description. This would allow the expert to appropriately apply the data, clearly understands its limitations and have a degree of flexibility in the exact manner in which it is applied; e.g., model development, egress calculations, model calibration, model validation, etc. For instance, raw data collected from an evacuation exercise, accompanied by a detailed description of the event, the procedures employed, the structure, the population involved, and data acquisition activities.

Detailed data-sets with limited background information possibly provide the greatest opportunity for misuse. An expert user may be able to apply the data to a range of target applications; however, with little background information they may not understand the appropriateness of such data for that specific application. In this situation, the likelihood of the problem occurring increases as the relevant expertise of the user decreases. For instance, video footage of an evacuation drill with little associated description. Detailed data could be extracted from this footage; however, the expert may not know whether the drill was announced or unannounced, which directly influences how the human response data can be used.

## Data Portal Development

To help address the issues highlighted, a NIST-funded project provided a detailed design for core components of an online *Data Portal* [1]. Primary in these elements are the *Data Acquisition Matrix* and a *Data Template*. The *Data Acquisition Matrix* provides detailed guidance on the data collection process: what factors should be considered by the data collector before, during and after the data-set has been collected? The *Data Template* provides guidance on the presentation of the data and the associated background information that allows informed assessment: what data and supporting information need to be provided for the data-set to be of use to a third party? A critical part in representing the data in an unambiguous manner is having access to as complete a data-set as possible. This completeness relates both to the data, the data collection methods employed, and the context under which the data-set was collected. This completeness depends on the data collection process itself, and cannot therefore be ensured purely by improved representation and access; i.e., purely through the use of *Data Template* alone. A key component of the *Data Portal* is therefore providing guidance on the data collection process to ensure that future data-sets are as complete and comprehensive as possible.

### *Data Acquisition Matrix*

Guidance has been developed, in the form of a *Data Acquisition Matrix*, to aid in the data collection process: to provide key reminders as to the elements of the event in question that should be documented. The guidance provided in the *Data Acquisition Matrix* ranges from the initial concept phases of the data collection process to the collection and analysis of the data. It is categorized according the stage of the data collection process and the component of the event being examined (a simplified version of this is shown in Fig. ). Each cell in the matrix leads to a resource (e.g., a set of documents), describing the component in question. Depending on the particular area of interest, this matrix leads to questions for the researcher to address, checklists, and/or guidance documents to refer to during the entire data collection process. This design was produced with potential online applications in mind.

| TIMELINE \ SCOPE | Procedure Pr | Response Re | Organization Or | Population Po | Objectives Ob | Structure St | Environment En | Data Acq. Da |
|---|---|---|---|---|---|---|---|---|
| Blueprint B | LINK TO QUESTIONS | LINK TO QUESTIONS | LINK TO QUESTIONS (See Table 9) | LINK TO QUESTIONS | LINK TO QUESTIONS | LINK TO QUESTIONS | LINK TO QUESTIONS | LINK TO QUESTIONS |
| Investigation I | LINK TO CHECKLIST | LINK TO CHECKLIST | LINK TO CHECKLIST | LINK TO CHECKLIST | LINK TO CHECKLIST | LINK TO CHECKLIST (See Table 10) | LINK TO CHECKLIST | LINK TO CHECKLIST |
| Preparation P | LINK TO DESCRIPTION OF ROLES | LINK TO DESCRIPTION OF ROLES | LINK TO DESCRIPTION OF ROLES | LINK TO DESCRIPTION OF ROLES | LINK TO DESCRIPTION OF ROLES | LINK TO DESCRIPTION OF ROLES | LINK TO DESCRIPTION OF ROLES | LINK TO DESCRIPTION OF ROLES |
| Execution E | LINK TO DESCRIPTION OF ROLES | LINK TO DESCRIPTION OF ROLES | LINK TO DESCRIPTION OF ROLES | LINK TO DESCRIPTION OF ROLES | LINK TO DESCRIPTION OF ROLES | LINK TO DESCRIPTION OF ROLES | LINK TO DESCRIPTION OF ROLES | LINK TO DESCRIPTION OF ROLES |
| Data $D_a^e$ | LINK TO DESCRIPTION OF ROLES | LINK TO DESCRIPTION OF ROLES | LINK TO DESCRIPTION OF ROLES | LINK TO DESCRIPTION OF ROLES | LINK TO DESCRIPTION OF ROLES | LINK TO DESCRIPTION OF ROLES | LINK TO DESCRIPTION OF ROLES | LINK TO DESCRIPTION OF ROLES |

**Fig. 6. Data Acquisition Matrix.**

Two acronyms have been developed to describe the scope of the guidance provided (PROPOSED) and the data collection timeline (BIPED). The scope of the data collection is categorized as follows: Procedure; Response; Organization; Population; Objectives; Structure; Environment; and Data Acquisition. The timeline is categorized as follows: Blueprint (planning what to do); Investigation (establishing specifically how to do it); Preparation (configuring the data acquisition elements); Execution (collecting the data); and Data (manipulating the data). The permutations of these two sets of factors (PROPOSED and BIPED) combine to address the key components of data acquisition. The nature of the guidance provided is sensitive to the stage of the data collection process and the factor being addressed. For instance, early on in the process (during the planning stages), questions are provided to prompt the user on issues that should be considered; later in the process (where an event has been decided upon) checklists are provided to remind the user of issues that need to be addressed as they prepare and collect the data; in the final (preparatory and execution) stages, guidance is provided on the roles that need to be adopted for these stages to be completed, and the activities associated with these roles.

## *Data Template*

The *Data Template* describes the information required for each of the data-sets provided to the portal – the data-set itself, the data acquisition methods and the associated description of the background event conditions. The *Data Template* evolved in response to the review of numerous data-sets and through the lessons learned during data collection activities and various feedback exercises [1]. Given the range of data sources examined, the scope of the general *Data Template* is well beyond the requirements of any specific data-set. The template is designed to capture as broad a range of information as possible. Therefore, the template is as flexible as possible, allowing information to be provided in summary form, in detail and in a range of numerical/graphical/descriptive formats (the section headings within the template are shown in Table 1). The template will be the basis of a searchable database of data-sets, but also as a placeholder for the data-set and associated information that is required. It is intended that the comprehensive nature of the template will act as a resource, providing sufficient context and detail of the data, and a motivating force during the data collection process encouraging collectors to gather the range of information necessary – to act in conjunction with the *Data Acquisition Matrix*.

**Table 1. Key sections of the Data Template**

| Section | Components |
|---|---|
| A. Background Information | [Reference information \| Organizations involved \| Data of data collection \| Review material \| Purpose] |
| B. Summary Information | [Variables examined \| Key terminology used] |
| C. Procedure | [Nature of event \| Procedure employed \| Procedure preparation \| Technology used \| Human resources used] |
| D. Structure | [Spatial characteristics] |
| E. Population | [Population characteristics] |
| F. Environment | [Environment conditions] |
| G. Data Processing | [Data collection methods \| Data extraction methods \| Data analysis methods \| Description of data presented] |
| H. Event Timeline | [Narrative timeline describing event] |
| I. Results | [Report results \| Key quotations \| Conclusions\| Theory development] |

The *Data Template* and the *Data Acquisition Matrix* have been designed to complement each other: the guidance provided in the *Data Acquisition Matrix* will allow the data collector to populate the template more completely. Similar assumptions and terminology are used throughout to reduce ambiguity. The goal is therefore to allow new data to be uploaded in the most complete form and for existing data-sets to be interrogated and employed in the most unambiguous and refined manner possible. This should benefit the presentation of data that is currently available and the collection of data that will become available in future.

## Conclusion

The availability of credible, accurate and comprehensive human egress data is critical for responsible engineering. Currently, much of the data available is incompletely described, inconsistently represented or difficult to get access to, making the misuse of data more likely. A design for a *Data Portal* has been produced that should help existing data-sets be described more comprehensively (identifying both what is included and excluded in the data), and aid future data collection efforts. As such, the portal should aid in the development of more accurate theories and in more responsible engineering activities.

## References

1. Gwynne, S.M.V.: Conventions in the Collection and Use of Human Performance Data. NIST GCR 10-928. National Institute of Standards and Technology (2010)
2. Kuligowski, E.D.: The Process of Human Behavior in Fire. Proceedings of the Human Behaviour in Fire Symposium, pp. 627-632. Cambridge, UK (2009)
3. Sime, J.: Human Behaviour In Fires Summary Report. CFBAC Report No.450. Portsmouth Polytechnic (1992)
4. Canter, D.: Overview Of Human Behaviour. In: Canter, D. (ed.), Fires and Human Behaviour (2nd Edition), pp. 205-234. Fulton (1990)
5. Quarantelli, E.L.: Panic Behaviour: Some Empirical Observations. In: Conway, D.J., Human Response to Tall Buildings, Community Development Series, 34, ISBN 087933268. Dowden, Hutchinson & Ross, Stroudsburg (1997)
6. Latane, B., Darley, J.M.: Group Inhibition of Bystander Intervention in Emergencies. Journal of Personality and Social Psychology. Vol. 10, No. 3, p. 218 (1968)
7. Wood, P.G.: Survey Of Behaviour In Fires. In: Canter, D. (ed.), Fires and Human Behaviour (2nd Edition), pp. 205-234. Fulton (1990)
8. Bryan, J.L.: Smoke as a Determinant of Human Behavior in Fire Situations. Department of Fire Protection Engineering, University of Maryland, College Park, MD (1977)
9. Gwynne, S.M.V., Kuligowski, E.D.: Simulating a Building as a People Movement System. Journal of Fire Sciences. Vol. 27, July, Sage Publishing (2009). DOI: 10.1177/0734904109102387

# Data Collection (Pedestrian)

# New Data for Human Performance in Planar Corridors

**A. Winkens[1], W. Klingsch[1], and A. Seyfried[2]**

[1]Institute for Building Material Technology and Fire Safety Science, University of Wuppertal
42285 Wuppertal, Germany
[2]Jülich Supercomputing Centre, Forschungszentrum Jülich GmbH
52452 Jülich, Germany
Corresponding author: winkens@uni-wuppertal.de

**Abstract** Design of escape routes by human performance data depends on flow rates given by e.g. the maximum of the flow-density relation. Although there are many different studies about this relation, even so for the qualitative shape there is still no agreement. To enhance the empirical database we performed experiments with up to 107 test persons under laboratory conditions to study local densities, speeds and flow in planar corridors. Due to new methods of digital image processing our analysis is based on data (trajectories) of very high accuracy. In this contribution we present an extract of our current studies concerning speed, density and flow in planar corridors. We also show effects on results by different measurement procedures. We used different sizes of measurement areas to determine the influence of this parameter. In particular the density where speed reaches zero due to overcrowding is sensitive to this variation.

## Motivation

There are several published studies of the flow density fundamental diagram which exhibit many quantitative and qualitative differences [9]. Additionally a majority of studies were made in the last century and there are just few current analyses, for example the studies by Nelson [3], Weidmann [4] and Predtetchenskii & Milinskii [5]. One can already see the serious discrepancies comparing these three studies with respect to their diagrams of flow-density relation. Actual there are new approaches to explain these discrepancies, e.g. cultural differences (see [6]) or demographic trends (see [7]). Thus we decided to perform new experiments under well-known conditions. Our aim was to eliminate some of these differences with the help of these investigations.

In this contribution we will show influences of chosen measurement areas on resulting values of speed, density and flow. Due to different corridor-widths the measurement area had to be adjusted, too. So, this analysis is another approach to explain differences in several studies of pedestrian movement. Our study also

R.D. Peacock et al. (eds.), *Pedestrian and Evacuation Dynamics*,
DOI 10.1007/978-1-4419-9725-8_6, 

represents an enlargement of the fundamental database for macroscopic data of pedestrian movement.

## Experimental Setup

The experiments were performed in November 2006 with up to 250 German soldiers in the barracks 'Bergische Kaserne Düsseldorf'. All together we had two main setups:

- planar corridors,
- bottleneck.

In this contribution we focus on the planar corridors. The course was arranged with tables at the straight sections and chairs at the curves. In addition we placed some screens for visual protection.

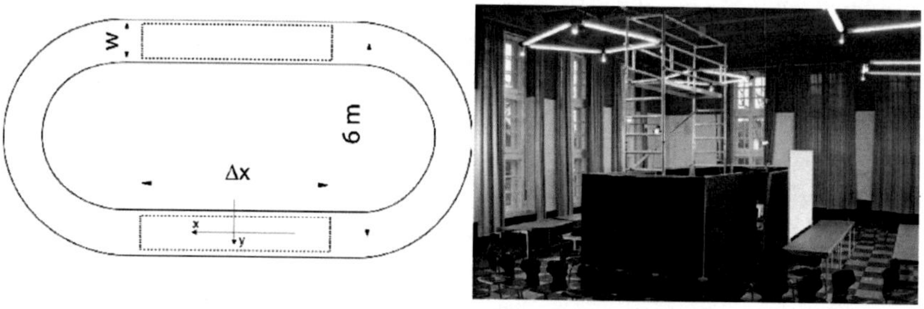


**Fig. 1. Experimental setup; left: sketch, right: picture**

The cameras for recording the experiments were placed under the ceiling at a height of 5.3 m. This height was needed to minimize the camera-caused distortions of pictures.

All together we performed 29 runs over three days with different corridor-widths and different numbers of people. We started at a width of w=0.7 m, where just single file movement was possible. The wider setups (w=0.85 m and w=1.0 m) also allowed movement of people side-by-side. To vary the density we changed the number of pedestrians inside the corridor from N=14 to N=107 subjects (see Table 1).

**Fig. 2. Snapshot of video analysis (including trajectories)**

Test persons were evenly distributed inside the setup and started to walk simultaneously after a start signal. They were instructed neither to run nor to push each other but to go at normal speed. A full overview of the whole series of experiments, we performed and a first analysis of these tests is given in [1].

**Table 1. Overview of performed experiments**

| Corridor-width $w$ (m) | Number of test persons $N$ (-) | Approx. global density $\rho_g(1/m^2)$ |
|---|---|---|
| 0.70 | 14, 17, 20, 22, 25, 28, 34, 39, 45, 56, 62, 70 | 0.5; 0.6; 0.7; 0.8; 1.0; 1.2; 1.4; 1.6; 2.0; 2.2; 2.5 |
| 0.85 | 20, 30, 40, 50, 55, 60, 65, 70, 75, 80 | 1.0; 1.5; 2.0; 2.5; 2.8; 3.0; 3.3; 3.5; 3.8; 4.0 |
| 1.00 | 24, 36, 48, 60, 72, 83, 107 | 1.0; 1.5; 2.0; 2.5; 3.0; 3.5; 4.5 |

## Methods of Measurements

We regulated the global density (see Table 1) by placing different numbers of test persons evenly inside the setup. For our measurements we focused on an area with length of $1.0 \text{ m} \leq \Delta x \leq 4.0 \text{ m}$ (see Fig. 1). Inside the center of this area (region of interest) we measured local density, local speeds and flow. Furthermore we wanted to study the influence of size and shape of the area. First we used an area with full corridor-width and a length of 2 m, so we had an adjusted area size for every width (areas $A_1$, $A_2$, $A_3$). We also considered a length of 4 m with full corridor-width (areas $A_7$, $A_8$, $A_9$). Then we fixed the area itself with three different floor spaces and adapted the length to the particular corridor-width (areas $A_{4,w}$, $A_{5,w}$, $A_{6,w}$). The detailed variations are listed below in Table 2.

**Table 2. Overview of measurement areas**

| | Area $A_j$ ($m^2$) | Width $w$ / length $\Delta x$ | Run ($w/N$) |
|---|---|---|---|
| $A_1$ | 1.4 | 0.7 / 2.0 | 0.7 / all |
| $A_2$ | 1.7 | 0.85 / 2.0 | 0.85 / all |
| $A_3$ | 2.0 | 1.0 / 2.0 | 1.0 / all |
| $A_{4,70}$ | 2.0 | 0.7 / 2.86 | 0.7 / all |
| $A_{4,85}$ | 2.0 | 0.85 / 2.35 | 0.85 / all |
| $A_{4,100}$ | 2.0 | 1.0 / 2.0 | 1.0 / all |
| $A_{5,70}$ | 1.5 | 0.7 / 2.14 | 0.7 / all |
| $A_{5,85}$ | 1.5 | 0.85 / 1.76 | 0.85 / all |
| $A_{5,100}$ | 1.5 | 1.0 / 1.5 | 1.0 / all |
| $A_{6,70}$ | 1.0 | 0.7 / 1.43 | 0.7 / all |
| $A_{6,85}$ | 1.0 | 0.85 / 1.18 | 0.85 / all |
| $A_{6,100}$ | 1.0 | 1.0 / 1.0 | 1.0 / all |
| $A_7$ | 2.8 | 0.7 / 4.0 | 0.7 / all |
| $A_8$ | 3.4 | 0.85 / 4.0 | 0.85 / all |
| $A_9$ | 4.0 | 1.0 / 4.0 | 1.0 / all |

The experiments were recorded with industrial video cameras, following a special image analysis with a software tool developed at Forschungszentrum Jülich. The resulting trajectories were used to study the movement parameters. For more information concerning the image analysis we refer to [2].

Several approaches to analyse the trajectories are used. Due to the varying dimensions of measurement areas there are resulting different values for the local density and speed. Density is measured by using the basic definition of density:

$$\langle \rho \rangle_{\Delta x} = \frac{N'}{b\Delta x}$$

where $N'$ stands for the number of pedestrians inside the particular floor space $b\Delta x$, which gives the size of the measurement area $A_j$.

Speed $v_i$ is measured for an interval $\Delta t$ of 0.5 s (= 13 frames). The corresponding speed $\langle v \rangle$ was calculated as follows:

$$\langle v \rangle_{\Delta x} = \frac{1}{N'} \sum_{i=1}^{N'} v_i \,.$$

So $\langle v \rangle$ is the mean value over all pedestrians $N'$ inside the measurement area $b\,\Delta x$. Flow was measured by counting the number of people $N$ crossing a line in a given time interval $\Delta t$:

$$J = \frac{N}{\Delta t}$$

Furthermore we calculated the flow using the hydrodynamic flow equation

$$J = \overline{\rho} \times \overline{v} \times b$$

respectively the specific flow

$$J_S = \overline{\rho} \times \overline{v}$$

where $\overline{\rho}$ and $\overline{v}$ are mean values of density $\langle \rho \rangle$ and speed $\langle v \rangle$ over time and $b$ is the actual corridor width. For a detailed analysis we considered a steady state of each run with length of 10 s (= 250 frames). Over this period we average our mean values of $\langle \rho \rangle$ and $\langle v \rangle$ over time for calculating flow. This was done due to large fluctuations we observed (see Fig. 6). The steady states were chosen manually by visual analysis of time series for the measurements of density and speed. So, $\overline{\rho}$ and $\overline{v}$ are calculated as follows:

$$\overline{\rho} = \frac{1}{250} \sum_{i=1}^{250} \langle \rho \rangle_i \qquad \overline{v} = \frac{1}{250} \sum_{i=1}^{250} \langle v \rangle_i$$

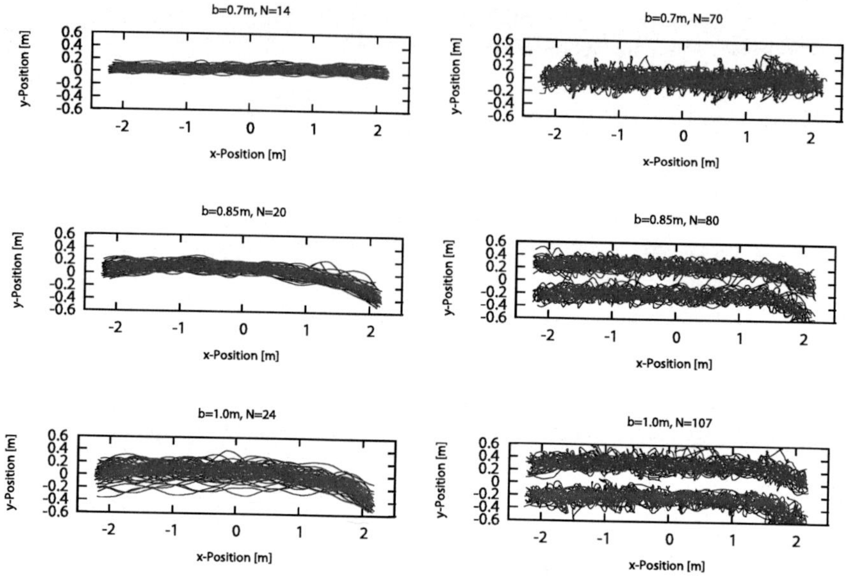


**Fig. 3. Examples of different trajectories (comparison of $N_{min}$ and $N_{max}$ for different widths)**

## Results

First it is obvious that trajectories for several runs show lane formation with increasing corridor width (see Fig. 3).

### *Measurement of flow*

By comparing the diagrams for *N(t)* of the different widths, where the flow is given by the shape of the curves, one can see that with increasing number of persons there appear more periods of stagnated movement.

Maximum flow for all tested widths occurs between values of $N = 20$ and $N = 25$. For higher values of N flow decreases again (see Fig. 4). Noticeable here is that the flows for $N = 20$, $N = 60$ and $N = 80$ are very similar at width b = 0.85 m as well as the flows for $N = 24$ and $N = 60$ at width b = 1.0 m.

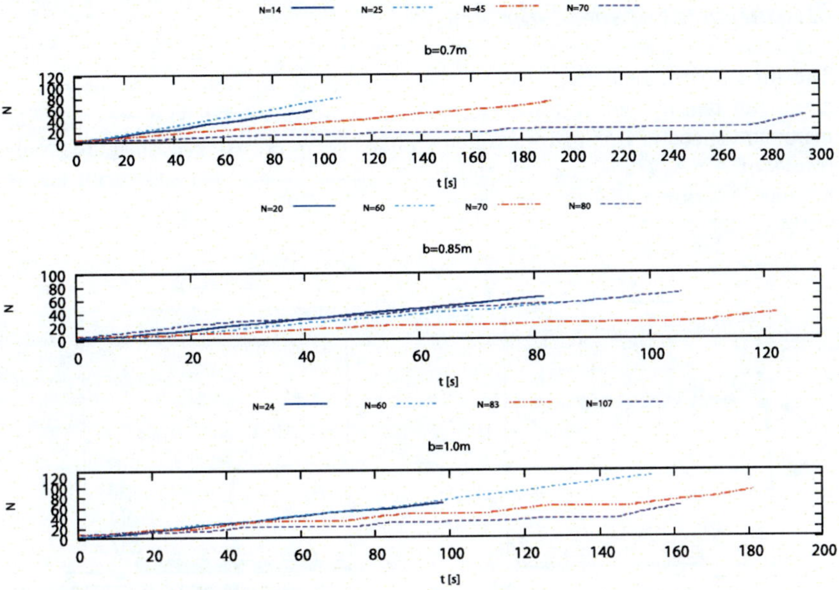


**Fig. 4. Number of people *N* crossing the center of observation field for different widths *w***

The simple flow calculation at the centre (x=0) of the particular observation area, using data from the whole run, shows values between 0.2 $s^{-1}$ and 0.8 $s^{-1}$ (right graph in Fig. 5). The results for measurement of flow using the flow equation with measured mean values of speed and density for areas $A_{6,70}$ ($rho_{ls}$), $A_{4,85}$ ($rho_l$) and $A_9$ ($rho_4$) are also shown below in Figure 5 (left graph). These three examples for different areas stand for the smallest, the biggest and a middle floor space.

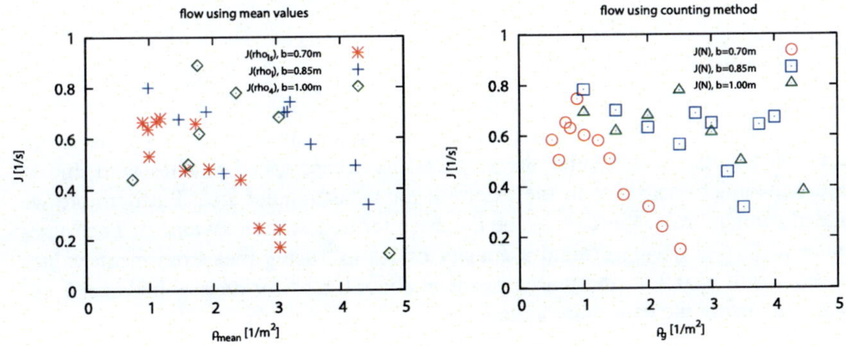


**Fig. 5. Flow J using mean values for speed $\overline{v}$ and density $\overline{\rho}$ (left) and estimated global density $\rho_g$ (right), both for different widths w**

## *Measurement of speed and density*

The time series show large fluctuations for density as well as for speed. The maximum density during all experiments is 8 m$^{-2}$ measured with area $A_9$. The maximum speed is 1.56 m/s measured with areas $A_1$, $A_4$, $A_5$, $A_7$. In figure 6 only results of $A_1$, $A_3$ ($rho_2$, $speed_{n2}$) and $A_7$, $A_9$ ($rho_4$, $speed_{n4}$) are shown for clarity reasons. Maximum density appeared in run 100/107, maximum speed in run 70/14.

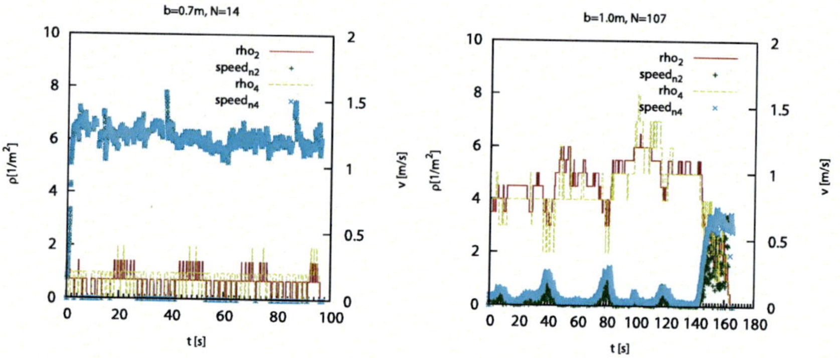


**Fig. 6. Time-series for ⟨ρ⟩ and ⟨v⟩ at run 70/14 (left) and run 100/107 (right)**

In Figure 7 the resulting speed-density relation is shown. For these mean values $\bar{v}$ and $\bar{\rho}$ we used the above named steady states of 10 s. The comparison of the left and right figure shows that larger measurement areas lead to higher densities where the speed reaches zero. Moreover it is noticeable that there are significant higher fluctuations for speed at lower densities but higher fluctuations for density at higher densities.

## Conclusion and Outlook

In our study we showed that the shape of measurement area has a serious influence on results. In particular the stop density where speed reaches zero due to overcrowding is sensitive to the extent of the measurement area. Furthermore we added some more actual data to the database for pedestrian movement. The maximum of flow $J$ occurs around a density of 1.6 m$^{-2}$ using measured mean values with a result of 0.9 s$^{-1}$ and with a result of about 0.8 s$^{-1}$ for density between 1 m$^{-2}$ and 2 m$^{-2}$ using the counting method.

A critical density where movement stops generally could not be found. In a solitary case we measured a density of 8 $m^{-2}$, where congestion shortly appears but we also observed congestion at runs with lower maximum densities, e.g. 5 $m^{-2}$. But this is also dependent on the measurement area, i.e. the measured density of 8 $m^{-2}$ was determined with area $A_9$ in run 100/107. For the same run the measured maximum density in $A_3$ amounts to 6.5 $m^{-2}$ as shown in figure 6.

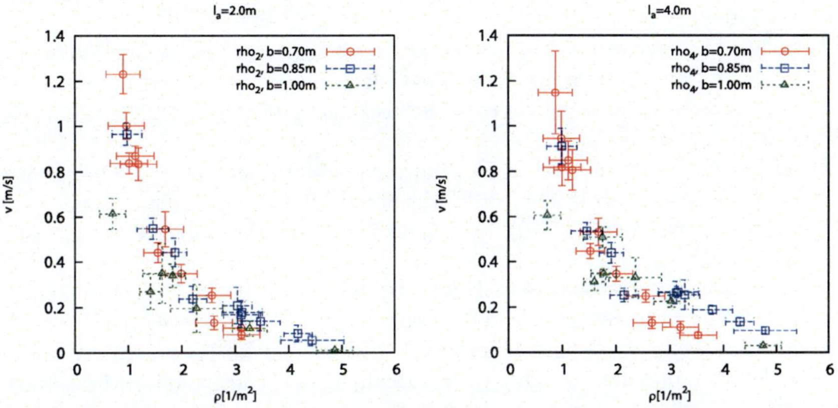


**Fig. 7. Speed-density relation obtained in areas $A_1$, $A_2$, $A_3$ (left) and $A_7$, $A_8$, $A_9$ (right)**

So it is very important for future studies to consider the measurement method. It would be helpful to find a standard method, which is generally accepted. But also when studying older research reports detailed information on how results are obtained is required to come to correct conclusions.

In our actual research project HERMES [8], which is funded by the German Federal Ministry of Education and Research, we will continue with these studies to get more into details. Furthermore the here presented results of this study have to be discussed more in detail, e.g. the detection of stop-and-go waves inside these closed-system experiments or methods of measurement for $\langle\rho\rangle$ and $\langle v\rangle$ as introduced in [1].

**Acknowledgements** The execution of experiments is supported by the German Research Foundation (Deutsche Forschungsgemeinschaft - DFG) through Grants No. KL 1873/1-1 and SE 1789/1-1. It is also supported by the German Government through their High-tech Strategy within the "Research for Civil Security - Protecting and Saving Human Life" programme of the Federal Ministry of Education and Research (BMBF). This support is gratefully acknowledged.

## References


1. Seyfried, A.; Boltes, M.; Kähler, J.; Klingsch, W.; Portz, A.; Rupprecht, T.; Schadschneider, A.; Steffen, B.; Winkens, A.: Enhanced empirical data for the fundamental diagram and the flow through bottlenecks. In: W. Klingsch & C. Rogsch & A. Schadschneider & M. Schreckenberg (Eds.), PED2008, **23**, 145-156 (2010)
2. Boltes, M.; Seyfried, A.; Steffen, B.; Schadschneider, A.: Automatic Extraction of Pedestrian Trajectories from Video Recordings. In: W. Klingsch & C. Rogsch & A. Schadschneider & M. Schreckenberg (Eds.), PED2008, **23**, 43-54 (2010)
3. Nelson, H. E. and Mowrer, F. W.: Emergency Movement. In: P. J. DiNenno & W. D. Walton (Eds.), The SFPE Handbook of Fire Protection Engineering (Third ed., pp. 3-367-3-380). Society of Fire Protection Engineers, Bethesda, MD (2002)
4. Weidmann, U.: Transporttechnik der Fussgänger; Institut für Verkehrsplanung, Transporttechnik, Strassen- und Eisenbahnbau; ETH Zürich; (1993)
5. Predtechenskii, V. M. and Milinskii, A. I.: Planning for Foot Traffic Flow in Buildings (Amerind Publishing, New Dehli, 1978). Translation of: Proekttirovanie Zhdanii s Uchetom Organizatsii Dvizheniya Lyuddskikh Potokov (Stroiizdat Publishers, Moscow, 1969).
6. Chattaraj, U.; Seyfried, A.; Chakroborty, P.: Comparison of pedestrian fundamental diagram across cultures. In: Advances in Complex Systems, Vol. 12, No. 3, 393-405. World Scientific Publishing Company (2009)
7. Pauls, J.: Demographic changes leading to deterioration of pedestrian capabilities affecting falls safety and crowd movement performance including facility evacuation. Submitted for Transportation Research Board Annual Meeting, Washington, DC (2008)
8. HERMES: Investigation of an evacuation assistant for use in emergencies during large-scale public events; http://www.fz-juelich.de/jsc/appliedmath/ped/projects/hermes
9. Schadschneider, A.; Klingsch, W.; Kluepfel, H.; Kretz, T.; Rogsch, C.; Seyfried, A.: Evacuation Dynamics: Empirical Results, Modelling and Applications. In: Meyers, R.A. (Ed.), Encyclopaedia of Complexity and System Science (3, 3142-3176). Springer (2009)

# Influence of Geometry Parameters on Pedestrian Flow through Bottleneck

**T. Rupprecht[1], W. Klingsch[1], and A. Seyfried[2]**

[1]Institute for Building Material Technology and Fire Safety Science, University of Wuppertal, Germany
[2]Jülich Supercomputing Centre, Forschungszentrum Jülich GmbH, 52425 Jülich, Germany
Corresponding author: rupprecht@uni-wuppertal.de

**Abstract** In pedestrian evacuations bottlenecks can be a crucial factor influencing the evacuation time. The main question involves the design of bottlenecks to enable unhindered pedestrian flow in order to optimize evacuation times. For better understanding of this problem, a set of experiments with pedestrians in different bottleneck-scenarios has been performed. The results enlarge the database and allow the testing of the basic assumptions of performance based egress design.

Within experimental series of evacuation trials the main focus is on two subjects of pedestrians flow dynamics: The fundamental diagram and bottleneck. In the following we present the bottleneck trials. The pedestrian experiments have been performed in well-controlled laboratory conditions. Special attention was given to the comparability of different experimental series because both variations of width and length of the bottleneck have an influence. Formation of lanes inside the bottleneck becomes less pronounced for wide bottlenecks, even for narrow bottlenecks the numbers of lanes don't interfere with the linear relations between flow and bottleneck width. Based on the length of the bottleneck the behavior of pedestrians is different, going through a tunnel or a door. Short bottlenecks show strong non-stationarity.

## Experimental Arrangement and Performance

To run this kind of experiment under well-controlled laboratory conditions, we used up to 230 soldiers. There were two main setups: planar corridors and bottlenecks. At PED 2008 in Wuppertal an overall outlook on these research series was shown [1]. Bottleneck effects are discussed now in detail in this contribution.

The setup of the bottleneck was realized by wooden frameworks, which were covered by non-transparent PVC-film. The tiles, which can be seen on the floor (Figure 1), have a size of 0.50 x 0.50 meters.

R.D. Peacock et al. (eds.), *Pedestrian and Evacuation Dynamics*,
DOI 10.1007/978-1-4419-9725-8_7, © Springer Science+Business Media, LLC 2011

**Fig. 1.** **Video capture as bird view**

Video equipment situated above the scene was used to record the experiments. Data processing was performed with PeTrack [2] yielding high precision trajectories. For comparability conditions like the start-up density of pedestrians, are held as constant as possible. Variation was restricted to one boundary condition per experiment. 24 experiments have been executed in total as well. Length and width of the bottleneck have been altered, as have different geometries of waiting area and distances of pedestrian starting positions. Variations are listed in Table 1.

**Table 1.** **Variations of experiments (standard size in bold)**

| Setup Condition | Variation |
|---|---|
| Bottleneck width w | 0.90, 1.00, 1.10, 1.20, 1.40, 1.60, 1.80, 2.00, 2.20, 2.50 m |
| Bottleneck length l | 0.06, 2.00, **4.00** m |
| Width of waiting area $w_c$ | 4.00, 5.00, 6.00, **7.00** m |
| Distance between waiting area and bottleneck d | 1.00, 2.00, 3.00, **4.00** m |
| Number of Pedestrians N | 50, 100, 150, 200, 230 |

**Fig. 2.** **Picture of a typical experimental run and the experimental setup**

A lot of experimental based flow values are published; however they are influenced by uncontrolled effects. The goal of our test series was to analyze as accurately as possible how the flow through bottlenecks is influenced by geometry parameters like width $w$ or length $l$. This will give answers to some basic questions, like whether the flow through the bottleneck increases linearly with the width as given by the equation $J = J_s * w$. Here $J$ is the flow of pedestrians, $J_s$ the specific flow and $w$ the width.

To ensure reliable results and to get general valid relations some special quality criteria for the experimental setup and data analysis are necessary. For instance it is important that test persons are as equal as possible, both mentally and physically, for the runs compared. By choosing soldiers for test persons and arranging sufficient breaks and feedback information provided a constant pool of test persons. Furthermore it is of crucial importance that the resulting values are independent of the starting conditions given by the setup of the experiment. A strong indication that the system is no longer influenced by the initial condition is the occurrence of stationary states. To analyze whether non-stationary states influence the results we choose different samples to calculate the flow and tested whether trends in the resulting relations vary.

## Trajectories and Lane Formation

In the following the trajectories are the subject of a more detailed analysis. Considering variation of the bottleneck width, the generation of paths by the trajectories can be recognized. For widths up to 1.10 m there is an accumulation of two paths on the right and left side next to the boundary. The persons walk arranged in a zipper-system through the bottleneck, already shown by Hoogedoorn [3]. If people were running close behind one another, the zipper-effect is disordered, small gaps are formed.

From a width of 1.00 m some single people start to walk in the middle of the bottleneck. From a width of 1.20 m a complete third path is formed. The development of the paths disappears from a width of 1.80 m. Just on the edges a forming of paths can be recognized. There arises the question of whether the formation of lane formation is caused by restrictions of potential paths near boundaries.

In front of the boundaries the cumulative trajectories show a funnel like shape of the space used. Also some passing lanes are visible. They pass the funnel and directly file at the entrance again. The probability to find a person at position x inside the bottleneck ($y \in [0$ m, $+ 4.00$ m$]$) is shown (Figure 3, right column). The minimal distance is 0.20 m for $w = 1.10$ m and increases to 0.30 m at a width of 2.00 m. This shows that in narrow bottlenecks the space is better utilized and the effective width is function of $w$.

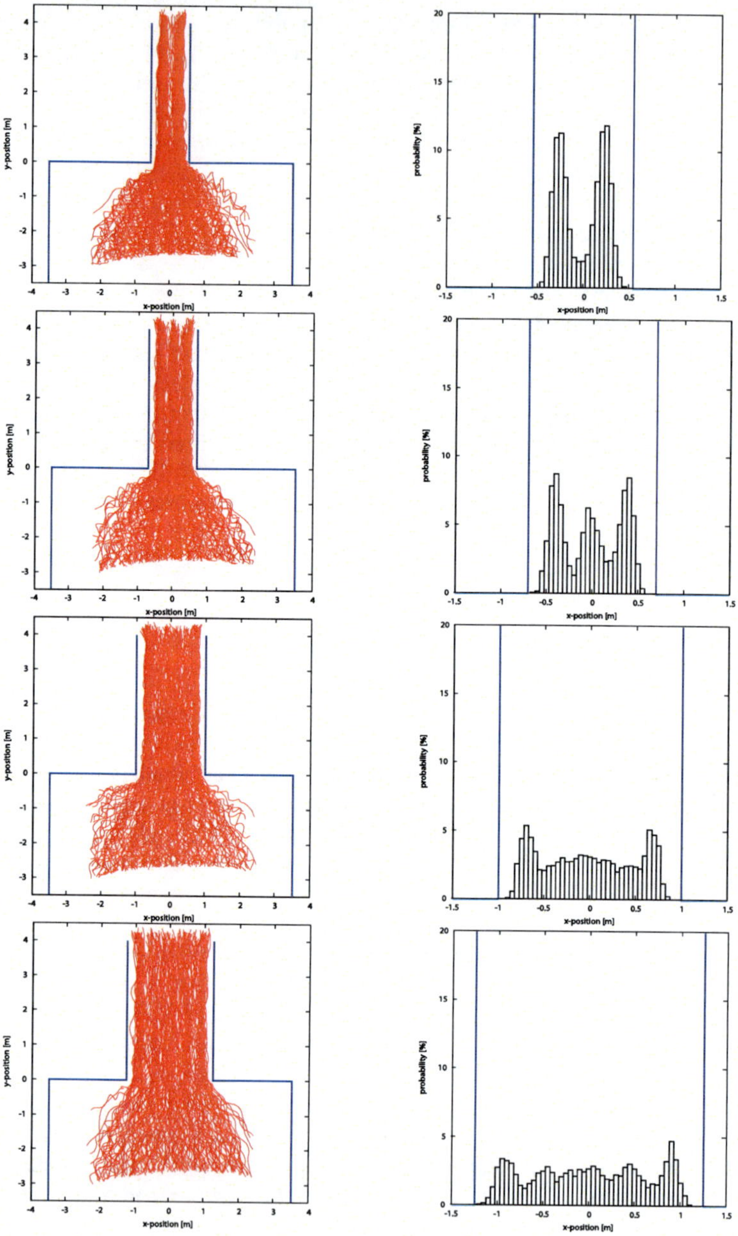


**Fig. 3.** **Trajectories (left), variation of bottleneck width *w*; 1.10, 1.40, 2.00, 2.50 m. Probability of finding a pedestrian at position *x* inside the bottleneck (right).**

Considering the variation of the length of bottleneck the people stream widens directly after the bottleneck for the short bottleneck ($l$ = 0.06 m). Here one must bear in mind, that the test persons had experience of the experimental setup after the first running. Trajectories at bottlenecks of length 2.00 m and 4.00 m show no significant differences (Figure 4).

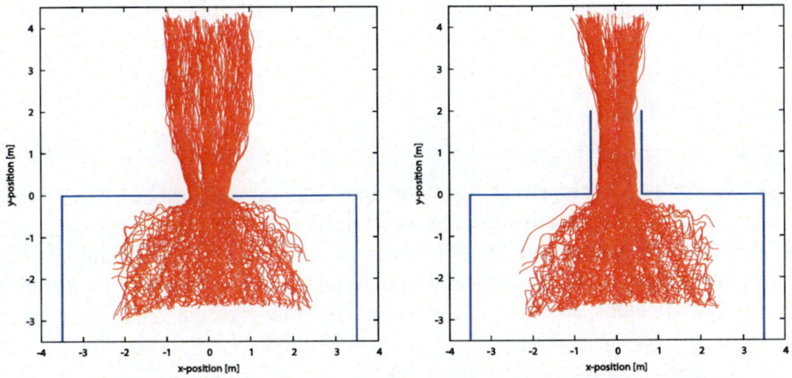


**Fig. 4. Trajectories, variation of bottleneck length; $l$ = 0.06 and 2.00 m**

Tests with a different number of persons show no significant differences in lane formation (Figure 5). But in all tests lane formation starts immediately along boundaries after entrance into the bottleneck.

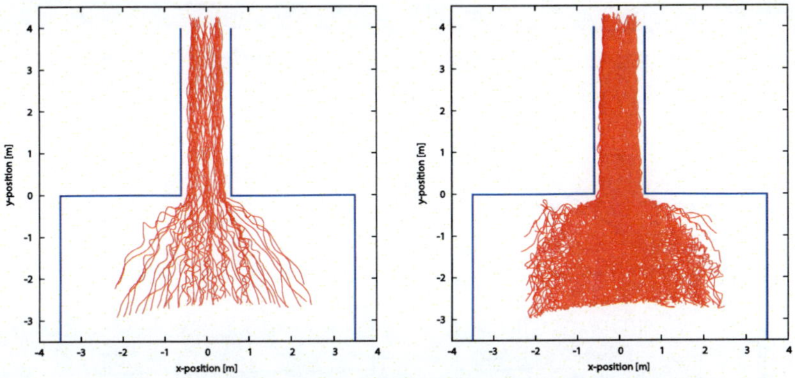


**Fig. 5. Trajectories, variation of pedestrian; $n$ = 50 and 230 people**

In another experimental series, the width $w_c$ of the corridor in front of the bottleneck has been reduced with the experiment for a waiting area of $w_c$ = 4.00 m a reduced angle of the funnel could be observed with again some passing lanes outside of the funnel.

Test results can be summarized qualitatively as follows: In front of the bottleneck a funnel is formed, with only few paths outside the funnel existing. Inside the bottleneck lanes are pronounced at the boundaries, with two paths forming for widths up to 1.10 m. From about 1.20 m a third path is created. For larger width the forming of lanes in between the exterior lanes diminishes. Moreover, lane formation starts already in front of the bottleneck (Figure 5, $N = 230$). This indicates that formation of lanes is a boundary effect.

## Pedestrian Flow in the Bottleneck

For determination of pedestrian flow the time interval $\Delta t$ has been determined, when a group of $\Delta N$ persons enter the bottleneck and when they have crossed a line, located at the entrance to the bottleneck (x = 0). To analyze the influence of non stationary states three different $\Delta N$ have to be taken into account: 1 – 120 persons, 41 – 120 persons, and 81 – 120 persons. The flow is calculated as follows:

$$J = \frac{\Delta N}{\Delta t}$$

Experiments, which do not include more than 120 people, were not included in the analysis[1]. The following persons-time diagrams are shown here as examples. Looking at these graphs, some improvements are shown in the pedestrian flow (Figure 6).

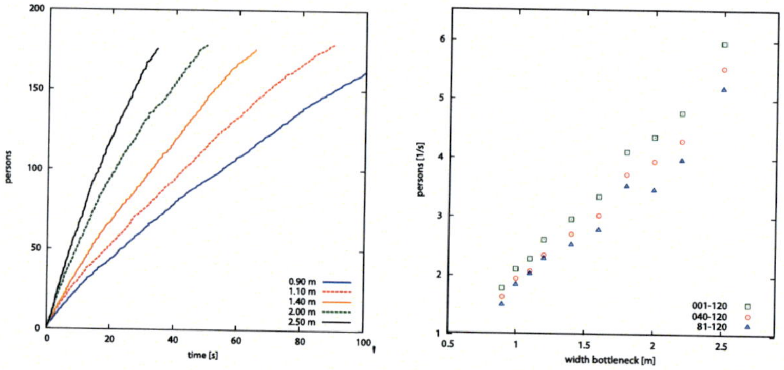


**Fig. 6. Variation of bottleneck width *w*, n-t diagram (left); flow-width diagram (right)**

[1] With one exception: At the experimental series with $w_c = 4.00$ m only 106 people participated. For comparison intervals of $\Delta N = 1$-100 persons, $\Delta N = 40$-100 persons and $\Delta N = 81$-100 persons are assumed.

As expected, pedestrian flow increases continuously with an increasing bottleneck width, with a specific flow of about 2 persons / (sec*m). In the first seconds of each run a slightly higher flow was measured for all widths. This also explains the slightly lower flow values for the interval of 81 – 120 persons in comparison to the interval of 1 – 120 persons.

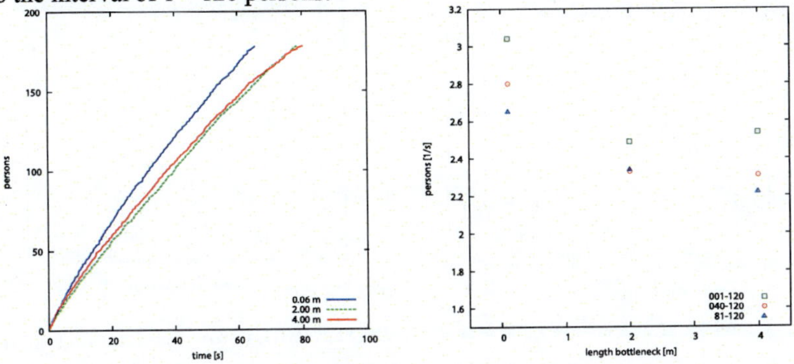


**Fig. 7. Variation of bottleneck length *l*; n-t-diagram (left), flow-width-diagram (right)**

An interesting phenomenon could be observed in the variation of the length of bottleneck. Pedestrian flows at a length *l* of 2.00 m and also 4.00 m are roughly on the same level; the pedestrian flow for the length of bottleneck of 0.06 m is significantly higher (0.06 m value is given by the thickness of the wood frame construction.). It must be considered that the subjects knew the experimental setup and that the length of bottleneck was visible. By screening of videos made in lateral view it could be recognized easily, that some of the subjects rotate immediately after bottleneck and withdraw from the main stream in a lateral direction. However it seems to be a regular behavior.

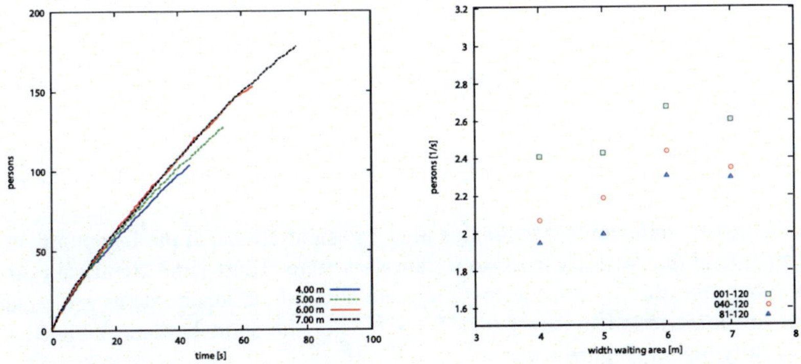


**Fig. 8. Variation of waiting area width $w_c$; n-t-diagram (left), flow-width-diagram (right)**

With variation of distance between waiting area and bottleneck only a slight increase of pedestrian flow is seen for increasing distances. This can be understood as a sufficient relation between *N, d, l* to avoid irregularities at beginning of evacuation.

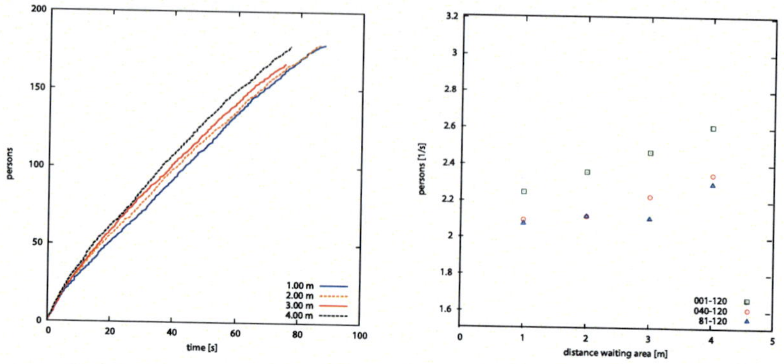


**Fig. 9. Variation of waiting area distance *d*; n-t-diagram (left), flow-width-diagram (right)**

Similar results were found for waiting area dimensions in front of the bottleneck: With increasing distance only small increases of pedestrian flow were seen (Figure 9).

## Local velocities in the bottleneck

Another important parameter in pedestrian dynamics is speed of individual persons. For comparability of the results speed is measured at the same place for each pedestrian, at the entrance of bottleneck. Speed is measured in principal direction, defined by

$$V_x = \frac{X_{(t_2)} - X_{(t_1)}}{t_2 - t_1}$$

with $t_2$-$t_1$ = 0.5 s. Lateral movement, caused by the movement of the head for example, is not considered here.

All diagrams indicate that velocities after transient effects at the beginning are all similar, and that velocity tends to a stationary state. There is no significant difference between the different tests. The velocities of stationary states are equal except for bottleneck lengths of (nearly) zero. For the short bottleneck speed is significantly higher (Figure 10).

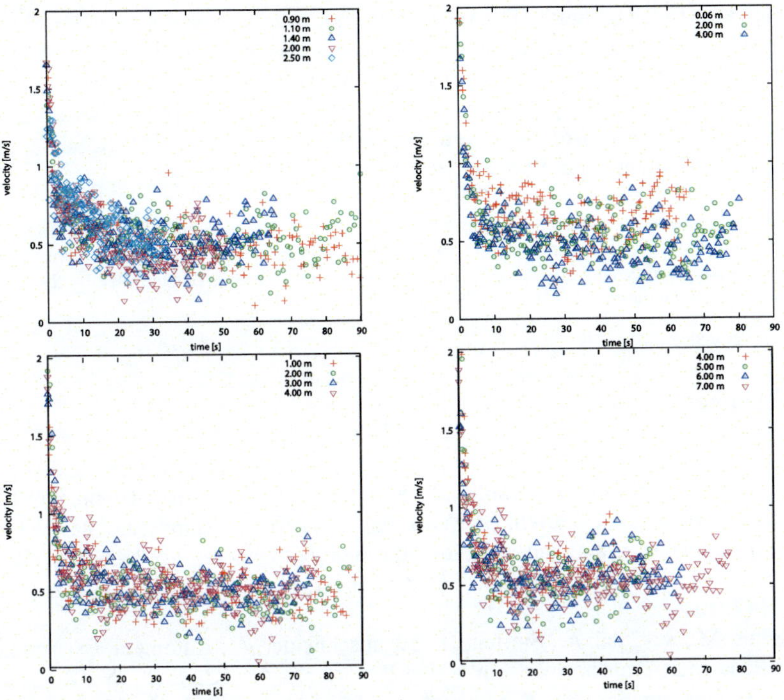


**Fig. 10. Pedestrian speed; variation of bottleneck width *w* (top left), length *l* (top right), distance *d* (bottom left) and width $w_c$ (bottom right)**

## Summary and Outlook

The experiments have been planned to investigate as precisely as possible the influence of defined parameters. As subjects we used male soldiers in the age of 20 to 30 years with a healthy constitution. This may influence the results. Therefore the applicability of these results to entirely different individuals or situations may be restricted.

The study offers some interesting results: There are strong indications that the formation of lanes inside the bottleneck is caused by movement restrictions at the boundaries. With increasing bottleneck width the flow increases linearly supporting the specific flow concept. However the value for the specific flow is influenced by non-stationary states and decreases if stationary states are considered only. Short bottlenecks, like doors, show a higher flow than long bottlenecks. But the difference depends again on non-stationary states. No clear trend is found for different widths of the waiting areas. The same is true of different distances from the waiting area to the entrance. The speed at the entrance shows transient effects but is same for all runs except the run with a short bottleneck length.

In general the examination of our experiments showed strong influences of transient or non-stationary states, which are not considered in performance based egress design, like Nelson [4] or Predtechenskii and Milinskii [5]. Influence of non-stationary states should be studied further.

**Acknowledgments** This study was supported by the German Government's high-tech strategy, the Federal Ministry of Education and Research (BMBF). Program on "Research for Civil Security - Protecting and Saving Human Life". Execution of experiments was supported by the German Research Foundation (DFG) KL 1873/1-1 and SE 1789/1-1. [6]

## References

1. Seyfried, A., Boltes, M., Kähler, J., Klingsch, W., Portz, A., Rupprecht, T., Schadschneider, A., Steffen, B., Winkens, A.: Enhanced empirical data for the fundamental diagram and the flow through bottlenecks. In: Klingsch, W., Rogsch, C., Schadschneider, A., Schreckenberg, M. (Eds.), PED2008, 23, 317-323 (2010)
2. Boltes, M., Seyfried, A., Steffen, B., Schadschneider, A.: Automatic Extraction of Pedestrian Trajectories from Video Recordings. In: Klingsch, W., Rogsch, C., Schadschneider, A., Schreckenberg, M. (Eds.), PED2008, 23, 317-323 (2010)
3. Hoogendoorn, S.P., Daamen, W.: Pedestrian Bahavior at Bottlenecks. Transportation Science. 39, 147-159 (2005)
4. Nelson, H.E., Mowrer, F.W.: Emergency Movement. In: DiNenno, P.J., Walton, W.D. (eds.), The SFPE Handbook of Fire Protection Engineering (Third ed.), pp. 3-367-3-380. Society of Fire Protection Engineers, Bethesda (2002)
5. Predtechenskii, V.M., Milinskii, A.I.: Planning for Foot Traffic Flow in Buildings. Amerind Publishing. New Dehli (1978) Translation of: Proekttirovanie Zhdanii s Uchetom Organizatsii Dvizheniya Lyuddskikh Potokov. Stroiizdat Publishers, Moscow (1969)
6. HERMES: Investigation of an evacuation assistant for use in emergencies during large-scale public events. http://www.fz-juelich.de/jsc/appliedmath/ped/projects/hermes

# Real-Time Video Analysis of Pedestrians to Support Agent Simulation of People Behavior

**O. Junker[1], V. Strauss[1], R. Majer[2], and N. Link[2]**

[1]Airport Research Center GmbH
52066 Aachen, Germany
olaf.junker@arc-aachen.de
verena.strauss@arc-aachen.de

[2]Vitracom AG
76133 Karlsruhe, Germany
majer@vitracom.de
norbert.link@vitracom.de

**Abstract** Online simulation coupled with real-time measurements of pedestrians in public buildings is a novel application which can be used to increase the security and safety of pedestrians within those buildings. To receive realistic forecasts it is necessary to update the simulation constantly against reality. The real-time video analysis can thereby support the simulation with the necessary data. This proposed system is largely supported by the state observer of the control theory.

Within a laboratory one room model pedestrian behavior is monitored and analyzed by video cameras. Evolving data is subsequently provided for the simulation where flow rates are additionally recorded. Thus, two resulting passenger flow rates are available which can be compared by the simulation controller. The actuators, receiving information from the controller, can afterwards trigger an appropriate action. Two different actuators have been realized in the model in order to bring the simulation passenger flow closer to the observed passenger rates: the velocity controller adjusts the walking speed of the passengers and the flow generator actuator has the ability to match the passenger generation rate.

Results show that the simulation passenger flow curve converges to the real passenger flow. As expected, the simulation curve follows the real passenger rate with a certain delay. Nevertheless, the simulation model appears to reflect the behavior of the persons in an appropriate way. Further investigations will show which additional instruments can be used to refine the simulation actor behavior.

## Introduction

People moving along a restricted spatial area represent a complex dynamic system due to their different objectives, capabilities and strategies, their mutual interactions and the geometric motion constraints.

Agent and flow based numerical simulation of the people motion system behavior is commonly used to plan traffic facilities. Also emergency and evacuation scenarios are simulated in order to derive pro-active measures to prevent critical

R.D. Peacock et al. (eds.), *Pedestrian and Evacuation Dynamics*,
DOI 10.1007/978-1-4419-9725-8_8, 

situations due to external impacts or internal fluctuations. The pre-requisite of such pro-activity is the detection of the evolvement of a dangerous situation in advance to trigger counter measures in time.

Another issue is the capturing of the traffic situation of large pedestrian traffic areas such as airports, train stations and other pedestrian traffic hubs or event locations. In these cases it is not possible to economically monitor the behavior of the people on the whole area with sensors. Measurements alone do also not allow to predict the evolution of the situation. On the other side, a simulation cannot provide any real-time information on the situation and the situation trend. This would not even be possible if the simulation gets the base information about the parameter configuration at the starting time: since the reality changes permanently simulation and reality would most probably drift apart very soon if the simulation does not continuously receive updated parameter data.

To address those problems we propose to combine the (fast-time) simulation of a pedestrian traffic model of the whole considered area with local sensor measurement at certain sub-areas. To avoid that simulation and reality drift apart a synchronization system will be adapted which can be seen as a control loop. Therein the simulation will constantly be calibrated and adjusted against the sensorial perception of reality. We expect to recover a complete and sufficiently precise real-time image of the current situation, which can be used as a starting point for situation prediction and the assessment of traffic control measures.

## Solution concept

On-line simulation integrated with real-time measurements of pedestrian traffic in public buildings like train stations or shopping centers is a novel approach which can be used to increase the security and safety of pedestrians within those buildings. Usually a model of the building structures and a course model of the people behavior are available, reflecting the prior knowledge about the situation. The model is capable to represent the microscopic state (cross product of the individual states of the people) of the system. Measurement could in principal capture the microscopic state in real-time. As mentioned before, the complete coverage of large areas is not feasible in most cases. Therefore we restrict ourselves to situations, where the people flow can be measured at only some dedicated points or areas. These values form the so-called observables of the system, which are only a few compared to the complete number of state variables. The problem of calculating the microscopic state from the observables is therefore underdetermined.

A solution consists in relating the microscopic state with the observables and adjusting the simulation parameters to produce the measured observables values.

## *General Approach*

The idea of using measurements to calibrate simulation has been followed by different research groups. In [8] measurement data is acquired off-line and used to calibrate the simulation system once before employment. In [5], [6] and [7] mesoscopic simulation systems are presented, which use online measurements to set the initial values of a simulation system or to set the actual start values of an online simulation system.

In this work, microscopic simulation and on-line measurement data are combined into an overall situation capturing system where the simulation parameters are permanently adjusted to optimally reflect the currently measured figures.

The capturing of the present microscopic state of a dynamical system by means of measuring only a few observables is addressed by so-called "observer models" in the control theory of linear systems. The observer consists of a virtual system, which runs parallel to the real system. As shown in figure 1, the real system is assumed to be linear and to obey the same dynamical state space model as the virtual system. This observer is named after its inventor Luenberger observer [1].

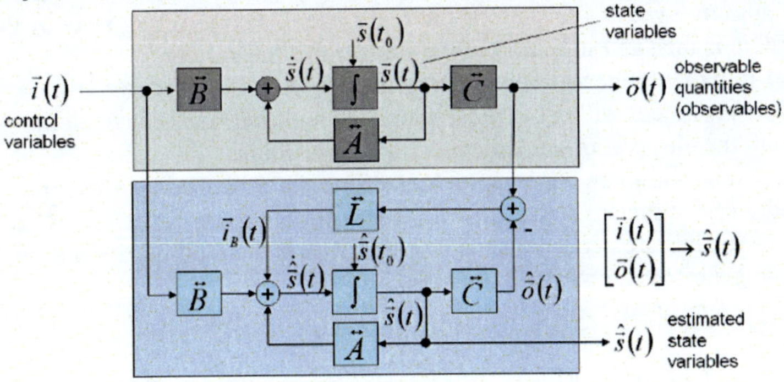


**Fig. 1. Linear observer model for the estimation of the system state from observable values**

This model is characterized by system matrix A, control matrix B and measurement matrix C. The system state is described by the state variables vector s. The evolution of s in time is determined by the system matrix (in the absence of external influence) and by external forces exerted by the control variables via the control matrix B. The measurement matrix C transforms the state variables into the observable quantities o, which can be measured with appropriate sensors.
The dynamical behavior of the system represented in Fig. 1 follows the following equations.

$$\dot{\vec{s}}(t)=\vec{\vec{A}}\cdot\vec{s}(t)+\vec{\vec{B}}\cdot\vec{i}(t);\quad \vec{o}(t)=\vec{\vec{C}}\cdot\vec{s}(t)$$

$$\dot{\hat{\vec{s}}}(t)=\vec{\vec{A}}\cdot\hat{\vec{s}}(t)+\vec{\vec{B}}\cdot\vec{i}(t)+\vec{i}_B(t);\quad \hat{\vec{o}}(t)=\vec{\vec{C}}\cdot\hat{\vec{s}}(t);\quad \vec{i}_B(t)=\vec{\vec{L}}\cdot\left(\vec{o}(t)-\hat{\vec{o}}(t)\right)$$

These equations can be summarized in one dynamical equation for the estimated state depending on the known input quantities and the observables by

$$\dot{\hat{\vec{s}}}(t)=\left(\vec{\vec{A}}-\vec{\vec{L}}\cdot\vec{\vec{C}}\right)\cdot\hat{\vec{s}}(t)+\vec{\vec{B}}\cdot\vec{i}(t)+\vec{\vec{L}}\cdot\vec{o}(t).$$

Matrix L must be chosen in such a way that the state estimation error governed by $\dot{\vec{s}}(t)-\dot{\hat{\vec{s}}}(t)=\left(\vec{\vec{A}}-\vec{\vec{L}}\cdot\vec{\vec{C}}\right)\cdot\left(\vec{s}(t)-\hat{\vec{s}}(t)\right)$ converges to zero.

The virtual system is given an estimated start state and is evolving parallel to the real system. The sensors measure the values of the real observables, while the virtual system model produces estimates of the observable values. If the real and the virtual system were in the same state, the difference of the real and estimated observables would vanish.

The main idea of the observer is to use this difference to create an additional external force applied to the virtual system via matrix L to drive the state of the virtual system towards the state of the real system. The resulting state of the virtual system is then an estimation of the state of the real system.

After initializing the virtual system with a roughly estimated initial state, the linear observer model will converge towards the real state and thus track the system state from the observation of some measurable quantities.

Pedestrians within an observation area cannot be described by a linear dynamical model of the system states. Instead, sophisticated multi-agent models or fluid dynamics models are needed. In the non-linear case, the multiplication of matrices turns into general vector-valued functions.

$$\dot{\vec{s}}(t)=\vec{F}(\vec{s}(t))+\vec{G}(\vec{i}(t));\quad \vec{o}(t)=\vec{H}(\vec{s}(t))$$

$$\dot{\hat{\vec{s}}}(t)=\vec{F}\left(\hat{\vec{s}}(t)\right)+\vec{G}\left(\vec{i}(t)\right);\quad \hat{\vec{o}}(t)=\vec{H}\left(\hat{\vec{s}}(t)\right);\quad \vec{i}_B(t)=\vec{J}\left(\vec{o}(t)-\hat{\vec{o}}(t)\right)$$

Despite this fundamental difference, we follow the basic idea of the observer model to adjust the model parameters such, that the simulation results and the measurements (observables) of people flow at dedicated positions have minimum difference. The force driving the estimated state quantities towards the real state quantities could be chosen to be proportional to the negative gradient of the squared difference between the measured and estimated observables $\vec{i}_B(t)=-\eta\frac{\partial\left(\vec{o}(t)-\hat{\vec{o}}(t)\right)^2}{\partial\vec{s}(t)}$ yielding an iterative adaptation scheme.

Extending the observer model to non-linear systems has already been considered before in chemistry [3] and production processes [2,4].

Within the state observer model two concurrent processes exist: (1) the real process with sensor systems giving observables values due to the real system state and (2) a pedestrian traffic simulation yielding an estimated system state.

The measured data allows to always use realistic start values for the current situation within the simulation and to continuously adjust the simulation parameters to minimize the difference between observed and simulated behavior. Since the simulated situation is always up to date based on the on-line real data the extrapolated forecast would increase in accuracy.

## *Measurement and Simulation*

Real-time video analysis can be used to perform such measurements and deliver people flow, speed and track data as observables of certain observed areas.

Depending on the video processing technology and mounting situation, also flow distribution data of a larger area or precise track data of a smaller area can be acquired. Tracks can be evaluated with counter lines in the image to generate the actual boundary conditions of an analyzed area by measuring the current in-flow and out-flow of people at entrances and exits from counting data.

The video analysis is performed by the Vitracom SiteView system, which is able to extract the trajectories, number and speed of people at 25 fps. From this information, general rules e.g. regarding people behavior would be derived.

For the simulation and visualization the CAST engine of ARC is used, which allows a discrete event simulation, that integrates modeling, simulation and visualization abilities in a common environment.

CAST was chosen because new simulation actors can easily be developed and integrated into the full environment thus gaining all the advantages of the class hierarchy ancestors quickly and efficiently.

Inside the CAST engine, every agent represents an actor (e.g. passenger, a vehicle or aircraft) which is able to react to the given situation according to its individual characteristics. According to the BDI idea [4], every agent thus has a specific knowledge of his goals. In addition, agents may also have detailed knowledge about their environmental situation.

Following this idea, sensors reading real and simulation data as well as actors making adjustments to other actors have been implemented in CAST.

### *Communication between reality and simulation*

The objective is to keep the simulation running parallel to reality, which is observed by the video analysis. In order to achieve this goal, the observables data gathered by the video cameras has to be delivered to the simulation. The interaction scheme –instantiating the observer control loop- is shown in figure 2.

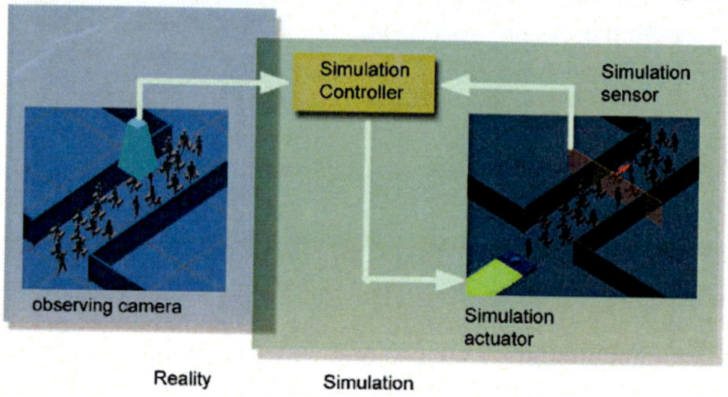


**Fig. 2. Adjustment between reality and simulation**

The observables of the real passenger flow captured by the video system are compared by the simulation controller with the data from the simulation sensor which extracts the same observables in the simulation model from the virtual agent flow. The simulation controller uses a simulation actuator to affect the simulation parameters in order to decrease the difference. This can be done with three different schemes. The respective effect on one traffic parameter (passenger flow rate in the example) is qualitatively shown in figure 3.

If the observed real passenger flow rate is higher than the current simulated flow rate it has to be increased. Increasing the simulation flow rate too fast by adjusting the passenger generation rate inappropriately will result in unrealistic, discontinuous passenger agent behavior (green curve). Using a certain attenuation factor (reflecting a realistic maximum passenger acceleration), the simulation curve will be slightly delayed against the real passenger flow and the simulation agents will additionally show a more realistic behavior.

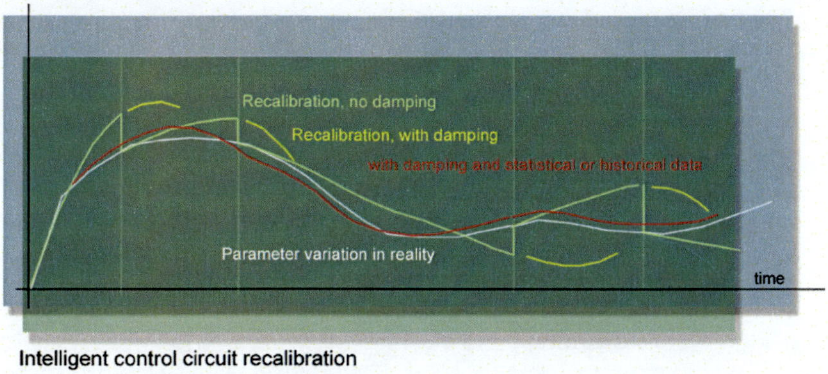


**Fig. 3. Intelligent control circuit recalibration**

If the simulation actuator is placed far apart from the sensor unit, a lag time will arise between the measurement itself and the time, where the action of adjusting simulation parameters is taken. This will result in unstable system behavior, which can be avoided by increasing the attenuation factor. Appropriate attenuation factor adjustment is a subject of ongoing research.

The most difficult part is to relate the observation results with actuator actions in order to bring the simulation closer to the reality. In the example above (fig. 3), the action that has to be performed - increasing the rate if it is too low - is obvious. For larger scopes, the real situation will never be covered completely but there is usually also more than one observer such as a camera sending data to the simulation. As a consequence, multiple simulation controllers as shown in fig. 2 might be employed having several data sources and even more than one actuator.

Future work will treat improvement measures. Stability can be improved by incorporating additional meta-information such as limits of certain traffic parameters which must not be exceeded. Precision improvement can be reached by sharing information among simulation controllers to benefit from the observations of the other controllers.

## Interface between measurement and simulation

To prove the validity of this analysis, a real scenario is set up with the proposed system using pedestrian flow as observables. Differences between the pedestrian flows regarding direction and speed can be measured and quantified for parameter optimization of the multi agent simulation. For measurement and simulation Site-View and CAST were used.

The first integration step was to define an interface to exchange data between the systems.

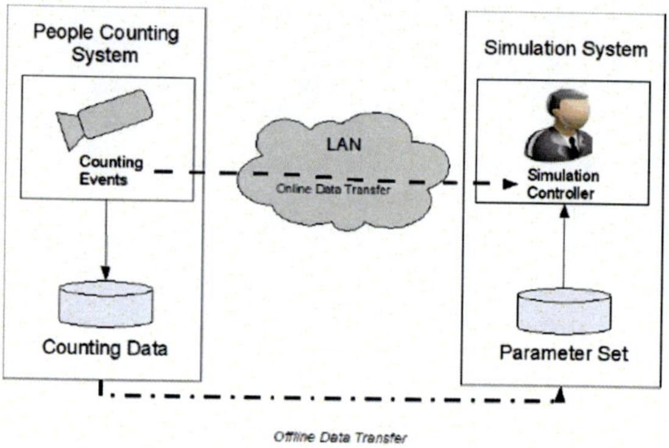


**Fig. 4. System layout and online/offline data transfer**

In figure 4 a technical overview of the system is given. For the experiments a file based offline exchange of data was implemented. Each trajectory which crossed a counting line, a time stamp and the direction are written into the CSV-file. Even though it is possible to exchange this data offline to run tests of both systems an asynchronous remote procedure call is currently implemented to exchange the data for each counting event. Therefore a standard network connection between the systems is needed, which is fast enough to exchange the data. The transfer has to be asynchronous because no system may wait for the other. If single data gets lost, this is no problem for each system in a technical matter since the parameter set of the simulation is given by historic data and "only" updated to the real situation. So this coupling of two existing and verified systems leads to more correct starting parameters of the simulation and to adapt better the real situation.

## Test scenarios

A laboratory sample situation was set up for the development and testing of the system. The sample setup only consists of one room with one entrance and one exit situation (fig. 6). The area of the room is about 15 square meters in size. In contrast to real world scenarios the whole area was completely observed for validation of the simulation results. Only the people crossing events at entrances and exits were transferred to the simulation, the frequency of which was affected by the speed of people.

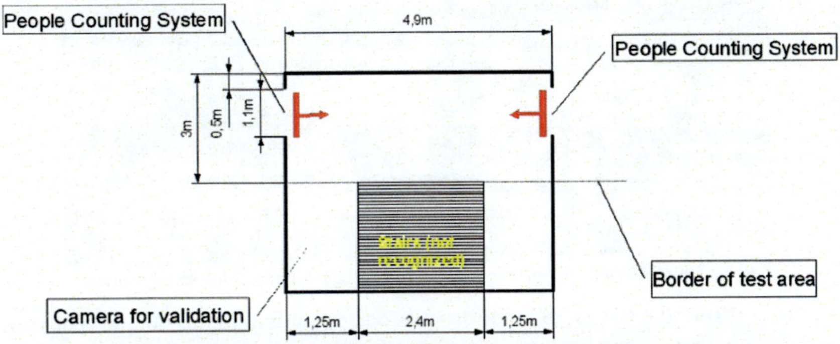


**Fig. 5. Laboratory setup for test data acquisition**

The experiments were about 20 minutes in total with no pre-defined number of participants. However 30 instructed persons (probands) wore a white cap for easier identification and for privacy reasons. Six behavioral scenarios were recorded for later analysis:

A. Unidirectional traffic

1. normal walking speed; test persons were told to behave as if they were on their way to the train station, few fast, few slow, many normal speed
2. hurried persons; as if leaving a show event, many leave fast, few normal, fewer slow
3. panic situation; everybody as fast as possible, only one exit used

B. Bidirectional traffic:

1. normal amount of traffic in one way, normal opposing traffic with normal speed
2. many walking in one way, with little opposing traffic
3. panic: everybody as fast as possible, both exits

In figure 6 some sample images are given to get an impression of the density and people moving in the example setup. The red lines in the left image show the trajectories of people and the green/red line is the virtual counting line with a hysteresis. The right image shows an overview of the test area, stairs are not included.

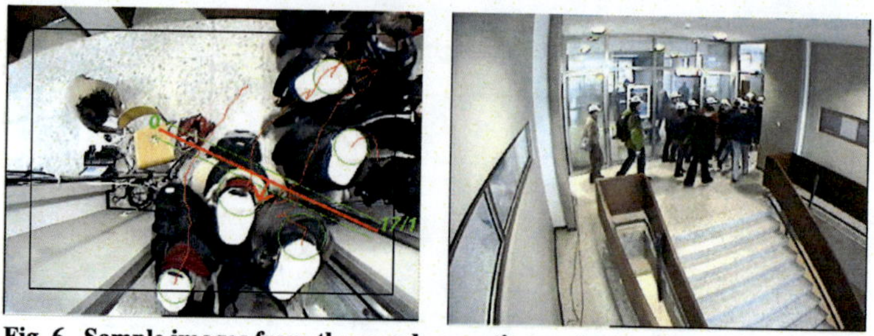

**Fig. 6. Sample images from the people counting system and overview camera**

Fig. 7 shows the CAST simulation setup corresponding to the sample scenario. The person agents have been given a plan which makes them walk bidirectionally through the investigated area in the middle of the model.

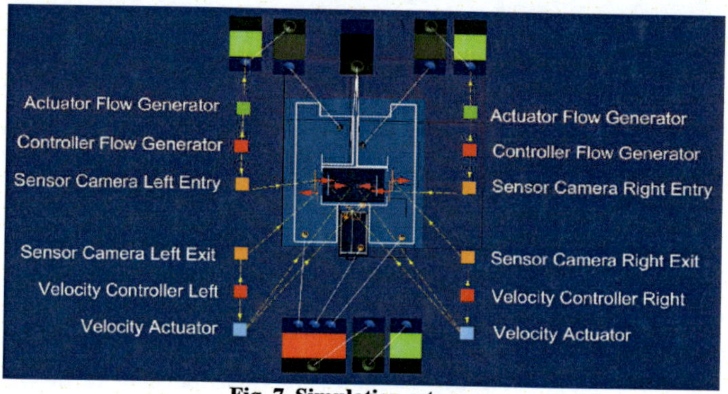


**Fig. 7. Simulation setup**

The simulation camera sensor provides imported real passenger flow data and can additionally perform passenger flow measurements inside the simulation. Thus, two resulting passenger flow rates are available which can be compared. Sending the resulting virtual and real flow rates to the controller (red cube in figure 9), the actuators (green) can trigger an appropriate action on the simulation actors to bring simulation closer to reality. Two different actuators have been realized in the model. First, the velocity controller, working on a spatial area increases or reduces the walking speed of the passengers thus either increasing or reducing the flow rate on the successive area to adjust the exit rate. Second, the flow generator actuator has the ability to increase or reduce the passenger generation rate. In summary, the sensors are monitoring dedicated areas of both the simulation model and reality and perform appropriate decentral actions in order to bring the simulation passenger flow close to the observed passenger rates.

For future real world system tests a video sensor system was installed at Cologne central train station, Germany, where the test area counts for about 200 square meters. Fig. 8 shows a screenshot of the corresponding simulation model.

## Results

The curves in figures 9 show the simulated and the observed passenger flow rates at the entrance (a) and the exit (b). In the diagrams the people flow is plotted against time. The plot comprises all six subsequent behavior scenarios A.1 to B.3 as described before which are marked in the charts. Only the flow in entrance-to-exit direction which is also representative of the opposite direction is shown. It can be seen that the simulated curve converges to the real curve. The simulation curve

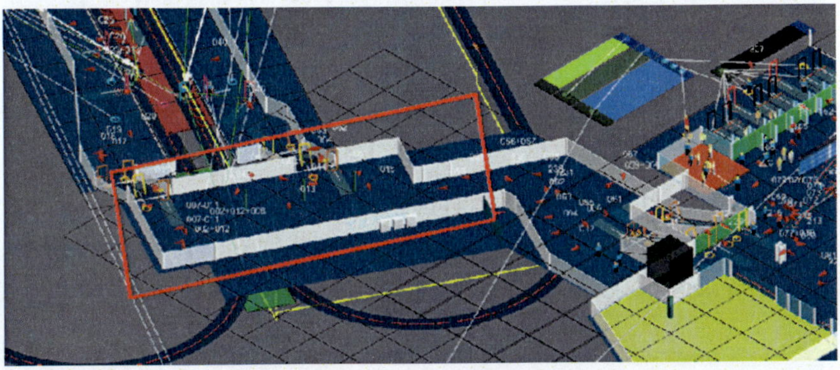

**Fig. 8. Screenshot of simulation model in the real world installation, red rectangle shows the observed area**

shows latency as discussed before.

The simulation controller used two actuators at the entrance, one adjusting the generation rate (based on the entrance flow sensor data) and the second the velocity of the agents (based on the exit flow sensor data). Comparing the curves the simulation model appears to reflect the behavior of persons in an appropriate way. Figure 9b shows higher latency and lower accuracy due to the higher distance between sensor and actuator.

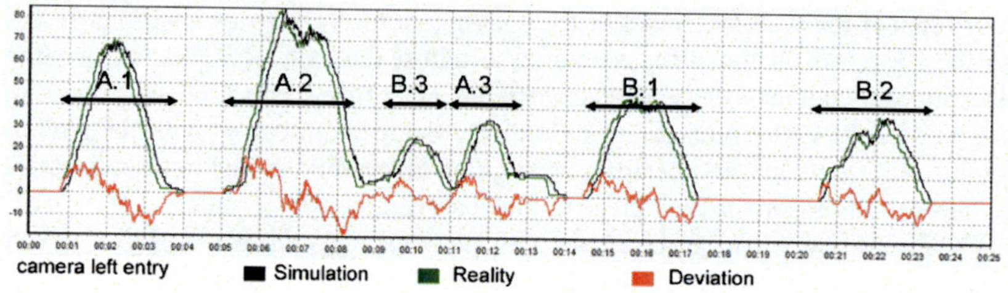


**Fig. 9a. Results entry: comparison between simulation and reality**

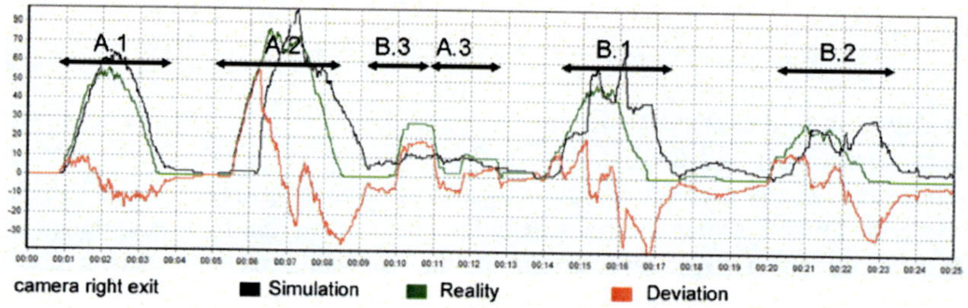


**Fig. 9b. Results exit: comparison between simulation and reality**

While the adjustment of the generation rate affects the simulation state directly with a small time delay the velocity adjustment of the agents based on the exit sensor data takes longer to take effect and the delay between the curves grows larger. These preliminary results show, that the model is capable to follow the observables by adjusting behavior parameters via the proposed controller. This changes the internal state of the simulation model according to the state of the real observed traffic situation. Further investigations will analyze the internal states dynamics, apply and assess the prediction capabilities and reveal which additional instruments can be used to refine the simulation actor behavior.

**Acknowledgements** This research was supported of the BMBF (German Federal Ministry of Education and Research) within the program "Research for Civil Security" under grant 13N9709. We also want to thank all our colleagues from VeRSiert, Airport Research Center and Vitracom.

## References

1. Luenberger, D.G.: Observing the State of Linear Systems. IEEE Transactions on Military Electronics, 74-80 (1964)
2. Istre, J.M., Zhang, Y.M.: Predictive generic model control for non-linear interval systems with application in arc welding. Int. J. Modelling, Identification and Control. Vol. 1, No. 2 (2006)
3. Lorenz, O.: Der Systembeobachter, ein Ansatz für komplexe chemisch-verfahrenstechnische Prozesse. Dissertation, TH Darmstadt (14.3.2003)
4. Sampaio, D.J.B.S, Zettel, D., Link, N., Peschl, M., Moscato, L.: Process Surveillance and State Sensing with Generic Model Parameter Estimation. AISTA 2004 in Cooperation with the IEEE Computer Society Proceedings, 15-18 November, Luxemburg, Germany (2004) ISBN: 2-9599776-8-8
5. Klügl, F.: Multiagentensimulation - Konzepte, Werkzeuge, Anwendung. Addison Wesley (April 2001) ISBN: 3-8273179-0-8
6. Hanisch, A., Tolujew, J., Raape, U., Schulze, T.: Online-Simulation für Personenströme in einem Frühwarnsystem. In: Hohmann, R. (ed.), Simulationstechnik, 17. ASIM-Symposium in Magdeburg, p. 221-226. SCS Int., Ghent (2003)
7. Hanisch, A., Tolujew, J., Meuschke, T., Schulze, T.: Datenkollektion zur online Simulation von Personenströmen. In: Schulze, T., Schlechtweg, S., Hinz, V., Proceedings Simulation und Visualisierung , pp. 27-38. SCS-European Publishing House (2004)
8. Hanisch, A., Tolujew, J., Richter, K., Schulze, T.: Online Simulation of Pedestrian Flow in Public Buildings. In: Chick, S., Sánchez, P.J., Ferrin, D., Morrice, D.J. (eds.), Proceedings of the 2003 Winter Simulation Conference, pp.1635-1641 (2003)

9. Daamen, W., Hoogendoorn, S.P.: Research on pedestrian traffic flows in the Netherlands. In: Proceedings Walk 21 IV, Portland, Oregon, United States, pp. 101-117 (2003)

# Free Walking Speeds on Stairs: Effects of Stair Gradients and Obesity of Pedestrians

**T. Fujiyama and N. Tyler**

Centre for Transport Studies, University College London
Gower Street, London, WC1E 6BT United Kingdom
Corresponding author: taku.fujiyama@ucl.ac.uk

**Abstract** Effects of stair gradients and obesity on walking speeds on stairs were empirically investigated. The participants included a group of elderly people (n=18) and a group of young people (n=15). They were asked to ascend/descend four staircases with different gradients, as well as to walk on a flat surface, at both normal and fast speeds. The study found the effects of stair gradients. When walking speeds on stairs are estimated, the gradient should be taken into account. In contrast, the study found no effect of overweight (or moderate obesity) on speeds.

## Introduction

Walking speed is an important element of the walking characteristics of pedestrians. In order to run a pedestrian simulation model, it is necessary to determine free walking speeds. Free walking speeds are defined as the speed at which a pedestrian wishes to walk when there are no obstructions [1]. There have been a good number of studies on walking speeds (on flat surfaces) and some studies have investigated the effects of personal characteristics, such as age and gender, on walking speeds (e.g. [2]). These studies have shown that elderly people walk more slowly than young or middle–aged people and that women's walking speeds are slower than those of men. In fact, differences in walking speeds according to age and gender have also been investigated by physiologists (e.g. [3]).

Although their number is far less than studies on flat surfaces, there have been a few studies on free walking speeds on stairs. Stairs are an important element of transport facilities, such as transport terminals or stations, and when pedestrian movements in a whole facility or building are simulated, stairs have to be included. Fruin [4] showed that walking speeds decline according to age and that women showed a lower value than men. Templer [5] suggested an equation which estimates walking speeds on stairs based on the dimensions of stairs (i.e. tread-depth and riser-height).

One issue that needs further investigation is how different are walking speeds on stairs from those on flat surfaces. In order to estimate walking speeds on stairs, many microscopic pedestrian simulation models use a "reduction coefficient", which is multiplied by walking speeds on flat surfaces. Knowledge of the differ-

R.D. Peacock et al. (eds.), *Pedestrian and Evacuation Dynamics*,
DOI 10.1007/978-1-4419-9725-8_9, © Springer Science+Business Media, LLC 2011

ence in walking speeds between stairs and flat surfaces can help simulation modellers use a better coefficient. Another issue can be the effects of obesity. Obesity is an emerging problem in many developed countries. Obese pedestrians could walk slower than ordinary pedestrians, and an increase in the number of overweight pedestrians could change the capacity of pedestrian flow. Therefore, this study examines the following questions:

- How different are walking speeds on stairs from those on flat surfaces?
- Does overweight affect walking speeds? If so, how much?

Note that the definitions of walking speeds on stairs are as shown in Figure 1. In this paper, speeds on stairs mean horizontal walking speeds unless specified.

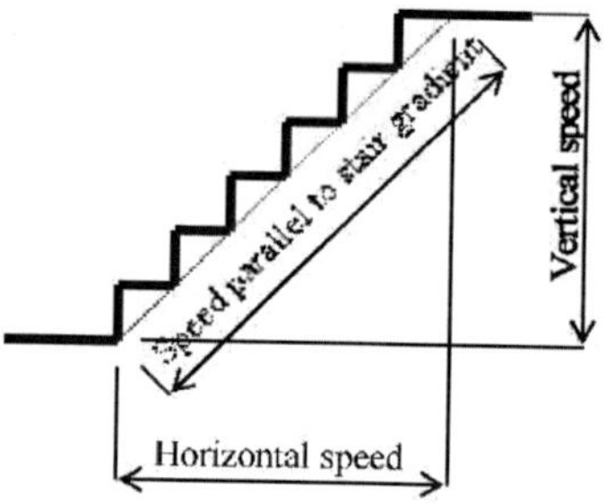


**Fig. 1. Definitions of speeds on stairs**

## Experimental data

An experiment was conducted on four staircases with different gradients in a university campus (Table 1). The details of the experiment were published elsewhere [6]. In the experiment, each participant was asked to ascend and descend a flight of each staircase at his/her normal and fast walking speeds, which means that we measured four patterns of speeds for each participant: namely the normal ascending speed, the normal descending speed, the fast ascending speed and the fast descending speed. Each participant was asked to repeat this exercise on four staircases with different stair-gradients as well as to walk at normal and fast speeds on a flat surface. The experiment included two groups of participants: one consisted of people aged 60 or over, whereas the other consisted of university students and employees. The characteristics of the participants are displayed in Table 2.

**Table 1. The dimensions of the staircases used in the experiment**

| | | Proportion | | | Total Length | | |
|---|---|---|---|---|---|---|---|
| Staircase No. | Number of Steps | Riser-height mm | Tread length mm | Stair-Gradient deg | Horizontal length m | Vertical Length m | Availability of handrail √ |
| 1 | 12 | 185 | 230 | 38.8 | 2.76 | 2.22 | √ |
| 2 | 12 | 175 | 250 | 35.0 | 3.00 | 2.10 | √ |
| 3 | 15 | 157 | 267 | 30.5 | 4.01 | 2.36 | √ |
| 4 | 9 | 152 | 332 | 24.6 | 2.99 | 1.37 | √ |
| Flat Surface | - | - | - | - | 8.00 | - | × |

**Table 2. The characteristics of the participants**

| | Elderly | Young |
|---|---|---|
| Sample number | 18 | 15 |
| (male sample) | 6 | 7 |
| (female sample) | 12 | 8 |
| Age (yrs) | 71 ± 5.9 | 34.5 ± 12.7 |
| Height (m) | 1.61 ± 0.72 | 1.74 ± 0.82 |
| Weight (kg) | 67.3 ± 11.7 | 66.4 ± 13.9 |

Results are given as mean ± SD.

## Results and Analysis 1: How walking speeds on stairs differ from those on a flat surface?

Figure 2 shows the percentages of normal walking speeds on stairs of both ascending and descending in proportion to those on a flat surface. For each walking pattern (ascending or descending) of each participant group (elderly or young), percentages were calculated as

$$p_k = \sqrt{v_{kn} / v_{fn}} \quad \cdots\cdots\cdots (1)$$

where $p_k$: percentage for staircase $k$

$v_{kn}$: walking speed on staircase $k$ of participant $n$

$v_{fn}$: walking speed on a flat surface of participant $n$

The figures suggest that percentages vary according to the stair gradient. Interestingly, a linear relationship can be seen between walking speeds and stair-gradients.

Figure 3 shows the percentages of ascending speeds (both normal and fast) on stairs in proportion to speeds on a flat surface. Note that for each pedestrian, normal ascending or descending speeds were compared to the normal walking speed on a flat surface, whereas fast walking speeds were compared to the fast speed on a flat surface. Figure 4 shows those of descending speeds. For ascending speeds at the normal speed, the elderly participants showed similar values to those of the young participants, whereas for descending this is the case only when the gradient was not steep.

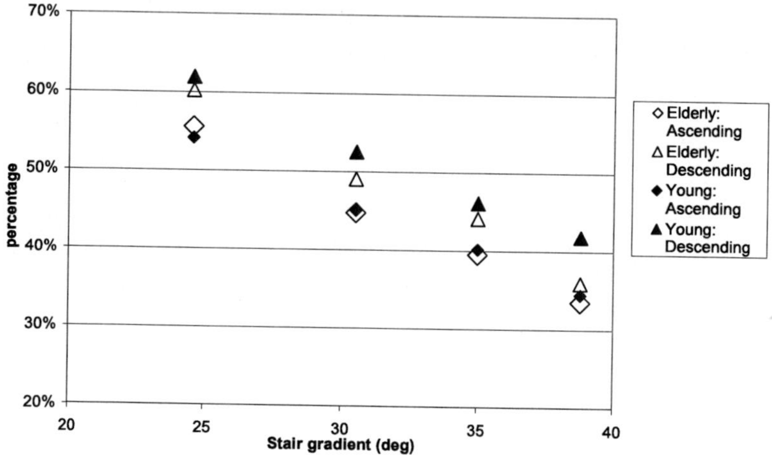


**Fig. 2. Percentage of normal walking speeds on stairs to those on a flat surface**

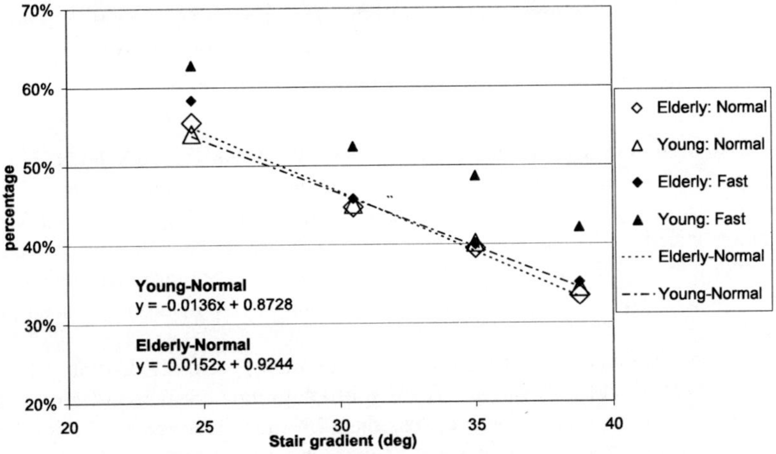


**Fig. 3 . Percentage of ascending speeds on stairs to those on a flat surface**

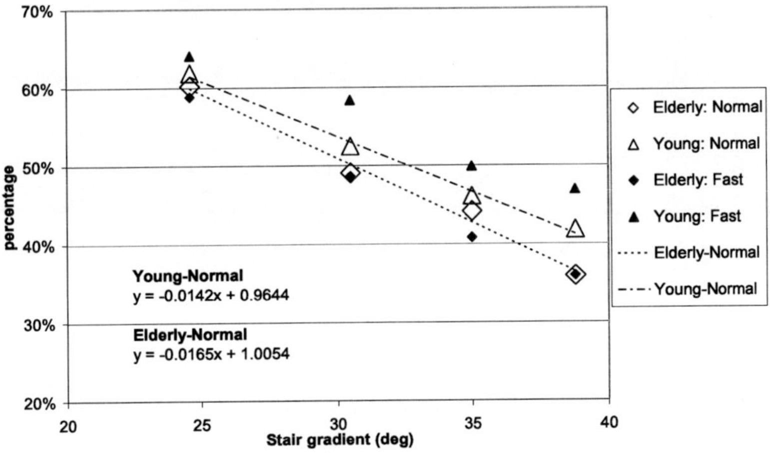


**Fig. 4. Percentages of descending speeds on stairs to those on a flat surface**

## Result and Analysis 2: Effects of overweight on walking speeds

We use Body Mass Index (BMI) as the index of overweight. BMI is calculated as follows.

$$B_n = {W_n}/{H_n^2} \cdots\cdots\cdots (3)$$

where $B_n$: Body Mass Index of participant *n* (kg/m²)
$W_n$: Weight of participant *n* (kg)
$V_n$: Height of participant *n* (m)

BMI is commonly used as an indicator of obesity. According to WHO [7], a person with a BMI of 25 or more is regarded as being overweight. Figure 5 shows normal ascending speeds on Staircase 2 by BMI, whereas Figure 6 shows fast ascending speeds. We had hypothesised that overweight has negative effects on walking speeds, which can be described as a negative proportionate relationship between BMI and walking speeds. However, such a relationship is not seen in the figures.

The participants of each group (elderly and young) were divided into two subgroups: Normal and Overweight. The "Normal" groups consisted of participants with a Body Mass Index (BMI) of less than 25, the "Overweight" having 25 or more. Table 3 shows the profile of each subgroup.

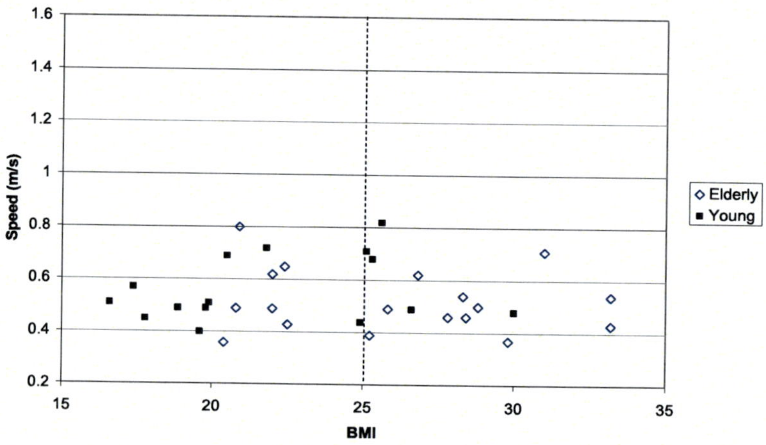


**Fig. 5. Relationships between BMI and normal ascending speeds on Staircase 2**

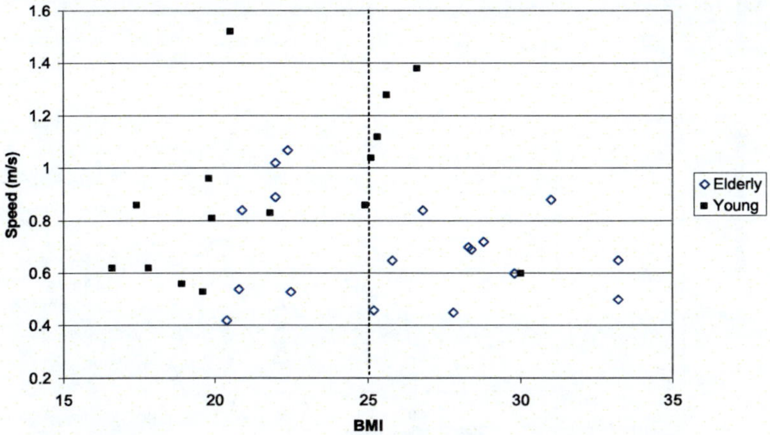


**Fig. 6. Relationships between BMI and fast ascending speeds on Staircase 2**

**Table 3. Profiles of subgroups**

| | Elderly | | Young | |
|---|---|---|---|---|
| | Normal | Overweight | Normal | Overweight |
| Sample size | 7 | 11 | 10 | 5 |
| Age (yr) | 72.6 ± 6.7 | 70.7 ± 5.6 | 34.4 ± 10.8 | 34.6 ± 17.2 |
| BMI (kg/m$^2$) | 21.6 ± 0.8 | 28.9 ± 2.7 | 19.7 ± 2.4 | 26.5 ± 2.0 |

Age and BMI are given as mean ± SD.

Figure 7 shows the average normal ascending speeds of each subgroup, whereas Figure 8 shows fast ascending speeds. In the figures, the elderly-overweight subgroup shows a lower value than the elderly-normal subgroup for all the staircases, whereas the young-overweight subgroup shows a higher value. Table 4 compares walking speeds between the subgroups within each group (elderly and young). The standard deviation of each entry is large, and a statistical significance is merely observed.

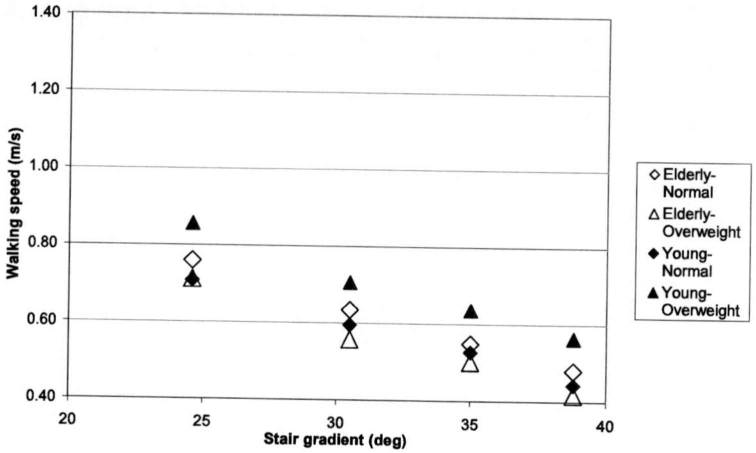


**Fig. 7. Average normal ascending speeds of each subgroup by stair gradient**

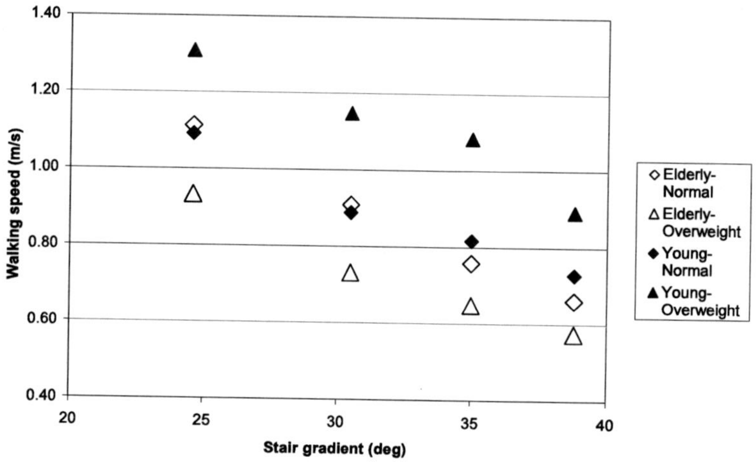


**Fig. 8. Average fast ascending speeds of each subgroup by stair gradient**

**Table 4. Comparison of walking speeds between subgroups**

| Patterns of speeds | Staircase No | Staircase Degree | Elderly Normal | Elderly Overweight | Elderly Sig. | Young Normal | Young Overweight | Young Sig. |
|---|---|---|---|---|---|---|---|---|
| Normal ascending | Stair 1 | 38.8 | 0.48 ± 0.15 | 0.42 ± 0.09 | *NS. p*= 1.13 | 0.44 ± 0.18 | 0.56 ± 0.11 | $p<0.05$ |
| | Stair 2 | 35 | 0.55 ± 0.15 | 0.50 ± 0.10 | *NS. p*= 0.82 | 0.53 ± 0.10 | 0.64 ± 0.15 | *NS. p*= 1.67 |
| | Stair 3 | 30.5 | 0.63 ± 0.15 | 0.56 ± 0.11 | *NS. p*= 1.25 | 0.60 ± 0.10 | 0.71 ± 0.19 | *NS. p*= 1.51 |
| | Stair 4 | 24.6 | 0.76 ± 0.22 | 0.71 ± 0.15 | *NS. p*= 0.56 | 0.71 ± 0.11 | 0.86 ± 0.23 | *NS. p*= 1.74 |
| Fast ascending | Stair 1 | 38.8 | 0.66 ± 0.23 | 0.57 ± 0.13 | *NS. p*= 1.02 | 0.73 ± 0.25 | 0.89 ± 0.23 | *NS. p*= 1.24 |
| | Stair 2 | 35 | 0.76 ± 0.26 | 0.65 ± 0.14 | *NS. p*= 1.17 | 0.82 ± 0.29 | 1.08 ± 0.30 | *NS. p*= 1.67 |
| | Stair 3 | 30.5 | 0.91 ± 0.24 | 0.73 ± 0.15 | *NS. p*= 1.89 | 0.89 ± 0.23 | 1.15 ± 0.31 | *NS. p*= 1.82 |
| | Stair 4 | 24.6 | 1.11 ± 0.30 | 0.93 ± 0.16 | *NS. p*= 1.69 | 1.09 ± 0.27 | 1.31 ± 0.38 | *NS. p*= 1.29 |
| Normal walking on the flat surface | | | 1.33 ± 0.24 | 1.30 ± 0.24 | *NS. p*= 0.25 | 1.35 ± 0.17 | 1.51 ± 0.13 | *NS. p*= 1.79 |
| Fast walking on the flat surface | | | 1.83 ± 0.31 | 1.64 ± 0.26 | *NS. p*= 1.35 | 1.81 ± 0.18 | 1.89 ± 0.07 | *NS. p*= 0.98 |

unit: (m/s)

## Discussions

**a) Stair gradients and walking speeds:** Figure 2 suggests a linear relationship between walking speeds on stairs and stair-gradients. The results are in contrast to the fact that many pedestrian simulation models use the same speeds for different stairs with different gradients. When walking speeds on stairs are estimated from speeds on flat surfaces using a reduction coefficient, the stair gradient should be taken into account. In Figure 2, the young group shows a large difference between percentages of ascending speeds and those of descending speeds, whereas the elderly group does not. When pedestrians that consist mainly of young people are simulated, it is recommended to use different coefficients for ascending and descending speeds. So, what value should be used? Readers can refer to the equations in Figures 3 and 4 for normal speeds.

**b) Effects of overweight :** In Figures 7 and 8, the elderly-overweight subgroup shows a lower value than the elderly-normal subgroup for all the staircases, whereas the young-overweight subgroup shows a higher value. One possible explanation for this is the components of each overweight subgroup. Because BMI is calculated by weight and height, the index does not distinguish between obese and muscular people. Some professional athletes are even categorised as "overweight" or "obese" [8]. On the other hand, muscle decreases with age [9]. It is speculated that the young-overweight subgroup included muscular people, whereas the elderly-overweight included fat people, and thus the trends are different.

So, do obese people walk more slowly than people of normal weight? We now focus on walking speeds of elderly people. From Figures 5 and 6, the relationship between BMI and walking speeds is not clear. In Figures 7 and 8, the elderly-overweight subgroup shows a lower value than the normal subgroup on all the staircases. Table 4 shows that a statistical significance is merely observed. These imply that there is no significant effect of overweight (because of obesity) on walking speeds. Existing studies have found no effect of obesity on walking speeds (e.g. [10]). There have been some studies that suggested such effects, but such studies have used a treadmill (e.g. [11]) or examined the speeds of obese people only and compared them with data of other experiments (e.g. [12]). Our result shows that, even though the average values of the elderly-overweight subgroup are lower than those of the elderly-normal subgroup, the differences were not significant compared to the variations of each subgroup. It is known that for each person, there is one walking speed that minimises the amount of energy required to walk, and his/her preferred walking speed is thought to correspond with this speed [10]. Browning and Kram [10] found that energy consumption in walking of obese people was not as great as physiologists had thought, and suggested that obese people learn to walk in a way that reduces the extra energy consumption due to obesity. Interestingly, Figures 7 and 8 show that differences in fast ascending speeds between elderly-normal and elderly-overweight subgroups were larger than those in normal ascending speeds. One possible reason for this is that because overweight people do not often walk fast, they are not familiar with ways to reduce the extra energy consumption in fast walking.

It should be noted that the overweight-elderly group in our experiment did not include any severely or morbidly obese person with a BMI of 35 or more, who falls into Class II or Class III obesity according to WHO [7]. One study estimated that around 8% of the population of the USA was Class II or Class III obese in 2005 [13]. Further research would be necessary to investigate walking speeds of severely and morbidly obese people on stairs.

## Conclusions

This study explored the effects of stair gradients and overweight on walking speeds on stairs. The study found that stair gradients affect walking speeds on stairs. In the estimation of walking speeds on stairs based on those on flat surfaces, it is recommended to take account of the stair gradient. The study did not find any significant difference between the walking speeds of normal and overweight (or moderately obese) participants. However, further research would be necessary on walking speeds of severely or morbidly obese people. In addition, differences in body size and potential differences in body movement of obese people from those of normal people could affect the capacity of pedestrian flow that includes obese people. Further investigation is necessary on this topic as well.

## References

(Websites and their contents were current on 31 January 2010)

1. Daamen, W., Hoogendoorn, S.P.: Free speed distribution - Based on empirical data in different traffic conditions. In: Waldau, N. et al. (Eds.), Pedestrian and Evacuation Dynamics 2005, pp. 13–25. Springer, Berlin (2007)
2. Wilson, D.G., Grayson, G.B.: Age-related differences in the road crossing behaviour of adult pedestrians. TRRL Laboratory Report 933. Transport and Road Research Laboratory, Berkshire (1980)
3. Bohannon, R.W.: Comfortable and maximum walking speed of adults aged 20-79 years: reference values and determinants. Age and Ageing. 26, 15-19 (1997)
4. Fruin, J.J.: Pedestrian Planning and Design, (Revised Edition). Elevator World, Inc., Mobile, AL (1987)
5. Templer, J.A.: The staircase: Studies of Hazards, Falls, and Safer Design. MIT Press, MA (1992)
6. Fujiyama, T., Tyler, N.: An Explicit Study on Walking Speeds of Pedestrians on Stairs." In: Proc. of the 10th International Conference on Mobility and Transport for Elderly and Disabled People, pp. 643-652 (2004)
7. World Health Organisation, BMI classification available at http://apps.who.int/bmi/index.jsp?introPage=intro_3.html
8. Body mass index. Wikipedia. http://en.wikipedia.org/wiki/Body_mass_index
9. Nair, K.S.: Aging muscle. The American Journal of Clinical Nutrition. 81, 953-963 (2005)
10. Browning, R.C., Kram, R.: Energetic cost and preferred speed of walking in obese vs. normal weight women. Obesity Research. 13(5), 891-899 (2005)
11. Melanson, E.L. et al.: Body mass index and sex influence the energy cost of walking at self-selected speeds. Medicine & Science in Sports & Exercise. 35(5), p. S183 (2003)

12.Mattsson, E. et al.: Is walking for exercise too exhausting for obese women?. International Journal of Obesity and Related Metabolic Disorders. 167, 845-850 (2001)
13.Sturn, R.: Increases in morbid obesity in the USA: 2000-2005. Public Health. 121, 492-496 (2007)

# Data Collection (Vulnerable Groups)

# Travel Along Stairs by Individuals with Disabilities: A Summary of Devices Used During Routine Travel and Travel During Emergencies

**G. E. Hedman**

University of Illinois at Chicago, Chicago, IL USA
Corresponding author: ghedman@uic.edu

**Abstract** Many devices exist which are designed to enable individuals with disabilities to travel between floors via staircases. The range of devices includes wheelchairs that descend and ascend stairs, platforms and chairs installed on the staircase, wheelchair carriers which transport the individual while occupying a wheelchair, and evacuation chairs used during emergencies.

A summary of commercially-available devices is presented, categorized by design type and usage. Specifications are provided for characteristics such as overall size, device weight, travel speed.

A summary of existing international standards and testing methods which apply to these devices is presented. A description of standards development work in the area of emergency evacuation chairs is also included.

## Introduction

As effective the effort to design both public and private spaces to maximize accessibility has been, stairs are a necessary component of the built environment. Although much consideration is being given to the use of elevators for evacuation, use of all means of evacuation must be maintained as a goal as a given incident and given building design may necessitate use the of stairways for evacuation by occupants.

There are many types of devices that have been developed to address the issue of stair travel by individuals with disabilities. These include special wheelchair designs, permanent attachments to wheelchair frames, wheelchair carriers, platform lifts and chairlifts installed directly on stairways and evacuation chairs. These approaches can be organized relative to design and role of the user (Figure 1).

R.D. Peacock et al. (eds.), *Pedestrian and Evacuation Dynamics*,
DOI 10.1007/978-1-4419-9725-8_10, © Springer Science+Business Media, LLC 2011

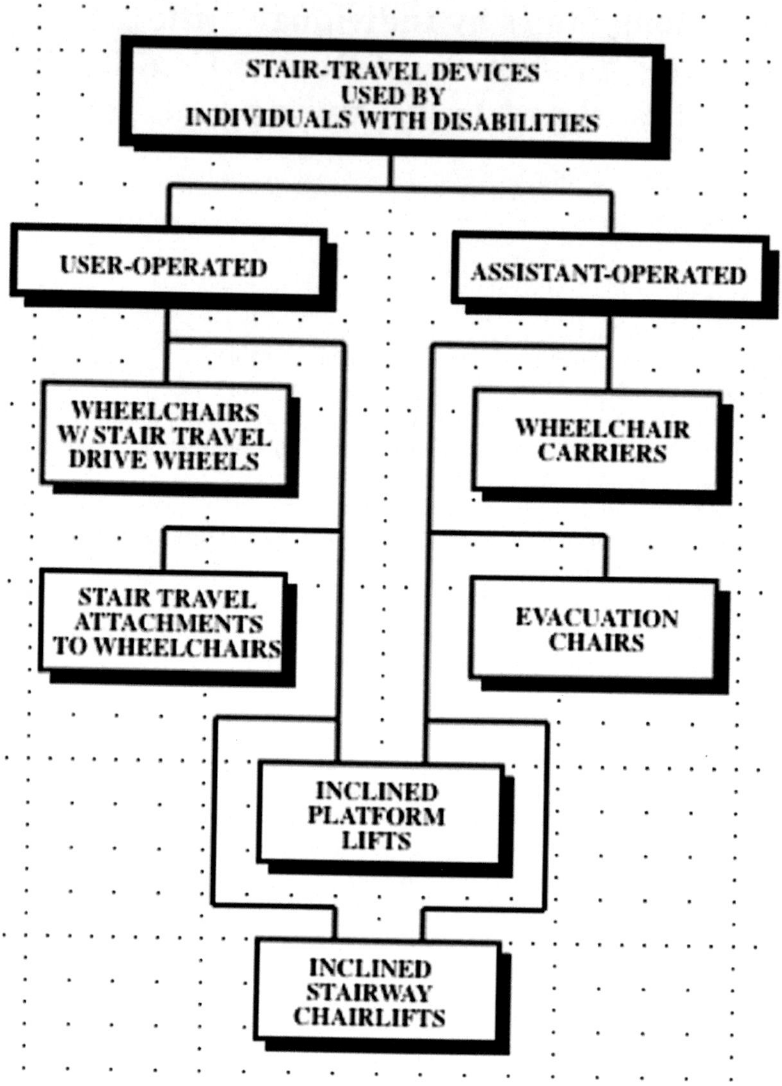


**Fig. 1. Stair travel device options relative to design and user control**

## Devices to Enable Travel along Stairs

Several different approaches to enable travel along stairs have been promoted, which include devices that are now common as well as devices which have not moved past the concept stage.

### *Wheelchairs & Wheelchair Attachments for Stair Travel (User-Controlled and Assistant-Controlled)*

One of the more intriguing approaches to enabling travel along stairways by individuals with disabilities involves wheelchair designs and / or adaptations to provide for this travel (Figure 2).

**Fig. 2. Examples of stair travel via wheelchair design (left, iBot 4000, Independence Technologies, LLC) and attachment to wheelchair frame (right, Hubbard prototype, University of California – Irvine).**

One of the goals of these designs is independent travel by the user. This may be achieved by motorized components or manual operation of the stair-traversing mechanism.

## *Wheelchair Carriers*

A category of portable devices exists, wheelchair carriers, which enable an individual with a disability to travel along stairways while seated in their wheelchair (Figure 3). These portable support frames are most commonly used by an individual that resides on an upper floor of an apartment building, where travel along a common stairway and landing system which has other apartment entrance doors present.

Use of the wheelchair carrier involves the securing of the support frame to the wheelchair frame and backrest frame, tilting the wheelchair / support frame backward, and careful operation of handle-mounted controls to guide the system down the stairs and to maneuver it at landings.

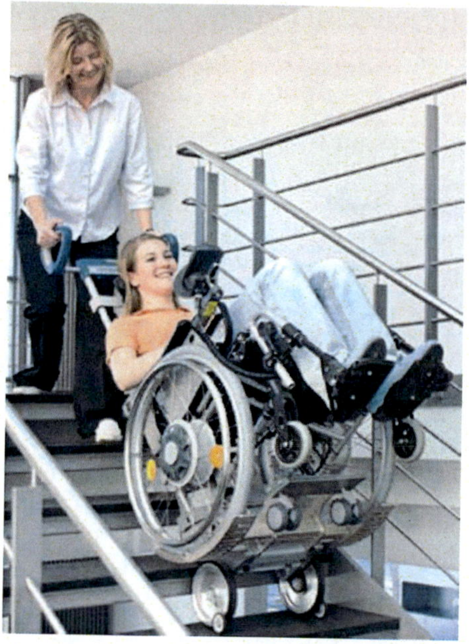

**Fig. 3. Wheelchair carrier (Frank Mobility Scalaport X7**

## *Platform Lifts and Stairway Chairlifts*

Where travel by an individual with a disability along an existing stairway is necessary, and that stairway can be adapted with an additional device to provide for travel, installation of an inclined platform lift or inclined stairway chairlift is common (Figure 4).

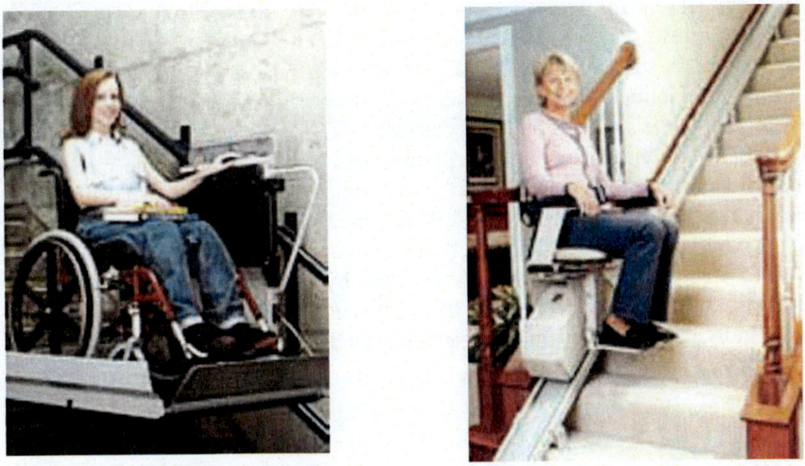

**Fig. 4. Inclined platform Lift (left) and inclined stairway chairlift**

Inclined platform lifts are platforms which support the user while seated in his or her wheelchair. The platform travels along a single track along one side, or along a double track, one located along each side of the stairway. Track sections are secured to the stairs, no attachment to side walls is necessary.

While installation of this device along both straight and L-shaped stairways is possible, the footprint of the platform during travel through turns makes installation along L-shaped stairways unfeasible except where the stairways and landings are relatively large (e.g., schools, libraries, churches). Additionally, even for installations along straight stairways, sufficient landing space must be present at the bottom landing must be sufficient to allow for the device to fully reach the lower level.

An important consideration regarding the possible use of inclined platform lifts is the opinion of the authority having jurisdiction regarding the clear space along the stairway for other occupants to utilize. Even when the platform is in the folded, stored position, the stairway clear width may be reduced. One possible solution, if space permits, is to extend the track section well away from the top and bottom landings, along the hallway or around the corner from the landing.

One benefit of the inclined platform lift is that the user is able to remain in their wheelchair during travel. This may be essential when the user has specialized wheelchair seating equipment, or if the user is at risk for injury if transfers to and from the wheelchair was necessary.

Inclined stairway chairlifts also operate along a track which is secured to the stairs. Via the modular track system, installation along straight, L-shaped, and even curved stairways is possible.

Supports present for users of inclined stairway chairlifts is usually limited to a one-piece footrest, seat belt, and bilateral armrests, although the installation of additional supports such as lateral trunk supports or a headrest is sometimes possible. Like the inclined platform lift stairway clear width issues, installation must allow for use of the stairway by other occupants.

Use of an inclined stairway chairlift requires two transfers to be made: one from the wheelchair to the chairlift at the first landing, one from the chairlift to the wheelchair at the destination landing. This is not an insignificant issue for many users of other mobility aids, such as canes, crutches, and walkers. Where wheelchairs are involved, transport of the wheelchair by others is necessary, or another wheelchair may need to be present at the destination landing. In private residences, individuals may position a backup wheelchair at the destination landing for use on that level.

## *Evacuation Chairs*

Several different models of evacuation chairs exist, which vary greatly in design, capacity, and cost. A recent survey of commercially-available evacuation chairs found a total of 34 stair descent devices, produced by 12 manufacturers worldwide[1].

The 34 devices were found to fall in to one of three design categories (Figure 5):

- Devices which use a track / belt system which contacts the stair nosings (14 devices).
- Devices which are supported or carried by one of more persons, and either roll along both the stair riser and tread, or are carried (19 devices).
- Devices which come in contact with the stairs, but use other designs (e.g., sled type, 1 device).

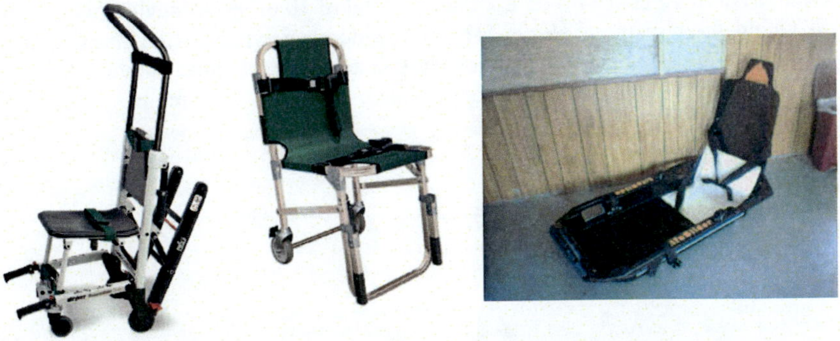

**Fig. 5. Examples of stair descent device design types. From left, Track/Belt (Stryker 6253), Carry Chair (Junkin JSA-800), and Other (LifeSlider).**

For stair descent devices of the track / belt design type, device key device capacities can be compared (Figure 6). Device weight ranged from 10 kg to 20.7 kg (22 lbs to 46 lbs). The maximum user weight of the devices ranged from 130 kg to 228 kg (287 lbs to 500 lbs), with most devices rated for 150 kg (330 lbs).

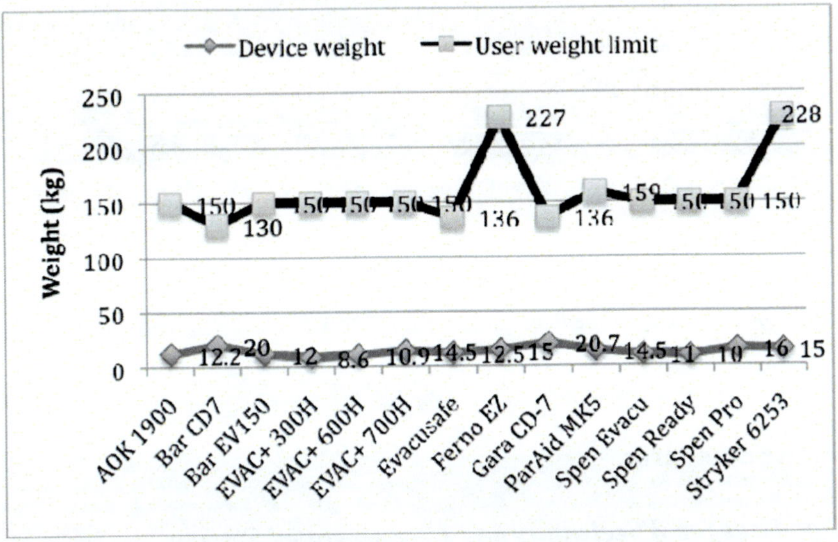


**Fig. 6. Stair descent device weights and user weight limits (Track / Belt type)**

For the same track / belt devices, overall width ranged from 42.6 cm to 58.4 cm (16.8 in to 23.0 in). For a stairway of width 112 cm (44 in), with a reduction in clear width of 10 cm (4 in) on each side, a total of 10 of the 14 devices would not fit within the remaining 46 cm (18 in) – wide clear width of the stairway half (Figure 7). If use of stair descent devices side-by-side with other individuals using the stairway is a goal, this would seem to have implications for minimum stairway width design.

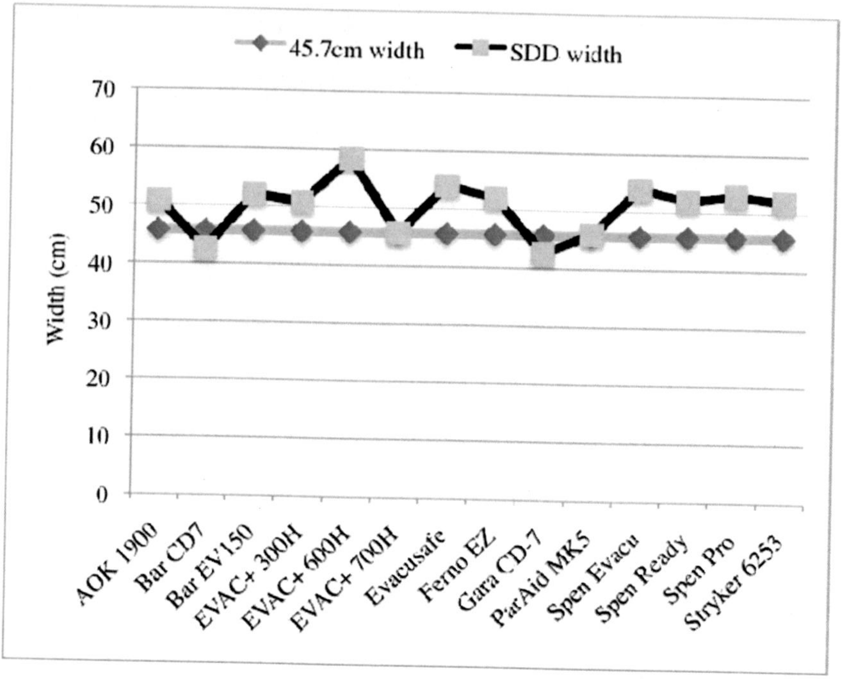


**Fig. 7. Stair descent device overall width (track / belt design).**

Costs for this group of devices ranged from 794 USD to 2,832 USD, excluding taxes and shipping.

## Device Standards

### *Standards for Wheelchairs and User-Operated Systems*

Whether stair travel is built in to the drive wheels of the wheelchair itself or is achieved via attachment of a separate mechanism, international standards exist via ISO 7176 – Part 24 (2005): Requirements and Test Methods for User-Operated

Stair-Climbing Devices[2]. This standard covers characteristics such as static and dynamic stability, static, impact, and fatigue strength, and step transition safety.

### *Standards for Assistant-Operated Stair-Climbing Chairs*

For wheelchairs equipped to travel along stairs, but whose operation requires an attendant, standards exist through ISO 7176 – Part 23 (2002): Requirements and Test Methods for Attendant-Operated Stair-Climbing Devices[3]. Most areas of standard ISO 7176 – Part 24 are covered, but concepts such as the methods for measuring speed and acceleration are introduced.

### *Standards for Stair-Climbing Wheelchair Carriers*

Use of Stair-Climbing Wheelchair Carriers brings additional issues, and these are addressed by ISO 7176 – Part 28 (2009): Requirements and Test Methods for Stair-Climbing Devices[4]. Ergonomics requirements and specifications on reach ranges of the assistant are provided.

### *Standards for Platform Lifts and Stairway Chairlifts*

In the U.S., coverage is provided by the American Society of Mechanical Engineers A18.1-2008 Safety Standard for Platform Lifts and Stairway Chairlifts[5]. Slight differences exist in the standards for each device in public vs. private locations, such as the maximum rated load for inclined platform lifts (320 kg (700 lbs) for public, 340 kg (750 lbs) for private residences). For any setting, speed of inclined platform lifts is limited to 0.15 m/s (30 ft/min) and speed for inclined stairway chairlift is limited to 0.2 m/s (40 ft/min).

In Canada, standards are in place via Canadian Standards Association (CSA) B355-09 Lifts for Persons with Physical Disabilities[6]. This standard limits travel speed of all devices traveling along stairways at 0.15 m/s (30 ft/min). Minimum weight capacities are specified for inclined platform lifts (180 kg, 397 lbs) and for inclined stairway chairlifts (110 kg, 243 lbs). No maximum weight capacities are specified, however the standard states that the rated load be increased when appropriate.

International standards exist via ISO 9386-2 (2000) Power-Operated Lifting Platforms for Persons with Impaired Mobility – Rules for Safety, Dimensions, and Functional Operation; Part 2: Powered Stairlifts for Seated, Standing, and Wheelchair Users Moving in an Inclined Plane[7]. This standard limits speed of travel of all devices to 0.15 m/s (30 ft/min). Minimum weight capacity is specified as 115 kg (254 lbs) for inclined stairway chairlifts and 150 kg (330 lbs) for inclined platform lifts. For public spaces (where the exact loading may not be known), a minimum weight capacity of 225 kg (496 lbs) is specified. In all situations, for all devices, maximum weight limit is specified as 350 kg (772 lbs).

## *Standards for Evacuation Chairs*

Evacuation chairs have gained an increased identity in life safety codes. The National Fire Protection Association Life Safety Code 101 notes overall goals of use[8], but no standards exist regarding the performance of these devices. Work has begun, however, regarding the development of standards for evacuation chairs. RESNA, the Rehabilitation Engineering and Assistive Technology Society of North America, in 2009 approved the formation of the Standards Committee on Emergency Stair Travel Devices used by Individuals with Disabilities (ESTD).

RESNA is an ANSI-accredited standards development organization, and the RESNA ATSB is the U.S. Technical Advisory Group for ANSI for the development of ISO standards pertaining to Assistive Technology and other products for individuals with disabilities. This established structure will enable RESNA and ANSI to participate in any future ISO discussions regarding evacuation chairs.

## References

1. Hedman, G.E.: Stair Descent Devices: An Overview of Current Devices and Proposed Framework for Standards and Testing. Proc Human Behavior in Fire, 4th International Symposium, Cambridge, UK (2009)
2. ISO 7176 - Part 24 - Wheelchairs: Requirements and Test Methods for User-Operated Stair-Climbing Devices. International Organization for Standardization. Geneva, Switzerland (2005)
3. ISO 7176 - Part 23 - Wheelchairs: Requirements and Test Methods for Attendant-Operated Stair-Climbing Devices. International Organization for Standardization. Geneva, Switzerland (2002)
4. ISO 7176 – Part 28: Requirements and Test Methods for Stair-Climbing Devices. International Organization for Standardization. Geneva, Switzerland (2009)
5. ASME A18.1: Safety Codes and Standards for Platform Lifts and Stairway Chairlifts. American Society of Mechanical Engineers. New York (2008)
6. CSA B355-09: Lifts for Persons with Disabilities. Canadian Standards Association. Mississauga, Ontario, Canada (2009)

7. ISO 9386-2: Power-Operated Lifting Platforms for Persons with Impaired Mobility – Rules for Safety, Dimensions, and Functional Operation; Part 2: Powered Stairlifts for Seated, Standing, and Wheelchair Users Moving in an Inclined Plane. International Organization for Standardization. Geneva, Switzerland (2000)
8. *NFPA Life Safety Code, 101-2009*. National Fire Protection Association. Quincy (2009)

# The Evacuation Training Problems of an Earthquake in China

**C. Lu**

School of Naval Architecture, Ocean and Civil Engineering
Shanghai Jiaotong University, Shanghai 200240, China
Chunxialu@sjtu.edu.cn

**Abstract** WenChuan Earthquake took place on May 12, 2008 in China. 69,227 were confirmed dead, but Shangzao junior high school was a miracle. There was no death or injured. It took over two thousand students and teachers only one minute and thirty-six seconds to evacuate safely from the two 5-floor buildings. One of the miracle resulted from their continuous evacuating training every semester. The paper summarized the evacuating experience of Shangzao junior high school firstly. Other trainings from high school to elementary school were analyzed after the earthquake. Be short of the guidance and organizations, there were some problems in the trainings. Speed should be focused as the No.1. Order is the most important especially in elementary school. Students should be told to run downstairs in two lines if the stairs were not wide enough. It was dangerous for students to put books on their heads while they ran down, because it led to tumble easily.

## WenChuan Earthquake and the miracle

WenChuan Earthquake took place on May 12, 2008 in China. 69,227 were confirmed dead, and 374,176 injured, with 18,222 listed as missing. The school, Shangzao junior high school was a miracle during this disaster because there was no death, even injured. All over two thousand students and teachers were safe. It took them only one minute and thirty-six seconds to evacuate safely from the two 5-floor buildings. The miracle resulted from two reasons. One was due to strengthening of the buildings; the other was that they insisted on the evacuating training once each term. It was amazing that the process, the pattern and the result of the evacuating during the “5.12” earthquake was the same as the training experience.

R.D. Peacock et al. (eds.), *Pedestrian and Evacuation Dynamics*,
DOI 10.1007/978-1-4419-9725-8_11, 

**Fig 1.** The teaching buildings of Shangzao junior high school after the earthquake

The miracle aroused not only the recognition from the governments, but also all schools from kindergartens to senior high schools. Most of the primary schools and middle schools have got on with earthquake evacuating training after the WenChuan Earthquake. The paper summarized the evacuating experience of Shangzao junior high school firstly. The evacuate trainings from other schools subsequently were described then. Some problems were analyzed and suggestions were put forward to help the evacuating training.

## Evacuating training of Shangzao junior high school

Shangzao junior high school arranged the evacuate training once each term. They have insisted the training system for 6 years. All the teachers and students joined together. They also have written down the training draft and perfected it continuously.

Students were told in advance there would be a evacuate training this week, but they did not know the certain day. The training would be arranged at a break time.

The following are their escape rules.

1. They never impose the speed. Order is the first. Only the order can guarantee the safety.
2. The number of students in each class is about 70. The students are seated 9-coloun 8-row. The students of fore 4 rows evacuate from the fore door. The students of latter 4 rows evacuate from the latter door. The students of each column are educated to evacuate from the appointed walkway.
3. The evacuating road of each class is fixed in advance.
4. Two classes will make use of one stairs together.
5. Students of each class must be in line.
6. Each line runs downstairs either along the wall or along the railing.
7. There are eight rows in each class. The front four rows escape from the front door, the back four rows from the back door.
8. The students have also been told who are in the first or the second floor should run fast in order not to build up the escape route, who are in the third or fourth floor should run slowly in order not to jam the crowd.

Unfortunately, all the training pictures and drafts were destroyed during the earthquake. The above rules were summarized by their teachers.

## The trainings of other schools after earthquake

Most of the elementary schools and middle schools have been on earthquake evacuate training after the WenChuan Earthquake in May or June. Be short of the guide book, there were lot of problems about these trainings which could be learned from the table 1 and pictures.

**Table 1 Evacuate trainings**

| School | age | The number of students and teachers | Evacuating time | Speed(n/m) |
|---|---|---|---|---|
| senior high school | 16-18 | 3000 | 5 minutes | 600 |
| senior high school | 16-18 | 2000 | 4 minutes | 500 |
| senior high school | 16-18 | 1400 | 2minuted 50 seconds | 494 |
| junior high school | 12-15 | 3000 | 4 minutes 15 seconds | 706 |
| junior high school | 12-15 | 2000 | 2 minutes 58 seconds | 665 |
| junior high school | 12-15 | 2000 | 3 minutes 17 seconds | 876 |
| elementary school | 6-11 | 1300 | 2 minutes 40 seconds | 488 |
| elementary school | 6-11 | 3500 | 3 minutes 3 seconds | 1147 |

| | | | | |
|---|---|---|---|---|
| elementary school | 6-11 | 1200 | 3 minutes | 400 |
| Kindergarten | 3-5 | 1000 | 4.5 minutes | 222 |

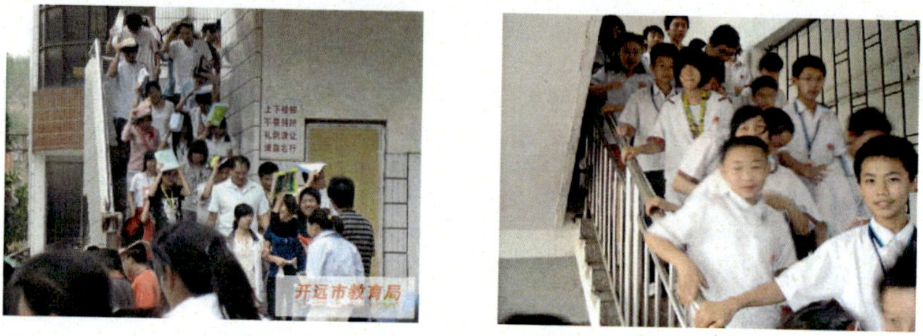


**Fig. 2**. Senior high school

**Fig. 3.** Junior high school

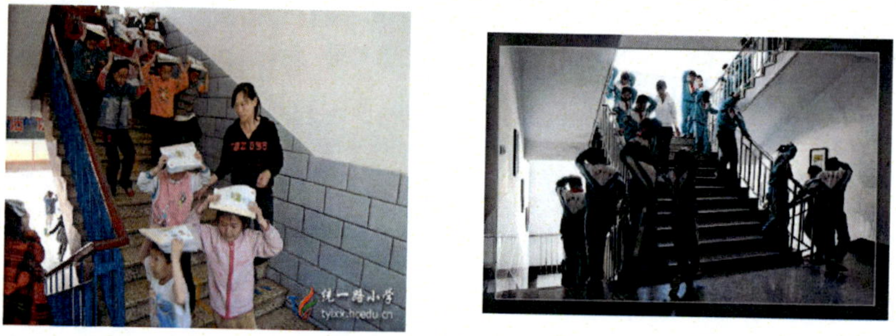


**Fig. 4.** Elementary school

**Fig. 5.** Kindergarten

Most of the trainings were not successful compared with Shangzao junior high school. It cost too much time. Not all could evacuate from the buildings within such time if they faced the real earthquake.

## The analysis of the trainings

The evacuating training is influenced by many factors, such as environment, psychology, age, gender and so on. These training data are analyzed in detail.

### *Speed Versus Age*

Most of people think that the older students have quicker response ability. The older the students are, the faster the evacuating speed is. The training result shows the relationship between speed and age is not taken for granted. The evacuating speed of junior high school is the fastest, next is elementary school. The speed of senior high school is only the third.

The older the students are, the laxer they are. It is why their evacuating speed is slow. From the records, we can see these following situations.

1. Some students of senior high school liked to talk and laugh while evacuating. Younger students walked faster without talking.
2. Younger students were more willing to stand by the instructions of teachers. Teachers were more powerful at their heart than that at older students' heart.
3. The order of younger students was better than that of older students. Some older students didn't evacuate as assigned lines.
4. Younger students trained much times. One elementary school trained four times in a week, while most older students trained only once because of their strenuous hard-working.

**Table 2. Evacuating speed versus age**

| School | Age | average speed(n/m) |
|---|---|---|
| senior high school | 16-18 | 531 |
| junior high school | 12-15 | 749 |
| elementary school | 6-11 | 678 |
| Kindergarten | 3-5 | 222 |

## *Two Lines Versus More Lines*

We got an astonishing result after studying the records. The students evacuated faster in two lines than in more ones especially when were young. It was justified much more by Shangzao junior high school.

The students are prone to fall down much more in the middle line. Younger students tumbled easily without supporting. Tumbling is the disaster while evacuating because it will result in trample accident.

We observed four evacuating trainings of a elementary school in a week. On the first day, students were encouraged to run downstairs fast. Students were crowded on the stairs. There were about 3 or 4 students on a step. 14 students tumbled because there was no order. Then students were arranged 3 lines to evacuate. 6 students tumbled. Finally, they evacuated in two lines along the wall or armrest. No one tumbled. The speed of the four trainings changed too much from the table 3.

Speed should not be emphasized during evacuating training especially in elementary schools. Pupils should also be educated to walk along the wall or armrest at break time to avoid tumbling. Sequence is the most important.

**Table 3. Four evacuating trainings of a elementary school**

| | 1st training | 2nd training | 3rd training | 4th training |
|---|---|---|---|---|
| Evacuating pattern | Free running | Free running | Three lines | Two lines |
| Evacuating speed(n/m) | 1112 | 807 | 624 | 558 |
| Pupils tumbling | 14 | 6 | 1 | 0 |

### *The Width Of Stairs*

Overcrowding is common in China. Most of the classes are over 50 students. There is not enough space to build stairs especially in big cities, as seen from the table 4.

**Table 4.** **The width of stairs**

| The width of stairs | 1.5m-2m | 2m-2.5m | 2.5m-3m | >3m |
|---|---|---|---|---|
| | 40% | 20% | 25% | 15% |

Most of the width of stairs in the elementary school are below 2 meters because most of these schools are located in centre city district. The stairs with 2 meters can hold three students on a step, but students in the middle without any support are easy to fall down especially in elementary schools. The speed of the middle line is also slower than that of both sides. The older the students are, the more steadily they run downstairs. The pupils should be encouraged not to run in the middle.

### *Protective Books on Head*

Most of the students are educated to put books or to clasp hands on their heads while running downstairs. Some experts think these books can protect students' heads from hitting by falling bricks. But it is not reasonable, maybe.

Firstly, it results in the low speed. Pupils are not used to running with these actions. They must pay more attention to the stairs. Most of them must watch the each stair. The speed reduces up to 20%.

Secondly, pupils cannot support wall with hands which leads to tumble easily as explained before.

Thirdly, the evacuating order cannot be kept especially when students are young.

## Conclusion

The paper firstly concludes the successful experience of Shangzao junior high school, and then investigates other school trainings. We have investigated about ten schools including elementary schools and middle schools. What impressed us best is that all these trainings are short of guidance. There are no organizations responsible for the trainings. Most of schools are crowded. Not all students have been educated to evacuate accurately.

# An Experimental Evaluation of Movement Devices Used to Assist People with Reduced Mobility in High-Rise Building Evacuations

**A.P.M. Adams and E.R. Galea**

Fire Safety Engineering Group,
University of Greenwich, London, SE10 9LS, UK
Corresponding author: E.R.Galea@gre.ac.uk

**Abstract** Evacuating people with reduced mobility (PRM) from multi-storey buildings can be a difficult task. A number of commercially available devices can be used to assist in moving the PRM to the ground however, there is little consistent data quantifying the relative performance of these devices. In this paper four commonly used assist devices, the Evac+Chair, Carry-Chair, Stretcher and Drag Mattress are used in a series of 32 evacuation trials designed to assess their performance. The trials involve moving a PRM from a wheelchair to the device, moving the PRM along a long corridor to a stair and then down 11 floors to the ground. The performance of the devices is then assessed in terms of travel speed on the flat and stairs, number of handlers required to operate and ease of overtaking by other stair users.

## Introduction

Evacuating people with reduced mobility (PRM) from multi-storey buildings can be a difficult task. In the UK, some high-rise buildings are equipped with fire fighter lifts which are designed to be operated in fire conditions and can be used to evacuate PRM. However, in most cases PRM are expected to remain within the building in a refuge or place of safety or can be assisted out of the building by fellow occupants. In some countries there is the expectation that the fire brigade will be able to rescue PRM located in refuges or places of safety. While the fire brigade (department) may be able to rescue PRM taking refuge in places of safety, there are several examples where this has tragically not been the case, for example the WTC [1], where many PRM left in places of safety were not able to be rescued. The recent Lakanal House fire [2] in the UK which claimed the lives of six residents who were trapped by smoke in their 14 storey apartment building also serves to demonstrate that the fire brigade may not always be able to rescue people seeking refuge in perceived places of safety. Indeed, in the UK, the Regulatory Reform (Fire Safety) Order 2006, which relates to places of employment, assembly, health care facilities, educational establishments, etc emphasises that it is a

R.D. Peacock et al. (eds.), *Pedestrian and Evacuation Dynamics*,
DOI 10.1007/978-1-4419-9725-8_12, 

managements responsibility to ensure that everyone can evacuate a building safely and that it is not acceptable to simply rely on the Fire and Rescue Services intervention to enable the safe evacuation of occupants [3].

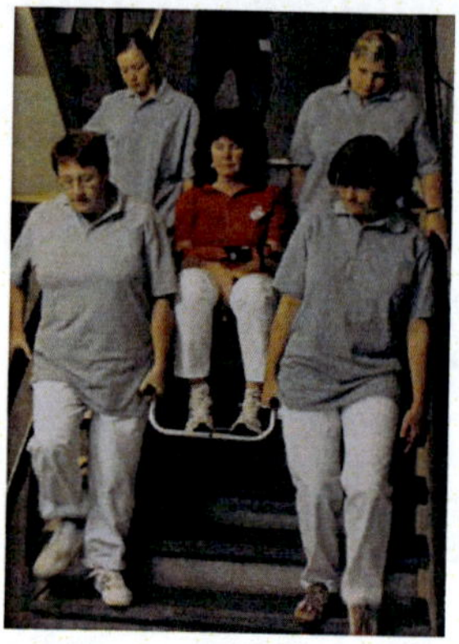

(a) (b)

**Fig. 1. (a) Female handling team with Carry-Chair assist device and female PRM (b) Female handling team with Evac+Chair assist device and male PRM**

While evacuating PRM from high-rise buildings may be a difficult task, an even more daunting situation involves the evacuation of hospitals and care facilities. The recent Royal Marsden Hospital fire in the UK [4] demonstrates that it may be necessary not simply to undertake a progressive horizontal evacuation of patients to places of relative safety, but to fully evacuate an entire hospital. The added complexity in hospital evacuations is due to a number of reasons including; the large number of occupants (patients) requiring assistance to evacuate, the (potentially) relatively small number of staff present to assist in the evacuation of patients (e.g. during night shifts), the need to have multiple staff to assist in the evacuation of single patients, the impact of fatigue resulting from the need for staff to make repeat trips, the time required to prepare patients for assisted evacuation and the potential blocking of stairs due to the assist teams carrying PRM delaying the evacuation of able body occupants.

In both high-rise buildings and hospitals, PRM may be evacuated using a number of different assist devices. Devices commonly used to assist in the evacuation of PRM include; Carry-Chair and Evac+Chair, see Figure 1. In addition to these two devices, in hospitals the drag mattress (with slide sheet) and the stretcher are also used, see Figure 2. While these assist devices are commonly used in both high-rise buildings and hospitals, there is little consistent data quantifying their relative performance or identifying the level of training required to safely and efficiently operate the devices. This includes issues such as, the relative ease (including number of required operators) in transporting the PRM to the assist device, the movement speed of the assist device on the flat and on stairs, the number of people required to operate the assist device, the impact that the device may have on the evacuation of others and the training required by device operators. It is thus difficult for safety managers to assess the relative merits of each device and more importantly, realistically plan how the device should be deployed in their buildings. A key recommendation from the recent Homeland Security Standards Panel of ANSI was that additional work is required specify standards for assist devices and their usage [5]. Furthermore, if these devices are to be represented within computer based evacuation models, it is essential that their performance is quantified.

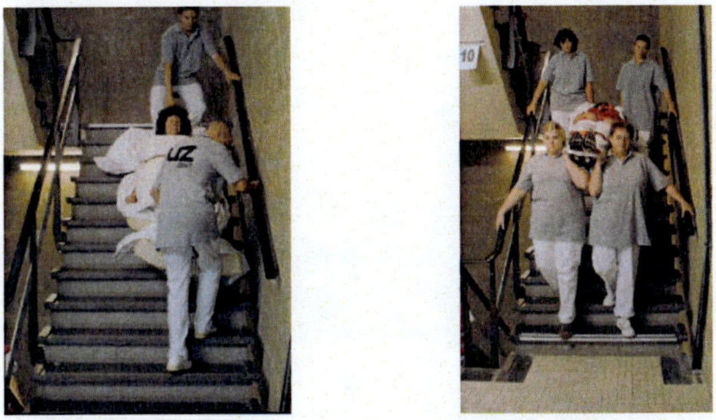

(a) (b)

**Fig. 2. (a) Female handling team with Drag Mattress assist device (with slide sheet) (b) Female handling team with Stretcher assist device**

This paper addresses these issues by presenting preliminary results from a series of experiments conducted by the Fire Safety Engineering Group (FSEG) of the University of Greenwich in collaboration with the Universitair Ziekenhuis (UZ) Gent (University Hospital of Gent) in Belgium measuring the performance of four commonly used assist devices (see Figures 1 and 2). The trials were designed by FSEG and conducted on the premises of UZ using UZ staff.

## Trial Plan, Building Layout and Data Collection Methodology

In total a series of 32 trials were undertaken over a two day period from 17 to18 September 2008. The trials were conducted in a 14 floor building of the University Hospital of Gent (see Figure 3) using trained staff from the UZ hospital. Four handling teams, two male and two female, used each of the four devices shown in

Figures 1 and 2. Two volunteers from the UZ acted as the PRM. All 18 staff (16 in the handling teams and 2 PRM) were highly trained in the use of the devices and in handling patients and were members of the UZ Manutentie Team. Using highly trained handlers removes the issue of training from the device performance analysis. While the two PRMs had different body weights, for consistency the weight of the two was made identical (i.e. 75 kg) through inserting lead weights into the pockets of the lighter PRM. Half the trials consisted of individual device trials while the other half consisted of trials in which a group of 24 people (students from UZ) were injected onto the stair from the 6$^{th}$ floor to investigate the ease with which other evacuees could pass the assist team with the PRM down the stairs.

**Fig. 3. Building in which evacuation trials undertaken**

For each trial the PRM was located in a room on the 11$^{th}$ floor of the building and was positioned in a wheelchair. At the sounding of the 'go' signal, the assist team would enter the room, move the PRM to the device, move the PRM out of the room into the corridor, travel 63.0 m along the corridor, pass through three sets of doors along the corridor, negotiate a left 90 degree turn into another corridor, move past the lifts (elevators), enter the stair case (see Fig. 4.) and descend 11 floors to the ground level, exit the stairwell, travel 5.0 m along the ground floor, exit the building and travel a further 32.1 m to an end point outside the building. The stairs were dog-legged, with a single flight down to half landing followed by another flight down to the next floor. The stairs were inclined at an angle of 34$^0$, were 1.4 m wide (handrail to handrail) and each flight had a drop of 2.1 m from floor to half landing and from half landing to the next floor, with the exception of the last flight which has a slightly shorter drop. The main landing on each floor measured 3.3 m x 2.1 m while the half landing measured 3.3 m x 1.4 m. The total travel distance down the stairs (as measured from the stair entry point on the 11$^{th}$ floor to the stair exit point on the ground floor taking a central path) was 169 m.

The progress of the handling teams was recorded using fixed and roaming video cameras and fixed observers with stop watches on each floor. The fixed video cameras were positioned on each floor and recorded the movement of the handling team down the stair and on the landings. The roaming video camera followed the handling team from the point that they first touched the PRM, transferred him/her to the assist device, moved them down the corridor onto and down the stair in each trial. In addition, at the end of each trial, the assist teams and the PRM completed a questionnaire. Separate questionnaires were administered to the handling team, the PRM and the group of people attempting to overtake the PRM.

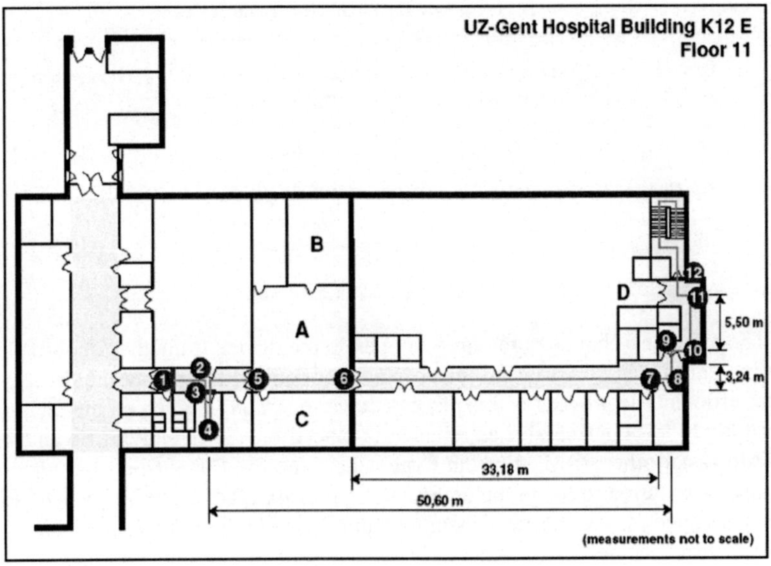


**Fig. 4. Floor plan of starting floor (floor 11)**

## Results and Discussion

Here we present preliminary results based on timings derived from the stopwatch data, some observations from video footage and some analysis of the assist team questionnaire data. Presented in Table 1 is a summary of the movement results for the corridor. This involves moving the PRM from their starting location to the stair entry point a total distance of 63.0 m. Along this route the PRM must pass through three sets of doors which were all originally closed. The results presented in Table 1 represent an average of eight trials per device and represent trials with both male and female assist teams and for the group and individual trials.

Clearly, the nature of the group trials did not have an effect on the corridor performance and so these performance results have been included in these averages.

As seen in Table 1, the Evac+Chair and the Carry-Chair are the fastest device, achieving an average speed of 1.5 m/s. The drag mattress is the slowest device achieving 0.9 m/s. The Evac+Chair and Carry-Chair are some 50% faster than the other two devices. However, all four devices produce reasonable speeds, with the Evac+Chair and the Carry-Chair producing speeds comparable to unobstructed fast walking speeds. It should be noted that in all these trials, the corridors where unobstructed by other evacuating individuals.

**Table 1. Average horizontal performance for 8 trials for each device**

| Device | Average Travel Time (sec) | Number of Handlers in Emergency | Average Speed (m/s) |
|---|---|---|---|
| Evac+Chair | 41 | 1 (+1 for doors) | 1.5 |
| Carry-Chair | 41 | 1 (+1 for doors) | 1.5 |
| Stretcher | 58 | 4 | 1.1 |
| Drag Mattress | 69 | 2 | 0.9 |

The number of handlers required to operate the device is another important operating parameter. Clearly the fewer handlers required to operate the device, the more efficient the process as this places a lower operating burden on the facility, is more likely to be successful as reliance is placed on fewer individuals and frees staff to assist other PRM. For the Evac+Chair and the Carry-Chair only a single handler is required to move the device along the corridor. However, when closed doors are encountered, a second handler can be used to open the doors. This is the method used in the trials. Alternatively, the handler pushing the device would need to stop, turn the device around and while holding the door open, pull the device through, then turn the device around and continue. While this is possible, it would clearly have a negative impact the horizontal travel speed performance. The Stretcher is not only one of the slowest of the devices, but as it requires four handlers, requires the most number of handlers to operate. It should be noted that the Drag Mattress was pulled by both operators along the corridor. As both the Stretcher and the Drag Mattress had operators in the front of the device, this proved relatively easy to negotiate closed doors however; the device did have to stop as the door was opened. In addition, both these devices occupy a large footprint when moving along the corridor. Compared to the other devices, which have a considerably smaller footprint, this may prove a disadvantage in crowded situations.

Presented in Table 2 is a summary of the movement results down the stairs. This involves moving the PRM from entry point of the stairs on the 11$^{th}$ floor down 21 flights of stairs to the exit point of the stairs on the ground floor, a total distance of 169.0 m. The results presented in Table 2 represent an average of four trials per device and represent trials with both male and female assist teams. Unlike the horizontal results presented in Table 1, the stair results are only for the individual trials. As seen in Table 2, the Evac+Chair is clearly the fastest device, achieving an average speed of 0.81 m/s. This is some 30% faster than the next fastest device, the Drag Mattress. However, all three other devices produce comparable speeds of approximately 0.58 m/s. The speed of the Evac+Chair on the stair is approximately half that of the same device on the flat. A significant difference between the performance of the devices on the flat and down the stairs is that when going down the stairs a number of stops were required. This was for a number of reasons including, resting the handlers, rotating the handlers or improving the handlers grip on the device. The number of stops for each device varied considerably as did the duration of the stop and contributed to the difference in performance. However, the Evac+Chair did not stop a single time during the descent.

**Table 2. Average vertical performance for 4 trials for each device**

| Device | Average Travel Time (sec) | Number of Handlers in Emergency | Average Speed (m/s) |
|---|---|---|---|
| Evac+Chair | 209 | 1 | 0.81 |
| Carry-Chair | 297 | 3 male or 4 female | 0.57 |
| Stretcher | 305 | 4 | 0.55 |
| Drag Mattress | 272 | 2 | 0.62 |

As with device performance on the flat, the number of handlers required to operate each device on the stairs varied between devices. Only the Evac+Chair required a single handler. It should be noted that while only a single handler is required, it is considered good practice to have a second handler in front of the device to reassure the PRM during the descent. During these trials the Evac+Chair was used in this way however, the second handler played no role in the stair descent. It is suggested in an actual emergency evacuation situation it would be possible to operate the device with only a single handler. The Drag Mattress required two handlers, one at the front and one in the rear. The handler at the rear assisted the descent in a number of ways such as, supporting the end of the mattress thereby reducing the jolting to the head of PRM during the descent, acting as a break so that the descent was controlled and assisting to turn the mattress on the landings (see Figure 2a). The Stretcher required the largest number of handlers, requiring four as on the flat.

The Carry-Chair proved to be the only device that was sensitive to the gender of the handlers. Using an all female handling team, the Carry-Chair required four operators, as shown in Figure 1a, while using an all male handling team the Carry-

Chair required three handlers. When using three handlers, the Carry-Chair would only be carried by two handlers, one at the front and one at the rear. When the carry team needed a rest, the third handler would relieve one of the carry team. In the four person female team, when the handlers needed a rest, they would rotate their location around the chair.

In an attempt to gauge the impact of the devices on other people simultaneously using the stairs a series of 16 group trials were also conducted. These involved a group of 24 people who enter the stairs on the 6$^{th}$ floor just after the device has passed their location and who attempt to overtake the device. From observing video footage of these trials it is clear that the Evac+Chair creates the least obstruction to other stair users. Other stair users are easily able to overtake the device on the stairs (see Figure 1b) as the device and its handler occupies a single lane on the 1.4 m wide stair. Other users can also get around the device on the landings. The Drag Mattress is the next best in offering least resistance to other stair users. The Drag Mattress can also be overtaken on the stairs (see Figure 2a) but does occupy more of the width of the stair than the Evac+Chair. However, the Drag Mattress is more difficult to overtake on the landings, requiring a greater turning circle than the Evac+Chair. This is particularly noticeable on the half landing which is not as deep as the main landing.

The Carry-Chair when operated by all female handlers does not provide an opportunity for overtaking on the stairs (see Figure 1a). Other stair users can only overtake when the handlers stop on the landing and allow other users to pass. When operated by all male handlers, the Carry-Chair can be overtaken on stairs. In this configuration, the Carry-Chair can also be overtaken on the landing if the handlers stop and let the other users by. The Stretcher cannot be overtaken on the stairs (see Figure 2b) and can only be overtaken if the handlers stop on the landing and let the other users by. It should be noted that these observations are specific to the stair configuration found in these trials. The stairs are particularly wide at 1.4 m and the landings are also quite wide. Stairs found in a typical office building can be somewhat narrower, for example, two of the stairs in the WTC (Stair A and C) were 1.1 m wide while the third stair (Stair B) was 1.4 m wide [1]. In addition, training of the handlers is also an important aspect to consider when assessing the obstruction caused by the devices. Handlers of devices such as the Evac+Chair and the Drag Mattress should be trained not to block the stairs handlers of all devices should be trained to allow others to pass on landings where possible.

The questionnaires provided an opportunity for the participants to express their opinion on a range of issues associated with the devices. The questions were in Flemish and generally used a five point Likert Scale. The questionnaire for the handlers consisted of 15 questions, some with multiple parts and a section at the end for comments. Question 2c asked the handling team to “Please rate this device on the physical effort to transport the PRM down the stairs (how demanding)”. Respondents could select from 1 (Very Difficult), 2 (Difficult), 3 (Neither Difficult nor Easy), 4 (Easy) and 5 (Very Easy). Each person in the handling teams were requested to complete the questionnaires for each of trials. As the size of the handling teams differed, the number of response also differed. For example, a Stretcher handling team consisted of four people and they undertake eight different trials and so there would be 32 replies to Question 2c for the Stretcher, whereas for the Drag Mattress, the handling team only consisted of two people and hence there would only be 16 replies to Question 2c. The responses for each device in each category was thus normalised by dividing by the total number of responses for that device. In response to Question 2c, 81.3% of the handlers responses classed the Evac+Chair in the Easy/Very Easy categories while none of the responses for the Evac+Chair were in the Hard/Very Hard category. In contrast, 88.6% (67.9% and 53.2%) of the responses classed the Drag Mattress (Carry-Chair and Stretcher respectively) in the Hard/Very Hard category. Clearly, the experienced handlers find that the Evac+Chair required the least effort of all the devices while the Drag Mattress required the greatest effort in descending 11 floors. Question 11 asked the handlers to, "Please rate this device on your level of discomfort from muscle soreness in the arms". Respondents could select from 1 (Very Much), 2 (Much), 3 (Neither Much nor Little), 4 (Little) and 5 (Very Little). In response to Question 11, 93.8% of the handlers responses classed the Evac+Chair in the Little/Very Little categories while 32.2% (32.2% and 21.4%) of the responses classed the Stretcher (Carry-Chair and Drag Mattress respectively) in the Little/Very Little categories. For the Evac+Chair, 0% of the respondents classed the Evac+Chair in the Much/Very Much categories. In contrast, 71.5% (48.4% and 39.3%) of the responses classed the Drag Mattress (Stretcher and Carry-Chair respectively) in the Much/Very Much categories. Clearly, all the experienced handlers found little muscle strain while using the Evac+Chair while the greatest muscle strain was experienced when using the Drag Mattress.

## Conclusions

A series of 32 evacuation trials assessing the movement capabilities of four different assist devices; Evac+Chair, Carry-Chair, Stretcher and Drag Mattress, have been successfully completed. The trials evaluated a number of performance criteria including; travel speed along a corridor and on stairs, number of handlers required, ease of overtaking by other stair users and subjective to questionnaires by handlers, PRMs and other stair users. Preliminary analysis presented in this paper is based on stopwatch timings, observations of video footage and questionnaire responses. These results allow different aspects of device performance to be assessed, providing building operators and safety managers a quantified basis upon which to make implementation decisions.

The results clearly show that in the hands of experienced handlers, the devices have significantly different performance capabilities. When travelling over 63 m of corridor, the Evac+Chair and the Carry-Chair are equal fastest (1.5 m/s), being some 50% faster than the other devices and requiring the least number of handlers. While descending 11 floors using the stairs, the Evac+Chair is the fastest device (0.81 m/s) being some 30% faster than the other devices and requiring the least number of handlers. The Evac+Chair also offered the least degree of obstruction to other stair users, enabling them to overtake both on landings and on the stairs.

It should be noted that these observations are specific to the stair configuration found in these trials. Furthermore, it again must be emphasised that the handlers used in these trials were professional staff of UZ, trained in the correct use of each devices. All the devices require that handlers are professionally trained in their use if they are to be used correctly and efficiently and in a manner that minimises the threat of injury to the handlers, the PRM and other stair users.

**Acknowledgments** The authors are indebted to a number of people who enabled this project to happen and who assisted in carrying out the work, in particular, Prof N Fraeyman of UZ who permitted the work to be undertaken at UZ, utilising both UZ buildings and staff, Mr Filip Buckens the coordinator of the UZ Manutentie Team and his 18 staff who donated their time to the project and did all the heavy lifting (and sitting), the 24 UZ student volunteers who ran up and down the stairs, the team of UZ administrative staff who contributed to the logistics of the operation, the 11 friends of Mr Adams who volunteered to assist with the stopwatch observations, the video camera team and finally, Ms Aoife Hunt, a PhD student from FSEG who assisted with the trials and who is currently undertaking a detailed analysis of the video footage as part of her PhD studies.

## References

1. Galea, E.R., Hulse, L., Day, R., Siddiqui, A., Sharp, G.: "The UK WTC 9/11 Evacuation Study: an Overview of the Methodologies Employed and some Analysis Relating to Fatigue, Stair Travel Speeds and Occupant Response Times." Proc 4th Int Symp on Human Behaviour in Fire, 13-15 July 2009, Robinson College, Cambridge, UK, pp. 27-40 (2009)
2. 'Fatal fire block's layout probed.' http://news.bbc.co.uk/1/hi/england/london/8135535.stm. Last accessed 26 July 2009
3. Fire Safety Risk Assessment, Means of Escape for Disabled people. HMG, Reference number 06 FRSD 03913(a) (2007)
4. Hospital fire was a 'wake up call.'http://news.bbc.co.uk/1/hi/programmes/file_on_4/7213205.stm. Last accessed 26 July 2009
5. Emergency Preparedness for Persons with Disabilities and Special Needs, Final Workshop Report, ANSI Homeland Security Standards Panel (2009)

# Evacuation Dynamics of Children – Walking Speeds, Flows Through Doors in Daycare Centers

**A. R. Larusdottir and A. S. Dederichs**

Department of Civil Engineering
Technical University of Denmark
2800 Kgs. Lyngby, Denmark
Corresponding author: and@byg.dtu.dk

**Abstract** The purpose of this study is to deliver new data and to bring attention to the subject of evacuation of children. Evacuation characteristics such as flow, densities and walking speeds are in focus. Current literature on evacuation is based mostly on studies on adults. Ten Danish daycare centers participated in full scale evacuation experiments where two age groups 0-2 years and 3-6 years were analyzed separately. The overall findings were as follows. Flow through doors, walking speeds and densities were age-dependent and differed strongly from the data in existing literature. The results showed higher walking speeds in spiral stairs when the children were familiar with the evacuation path. Higher person densities and faster flow through doors were obtained among the children than found in the current literature on adults. Children in the younger age group were generally slower than the older children. The children walked slower in horizontal plane than adults, however they were keen to run during the evacuations, in the latter case their travel speed increased and exceeded the adults'. Since the evacuation characteristics of children differ in many ways from those of adults, nowadays models badly comprehend the evacuation behavior children.

## Introduction

An understanding of human evacuation dynamics and performance are important when designing new buildings and applying performance-based codes in order to reduce the risk of exposing occupants to critical conditions in case of fire. The existing literature provides a number of case studies of real fire incidents as well as experiments concerning fire and evacuations [1-3]. These studies provide information on walking speed in horizontal plan and in stairs and flow through doors as well as other aspects relevant for evacuation [4]. The results are applied in models and introduced in simulation programs [5] for prediction of the evacuation process. The majority of previous studies deal with the evacuation behavior of

R.D. Peacock et al. (eds.), *Pedestrian and Evacuation Dynamics*,
DOI 10.1007/978-1-4419-9725-8_13, 

adults where users are expected to be able to bring themselves to safety in case of an evacuation. These include studies on evacuation dynamics in private, public, and commercial buildings with different occupational hours, where the evacuees are either familiar or unfamiliar with the building [1-9]. However in recent years studies have focused on a broader population for experiments and models. For instance by considering people with disabilities and other groups that might require assistance during an evacuation [9-13].

However, very few studies can be found which provide information on evacuation dynamics and behavior of children [1, 9,14] and embrace buildings with mass stay of children, such as daycare centers. One of these is a study on daycare centers performed in Kobe, Japan in 1985 [15]. The focus of the research was on evacuation using stairs and slides. A new Russian study from 2009 also investigates evacuation of daycare centers for children [16]. The focus of that study is on human behavior and travel parameters. Hence, there is very little existing data on walking speeds in a horizontal plane and on stairs as well as the flow through doors and densities of children, especially for children at this young age.

## Method

The present study investigates the evacuation dynamics of children and provides experimental data on flow through doors as well as travel speeds for children. Ten Danish daycare centers participated in full scale evacuation experiments. Two daycare centers performed the experiment twice and one performed the experiment three times involving a total of 1017 persons, where of 173 evacuated twice and 67 three times. The experimental period was from March to May and November 2009. Danish daycare centers host two age groups, "younger children" aged between 6 months and 2 years and "older children" aged 3-6 years.

The motion of 71 children was considered when collecting the data for movement speed; 805 counts (mainly children) contributed to the data collection on the flow and 66 persons (mainly children) to the data on spiral stairs. More data would be necessary to prove statistical validity. Hence the results presented in this work are not general, but indicate trends.

A total of fourteen full scale evacuation experiments, in the form of fire drills, were performed and data was collected using video cameras. In all of the daycare centers the subject of a fire drill had been discussed in a staff meeting. The children were less prepared than the adults but in most of the daycare centers the upcoming fire drill had been mentioned and explained to the older children, but without indicating the specific day.

Each of the experiments had a similar course of action. After arriving at the daycare center and talking to the contact person (typically the leader or a safety person[1]) the cameras were set up, focusing on the exits. Shortly before the fire drill the cameras were turned on, one by one. Then a signal was given to start the fire drill, fire alarms were used where possible otherwise an adult started a verbal warning process. The evacuation started and all children and adults evacuated to the outside and gathered at the previously determined assembly point. When all of the children had been accounted for, an "all clear" signal was given. After the drill, most daycare centers chose to discuss the fire drill with the children, which worked well and helped the children to process this experience.

The recorders were partially exposed and out of reach for the children. The dimensions of fixed points in the rooms were taken. The films were analyzed for walking speed in horizontal plane and spiral stairs and flow through doors as well as behavioral pattern.

## Results

In the following three subsections the results of the previously described experiments are introduced. The children are on a daily basis divided into the two previously mentioned age groups, which made it possible to compare results between the two ages.

### *Travel speed – horizontal*

Figure 1 shows the percentage of children moving at a certain speed interval. The speed is shown in intervals of 0.20 m/s (vertical axis) and is measured at low person densities (< 0.5 person/m$^2$) where the children move independently. The travel speeds are differentiated into four series: walking of younger and older children and running of younger and older children. As can be seen in figure 1, more than 78 % of the younger children have a walking speed of 0.41-0.80 m/s and more than 66 % of the older children have a walking speed in the range of 0.61-1.00 m/s. It is also clear that younger children move slower than the older ones, regardless of whether they are walking or running. The average walking speeds were 0.60 m/s and 0.84 m/s for the younger and older age group respectively and 1.14 for younger and 2.23 for older, when the children ran.

Common overall values used for adults' average walking speed in a horizontal plane with low person densities is 1.2 m/s -1.3 m/s [17, 18]. The range for this speed is marked in figure 1 as a hatched background. Comparison of the children's

---

[1] A member of the staff which together with the leader of the daycare center takes care of relevant safety issues. Also known as health and safety officer.

movement speeds to these values from literature concerning adults, shows that the children generally move slower, except when the older children run.

It should be mentioned here that not all children in the younger age group were able to walk by themselves; hence fewer measurements were available for that age. The travel speed of children carried by adults or holding an adult's hand was excluded in this work, since it is likely that in that situation the adult controls the travel speed more than the child.

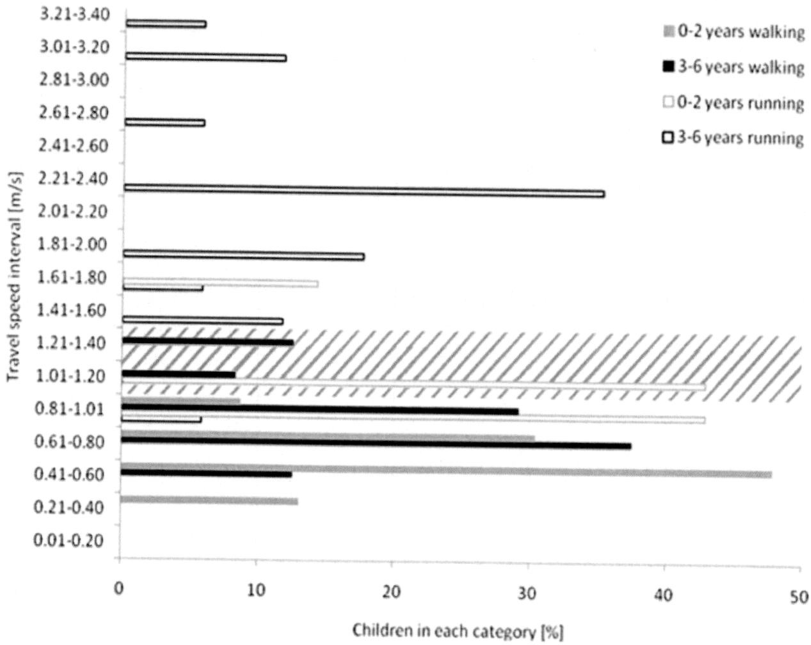


**Fig. 1. Travel speed (intervals of 0.20m/s) for differentiated age groups 0-2 (grey) and 3-6(black) years running (empty box) and walking (filled box). The background of the two intervals including the most commonly used average walking speeds for adults, is hatched**

## *Travel speed – stairs*

The data on stairs only includes the older age group since the younger children were in all cases located at ground level. The study involves three spiral stairs. Width of the steps, the slope of the walking path (defined 0.25 m from the wider end of stair), the average travel speed and the standard deviation of the speed of each stair is shown in table 1. The travel speed is defined as the movement along the slope of the stair.

**Table 1. Results on spiral stairs**

| Stair | Width (m) | Slope (°) | Average speed (m/s) | Standard deviation (m/s) |
|---|---|---|---|---|
| Stair 1 | 0.80 | 33 | 0.58 | 0.31 |
| Stair 2 | 0.87 | 33 | 0.38 | 0.07 |
| Stair 3 | 0.91 | 30 | 0.13 | 0.06 |

Stair 1 is an internal stair used by the children every day. It also has an extra handrail at an appropriate height for the children. Stair 2 is a spiral stair in a square stair case in an old house. The children do not use the stair regularly and the stair only has an inconvenient handrail in the center. Stair 3 is a typical metallic external fire escape, where the steps are see-through. The children had never used the stair before the experiment and the handrail is high and hard to reach for the children.

Although the three stairs have similar dimensions, there was remarkable difference in the average travel speed on the stairs. This can be linked to the difference in the children's familiarity of the stairs and the stairs' design, which is roughly described above.

Based on experiments on spiral stairs, involving adults [5] a general value for the average travel speed of 0.5 m/s is suggested, nondependent on the width of the stair [18].

As seen in table 1 the children's average travel speed on stair 1 exceeds the suggested average travel speed for adults on spiral stairs. It can however also be seen that the standard deviation is high for stair 1. In fact when looking at the travel speed of individual children it was noticed that the first few children moved the fastest and as more children entered the stair the travel speed decreased. The fastest measured speed on stair 1 was 1.4 m/s and the slowest speed measured on the same stair was 0.25 m/s. A likely explanation of this is that the increased person density on the stair affected the travel speed. Another possible factor is that the more hesitant children waited as long as they could to enter the stair, and they did not move as fast as the ones who were eager to evacuate.

The high movement speeds achieved on stair 1 and the slow speeds on the other two stairs strongly indicates that familiarity with the evacuation route leads to a faster evacuation, which was in fact one of the main conclusions of Murozaki's and Ohnishi's study mentioned earlier [15].

It was obvious from the recordings that stair 3 caused the most problems for the children, leading to an extremely low travel speed. Apart from being insecure about walking on the stair and using the vertical bars in the handrail for support, the children were also curious and stopped to look around since this was a totally new environment for them.

## *Flow*

Figure 2 shows the flow through doors for older children (age group 3-6 years, □), compared to younger children (age group 0-2 years, ○) and a mixed group (0-6 years, Δ) and the data from Nelson and Mowrer [17] as reference. The data in figure 2 includes a few adults, accompanying the children, but the majority of the people are children. Common ratios between children and adults was 3-4 children per adult in the younger age group and 6-10 children per adult for the older age group at the time of the evacuations. However in the flow measurements for the older children only a few adults are included as the adults typically waited until last to evacuate, and were not a part of a flow.

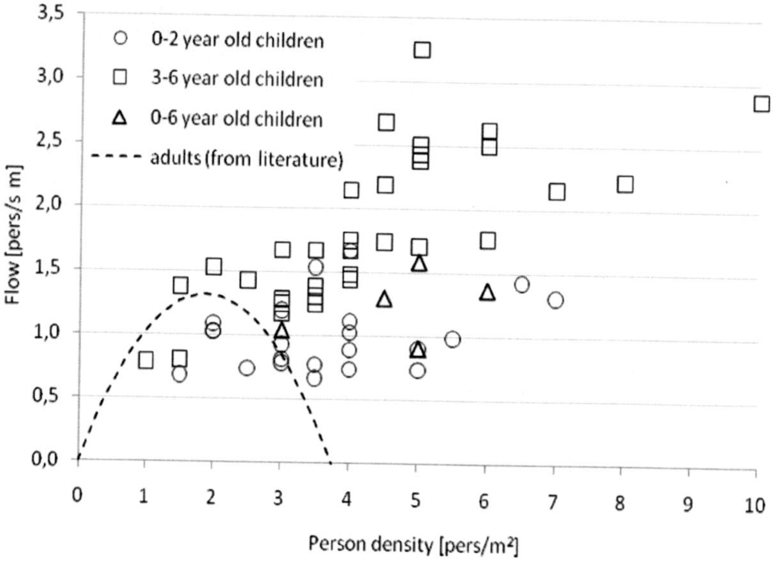


**Fig. 2. Flow through doors, measured during evacuation experiments, with respect to person density**

It can be seen in figure 2 that the flow increases with increasing person density. Again a clear difference can be seen between the two age groups analyzed, as the older children generally move faster than the younger ones, especially at high person densities. This difference was expected as the younger children have a lower walking speed. Other factors that might contribute to this difference are that the younger children were more hesitating, as they did not fully understand what was going on and there were more adults among the young children than in the measurements of the older children.

Compared to the flow curve found in literature, the flow does not stop at a density of 3.8 pers/$m^2$ as suggested for adults but it keeps growing with higher person densities. This may be explained by the size of children. Another reason could be that the children know each other and are comfortable being close to each other, whereas adults generally need more personal space.

The data obtained does not show a clear pattern or a peak. This could be explained by a lack of data especially at person densities higher than 6 pers/$m^2$. The highest person density obtained naturally, during the evacuation experiments, was 8 pers/$m^2$. In the one measurement where the density was 10 pers/$m^2$ the children were instructed to stand in a crowd by the door and then walk through when given a signal. As figure 2 shows, this resulted in a flow of 2.9 pers/s m, but a descent in the flow at such high person density was expected.

Furthermore it could be observed that the children had no problem passing the doors two side by side, even where the doors were only 0.6-0.7 m wide. Concerning the effective width of the doors, it could be seen from the recorded material that the whole free width of the door was used when needed, so the door width was not reduced in the flow calculations, as if there was a boundary layer.

It was found that the two age groups (younger children aged 6 months - 2 years and older children aged 3 - 6 years) vary from each other when it comes to behavior, travel speed and flow through doors. An example of the behavioral difference is that very few of the younger children ran during the evacuations (about 5 %) (not shown), however it was common among the older children to run to the exit (about 40 %). In some cases the children did not have the opportunity to run due to crowd or orders from the teachers on staying in line. It should also be noted that not all of the younger children were yet physically able to run, which also affects these numbers.

## Conclusion

This study presents new data concerning the evacuation dynamics of children. Fourteen evacuation experiments in daycare centers in Denmark were performed, evacuation times were measured and video films were analyzed. The experiment gives new information on flow through doors, walking speeds in a horizontal plane and in stairs.

The experimental study indicates that evacuation characteristics of children, concerning travel speeds on horizontal planes and down spiral stairs as well as flow through doors, differ from the data in literature which is focused on adults. The travel characteristics of children are age-dependent since the results show a clear difference between the two age groups analyzed, younger children: aged 6 months - 2 years and older children: 3 - 6 years old. They also depend on how familiar the children are with the path of escape.

When looking at travel speed in horizontal planes it was found that children move slower than adults and that younger children move slower than older ones. Walking children were slower than the predicted average walking speed usually applied for adults. However, the adult walking speed is exceeded by the running speeds of younger and older children. Hereby it has to be accounted for that the older children were running in a higher frequency than the younger children.

It can be concluded that the walking speed for children deviates from the data obtained and applied from literature.

The results on the travel speed on spiral stairs indicate that familiarity with the stair and the design of the stair greatly affects the speed. In the only stair which the children used on a daily basis and which had a special handrail for the children, the average travel speed exceeded the data from the literature on adults. The two other stairs had average travel speeds which were much lower than the speeds obtained from experiments with adults.

Results regarding flow indicate that flow of children through doors is generally higher than the reference data on adults found in existing literature. Higher densities were obtained, without stopping the flow, and in fact the flow increased at high densities. A clear pattern or a peak was not found, perhaps due to limited measuring points at densities higher than 6 pers/m$^2$. The older children reached both higher person densities and higher flows than the younger children did.

Nowadays evacuation models badly comprehend the behavior of children This suggests that children are less safe in buildings than adults. In order to describe the evacuation of mixed populations the evacuation characteristics of children need to be accounted for. More data and models are needed for further understanding on the subject.

## References


1. Proulx, G.: Evacuation time and movement in apartment buildings. Fire Safety Journal. **24,** 229-246 (1995)
2. Proulx, G.: Movement of People: The Evacuation Timing. In: DiNenno, P.J., Walton, W.D. (Eds.), The SFPE Handbook of Fire Protection Engineering Third ed., pp. 3-341-3-366. Society of Fire Protection Engineers, Bethesda, MD (2002)
3. Nilsson, D.: Exit choice in fire emergencies. Ph.D. thesis, Department of Fire Safety Engineering and Systems Safety. Lund, Sweden (2009)
4. Frantzich, H.: En modell för dimensionering av förbindelser för utrymning utifrån funktionsbaserede krav. Tech. rep., Department of Fire Safety Engineering. Lund University (1994) (In Swedish)

5. Frantzich, H.: Utrymningsvägars fysiska kapacitet, sammanställning och utvärdering av kunskapsläget. Tech. rep. Department of Fire Safety Engineering, Lund University (1993) (In Swedish)
6. Prolux, G., Reid, I.M.A.: Occupant behavior and evacuation during the Chicago cook county administration building fire. Fire Protection Engineering. **16,** 283-309 (2006)
7. Steffensen, F.B.: Brand og brandsikkerhed i forsamlingslokaler et fælles ansvar. Technical report, Københavns Brandvæsen (May 2000) (In Danish)
8. Ravn, K.: Undersøgelse af Reaktions- og Beslutningstid. Master thesis, Technical University of Denmark (2008)
9. Bryan, J.L.: Behavioral Response to Fire and Smoke. In: DiNenno, P.J., Walton, W.D. (Eds.), The SFPE Handbook of Fire Protection Engineering, Third ed., pp. 3-315-3-341. Society of Fire Protection Engineers, Bethesda (2002)
10. Shields, T.J., Smyth, B., Boyce, K.E., Silcock, W.H.: Towards the prediction of evacuation behaviours for people with learning difficulties. Facilities. Vol.17, Issue. 9-10, pp. 336-344. NBC, University Press (1999)
11. Nisser, M.: Utrymningssäkerhet för rörelsehindrade. Tech. rep., Räddningsverket. Sverige (2001) (In Swedish)
12. Gwynne, S.M.V., Boswell, D.L., Proulx, G.: Understanding the Effectiveness of Notification Technologies in Assisting Vulnerable Populations. Journal of Fire Protection Engineering. Vol. 19, pp. 31-48 (2009)
13. Bruck, D., Thomas, I.: Comparison of the Effectiveness of Different Fire Notification Signals in Sleeping Older Adults. Fire Technology. Vol. 44, pp. 15-38 (2008)
14. Bruck, D.: Non-awakening in children in response to a smoke detector alarm. Fire Safety Journal. Vol. 32, Issue. 4, pp. 369-376 (1999)
15. Murozaki, Y., Ohnishi, K.: A study of fire safety and evacuation planning for pre-schools and day care centers. Memoirs of the Faculty of Engineering, Kobe University. **32**, 99-109 (1985)
16. Kholshevnikov, V.V., Samoshin, D.A., Parfenenko, A.P.: Pre-School and School Children Building Evacuation. State Moscow University of Civil Engineering, Academy of State Fire Service of Russia, Moscow, Russia (2009)
17. Nelson, H. E, Mowrer, F.W.: Emergency Movement. In: DiNenno, P.J. , Walton, W.D. (Eds.), The SFPE Handbook of Fire Protection Engineering (Third ed.), pp. 3-367-3-380. Society of Fire Protection Engineers, Bethesda (2002)
18. Brandskyddshandboken, Rapport 3134, Department of Fire Safety Engineering, Lund University (2005) (In Swedish)

# Data Collection (Transport)

# Evacuation Analysis of 1000+ Seat Blended Wing Body Aircraft Configurations: Computer Simulations and Full-scale Evacuation Experiment

**E. R. Galea, L. Filippidis, Z. Wang, P. J. Lawrence and J. Ewer**

Fire Safety Engineering Group, University of Greenwich
London, SE10 9LS, UK
Corresponding author: E.R.Galea@gre.ac.uk

**Abstract** Blended Wing Body (BWB) aircraft with around 1000 passengers and crew are being proposed by aircraft manufacturers. This type of aircraft configuration is radically different from conventional tube type passenger aircraft and so it is essential to explore issues related to both fire and evacuation for these configurations. Due to both the large size and the unusual nature of the cabin layouts, computer simulation provides the ideal method to explore these issues. In this paper we describe the application of both fire and evacuation simulation to BWB cabin configurations. The validity of the computer evacaution simulations is also explored through full-scale evacuation experiments.

## Introduction

Very Large Transport Aircraft (VLTA) pose considerable challenges to designers, operators and certification authorities. Capable of carrying more than 800 passengers, the A380 may be considered a VLTA however; it is nevertheless a conventional aircraft configuration and so falls within the realms of past operations and certification experience. The aviation industry's drive for increased efficiency is leading to the consideration of less conventional designs and even greater passenger capacity, such as the Blended Wing Body (BWB or Flying Wing) passenger aircraft.

BWB designs being considered by the EC Framework 6 project NACRE (New Aircraft Concepts REsearch) are capable of carrying in excess of 1000 passengers on a single deck with 20 exits and eight longitudinal aisles. Furthermore, BWB layouts will mean that cabin crew at exits will not be able to assess the situation at opposite exit locations making redirection of passengers difficult. Indeed, the restricted and complex visual access and complex spatial connectivity offered by these aircraft configurations make wayfinding by passengers and redirection by cabin crew difficult and challenging. The industry standard evacuation certification regulations [1,2] require the aircraft manufacturer to demonstrate that the maximum

R.D. Peacock et al. (eds.), *Pedestrian and Evacuation Dynamics*,
DOI 10.1007/978-1-4419-9725-8_14, 

complement of passengers and crew can be evacuated from the aircraft within 90 seconds through half the normally available exits. The BWB concept represents a significant departure from conventional aircraft design and as a result there are many challenging questions that need to be addressed. How long would it take to evacuate a BWB aircraft with around 1000 passengers and crew? How long would it take an external post-crash fire to develop non-survivable conditions within the cabin of a BWB aircraft? Is it possible for all the passengers to safely evacuate from a BWB cabin subjected to a post-crash fire?

These questions are explored in this paper through computer simulation and experimental analysis. As part of project NACRE, a specially modified version [3] of the airEXODUS aircraft evacuation model [4] was used to explore evacuation issues associated with BWB aircraft. In addition, a series of full-scale egress trials were conducted using a specially constructed BWB mock-up to verify key airEXODUS predictions. To simulate the fire, the SMARTFIRE [5] Computational Fluid Dynamics (CFD) software was used. Finally, the results from the fire simulation and the evacuation simulation were linked to investigate the evacuation in the presence of the developing fire. The results from these evacuation and fire simulations along with the results from the experiment are briefly presented in this paper.

## airEXODUS and SMARTFIRE Simulation Models

The airEXODUS evacuation model is used to perform the evacuation simulations presented in this paper. airEXODUS [4,6] is designed for applications in the aviation industry including, aircraft design, compliance with 90-second certification requirements, crew training, development of crew procedures, resolution of operational issues and accident investigation. Within the software, parameters such as aisle walking speeds, passenger exit hesitation times, exit opening times etc are derived from the industry standard certification trials. Cabin crewmembers can also be represented and require an additional set of attributes such as, range of effectiveness of vocal commands, assertiveness when physically handling passengers and the extent of their visual access within the cabin. The atmospheric conditions generated by the fire such as heat, radiation, smoke and toxic fire gases are derived from the SMARTFIRE CFD fire model [5]. The impact that these hazards have on the exposed population is determined using the Fractional Effective Dose (FED) and Fractional Irritant Concentration (*FIC*) concept [6,7]. These models consider the toxic, irritant and physical hazards associated with elevated temperature, thermal radiation, HCN, CO, $CO_2$, low $O_2$, HCL, HBr, HF, $SO_2$, $NO_2$, Acrolein and Formaldehyde and estimates the time to incapacitation. Finally, when a passenger moves through a smoke filled environment their travel speed is reduced according to the experimental data of Jin [8]. To address issues associated with BWB cabin configurations, the airEXODUS evacuation model was modified in three specific areas:

- A novel scheme for passenger navigation was introduced based on wayfinding techniques used in the buildingEXODUS evacuation model.
- A modified model for passenger aisle swapping behaviour was introduced more appropriate for the BWB layout.
- A modified model to simulate cabin crew redirection procedures in BWB aircraft.

A research version of the SMARTFIRE V4.1 [5] software is used to perform the fire simulations in this study. The fire simulation model incorporated a range of sophisticated sub-models. A flame spread model including three ignition criteria [5] is used to generate gaseous fuel at the interior burnable surfaces. A toxicity model based on local equivalence ratio [9] is used to calculate the generation and spread of fire gases within the cabin. The calculation of smoke optical density utilises the mass optical density. Finally, the parallel version of SMARTFIRE is used to simulate the large-scale fire scenarios. The fire model has been validated by successfully reproducing the C133 fire test conducted by the US Federal Aviation Administration [10].

## BWB Configuration

As part of project NACRE many BWB configurations were under investigation. In this paper we consider configuration FW1-1-1. The FW1-1-1 configuration is the base case from which all other NACRE BWB variants are generated. The FW1-1-1 configuration consists of 1020 passengers in a single class configuration, 25 cabin crew and 20 floor level Type-A exits (see Fig. 1). The exits on the left side of the aircraft are numbered L1, L2, up to L10 going anti-clockwise from from the front to the rear of the aircraft as shown in Fig. 1.

## Evacuation Model Predictions

As airEXODUS is a stochastic model, the agents will not necessarily make the same decisions if the simulation is repeated, it is thus necessary to run the model several times for each scenario. For the results presented here, the model was run 10 times. The scenario considered here was a standard evacuation certification case where the exits on one side of the aircraft are considered unavailable. Thus of the 20 exits, 10 were made available on the left side of the aircraft. A standard opening time of 11.1 sec was used for each of the Type-A exits.

Also, note that the times specified in this paper refer to out of aircraft times and not on-ground times as exit slide configurations have not yet been determined. For the above scenario the out of aircraft times ranged from 80.6 sec to 92.8 sec with an average of 85.9 sec. While the minimum and average egress times are well under 90 sec, we note that the maximum evacuation time is some 3 sec over the maximum permitted time. It should also be recalled that these times represent out of aircraft times and not on ground times which may be some 3 sec longer.

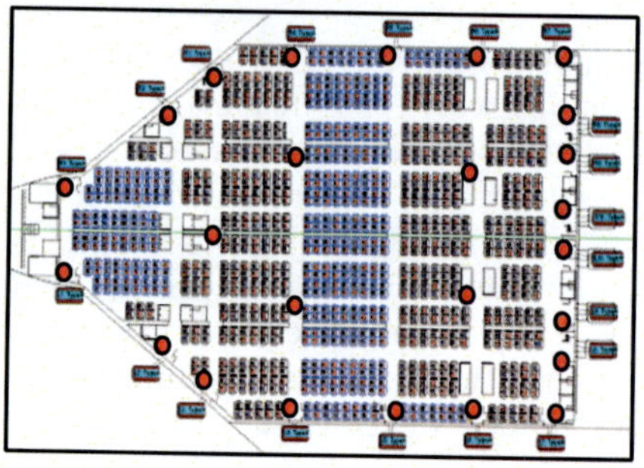

**Fig. 1. Cabin layout for FW1-1-1 showing location of cabin crew (circles) and exits (blue rectangles)**

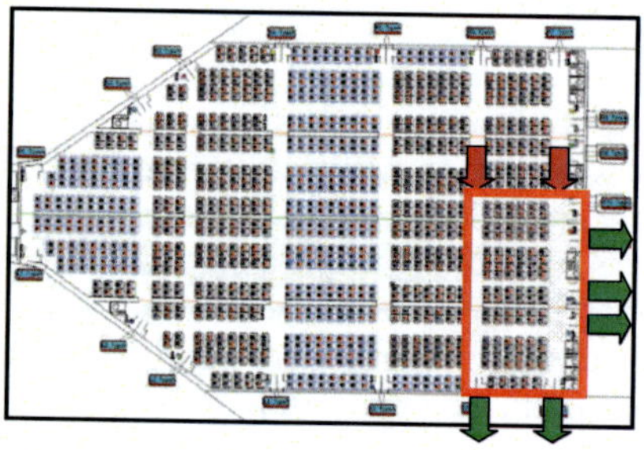

**Fig. 2. Section of full-scale cabin represented within the experimental mock-up**

From the predicted exit usage results (see Fig. 3) it is evident that the exits located at the south east corner of the cabin experience very low passenger usage. The worst offenders are the corner exits L7 and L8 with an average of 30 and 56 passengers using these exits respectively (i.e. the two exits in the bottom right corner of Fig. 1). The passenger exit usage results also indicate that exits L2, L3, L4 and to a lesser extend L5 are over-utilised. There is a clear trend that the exit capacity in the rear corner of the cabin cannot be fully utilised. This is thought to be for several reasons, firstly, to utilise L7 and L8 requires passengers to by-pass other functioning exits. Secondly, the location of these exits in the corner of the cabin means that they have a small natural catchment area of passengers for which these exits are their closest exits. Finally, the physical location in the corner provides poor visual access within the cabin. As a result it is difficult to reduce the heavy congestion in cross aisles 2-5 and the heavy usage of the forward exits (i.e. L2 to L6). If we consider the ratio of the time wasted in congestion to the time spent in evacuating we find that in the average simulation, passengers spend on average 40% of their personal travel time caught in congestion. This indicates that a significant amount of time is lost to congestion in this scenario.

This trend in exit usage has been observed in all of the numerical predications for the various configurations examined. While the results appear to be consistent and plausible, it was not clear if this was an artefact of the numerical simulation or if it was a realistic result. In particular it was not clear if the crew redirection model and the passenger navigation model were producing realistic predictions. To investigate this further it was necessary to undertake experimental evacuation trials.

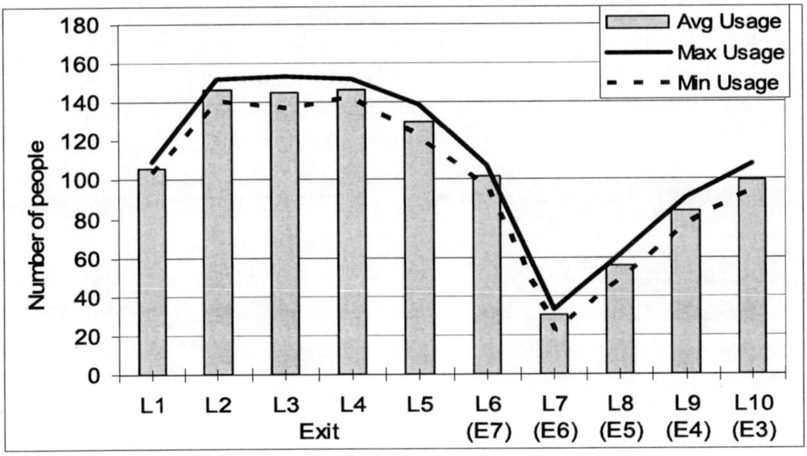


**Fig. 3. Predicted average exit usage for the 10 exits, L1 to L10**

## Large Scale Evacuation Trials

The purpose of the experimental programme of work was to observe and quantify the evacuation behaviour and performance of passengers and crew in novel BWB configurations and validate the computer simulations. Conducting full-scale trials involving over 1000 people was prohibitively expensive and impractical and so it was decided to undertake full-scale trials using a portion of the BWB cabin. Furthermore, given the concern over the modelling of the rear part of the cabin, the trials focused on this part of the cabin (see Fig.2). The key issue of interest was identifying whether participants would redirect and bypass a usable exit while trying to evacuate. To accurately represent this behaviour within the mock-up it was estimated that 380 people would need to be utilised in the mock-up of this area. Note that in order to measure whether occupants are willing to bypass a usable exit there was no need to have all the test subjects seated within the mock-up. In total some 88 participants would be seated in the mock-up and 146 participants would be brought into the mock-up via the two cross aisles feeding the mock-up section (see Fig.2).

The cabin mock-up was constructed at Cranfield University who also recruited the trial participants under contract to the University of Greenwich. A series of four trials were conducted over two days with two groups of participants, 375 participants on the first day and 358 participants on the second day. Trials considered full and partial partitions, additional crew and a repeat of the full partition trial. The participants were aged between 20 and 50 and each cohort of participants was used in all four trials on each day. Data from the trials was collected using some 12 internal fixed mounted cameras (see Fig. 4) and five external fixed mounted cameras. It is important to note that the trials were conducted in non-competitive conditions similar to those found in certification trials. Only the results from trial 1 session 1 are discussed here (trial with full partitions) however, these results are indicative of the findings from all the trials.

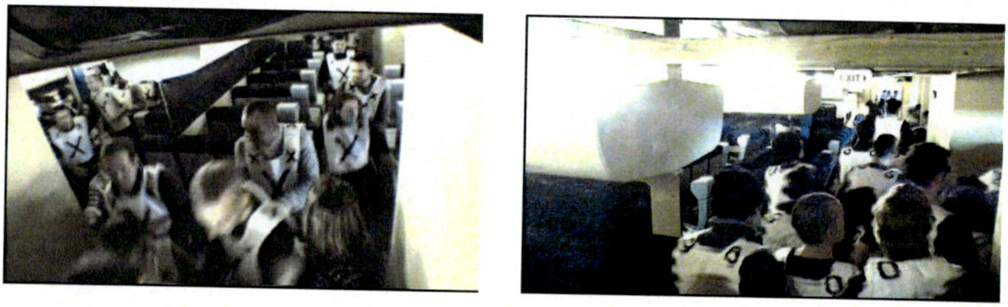

**Fig. 4. View from Cameras 9 and 12 during Trial 1 Session 1**

In comparing the exit locations used in the full-scale aircraft (and in the computer model) with those in the experimental mock-up, the designation L1 – L10 are used to represent the exits on the left side of the aircraft. In the mock-up, an E designation is used to describe the exits in the experiment. The link between the exits used in the experimental mock-up and simulation is as follows: L6 – E7, L7 – E6, L8 – E5, L9 – E4, L10 – E3 (see Fig.2).

A significant observation to emerge from the trials is that the exit usage distribution predicted by the airEXODUS software (see Fig.3) is reflected in the results found in the experimental trial (see Fig. 5). In particular, the corner exit E6 (L7) is the most underutilised exit while the first back exit that the participants encounter, E3 (L10) is heavily used. There is a gradual decline in the number of people using the next exits along (E4 (L9) and E5 (L8)) culminating in the minimum exit usage for E6 (L7) in the corner. The number of people using the next exit (E7 (L6)) then increases significantly. It should be noted that the modelling results depicted in Fig. 5 represent an average over 10 simulations while the experimental trial results represent the observations from a single trial. There is expected to be significant variation in exit usage for repeat trials which is not reflected in the trial results. This explains some of the differences between the predicted and measured exit usage values. It should also be noted that in the simulations there is a supply of passengers along the longitudinal aisles closest to the L6 (E7) exit that will also feed the exit. This will also contribute to the slighter higher number of people predicted to use the L6 (E7) exit.

The exit by-pass that was noted in the trials is also of interest. If we consider the stream of people coming down the cross aisle closest to the rear three exits (145 participants) we note that 39.3% by-passed the first exit (E3), 6.9% by-passed the second exit (E4), 2.1% by-passed the third exit (E5) and no one by-passed the forth exit (E6). In comparison, airEXODUS predicts that 41.0% of the passengers will by-pass the first exit which is in good agreement with the experimental findings. We note that while just over a third of the participants are prepared to by-pass one exit, very few will by-pass more than one exit.

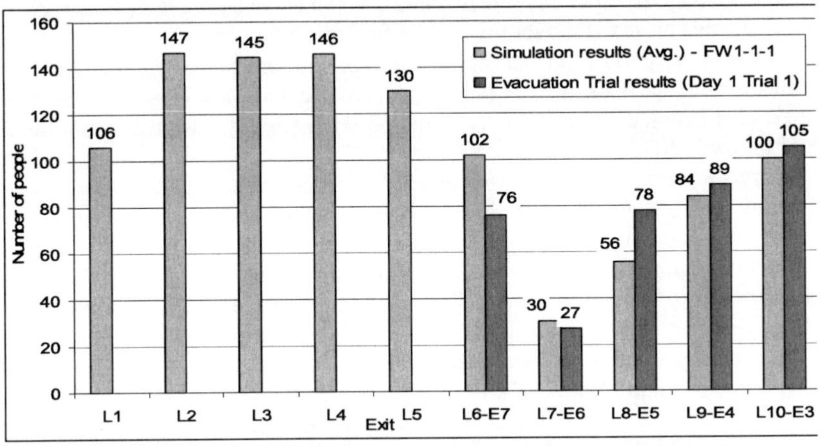


**Fig. 5. Comparison of exit usage between modelling predictions for full cabin and experimental results for cabin section**

## Fire Model Predictions

In a post-crash aircraft fire, the fire is typically initiated outside the cabin usually due to a fuel spill. The fire then attacks the aircraft cabin gaining entry via ruptures to the fuselage due to impact damage, or burn through and ignites the interior materials. In the NACRE simulations, the external fuel fire source is located on the right side of the aircraft. Six different fire scenarios were investigated, all of which involved opened exits on the left side of the cabin during the entire fire simulation. Here we report the results of Scenario 3, with the wide cabin rupture, equivalent to three Type-A exits. The external fire had dimensions of 5.2 m long by 2.5 m wide and the fire reached a maximum heat release rate of 18 MW after 8 sec and burnt at this maximum rate for 10 minutes. The computational mesh used for the NACRE simulations consisted of approximately 650,000 cells. A parallel cluster consisting of seven processors was used for the simulations. This reduced the run time from 425 hours on a single processor to around 70 hours for a single 480 second fire simulation.

At flashover, the fire very rapidly changes from being localised to engulfing the entire volume. An important outcome of this analysis is that flashover is not observed within the first 480 sec, which is much longer than the certification requirement of 90 sec. The combustion behaviours over the entire simulation time do not display the rapid increase in values, which is the hallmark of flashover.

The seats close to the fuel fire are the first cabin fixture to be ignited. Later, the fire spreads to portions of the seats in front of and just behind the initially ignited seats. At 480 seconds, the fire mainly remains localised and confined to seats and overhead materials in the vicinity of the rupture. Clearly flashover is not the factor that will drive survivability in this type of scenario. Predicted (interior) HHRs reach a local maximum at approximately 60 sec. At 60 sec, severe fire hazards are mainly confined within the immediate vicinity of the rupture at head height (1.7 m above the floor). Within the lower layer (0.5 m above the floor), fire hazards such as temperatures and toxic gas concentrations are at very low levels in the vicinity of the rupture however, radiation fluxes are at untenable levels. After 80 sec the hot fire gases have spread throughout the cabin section closest to the rupture. Temperatures at head height are around 100$^{o}$C through most of the section. Hot fire gases begin to spill into the next cabin section with temperatures around 60$^{o}$C in parts of the third longitudinal aisle. The atmospheric conditions in most of the cabin at around 90 seconds appear to be survivable. Only conditions in the cabin section immediately adjacent to the rupture pose a threat to life.

In order to analyse the likely impact of fire hazards on the evacuating passengers, the NACRE cabin is divided into 67 zones for data output from the fire simulations. The fire hazard data in the upper layer (1.5 m to 2 m) and lower layer (0.3 m to 0.8 m) within each zone is a weighted average of variable values of all cells within the layer. This data at each time step is then exported to airEXODUS and used in the evacuation simulation, exposing the population to the evolving fire hazards. Presented in Fig. 6 are the predicted radiation fluxes at Zone 2 and 61.

Zone 2 is in the section of longitudinal aisle immediately opposite the cabin rupture and hence the external fuel fire while Zone 61 is in the section of cross aisle adjacent to exit L4 on the opposite side of the cabin to the fire. As seen in Fig.6, the radiation fluxes in both the upper and lower layers of Zone 2 reach hazardous levels of 10 $kW/m^2$ just before 10 sec. The local CO concentrations peak at approximately at 60 sec, which is 50 sec after the radiation flux reaches critical values. This demonstrates that in the vicinity of the rupture, radiative flux is the key threat to survivability in Zone 2. In Zone 61 we note that the radiative fluxes and CO values are near ambient values up to 90 sec after ignition and pose no threat to the passengers. The same conditions exist in the zone opposite L5. Thus conditions at two heavily used exits pose no threat to the passengers.

As with the case without fire, the evacuation simulation was run 10 times. This produced an average evacuation time of 89.3 sec compared with 85.9 sec without the fire. This modest increase in evacuation time is due to the presence of smoke within the cabin which reduces visibility and reduces travel speeds. While there is only a modest increase in evacuation times there are 12 predicted fatalities in this simulation. All 12 fatalities occur in the immediate vicinity of the rupture and all the fatalities are a result of exposure to radiative heat. The fatalities occur between 8 and 34 secs from the start of the simulation, with three fatalities occurring within the starting location and nine fatalities occurring in the aisle adjacent to the starting location. Given these conditions, it is felt that these fatalities are unavoidable, given their starting location and proximity to the fire.

In addition to the predicted fatalities, some 25 passengers are predicted to be injured due to heat exposure. Of these, 3 passengers are considered to have serious life threatening injuries. None of the survivors suffers from serious exposure to the toxic fire gases however, most of the survivors suffer from light exposure to HCl.

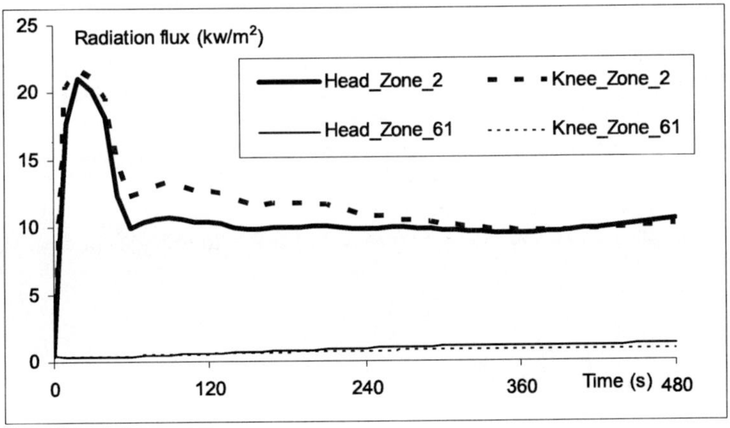


**Fig. 6. Predicted radiation fluxes in Zone 2 and 61**

## Conclusions

The airEXODUS evacuation simulation suggests that the NACRE BWB with 1045 passengers and crew can be evacuated within 80.6 sec to 92.8 sec with an average of 85.9 sec. Improved performance can be expected by better utilisation of the rear, and in particular the corner cabin exits. This may be achieved through improved passenger familiarisation with the cabin layout and improved visual access. However these times represent out of aircraft time and not the on-ground time as required by current regulation.

Experimental data from full-scale evacuation trials support the appropriateness of the passenger exit selection behaviour implemented within the airEXODUS evacuation model and suggest that it is suitable for these types of applications. The experimental trials also support the overall findings of the numerical simulations. The experimental results highlight the importance of situational awareness and visibility in navigating a successful exit path within the complex layout of the BWB. Improving the passenger's knowledge of the cabin layout and the location of the exits and providing them with good visual access of the exits and aisles will be essential in achieving an efficient evacuation of complex BWB configurations.

Fire simulations suggest that the BWB cabin exposed to an 18 MW post-crash external fuel fire via a large cabin rupture does not flashover within the first 480 sec. This suggests that, unlike conventional tube style aircraft, flashover is not the primary factor driving passenger survivability. When the SMARTFIRE fire simulations are linked to the airEXODUS evacuation simulation, thereby exposing passengers to the developing fire, the average evacuation time increases to 89.3 sec. In addition, some 12 fatalities and 3 serious injuries are predicted. All the fatalities and injuries are the result of exposure to radiative heat and all are initially located in the immediate vicinity of the rupture. Smoke and toxic gases are not considered a serious threat in these scenarios. Given the location of the fatalities and the severity of the fire conditions, it is felt that these fatalities are unavoidable and are not inherently due to the cabin architecture.

Ultimately, the practical limits on passenger capacity and aircraft design are not based on technological constraints concerned with aircraft aerodynamics but on the ability to evacuate the entire complement of passengers and crew within agreed safety criteria. This work has demonstrated that the NACRE BWB configuration has the potential of satisfying such safety criteria and is arguably capable of providing an equivalent or better level of safety to today's conventional aircraft.

## References


1. Title 14, Code of Federal Regulations (14 CFR). Federal Aviation Regulations, Washington, USA (1999)
2. JAR Section 1 Part 25.807 Large Aero planes: Subpart D Design and Construction. As published in Joint Aviation Requirements (Change 15) ( 2001)
3. VELA: Very Effecient Large Aircraft. University of Greenwich. http://fseg.gre.ac.uk/fire/VELA.html.
4. Galea, E.R., Blake, S., Gwynne, S., Lawrence, P.: The use of evacuation modelling techniques in the design of very large transport aircraft and blended wing body aircraft. The Aeronautical Journal. Vol 107, 1070, pp 207-218 (2003)
5. Jia F., Patel, M.K., Galea, E.R., Grandison, A., Ewer, J.: CFD Fire Simulation of the Swissair Flight 111 In-flight Fire – Part II: Fire Spread within the Simulated Area. The Aeronautical Journal. Vol 110: 303-314 (2006)
6. Galea, E.R., Owen M., Lawrence, P.: Computer modelling of human behaviour in aircraft fire accidents. Toxicology. 115, (1-3), pp. 63-78 (1996)
7. Purser, D.A.: Toxicity Assessment Of Combustion Products. In: DiNenno, P.J. (ed.), The SFPE Handbook Of Fire Protection Engineering (3rd Edition). Drysdale (2002)
8. Jin, T.: Visibility Through Fire Smoke. Report Of Fire Institute Of Japan #42 (September 1976)
9. Galea, E.R., Jia, F., Wang, Z.: Predicting toxic gas concentrations resulting from enclosure fires using local equivalence ratio concept linked to fire field models. Fire and Materials. Vol 31, Issue 1, pp 27-51 (Jan/Feb 2007)
10. Wang Z., Galea, E.R., Jia F.: A computational study of the characteristics of aircraft post-crash fires. Proc Int Aircraft Fire & Cabin Safety Conf, Oct 29 – Nov 1, 2007. Atlantic City, USA (2007)

# Collection of Evacuation Data for Large Passenger Vessels at Sea

**E.R. Galea[1], R.C. Brown[1,2], L. Filippidis[1], and S. Deere[1]**

[1] Fire Safety Engineering Group, University of Greenwich, London, UK
[2] Offshore Safety and Survival Centre, Fisheries and Marine Institute, Memorial University, St. John's, Canada
Corresponding author: E.R.Galea@gre.ac.uk

**Abstract** In the past decade, significant effort has gone into the planning and execution of full-scale sea trials in an attempt to improve, calibrate and validate existing evacuation models for passenger ships. In September, 2009 two assembly exercises were conducted at sea onboard the RO-PAX ferry SuperSpeed 1 by team members of the EU-funded project SAFEGUARD. The exercises were conducted with passengers during routine sailings between the ports of Kristiansand, Norway and Hirtshals, Denmark. Between both trials, a total of 1,769 passengers were assembled, on day one, 902 passengers and on day two 867 passengers. As part of the data collection exercise, passenger response time data was collected – using video cameras – and passenger movement data was collected using a novel infrared (IR) based position logging system. This paper briefly describes the development and testing of the data acquisition system and briefly discusses preliminary results.

## Introduction

Understanding how people behave in emergency situations within maritime settings is vital if we are to; design and develop evacuation efficient vessels and evacuation procedures, train crew in the management of evacuation situations, develop reliable ship evacuation models and regulate the design and operation of vessels. An essential component of this understanding is the collection and characterisation of human performance data. Unfortunately, little data relating to passenger response time or full-scale validation data in maritime environments exists. In the first International Maritime Organisation (IMO) document to specify protocols for the use of ship evacuation models in the analysis and certification of passenger ship design, IMO MSC Circ. 1033 [1], an arbitrary uniform random distribution was set to represent the response time behaviour of passengers. It has been shown that this is unrepresentative of actual passenger response time and liable to produce incorrect or misleading conclusions concerning the suitability of ship design for evacuation [2]. As part of the EU Framework V project FIRE EXIT [3], passenger response time data was collected for a passenger ship at sea [3, 4]. This

R.D. Peacock et al. (eds.), *Pedestrian and Evacuation Dynamics*,
DOI 10.1007/978-1-4419-9725-8_15, 

data was accepted by the IMO and used in the formulation of IMO MSC Circ. 1238 [5], the modified protocols for passenger ship evacuation analysis and certification. However, the response time data produced by FIRE EXIT [3,4] related to only a single passenger vessel. As such the data cannot be considered representative of passenger ships in general. The IMO Fire Protection (FP) Sub-Committee in their modification of MSC Circ. 1033 at the FP51 meeting in February 2007 [6] invited member governments to provide, "...further information on additional scenarios for evacuation analysis and full scale data to be used for validation and calibration purposes of the draft revised interim guideline". To this end, project SAFEGUARD was proposed and successfully funded through the EU framework 7 programme. The project aims to address this IMO requirement by providing relevant full-scale data and proposing and investigating additional benchmark scenarios that can be used in certification analysis. Six full-scale data sets will be collected as part of SAFEGUARD - two trials on each of three different types of passenger vessels.

This paper concentrates on the first two data sets collected on the first vessel - a large RO-PAX ferry operated by Color Line AS called SuperSpeed 1. The vessel can carry approximately 2000 passengers and crew and over 700 vehicles. It operates on the route between Kristiansand in Norway and Hirtshals in Denmark, a trip of 3 hours and 15 minutes. The ship contains a mixture of spaces spread over three decks including; business and traveller class seating areas (airline style seating), large retail and restaurant/cafeteria areas, bar areas, indoor and outdoor general seating areas and general circulation spaces. The ship has four assembly stations, three located on Deck 7 (assembly stations A, B and C) and one located on Deck 8 (assembly station D). Assembly stations B and C are located on the outer decks while assembly stations A and D are internal.

Three types of data sets were collected in each trial. The first consisted of response time data collected using video cameras positioned throughout the vessel. Some 30 battery powered mini digital video cameras were used to collect the response time data (see Figure 1). The cameras were placed at strategic locations throughout the vessel to record not only the time at which the passengers respond, but also the nature of the activities that they were involved in at the time. The second type of data collected comprised validation data for ship based evacuation models. This consisted of starting locations of passengers, arrival time at the designated assembly locations and the paths taken by the passengers from the start location to the assembly location. This data was collected using a novel data acquisition system consisting of 30 Infra-Red (IR) beacons, each emitting unique IR signals and data logging tags worn by each passenger (see Figure 2). The third type of data consisted of a questionnaire completed by each of the participants.

**Fig. 1. Example video camera mounting location (left, circled) and mounting equipment with camera (right)**

**Fig. 2. IR field generating beacon (left) and IR logging tag (right)**

## Data Collection Methodology

Here we describe the process of collecting the validation data. Previous efforts at collecting comprehensive full-scale ship evacuation validation data have been less than successful due to the complication of the associated data analysis. Previous efforts have attempted to use video footage to manually track individuals through the vessel [3]. However, tracking individuals through the complex layout of a large passenger vessel is extremely time consuming. Depending on the complexity of the structure, the analyst may have to track an individual through tens of different video camera locations. Attempting to track a handful of individuals this way can be extremely tedious and prone to error. In the case of SuperSpeed, tracking almost 1000 passengers across three large decks would have been unthinkable! Automated video tracking systems also have problems as they require a 'birds eye' view of the targeted individuals if they are to accurately monitor an individual's progress through a particular location, making installation of the video equipment difficult due to the low head room often found on ships [7]. Furthermore, the problem still persists of tracking individuals passing through many different camera locations.

A comprehensive investigation was undertaken of technologies which may be useful in addressing this problem. This identified two specific technologies - passive radio frequency identification (RFID) and infra-red (IR) position logging.

Both systems rely on similar underlying concepts - devices are mounted throughout the structure that generate uniquely identified radio frequency or IR fields or "gates" and passengers wear a device that allows for their unique identification as they move throughout the structure and pass through each gate. In this way, if the structure is instrumented with a sufficient number of gates, then as a person moves around the structure, their tag either logs or permits the logging of the ID for the gates that were passed and at what time (see Figure 3).

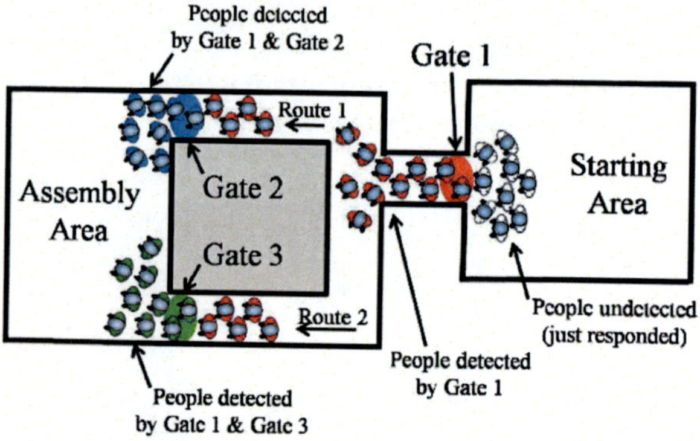


**Fig. 3. Example of tracking system following individuals from a starting area to an assembly area or exit point via two different routes**

Both systems rely on the population agreeing to wear a tag for the purposes of the trial. As the trial may take place at any time, the participant population must be prepared to wear the tag for an extended period of time (possibly all day/night) and so it must not interfere with normal operations, be comfortable, and if possible must blend in with their normal attire. For example, attaching the tags to a hat or cap, while ideal for detection, would not be acceptable.

There are fundamental differences between the RFID and IR systems, particularly the way in which the gates are generated and the way in which communication takes place between the tag and gate for logging of position and time. The passive RFID technology examined (see Figure 4) relies on a pair of antennae that generate a radio frequency (RF) field (*gate*) with sufficient power to energise tags that enter it. The tags use the RF energy from the field to wirelessly transmit a signal to the receiving antennae (the same antennae that generate the RF field) which then sends the information to a processor that logs the tag's unique ID and the time at which the signal was received. For this system, the RFID tag acts as a passive device with no built-in power source and no data storage capacity. The data for people tracked is stored either by the component that generates the RF field or an attached computer. The main difficulty with this technology is that human bodies attenuate RF signals, sometimes in an unpredictable manner, thus making the placement of RFID antennae and tags of critical importance. In large crowds, especially, data can be lost creating inaccuracies in the validation dataset. In a recent series of evacuation trials using RFID [8,9], it is suggested that read rates "will be better than 50% if proper alignment and measurement power of reader antennas is found through experiments" in crowded situations [8]. The success rate is also critically dependent on how the tag is worn, whether it is in contact with the skin, near metal objects or concealed by clothing. In addition, RFID systems tend to be bulky for temporary applications and logistics of setup becomes time consuming due to the need to carefully run cables and ensure suitable antennae orientations. Further, while RFID tags tend to be inexpensive in large quantities required for test series, they can be damaged relatively easily jeopardising the quality of the dataset. Considerable effort is required to select the correct tag and form factor to ensure the highest read-rate possible.

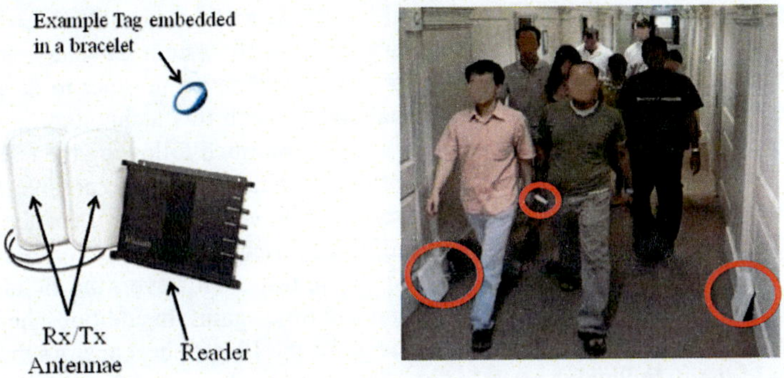


**Fig. 4. Example RFID system tested (left) and example of field test (right)**

A series of tests was undertaken of a representative RFID system, first in a corridor at the University of Greenwich and then on-board the SuperSpeed vessel. The RFID system was manufactured by Alien Technology Corp. and consisted of an Alien model ALR-8800 reader and a pair of Alien model ALR-8610 circular polarised multistatic antennae (i.e. capable of both generating the RF field and receiving tag transmission data for logging). This system was designed to operate in the European UHF band in the 865.7 - 867.5 MHz range with power levels of 2W ERP and is compliant with European radio regulations. The system was designed to read EPC Class 1 Generation 2 UHF tags. In addition to the RFID system, various types of EPC Class1, Gen2 UHF tags were purchased in different form factors, specifically; peel and stick labels, silicone rubber wristbands and plastic wrist/ankle hospital-style bands. The corridor tests consisted of some 12 people walking down a 1.89 m wide corridor of which all 12 were wearing the RFID tags. Trials with both wrist and ankle bands were undertaken. Subjects were asked to walk together past the antennae as a group keeping their speed, position and group density consistent from test to test (see Figure 4). Tag reads were stored on a computer for each test. The maximum read rate was 75% for tests where subjects were permitted to walk normally (i.e. with arms swinging by their side). One test case was conducted where the subjects were asked to fold their arms, thereby shielding the tags somewhat. For this test, it was found that read rate decreased significantly to 17%. A series of trials was also conducted on-board the SuperSpeed. Unfortunately, the trials on-board SuperSpeed were conducted using only nine people (equal to the number of tags) in an open space and so do not represent a reliable test under crowded situations, nevertheless, these trials returned an average read rate of 86%. The ship trials demonstrated that the RFID system could work within the confines of the metal environment of a passenger ship. The corridor tests suggest that read rates of up to 75% can be achieved using the RFID technology in crowded situations.

The IR technology examined (see Figure 2) relies on a beacon that generates an IR light field (*gate*). As a tagged individual passes through the field, IR light sensors in the tag detect the IR light and log its ID and the time at which it was detected in the tag's own internal memory. For the IR system, no data is transmitted from the tag. Following the test, tags must be retrieved in order to determine the occupant's route data. The main disadvantage with this technology is that occupant route data is not collected unless tags are returned following the test. In addition, the IR tag is also considerably more expensive than the passive RFID tag, costing approximately 15 times more. This is very different than the (passive) RFID tags which, due to their low individual price and inability to log route data, do not need to be returned at the end of the tests. However, the IR tags, if collected after the trial, can be reused time and time again. In addition, the disadvantage of the IR tag turns into an advantage for the IR gate beacons. As the path history of the individual is recorded on the tag itself, this simplifies the nature of the IR beacons, reducing their cost (by a factor of approximately 60), as well as their size and power requirements compared to the RFID system. This allows many IR beacons to be placed through the structure, thus allowing for more granular definition of occupant routes. In addition, the beacons are much easier to set-up, greatly simplifying instrumentation setup and due to their reduced size, simplifying logistics involved in transporting the equipment to the test site.

The IR system was manufactured by RFID Centre Ltd. A modified version of their *TagMobile* system was employed which includes IR generating beacons and logging tags which are hung around the neck using a lanyard. The RFID Centre worked with FSEG to modify this system to make it more appropriate for use in evacuation applications. This involved a redesign of the standard IR tag. The modified IR system was put through a similar series of tests to the RFID system. In corridor tests at the University of Greenwich, the beacon was mounted on one side of the corridor at a height of 2.13 m above the floor, facing perpendicular to the opposite wall. A total of ten tags were used and a group of 23 individuals was formed with the ten tagged subjects mixed throughout. The group was instructed to walk past the beacon keeping speed, position and group density consistent from test to test (see Figure 5). Subjects were asked to raise their hand when the tag indicated a successful read. These tests always returned a 100% read rate. Ship board tests using the IR system involved 10 test subjects wearing the IR tags. These tests returned an average of 93.5% successful read rate however, as with the

RFID trials on the ship, these trials were not under heavily crowded situations and so were not considered representative of the intended application. However, the tests demonstrated that the system could work in a ship board environment.

**(a)** **(b)**

**Fig. 5. IR corridor test (a) tagged people (down arrows), IR beacon out of view (up arrow) and IR tag (circled) (b) tagged people raising hands when tag is detected with IR beacon out of view (circled)**

Based on the results generated in the various trials, it was concluded that the IR-based tracking system was better able to accurately track large numbers of individuals in high density crowds. While the RFID system provided reasonable read rates, performance of the IR system was superior in all cases. The success of the IR system is due in part to it not relying on direct line-of-sight between the beacon and the detector, the IR beams being readily reflected from surfaces and not distorted by human bodies. Furthermore, considering the logistical issues associated with using the two systems, the IR system was considered more versatile being; easier for setup and knock-down, not requiring external power supplies or cables, being relatively cheap to add additional gates and easier to transport. On completion of the trial, data is transferred from each retrieved tag to a computer via an IR reader. Software developed by FSEG then reads the tag information and, for each tag, identifies when the participant passed each numbered beacon and, for this application, when they arrived at the assembly station. Travel times, average travel speeds and levels of congestion can then be determined and associated with each beacon and hence beacon location. The data transfer process and path extraction is quick and reliable.

## Results and Discussion

Both full-scale sea trials were conducted on the SuperSpeed's Kristiansand to Hirtshals crossing. While the passengers were aware that they would be involved in an assembly trial, they did not know at what time on the 3 hours and 15 minute crossing the trial would take place. Participation in the trials was not compulsory and children under the age of 12 were not permitted to take part. Passengers were given the IR tags as they boarded, together with an information sheet and asked to wear the IR tag at all times while on the vessel. The trials consisted of the ship's Captain sounding the alarm and crew moving the passengers into the designated assembly areas. The assembly trials were successfully conducted on the 4th at 08:20 and 5th at 08:19 of September 2009. On day 1 (day 2) there were 1431 (1349) passengers on board of which 1296 (1243) were eligible to participate in the trial. In total some 1170 (1192) IR tags were issued (some passengers refused to participate) of which 902 (867) passengers with IR tags participated in the trial in some capacity. Of these, a number of passengers who had been issued IR tags changed their minds and decided not to participate and so handed back their tags. Others with tags simply refused to participate while others responded, but then did not immediately move off to the assembly stations and so were not counted in the assembly total. The 902 (867) participating passengers represents 70% (70%) of the passengers on board and 77% (73%) of the issued tags. In total 13 (0.5%) of the tags were lost together with 60 lanyards. The first assembly trial was completed in 12 min while the second assembly trial was completed in 10 min. The end of each assembly trial was determined by the Captain.

Here we consider some data extracted from the second trial. In 10 min some 841 passengers were assembled, this included 345 passengers who were already in the assembly stations and 496 passengers who made their way to the assembly stations. Depicted in Figure 6 is the assembly curve determined from the data collected by the IR tags. It displays the characteristic shape of a typical arrival curve. Note the passengers who assembled after 10 min (26 passengers) were not considered to be participating in the trial. A similar curve is available for each unique assembly station.

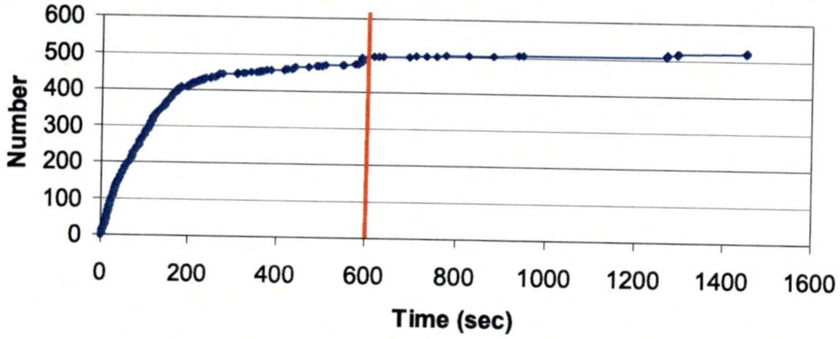


**Fig. 6. Total assembly curve for trial 2**

Presented in Figure 7 is a typical path reconstructed from the tag information for a passenger. The passenger was located in the restaurant area and exited the restaurant, passing beacon 14 69 secs after the sounding of the alarm, exiting the entrance to the dining area, passing beacon 12 89 secs after the alarm and entering Assembly area D (end point), passing beacon 11 103 secs after the sounding of the alarm. This type of analysis will enable an accurate data set to be assembled for model validation purposes, providing details of the routes taken by individual passengers, which assembly stations were used and associated assembly times. The collected data can also be used to generate approximate average travel speeds and in some cases population densities.

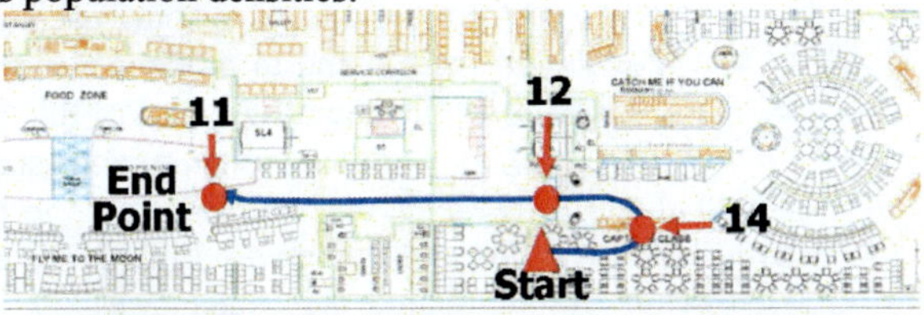


**Fig. 7. Path adopted by a passenger during assembly trial 2**

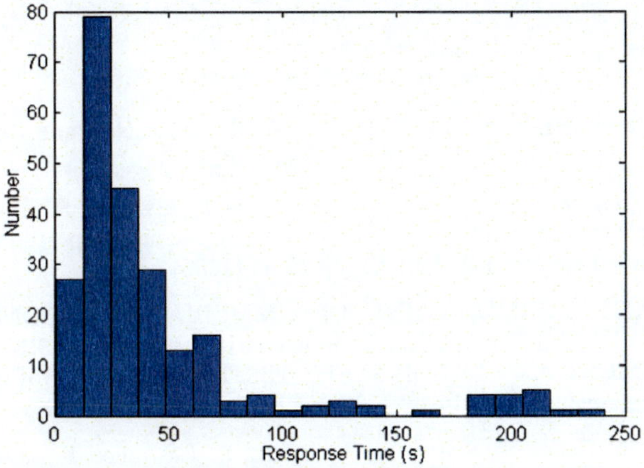


**Fig. 8. Response time distribution derived from both trials for population in the bar region**

The travel speeds are only approximate as the size of a gate region, which is tuned as required for the specific location, can be between 4 m and 11 m wide. The population densities are determined using information concerning the number of people entering and leaving each gate region over a period of time. The accuracy of this measurement is improved if the flow is uni-directional and if densities are not extreme. The population response times are being determined by analysis of the video footage. This analysis is not only determining the response times but also the nature of the activities that the population was engaged in during the response phase, including determining the number of information and action tasks completed by each individual during the response phase. Shown in Figure 8 is the response time distribution for people located in the bar area. It displays the typical log-normal shape with a mean of 45.2 secs and a standard deviation of 48.6 secs.

## Conclusions

Two assembly trials have been successfully conducted at sea involving a total of 1769 passengers. The data collected from these trials is being used to formulate a validation data set for ship evacuation models and to produce a more representative response time distribution for ship applications. As part of this project a new technology to track individuals through a complex structure, consisting of IR tags and beacons, was developed and tested. The technology is able to reliably track large numbers of people through complex structures, is relatively easy to setup and enables the rapid extraction of individual person trajectories.

## References

1. IMO, Interim Guidelines for Evacuation Analyses for New and Existing Passenger Ships. IMO MSC/Circ 1033 (6 June 2002)
2. Deere, S., Galea, E.R., Lawrence, P., Filippidis, L., Gwynne, S.: The impact of the passenger response time distribution on ship evacuation performance. Int J of Maritime Eng, Vol 148, Part A1, pp 35-44 (2006)
3. FIRE-EXIT: FORMULATION OF IMMEDIATE RESPONSE AND EVACUATION STRATEGIES THROUGH INTELLIGENT SIMULATION ASSESSMENT AND LARGE-SCALE TESTING. University of Greenwich. http://fseg.gre.ac.uk/fire/fire_exit_project.html (2009)
4. Galea, E.R., Deere, S., Sharp, G., Filippidis, L., Lawrence, P., and Gwynne, S.: Recommendations on the nature of the passenger response time distribution to be used in the MSC 1033 assembly time analysis based on data derived from sea trials. Int J of Maritime Eng, Vol 149, Part A1, pp 15-29 (2007)
5. IMO MSC Circ 1238 (2007)
6. FP 51/WP.3 (8 Feb 2007)
7. Boltes, M., Seyfried, A., Steffen, B., Schadschneider, A.: Automatic Extraction of Pedestrian Trajectories from Video Recordings. Proc Pedestrian and Evacuation Dynamics 2008, pp. 43-54. Springer (2010)
8. Hostikka, S., Paloposki, T., Rinne, T., Saari, J.M. Korhonen, T. and Heliövaara, S.: Experimental Observations of Evacuation Situation. VTT Technical Research Centre of Finland (September 2007)
9. Rinne, T., Hostikka, S., Paloposki, T., Korhonen, T., Saari, J.M., Heliövaara, S.: Application of RFID and Video Imaging on Evacuation Observations in Offices and Public Buildings. In: Karlsson, B. (ed.), Proceedings, 9th IAFSS Symp, pp. 329-340 (2008). DOI:10.3801/IAFSS.FSS.9-329

# Experimental Research on Investigation of Metro Passenger Evacuation Behaviors in Case of Emergency

**H. Li[1,2], Z. Maohua[1], S. Congling[1], S. Jiehong[1,2], C. Haicheng[1] and X. Qiaoxiang[3]**

[1]China Academy of Safety Science and Technology, State Administration of Work Safety, 17 Huixin Xijie, Chaoyang District, Beijing, 100029 P. R. China
[2]University of Science and Technology Beijing, Beijing, 100083, P. R. China
[3]Nanjing Metro Co., Ltd. Nanjing 210008 Jiangsu, P. R. China
Corresponding author: Zhong Maohua: mhzhong@chinasafety.ac.cn

**Abstract** As traffic problems in major and medium cities of China are getting increasingly serious and the demands for public transit are rising dramatically, the construction of subways and light-rail transit systems is imperative in the construction of infrastructures in various major cities and is also the symbol of modernized metropolitans. The total length of subway to be built and completed in China by 2010 is 1200 km. However, the social influence would be great once the emergency occurs in metro. To realize and grasp the characteristics of evacuation behavior of passengers in metro emergency is the precondition for the operation enterprise to constitute effective planning of emergency evacuation and conduct the passengers' evacuation to safe area reasonably and quickly. The discrepancy in economy, culture, population and number of operational subway routes in different cities are bound to result in different passenger evacuation characteristics in case of any subway emergency. In this article, by questionnaire survey method and statistical analysis method, an investigation was conducted on the safety awareness and safe evacuation behavior of subway passengers in three different regions in China, namely Beijing (BJ),Nanjing (NJ) and Guangzhou (GZ), based on which the characteristics of subway passengers evacuation behavior in these cities were analyzed.

## Introduction

Subway is a heavy-duty urban transit means. During its operating period, there are a huge amount of passenger streams in the substations, and if any emergency such as fire, etc. occurs, it is very likely to cause mass injuries and casualties. Therefore, it has become an issue of common concern of the subway operators and relevant departments of the government as to how to formulate an effective emergen-

R.D. Peacock et al. (eds.), *Pedestrian and Evacuation Dynamics*,
DOI 10.1007/978-1-4419-9725-8_16, 

cy evacuation plan for such kind of densely populated venues as subway stations, and as to how to reasonably and quickly organize the subway passengers to evacuate to a safe area in case of emergency. To formulate an emergency subway evacuation plan, we must first get to know the passengers' behavioral characteristics under any emergency incident.

Many researchers have conducted a great deal of researches on the human behavior under emergency incidents. The traditional evacuation theoretical research began in the early 1930s covered a series of large-scale public buildings constructed at that time, including railway station, subway, theatre, supermarket and government office building, etc. Relevant researches conducted between the end of the 1970s and early 1990s mainly focused on human behavior in fires, and also strengthened the research on the process of evacuation from different types of buildings in case of fire. An investigation and analytic research was conducted by Liu Kehui on the subway passenger safety awareness and behavioral status in BJ [1]. In addition, relevant simulated researches and analysis were conducted by Zhong Maohua, Shi Congling and other researchers on the personal evacuation behavior in case of different types of building fires [2,3].In this paper, we will carry out an investigation and analytic research on the passenger evacuation behavior and psychological characteristics in case of subway emergency in different cities in China by employing questionnaire survey method.

## General Information of subway in three cities

For this paper, the objects of investigation are respectively selected from the three cities with operational subway routes, namely, BJ, NJ and GZ, respectively located in the northern part, central part and southern part of China.

### *GDP and population*

GDP and total population of BJ, GZ and NJ from 2001 and 2009 are shown in Fig. 1, which indicate that GDP and population of BJ rank the first among the three cities, followed by GZ. Economic development is one of the major factors supporting urban progress, stimulating urban social activities and affecting residents' trips. It is quite apparent that the trip rate is closely related to the degree of richness of the citizens in such a manner that the higher the urban economic level, the higher the citizens' trip rate and the more opportunities for the citizens to choose subway as their means of transportation. Meanwhile, the intensity of investment on subways is also closely related to the economic development. Due to the economic development and degree of mobility, there exists a close relation between the total trip rate and private motorized trap rate. Therefore, the trend of GDP growth is of great significance for the development of subways.

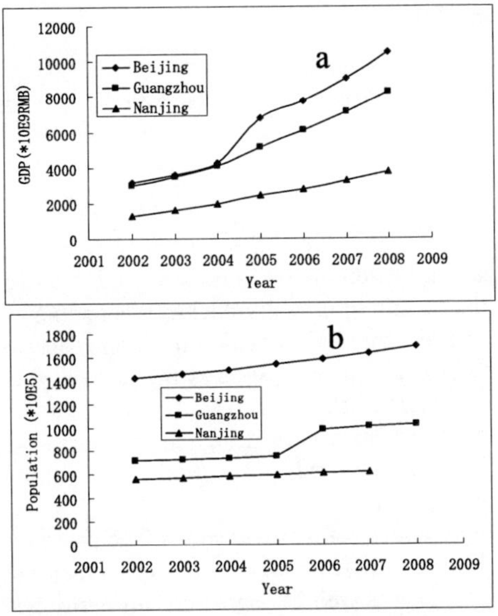


**Fig. 1. Comparison of GDP and population among the three cities. (a) GDP, (b) population**

The growth of population is bound to result in the expansion of city size as well as the convergence of population in outskirts of a city, thus, affecting the demand for urban public transit, and increasing the need for intensifying construction of urban subway routes in order to address people's traveling needs.

## *Operational subway lines*

As of May 2009, subway lines already put into operation in BJ are Line 1, Line 2, Line 13, Ba-Tong Line, Line 5, Phase I of Line 10, Airport Line and Olympics Branch Line, with a total operational length of 200km. In NJ, the subway line already put into operation is Line 1 with a total operational length of 22km. Subway lines put into operation in GZ include Line 1, Line 2, Line 3 and Line 4, with a total operational length amounting to 116km.

## Research methods

### *Questionnaire design*

According to the passenger evacuation characteristics in case of subway emergency, survey questionnaire covering three aspects [2] is designed, which mainly includes:

1. **Investigation of personal information about objects of investigation:** Human being's behavioral reactions are closely related to social and cultural environments while the quality of an individual as a constituent of the society even has a closer relationship with its behavioral reaction. Therefore, the personal information of the objects of investigation is the important original information for studying its evacuation behavioral reaction in case of emergency. The passenger personal information survey designed for this article mainly includes age, sex, educational background, occupation and dwelling condition, etc. of the objects of investigation.
2. **Investigation of passenger safety awareness:** The passenger safety awareness survey designed for this article mainly includes two major aspects, namely, the subway safety awareness as well as degree of familiarity with the subway safety facilities. The main contents thereof are shown in Table 1.

**Table 1 Contents of survey on passenger safety awareness.**

| No. | Questionnaire | Choice |
|---|---|---|
| Q1 | How do you care about the subway safety? | A1 (1) Very much; A1 (2) Fairly; A1 (3) Generally; A1 (4) Seldom; A1 (5) Not at all. |
| Q2 | Have you ever received any promotion of subway safety knowledge? | A2(1)Yes, I have; A2(2) No, I haven't. |
| Q3 | What do you think is the best solution in case of any subway emergency (such as fire, explosion, power outage and terrorist attack, etc.)? | A3(1) Watch out calmly and obey command; A3(2) Seek help; A3(1) Try to find a way to evacuation. |
| Q4 | How do you know about the functions of safe evacuation signs? | A4 (1) Very much; A4 (2) Fairly; A4 (3) Generally; A4 (4) Not much; A4(5) Not at all. |
| Q5 | Do you know if there is any alarm device in the subway cars? | A5(1) Yes, I do I know their exact locations; A5(2) Yes, I do, but I don't know their exact locations; A5(3) No, I don't. |

Notes: QN(N=1,2...) in the table means the N-th question: AN(i) (N=1,2..., i=1,2...) means the i-th answer to the N-th question.

(3) Survey on passenger evacuation behavior

The passenger evacuation behavior survey designed for this article mainly covers two aspects, ie., the evacuation behavior in the case of emergencies occurred in different areas as well as the evacuation behavior in the course of passenger evacuation and evacuation, and the main contents thereof are shown in table 2.

**Table 2 Contents of survey on passenger evacuation behavior.**

| No. | Questionnaire | Choice |
|---|---|---|
| Q6 | How would you evacuation in case of any emergency occurred in the subway station (such as fire, explosion, power outage and terrorist attack, etc.)? | A6(1) Obey the command; A6(2) Follow the main passenger stream; A6(3) Evacuation by selecting the evacuation passage with the least passengers; A6(4) Evacuation by selecting the fastest evacuation passage. |
| Q7 | Do you know how the passengers should evacuation when the train stops in the running tunnel due to a fire? | A7(1) According to the location of train fire, evacuation toward the direction of fresh air along the emergency platform inside the tunnel by following the broadcasted reminder; A7(2) Try to find a solution by myself; A7(3) I don't know how. |
| Q8 | How would behave when there is a fire in the subway hall of an overhead station? | A8(1) Wait for rescue at the upper platform level; A8(2) Evacuation from the platform level toward the lower subway hall where the fire starts, and evacuate from the station as fast as possible; A8(3) Jump off down to the track area and evacuation toward the overhead viaduct |
| Q9 | How should the emergency door be opened in case of emergency? | A9(1) Press the push bar to release the lock on the emergency door and then push the door open toward the platform by following the instruction broadcasted by the train driver, and get off the train; A9(2) Find a solution by myself; A9(3) I don't know how. |
| Q10 | Would you choose to evacuation the platform by using the escalator or the stair in case of any emergency occurred in the subway station (such as fire, explosion, power outage and terrorist attack)? | A10(1) Escalator; A10(2) Stair; A10(3) Uncertain. |
| Q11 | When passengers are evacuating in case of any emergency occurred in the subway station (such as fire, explosion, power outage and terrorist attack, etc.), how would you usually do if you find a major congestion on the escalator or the stair ahead of you? | A11(1) Choose the escalator or stair at the other end of the platform; A11(2) Continue to wait at the same place; A11(3) Estimate the time needed for getting to the evacuation stair at the other end according to the degree of congestion before making any decision. |
| Q12 | During evacuation in case of any emergency occurred in the subway station (such as fire, explosion, power outage and terrorist attack, etc.), how would you behave if any passenger is found to have fallen off ahead of you? | A12(1) Try to avoid stumbling on the fallen passenger; A12(2) Evacuation myself by directly jumping over; A12(3) Help the passenger stand up. |
| Q13 | the evacuation passages of a subway station can evacuate all passengers from the platform where the fire occurs within 6 minutes. If you are not evacuated after 6 minutes and there is a huge amount of smoke converged over the platform at that time, what would you do? | A13(1) Continue to evacuate from the same exit without any panic; A13(2) Panic, resulting in extreme behavior such as pushing and shoving, stampede, etc.; A13(3) Try to find another exits in the smog. |

## *Data sources*

For this article, some 240 survey questionnaires were distributed randomly for Line 5 of BJ Subway, Line 1 of NJ Subway and Line 4 of GZ Subway, as a result of which 230 effective questionnaires were reclaimed. Specific details are shown in Table 3.

**Table 3 Sources of survey questionnaire data.**

| City | Time | Venue of survey | Type of station | Number of questionnaires issued | Number of valid questionnaires |
|---|---|---|---|---|---|
| BJ | May 14, 2009 | Huixin Xijie Beikou Station of Line 5 | Underground | 80 | 80 |
| NJ | Jan 13, 2009 | Xinjiekou Station and NJzhan Station of Line 1 | Underground | 100 | 97 |
| GZ | Dec 25, 2008 | Shiqi Station of Line 4 | Overhead | 60 | 53 |

## *Analysis method*

This paper utilizes histogram for statistics and analysis of the investigative data. Longitudinal coordinate of the histogram represents the percentage of persons, the lateral coordinate represents the city under investigation, and the legend represents different contents of investigation.

$$\mathrm{P}_N^i = \frac{K_N^i}{K}$$

where: $\mathrm{P}_N^i$ represents the percentage of persons choosing the i-th answer to the N-th question in the survey questionnaire; $K_N^i$ represents the number of persons choosing the i-th answer to the N-th question in the survey questionnaire; $K$ represents the effective number of persons surveyed on subway of a city (ie., the effective number of survey questionnaires reclaimed for subway of a city).

## Result analysis and discussions

### *Basic information about object of investigation*

- Sex: Percentage of male is relatively high among the objects of investigation in BJ and NJ while the percentage of female surveyed is relatively high in GZ, see Table 4.
- Age: Most of the objects of investigation in all three cities are around 30 years of age, and there are very few objects of investigation older than 60, which indicates that most of the subway commuters are employees on the way to/from work while other passengers are relatively few.
- Educational background: of the objects investigated, 70% or so are graduates of junior college or senior college while postgraduates are very few.
- Place of dwelling: 60% of the objects investigated in BJ and NJ live in the urban area while about 50% of the objects investigated in GZ live in the suburb.
- Period of dwelling: of the three cities, more than 50% of the objects investigated live in their respective areas for more than 5 years.

**Table 4 Analysis of basic information about the objects of investigation.**

| Characteristics | BJ (%) | NJ (%) | GZ (%) | |
|---|---|---|---|---|
| Sex | Male | 56.00 | 55.00 | 42.00 |
| | Female | 44.00 | 45.00 | 58.00 |
| Age | 15-30 | 75.00 | 60.82 | 84.90 |
| | 30-60 | 21.25 | 38.15 | 15.10 |
| | >60 | 3.75 | 1.03 | 0 |
| Educational background | High school or under | 27.50 | 34.02 | 24.53 |
| | Junior college, senior college | 67.50 | 64.95 | 71.70 |
| | Postgraduate or higher | 5.00 | 1.03 | 3.77 |
| Place of dwelling | Urban area | 58.75 | 67.01 | 30.19 |
| | Suburb | 27.50 | 22.68 | 47.17 |
| | Junction between urban area and suburb | 13.75 | 10.31 | 22.64 |
| Period of dwelling | <0.5 | 11.25 | 14.43 | 13.21 |
| | 0.5-5 | 28.75 | 23.71 | 32.74 |
| | >5 | 60.00 | 61.86 | 51.05 |

## *Passenger safety awareness*

### Cognitive degree of subway safety

**Degree of concern about subway safety (Fig. 2a):** In the three cities, about 80% of the passengers care about the subway safety. About 10% of the passengers in BJ and GZ showed insufficient concern about the subway safety while this percentage in NJ reached 20% or above. By comparing GDP and operational subway mileages of these three cities, the following can be established:

1. The improvement of urban economic level increased the citizens' trip rate as well as their opportunities of taking subway. Meanwhile, as the economic level in cities further improves, the urban area becomes bigger and bigger. The too high a housing price in the center area of the cities results in a large convergence of working classes in the suburbs who frequently take subway on their way to/from work places. Therefore, the higher the economic level in a city, the higher the quality of the citizens, and the higher degree of concern about the subway safety.
2. As the number of operational subway lines increases, passengers tend to have more opportunities in taking subways. Therefore, the more the number of operational subway lines, the higher the citizens' concern about the subway safety
3. Since 38.15% of the objects investigated for NJ Subway are aged between 30 and 60, which is obviously higher than 21.25% in BJ and 15.10% in GZ, indicating that the elderly are not as concerned about the subway safety.

**Degree of popularity of subway safety (Fig. 2b):** About 60% of the passengers in BJ and NJ have received education about the subway safety while 51% of the passengers in GZ have not received any such education. The analysis with respect to the place of dwelling of the objects investigated in these three cities indicates the following:

1. The subway safety education has effectively covered the passengers in the urban areas of the three cities while the effort of education of the passengers in the suburbs of the cities is insufficient.
2. Since there are many migrant workers in GZ who live in there not for a long time, this is bound to affect the degree of concern and education about the subway safety.

3. **Safety awareness in case of emergency (Fig. 2c):** More than 70% of the passengers in BJ and GZ are likely to observe calmly and obey the commands in case of emergency while only 60% of the passengers in NJ would do so; about 20% of the passengers in all three cities would need help in case of emergency; of the passengers who would find a solution to evacuation by themselves, NJ accounts for 16% while this percentage is under 10% in the other two cities.

4. It can be seen from the above that as the number of operational subway lines increases, passengers tend to experience more subway emergencies and receive more educations thereof, and they would be more cool-headed when it comes to dealing with emergencies.

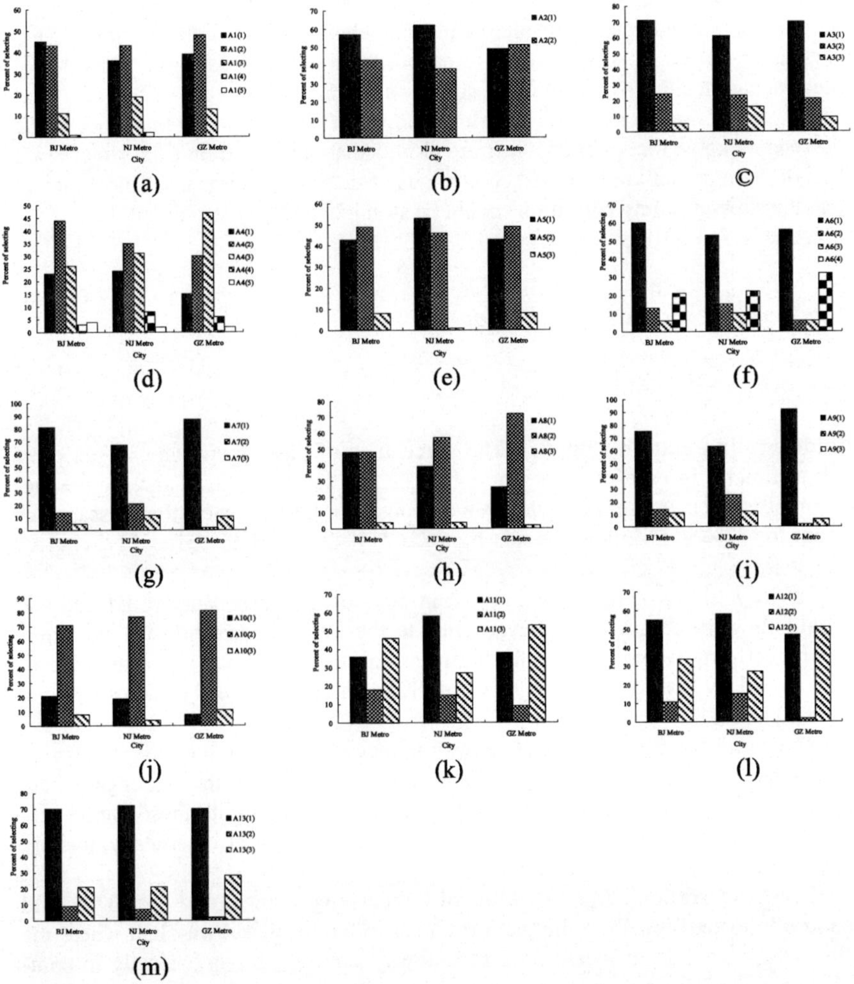


**Fig. 2.** Comparative analysis according to the survey results. (a) Q1, (b) Q2, (c) Q3, (d) Q4, (e) Q5, (f) Q6, (g) Q7, (h) Q8, (i) Q9, (j) Q10, (k) Q11, (l) Q12, (m) Q13.

### Degree of familiarity with safety facilities

**Degree of familiarity with safe evacuation signs (Fig. 2d):** More than 60% of the passengers in BJ and NJ know about the safe evacuation signs provided in the subway system, while this number is GZ is only 45%, where more than 50% of the passengers are not sure about or don't know at all the purpose why the signs are provided. The analysis of the place of dwelling of the objects investigated in the three cities indicates that safety education should be strengthened for passengers living in the suburbs in order to enable them to understand the functions of the evacuation signs provided in the subway system.

**Degree of familiarity with the alarm devices (Fig. 2e):** More than 90% of the passengers know that alarm devices are provided aboard the train cars. However, a half of them do not know the exact locations of these alarm devices. This indicates that the subway safety education should be strengthened in every detail.

## *Passenger evacuation behavior*

### Evacuation behavior in different areas

**Underground station (Fig. 2f)**: The percentage of passengers who would obey the command in case of emergency is basically the same in all three cities, ie, about 60% each. Passengers who would choose the fastest evacuation passage account for 32% in GZ, whereas this figure is 21% and 22% for BJ and NJ respectively.

**Subway tunnel (Fig. 2g ):** More than 80% of the passengers in BJ and GZ would obey the command and evacuation to the fresh air direction via the emergency platform, whereas in NJ, this figure is only 61%. In NJ, 21% of the passengers would try to find a solution by themselves, and this figure is obviously higher than BJ and GZ.

By comparing the operational subway mileages of these three cities, it can be found that since there are few operational subway lines in NJ, the passengers there experience less emergencies and lack experience in dealing with these. Some passengers would try to find a solution by themselves in case of emergency, and are not used to obeying to the command.

**Overhead station (Fig. 2h):** Most of the passengers in BJ, NJ and GZ would choose to evacuation from the platform level to the lower subway hall where the fire occurs so as to evacuate from the subway station. This is basically in compliance with China's quick subway evacuation plan in case of emergency. However, it can also be seen from the figure that 72% of the passengers have chosen "Quickly evacuation out of the station by escaping from the platform level to the lower subway hall", which is obviously higher than 48% for BJ and 57% for NJ.

With the type of subway stations under the investigation taken into account, it shows that a certain percentage of subway passengers show a certain pattern, and this pattern has a certain impact on the tendency of their subway safety concerns. For example, the subway stations investigated in GZ are overhead stations, and the passengers are obvious know better how to deal with the emergencies occurred in the overhead stations than the passengers of the underground subway stations in BJ and NJ.

**Evacuation behavior in the course of evacuation**

**Opening of emergency door (Fig. 2i):** 75% of the passengers in BJ and 92% of the passengers in GZ would obey the commands and handle calmly in case of any emergency whereas this figure is 63% in NJ. 25% of the passengers in NJ would choose to find a solution by themselves, and this figure is obviously higher than in BJ and GZ. By comparing the operational subway mileages of these three cities, it can be found that since there are few operational subway lines in NJ, the passengers there experience less emergencies and lack experience in dealing with these.

**Selection of escalator or stair (Fig. 2j):** More than 70% of the passengers in all of the three cities have selected evacuation via stairs in case of emergency. About 20% of the passengers in BJ and NJ have selected the use of escalators whereas this figure is 8% in NJ.

**Evacuation behavior in case of congestion (Fig. 2k):** About 605 of the passengers in NJ would evacuation via the escalator or stair at the other end of the hall in the case of any congestion on the escalator or stair at the present end, whereas this figure is about 40% for BJ and GZ. About 50% of the passengers in BJ and GZ would decide whether to choose the escalator or stair at the other end to evacuation after an analysis and comparison. This indicates that passengers in BJ and GZ are more cool-headed in dealing with emergencies.

**Evacuation behavior when somebody falls down (Fig. 2l):** In all of the three cities, more than 50% of the passengers would try to avoid stumbling over somebody who has fallen down. Some 51% of the passengers in GZ would choose to help the passenger stand up in such case, which figure is obviously higher than in BJ and NJ. It can be seen from Table 4 that under the influence of traditional education, young people in China tend to help others out of harm's way together in case of emergency.

**Evacuation behavior in case of failure to evacuate after 6 minutes (Fig. 2m):** In the three cities, about 70% of the passengers would continue to evacuation from the same exit in the case of failure to evacuate within 6 minutes without resulting in any panic. Under the same situation, more than 20% of the passengers in all of the three cities would try to find other exits to evacuation.

## Conclusions

The investigation and analysis of the safety awareness and evacuation behavior of the subway passengers in BJ, NJ and GZ indicates that the safety of subway passengers' evacuation behavior is closely related to the operational subway mileage, citizen trip rate, safety education and popularity as well as passenger quality among other factors. These specifically lie in the following:

- As the urban economy grows and the citizens' trip rate increases, the citizens have more opportunities of engaging subways, and their safety awareness improves gradually.
- As the urban economy grows and people's quality improves, people will care more about the subway safety and will more standardize their safety behavior.
- With the increase of operational subway lines and operating time in a city, the passengers' subway safety awareness is improving as they experience more emergencies. As a result, they will be more cool-headed and rational in dealing with the emergencies.
- With respect to safety education, the popularity in the center area of a city is better than in the suburb, and the popularity among young people is better than among the elderly. Therefore, safety education and training should be strengthened among passengers who live in the suburbs and among the elderly.

**Acknowledgments:** The authors would like to thank National Natural Science Foundation of China for supporting this research under contract 50704027, 50674079 and 70833006, and "11th Five" National Key Technologies R&D Program under contract 2008BAB37B05-05.

## References

1. Liu, K.H.: Analysis and countermeasures study on the status quo of beijing subway passenger's safety awareness. Master Thesis, Beijing University of Chemical Technology (2007)
2. He, L., Zhong, M.H., Shi, C.L., Shi, J.H.: Investigation on evacuation behavior of passengers in metro emergency. Journal of Safety Science and Technology. 5(1), 53-58 (2009)
3. Jin, R.F., Gan, C.X., Chai, S.J., Sun, X., Li, J., Xia, S.L.: A cross-section study on workers safety consciousness attitude and behavior in chemical industry. Industrial Health and Occupational Diseases. 29(2) , 92-95 (2003)

# Modelling Random Taste Variations on Level Changes in Passenger Route Choice in a Public Transport Station

**I. Zeiler, C. Rudloff, and D. Bauer**

Austrian Institute of Technology
Department Mobility, Business Unit DTS.
Giefinggasse 2, A-1210 Wien, Austria.

**Abstract** In large stations of public transportation high crowd densities can lead to potential safety risks and to unnecessary delays. To assess the *actual* capacity of potential bottlenecks a deeper understanding on the route choice of pedestrians is of great importance. This paper investigates the factors that influence the route choice of pedestrians when facing a stair/escalator combination in a major Austrian train station. We employ random utility models on data sets of revealed and stated preferences. In particular we investigate the potential for heterogeneities in taste by employing *mixed logit models*. The results show that, first, crowding is an important factor for route choice, second, that the application of mixed logit models is appropriate and, last, that the use of both revealed and stated preference data adds valuable information.

## Introduction

Multi-modal interchange nodes of public transportation systems play an important role in the daily commute of millions of passengers. To allow a safe and efficient flow of passengers the design of an interchange node has to meet a number of requirements, most important among which is providing sufficient capacity for the expected demand. Hence, the key elements of the infrastructure have to be dimensioned such that high crowd densities and therewith unnecessary delays or potential safety risks can be avoided. In particular the *actual* demand during normal operation and during peak hours is of great interest.

In the last decade simulation models of pedestrian behaviour have become an increasingly popular planning tool for assessing the quality and functionality of transportation infrastructures in the planning stage. While the operational level of these models is well developed (e.g., replication of fundamental diagrams) seemingly most programs use only a simple model for route choice. The most common underlying assumptions are that individuals minimize walking distance or walking time. In certain situations this may lead to a distortion in the route choice and thus to a faulty allocation of crowds. Furthermore, in choice situations pedestrians are

R.D. Peacock et al. (eds.), *Pedestrian and Evacuation Dynamics*,
DOI 10.1007/978-1-4419-9725-8_17, 

usually assumed to be homogenous in both taste and attributes. These simplifications may sometimes be justified by application, however, a deeper understanding of how route choices are made and of the factors influencing these choices are crucial for advancing predictive models of pedestrian behaviour.

A very common route choice situation occurs when pedestrians face a level change. Most often two or more options are available, e.g. stairs, escalators or elevators. Moreover, such facilities are often potential bottlenecks of public transportation nodes. So, a better understanding on the route choice in such situations helps to assess the *actual* capacity of these facilities and is a step towards a more realistic replication of pedestrian behaviour in simulation models.

However, literature on the route choice when changing levels within a building is only sparsely available. Two notable contributions on this topic are [1] and [2]. In more detail, [1] found that besides walking time and walking distance the necessity to climb a grade is a determining factor. More generally [2] investigated the impact of different facilities for level changes on route choice in two Dutch railway stations. The investigated facilities were ramps, stairs and escalators; the model employed for describing these discrete choices was a standard logit model-which assumes homogeneity in taste.

In this paper we build on existing literature by contributing in both substance and method. Substantially, we investigate how *"dynamic"* factors, e.g. crowding, and personal factors, e.g. age or carrying luggage, influence the choice between descending, adjacent escalators and stairs. Furthermore, we investigate the potential for heterogeneities in taste in these factors by employing *mixed logit models*.

Methodologically, we estimate route choice models on both revealed preference data (RP) and stated preference data (SP) as well as a joint model. SP data has the advantage that hypothetical situations can be included, introducing variability in the aspects underlying the choice and hence contributing to the identifiability of factors influencing the choice process. At the same time collecting RP data enhances realism of the answers obtained. Model estimation using a combination of RP and SP data are reported in economics [3] and in transport related issues [4]. There is also a growing body of literature on the particular problems of estimating these joint models using different simulation based methods (see e.g., [5], [6]). The application of the modelling methodology to pedestrian route choice decisions, however, appears to be an innovative feature of this paper.

This paper is structured as follows. We begin with describing the extensive survey undertaken for collecting the data. We then present the methods and the analysis. Lastly, we describe our results and present conclusions.

## Empirical Study

We collected data on pedestrian's route choice behaviour by undertaking an extensive survey in a major Austrian train station, the "Westbahnhof". During the survey 200 persons were observed and interviewed immediately after using a highly frequented escalator/stair combination (see Figure 1). This combination was a key element of the train station, since it was used by a large share of the passengers coming from the platforms.

The sample consisted of 106 men and 94 women; the youngest subject was 15 years old, the oldest 88 years old. Of the participants 7.76% were between 15 and 19 years, 27.14% between 19 and 30 years, 46.73% between 31 and 60 years, and 17.66% older than 60 years. A detailed description of the survey and the data can be found in [7].

**Fig. 1. Screen shot of the examined stair/escalator combination. The right escalator leads down towards the exit of the train station. Passengers coming from the platforms arrive from the left hand side of the picture.**

During the survey different types of data were collected in three ways.

- To collect RP data and personal factors, an investigator standing at the bottom of the stairs recorded the choice of a particular pedestrian and then asked a number of questions regarding their personal attributes.
- The area in front of the escalator was videotaped during the whole survey. That way the exact number of persons queuing in a predefined area could be determined for each of the respondents' choice situations.
- For collecting SP data, each pedestrian was shown six different video-sequences and was asked which choice they would have taken in each situation. To make the choice situations as realistic as possible the sequences showed approaches to the very same stair/escalator combination from a first-person perspective and with different degrees of crowding. The degree of crowding was quantified in the numbers of passengers queuing in a predefined area in front of the escalators.

Summing up, we collected two "dynamic" factors, which were the number of persons queuing with luggage and the number of persons queuing without luggage at the time the choice was taken.

The personal factors we collected were the following. Besides age and gender we collected information on trip purpose (work, business, travel, education, shopping), frequency of visit (six levels between first time visit and daily visit), self-reported walking speed (18.32% slow, 13.2% normal, 68.5% fast), and level of education. Furthermore, the subjects were asked for the reasons for their choice. The answers included luggage, haste, crowding ("too many people waiting"), saving time, convenience, health issues, free space ("no people waiting"), tiredness, and miscellaneous.

## Methodology

We modelled the choice between escalator and stairs by discrete choice models (see e.g., [3]), which belong to the family of random utility models (RUM). In a RUM, a decision maker $n$ gains from each possible alternative $j$ a certain utility $U_{nj}$ and, under the assumption of rational choice behaviour, chooses the alternative with the highest utility. However, due to modelling inaccuracies and the inability to observe all underlying factors only a portion $V_{nj}$ of the utility is observable; the remaining part, $\varepsilon_{nj}$, is modelled as a random variable.

In our context, individuals face a *binary* choice only between the two options, stairs and escalator. Since the alternative yielding the higher utility is chosen, the probability of an individual choosing the stairs can be modelled as depending on the *difference* of the respective utilities. Hence, when the alternatives are labelled by $e$ for "escalator" and $s$ for "stairs" it is sufficient to consider $\underline{U}_n=U_{ns}-U_{ne}$. When $\underline{U}_n>0$ (that is $U_{ns}>U_{ne}$) holds, individual $n$ takes the stairs.

The observable part of utility $\underline{U}_n$ is modelled as $\underline{V}_n=\beta' x_n$, where $x_n$ denotes a vector of the *differences* of the attributes of the alternatives for the $n$-th individual and $\beta$ is the vector of parameters to be estimated.

To capture the average effect of all unobserved factors not included in the model, alternative specific constants are introduced. When the random part of the utility is rewritten as $\underline{\varepsilon}_n=c+\eta_n$ with $\eta_n$ a random variable with mean zero, $c$ is the alternative specific constant for taking the stairs. It is estimated together with all other parameters. This yields $\underline{U}_n=\underline{V}_n+c+\eta_n$.

Using the **standard logit formula** the probability of an individual $n$ choosing the stairs can then be derived as

$$P(n\ chooses\ stairs)=\frac{\exp(\underline{V}_n+c)}{1+\exp(\underline{V}_n+c)} \tag{1}$$

using the binary nature of the choice situation.

Standard logit, however, has its limitations as it assumes homogeneity in taste across all individuals. In particular, each individual uses the same parameter $\beta$ in order to measure the utility. This limitation can be overcome by using **mixed logit models**, which are highly flexible models that can approximate any random utility model [9]. Here, instead of assuming fixed parameters the vector $\beta$ is allowed to be random following a distribution depending on some underlying parameters.

Two more advantages of the mixed logit specification are that it alleviates the independence of irrelevant alternatives (IIA) restriction by allowing for a more flexible specification of the correlation structure of the random utility term across alternatives. Furthermore, it accommodates the correlation structure for repeated choice situations in general and in particular those present in SP data sets (cf. [3]).

Due to the distributed parameters, the choice probability is the integral of the adapted logit formula over the density of $\beta$ following a pre-defined distribution with probability density function $f$ with parameters $\theta$ (e.g. $f$ denotes the Gaussian distribution where $\theta$ contains the mean and the variance),

$$P(n\ chooses\ stairs) = \int L_n(\beta, x_n) f(\beta \mid \theta) d\beta\,,$$

where $L_n(\beta, x_n)$ is the logit probability (1) evaluated at parameters $\beta$. The parameters to be estimated are $\theta$.

## *Combining RP and SP data*

When estimating on a data set that consists of both SP and RP data, it cannot be assumed that the random part of the utility has same variance and mean for the two types of preference data. In particular, when looking at real choice situations a large number of unobserved factors, like tiredness, can influence the decision maker's choice. Thus, when answering to a stated preference survey, the decision maker is asked to treat all the decisions similarly with respect to factors not mentioned in the survey. Consequently the variance of the random part of the utility is likely to be smaller than for the RP data.

To include these differences in random utility the RUM has to be changed (see [3]). When $t$ is a stated preference situation the utility difference is defined as $\underline{U}_{nt}=\beta' x_{nt}+c^s+\underline{\varepsilon}_{nt}$ and when $t$ is a revealed choice situation as $\underline{U}_{nt}=(\beta/\lambda)' x_{nt}+c^r/\lambda+\underline{\varepsilon}_{nt}$.

Here $\lambda$ scales the utility such that the variance of stated and revealed preference situations is equal and $c^{r/s}$ are the different alternative specific constants for revealed and stated preferences. The parameters $\theta$, $\lambda$ and $c^{r/s}$ can now be jointly estimated by maximizing the corresponding likelihood function.

## *Specific Modelling Approach*

In all models estimated we considered the same set of explanatory variables being gender, age, self-reported walking speed, frequency of visit, carrying luggage, number of persons queuing without luggage, and number of persons queuing with luggage. For numerical reasons we divided age by 10. Self-reported walking speed was encoded as 0 for "slow", 1 for "normal" and 2 for "fast". We distinguished six different levels for the frequency of visit, encoded as 0 for "daily", 1 for "several times a week", 2 for "several times a month", 3 for "1-2 times a month", 4 for "infrequently" and 5 for "first time visit". All other variables were encoded by their actual number.

For analyzing the data we chose the following approach. We firstly estimated standard logit models separately on the RP and the SP data. Secondly, we tested whether distributed parameters enhance the estimate by applying mixed logit models. Thus, we tested whether there is heterogeneity in taste in the sample. Finally, we used the combined data set to estimate mixed logit models for finding out whether the variance in choice is different on the two data sets.

This last model combined the advantages of both SP and RP data. These were the consideration of real choices, and the chance to choose a design for the SP questions that emphasized the attributes of interest (crowding).

**Table 1. Estimation results from standard logit and mixed logit for the SP data only. Heterogeneities in taste could be found in the parameter corresponding to age. LL(θ) gives the value of the log-likelihood function at convergence.**

| SP Data | Standard Logit | | Mixed Logit | | |
|---|---|---|---|---|---|
| | *Estimate* | *t-Ratio* | *Estimate* | *t-Ratio* | *Distribution* |
| *c* | -0.7061 | -18.581 | -2.2673 | -2.7310 | - |
| Age | -0.1751 | -4.607 | μ: -0.0608 | -0.3447 | Normal |
| | | | σ: 0.8096 | 9.0661 | |
| Speed | 0.4470 | 5.4981 | 0.7060 | 2.3557 | - |
| #P w Luggage | 0.3672 | 6.0795 | 0.7489 | 7.8832 | - |
| #P w/o Luggage | 0.1962 | 7.1087 | 0.4031 | 8.9181 | - |
| LL(θ) | -703.1074 | | -528.2375 | | |

## Results of the Different Models Estimated

When looking descriptively at the data two points stand out. First, out of the 200 persons observed 182 used the escalator and 17 used the stairs (1 observation missing) corresponding to 8.54%. This is contrasted by the stated preferences where the subjects stated in 741 out of 1164 choice situations (5 observations missing) to take the stairs. This corresponds to 63.66%.

Second, each subject in the sample that carried luggage took the escalator. So, an estimation of the influence of this variable yields a probability of 1 for taking the escalator when carrying luggage. In order to avoid a distortion of the results these subjects were removed from the sample.

## *Comparing Stated and Revealed Preferences*

When estimating discrete choice models on the RP data, only the number of persons queuing with luggage influenced the choice significantly. The corresponding parameter was estimated at 1.7168 (t-ratio 4.2973), the alternative specific constant was -2.8295 (-7.8488). Using distributed parameters did not improve the results significantly, so the influence of this factor was homogeneous within the tested sample.

The situation is different when looking at the SP data. When estimating a standard logit model four factors were significant, age, walking speed, number of persons queuing with luggage and without luggage. Table 1 provides all estimates and the corresponding t-ratios. Here, using a mixed logit model that allows normally distributed parameters for the effects of age increased the value of the log likelihood by 174.9 points and thus improved the model significantly. The estimated distribution had a mean of -0.0512 and a standard deviation of 0.8088, which results in a share of 52.8% persons with a negative parameter.

When comparing the results, we see that age and self-reported walking speed seemed to influence the *self-assessment* of the passengers interviewed, even though they apparently had no influence on the actual, revealed choices. Further differences can be found when directly comparing the different probabilities yielded from the estimates.

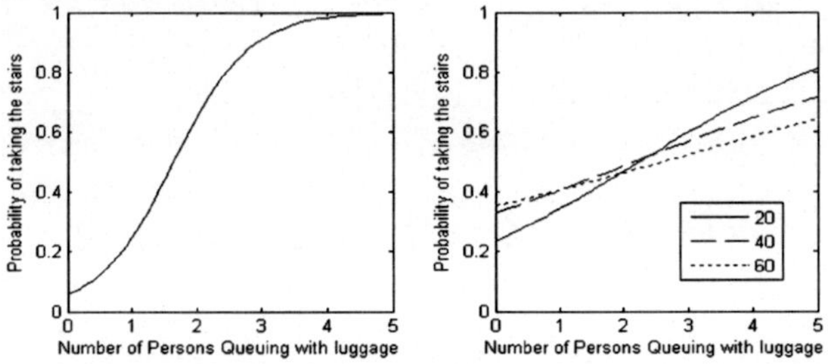


**Fig. 2. Probability of taking the stairs as function of the number of persons queuing with luggage, estimated for the separate revealed (left plot) and stated preferences (right plot) data sets. Self reported speed has been set to normal; number of persons queuing without luggage is zero. Age has a significant influence in the stated preferences only and moderates the effect of the persons queuing with luggage.**

Figure 2 shows the probability of taking the stairs for different numbers of pedestrians queuing with luggage. The figure shows the results from both revealed preferences (left) and stated preferences (right). For making the probabilities comparable in both models normal walking speed was assumed. For the SP model we set the number of persons queuing without luggage to zero. Furthermore, in the SP model we considered persons of different age, namely 20, 40 and 60.

The first result to be pointed out is that when having an empty escalator in the RP model the probability of taking the stairs approximately 0.1. The same probability in the SP data is between 0.21 for younger pedestrians and 0.39 for the elderly. So, without any crowding the models yield very different probabilities.

The second interesting phenomenon is that in the estimated SP model age does not have a monotonous effect on the probability of taking the stairs. If it were true that younger persons are generally more likely to take the stairs, the three different curves would not change their order. Here, however, the curves intersect at some point, so older subjects are more likely to take the stairs when there is no crowding. In crowded situations the opposite is true according to the model.

**Table 2. Estimation results from standard logit and mixed logit for the combined SP and RP data. LL(θ) gives the value of the log-likelihood function at convergence.**

| | Standard Logit | | Mixed Logit | | |
|---|---|---|---|---|---|
| | *Estimate* | *t-Ratio* | *Estimate* | *t-Ratio* | *Distribution* |
| $c^r$ | -2.3982 | -3.1714 | -7.6442 | -3.4806 | - |
| $c^s$ | -0.7609 | -2.6365 | -1.7785 | -2.3014 | |
| λ | 0.9224 | 2.5587 | 2.7347 | 2.7523 | |
| Age | -0.1654 | -4.3412 | μ: -0.264 | -1.7449 | Normal |
| | | | σ: 0.7994 | 9.0532 | |
| Speed | 0.4549 | 5.7655 | 0.689 | 2.3932 | - |
| #P w Luggage | 0.3774 | 6.4845 | μ: -0.2655 | -2.0129 | Lognormal |
| | | | σ: 0.3154 | 3.1477 | |
| #P w/o Luggage | 0.1949 | 7.3271 | 0.4079 | 8.9846 | - |
| LL(θ) | -767.8122 | | -585.3091 | | |

## *Analyzing the Combined Data Set of Stated and Revealed Preferences*

The analysis of the **combined data** set yielded the following results. As in the estimate on SP data only, the four significant factors are numbers of persons queuing with or without luggage, age and self reported walking speed. Table 2 provides all estimated values and the corresponding t-ratios.

Since the unobserved factors (e.g., tiredness) are generally different for RP and SP data a scaling factor $\lambda$ for normalizing the variance of the errors was estimated along with the parameter vector. Here, the factor $\lambda$ scales the variance of the unobserved utility for RP situations. In particular, the variance for the unobserved factors of the RP situations is $\lambda^2 \text{var}(\varepsilon_{nt})$. This means that the variance for the RP data was roughly seven times that of the SP data, i.e. there was a higher variance in unobserved factors for revealed choice situations than for the ones in the SP survey.

So, the estimation of mixed logit models improved the results significantly, as well. Hence, our model showed heterogeneities in taste. Best results were achieved when two of the coefficients were mixed. First the coefficient corresponding to age is normally distributed with mean -0.264 and standard deviation 0.7994. Second the coefficient for the number of persons queuing without luggage has a lognormal distribution with mean 0.8038 and standard deviation 0.2582. Thus, the influence of age is negative for taking the stairs for about 63% of the passengers, and positive for about 37% of the passengers.

## Conclusions

In this paper we investigated the route choice problem of choosing between one of adjacent stairs or escalators in a public transportation infrastructure. We determined the relevant factors by estimating a series of random utility models on both RP and SP data. From our results we can draw the following conclusions.

When considering the actual choices of the observed individuals, the only significant factor was the number of persons queuing with luggage. This matches the results of [8] who found that crowdedness is an important factor in route choice. Surprisingly, age or self-reported walking speed did not play a role. So, for assessing the capacity of stairs/escalator combinations one should consider the share of pedestrians carrying luggage, since they have a large propensity to take the escalator *and* influence the route choice of other passengers at the same time.

A comparison with the SP models yields interesting insights into the self assessment of the subjects. Age has a significant influence on the choice. However, for a share of 47.2% a higher age increased the likelihood to take the stairs. A possible explanation is that at a certain age factors like physical fitness could gain more importance. For example, some elderly people might view taking the stairs as exercise, using them without letting exogenous factors influence their decisions. For others, taking the stairs might not be possible due to a reduced agility. The assumption that younger persons react more sensitive to crowding is an explanation for the differences that can be observed in Figure 2. In particular, the probability for taking the stairs shows less dependence on crowding for older people than for young ones. In any case the influence of age is heterogeneous, so presuming that older persons are less likely to accept physical effort is misleading.

Self-reported walking speed and numbers of person queuing had an influence in the expected direction. Fast walkers are more likely to state to take the stairs than persons walking slowly. The number of persons queuing influences the subjects towards choosing the stairs, as well.

The estimation of discrete choice models on a combination of RP and SP data allowed us to emphasize the attribute of interest, namely crowding, while still considering the actual choices made. The model shows that the variance in choice is larger for the RP data. This confirms that we successfully reduced the influence of unobserved factors (for instance tiredness) by including stated preferences.

However, there is a potential bias in the combined model, since the model implies differences in the frequency of choices in the revealed and stated preference data. In particular, pedestrians stated more often that they would take the stairs than they actually did according to the RP data. On the RP data the model estimated on that data shows an average of 11% of persons taking the stairs. When using the SP model on the RP data this share increases to 33% of people. This shows that people severely overestimated their own willingness to take the stairs compared with their actual behaviour. Applying the jointly estimated SP/RP model to the RP data deals with that problem by estimating two separate alternative specific constants, one for the SP and on for the RP data. This leads to a less biased model and demonstrates that joint estimation combines the advantages of RP and SP data. The bias introduced by using SP data is counteracted by the RP data.

While this shows the importance of using RP data for estimating choice models, it does not mean that SP surveys are less valuable. The SP survey added variance to the input data which can give a much better estimation of the relative importance of the parameters in the choice situation the decision maker faces and leads to significant parameters for more explanatory variables.

Another interesting point is that in the joint model estimation age is estimated to have a negative influence on the utility of the stairs for a share of about 63% of the passengers, and a positive influence for about 37%. This circumstantiates that age might be an imperfect proxy for the readiness to invest a physical effort to bridge a level change.

In summary, this paper investigated the significant factors for taking stairs and explored the potential for heterogeneity in taste in these factors. The results show that, firstly, crowding is an important factor for route choice, secondly, that the application of mixed logit models is appropriate and, lastly, that the use of both revealed and stated preference data adds valuable information.

**Acknowledgments** The authors thank Clarissa Knehs for her dedication to the project and the time spent for evaluating hours of video footage. Financial support was granted by the Austrian Ministry for Traffic, Innovation and Technology (BMVIT) within the project "mPed" which is gratefully acknowledged.

## References

1. Lam, W.H.K., Cheung, C., Lam, C.F.: A study of crowding effects at the Hong Kong light rail transit stations. Transportation Research Part A: Policy and Practice. **33**, 401-415 (1999)
2. Daamen, W., Bovy, P.H.L., Hoogendoorn, S.P., Van de Reijt, A.: Passenger route choice concerning level changes in railway stations. Proceedings of the TRB Annual Meeting, 1 -18. National Academy Press, Washington, DC (2005)
3. Hensher, D.A., Louviere, J.J., Swait, J.D.: Combining sources of preference data. Journal of Econometrics. **89**, 197-221 (1999)

4. Börjesson, M.: Joint RP-SP data in a mixed logit analysis of trip timing decisions. Transportation Research E. **44**, 1025-1038 (2008)
5. Train, K.: Discrete choice methods with simulation. Cambridge University Press (2003)
6. Bhat, C.R., Castelar, S.: A unified mixed logit framework for modeling revealed and stated preferences: formulation and application to congestion pricing analysis in the San Francisco Bay area. Transportation Research B. **39**, 593-616 (2002)
7. Knehs, C.: Untersuchung und Modellierung des Routenverhaltens von Fußgängern in Bahnhöfen. Master thesis, TU Wien (2010)
8. Seneviratne, P., Morall, J.F.: Analysis of factors affecting the choice of route of pedestrians. Transportation Planning and Technology. **10**, pp. 147-159 (1985)
9. McFadden, D., Train, K.: Mixed MNL models of discrete response. Journal of Applied Econometrics. **15**, 447–470 (2000)

# Data Collection Methods

# Extended Range Telepresence for Evacuation Training in Pedestrian Simulations

**A. Pérez Arias[1], U. D. Hanebeck[1], P. Ehrhardt[2], S. Hengst[2], T. Kretz[2], and P. Vortisch[2]**

[1]Intelligent Sensor-Actuator-Systems Laboratory (ISAS)
Institute for Anthropomatics, KIT, Germany

[2]PTV Planung Transport Verkehr AG
Stumpfstraße 1, D-76131 Karlsruhe, Germany

Corresponding author: antonia.perez@kit.edu

**Abstract:** In this contribution, we propose a new framework to evaluate pedestrian simulations by using Extended Range Telepresence. Telepresence is used as a virtual reality walking simulator, which provides the user with a realistic impression of being present and walking in a virtual environment that is much larger than the real physical environment, in which the user actually walks. The validation of the simulation is performed by comparing motion data of the telepresent user with simulated data at some points of the simulation. The use of haptic feedback from the simulation makes the framework suitable for training in emergency situations.

## Introduction

Telepresence allows visiting remote or virtual places with a high degree of realism and offers new tools for human motion understanding. Applications of Telepresence include teleoperation, military training, driving and flying simulators, etc.

The feeling of presence is achieved by visual, acoustic, and haptic sensory information recorded from the remote environment and presented to the user on an immersive display. The more the user's senses are involved, the better the immersion in the target environment is. In order to make use of the sense of motion, which is especially important for human navigation and way-finding, the user's motion is tracked and transferred to the avatar in the target environment. This is known as Extended Range Telepresence, and enables the user to make use of proprioception, the sense of motion, to navigate the avatar intuitively by natural walking, instead of using devices like joysticks, keyboards, pedals, or steering wheels. To allow exploration of an arbitrarily large target environment while moving in a limited user environment, we developed Motion Compression [1]. By preserving

R.D. Peacock et al. (eds.), *Pedestrian and Evacuation Dynamics*,
DOI 10.1007/978-1-4419-9725-8_18, 

the walked distances and the turning angles, Motion Compression transforms the path in the target environment into a feasible path in the user environment, and guides the user on his path.

Fig. 1 shows the user interface in the Extended Range Telepresence system.

Through the egocentric view, this framework provides a first person evaluation of the simulation. The user is not passively looking at the simulation, but he feels present in the simulation and can interact with other pedestrians. Extended Range Telepresence also allows the evaluation of pedestrian models based on gathered real user data. These experiments in the virtual environment are not only cheap to set up, but are also quick to evaluate, as all position data of the user is available anyway.

An application of our framework with particular focus on gaining spatial knowledge is the training of evacuations, where people are trained to find the way out of buildings, ships, or planes, so that planners could check how intuitively occupants find their way out before beginning with the real-world construction.

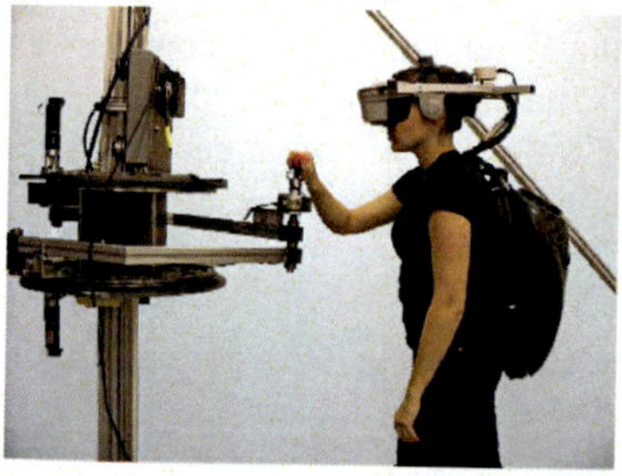

**Fig. 1. User interface in the Extended Range Telepresence system**

This work aims at employing Extended Range Telepresence first as a tool for evaluating and calibrating pedestrian simulations and second to train people in emergency evacuations using the validated simulations. In order to increase the degree of immersion and to present additional information to the user, the use of haptic information is also explored. Finally, a route choice scenario is examined and the potentials of the system in various fields of application are discussed.

## Extended Range Telepresence with Motion Compression

In order to allow for exploration of an arbitrarily large target environment while moving in a limited user environment, Motion Compression provides a nonlinear transformation between the desired path in the target environment, the target path, and the user path in the user environment. The algorithm consists of three functional modules (see Fig. 2).

First, the *path prediction* gives a prediction of the desired target path based on the user's head motion and on knowledge of the target environment. If no knowledge of the target environment is available, the path prediction is based completely on the user's view direction. Second, the *path transformation* transforms the target path into the user path in such a way, that it fits into the user environment. In order to guarantee a high degree of immersion, the user path has the same length and the same turning angles as the target path. The two paths differ, however, in path curvature. The nonlinear transformation found by the path transformation module is optimal regarding the difference of path curvature. Fig. 3 shows an example of the corresponding paths in both environments. Finally, the *user guidance* steers the user on the user path, while he has the impression of actually walking along the target path. It benefits from the fact that a human user walking in a goal oriented way constantly checks for his orientation toward the goal and compensates for deviations. By introducing small deviations in the avatar's posture, the user can be guided on the user path. More details can be found in [1, 2].

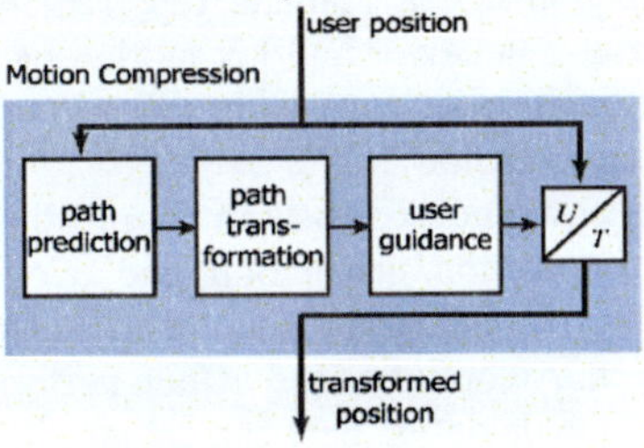


**Fig. 2. Motion Compression**

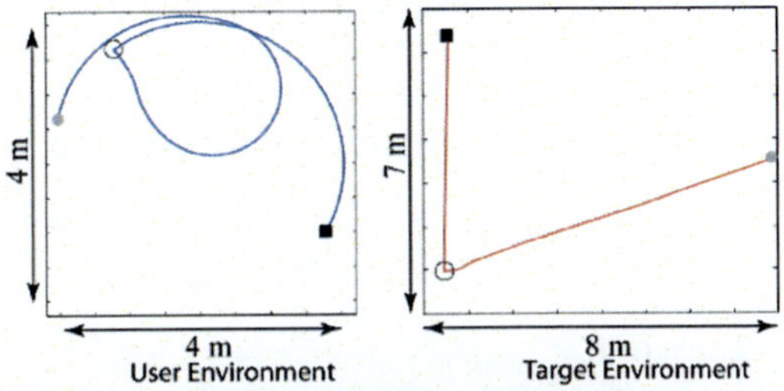


**Fig. 3. Path in both environments**

## Evaluation of the Pedestrian Simulation

A common problem in pedestrian simulations is the lack of available data for the calibration of pedestrian models [10]. Not all kinds of real experiments can be replaced by virtual experiments – for example high density situations with many contacts would be difficult – but Extended Range Telepresence should provide a suitable framework for gathering data, for example, for route choice issues and obviously in scenarios that involve some risk for the test participants. Furthermore, Extended Range Telepresence represents a systematic evaluation tool, since all data is logged, while trajectory extraction of real experiments takes quite some effort [11-13].

First, it is necessary to specify reproducible test set-ups in the proposed framework. To do this, it is required to compare data from real experiments with data from experiments in the virtual reality, in order to identify the situations in which the behavior of the user in the Extended Range Telepresence system differs most from real situations. We expect that the user's behavior will be more realistic if the quality of the virtual scenario increases. For this reason, we intend to investigate, how haptic information and a realistic representation of the pedestrians' morphology affect the user's behavior and his motion parameters (e.g., position, velocity, distance to next pedestrian, etc.). This will allow us the definition of reasonable criteria to assess test results.

The evaluation of the simulation consists in augmenting the simulation with a real user. The user's state (position, orientation, and velocity) is used to initialize the simulation, and at every time step the state of the telepresent user and the simulated state are compared. An optimization is then performed in order to find the parameters that minimize the deviation between actual and simulated state. Global parameters of the simulation that affect the route choice and the acceptance of the simulated traffic flow will also be evaluated.

Finally, owing to the fact that unexpected and dangerous escape situations exclude real-life experiments [7,14], it is desired to find suitable parameters on escape panics, which permit to use the pedestrian simulation also in evacuation scenarios.

## Evaluation Framework

A connection of Extended Range Telepresence and the VISSIM [3] pedestrian simulation was implemented [4] so that the scene shown to the user is populated with pedestrians from a VISSIM simulation.

The framework was extended in order to introduce haptic information into the simulation, since it is indispensable for increasing the degree of immersion and to present additional information to the user, e.g., by displaying the contact force against other pedestrians or by signalizing a forbidden or too dangerous path, which must be avoided. A haptic interface especially designed for Extended Range Telepresence provides haptic information from the target environment [5]. The data flow in the framework is shown in Fig. 4.

The user's posture is tracked and fed into the system. Every time an update of the user's position is available, the Motion Compression algorithm is executed, and the transformed avatar's posture is calculated and sent to the simulation. The simulation constantly captures live images from the avatar's view, which are sent to the user and presented on a head-mounted display, so that the user has the impression of walking, without any restriction, in the simulation.

The simulation also calculates the resulting force acting on the user, which results from the contact with obstacles and other pedestrians, and sends this to the Motion Compression server, which transforms the force from the target to the user environment. The resulting force vector has the same magnitude and relative direction to the user path in both environments. The transformed force vector is sent to the haptic interface that displays the force on the user's hand.

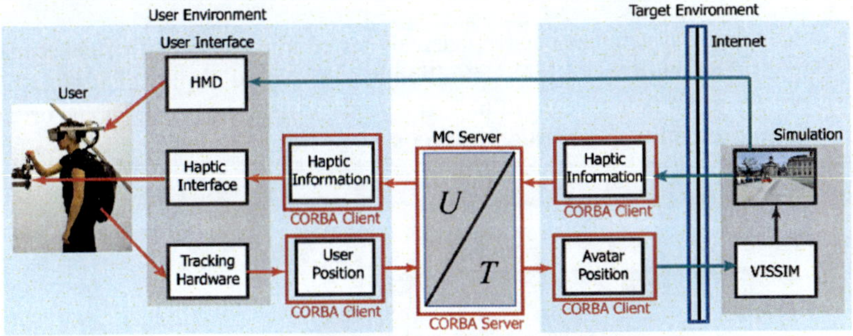


**Fig. 4. Data flow in the proposed framework**

## Use of Haptic Information

The use of haptic information is indispensable for increasing the sense of presence so that the user behaves realistically when immersed in the simulation. When the simulation is augmented by a real user, the simulated pedestrians react to him and try to avoid him. However, without physical forces displayed on the haptic interface, it would be difficult for the user to go ahead into the crowd without walking through simulated pedestrians.

The physical force acting on the user is due to the contact with other pedestrians and with borders of buildings, walls, obstacles, etc. This force can be obtained by computing the repulsive effects with other pedestrians and obstacles when a collision takes place in the direction of motion. The pedestrians touch each other if their distance is smaller than the sum of their radii. In this case, the interaction force is the sum of two forces [7]: a normal force counteracting body's compression and a sliding friction force impeding relative tangential motion. The interaction force with the walls is treated analogously.

Since the Social Force Model [6-9], which models the behavior of a pedestrian in a crowd, does not draw a clear distinction between physical and psychological forces, we apply on the real user, in a first approach, the resultant force calculated by the Social Force Model. As the user sees the simulated pedestrians and the walls in the display (and as the forces in the model are mediated by the senses and the psyche), by including all forces in the resultant force displayed by the haptic interface, the psychological forces are in a way doubled. However, since the physical forces are much higher than the psychological forces during contact, the resultant social force can be regarded as an approximation of the interaction force, just when a collision takes place. Otherwise the applied force is zero.

## Evaluation Scenario

The implemented system has been extensively tested in a number of virtual scenarios (Fig. 5). The setup uses a high-definition head-mounted display of 1280x1024 pixels per eye and a field of view of 60°. The user's posture, i.e., position and orientation, is estimated by an acoustic tracking system that provides 50 estimates per second [15].

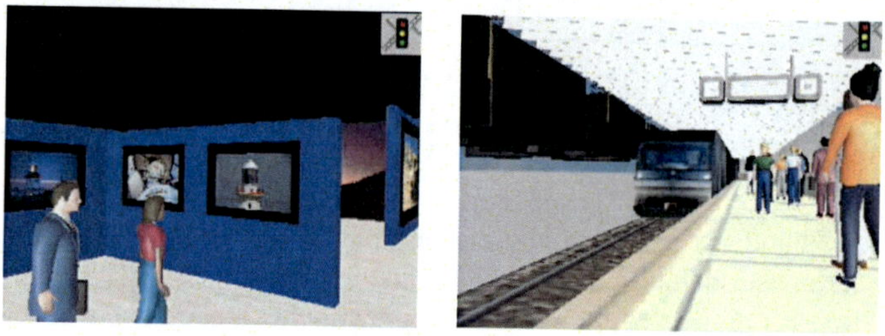

**Fig. 5. Impression of tested scenarios**

A simple route choice scenario has been selected to illustrate the proposed approach (Fig. 6). Such a scenario is suitable to calibrate VISSIM's simulation module that allocates the pedestrians to the gates. A rigorous calibration of the simulation would require the comparison of virtual with real experiments, as explained in the previous section but we assume here that in such a scenario the user in the Telepresence system behaves realistically.

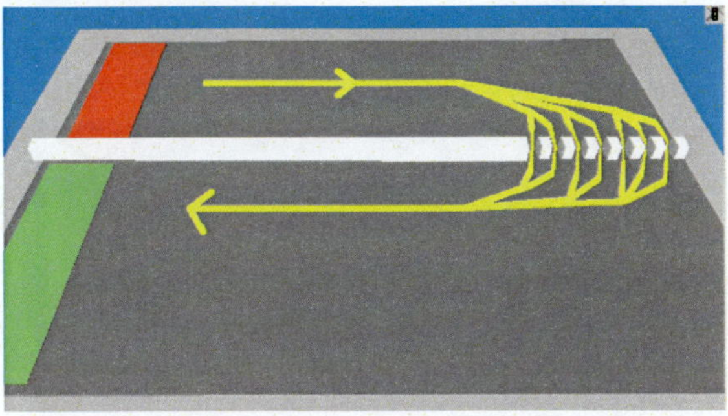

**Fig. 6. Schematic view of route choice scenario used to evaluate VISSIM**

In this scenario, pedestrians start walking on the red surface, walk across one gate, and disappear on the green surface. In each trial, 150 pedestrians were simulated. The task completion time, the covered distance, and the chosen gate were recorded and compared with those of the test participants.

Fig. 7 shows the trajectories of pedestrians and participants.

The framework allows performing reproducible tests that can easily be evaluated. The simulation first calculates a realistic distribution of pedestrians over the gates. Most of pedestrians take the closest gate, as expected, and the others take farther gates in decreasing number. By monitoring together the behavior of real and simulated pedestrians, it is possible to compare their distributions.

It is notable, that most pedestrians in the simulation choose the first gate, whereas three of five test participants take the second one. In the last path segment, the trajectories of pedestrians and participants differ considerably. This is due to the fact that pedestrians in the simulation do not use the entire available surface, but this is not relevant for the experiment. The observation that a real user considers not only the distance but also the availability of the infrastructure and the waiting time in order to choose a farther gate is also in agreement with real observations.

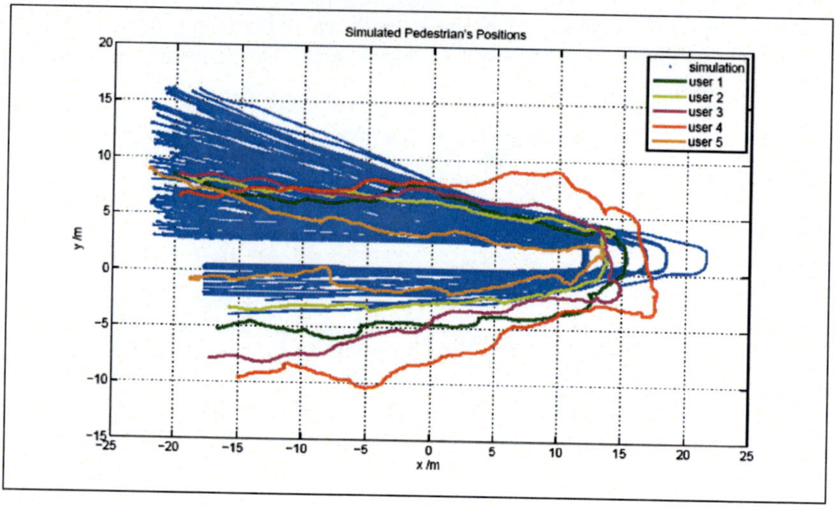


**Fig. 7. Trajectories of simulated pedestrians and test participants**

The calibration process must be iterative. Test participants behave at the beginning differently from the simulated pedestrians. Once the simulation has been so adjusted that the distribution of simulated pedestrian matches the distribution of test participants, another calibration run must be performed. This procedure, which is known as *dynamic assignment*, must be repeated until it converges. In traffic assignment, there is one iteration method, in which the convergence is demonstrated. It consists in regarding at the iteration $N+1$ the past $N$ iterations with a weighting factor $1/N$. However, this method converges very slowly and needs a lot of experiments. Therefore, a simplified method is used, which just considers the last iteration (iteration $N$) with a weighting factor $W$ and the iteration before (iteration $N$-$1$) with a weighing factor $1$-$W$.

## Conclusions and Future Work

Extended Range Telepresence offers an appropriate tool for the evaluation and the calibration of pedestrian simulations, by having one real person walking through a crowd of simulated agents. The main benefit of the proposed framework is that experiments with a real person walking in a virtual environment are cheap and easy to evaluate, since the position of the test person is available at any time during the experiment. Haptic information is used to enhance the realism of the simulation and therefore, the validity of the given evidence.

Currently, it seems very difficult to give a highly immersive and realistic impression for high-density situations, when contacts to other pedestrians occur frequently and from all sides. But the system might prove very useful for virtual experiments on topics where currently only few data are available: route choice [16-19], behavior as participants of road traffic (in VISSIM, vehicles can easily be included to virtual experiments) [20], and behavior in certain dangerous situations [21, 22], which anyway are not directly and fully realistically accessible by experiment. Further work must be done in defining an evaluation metric to prove that the experiments in the Telepresence system are as valid as real experiments for the purpose of improving pedestrian simulations in specific situations, especially in evacuation scenarios.

## References


1. Nitzsche, N., Hanebeck, U.D., Schmidt, G.: Motion Compression for Telepresent Walking in Large Target Environments. Presence. 13(1), 44-60 (2004)
2. Rößler, P., Hanebeck, U.D., Nitzsche, N.: Feedback Controlled Motion Compression for Extended Range Telepresence. In: Proceedings of IEEE Mechatronics & Robotics (MechRob 2004), Special Session on Telepresence and Teleaction, 1447-1452 (2004)
3. PTV Planung Transport Verkehr AG – VISSIM 5.20 User Manual (2009)
4. Arias, A.P., Kretz, T., Ehrhardt, P., Hengst, S., Vortisch, P., Hanebeck, U.D.: A Framework for Evaluating the VISSIM Traffic Simulation with Extended Range Telepresence. In: Proceedings of the 22nd Annual Conference on Computer Animation and Social Agents (CASA 2009), 13-16 (2009)
5. Arias, A.P, Hanebeck, U.D.: A Novel Haptic Interface for Extended Range Telepresence: Control and Evaluation. In: Proceedings of the 6th International Conference on Informatics in Control, Automation and Robotics (ICINCO 2009), 222-227 (2009)
6. Helbing, D., Molnar, P.: Social Force Model for Pedestrian Dynamics. Phys. Rev. E, 51, 4282-4286 (1995)
7. Helbing, D., Farkas, I., Vicsek, T.: Simulating Dynamical Features of Escape Panic. Nature. 407, 487-490 (2000)
8. Johansson, A., Helbing, D., Shukla, P.K.: Specification of the Social Force Pedestrian Model by Evolutionary Adjustment to Video Tracking Data. Advances in Complex Systems. 10(4), 271-288 (2009)

9. Helbing, D., Johansson, A.: Pedestrian, Crowd and Evacuation Dynamics. In: Meyers (ed.), [24], 16, 6476-6495 (2009)
10. Schadschneider, A., Klingsch, W.W.F., Klüpfel, H., Kretz, T., Rogsch, C., Seyfried, A.: Evacuation Dynamics: Empirical Results, Modeling and Applications. In: Meyers (ed.), [24], 5, 3142-3176 (2009)
11. Boltes, M., Seyfried, A., Steffen, B., Schadschneider, A.: Automatic Extraction of Pedestrian Trajectories from Video Recordings. In: Klingsch et al (eds.), [23], 43-54 (2010)
12. Bauer, D., Kitazawa, K.: Using Laser Scanner Data to Calibrate Certain Aspects of Microscopic Pedestrian Motion Models. In: Klingsch et al (eds.), [23], 83-94 (2010)
13. Johansson, A., Helbing, D.: Analysis of Empirical Trajectory Data of Pedestrians. In: Klingsch, W.W.F. et al (eds.), [23], 203-214 (2010)
14. Rogsch, C., Schreckenberg, M., Tribble, E., Klingsch, W.W.F., Kretz, T.: Was it Panic? An Overview about Mass-Emergencies and Their Origins All Over the World for Recent Years. In: Klingsch, W.W.F. et al (eds.), [23], 743-755 (2010)
15. Beutler, F., Hanebeck, U.D.: Closed-Form Range-Based Posture Estimation Based on Decoupling Translation and Orientation. In: Proceedings of the IEEE International Conference on Acoustics, Speech, and Signal Processing (ICASSP 2005), 989-992 (2005)
16. Kretz, T.: Pedestrian Traffic: on the Quickest Path. JSTAT. P03012 (2009)
17. Gräßle, F., Kretz, T.: An Example of Complex Pedestrian Route Choice. In: this PED2010 Proceedings Book (2010)
18. Kretz, T.: Applications of the Dynamic Distance Potential Field Method. In: Traffic and Granular Flow '09 (in press)
19. Kretz, T.: The Use of Dynamic Distance Potential Fields for Pedestrian Flow around Corners. In: Proceedings of the First International Conference on Evacuation Modeling and Management (ICEM 2009) (2009)
20. Bönisch, C., Kretz, T.: Simulation of Pedestrians Crossing a Street. In: Traffic and Granular Flow '09 (in press)
21. Kretz, T., Mayer, G.: Behavior and Perception-Based Pedestrian Evacuation Simulation. In: this PED2010 Proceedings Book (2010)
22. Kobes, M., Oberijé, N., Duyvis, M.: Case Studies on Evacuation Behaviour in a Hotel Building in BART and in Real Life. In: Klingsch, W.W.F. et al (eds.), [23], 183-201 (2010)
23. Klingsch, W.W.F., Rogsch, C., Schadschneider, A., Schreckenberg, M.: Pedestrian and Evacuation Dynamics 2008. Springer-Verlag, Berlin, Heidelberg (2010)
24. Meyers, R.A.: Encyclopedia of Complexity and Systems Science. Springer, Science + Business Media, New York (2009)

# Proof of Concept: Use of Eye-Tracking to Record How People Use Exit Signage

**R. C. Till[1] and J. S. Babcock[2]**

[1]John Jay College of Criminal Justice
New York, NY 10019 USA
[2]Positive Science, LLC
New York, NY 10128 USA

Corresponding author: rtill@jjay.cuny.edu

**Abstract** To date, there is little research concerning selection of exit paths in emergency and non-emergency situations. Whether an individual takes their cues from others, follows signs, uses some other unknown method or some combination of the three when navigating unknown structures needs clarification. We used a wearable eye-tracking system to record where subjects look when they are exposed to new environments to determine how they use their visual awareness to navigate, and whether they are using exit signage to do so. The system provides a video record of the subject's point-of-view. This knowledge could ultimately help improve evacuation times and save lives during emergencies by allowing for better design of egress systems and/or models, particularly those where a subject has no prior knowledge of the layout of the structure, such as an office building, subway/rail station, or an airport.

## Introduction

In 2004, Fahy [1] identified "exit choice decisions, which determine travel paths and affect travel times" as a type of data needed for evacuation models that we currently lack. She also stated "videotaped observations of actual evacuations are ideal, since they show exactly what people did, and the elapsed time can be calculated directly from the tape".

How do pedestrians behave when searching for the nearest exit? Do they take cues from others, follow signs, use some other unknown navigation method, or a combination of these to exit a building?

Fruin used external and overhead movie cameras to record and extract pedestrian flows [2]. More recently others have used various sensor technologies to model the flow of individuals [3, 4]. Eye tracking has also been used to determine how pedestrians avoid collisions [5, 6]. Despite these efforts, we know little about what observers actually look at during the exit or evacuation process.

R.D. Peacock et al. (eds.), *Pedestrian and Evacuation Dynamics*,
DOI 10.1007/978-1-4419-9725-8_19, 

As suggested by Fahy, capturing a video record from the perspective of the observer provides an "ideal" output format. Eye-tracking software goes a step further, supplying a marker on the video scene indicating where the subject is looking.

We used a wearable eye-tracking system to examine where people looked when exposed to a new environment to determine how they use visual awareness to navigate, and whether they are using signage to do so. The paper presents a "proof of concept", suggesting that eye-tracking can be a useful tool in studying pedestrian and evacuation dynamics.

## *Use of Exit Signage*

It is a generally accepted principle in fire protection that in an emergency most people will first attempt to leave a building the same way they entered [7, 8]. However, this rule does not apply in the following situations:

- *mass transit stations with which disembarking passengers are unfamiliar, but where they may be forced to disembark in an emergency.*
- *high-rise buildings where elevators are used as the primary means of occupant entry to a given floor, but where stairs must be used as a means of egress in an emergency (as is currently prescribed in the US for example).*
- *airports with which disembarking passengers are unfamiliar.*
- *any situation where the means that a person used to gain entry to a facility is no longer available to them.*

In mass transit stations people enter at track level, and may be unfamiliar with the exact path to the street that they need to follow. In high-rise buildings people enter the higher floors using elevators. However, during emergencies people are directed or inclined to use stairs that they may have never used before finding themselves in locations that are unfamiliar. Upon arrival at a new airport, way finding can also be challenging. Difficulties are compounded by the fact that travelers need to retrieve baggage and then find an appropriate exit.

There are other situations where the means of entry may be blocked by fire or may not have the capacity in an emergency situation to provide adequate egress. We suggest that eye-tracking technology can be used to assist in the development of better egress designs and also assist with exit (or informational) sign placement. Although not the immediate goal of this paper, we hope to pair eye-tracking technology with other pedestrian tracking approaches to improve computer simulated modeling.

### *Gaze During Natural Behavior*

Portable eye-tracking technology has allowed researchers to study the natural behavior of viewers outside of the laboratory. The average person makes roughly 150,00 eye movements a day [9]. Except for an occasional eye-to-eye glance at a stranger, we are rarely conscious of the eye's point of gaze. However, recording where people look can offer details about the viewer's attention that occur subconsciously. For example, researchers have studied observer's gaze behavior during natural tasks such as driving a car [10], making tea [11] or a sandwich [12], washing hands [13], playing table tennis [14] and cricket [15], and walking through a room cluttered with obstacles [16]. Conclusions from this body of research show that where people fixate is closely linked to the attentional demands of the task and that wearable eye-tracking is a good tool for monitoring the thought process.

Eye movements are an especially rich source of information for research in the motion of pedestrians. Despite decades of building code development in which the location of exit signs is specified, we know little about how natural vision influences these specifications or how they may be improved. A fundamental question remains unanswered: Where do people look when they are trying to exit a structure that they may or may not be familiar with?

## Eye-tracking Hardware and Software

***Headgear*** - For this experiment we used a wearable eye-tracking system from Positive Science [www.positivescience.com] to record gaze while the subject walked to a nearby ground floor exit to take a picture of it. Figure 1 shows a person wearing the eye-tracking headgear and the components inside the backpack. The headgear consists of two miniature cameras. One camera records the participants' eye and a second camera, facing outward, records the subject's point of view. An infrared emitting diode (IRED) illuminates the eye allowing the software logic to track the pupil center and the corresponding glint from the IRED. Video output from the headgear is captured on two external SD cards from the camcorders in the backpack. The videos are then loaded into the software post-capture for gaze analysis (we call this "Offline" mode).

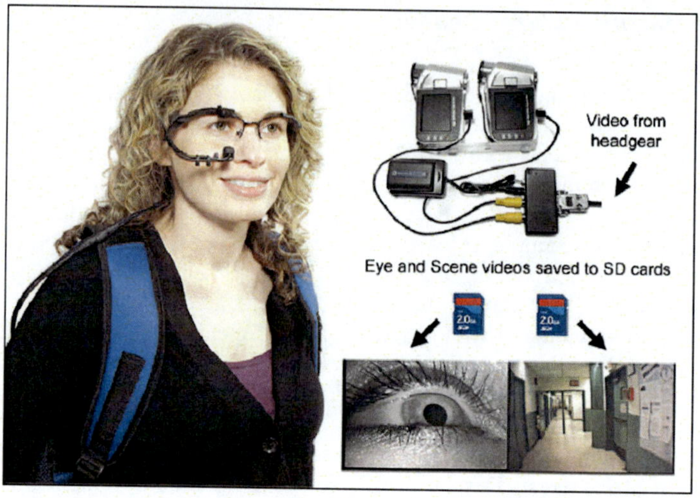


**Fig. 1. Left) eye-tracking headgear Right) components inside backpack**

***Software*** - Yarbus 2.0 eye-tracking software [positivescience.com] was used to compute subjects' gaze. The eye window (upper left) tracks the pupil center and the center of the IRED specular reflection. Figure 2 shows the layout of the "Offline" user interface. The software loads and synchronizes eye and scene videos. A calibration procedure creates the mapping of raw eye position onto the scene video. The output movie renders the subjects' field of view with a crosshair superimposed to indicate gaze location. The average accuracy of the track was measured to be less than 2 degrees (measured over a 30 x 28 degree field of view).

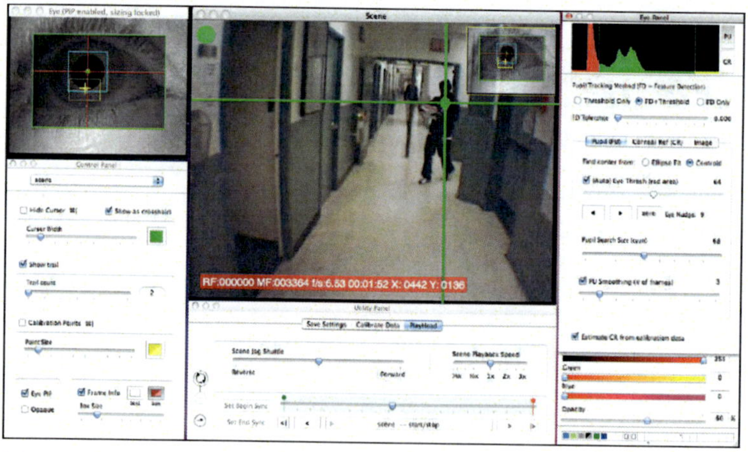

**Fig. 2. Yarbus eye-tracking software with gaze position (crosshair)**

## Method

Four males and six females participated in the experiment. Five subjects were considered "experts" because they were recruited from faculty or staff that regularly work in the building. The other five subjects were deemed "novices" because they had never been to the building before. Novice subjects were brought to a conference room from an elevator that was far from the *start area* (room 3531). This route was used so that novices did not see the "experimental" exit location. Experts were simply told to report to the start room, usually coming from their nearby office.

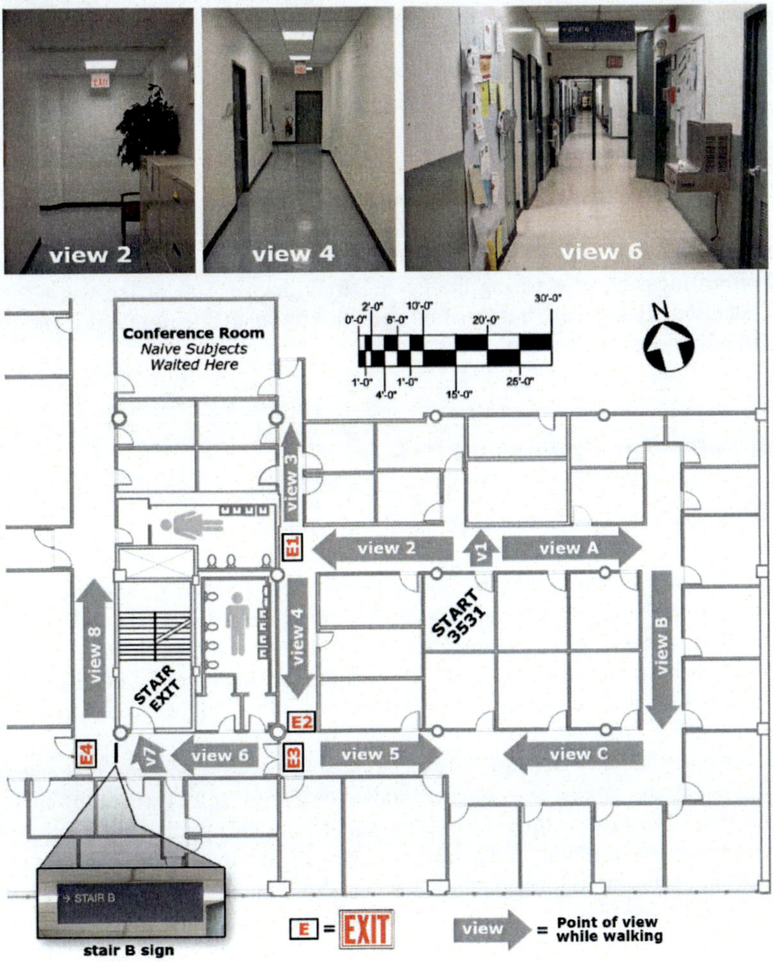


**Fig. 3. Floor plan with video event coding references**

The equipment was fitted to the all subjects by having them put on the eye-tracking headgear and then the backpack containing the DVR recorders. For each participant the experimenter performed a calibration procedure so that raw eye position could be mapped to the corresponding location in the scene camera. To calibrate, subjects followed a marker in the experimenter's hand with their eyes while holding their head steady. Two camera flashes were used to sync the begin/end frames in the eye and scene movies. Synchronization was done later in software. Once calibration was complete, the following statement was read to the subject:

> *"Your goal is to locate a doorway that will take you outside of the building - i.e. you should be able to see daylight or a marked way out of the building - and take a picture of that location for documentation.*
>
> *Do not actually leave the building, and avoid passing through any turnstile, as this will require us to re-check you in through the security desk. Also, do not make use of any elevators.*
>
> *Remember, your goal is to simply "find" a way out of the building and document it. Once you have taken the picture, return to this office and the test will be complete."*

The goal of having subjects take a picture of the ground floor exit was to distract them from focusing on wearing the eye-tracking device. Use of the word "exit" was intentionally avoided, but in some cases it was inadvertently used if the subject asked for clarification of the instructions.

## *Data Collection, Coding and Reduction*

Video frames were manually scored using InqScribe movie transcription software [www.inqscribe.com] according to the scene views shown in Figure 3. While all test subjects descended to the first floor and recorded their picture there, data was only scored to the top of the staircase, as both the Life Safety Code (NFPA 101) and the ICC codes would consider the subjects as being safely "in an exit" once they were behind the staircase door. The results can therefore be generalized to any office floor with a similar configuration, including floors located in a high rise.

***Event-based Coding Scheme*** - The coding scheme was based on two categories: 1) where the subject was looking, known as a scan, and 2) where the subject was moving. Most of the time if a subject was moving in a particular direction, he or she was usually scanning in that direction as well. An example of a novice subject (test subject 5) with corresponding event codes is summarized in Table 1.

While this table summarizes an entire procedure from the start, the overall task time was the time to move to view 1 (when the start room was exited) subtracted from the time to touch the stairwell exit door's push bar, in this case 21.634 seconds.

**Table 1. Event codes – test subject 5 - novice**

| Subtask | Time (s) |
|---|---|
| Begin | 0.000 |
| Begin Calibration | 3.967 |
| End Calibration | 19.833 |
| Move View 1 | 27.933 |
| Scan View A | 30.233 |
| Scan Back to Office | 31.167 |
| Move View 2 | 31.933 |
| Scan Towards North | 33.500 |
| Scan View A | 33.900 |
| Scan View North | 34.233 |
| Move View 2 | 34.733 |
| Scan Exit 1 | 35.267 |
| Scan Exit 1 End | 37.633 |
| Move View 4 | 38.600 |
| Scan Exit 2 | 39.167 |
| Scan Exit 2 End | 40.967 |
| Scan Exit 3 | 42.300 |
| Scan Exit 3 End | 42.533 |
| Move View 6 | 43.833 |
| Scan Stairwell B Sign | 44.300 |
| Scan Stairwell B Sign Again | 45.100 |
| Move View 7 | 45.533 |
| Scans Pedestrian | 47.046 |
| Scans other pedestrian exiting stair B | 47.227 |
| Touches Door B push bar before door closes | 49.567 |

Screen captures of event code views are shown in Figure 4. The cross hair indicates the subject's gaze.

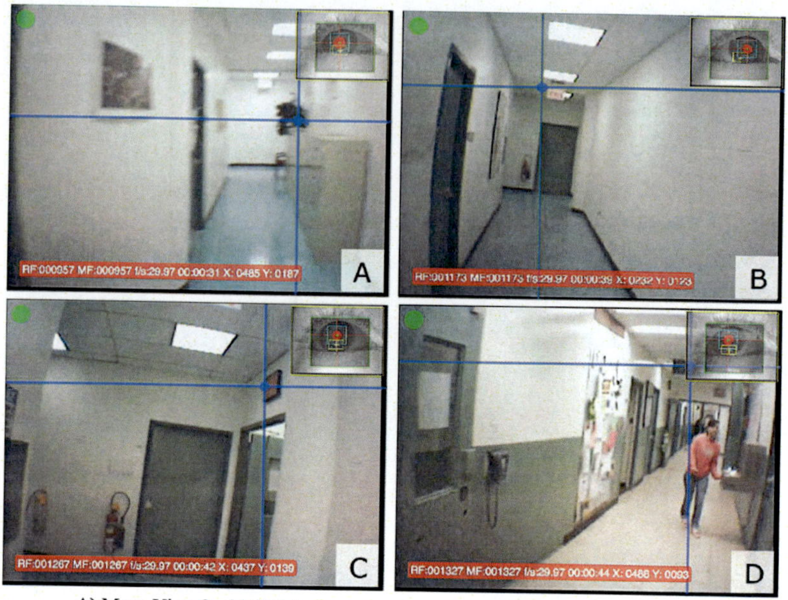


**A)** Move View 2 - 31.93 (s)
**B)** Scan Exit Sign 2 - 39.17 (s)
**C)** Scan Exit Sign 3 - 42.30 (s)
**D)** Scan Stairwell B Sign - 44.30 (s)

**Fig. 4. Event code views for novice subject 5 (time in sec)**

## Results

Results are summarized in Table 2. The average time for the experts to get to the exit door was 21.3 seconds and for novices it was 31.3 seconds. Naturally more test subjects would provide more significant results, and since we have human subjects performing a task, a normal distribution of results is unlikely [18].

***Experts*** – Video analysis revealed that experts performed little or no scanning of signage. No "stop and scan behavior" was observed in any of the expert users. Move and scan" behavior was noted. Scanning in directions other than the direction of motion sometimes occurred (i.e. large head rotations while walking), however in these cases it was to acknowledge a colleague or to avoid a potential collision with people moving through an intersection.

***Novices*** – Some interesting observations can be made about novice observers. The observers that used exit signs had performance times that were as fast as the experts. This was apparent in subject 5, who made extensive use of exit signage, and achieved performance that was actually better than one of the experts. This extensive use of signage by subject 5 is shown in Figure 4. Subjects 4 and 6, both novices, had task times that exceeded 40 seconds. Each made almost no use of signage, seeming to rely on a "brute force" method for finding the exit, possibly because of their own *learned irrelevance* of all exit signs [7]. Their task time seems to have suffered greatly because of this.

**Table 2. Test subject – designation and time to perform task**

| Test Subject Number | Designation | Gender | Time (s) |
|---|---|---|---|
| 2 | Expert | Male | 19.800 |
| 7 | Expert | Male | 20.067 |
| 8 | Expert | Female | 21.134 |
| 5 | Novice | Female | 21.634 |
| 1 | Expert | Male | 21.951 |
| 10 | Novice | Female | 23.333 |
| 3 | Expert | Female | 23.655 |
| 9 | Novice | Female | 26.600 |
| 6 | Novice | Male | 42.266 |
| 4 | Novice | Female | 42.666 |

## Conclusions and Recommendations

The goal of this experiment was to show that wearable eye-tracking technology could be used as a tool to better understand pedestrian dynamics, and that collection of event-based data from video scenes could be enhanced by knowing what observers look at during the task of exiting.

Analysis of eye-tracking video was useful in distinguishing between the actions of novice subjects and expert subjects as they searched for an exit location. As might be expected, experts did not make eye movements to exit signage at all. Novice subjects that did look at exit signage showed an advantage in task times over those subjects that did not. Given our small data set, a larger study with more subjects, and possibly extending the eye-tracking experiment to more locations (i.e. an airport, subway, high rise etc.) could provide more relevant statistics regarding where users get information when exiting a building that they are unfamiliar with. Because many pedestrian and evacuation models rely on assumptions of what people do, our future goal is to use eye-tracking technology to show what viewers actually do and what they look at so that we can use this information to make better egress models.

## References


1. Fahy, R.: Available Data and Input into Models. Paper Presented at the Workshop on Building Occupant Movement During Fire Emergencies, Session 4.3. Gaithersburg (2004)
2. Fruin, J.J.: Pedestrian Planning and Design, (Revised Edition). Elevator World, Inc., Mobile (1987)
3. Liu, X., Song, W., Zhang, J.: Extraction and Quantitative Analysis of Microscopic Evacuation Characteristics Based on Digital Image Processing. Physica A. **388**, 2717-2726 (2009)
4. Boltes, M., Seyfried, A., Steffen, B., Schadschneider, A.: Automatic Extraction Of Pedestrian Trajectories From Video Recordings. In: Pedestrian and Evacuation Dynamics (In Press). Springer-Verlag, Berlin (2008)
5. Fujiyama, T.: Investigating Density Effects on the "Awareness" Area of Pedestrians Using an Eye Tracker. CTS Working Paper (2006/4), Centre for Transport Studies, University College, London (2006)
6. Jovancevic, J., Sullivan, B., Hayhoe, M.: Control of Attention and Gaze in Complex Environments. Journal of Vision. **6,** 1431-1450 (2006)
7. McClintock, T., Shields, T.J, Rutland, T.R., Leslie, J.: Dishabituation and Stimulus Equivalence Could Make All Emergency Exits Familiar. Journal of Applied Fire Science. **9**(2), 125-134 (2000)
8. Sime, J.D.: Crowd Psychology and Engineering. Safety Science. **21,** 1-14 (1995)
9. Abrams, R.A.: Planning and Producing Saccadic Eye Movements. In: Rayner, K. (ed.), Eye Movements and Visual Cognition: Scene Perception and Reading, pp. 66. Springer-Verlag, NY (1992)
10. Land, M.F., Lee, D.N.: Where We Look When We Steer. Nature. **369**, 742-744 (1994)
11. Land, M., Mennie, N., Rusted, J.: The Roles of Vision and Eye Movements In the Control of Activities of Daily Living. Perception. **28**, 1311-1328 (1999)
12. Hayhoe, M.M., Shrivastava, A., Mruczek, R., Pelz, J.B.: Visual Memory and Motor Planning In a Natural Task. Journal of Vision. **3**, 49-63 (2003)
13. Pelz, J.B., Canosa, R.: Oculomotor Behavior And Perceptual Strategies In Complex Tasks. Vision Research. **41**, 3587-3596 (2001)
14. Land, M.F., Furneaux, S.: The Knowledge Base of The Oculomotor System. Philosophical Transactions of the Royal Society of London A. **352**, 1231-1239 (1997)
15. Land, M.F., Mcleod, P.: From Eye Movements To Actions: How Batsmen Hit The Ball. Nature Neuroscience. **3**, 1340-1346 (2000)
16. Franchak, J.M., Adolph, K.E., Gabelman, L., Babcock, J.: Visual Guidance of Locomotion In Children: Navigation From The Periphery. Paper presented at the 2009 meeting of the Society for Research in Child Development (2009)

17. Law, A.M., Kelton, W.D.: Simulation Modeling and Analysis. McGraw Hill, Boston (2000)

# Measurement Techniques for Unannounced Evacuation Experiments

**D. Nilsson and H. Frantzich**

Department of Fire Safety Engineering and Systems Safety
Lund University, Box 118, SE-22100 Lund, Sweden
Corresponding author: daniel.nilsson@brand.lth.se

**Abstract** There are many available measurement techniques for documenting people's movement patterns, but not all of them are appropriate for unannounced evacuation experiments. An unannounced evacuation requires that participants are not influenced beforehand. In addition, experiments are performed in all types of buildings and low ceiling height is often a problem. This paper describes three techniques that can be used for unannounced evacuation experiments, namely (1) filming from above – cameras with wide angle lenses, (2) triangulation with two cameras, and (3) distance measurement with a laser scanner. The description is based on the results from a research study in which the three measurement techniques were tested and evaluated. The study also involved collection of data in unannounced evacuation experiments.

## Introduction

Many previous studies have examined the use of measurement techniques for documenting people's movement patterns [1,2,3,4,5]. Examples of proposed techniques include video recording [2], infra-red sensors [3] and radio frequency identification (RFID) [4]. Many of the techniques require that specific conditions are met in order to ensure high validity of the measurements. For example, the use of RFID requires that antennas are placed at appropriate locations and that tags are somehow planted in the crowd. A very common data collection technique is to film the movement of people from above with ordinary video or CCTV cameras. In order to get the best accuracy with this technique the cameras should be placed relatively high above the floor and point directly downwards. This basic method has been used numerous times to collect valuable data, such as data on the movement of people in open spaces during the Hajj [6].

Many existing measurement techniques require that specific conditions are met, e.g., that cameras are placed relatively high above the floor. These types of necessary preconditions may not always be possible to achieve in unannounced evacuation experiments, i.e., experiments where the participants are not informed about the evacuation beforehand. It is argued that an unannounced evacuation experiment is the best ways to get valid data because participants are not aware that they

R.D. Peacock et al. (eds.), *Pedestrian and Evacuation Dynamics*,
DOI 10.1007/978-1-4419-9725-8_20, © Springer Science+Business Media, LLC 2011

are taking part in an experiment [7]. From the viewpoint of the participants the situation is potentially a real emergency and they therefore act as they would have done in a real situation, e.g., similar walking speeds and movement patterns.

Unannounced evacuation experiments are most often performed in real buildings and very seldom in artificial laboratory environments. This means that the design of the building will influence the type of measurement techniques that can be used. One of the biggest difficulties is low ceiling height, which is a problem in almost all types of buildings. Low ceiling height makes it difficult to use the most basic method of filming the crowd from above. Instead it is often necessary to film the crowd from an angle. It is possible to get useful trajectory data even if the camera is not placed directly above [8], but these techniques are in some cases associated with big measurement errors [9].

Another limitation of unannounced evacuation experiments is that the participants can not be influenced beforehand. If they are influenced in any way, e.g., tagged or marked, they may suspect that something is going to happen, which would lead to low validity. Measurement techniques that rely on manipulation prior to the experiment, e.g., giving transmitters to people (RFID), will therefore not be appropriate for unannounced evacuation experiments.

The discussion above clearly illustrates that there is a need to develop and test techniques that can be used in unannounced evacuation experiments. In this paper, three techniques that were included in a study[1] at Lund University are described. The objective of the study was to explore the use of different measurement techniques for documenting people's movement patterns and to collect trajectory data using selected techniques. This paper describes the study and the three tested techniques, but collected data are not presented. The data will, however, be made publicly available at www.brand.lth.se/research in summer 2010.

## Measurement techniques

The research study at Lund University began in autumn 2007 and was completed in spring 2010. One of the objectives of the study was to test and evaluate measurement techniques for unannounced evacuation experiment, but also to collect data. Experiments were performed both to calibrate the tested measurement techniques (calibration experiment) and to collect data (data collection experiment). A complete list of all experiments that were performed in the study can be found in Table 1.

The first step of the study was to make an inventory of existing measurement techniques that could potentially be used in unannounced evacuation experiments. The two most important selection criteria were (1) that the technique would not require that the participants were influenced before the experiment, and (2) that

[1] The study was funded by Brandforsk, Sweden.

the technique would be possible to use in spite of low ceiling height (ca 2.4 to 3.5 meters – standard ceiling heights in Swedish buildings). Based on the inventory three measurement techniques were selected for testing:

- Filming from above – cameras with wide angle lenses (*Standard method*)
- Triangulation using two cameras (*Triangulation*)
- Measurement of distance to people with a laser scanner (*Laser scanner*)

Only the actual measurement technique was evaluated in the study, and limited consideration was taken of the data acquisition process, i.e., how the data can be effectively collected and stored. Other studies have focused on such aspects as automatic data collection from video footage (see for example [1] and [6]), but this was not the focus of the present study. However, the three studied measurement techniques should ideally be combined with these types of automatic data collection methods to reduce the work load for the researcher.

**Table 1. Experiments in the study at Lund University (autumn 2007 – spring 2010)**

| No. | Location | Year | Type of experiment | Description | Tested technique(s) |
|---|---|---|---|---|---|
| 1 | School of Economics and Management, Lund University, Lund, Sweden | 2008 (Apr) | calibration, data collection | Movement on spiral stairs (balcony to ground level); *unannounced evacuation* | standard method, triangulation |
| 2 | IKEA, Helsingborg, Sweden | 2009 (May) | calibration, data collection | Movement around a pillar and a display (furniture); *unannounced evacuation and normal use* | standard method, laser scanner |
| 3 | IKEA, Malmö, Sweden | 2009 (Jun) | data collection | Movement around a pillar and a display (furniture); *unannounced evacuation and normal use* | standard method, laser scanner |
| 4 | Civil engineering building, Lund University, Lund, Sweden | 2009 (Oct) | data collection | Movement through an opening; *unannounced evacuation* | standard method, triangulation |
| 5 | Civil engineering building, Lund University, Lund, Sweden | 2009 (Oct) | calibration | Laboratory experiment to determine the accuracy; *no participants* | standard method, triangulation |
| 6 | Civil engineering building, Lund University, Lund, Sweden | 2009 (Nov) | calibration, data collection | Movement through an opening (width varied); *announced evacuation (laboratory experiment)* | standard method, laser scanner |

### *Standard method*

Filming from above (called the standard method in this paper) is one of the most common measurement techniques. The technique is particularly useful if the cameras can be placed relatively high above the floor. However, problems arise when the ceiling height is low, which is the case in most buildings. For these types of settings it is often necessary to use an array of cameras and also to use optical lenses with short focal length (wide angle lenses). An example of images from an array of cameras is shown in Fig. 1. The images in Fig. 1 were taken during the initial calibration before the experiment at IKEA in Helsingborg, Sweden (experiment 2). By combining the images from the different cameras it was possible to trace the walking path of each participant in the experiment.

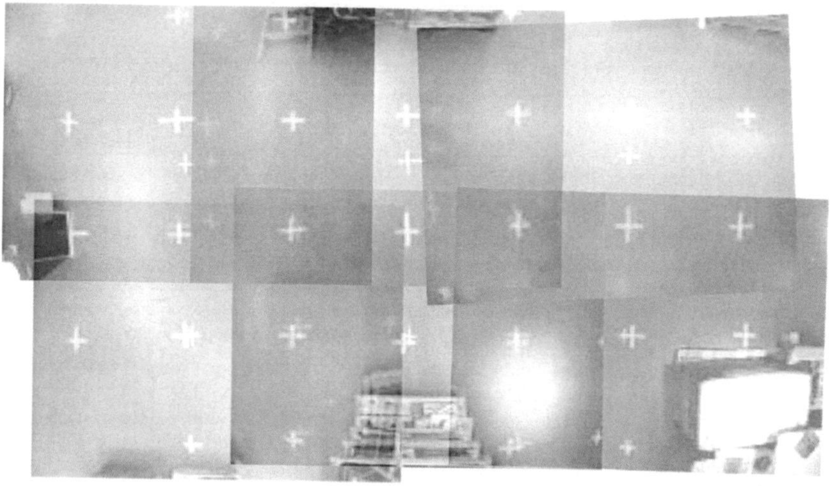

**Fig. 1. Images from an array of cameras at IKEA in Helsingborg, Sweden (experiment 2)**

As mentioned previously, the standard method often requires the use of wide angle lenses if the ceiling height is low. The cameras that were used in the study were therefore equipped with a wide angle lens in the experiments (see Fig. 2). The main benefit of using wide angle lenses is that a bigger area can be covered by the camera, but the big disadvantage is that the image becomes distorted. Fig. 2 shows the distorted image captured by a camera with a wide angle lens when filming a 2D grid with cells of equal size (0.10 m x 0.10 m).

In the experiments where the standard method was used (experiments 1 to 6) a grid was first marked on the floor. This grid was then filmed for a short time period before it was removed. In the analysis of the video footage the grid was then used to determine the position of each participant as he or she moved through the area of interest. This was a relatively simple task for the experiments at IKEA (experiments 2 and 3), but was much more difficult for the other experiments (experiments 1, 4 and 6). The reason for this was that the population density was low in the experiments at IKEA which made it easy to determine exactly where the people were standing (called original approach, see Fig. 3). The density was much higher in the other experiments which made it difficult to see exactly where people were standing. An alternative approach was therefore used for cases with high population densities (see Fig. 3).

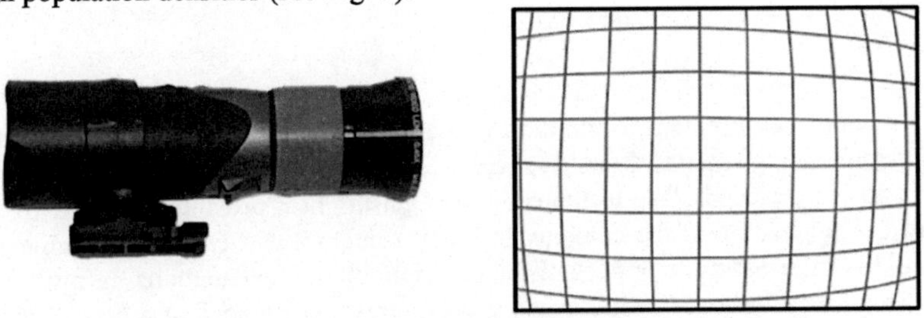

**Fig. 2. Camera used in the experiment (left) and the distortion of the image with a wide angle lens (right)**

The first step of the alternative approach was to determine exactly where the camera was pointed in the experiment. The direction of the camera was calculated based on the grid that was marked on the floor in the beginning of the experiment. Once it had been determined exactly where the camera was pointed it was also possible to calculate the vector from the camera to a person, namely to the centre of the upper torso of the person (between the shoulders). In these calculations the distortion of the image (see Fig. 2) was taken into account.

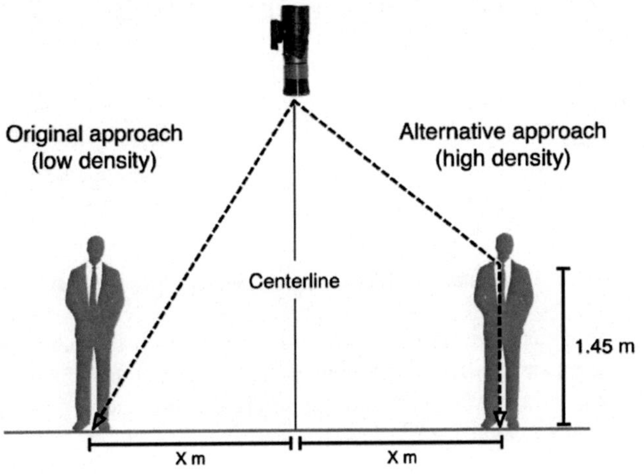


**Fig. 3. The approach used to determine the position of each participant for low densities (experiment 2 and 3) and high densities (experiment 1, 4 and 6)**

Finally, it was determined where the vector crossed the horizontal plane at the height of 1.45 meter, which approximately corresponds to the shoulder height of a person (Fig. 3). The point of intersection provided a reasonable estimate of the position of the person. This approach is similar to the calculation procedure used in the computer program called Persias [8].

## *Triangulation*

The second measurement technique that was evaluated in the study involved triangulation using images from two video cameras. In the experiments the two cameras were placed close to the ceiling to get the best possible overview of the crowd. The position of the cameras and their direction was carefully documented since this information is important for the triangulation calculations. In most cases, wide angel lenses were used and the cameras were placed at a height of approximately 2.3 to 2.5 meters. Fig. 4. shows the location of the cameras in one of the experiments at Lund University (experiment 4).

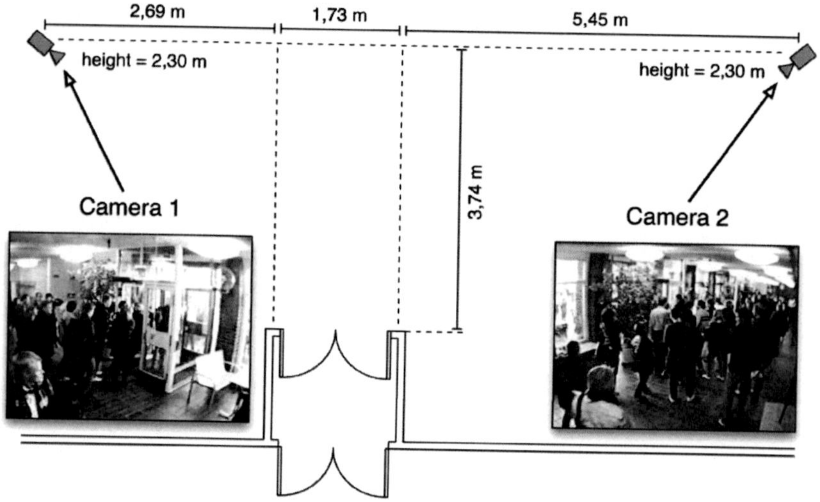


**Fig. 4. Triangulation with two video cameras (experiment 4)**

In order to find the position of a person the person was first identified in the images from both cameras. The line that extended from each camera to the person (the head of the person) was then identified, taking the distortion of the image into account. Finally, the person's position was determined by finding the intersection between the two lines. In most of the experiments, triangulation generated results with reasonable accuracy. The main advantage of the technique is that it does not only provide the position (x- and y-coordinates) but also the height of the person (z-coordinate). However, a serious disadvantage is that it is often difficult to correctly identify a person in the two images, which is partly due to the fact that people are often obscured by others.

## *Laser scanner*

The third measurement technique involved measurement of the distance to people with a laser scanner. The scanner that was used in the study was a LMS221 from SICK [10]. LMS221 performs measurements of the distance in a plane (2D) and the maximum scan frequency is 75 Hz. The field of view is 180° and the angular resolution can be set to 0.25°, 0.5° or 1.0°. LMS221 uses the time of flight of laser light (infrared "eye-safe" laser) to calculate the distance. The laser light (pulse) is sent from the scanner via a rotating prism and is then reflected back to the scanner when it encounters an object. The output from LMS221 can be presented as a scan line that displays the distance to obstacles, e.g., people (see Fig. 5).

The laser scanner was used in two different ways in the study. In the experiments at IKEA (experiments 2 and 3) the scanner was fitted in a pillar and scanned along a horizontal plane (Fig. 5). The height of the plane was 1.3 meters above the floor, which corresponded to the height of the torso for most of the adult participants. In the experiment at Lund University (experiment 6) the scanner was instead attached to the ceiling and scanned along a vertical plane in the door opening. A similar approach has been applied by Walkow [5] who used two scanners to also get information about the speed and direction of travel.

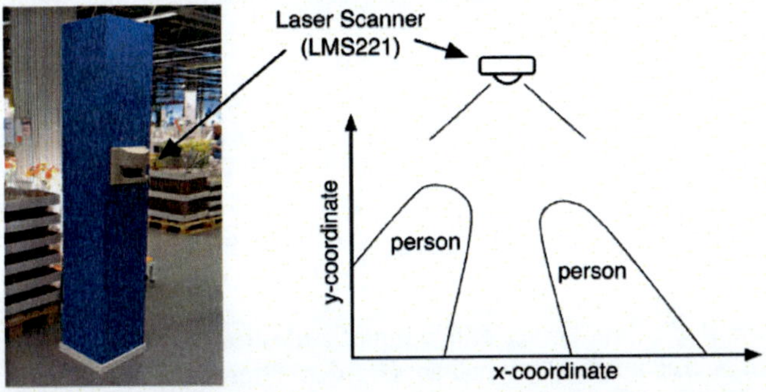


**Fig. 5. LMS221 in a pillar at IKEA (left) and a distance profile from the scanner (right)**

The laser scanner generated highly accurate data. One limitation of the approach used at IKEA (horizontal plane) is that it can only be used for low densities. Also, the approach that was used in the experiment at Lund University (vertical plane) needs to include an additional scanner to provide more useful data.

## Conclusions

All three measurement techniques were able to estimate people's position with reasonable accuracy. However, all techniques were also associated with limitations that need to be considered. Difficulties typically arise when the population density is high, since people are easily obscured by others. This problem can partially be addressed with multiple cameras/scanners, but this option was not explored in the present study. Based on the study it is concluded that all the tested techniques can be used in unannounced evacuation experiments. Measurement of the distance with laser scanners is the technique that is believed to have the greatest potential and therefore should be developed further. More specifically, it should be investigated if and how may scanners can be used together to track walking paths and to collect trajectory data.

## References


1. Boltes, M., Seyfried, A., Steffen, B., Schadschneider, A.: Automatic extraction of pedestrian trajectories from video recordings. Abstracts - Pedestrian and Evacuation Dynamics 2008. pp. 29-30 (2008)
2. Meysel, F., Reulke, R.: Automated multi-camera pedestrian dynamics acquisition and analysis. Abstracts - Pedestrian and Evacuation Dynamics 2008. pp. 33-36 (2008)
3. Clayton, S., Urquhart, N., Kerridge, J.: Real-time pedestrian tracking using an array of low-cost infrared sensors. Abstracts - Pedestrian and Evacuation Dynamics 2008, pp. 15-16 (2008)
4. Rinne, T., Hostikka, S., Paloposki, T., Korhonen, T., Saari, J., Heliövaara, S.: Application of RFID and video imaging on evacuation observations in offices and public buildings. Fire Safety Science – Proceedings of the 9th International Symposium, Karlsruhe, Germany, pp. 329-340 (2008)
5. Walkow, K.F.: Empirische Querschnittsdaten von Fußgängerströmen: Erfassung und Analyse, Vervendung in der Verkehrsmodellierung und ökonomische Bedeutung, Fakultät Wirtschaftwissenschaften. Technische Universität Dresden. Dresden, Germany (2006)
6. Johansson, A.F.: Data-Driven Modeling of Pedestrian Crowds. Faculty of Traffic Sciences. Technische Universitäat Dresden, Dresden, Germany (2009)
7. Nilsson, D.: Exit choice in fire emergencies - Influencing choice of exit with flashing lights. Department of Fire Safety Engineering and Systems Safety, Lund University, Lund, Sweden (2009)
8. Thompson, P.: Simulex: simulated people have needs too. Workshop on Building Occupant Movement During Fire Emergencies. National Institute of Standards and Technology, Gaithersburg (2004)

9. Nilsson, J., Petersson, R.: Utvärdering av videoanalysmetoder för gångvägar med tillämpning på hörn [Evaluation of video analysis techniques and applied to movement around corners], Report 5256. Department of Fire Safety Engineering and Systems Safety, Lund University, Lund, Sweden (2008)
10. Sick, Laser Measurement Technology – LMS2xx / LMS221 / Outdoor / Mid-Range – LMS221-30106, Online Data Sheet. From: https://www.mysick.com/. Retrieved 8 February 2010

# Automation of Pedestrian Tracking in a Crowded Situation

**S. Saadat and K. Teknomo**

Ateneo De Manila University, Philippines
Corresponding author: samansad@gmail.com

**Abstract** Studies on microscopic pedestrian requires large amounts of trajectory data from real-world pedestrian crowds. Such data collection, if done manually, needs tremendous effort and is very time consuming. Though many studies have asserted the possibility of automating this task using video cameras, we found that only a few have demonstrated good performance in very crowded situations or from a top-angled view scene. This paper deals with tracking pedestrian crowd under heavy occlusions from an angular scene.

Our automated tracking system consists of two modules that perform sequentially. The first module detects moving objects as blobs. The second module is a tracking system. We employ probability distribution from the detection of each pedestrian and use Bayesian update to track the next position.

The result of such tracking is a database of pedestrian trajectories over time and space. With certain prior information, we showed that the system can track a large number of people under occlusion and clutter scene.

## Introduction

Measurement of individual pedestrian trajectories, which is part of microscopic pedestrian field of study, is useful in a variety of situations. In business environments, information about customer's movement derives better knowledge on shopping behavior and preferences. In crowded locations (such as sports venues, public transport station, religious events), pedestrians' movement influences safety concerns.

One major approach to obtain pedestrian trajectories is to use video camera and track each individual pedestrian from the video camera from the time he/she is detected on the scene until he/she leaves the scene. However, given the normal video frame rate of 25-30 images per second, to track hundreds of individual pedestrians manually would requires herculean effort despite its accuracy. Better approach is to automate the tracking system.

While such automation in tracking in video camera is not new in the literatures of computer vision (e.g. see [1], [2], and [3] for good survey on the field), most of them are performed only for a relatively few pedestrians. Relatively few recent papers (e.g. [4]), however, reported to track pedestrian on a crowded situation in

R.D. Peacock et al. (eds.), *Pedestrian and Evacuation Dynamics*,
DOI 10.1007/978-1-4419-9725-8_21, © Springer Science+Business Media, LLC 2011

short-medium range sometime up to maximum distance of 8 meters from the camera. Most of the person detection technique works better in a well controlled environment such as laboratory setting with a small amount of people [2] whereas other techniques require more than one camera with overlapping fields of view in order to cope with the pedestrian segmentation problem in the a dense crowd [5].

In the real environments, while walking, the people form groups and then separate from one another, and more over they have shadows [1] and the object, especially, people undergo a change in their shapes while moving and their motion is not constant that cause the tracking is a difficult problem [3].

This paper describes our attempt to automate such tracking. Tracking pedestrians in a crowded situation is a hard problem because pedestrian does not move as a rigid body. Human body goes through a large range of variation during walking. In crowded situation, due to the angle of the camera, the bodies of pedestrians which are farther from the camera are occluded by other nearer pedestrians. Furthermore, in outdoor scene, the lighting condition is uncontrolled and it may create shadow that hinders correct detection of pedestrians.

Our goal is to track all pedestrian's trajectories in the scene and save the data of tracking as a NTYX table where, N is the pedestrian number, T is time in video, and X and Y are the coordinate position of the pedestrian in the image that readily converted into scene coordinate.

Our automated tracking system consists of two modules that perform sequentially. The first module is called pedestrian detection. The second module is pedestrian tracking. Therefore, this paper is organized as follow. First, we explain the two modular part of our system: pedestrian detection and pedestrian tracking. Then, we discuss the result of our system compared to the ground truth data. Finally we conclude the paper.

## Pedestrian Detection

The first module detects moving objects as blobs. For this module we have two options:

- *Without prior information*: this option creates generic background image and then subtracts it from the video data to obtain the foreground scene.
- *With prior information*: as we as the pedestrians to wear red hat, we only detect the red hat based on color.

We employ background modeling to create background image when the pedestrian detection system run without prior information. The background modeling is also dealing with the difficulty to detect the shadow and part of the background. We assume the background model follows Gaussian distribution, thus we compute the average and variance of the image sequence using recursive time average

$$\mu_t = \frac{t-1}{t}\mu_{t-1} + \frac{I_t}{t} \tag{1}$$

$$\sigma_t^2 = \alpha\sigma_{t-1}^2 + (1-\alpha)(I_t - \mu_t)^2 \tag{2}$$

The background image is updated every time using the following formula

$$B_t = \mu_t + \eta\sigma_t \tag{3}$$

The foreground image is obtained through background subtraction and thrsholded to get the binary image.

$$F_t = (I_t - B_t) > \phi \tag{4}$$

The second option does not employ any background modeling because we have prior information that all the pedestrians we track wear the same red color of hats. In this case, our goal is to extract only the red hat information by removing other unnecessary color information. Subtracting green channel from the red channel will produce color of high reddish value and very low greenish value at the same time. That is equivalent to the range of violet to red. Since we also have prior knowledge that the road color was dark, the Blue channel does not carry much information to be subtracted from the red.

Therefore our object detection algorithm is simply a subtraction of green channel from the red channel and put some threshold to make them binary image.

$$F_t = (R_t - G_t) > \phi \tag{5}$$

Higher threshold value produces cleaner blob but it will also remove small blobs. After a sequence of morphological image operations (dilate, open and close) to clean the noise, we get the blobs ready for pedestrian tracking.

## Pedestrian Tracking

The tracking procedure begins when the moving objects on the scene have been detected as blobs. These binary blobs are used as masks to the color image to obtain the color blobs. Utilizing these blobs, we can find the basic features such as color histogram, area, and center of gravity. Our tracking algorithm is based on the Bayesian update and the probabilities are computed for each blob on each frame.

In each set of computation, we consider three consecutive frames (i.e. frame t-2 , t-1 and current frame t).

All of these features of the blobs are stored into a multidimensional feature matrix for each two consecutive frames. Each column in the matrix indicates a blob in the lower frame number (i.e. frame t-2 in matrix of t-2 and t-1, or frame t-1 in matrix t-1 and t), and each row represents a blob in the higher frame number (i.e. frame t in matrix t-1 and t). The other dimensions of the matrix are used to store each features.

Then, these features are thresholded to get binary multidimensional matrix. The entries of the binary matrix represent possibilities that a blob in one frame will become the equivalent blob in the next frame because they represent the same pedestrian. Equivalent blobs can either exactly the same blob, or merge with other blobs, or separated into several blobs. It is the task of tracking system to detect these possibilities through the computation of probabilities.

The probability of each blob is computed based on derived features from the feature matrix. We have four derived features namely entropy difference, movement angle, speed and distance. The derived features are explained as follow.

One unique feature of our algorithm is the usage of entropy of each blob. The entropy is computed based on the color histogram of the blob. To calculate the entropy we use the following equation.

$$E_c = -\sum \frac{H_c}{A_c} * \log(\frac{H_c}{A_c}) \tag{6}$$

Where, H is the color histogram of the blob when there is non-zero value (complete black) and subscript c is the index of three color channels and A is the area of the blob in color channel. We have three values in the entropy of each blob representing each color channel and finally we sum the entropy of the three channels to get entropy of the blob. Once entropy of the blob is computed, we can obtain the entropy difference by subtracting the entropy of the blob on the higher frame number from the entropy of the lower frame number, and then take the absolute value.

$$\Delta E = |E_h - E_l| \tag{7}$$

The next derived features that will be used to compute probability value are movement angle, speed and distance. Given the center coordinate of the blob X, we can compute the directional angle and speed (distance between two blobs in three consecutive frames) as suggested by [6].

$$\theta = w_1\left(1 - \frac{\overline{X_{i,t-2}X_{j,t-1}}\,\overline{X_{j,t-1}X_{k,t}}}{\|\overline{X_{i,t-2}X_{j,t-1}}\|\|\overline{X_{j,t-1}X_{k,t}}\|}\right) \tag{8}$$

$$v = w_2\left(1 - 2\frac{\sqrt{\|\overline{X_{i,t-2}X_{j,t-1}}\|\|\overline{X_{j,t-1}X_{k,t}}\|}}{\|\overline{X_{i,t-2}X_{j,t-1}}\| + \|\overline{X_{j,t-1}X_{k,t}}\|}\right) \tag{9}$$

The subscripts i, j, k indicate index of the blobs at time t-2, t-1 and t consecutively and w1 and w2 are weight parameters. The next derived feature is distance between blobs between two blobs in two consecutive frames.

The values of the derived features indicate the dissimilarity of two blobs, means if the values are zero then they are more similar and higher values means less similarity. Therefore we need to normalize the derived features to represent similarity. The normalization of the derived features is using Eq. (10) to make slow decrease for small value of the derived feature and then rapidly decrease if the value of the derive features is large.

$$f_{l,h,m} = \sqrt{1 - (f_{l,h,m} / \varphi_m)^2} \tag{10}$$

Having all the derived features that stored into multidimensional matrix, we can flatten the multidimensional matrix into 2-dimensional matrix by setting a linear combination with the weight of each feature. Then we also put threshold for each feature to obtain a binary matrix.

$$\sum_m \omega_{l,h,m} f_{l,h,m} < \varphi \tag{11}$$

For each two consecutive frame, we have one binary matrix that we called possibility matrix. Based on the possibility matrix, we can derive a tree structure of possibilities that a blob in one frame is equivalent to another blob in other frames.

Figure 1 illustrates a hypothetical probability tree with all the necessary naming convention of the probabilities. The probability of the second and third level of the probability tree is computed as follow

$$P_{i|j} = \frac{\sum_m \omega_{i,j,m} f_{i,j,m}}{\sum_i \sum_j \sum_m \omega_{i,j,m} f_{i,j,m}} \tag{12}$$

$$P_{k|i,j} = \frac{\sum_m \omega_{j,k,m} f_{j,k,m}}{\sum_j \sum_k \sum_m \omega_{j,k,m} f_{j,k,m}} \tag{13}$$

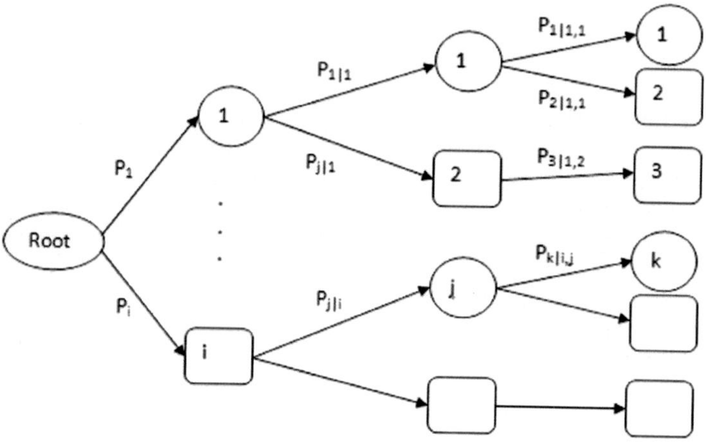


**Fig. 1. Probability tree name convention**

After calculating all the probabilities, we calculate the posterior probability using Bayes Theorem for each leaf in tree as follow

$$P_{i|j,k} = \frac{P_{j|i} P_{k|i,j} P_i}{\sum_i \sum_j \sum_k P_{j|i} P_{k|i,j} P_i} \tag{14}$$

The prior probability is computed based on the update of features on the second level of the probability tree

$$P_i \leftarrow P_j = \sum_i P_{j|i} \tag{15}$$

First we find the maximum posterior probability in possibility matrix. Once we found match blob, then we make the row and the column of the matched blob into to zero. The procedure is repeated until the possibility matrix becomes a zero matrix.

## Results and Analysis

The result of our detection system without prior information is relatively good to detect pedestrian in not so crowded situation as illustrated in Figure 3 with only 80% accuracy. The top left image is the original image of the video scene. The blobs are shown on the top right and in the bottom left shows the detected pedestrians. The left bottom show the trajectories tracking results in the image coordinates.

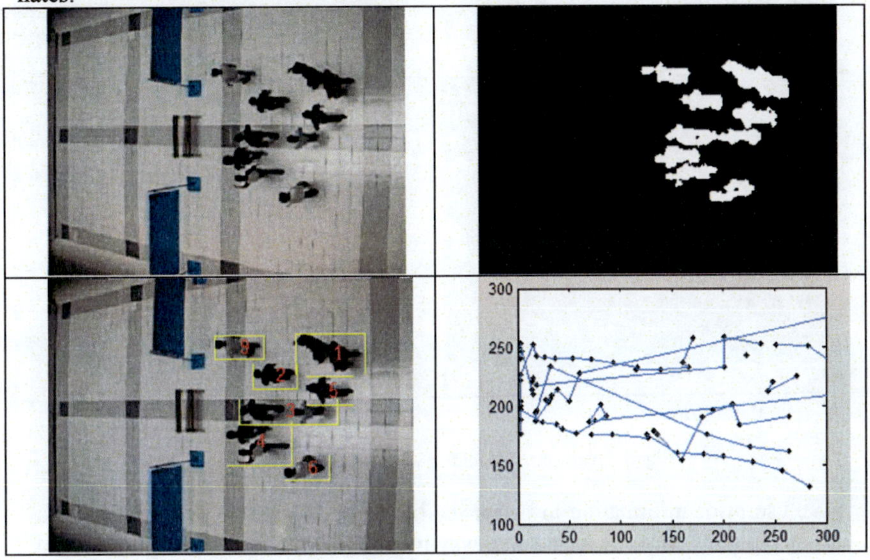


**Fig. 2. Example of tracking without prior information**

For crowded scene, the tracking without prior information produces low level of accuracy below 40%.

The crowded scene as illustrated in Figure 3 was taken during pedestrian experiment in Ateneo de Manila University, Philippines for the study of microscopic pedestrian. The video was taken from the third floor of a building (about 10 meter height) at unknown angle. The camera was not calibrated.

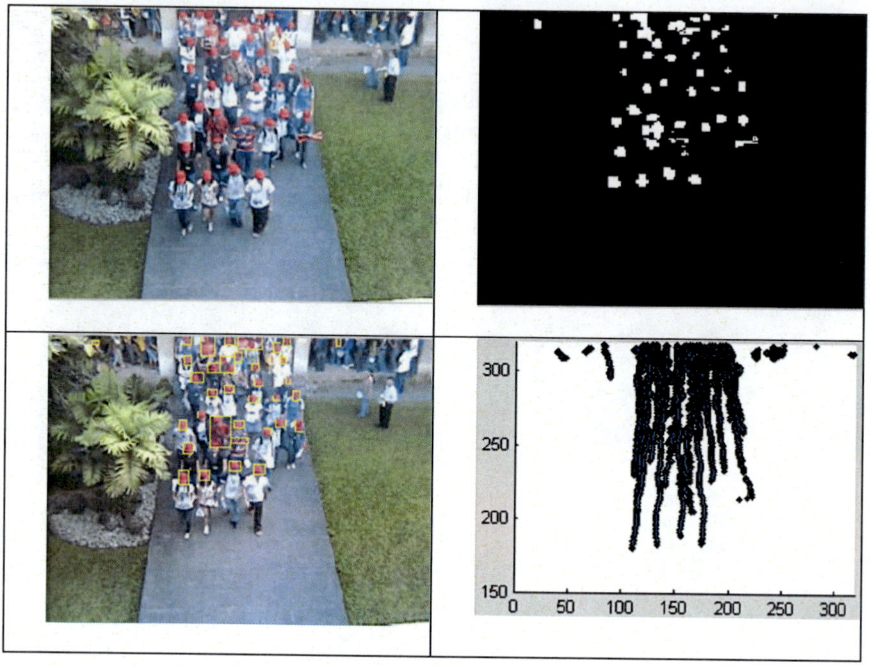


**Fig. 3. Example of tracking with prior information**

We use prior information to detect the blobs, our algorithm was able to track with much higher result and in some way more than 90% and the algorithm was able to find lost pedestrian even after 10 frames of occlusion.

This algorithm can track pedestrian very well because the detection part was almost perfect while the Bayesian update make the algorithm to be get better result over the longer time of tracking. Each frame that a pedestrian track it will increase its posterior probability and it eventually produces a better result. At the same time, however, our algorithm needs a very large amount of memory to save all blobs images and the computational speed becomes very slow.

## Conclusion

Tracking pedestrian in a crowded situation is a hard problem due to occlusion and clutters. We attempted to solve such problem by employing uncalibrated video camera from a top view angle. We compare two options of detections: one without prior information and the other using prior information about the scene. Then, we use the same tracking algorithm to track them. It is showed that prior information significantly increases the accuracy of detection.

We also presented a new algorithm to calculate the probability of tracking system using Bayesian Update in a probability tree. It has been shown that the tracking system can detect more than 90% of the pedestrian even in crowded situation and even when they move zigzag.

Improvement of such system for both part detection and tracking to have faster and more reliable system would be subject to further research study.

**Acknowledgments** This project supported by Pedestrian Research Group of Ateneo De Manila University

## References

1. McKenna, S.J.: Tracking Groups of People. Computer Vision and Image Understanding. Vol. 80, pp. 42–56 (2000)
2. Kelly, B.A.P.: Pedestrian Detection and Tracking Using Stereo Vision Techniques. PhD. Dissertation, Dublin City University, Dublin (2007)
3. Omar, J., Mubarark, S.: Automated Multi-Camera Surveillance: Algorithm and Practice. Springer, New York (2008)
4. Kelly, P., O'Connor, N.E., Smeaton, A.F.: Robust pedestrian detection and tracking in crowded scenes. Image and Vision Computing, Vol. 27, pp. 1445-1458 (2009)
5. Eshel, R., Moses, Y.: Homography Based Multiple Camera Detection and Tracking of People in a Dense Crowd. IEEE Conference on Computer Vision and Pattern Recognition (2008)
6. Chetverikov, D., Verestoy, J.: Tracking Feature Points: a New Algorithm. Computer and Automation Research Institute (1998)
7. Chetverikov, D., Verestoy, J.: Motion Tracking of Dense Feature Point Sets. Computer and Automation Research Institute (1997)
8. Masoud, O., Papanikolopoulos, N.P.: Robust Pedestrian Tracking using a model-based approach. Artificial Intelligence, Robotics, and Vision Laboratory (1997)
9. Saadat, S., Heng. S., Teknomo, K.: Automation of Pedestrian Tracking in a Crowded Situation using Angled Uncalibrated Camer. Proceeding of the 3rd Engineering Research and Development for Technology (ERDT) Conference, Metro Manila, Philippines, pp.119-122 (2009)

# Theory for Models

# Pre-Warning Staff Delay: A Forgotten Component in ASET/RSET Calculations

**S.M.V. Gwynne[1], D. Purser[3], and D.L. Boswell[2]**

[1]Hughes Associates UK, [2]Hughes Associates, Inc., [3]Hartford Environmental Research
Corresponding Author: sgwynne@haifire.com

**Abstract** In this article, the Pre-Warning (warning time) concept is developed: the time between an incident being noted by a member of staff (either directly or indirectly) and the raising of a general alarm. This represents the potential delay in staff response as they interpret their provision of cues and respond; a delay that may be procedural and/or cognitive. The theoretical basis of this concept is discussed, examples of incidents involving this factor are described, and data is examined from experiments and incidents to quantify the extent of the impact and the effect of this concept upon the ASET/RSET calculation. Examples of how Pre-Warning delay can influence RSET will be presented, along with a discussion of those procedures that are particularly susceptible to the delay and suggestions as to how this might be remedied.

## Introduction

Increasingly, a performance-based approach is adopted to demonstrate the safety level of a structure and the procedures employed in response to a fire. This requires a comparison between the Available Safe Egress Time (ASET - the time between the ignition of a fire and the time at which tenability criteria are exceeded), and the Required Safe Egress Time (RSET - the time between ignition of a fire and the time at which occupants in a specified space in a building are able to reach a place of safety) [1,2,3]. Similar comparisons can be made for other types of events such as security incidents. It is therefore critical to include the appropriate components when comparing these two calculations. Various methods are used to perform this comparison including engineering calculations and computational tools. This article addresses a potential limitation in the typical RSET calculation – the exclusion of a sufficient representation of the time for notification. This exclusion might lead to the RSET value being underestimated in certain situations. Critically, this limitation may misrepresent the effectiveness of procedural variants during RSET comparisons.

R.D. Peacock et al. (eds.), *Pedestrian and Evacuation Dynamics*,
DOI 10.1007/978-1-4419-9725-8_22, © Springer Science+Business Media, LLC 2011

## Establishing RSET – Engineering Timeline

Currently, any performance-based analysis of a structure requires the comparison of the Available Safe Egress Time (ASET) and Required Safe Egress Time (RSET) for a particular scenario. The RSET calculation is formed from a number of sub-components, relating to the procedural and evacuee performance. The RSET value can be formulated as follows [4]

$$RSET = t_d + t_n + t_{p\text{-}e} + t_e \tag{1}$$

where $t_d$ is the time from fire ignition to detection, $t_n$ is the time from detection to notification, $t_{p\text{-}e}$ is the time from notification until evacuation commences and $t_e$ is the time from the start of purposive evacuation movement until safety is reached. Typically, $t_d$ and $t_n$ are determined according to the technological resources employed (e.g., manufacturer's guidance on sensor response) and often do not include any reference to staff activities.

It is apparent that the first two components of the RSET calculation ($t_d + t_n$) are procedural. As such, they may be formed from technological or human resources; i.e., that staff or equipment may be employed to detect an incident or initiate evacuation. The selection of the procedural resources employed will be influenced by a number of factors (e.g., the cost, the infrastructure, the nature of the occupancy, the expected incident scenarios, etc.); however, it is possible that staff will be involved in the detection of the original incident and the raising of the alarm. This involvement may have a direct impact upon the time for information to be provided to the target population and thus, depending on the nature of the procedure, may be a prominent factor in the RSET calculation.

The RSET timeline can be broadly categorized into the fire development phase until detection (detection time), the notification phase (also known as alarm or warning time), the pre-response phase, and the response phase (see Fig. 1). Depending on the scenario, each of these phases can have a significant impact on the time for the target population to reach safety. However, it is contended here that the notification phase is often simplified; therefore overlooking key factors that, given the scenario, may have a significant impact on the RSET time. For this reason, a Pre-Warning component should be introduced into this notification phase to account for the staff decision-making and response activities. *In effect, the notification phase could be better represented by including the accumulation of individual staff decision-making processes and subsequent activities.*

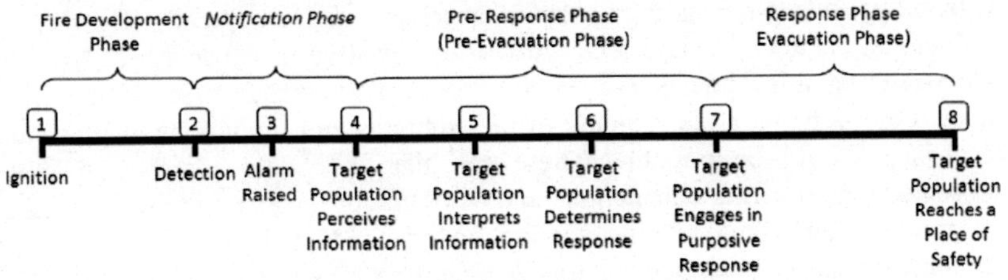


**Fig. 1. Modified representation of the RSET components**

*Key Point: The notification phase can be influenced by staff activities, and can therefore be influenced by the individual decision-making process.*

## The Human Element in Pre-Warning

In this article, the Pre-Warning (warning time) concept is suggested: the time between the fire being noted by staff and the eventual raising of a public alarm; i.e., notifying the target population that there is an incident and that they need to respond. This falls within markers (2) and (4) in **Fig. 1**. Often, with the RSET calculation, this time period (between (2) and (4)) represents the performance of technological devices (e.g., sensors, alarms, etc.). In reality, depending on the nature of the procedure itself, it may also include a significant human element. This includes the potential delay in staff response as they react to the provision of cues and determine a response– a delay that may be procedural and/or cognitive – and then enact that response.

The human element in warning times arises in any situation in which the behavior of an individual person intervenes between detection of a fire (or comparable event) and the notification of a fire to the affected occupant population. This may involve situations such as those involving simple manual alarm call point systems, those where confirmation is required, or those where automatic detection triggers a pre-alarm to designated security staff, who are then required to instigate appropriate warnings [5]. The need to determine the times required for persons first discovering a fire or staff receiving a pre-alarm to engage in the activities leading to the provision of warnings to affected occupants is recognized in fire safety engineering standards [1,2], but limited guidance is available on the behavioral parameters involved, how they should be managed and how they can be quantified in a design context. This concept is expanded here.

In reality, the Pre-Warning delay may consist of a combination of cognitive processes (processing information and then determining an action) and responses (the performance of the resultant action) that are much more significant than the equivalent technological performance times. This article focuses on the cognitive process given that the nature of this process is reasonably consistent between different actions, whereas the actions themselves may vary greatly. However, in both cases, the actual delays incurred will be sensitive to the scenario. By better understanding this phenomenon, a more accurate representation of the RSET process and a better understanding of the strengths and limitations of procedural responses should be possible. These can then be better represented in the subsequent engineering calculations and computational tools employed, allowing for more credible and reliable analysis to be conducted. *Key Point: In some circumstances, a component may be missing from the traditional RSET calculation related to the time for staff to go interpret and respond to the information available to them.*

## Cue Perception and Decision-Making

Recent developments in our understanding of human performance in fire have been characterized by a number of key elements [4,5]:

- Panic-based evacuee response is not inevitable and is indeed rare.
- People do not necessarily panic should information be provided.
- The time to reach safety is not necessarily dominated by physical factors, but can be significantly influenced by the individual's decision-making process.
- An individual is sensitive to the information around them.
- An individual will process this information in accordance with their abilities, attributes, experience, expertise, the situation, the surrounding environment (physical, social and procedural) and their objectives to formulate a response given the time available.
- Physical, social, and psychological attributes influence performance (and the decision-making process) and these attributes vary across a population.

The speed with which the individual arrives at decisions (i.e., interpret external information, integrate it with existing understanding, and arrive at an appropriate response) will be highly influenced by their expertise, experience, the situation faced and the nature and content of the information available; however, the decision-making process will broadly be the same. Critically, both staff and the target population (i.e., those occupants that need to respond) will pass through this process during any decision-making process. Training can help make this process more efficient and even make it more likely to arrive at an appropriate outcome; however, it does not alter or remove the process entirely. In the context of this article, it may make this process shorter, but it will still exist and delay the final action. *Key Point: Irrespective of the individual involved, they will be subject to the same general decision-making process when faced with a change in their surrounding conditions.*

## Examples of the Impact of the Staff Decision-Making Process

Pre-Warning is not simply a theoretical issue; it has been observed in real incidents and in drills. Several examples are discussed where the delay produced due to the staff decision-making process affected the notification time.

- *Nagasikiya Store Fire, Amagasaki City, Japan [6]* - A fire started on the fourth floor of a store and spread throughout the floor in approximately ten minutes. The fire was detected by a smoke alarm system at approximately 12:32. The signal was relayed to the emergency officer on the fifth floor who then confirmed the signal with staff on the fourth floor. The fourth floor staff commenced the evacuation of their floor. The emergency officer attempted to contact a senior member of staff via a coded messaged through the PA system. At 1238 the emergency officer on the fifth floor informed staff to commence a general evacuation, by which time the conditions had deteriorated precluding safe egress. Fifteen people died all of whom were initially located on the fifth floor.
- *Unannounced Evacuation Drills from Retail Stores, Ulster, UK [7]* - Five unannounced evacuation drills were conducted in four department stores. It took the customer population on average 30.3 seconds to respond to the call to evacuate (ranging from 1 to a 100 seconds) [7,8]. Samochine et al examined the staff delay in evacuating the customers and found that it took staff an average of 18.1 seconds (ranging from 2 to 57 seconds) to initiate the evacuation of the customers. The first staff actions were established: 1% ignored the alarm, 61% waited or sought more information, 29% evacuated themselves or others, with the rest leaving their immediate area. Here, it is apparent that many of the first actions were to aid the staff decision-making process. 79.5% of the staff observed had an impact on customers' behavior; indeed, 47% of customers received their first cue from staff.
- *Staff training and response studies, Tokyo, Japan [9,10]* - A detailed study of emergency staff behaviors and of the associated time distributions was carried out in Tokyo Fire Department as part of their training program. During simulated fire emergencies, teams received a fire pre-alarm in a control room and then investigated and responded to a fire on the 22nd floor of an office building. During these studies, involving 222 separate team responses, the times from pre-alarm to emergency announcement were characterized: on average, 120 seconds to arrive on scene; 42 seconds to confirm the incident; 71 seconds to evacuate the floor and/or 167 seconds to evacuate the structure as required. This resource provides a useful pre-warning time database for the scenario studied.

## Understanding the Impact of Staff Decision-Making

It is now widely accepted that a key component in the time to reach safety is the pre-response time (see Fig. 1): the time between the target population being notified (or becoming aware) of an emergency and them initiating purposive egress movement toward a place of safety [5]. It is therefore broadly accepted that the response of the occupant population to the provision of cues indicating an incident is important. Depending on the nature of the occupancy, the scenario, the procedure and the population, this pre-response time can represent the majority of the time to reach safety; e.g., in hotels where the occupants may be asleep. This is dependent upon the performance of pre-response activities (i.e., action prior to a purpose attempt to reach safety) and the cognitive process that assesses the information and the nature of the situation (i.e., that establishes an individual's situation awareness and then determines how they should respond)

Traditionally, the RSET calculation progresses from the time of ignition, to the time for detection, to the time of warning and then to the time for occupant response: occupants perceiving a cue, interpreting it, deciding to act and then moving to a place of safety (although this is simplified during the engineering calculations).

The extent to which human factors affect notification time depends upon the type of occupancy and in particular on the detection and warning systems used. In this context three basic types of system have been previously recognized, which have been classified into three performance levels in relation to the time required for notification its reliability [3]:

- **Level A1** alarm system consists of automatic detection throughout the building, activating an immediate general alarm to occupants of all affected parts of the building.
- **Level A2** alarm system consists of automatic detection throughout the building providing a pre-alarm to management or security, with a manually activated general warning system sounding throughout affected occupied areas and sometimes a general alarm after a fixed delay if the pre-alarm is not cancelled
- **Level A3** alarm system consists of local automatic detection and alarm only near the location of the fire or no automatic detection, with a manually activated general warning system sounding throughout all affected occupied areas.

For a Level A1 system, once ignition has been detected the alarm is automatically raised and evacuation can begin immediately, so the notification time is effectively zero. For a Level 2 system, the notification time depends at least partly upon the behavioral response of the staff. For a Level 3 system, the notification time depends upon the characteristics and behaviors of the person (or persons) discovering the fire. These concepts are expanded upon here.

For instance, in a small office building may have a Level 1 system such that once a fire is detected by a (functioning) smoke detector the alarm will sound. In this instance, the delay in notification will be due to technological performance alone. However, in a domestic residence, automatic detection may trigger only a local alarm, which may be heard by only one nearby person, so that notification of other occupants then depends upon the behavior of this individual. In such situations notification times may be long and unpredictable, presenting significant difficulties for RSET estimations. In more complex structures/scenarios (such as transport terminals, hospitals, large hotels, high-rise structures, etc.), more sophisticated emergency procedures involving staff decisions/activities are often essential and put into practice to combat the complexities faced, to minimize disruption and/or to minimize security issues. These procedures often require staff to interpret information, the notification of staff prior to the notification the general population (often referred to as private mode) and/or the (manual) confirmation of the incident by the staff prior (pre-signal) to the general population being notified. Pivotal in this type of procedure is the reaction and performance of the staff. It should be emphasized that this article is in no way suggesting that staff decision-making should be removed or minimized from emergency procedures. In many situations, this decision-making enables the procedure in place to function.

Although it has been identified as a key component of warning time within RSET [3], the time for the staff to go through the decision-making process is often omitted. Time for decision making is often only (crudely) represented for the general population (i.e., the evacuating occupants). Although staff may be better trained and have more experience (both of which will influence the extent of any delay), they will go through an analogous process to that of a non-expert evacuee who perceives new information.

Staff decision-making may occur in a number of situations; e.g., in discovering a fire, interpreting cues first hand and recognizing them as indicating a real incident, or being notified of an incident (i.e., via the notification system at the fire panel, by other members of staff, by members of the occupant population, etc.). In these instances, staff members will need to perceive, interpret, decide on an action, and then perform the action – in much the same way as other occupants; all of which may take time. Although trained staff may be expected to go through this process more quickly, they would still have to go through it, delaying their response and in turn, the response of the general population of the building. The time during the notification period associated with staff response can be characterized as

$$\sum_{1}^{\#}(t_{dm}+t_{a}) \tag{2}$$

where # is the total number of decision making processes, $t_{dm}$ is the time associated with a decision making process, and $t_a$ is the time associated with the subsequent action (which can approach zero in some instances). The discussion during this article focuses primarily on the $t_{dm}$. *Key Point: Staff activities are just as reliant upon the decision-making process as non-staff, and will then incur a delay, albeit potentially at a reduced level.*

## Procedural Variants

The procedural response to an incident will be dependent on the resources available, the type of occupancy and target population, the expected incident scenarios and the structure. Depending on the nature of the required response, procedures may require the immediate evacuation of the full population or require some pre-response management in order to ensure success. This management may involve confirming the nature of the incident (e.g., that is real and warranting response), determining the areas in the structure that require action (e.g., establishing which zones need to be evacuated in response to the incident), and determining when these zones should be evacuated (e.g., staging the response of zones in order to avoid egress routes being overloaded).

Broadly speaking, the procedure will require the detection of the incident and the notification of the population that there is an incident. However, the exact nature of these two actions is sensitive to a number of procedural attributes:

- Whether detection is through a sensor or a person (a member of staff or the public);
- Whether the signal detection automatically triggers a public alarm or provides a private signal to staff (e.g., whether the signal sets off an alarm or initiates a management procedure that is not immediately apparent to the occupant population);
- Whether the information is relayed to a central command or whether it is localized (e.g., an activated detector sets off a local alarm where the signal is not shared to staff or the occupant population outside of this area);
- Whether the report of the incident needs confirmation (e.g., given the propensity or the consequence of false alarms, that a detection signal must be confirmed prior to a public alarm);
- Whether the confirmation requires manual observation or whether remote observation is possible
- Whether a confirmed incident leads to a full evacuation or requires a managed response (e.g., zoned, staged, partial, etc.)
- Whether the pre-alarm can be cancelled or defaults to a general alarm after a fixed delay
- Whether activation of more than one detector automatically activates a general alarm

*Key Point: The extent of the impact of the decision-making process of staff activities will be dependent on the procedure employed.*

## Representing Staff Decision-Making

The theoretical impact of staff decision-making upon RSET can be estimated according to the nature of the procedure response and the number of decision-making activities involved. This type of approach may inform the development of engineering and regulatory guidance. Three main categories are considered from the list presented in the previous section:

- the mode of detection (**Detect.**) – [Sensor (Sn) | Staff (St) | Public (Pu)]
- the mode of dissemination (**Dissem.**) – [General Alarm (GA) | Local Alarm (LA) | Private Alarm (PA)]
- the mode of confirmation (**Confirm.**) – [None Required (N) | Remote Observation (R) | Manual (M)]

The additional activities required by staff are too numerous to list. However, they may certainly impact on the overall Pre-Warning time. Table 1 provides examples of the number of staff and public decision processes associated with several of the procedural responses. There is no attempt to quantify the expected time of these specific processes as this will be highly dependent upon the specifics of the situation and the individuals involved. However, it is apparent that the different category permutations require a different number of decisions each of which has the possibility of extending the time that it takes to eventually notify the general population of the response required of them. It is certainly not the case that procedures should be discarded because of the number of decision-making processes required within them. Indeed, it is the very ability of procedures to be sensitive to the incident and react accordingly that safeguards against unnecessary disruption or inappropriate response in complex situations. However, the potential for these decisions being required in order for the inclusion of legitimate procedural responses should be represented within the engineering process: within RSET. It should also be noted that we have deliberately underestimated the theoretical impact of the decision-making process upon RSET. In many instances, confirmation and reporting may be even more of an iterative process as more information is required and processed. In addition, no assessment is made here of the actions that staff need to support a procedure; e.g., movement, operating communication devices, etc. that follow on from the decision making processes. This will again add to the RSET value and should then form part of the *Pre-Warning* consideration.

**Table 1. Number of decision processes given the incident response**

| Detect | Dissem. | Confirm | Description | Staff | Public |
|---|---|---|---|---|---|
| Sensor | Gen.Al | None | No human decision-making | [0] | [0] |
| Sensor | Pr.Al | Manual | Automatic Detection. Private signal sent to control [1]. Confirmation required through manual observation [1], reported to control [1] prior to a) public announcement, or b) private announcement staff to evacuate [1]. | a) [3] or b) [4] | [0] |
| Public | Loc.Al. | None | Member of Public activates [1] Local Alarm. Local staff interpret signal, decide [1] to a) activate General Alarm, b) notify control who [1] activate General Alarm or c) communicate privately with staff to evacuate [1]. | a) [1] or b)[2] or c) [3] | [1] |

* Decision Processes Prior to Notification

From the material examined, it is apparent that the staff decision-making process required as part of any emergency procedure can affect the notification time. This delay is further extended (more intuitively perhaps) by the actions then required of the staff; however, the nature of this delay will be highly sensitive to the actions required of the staff, whereas a broadly applicable baseline for the decision-making process may be developed. Estimates are often used in the RSET calculation in order to represent the pre-response times of the general population based on the nature of the notification system, population status, etc. [1,2]. It is suggested that a similar approach be adopted for the staff decisions required in a procedure. For instance, it may be reasonable to assume a staff decision-making process takes 50% of a non-staff member, given the findings of Samochine. Therefore, this could be factored into the RSET equation: every time a staff decision is required given the procedure in place (see Table 1), then a suitable delay (50% of the equivalent non-staff delay) should be incurred (see Equation 2) and accounted for in the RSET equation (see Equation 1). Similar staff activity times could also be included in a manner similar to the evacuee response calculations (e.g., calculated according to expected distances to be covered and the specific actions to be performed, etc.).

## Conclusion

The notification time is not simply dependent upon the performance of technology. There are a number of legitimate scenarios that require managed procedures to ensure a degree of flexibility and rigor in the procedure employed. However, these procedures, although essential, may have an inherent set of delays that are currently not addressed in the RSET calculations. It is the very strength of these procedures (i.e., that the employ the information available to help manage the procedure in a specific manner) that requires individual staff members to make decisions, which delay the overall notification process. These decisions are vital; however, they need to be accounted for in the RSET calculations. This can be achieved through the Pre-Warning concept, where the notification phase is extended according to the estimated number of decisions required during a particular procedure.

## References


1. The application of fire safety engineering principles to fire safety design of buildings – Part 6: Human factors: Life safety strategies – Occupant evacuation, Behaviour and Condition, BS PD 7974-6. British Standards Institution, London, UK (2004)
2. Fire-safety engineering – Technical Information on Methods for Evaluating Behaviour and Movement of People, ISO TR16738 (2009)
3. Purser, D.A.: ASET and RSET: Addressing Some Issues in Relation to Occupant Behaviour and Tenability. FIRE SAFETY SCIENCE – Proceedings of the 7th International Symposium On Fire Safety Science, 16 – 21 June 2002, Worcester Polytechnic Institute – Worcester Massachusetts, USA, pp. 91-102. International Association for Fire Safety Science (2003)
4. Nelson, H.E., Mowrer, F.W.: Emergency Movement. In: DiNenno, P.J., et al (eds.), The SFPE Handbook of Fire Protection Engineering, (3rd edition), pp. 3-367-3-380. National Fire Protection Association, Quincy (2002)
5. Bryan, J.: A Selected Historical Review of Human Behavior in Fire. Fire Protection Engineering. 16, pp 4-10 (Fall 2002)
6. Sekizawa, A., Kakegawa, S., Ebihara, M.: Review of A Real Multi-story Store Fire by Applying Evacuation and Smoke Movement Interactive Simulation Model. Fire Safety Science. 9: 477-488 (2009). doi:10.3801/IAFSS.FSS.9-477
7. Samochine, D.A., Boyce, K., Shields, J.: An Investigation into Staff Behaviour in Unannounced Evacuations of Retail Stores – Implications for Training and Fire Safety Engineering. Fire Safety Science. 9: 519-530 (2005). doi:10.3801/IAFSS.FSS.9-477
8. Shields, T.J., Boyce, K.E.: A Study of Evacuation from Large Retail Stores. Fire Safety Journal, Elsevier Science Ltd. 35, pp. 25-49 (2000)
9. Ebihara, M., Notake, H., Yashiro, Y.: Fire risk assessment method for building consideration of actions of security staffs by using an idea of fire phase. Proceedings of the First International Symposium on Human Behaviour in Fire, 31 August – 2 September, Belfast, UK, pp. 421-428 (1998)
10. Ebihara, M., Notake, H., Yashiro, Y.: Study of the security staff's action taken in the event of a building fire. Proceedings of the 2nd International Symposium on Human Behaviour in Fire, Massachusetts Institute of Technology, 26-28 March, Boston, USA, pp. 341-348. Interscience Publications, UK (2001)

# Fundamental Diagrams for Pedestrian Networks

**S.P. Hoogendoorn, M.C. Campanella, and W. Daamen**

Delft University of Technology, Transport & Planning
Stevinweg 1, 2628CN Delft, the Netherlands
Corresponding author: s.p.hoogendoorn@tudelft.nl

**Abstract** This paper explores the concept of the Network Fundamental Diagram for two-dimensional pedestrian networks. In doing so, we investigate if the average performance of the network can be described as a function of the average density or accumulation of the network, in line with recent findings on discrete vehicular traffic networks. We show that this is indeed the case, by considering data from walking experiments and from simulation using a microscopic pedestrian model. It turns out that the shape of the diagram is determined by a number of factors, such as the shape and size of the area, its use and its function, and the composition of the pedestrian flow. Finally, we propose several applications of the diagram for pedestrian networks.

## Introduction

Recent work of [1] has shown that for (homogeneous) traffic networks, aggregate network dynamics can be described by a single relation: the Network Fundamental Diagram (NFD). This NFD relates average flow variables such as average performance (weighted flow), flow, or speed, to aggregate network flow variables, such as the average density or number of vehicles in the network (accumulation). The NFD has a strong significance, not only from the viewpoint of performance identification of networks, but also from the viewpoint of network traffic management (e.g. perimeter control, see [1,2]).

In this paper we set out to establish a similar relation for pedestrian flows, characterised by their two dimensions. In doing so, we investigate whether the average performance of the area (building, railway station, mall, etc.) can be explained sufficiently well by the total number of pedestrians in the area. To establish this, we analyze both empirical data from pedestrian experiments and synthetic data generated by a microscopic pedestrian flow model.

It turns out that the NFD can indeed also be established for walking areas. In doing so, we are able to identify the network load, which will be critical from a performance point of view. In other words, we can pinpoint a critical accumulation $N_{crit}$ that describes the boundary accumulation after which the network performance will start to deteriorate with increasing accumulation. For some specific

R.D. Peacock et al. (eds.), *Pedestrian and Evacuation Dynamics*,
DOI 10.1007/978-1-4419-9725-8_23, © Springer Science+Business Media, LLC 2011

areas, the performance does not deteriorate but will remain constant until some even higher accumulation is reached.

The notion of the NFD for pedestrian flows has a strong significance, for instance for determining the Level-of-Service (LoS) for an area, from the perspective of aggregate flow modelling (coarse dynamics), and from the perspective of crowd-control. All these aspects will be presented briefly in the paper.

## Network Fundamental Diagram Basics

The NFD shows that as long as the accumulation is less than a critical level $N_{crit}$, the production, defined as the number of pedestrians exiting the network per time unit, increases with increasing accumulation. However, when the accumulation is larger than the critical accumulation, the network production will start to drastically deteriorate. For the example shown in Fig. 1, the production will even become equal to zero. The fact that the network performance degenerates with increasing density is a *characteristic and fundamental property of traffic networks*. Other types of networks do not show this degradation.

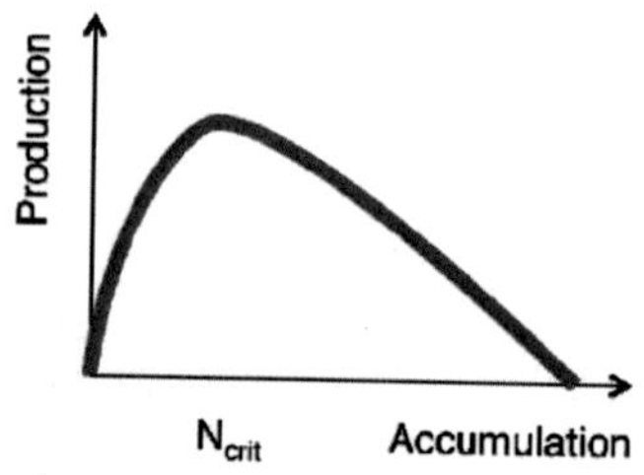


**Fig. 1. Example of Network Fundamental Diagram showing the relation between production (arrival rate per unit time) and accumulation (number of pedestrians in network).**

There are two main causes for the production degeneration:

1. Spill-back effects causing grid-lock;
2. Capacity drop or 'faster-is-slower' effect.

The first is probably the most important of the two. It describes the fact that when considering multi-destination flows, queues in one direction may severely hamper flow conditions in the other direction. Fig. 2 illustrates this: the high density queue (blue) has only limited penetrability and pedestrians walking into the perpendicular direction (orange) have very limited possibilities to pass through the queue.

The second cause is the capacity drop (or the 'faster is slower effect'). For pedestrian networks, this describes the fact that with increasing pressure from behind, the bottleneck capacity may be reduced substantially (see [3]). This in particular applies to situations where the pedestrians are eager to get out of a room (so-called 'faster-is-slower' effect).

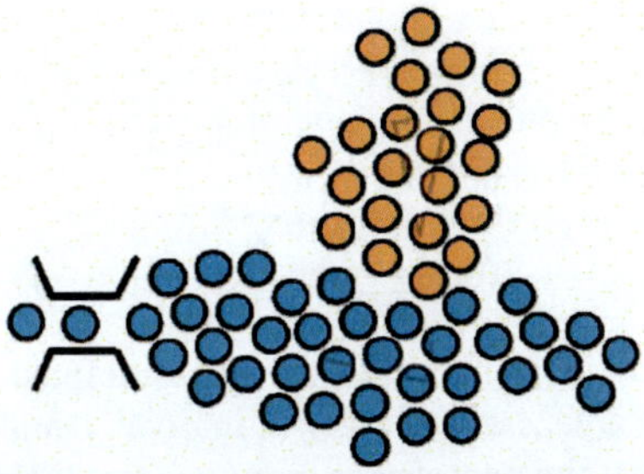

**Fig. 2. Example of spillback of pedestrians passing bottleneck (blue, from right to left) hampering pedestrians in other direction not having to pass the bottleneck (orange, top to bottom)**

In the remainder of the paper, we will show some examples of NFDs from both experimental data and simulated data, for different types of networks (from simple to more complex). In doing so, we show that NFD's can indeed be sensibly defined for pedestrian networks.

## Definition of flow variables

Before deriving the NFD, let us first define the relevant variables accumulation, average (network) speed, production and performance for two-dimensional pedestrian flows.

For the definition of the variables, we will use the generalized definitions of Edie. Let us consider a general area $A$ in which pedestrians $\alpha$ are present. The area can have a general shape. The locations of the pedestrians are denoted by $\mathbf{x}_\alpha$; the velocities of the pedestrians $\alpha$ at time instant $t$ will be denoted by $\mathbf{v}_\alpha(t)$.
Let us now consider a period $T = [t_0,t_1]$. Let $\mathbf{d}_\alpha$ be the distance vector defined by:

$$\mathbf{d}_\alpha = \mathbf{x}_\alpha(t_1) - \mathbf{x}_\alpha(t_0)$$

Furthermore, let $\tau_\alpha$ denote the time spent by pedestrian $\alpha$ in region $A$ during period $T$.

Let us now start by defining the generalized velocity over area $A$ and period $T$. We have:

$$\mathbf{u} = \frac{\sum_{\alpha \in A} \mathbf{d}_\alpha}{\sum_{\alpha \in A} \tau_\alpha}$$

This definition is intuitively correct: the average velocity for all pedestrians in area $A$ during period $T$ is given by the total distance travelled by the pedestrians (also referred to as the production) and the total time spent by all the pedestrians. In the remainder, we will also use the generalized speed $U$ defined by:

$$U = U_{\mathbf{e}} = \frac{\sum_{\alpha \in A} \mathbf{d}_\alpha \cdot \mathbf{e}}{\sum_{\alpha \in A} \tau_\alpha}$$

where **e** denotes the unitary direction vector along which the speed is determined.

The generalized density $k$ is defined as follows:

$$k = \frac{\sum_{\alpha \in A} \tau_\alpha}{(t_1 - t_0) \cdot \|A\|} = \frac{\sum_{\alpha \in A} \tau_\alpha}{(t_1 - t_0) \cdot n} \cdot \frac{n}{\|A\|}$$

The equation shows that the generalized density is defined by the number of pedestrians $n$ that were in area $A$ at any time in interval $T$ multiplied by the fraction of the period length $T$ that the pedestrians spent in area $A$. Note that if $(t_1 - t_0) = 1$ than $\tau_\alpha = (t_1 - t_0)$ and thus:

$$k = \frac{n}{\|A\|}$$

which is the traditional (instantaneous) definition of the density at time instant $t_0$.

The generalized flow $\mathbf{q}$ is now defined by the relation $\mathbf{q} = k \cdot \mathbf{u}$. The average absolute flow $Q$ is defined via the relation $Q = k \cdot U$.

The production $F$ is defined as the number of pedestrians that leave the network per time unit.

## Pedestrian Network Fundamental Diagrams examples

The Network Fundamental Diagram (NFD) relates the different pedestrian flow variables defined in the preceding section to each other. In doing so, it describes the average area (or continuous network) performance as a function of the average density (or number of pedestrians in the area). More specifically, we will focus on two relations:

- Relation between absolute generalized flow $Q$ and the number of pedestrians $n = k \cdot \|A\|$ (referred to as the accumulation in the remainder) in the area $A$;
- Relation between generalized speed $U$ and the accumulation $n$ in the area $A$.

Next to this, we will look at the relation between the production $P$ and the average flow $Q$; just as for car traffic networks, it turns out that this relation is approximately linear ($Q = \phi \cdot F$).

**Narrow bottleneck experiment.** The first situation that we will consider is a simple bottleneck scenario. The data stem from a large scale walking experiment in which we analyzed the pedestrian walking behavior in case of a narrow corridor of 1 m width. Note that in the data, the positions $\mathbf{x}_\alpha(t_k)$ are known for time instants $t_k$

= $0.1k$. Fig. 3 shows several pedestrian trajectories. For a detailed description of the experiments, see [3].

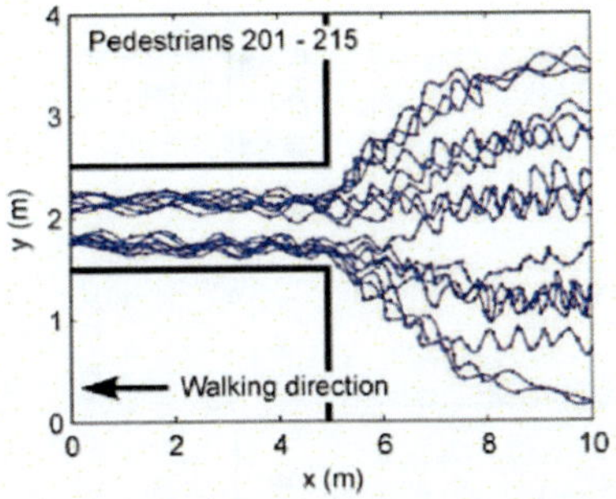


**Fig. 3. Example set of trajectories through narrow bottleneck (from [4]).**

Fig. 4 shows the different relations established from the narrow bottleneck experiment, with $|A| = 25$ m$^2$ and $|T| = 30$ s. Fig. 4a shows that when the accumulation in the area increases, the average speed decreases (as is expected). Fig. 4b shows that the generalized flow first increases when the accumulation increases, but once a critical accumulation level is reached (12.5 pedestrians) the generalized flow becomes a constant value of approximately 1.2 Ped/m/s, which is the capacity of the bottleneck. Fig. 3c shows this relation for the outflow of the area, while Fig. 4d shows the relation between generalized flow and outflow.

For this single bottleneck scenario, no performance degradation is expected when the network accumulation is overcritical. Since there is no spillback or gridlock effect expected, the capacity of the narrow corridor is the limiting factor. Hence, even if congestion becomes very severe, the flow is equal to the bottleneck capacity. Only in some cases, where the pressure due to the upstream pedestrians causes 'arc-formation' in front of the bottleneck (the so-called 'faster-is-slower' effect) we would expect a reduction in the bottleneck capacity. This would yield an observable reduction in the performance that would become visible in the NFD as a decrease of the average flow or outflow with increasing accumulation. The fact that in the experimental results shown, such a reduction is not visible shows that there was no observable faster-is-slower effect for this situation.

Also note the strong relation between the outflow and the generalized flow. It turns out that for this specific situation, $Q(t) = 0.3635 \cdot F(t)$, where $F$ denotes the outflow from the area (the production) in Ped/s.

Finally, note that although for the perspective of the network the performance does not decrease, the average speed does decrease. In fact, while we see that at $N_{crit} = 12.5$ Ped the flow becomes constant, the average speed reduces from $U = 0.8$ m/s at $n = 12.5$ Ped to $U = 0.4$ m/s at $n = 40$ Ped. So the time to traverse the area will be longer when the accumulation increases.

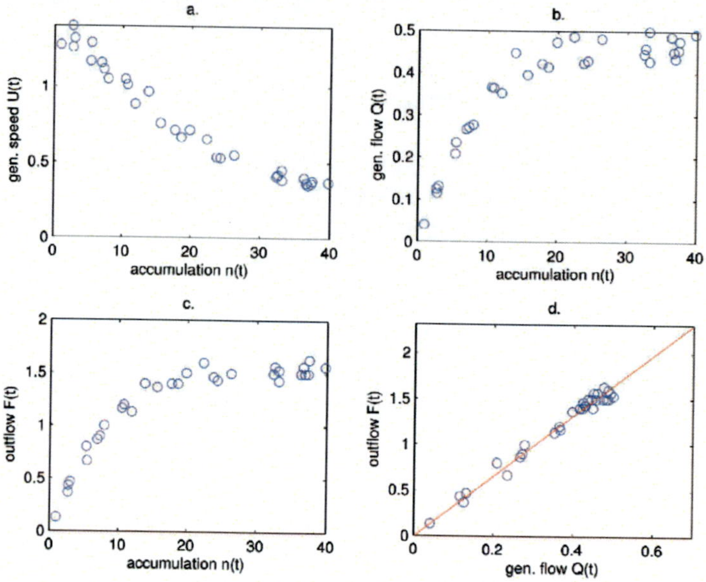


**Fig. 4. Network Fundamental Diagram for narrow bottleneck experiment described in [4].**

**Wide bottleneck NFD.** Fig. 5 shows similar results, but in this case for the wide bottleneck experiment, with $|A| = 30$ m$^2$ and $|T| = 30$ s. Also here, the capacity of the 2 m wide bottleneck determined the maximum generalized flow due to the lack of spillback and gridlock effects. It is remarkable to observe that the relation between the outflow and the generalized flow is nearly the same as in the narrow bottleneck case, namely $Q(t) = 0.3030 \cdot F(t)$. The critical accumulation seems to be much larger, about 25 pedestrians. This is due to the bottleneck capacity and the different geometry of the area compared to the narrow bottleneck.

**Pedestrian Network Fundamental Diagram of a complex area.** In this section the NFD of a complex walking area, namely the central hall of the Schiphol Airport in Amsterdam also known as Schiphol Plaza, is presented and discussed. The simulations were run using the microscopic pedestrian simulation tool Nomad [4].

The Schiphol simulation was based on a realistic scenario developed to simulate one day of pedestrian traffic in the Plaza area. The simulation started with low inflows at 5:00 in the morning and presented a steady increase until 8:00 when a large inflow indicated the beginning of the peak traffic in the area. At this time of the day the inflows mainly originated in the exits of the train platforms inside the court. The original OD tables were kept, but the pedestrian inflow values were multiplied by three. This extreme inflow formed three gridlocks. The biggest congestion was caused by the gridlock in the centre court of the Plaza (Fig. 6). During the congested period, pedestrians walking in areas adjacent to the congestion could walk freely.

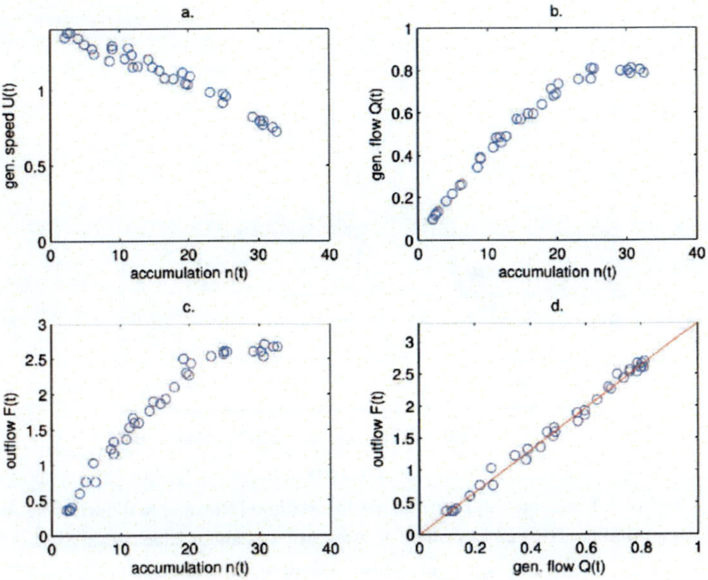


**Fig. 5. Network Fundamental Diagram for wide bottleneck [4].**

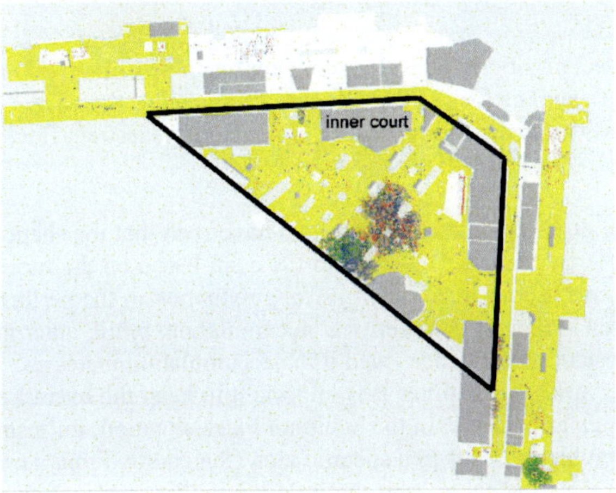


**Fig. 6. Overview of Schiphol Plaza Area.**

Fig. 7 shows the resulting NFD of the court part of Schiphol Plaza. We clearly see that there is a peak in the generalized flow at approximately 2000 pedestrians. When the number of pedestrians in the area increases further, the performance will start to reduce.

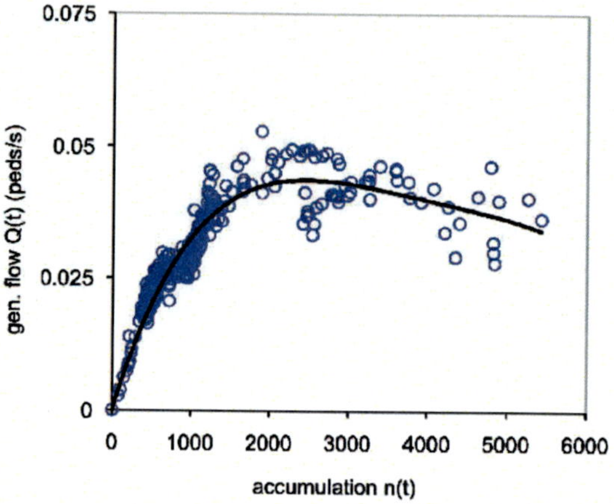


**Fig. 7. Network Fundamental Diagram for the Schiphol Plaza area (court). The figure clearly shows that the NFD has a critical accumulation of about 2200 pedestrians. After this point, flow (and production) deteriorates.**

## Example applications

Let us in this section briefly consider some applications of the NFD. Specifically we will consider applications to crowd control, modeling the coarse dynamics in pedestrian networks, and judging the performance of pedestrian flow models (validation).

**Crowd control.** In the previous sections, we have seen that the shape of the NFD can be very different from the one area to the next, based on the function, design, and use of the area. Some areas show a strong reduction in the performance when the accumulation surpasses some critical accumulation, while other areas show a much more constant performance, even if the accumulation increases.

For crowd control applications, it is important to keep the overall performance of the area as high as possible. In the Schiphol Plaza situation, for instance, we see that crowding problems occur in a specific area (the court). From a crowd-control perspective, it seems logical to limit the inflow into that area such that the performance stays at a high level.

The following control law will keep the performance of the area at the maximum level:

$$q_{\max}(t) = q_{\max}(t-1) - \alpha \cdot (n(t) - N_{crit})$$

This equation shows that when the accumulation surpasses the critical value $n'_{crit}$, the maximum inflow $q_{max}$ into the area will be reduced. The maximum inflow will be increased again when the accumulation is less than the critical accumulation.

**Coarse dynamics modeling.** As a second application example of the NFD, we consider dynamic modeling. The basic concept is that we can divide the considered area A we want to model into disjoint subareas $A_j$ with $j = 1,...,m$. For each of these subareas, we can establish NFDs.

Suppose that for a specific area $A_i$, we are only interested in the coarse dynamics. In this case, we can use the NFD to describe the outflow / number of completed trips. Let $F_i(n_i(t))$ again denote the outflow of the area as a function of the accumulation $n_i(t)$ in the area. For the accumulation $n_i(t)$ we can then establish the following dynamic equation:

$$dn_i = \left(\sum_j f_{ji}(t) - F_i(n_i(t))\right)dt$$

where $f_{ji}(t)$ denotes the outflow from area $A_j$ to $A_i$. Now, we let $\varphi_{ij}(t)$ denote the share of the outflow from area $A_i$ to $A_j$. We can then write:

$$f_{ij}(t) = \phi_{ij}(t) \cdot F(n_i(t))$$

**Pedestrian simulation model validation.** The final application we will illustrate is the application to model calibration and validation. The NFD of a specific area *A* describes the coarse pedestrian flow behavior within this area. It summarizes the main flow characteristics of the area. When calibrating or validating a pedestrian flow model, the ability of a model to replicate the empirical NFD is essential for realistic model outcomes. This holds for pedestrian flow models, but equally for vehicular traffic models.

## Conclusions and future work

In this paper, we have investigated the existence of the Network Fundamental Diagram for pedestrian flows. This is achieved by considering both experimental and microscopic simulation scenarios.

Based on these investigations, it is concluded that NFDs can be meaningfully defined for two-dimensional pedestrian flow areas and that the concept of the NFD can be extended to flows in two-dimensional areas.

The shape of the NFD depends on the shape of the area and the size and shape of the obstacles therein, the use of the area and consequently on the origin-destination matrix. Furthermore, we expect that other factors, such as the composition of the flow (trip purpose, gender, age distribution), and psychological factors (haste) will have an effect as well.

We have seen that based on the shape of the NFD we can see at which accumulation the performance of the area will start to reduce. For some areas, it turns out that only for very high accumulation levels, the performance will start to reduce. In designing crowd control strategies, it turns out that these areas could be designated as (temporary) storage areas for pedestrians, while other areas that show a strong performance decline with increasing accumulation levels, should be kept sufficiently empty.

Future work is aimed at gaining more insights into the relation between the shape of the NFD and the factors considered to be of substantial impact. This could be done in a simulation environment, such as the microscopic simulation model Nomad, but preferably this is done in an empirical setting. The data needs for such an experiment are however likely to be demanding, in particular for larger areas, and advanced data collection techniques will be required.

## References


1. Daganzo, C.F., Geroliminis, N.: *An analytical approximation for the macroscopic fundamental diagram of urban traffic*. Transportation Research Part B. Volume 42, pp. 771-781 (2008)
2. Geroliminis, N., Daganzo, C.F.: *Existence of urban-scale macroscopic fundamental diagrams: Some experimental findings*. Transportation Research Part B. Volume 42, pp. 759-770 (2008)
3. Hoogendoorn, S.P., Daamen, W.: *Pedestrian behavior at bottlenecks*. Transportation Science. 39(2), pp. 147-159 (2005)
4. Hoogendoorn, S.P., Bovy, P.H.L.: *Normative pedestrian behaviour theory and modelling*. In: Taylor, M.A.P (ed.), Transportation and traffic theory in the 21st century. pp. 219-245. Elsevier Science, Oxford: Pergamon (2002)

# Modeling of Human Behavior in Crowds Using a Cognitive Feedback Approach

**Y. E. Papelis[1], L. J. Bair[4], S. Manepalli[1], P. Madhavan[2], R. Kady[3], and E. Weisel[4]**

Virginia Modeling Analysis & Simulation Center [1]
Dept. of Psychology[2]
Dept. of Engineering Management and Systems Engineering[3]
Old Dominion University
Norfolk, VA 23529
WernerAnderson, Inc.[4], Gloucester, VA 23061
Corresponding author: ypapelis@odu.edu

**Abstract** We present a real-time agent-based approach to modeling crowd behavior that is based on complementary psychological and engineering principles. The application focus is for developing realistic models that address not only the physical but also the psychological aspects of crowd behavior. Our approach to modeling the psychology of a crowd is based on the principle of emotional reflection. According to this principle, our emotions are evoked in response to our perception of other people's emotions; hence emotions propagate through a crowd as a result of each person's perception of other crowd member's emotions as well as external factors. The emotional model is coupled with a movement model that is based on the social forces formulation, but with parameters modified to represent the current emotional state of each crowd member. We present the model along with results of how different emotional levels can affect the movement dynamics of crowds.

## Introduction

As crowd and evacuation models are increasingly becoming a part of an innovative approach to assess public safety, model developers are faced with the challenge of demonstrating that their models accurately represent human physical abilities and behaviors during emergency conditions. As sophisticated crowd and evacuation models continue to emerge [1-3], quantifying and predicting people behavior remains a fundamental element in estimating the required time to reach safety. At the same time, there is little, if any research on the quantification of how emotions affect behavior during crowd movement. The purpose of our research is to incorporate an emotional component to crowd models and link the emotional component to the movement model.

R.D. Peacock et al. (eds.), *Pedestrian and Evacuation Dynamics*,
DOI 10.1007/978-1-4419-9725-8_24, 

Conventional psychology provides numerous qualitative trends on how stimuli are processed and how they affect our emotional state. For example, within a crowd, physical cues such as emotional expression and race are quickly detected by perceivers [4]. In a large crowd it may be difficult or impossible to see every crowd member; therefore perceivers often base their perception and responses on their impressions of a subset of the crowd members. Research on face perception has shown that perceivers orient toward threatening stimuli, and angry faces tend to 'pop-out' and lead to reciprocal responses [5]. Similar effects have been found for other negative emotions such as fear and sadness, wherein negative faces typically capture attention and instigate responses [6].

This paper describes a quantitative crowd model that is built on the aforementioned principles, and demonstrates the feasibility of linking a movement model that is based on the principle of emotional reflection. The model utilizes stimuli generated by the perceived emotions of other crowd members in order to affect the ongoing emotional state of each crowd member. We use a physics-based perception model to capture realistic propagation of emotions. In this paper, we first describe the model then demonstrate a short example of its operation.

## Model Description

The basic architecture of the model is illustrated in Figure 1. Here, we use the term cognitive and psychological model interchangeably. The cognitive portion of the model is responsible for simulating the mental state of each individual crowd, whereas the physical portion of the model includes both the movement and perception simulation.

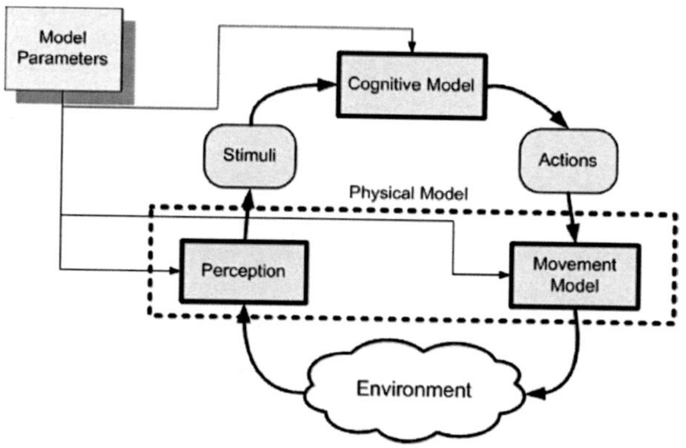


**Fig. 1. Overall crowd modeling architecture**

The cognitive and physical models are linked through actions and stimuli. Actions reflect the visible behavior of each crowd member, which may or may not directly relate to actual movement. Stimuli reflect the inputs to the cognitive model and capture the propagation of emotions among crowd members. Finally, all interactions are parameterized, allowing the incorporation of a wide range of cultural, age, gender and similar factors into the overall model operation.

### *Notation and Model Formulation*

As a matter of notation, the vector variable $Q$ represents the emotional state, and the lower letter $q_i$ represents a specific element of the vector $Q$, with $1 \leq i \leq N$. Stimuli belong to a set $S = \{s_1, s_2, \ldots, s_k\}, 1 \leq k \leq K$ , each member representing one of possible K actions. We use the notation $s_i^j$ to indicate that crowd member $j$ receives stimulus $i$.

Our approach for this crowd model utilizes a 7 member state vector (N=7) which consists of real numbers whose value is normalized to the interval [0, 1]. Three elements are used for basic psychological state, namely fear, anger and joy (emotional fullness). Surprise and pain are used to capture transient psychological states that also have a physical component. Final, injury represents an aggregate state of reduced physical capacity.

Stimuli are the primary inputs to the model and are treated as messages that are broadcast by each crowd member to the surrounding environment according to the emotions that the crowd member is currently experiencing.

The model employs three types of stimulus propagation: visual, aural and direct. Visual delivery implies that the recipient of the stimulus must be able to see the originator. Similarly, aural delivery implies that the recipient of the stimulus must be able to hear the originator. The direct propagation method is reserved for programmatic injection of stimuli into the simulation. Because each emotional state exhibited by a crowd member can propagate visually and aurally, each emotion can generate at least two types of stimuli, one per propagation type. For example, in the case of pain, the two stimuli would be "See-Pain" and "Hear-Pain".

### *Emotional Model*

For the implementation discussed here, the model makes the following assumptions.

- The effect of stimuli on each emotional state is independent of the values of other emotional states.
- The effect of a continuously received stimulus diminishes.

- Lacking any stimulus, emotional states diminish to a baseline.
- The effect of multiple stimuli on a specific emotional state is not linear; in fact, the model assumes that the dominant stimulus is the only one that affects the emotional state.

Figure 2 depicts the resultant model. The time decay component diminishes the value of each state over time, provided no stimulus is received. We use an exponential function with a parameter that controls the rate of decay. When one or more stimuli affect the state, the change is dominated by the effect of the dominant stimulus according to the effect computation model. The time decay computation is as follows:

$$q_{t+dt} = q_t e^{-\frac{\lambda}{t-t_u}}$$

The parameter $\lambda$ is the decay rate, and $t_u$ is the most recent time that a stimulus affected the state.

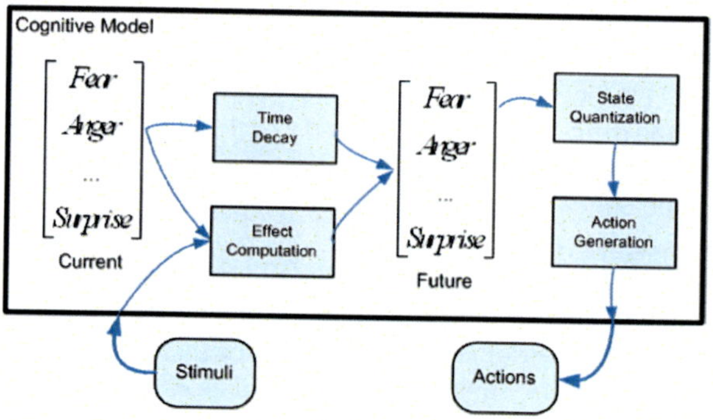


**Fig. 2. Cognitive crowd model**

The effect of stimuli on elements of the state vector is formulated according to the following equation:

$$q_{t+dt} = q_t + \max_{S} \left| s_i k_1 e^{\frac{-k_2}{t-t_0}} \right|$$

In this equation, the set $S$ represents all the incoming stimuli and $s_i$ represents the intensity of a specific stimulus. The parameters $k_1$ and $k_2$ are specific to the stimulus-state pair, and $t_0$ represents the time on which the specific stimulus first became available.

Note that the effect of stimulus $i$ on emotional state $j$ requires a set of three parameters, $k_1$, $k_2$ and $\lambda$. The sign of $k_1$ governs if the stimulus enhances or diminishes the emotional state and its amplitude scales the interaction. The $k_2$ parameter controls the diminishing effect of a stimulus when it is continuously applied over time.

Ultimately, the emotional state of a crowd member governs its actions. The model uses a two-step approach for mapping the emotional state vector into actions. The first step quantizes the values of the states, thus providing a finite set of states that need to be converted into actions. Once the states have become finite, a lookup function maps the state onto a set of actions, which are then implemented by the physical model. Note that not all actions involve locomotion; however all actions can be perceived by other characters and affect their emotional state through the generic stimuli mechanism.

### *Movement Model*

The movement model uses two complementary sub-models to generate realistic behavior that is also related to the emotional state. At the lower level, Helbing and Molnár's social forces model [7] is used to move individual crowd members. The model determines the motion of a member seeking a certain destination at certain period of time, which requires desired direction and velocity, and a repulsive effect which is the influence of a member on others or that provided by a boundary. In addition, a flow field, similar to the artificial potential field approach [8], is associated with each action that involves movement.

The social forces model has been covered extensively; we briefly present the relevant mathematical equations here. In this model, the movement of an entity is controlled by forces acting on the individual. The resultant acceleration is modeled by the Newtonian equation: $a = \frac{1}{mass} F, |a| < a_{max}$. The total force is the sum of the self force $F_s$, people force $F_p$, and wall force $F_w$.

The self-force models the attraction of an individual towards their desired goal and is modeled by the equation: $F_s = \frac{mass}{\tau}[v_d d - v(t)]$

Here, $v_d$ represents the desired velocity, d is the desired direction vector and v(t) is the actual velocity of the individual.

The people force represents the tendency of people to maintain a certain distance from each other, and is modeled by the equations:

$$F_p = \sum_{j=1, i\,!=j}^{P} f_{ij}$$

$$f_{ij} = \left[ A e^{-\frac{d_{ij}}{B}} + k_1 g(d_{ij}) \right] n_{ij} - k_2 g(d_{ij}) \Delta v_{ij}^{t} t_{ij}$$

The, $f_{ij}$ force is applied between crowd members i and j. The force has two components, a repulsion component and a friction component. The friction component only exists when the distance between the crowd members, $d_{ij}$ is below a threshold. The function g(i) returns 1 if i is positive and 0 otherwise and facilitates the inclusion of the friction component in the model. $\Delta v_{ij}$ is the difference in velocity between the two crowd members. Finally, the parameters A and B control the amount of force generated based on the distance between two members, while $k_1$ and $k_2$ are values that scale the forces.

The wall force represents the force that repels individuals to move away from walls and other physical obstacles. Its formulation is similar to the people force and is not repeated here for brevity.

The social forces model provides reasonable movement as far as the interactions between individuals and structures, but requires a desired destination in order to produce the *d* term in the self-force equation. The model itself does not contain any path finding or route selection. Furthermore, it is typical to maintain the model parameters fixed for the duration of the simulation. We have modified the social forces model to address two requirements: action specific navigation and dynamic changes to the movement parameters to reflect the emotional state of an entity.

The first modification involves using a flow field that provides the desired direction of movement based on the location of the entity. This desired direction is used in the self-force equation, thus guiding the entity around obstacles or through complex structures. See Figure 3 as an example of the flow field that would guide individuals to exit a room.

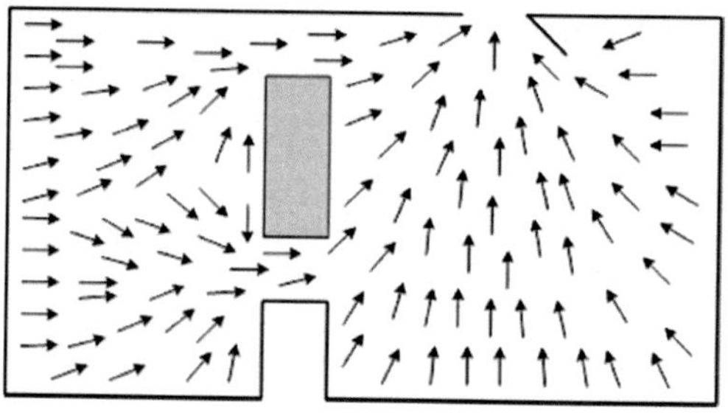

**Fig. 3. Use of flow fields for density velocity**

The second modification involves the dynamic change of the social forces parameters based on the emotional state of the individuals. Specifically, the people-force and wall-force equations contain four parameters each, $\{k_1, k_2, A, B\}$ that scale the forces and control the distance-based drop-off of the force magnitude. In our model, one set of such parameters is associated with each action, and multiple actions are used for locomotion, each reflecting different emotional states. For example, there can be four 'movement' actions: [Seek Exit Normal, Seek Exit Hurried, Seek Exit Fearful, Seek Exit Panicked]. Each of these actions utilizes the same flow field but has different parameter values; for example, the desired velocity for "Seek Exit Hurried" is higher than "Seek Exit Normal". Similarly, the force scale is smaller for "Seek Exit Fearful" than for "Seek Exit Hurried". As the emotional state of the crowd evolves during the simulation, crowd members are assigned actions that reflect their emotional state resulting in different movement.

## Example Application

The following scenario demonstrates how changes in the movement model parameters can affect the dynamics of pedestrian evacuation. Due to size limitations, we provide an example that focuses on the change in evacuation dynamics based on variation in the movement model; in an operational scenario, such changes would be dynamic and generated in response to the crowd emotional changes. The scenario involves pedestrian flow through a 3-meter wide corridor. A smooth entry way funnels pedestrian traffic into the corridor.

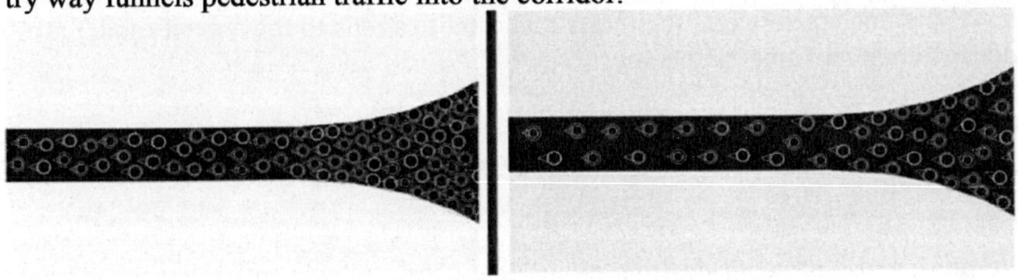

**Fig. 4. Pedestrian flow at B=0.4 and B=0.2.**

We are interested in the change in pedestrian flow as it depends on the variation of the model parameters. The model is run several times, with the people-force B parameter ranging between the values of 0.15 to 0.5. On one extreme of this range (0.15) the force diminishes rapidly and people will move in close proximity; on the other extreme of this range (0.5), the force balances the wall force created a singular column of travel. To help visualize the effect, Figure 4 depicts a snapshot of the simulation for a setting of B at 0.2, and 0.4. In both cases, travel is towards an exit located to the left of the corridor. As expected, the density and consequently the flow are significantly different, both at the funnel entry point as well as during steady state inside the corridor. The model can map a continuous value of B and represent varying emotional states driven by the cognitive model though emotional reflection.

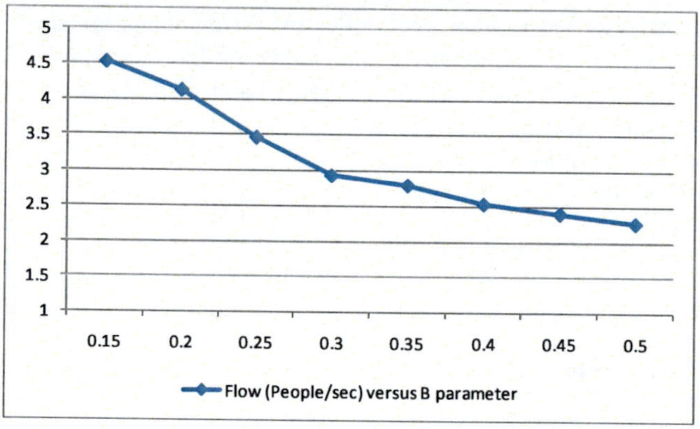


**Fig. 5. Pedestrian flow versus people-force B parameter**

Figure 5 provides a plot of the flow, represented in pedestrians per second, for the values of B used in this simulation. The curve asymptotically approaches a maximum flow which in this case would depend entirely on the programmed speed of each pedestrian. The sensitivity of flow to the just one of the movement model parameters supports the thesis that modifications to movement model provide different movement dynamics.

## Conclusion

We presented a crowd/pedestrian model that uses a movement and an emotional model in order to capture the effects of crowd psychology on movement dynamics. The formulation was described and results were provided to demonstrate the feasibility of linking the cognitive and physical models in order to provide an integrated model that can be used in a wide range of applications. The complexity of modeling people behavioral characteristics such as emotions demand not only a vast amount of input data to improve the accuracy of the models, but also a systematic comparison between experimental data and model predictions to provide more useful information to end users. This is part of our future research plan.

## References

1. Buckmann, L.T., Leather, J.: Modeling Station Congestion the PEDROUTE Way. In: Traffic Engineering and Control. Vol. 35, 373-377 (1994)
2. Klupfel, H., Meyer-Konig, T.: Characteristics of the PedGo Software for Crowd Movement and Egress Simulation. In: Galea, E.R. (Ed.), Proceedings of

the 2nd International Conference in Pedestrian and Evacuation Dynamics (PED), University of Greenwich, UK, pp. 331-340 (2003)

3. Bensilum, M., Purser, D.: GridFlow: An Object-oriented Building Evacuation Model Combining Pre-movement and Movement Behaviours for Performance-based Design. In: Evans, D. (ed.), 7th International Symposium on Fire Safety Science,Worcester Polytechnic Institute, MA, pp. 941-952 (2002)
4. Ito, T.A., Urland, G.R.: "The influence of processing objectives on the perception of faces: An ERP study of race and gender perception." Cognitive, Affective & Behavioral Neuroscience. 5, 21-36 (2005)
5. Juth, P., Lundqvist, D., Karlsson, A., Ohman, A.: "Looking for foes and friends: Perceptual and emotional factors when finding a face in the crowd." Emotion. 5, 379-395 (2005)
6. Eastwood, J.D., Smilek, D., Merikle, P.: Negative facial expression captures attention and disrupts performance. Perception and Psychophysics. 65, 352-358 (2003)
7. Helbing, D., Molnar, D.: Social Force Model for Pedestrian Dynamics. Physical Review E. 51, 4282-4286 (1995)
8. Khatib, O.: Real-time Obstacle Avoidance for Manipulators and Mobile Robots. In: Proceedings of the IEEE International Conference on Robotics and Automation, St. Louis, MO, 500 – 505 (1985)

# How Do People with Disabilities Consider Fire Safety and Evacuation Possibilities in Historical Buildings?

**S. Bengtson, L. Kecklund, E. Siré, K. Andrée and S. Willander**

Brandskyddslaget AB, MTO Säkerhet AB, Elena Siré Architect AB
Stockholm, Sweden
Corresponding author: staffan.bengtson@brandskyddslaget.se
lena.kecklund@mto.se
arch@elenasire.se

**Abstract** The purpose of the study was to explore how evacuation safety in historical buildings can be improved for people with various disabilities. Accounts of real experiences on how well evacuation routes in historical buildings are adapted to people with different types of impairments, as well as suggestions for safety enhancing measures, were collected by the use of focus group interviews. Some examples of problems reported when evacuating from historical buildings were, level differences on the way to and in evacuation routes, problems concerning orientation and problems with detecting the alarm signal. People had different needs depending on their type of impairment. This means that safety in evacuation routes must be given much more focus when improving accessibility in historical buildings. Measures must be taken to address the different needs of people with different types of impairments. The results can be used for the whole range of professions taking part in the project of planning refurbishments of historical buildings.

## Introduction

We, as members of society, differ in many ways. For example due to age differences or impairments, our ability to perceive and process information as well as our ability to move around varies. In the OECD countries the population in general is getting older. Also, people with disabilities can be expected to be more active in society in the future. A comparative study, carried out between 1982 and 2006 by The Swedish Arts Council, shows a tendency for elderly people to attend culture events more frequently [1]. The functional capabilities of elderly people decrease over time. It is also common that elderly people have several functional disabilities at once, for example, their eyesight has decreased, their hearing is impaired and they have difficulties finding their way. The needs of the elderly are

R.D. Peacock et al. (eds.), *Pedestrian and Evacuation Dynamics*,
DOI 10.1007/978-1-4419-9725-8_25, 

therefore often quite similar to the needs that younger disabled people have concerning information and accessibility with mobility aids. The number of people with activity limitations in Europe is around 10 – 20 % of the population [2]. This means that Sweden for example has approximately 1 – 1.5 million inhabitants suffering from various kinds of impairments.

Many countries have regulations to ensure accessibility to public buildings as well as to private homes and apartment buildings for people with different types of impairments and different user needs. Of course access to such buildings also requires the provision of safe evacuation routes which means that the design, technical functions of buildings as well as the organization of an evacuation must be adapted to the needs of different user groups.

Older buildings of significant cultural value represent a cultural heritage which is irreplaceable. When the demands of today necessitate changes to buildings it is of the utmost importance that these changes are carried out with great care. There are several problems related to existing constructions, combined with the cultural and historical value of the buildings, which makes it both difficult and sometimes even illegal to carry out required measures to improve evacuation safety.

A systematic compilation of how people with activity limitations experience fire safety, their risk perception regarding public environment and their practical experience of evacuation situation is missing today. To explore these issues a focus group study was carried out in Stockholm, Sweden in 2004. A brief summary of the results from this study has previously been published in *Safe evacuation for all – in cultural buildings* [3].

## Research Aim

The purpose of the study was to explore how emergency evacuation safety in historical buildings can be improved for people with activity limitations. The study was set out to investigate how well people with different kinds of impairments (mobility, visual and/or auditory) consider evacuation possibilities from historical buildings in general.

The study was designed to explore risk perception and previous experiences, as well as suggestions for measures, to improve evacuation safety in historical buildings for people with various disabilities.

The study was carried out with an explorative approach to increase understanding and knowledge amongst decision makers as well as several professional groups involved in the project and planning process when it comes to designing buildings so that people with impairments will be able to have and perceive a good evacuation safety level. People with activity limitations should not be exposed to a greater individual risk than a person without any impairment.

## Method

In order to gather as much information as possible on the subject a qualitative methodology was used. Accounts of real experiences on how well the design of historical buildings are adapted to people with different types of impairments, and thus the need for different safety measures, were collected by the use of focus group interviews. In addition suggestions for improvements were also collected.

A focus group study is a group discussion with the aim to gather people with different perspectives to discuss a given theme, in this study evacuation for people with activity limitations, with focus on evacuation from historical buildings. In a focus group study the subjects are chosen on the basis of the information they can contribute with on the matter, i.e. people with great experience and knowledge on the topic are selected. Through this approach it is possible to increase the knowledge on the theme but it is not possible to draw any conclusion on the matter that is valid for the entire population. [4]

Focus groups with people with different impairments were conducted in Stockholm, Sweden in 2004. The participants were recruited with help from a reference group, consisting of people from different disability organizations. They suggested other people who had different impairments, and had previously shown active interest in these matters, to participate in the study.

The data were collected through four focus group interviews. Each group consisted of 4-6 people with the same kind of impairment. There were four different impairments included in the study; mobility impairment – non-wheelchair users, mobility impairment – wheelchair users, visual impairment and auditory impairment. The group interviews were each lasting for approximately two hours and were led by an interview controller. Three of the interviews were audio recorded and was afterwards transcribed. The focus group interview, conducted with people with hearing impairment, was summarized through taking notes instead of using audio recording.

A requirement for the participants of the study was that he or she had to be a frequent visitor of historical buildings. The age range of the participants varied between 20-70 years. In total, the focus groups had 20 participants, nine males and eleven females participated. It was an even distribution of age and gender in each focus group.

After conducting the focus group interviews, each transcript was analysed using content analysis to be able to draw any conclusions from the data. Content analysis is used to break down interviews and take out major themes from them [5].

## Results

Several of the participants have had previous negative experiences from fire drills and evacuation situations. The bad experiences were mainly associated with insufficient organization, even if evident lacks in the built environment in the majority of the situations were the reason that evacuation of people with activity limitations was a problem in the first place. Since the environment is not adapted to the needs of people with impairments the exercise control tends to improvise during the drill, which according to the participants had not worked out in any of the drills they had taken part in.

It was highlighted in all focus groups that a person with activity limitations takes a greater risk when visiting a public building than what people without any impairments do. This matter was especially discussed in the group of people with visual impairment, where it was emphasized that a person with limited vision has to be brave and tough to visit public places. The participants also stressed that people with activity limitations are used to get into difficult situations and to handle them and hence in general are calmer than people without any impairment when it comes to handling tough and stressful situations. The participants described a couple of stressful situations where they had acted more calmly than their assistant or friend.

Several of the participants (especially the wheelchair users and people with visual impairment) expressed that they in great extent are dependent on help from others to evacuate from a historical building. The participants believed that people in general are very helpful when someone with an impairment is asking for help, but a concern regarding if people are as helpful in an emergency as they are in a normal situations was expressed. Some of the participants stressed that they were expecting to get help from their personal assistant* and that they felt safe with him or her. Others emphasized that they could only really rely on themselves in an emergency situation, why they found it better not to visit a building if the building was considered difficult.

In the two groups of people with mobility impairment the importance of getting valid information about a building's accessibility and emergency evacuation safety before a visit was emphasized. According to the participants this also applies to buildings in which it is considered hard to get around. In one of the groups, the experience was that historical buildings generally are difficult to move around in.

The participants' convictions were that staff in general has hard to understand difficulties related to people with different impairments and hence how the building should be designed to meet their needs. Several of the group members believed the best way to get appropriate information about a building's accessibility and emergency evacuation safety was instead to question someone who had already visited the building.

---

* In Sweden people with activity limitations are entitled to have a personal assistant.

The common opinion amongst the mobility impaired was that one can never trust that an assigned evacuation route is accessible before you have inspected the route yourself. Since it is very common that doors in evacuation routes in historical buildings in Stockholm have alarms, it is not possible to try the evacuation routes out in advance. Hence the participants preferred to use the same way to evacuate as they used when entering the building. The participants with visual impairment stated that they also normally try to use the same way to evacuate as they entered, since evacuation routes are considered uncomfortable to use, and that help from others is normally required to be able to evacuate.

## *Difficulties Related To the Built Environment*

People with different kinds of impairments have different problems and needs when it comes to the built environment; hence this section describes problems in the built environment by impairment.

### Mobility Impairment (Non-Wheelchair Users and Wheelchair Users)

Stairs in evacuation routes makes it hard for people with mobility impairment to evacuate, something that was stated in both focus groups. People with mobility impairment, who are not in need of a wheelchair, often suffer from problems with balance which makes it hard for them to descend stairs. Especially spiral staircases were considered problematic and a common fear amongst the participants of falling down the stair when using a spiral staircase was accentuated. To make it easier for people with limited mobility to descend stairs it is important that there are handrails to hold on to.

In both groups it was highlighted that it in general is considered hard and heavy to open doors, especially fire doors. The participants remarked that it is hard for them to grab the opening devices, if at all possible to reach. To manage a two-handed grip is almost impossible to them. The wheelchair users considered the best solutions for opening doors were to drive right on to them or to use an electrical opening device to open them.

### Visual Impairment

To people with visual impairment orientation is crucial. To improve the evacuation safety for visually impaired people it is important that the environment is designed to be easily read and understood. The participants pointed out that it is often easier to get orientated in old buildings, though these are generally more logically planned than newer ones. It was expressed that it is important to find a staircase in the event of an evacuation. A staircase symbolises an important part of the way to safety. The belief among the participants was that if you are able to find the stair it is often possible to get out of the building. One of the participants expressed that when you have found a staircase it is mainly about keeping count of how far you have to go before reaching the entrance level.

The group of visually impaired pointed out that spiral staircases is inappropriate for them to use, since it is hard to choose the right side (the wide side) of the stair when you can not see. It was emphasized that it is important for people with visual impairment to know which part of the staircase is wide and which part is narrow, because if they choose the wrong side it is possible to fall right down the staircase. To facilitate the understanding the staircase should be carried out with the handrail located at the wide side of the steps and each step should have the same height. It would make it easier for visually impaired people if a standardization of the design of staircases would be set.

The participants also described the difficulties they have with opening doors and windows in escape routes and manoeuvre opening devices. It could be hard to understand how to manoeuvre opening devices, especially when more than one grip is needed to open the door or when the door is carried out with more than one opening device.

### Auditory Impairment

The main problem for people with auditory impairment is to perceive the warning signal to get awareness of the fire or other cause of evacuation. Since people suffering from a hearing disability do not have any problems moving around in a building, no special requirement was mentioned regarding the built environment in this group.

## *Safety Improvements*

To be able to receive appropriate information about a building's accessibility and emergency evacuation safety before a visit, it was suggested that this information could be available on the building's home page. In the group of people with visual impairment, the desire to receive information about the building's emergency evacuation safety at arrival was highlighted. In one of the groups of people with mobility impairment a suggestion was presented that a safety folder could be given to people with activity limitations on arrival.

Level differences in the evacuation routes have to be compensated for to enable people with mobility impairment (especially wheelchair users) to evacuate on their own. The suggestion to establish fire protected elevators was brought up by the participants and was generally considered to be the best solution to level differences. The opinion of using an area of rescue assistance in case of emergency varied between the participants. One of the participants expressed that she would personally not be able to stay for an extended period of time in a separate fire compartment in the event of fire, while others were rather positive about the idea. Area of rescue assistance was a matter also discussed amongst the participants with visual impairment. The opinion about areas of rescue assistance was turned out to be more individually dependant than dependant on kind of impairment. To be carried out of a building was considered acceptable if it was a real emergency. It was emphasized though that different people have very different possibilities of being carried out. The group of wheelchair users expressed that it feels safer to be able to stay in the wheelchair at all times. If one has to leave the chair it means that the possibilities to be able to manage the situation on one's own are strongly reduced. It was also considered appropriate that the staff of a historical building is trained in lifting technique.

People with visual impairment are in need of getting help with orientation. Since it is very important to people with limited vision to find a staircase in the event of evacuation it was considered reasonable that it should be a requirement that the way to a staircase can be easily understood. Heading in the wrong direction during evacuation was a major concern to the group of visually impaired. It was suggested that the fire alarm could be supplemented with a spoken message informing about the way to an emergency exit or a stairwell. A beeping signal was not considered to be a good alternative to a spoken message since this signal was considered easy to misinterpret. It could both be perceived as information of the fire location or the evacuation route.

A common opinion was that it is important that the staff, regular as well as temporary, have adequate knowledge about the building's accessibility and evacuation safety to be able to give appropriate information to visitors and to assist people with activity limitations during evacuation. It was considered appropriate that staff working in a historical building should be educated and get the opportunity to practice evacuation of people with impairments. This will contribute to a sufficient knowledge of the building's emergency evacuation safety and how it is adapted to meet the needs of people with different impairments.

**Table 1. Overview of a comparison of the results**

| | Mobility impairment | Visual impairment | Auditory impairment |
|---|---|---|---|
| General problem | Movement towards an exit | Orientation – find the way to safety | Perception – getting awareness of the fire |
| Difficulties in the built environment | Level differences, especially in evacuation routes<br><br>High thresholds and obstacles in evacuation routes<br><br>Spiral staircases and no handrails<br><br>Hard to open doors and complicated if a two handed grip is needed | Small and high located evacuation signage<br><br>No contrasts in the built environment<br><br>Spiral staircases<br><br>Hard to find the opening device if there are more than one, and understand how to use it | Evacuation alarm without flashing lights and low-frequency sound |
| Safety improvements | Receive information about emergency evacuation safety before a visit<br><br>Receive a safety folder at arrival<br><br>Establish fire protected elevators<br><br>The opinion of using an area of rescue assistance in case of emergency varied between the participants<br><br>The staff should have adequate knowledge about the buildings' evacuation safety | Receive information about emergency evacuation safety at arrival<br><br>Evacuation alarm supplemented with a spoken message informing about the way to an emergency exit or a stairwell<br><br>The staff should have adequate knowledge about the buildings' evacuation safety | Evacuation alarm supplemented with flashing lights and a low-frequency sound<br><br>The staff should have adequate knowledge about the buildings' evacuation safety |

## Discussion

Some major groups of problems associated with each kind of impairment were elucidated during the focus group interviews. There are many features in the built environment which constitutes problems to people with activity limitations. There are also several possible technical measures to compensate for level differences and other difficulties. The problem is that many of these potential measures will, when carried out, have a great impact on the building and when considering historical buildings several of these might not be possible or even prohibited to perform due to the affect they might have on the building's cultural and/or historical value. Therefore it is important to develop and provide measures to improve evacuation safety that will not have an extensive affect on the building.

The study showed that the participants' negative experiences from previous fire drills were mainly associated with insufficient organization, even if evident lacks in the built environment in the majority of the situations were the reason that evacuation of people with activity limitations was a problem in the first place. The participants also stated that they normally choose to use the same way to evacuate as they used to enter the building, since they could never trust an assigned evacuation route they have not tried out.

Both of these examples demonstrate the importance of a well-working organisation to improve evacuation safety for people with activity limitations. Also measures should be taken to improve the organization of the evacuation. This is especially important in buildings where it is hard to carry out technical measures to meet the needs of people with various disabilities, often historical buildings. It is therefore important to consider the interaction between people - the user groups, the built environment and technical and organizational measures in order to improve emergency evacuation safety for different user groups. This means that professionals from different fields of expertise, e.g. architecture, behavioral science and fire safety engineering, should work together to create provisions for safe environments which considers an effective interaction between humans, technology and organization (MTO).

It is hard to receive the same emergency evacuation safety for people with activity limitations as for people not suffering from any impairments, consequently people with activity limitations expose themselves to greater risks when visiting public places. Hence it is important that people with various disabilities should be able to make well-founded choices regarding if they are willing to expose themselves to the risk or not. The participants' were of the opinion that it seems very important to the staff to be able to tell visitors that the building is accessible and they accentuated that it is important to make the staff think differently in these matters.

This study constitutes a pre-study of people with activity limitations' experiences, needs and wishes regarding evacuation safety in historical buildings. It is important to make great efforts to improve evacuation safety for people with activity limitations. There are several areas of research to be worked on before one could establish an appropriate level of adaptation of historical buildings. Some examples are offered below.

- To make a survey involving a larger population regarding people with activity limitations' needs and experiences of emergency evacuation safety in public buildings.

- Develop methods to include safety improvements for these groups in the design and construction process and to increase knowledge of what this involve is a part of the designing process
- Evaluate new measures in a real user environment

## References

1. Swedish Arts Council: New Cultural Habits, Cultural habits in Sweden in a 30 years perspective 1976-2006 (2008)
2. Boström, G.: Health on equal terms. Swedish National Institute of Public Health, 2008:17 (2008) (title translated from Swedish)
3. Siré, E. et al.: Safe Evacuation for all – in cultural buildings. AB Svensk byggtjänst and the authors (2006)
4. Kreuger, R.A.: Developing Questions for Focus Groups. Sage Publications, USA (1998)
5. Breakwell, G.M.: Interviewing. In: Breakwell, G.M., Hammond, S., Fife-Sachaw, C. (eds.), Research methods in Psychology, pp. 230-243. Sage, London, UK (1995)

# General Model Development

# Risk Minimizing Evacuation Strategies under Uncertainty

**G. Lämmel[1] and H. Klüpfel[2] and K. Nagel[1]**

[1]Technische Universität, ILS, 10623 Berlin, Germany
[2]TraffGo HT GmbH, Bismarckstraße 142a, 47057 Duisburg, Germany

**Abstract** This paper presents results on the simulation of the evacuation of the city of Padang with approximately 1,000,000 inhabitants. The model used is MATSim (www.matsim.org). Three different strategies were applied: shortest path solution, user optimum, system optimum, together with a constraint that moves should reduce risk whenever possible. The introduction of the risk minimization increases the overall required safe egress time (RSET). The differences between the RSET for the three risk minimizing strategies are small. Further quantities used for the assessment of the evacuation are the formation of congestion and the individual RSETs (in comparison with the available SET).

## Introduction: Safety, Risk, and the Need for Simulation

Safety is a basic need for individuals and societies. Safety can be roughly defined by: existing risk < acceptable risk. It can also be discriminated from security by dealing with non-intentional threats. In this paper, the potential threat is a natural hazard: a submarine earthquake in the Indian Ocean causing a Tsunami wave hitting the coast of Sumatra, Indonesia and the city of Padang. The risk, and consequently also the safety if the acceptable risk is specified can be quantified based on the following formula:

$$R = \int D \cdot (1 - C) \cdot P(t) dt \qquad (1)$$

The damage is denoted by $D$, the coping capability by $C$, and $P(t)$ is the probability of the wave reaching the coast. The criterion usually applied to assess a risk is: $R <$ acceptable risk. Please note that there is always a residual risk ($RR>0$), which cannot be reduced by technical or management means. In case of a tsunami, the physical safety or lives of people are at risk. Evacuation is one means in ensuring the safety, especially to avoid the risk and threat to human life. Evacuation reduces the damage. Another strategy would be to build tsunami safe buildings which would increase $C$. This is beyond the scope of this paper. We focus on the evacuation.

The condition for a safe egress is RSET < ASET, where ASET is the available safe egress time and RSET is the required safe egress time. In this paper, we

R.D. Peacock et al. (eds.), *Pedestrian and Evacuation Dynamics*,
DOI 10.1007/978-1-4419-9725-8_26, 

present the calculation of RSET (based on a microscopic multi-agent simulation). ASET is provided by inundation simulations that show the consequences of an earthquake off-shore the island of Sumatra (Indonesia) for the coastal city of Padang. The overall egress time is one major criterion for assessing an evacuation plan. Such a plan addresses – among many other issues – evacuation routes for the endangered population. There are many models that find optimal routing strategies (i.e. minimizing RSET) for a given road and walkway network. In the case of large-scale inundation, the network changes with time. Links or edges (i.e. roads or lanes) become impassable due to flooding. The evacuation simulation based on a dynamic network works only as long the advance warning time is known beforehand, though. When this is not the case, the optimal routing strategy might increase the risk for some persons on some stretch of way. This issue is addressed in the next section on utilities of evacuation strategies. Implementation details are given in section 3, experimental results discussed in section 4. The paper concludes with a discussion of the simulation results (section 5) and a conclusion and recommendations (section 6).

## Utility of an Evacuation Strategy

The utility of an evacuation path often depends on uncertain aspects. One uncertain aspect is the advance warning time $\tau_{warn}$. We assume that $\tau_{warn}$ follows an unknown probability distribution with, for this section, $P(\tau_{warn} > 0) = 1$, i.e. there is always a warning *before* the event. Let us consider a situation with two different evacuation paths $p_0$ and $p_1$; $p_0$ does not depend on the advance warning time $\tau_{warn}$ but has considerably longer travel time than $p_1$. The path $p_1$ first leads "towards danger" for a time period $T$ before it leads to safety, i.e. when the warning time is too short one cannot take it. An example is a bridge close to the shore heading to a safe area. If an evacuee takes $p_1$, she moves towards the shore (danger) in order to reach the bridge. As a result the utility of $p_1$ depends on $\tau_{warn}$. The utility for $p_0$ and $p_1$ can be formulated as follows:

$$U(p_0 \mid \tau_{warn}) = \begin{cases} -\infty & \text{if} \quad \tau_{warn} \le 0 \\ -t_{travel}(p_0), & \text{otherwise} \end{cases} \tag{2}$$

$$U(p_1 \mid \tau_{warn}) = \begin{cases} -\infty & \text{if} \quad \tau_{warn} \le T \\ -t_{travel}(p_1), & \text{otherwise} \end{cases} \tag{3}$$

where $t_{travel}(p_1)$ denotes the travel time for path $p_1$. Taking the information of the probability distribution for $\tau_{warn}$, we can calculate the expectation value for each utility:

$$E(U(p_0 \mid \tau_{warn})) = E(U(p_0)) = t_{travel}(p_0) \tag{4}$$

$$\begin{aligned} E(U(p_1 \mid \tau_{warn})) = \; & P(\tau_{warn} < T) \cdot (-\infty) \\ & + (1 - P(\tau_{warn} < T)) \cdot t_{travel}(p_1) = -\infty \end{aligned} \tag{5}$$

Based on these expectation values, risky evacuation paths $p_1$ are banned in the remainder of this paper, as long as a non risky solution $p_0$ exists. If no risk-free path exists, then the solution with the lowest risk should be chosen. Implementation details are given in the following section.

## Routing strategies

In this section we discuss three different routing strategies. The most straightforward approach to an evacuation problem is the shortest path solution, where every evacuee takes the shortest path to safety. The Dijkstra shortest path algorithm [1] finds the shortest path in a weighted graph from one node to any other. The weights for a link are defined by a time- and/or distance-dependent cost function. The algorithm relies on the information about the free-flow travel time $\tau_a$ for every link $a$. Algorithm 1 shows the shortest-path routing logic.

The shortest path solution does not take congestion into consideration, though. In reality, the link travel time depends on the level of congestion. In the underlying traffic flow simulation every link has a specific flow capacity; if this capacity is exceeded, congestion occurs and increases the link travel time. Since the demand on a link is not constant over time the link travel time is time dependent. There are different optimization approaches to find better solutions than the shortest path solution. In this paper we discuss the Nash equilibrium (NE; = user optimum) and the system optimum (SO).

**Algorithm 1.** Shortest path routing

1. initialize $\tau_a$ with the free-flow travel time for all links
2. calculate routes based on link costs $C_a = \tau_a$

The NE is named after John Forbes Nash and describes a state in a competitive two or more player game where no player can gain by unilateral deviation from her current strategy [2]. In the evacuation context the NE describes a state where no evacuee can improve her evacuation performance by unilateral deviation from her current evacuation route (user optimum). In most (but not all) evacuation situations, the NE leads to a shorter overall evacuation time than the shortest path solution. In the NE nobody has an incentive to change his path. It is therefore a solution that can be reached by appropriate training. In multi-agent simulations the solution can move towards the NE through iterative learning [3, 4]. An iterative learning algorithm starts with a given starting solution and tries to improve it through trial and error. Learning means re-planning agents paths. The learning algorithm uses a cost function based on travel times. Formally, the real-valued time is divided into $K$ segments ("bins") of length $T$, which are indexed by $k$=0,...$K$-1. The time-dependent link travel time when entering link $a$ in time step $k$ is denoted by $\tau_a(k)$. Implementation details are given in [10]. Algorithm 2 shows the Nash-equilibrium routing logic.

**Algorithm 2.** Nash equilibrium routing

1. initialize $\tau_a(k)$ = free-flow travel time for all links $a$ and time steps $k$
2. repeat for many iterations:
   (a) recalculate routes based on link costs $C_a(k) = \tau_a(k)$
   (b) load vehicles on network, obtain new $\tau_a(k)$or all $a$ and $k$

The SO can be achieved by applying a similar learning algorithm as for the NE approach. The only difference is that for a SO, the travel time based on which agents evaluate their routes needs to be replaced by the marginal travel time [5]. The marginal travel time of a route is the amount by which the total system travel time changes if one additional evacuee takes that route. It is the sum of the cost experienced by the added evacuee and the cost imposed on other evacuees. The latter is denoted here as the "social cost" ($C^S$). Implementation details for the approximated system optimum (SO) in the evacuation context are discussed in [6]. An application of this result to a system optimal route assignment requires to calculate $C_a^s(k)$ for every link $a$ and entry "time bin" $k$ in the network and to add this term to the time-dependent link travel time that is evaluated in the route re-planning of every agent. Algorithm 3 outlines the arguably most straightforward implementation of this approach.

**Algorithm 3.** System optimum approach

1. initialize $C_a^s(k) = 0$ and $\tau_a(k)$ with the free-flow travel time for all links $a$ and time steps $k$
2. repeat for many iterations:
   (a) recalculate routes based on link costs $C_a(k) = \tau_a(k) + C_a^s(k)$
   (b) load vehicles on network, obtain new $\tau_a(k)$and $C_a^s(k)$ for all $a$ and $k$

## Risk costs

In this section we propose a strategy that allows only risk-decreasing moves, as long as such moves exist. This approach is similar to the system of priority levels proposed by Hamacher and Tjandra [7]. A move is defined as risk-decreasing if it increases the evacuee's distance to the danger. Inside the endangered area the distance describes the temporal distance. For inundation scenarios this means that the evacuee's position before the move will be flooded earlier than the position after the move. But even people outside the area directly affected should keep some distance to the danger. This is important because otherwise those people could block evacuees from leaving the endangered area. Therefore we propose an additional buffer around the endangered area that also has to be evacuated. Within this buffer a move is defined as risk-decreasing if it increases the evacuee's spatial distance to the danger. In general, some evacuation paths might always be risk decreasing others not. In our simulation, the only decision points are at nodes. As soon as an evacuee has entered a particular link she has to travel along that link until the next node. Therefore we calculate risk levels for nodes. If a link leads from a node with lower risk to a node with higher risk than that link will be charged an additional penalty. This is achieved by adding a risk cost term $C^r$ to the cost function $C$. The cost terms for algorithms 1, 2, 3 are thus extended by the static risk cost $C^r$.

The cost term for the shortest path routing and the NE approach is now:

$$C_a^k = \tau_a(k) + C_a^r \tag{6}$$

for the NE; for the system optimal approach it is

$$C_a(k) = \tau_a(k) + C_a^s(k) + C_a^r . \tag{7}$$

The risk cost for link $a$ connecting nodes $(i, j)$ with risk levels $r_i$ and $r_j$:

$$C_a^r = \begin{cases} l_a \cdot penalty & \text{if} \quad r_i < r_j \\ 0, & \text{otherwise} \end{cases} \tag{8}$$

where $l_a$ is the length of link $a$ and $penalty$ is a constant that has to be chosen so that the cheapest risk increasing path is more expensive then than the most expensive risk decreasing path. In the underlying scenario, $penalty$ has been set based on a heuristic estimate to 30 hours per 100 meters. We conducted experiments for each of the three routing strategies (shortest path, user optimum, system optimum) with risk avoidance and compared them to the NE (user optimum) approach without additional risk costs. The results of the experiments are given in the following section.

## Simulation results

The risk minimizing routing strategy has been investigated through the application to a real world evacuation scenario namely the evacuation of the city of Padang in the case of a tsunami warning. Padang is located at the West Coast of Sumatra Island and is exposed to earth quakes triggered tsunamis (see, e.g. [8]). The city has more then 800 000 inhabitants, where several hundred thousand are living in the endangered area. The geographical data, socio-economic profile and expected inundation scenarios for the city have been discussed in many of our previous publications (see, e.g. [9]). The simulations have been performed in the MATSim simulation framework. The MATSim simulation frame work and its adaptation to pedestrian evacuation simulation have also been discussed broadly (see, e.g. [4,10]). We conducted four runs to investigate the risk minimizing strategy. *Run 1* implements the NE approach (w/o risk costs), *Run 2* implements the risk minimizing shortest path solution, *Run 3* the risk minimizing NE approach and *Run 4* the risk minimizing SO approach. The synthetic population is the same for all runs. It consist of 277 299 agents. The number of agents and the initial distribution corresponds to the real population of Padang.

The overall run-time for *Run 3* was 12 hours and 21 minutes. For *Run 4* the overall run-time was 26 hours and 4 minutes. This demonstrates that the pedestrian flow model can deal with large-scale scenarios. Some visualizer screen shots of the first 30 minutes are shown in fig. 1. The agents are colored according to their evacuation time: green indicates fast, red slow escape. It is shown that *Run 1* (left column in the figure, reference case) performs best. In the visualizer snapshots, no major differences between the three risk minimizing runs (2 to 4) can be identified. In *Run 2* there are many red colored agents in the northern part of the city, indicating a longer evacuation time. *Run 3* and *Run 4* are almost identical. Based on the screen shots alone, no advantage of the risk minimizing approach (compared to the reference case) can be identified. A detailed examination of the results shows the advantage of the risk minimizing approach, though.

In fig. 2 (right) there are two visualizer screen shots of the Siti Nurbaya Bridge each taken after 5 minutes The left part shows *Run 1* (reference case) and at the right part *Run 3* (risk minimization). In *Run 1* agents cross the bridge towards the mountains in the south. This strategy would be a good strategy if the advance warning time was known to be long enough; otherwise, if the wave arrived earlier than expected, this strategy would be disastrous. In contrast in *Run 3* the agents avoid the bridge and move away from the river (and from the danger).

As a consequence of risk minimization, many agents in the northern part of the city do not have enough time to evacuate (RSET < ASET). This fact is also reflected in the evacuation curves. Fig. 3 (left) shows the evacuation curves for the four runs discussed The simulation results show that the risk minimization does in our scenario comprise agents which do not have enough time for evacuation.

**Fig. 1.** From left to right: Nash equilibrium without risk, risk minimization (rm) for shortest path, Nash (user optimum), and approximated system optimum. Agents with RSET > ASET are shown in red. For all three risk minimizing strategies (column 2 to 4), namely shortest path rm, Nash rm, and approximated system optimum rm, the results are similar. The time is (from top to

bottom): 1 minute, 15 minutes, and 30 minutes after the alarm. Please note that the warning time is the time between the alarm and the wave reaching the coast.

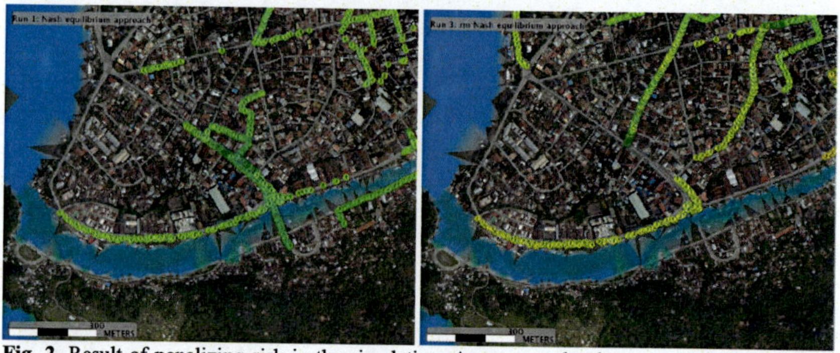


**Fig. 2.** Result of penalizing risk in the simulation: Agents use the dangerous bridge in the left case, but avoid it if its usage is more costly. In the right case, crossing the bridge is costly ($r_j > r_i$ in eq. (8), i.e. node *j* will be flooded earlier than node *i*).

Note that the evacuation curves for *Run 3* and *Run 4* almost coincide. Therefore, the risk costs are an additional constraint pushing NE and SO solution towards each other. There are still fundamental differences between both approaches, however. These differences are indicated by the area where ASET<RSET (available time<required time), e.g. the area in the north.

Therefore, this area is highly endangered and an alternative strategy is required. A first step is the detailed analysis of this area. Therefore a GIS based analysis of the number of endangered agents has been performed (fig. 3, right). This analysis demonstrates the differences between *Run 3* (NE) and *Run 4* (SO). Even if there are no big differences in the spatial distribution of the endangered agents for *Run 3* and *Run 4*, the figure shows one important aspect of the social cost optimization: The number of endangered agents further inland is higher in *Run 4* than in *Run 3* and vice versa at the coastline. The agents starting near the coast gain through the social cost optimization and the agents further inland lose. The latter make space for others in order to reduce their own social costs. Such behavior reduces the average evacuation time and also increases the number of rescued agents. The price, however, is the organized "sacrifice" of some. In reality, this is does not seem to be an option and therefore we argue in support of the NE approach.

## Summary and Discussion

We have presented simulation results for the coastal city of Padang. The results are based on the MATSim multi-agent simulation framework (www.matsim.org) adapted for evacuation simulation. The introduction of risk costs (eq. (8)) increases the overall evacuation time (RSET) considerably. This result holds irrespective of the routing strategy. Three different routing strategies have been investigated: shortest path, Nash equilibrium (NE; = user optimum), and system optimum (SO). The SO is realized by introducing social costs on each link; those social costs do not represent an intrinsic motivation but have to be enforced externally, i.e. they depend not only on the agent's own travel time but also on the travel time of others. The NE approach leads to a shorter overall evacuation time than the shortest path strategy. In the simulation, the NE is approximated by iterative simulation runs with re-planning based on the results of the previous run.

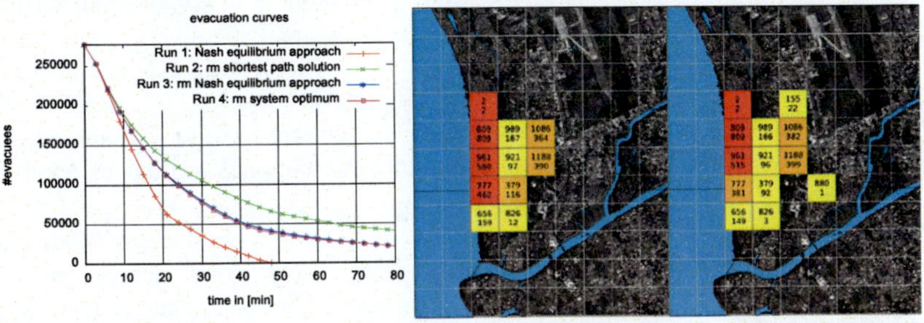


**Fig. 3.** Left: Evacuation curves of the four runs. Right: GIS analysis of the highly endangered area on a 300 meter grid. From left to right: Run 3 (rm for NE) and Run 4 (rm for SO). The colors of the squares describe the percentage of agents for whom ASET<RSET (red: ≥50%, orange: ≥25% and yellow: <25%). The numbers in the squares denote the total number of agents departing from that square (top) and the total number of agents with ASET<RSET.

## Conclusion and Recommendations

In the case of an evacuation, re-planning as described in the previous section does not take place. An evacuation is – other than commuter traffic – a singular event. This distinguishes evacuation simulations from traffic simulations. On the other hand, the results of the evacuation simulation are intended to provide a basis for an evacuation plan which has to be implemented by local authorities and groups. And for a recommendation, a stable and acceptable solution is required. Such a solution, in terms of evacuation paths, is provided by the NE approach: no single agent can gain by unilateral deviation and there is no gain based on the loss of someone else. Of course, the evacuation paths have still to be communicated to the population, e.g. by signage, the authorities, police, and volunteer groups. One of those groups is KOGAMI (kogami.multiply.com), which provides training for the local population to prepare for Tsunamis. The recommendations of KOGAMI turn out to be most similar to the NE solution for the risk minimization case. One prominent example to illustrate this is the bridge shown in fig. 2. In one sentence, the strategy derived from the simulations could be summarized as “avoid bridges and keep away from the water”. But the simulation allows more detailed analysis, including how much one loses through the “risk minimization” approach, or the possibility to evaluate changes to the scenario such as capacity expansions, shelters, or changes to the whereabouts of people when the evacuation starts.

**Acknowledgements** This project was funded in part by the German Ministry for Education and Research (BMBF), under grants numbers 03G0666E ("last mile") and 03NAPI4 ("Advest").

## References

1 Dijkstra, E.: A note on two problems in connection with graphs. Numerische Mathematik. 1:269–271 (1959)

2 Nash, J.: Non-cooperative games. Annals of Mathematics. 54(2):286–295 (1951)

3 Gawron, C.: An iterative algorithm to determine the dynamic user equilibrium in a traffic simulation model. International Journal of Modern Physics C. 9(3):393–407 (1998)

4 Lämmel, G., Klüpfel, H., Nagel, K.: The MATSim Network Flow Model for Traffic Simulation Adapted to Large-Scale Emergency Egress and an Application to the Evacuation of the Indonesian City of Padang in Case of a Tsunami Warning, Chapter 11, pages 245–265. Emerald Group Publishing Limited (2009)

5 Peeta, S., Mahmassani, H.S.: System optimal and user equilibrium time-dependent traffic assignment in congested networks. Annals of Operations Research. 60:81–113 (1995)

6 Lämmel, G., Flötteröd, G.: Towards system optimum: Finding optimal routing strategies in time-dependent networks for large-scale evacuation problems. In: Mertsching, B., Hund, M., Aziz, Z. (eds.), LNCS (LNAI), 5903, pages 532–539. Springer, Berlin (2009)

7 Hamacher, H.W., Tjandra, S.A.: Mathematical modeling of evacuation problems: A state of art. Technical report. Fraunhofer ITWM (2001)

8 Natawidjaja, D.H., Sieh, K., Chlieh, M., Galetzka, J., Suwargadi, B.W., Cheng, H., Edwards, R.L., Avouac, J.P., Ward, S.N.: Source parameters of the great Sumatran megathrust earthquakes of 1797 and 1833 inferred from coral microatolls. Journal of Geophysical Research. 111 (2006)

9 Taubenböck, H., Goseberg, N., Setiadi, N., Lämmel, G., Moder, F., Oczipka, M., Klüpfel, H., Wahl, R., Schlurmann, T., Strunz, G., Birkmann, J., Nagel, K., Siegert, F., Lehmann, F., Dech, S., Gress, A., Klein, R.: "Last-Mile" preparation for a potential disaster – Interdisciplinary approach towards tsunami early warning and an evacuation information system for the coastal city of Padang, Indonesia. Natural Hazards and Earth System Science. 9(4):1509– 1528 (2009)

10 Lämmel, G., Grether, D., Nagel, K.: The representation and implementation of time-dependent inundation in large-scale microscopic evacuation simulations. Transportation Research Part C: Emerging Technologies. 18(1):84–98 (February 2010)

# Methods for Improving Efficiency of Queuing Systems

**D. Yanagisawa[1,2], Y. Suma[1], Y. Tanaka[1], A. Tomoeda[3,4], K. Ohtsuka[4], and K. Nishinari[4,5,1]**

[1]Department of Aeronautics and Astronautics, School of Engineering, The University of Tokyo, Hongo, Bunkyo-ku, Tokyo, 113-8656, Japan.
[2]Research Fellow of the Japan Society for the Promotion of Science, Kojimachi, Chiyoda-ku, Tokyo 102-8471, Japan.
[3]Meiji Institute for Advanced Study of Mathematical Sciences, Meiji University, Tama-ku, Kawasaki-city, Kanagawa, 214-8571, Japan.
[4]Research Center for Advanced Science and Technology, The University of Tokyo, Komaba, Meguro-ku, Tokyo, 153-8904, Japan.
[5]PRESTO, Japan Science and Technology Agency, Sanbanmachi, Chiyoda-ku, Tokyo 102-0075, Japan.

Corresponding Author: tt087068@mail.ecc.u-tokyo.ac.jp

**Abstract** We have considered the methods for improving efficiency in queuing systems by theoretical analysis and experiments. First, a queuing system which has plural service windows is studied. There are mainly two kinds of systems which are a parallel-type queuing system and a fork-type queuing system. Queuing theory is often used to analyze these queuing systems; however, it does not include the effect of walking distance from the head of the queue to service windows; thus, a walking-distance introduced queuing theory is investigated. By using this model, we have discovered that the suitable type of system changes according to the utilization of the system. We have also verified that when we keep one person waiting at each service window in the fork-type queuing system, the waiting time dramatically decreases. Secondly, we consider queuing systems in amusement parks. Plural people waiting in the queue move to get on a roller coaster at the same time; therefore, the efficiency of the system is improved by shortening the moving time. The result of the experiments indicates that the moving time decreases if people keep walking in the queue to start instantaneously.

## Introduction

Pedestrian dynamics has been studied vigorously over last decades and many successful models have been developed [1,2]. There are two major topics focused on by many researches in pedestrian dynamics. One is one dimensional flow [3],

R.D. Peacock et al. (eds.), *Pedestrian and Evacuation Dynamics*,
DOI 10.1007/978-1-4419-9725-8_27, © Springer Science+Business Media, LLC 2011

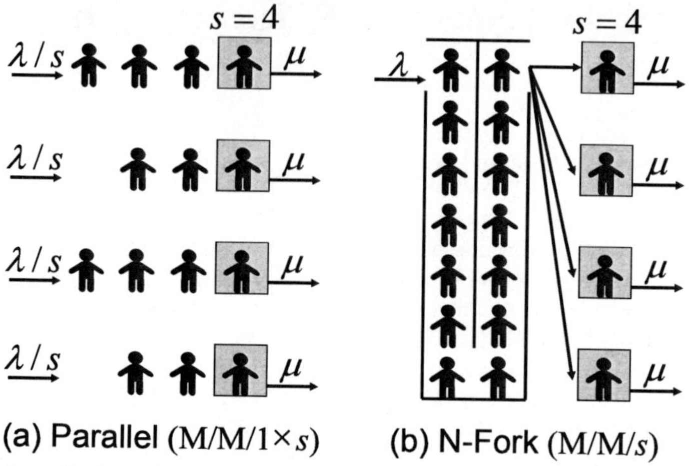


**Fig. 1. Schematic views of queuing systems in the case that the number of service windows $s = 4$. (a) Parallel (M/M/1 × 4). (b) N-Fork (M/M/4), which is an ideal fork-type queuing system. A person at the head of the queue moves to the service windows instantaneously when one of them becomes vacant.**

which is very important to investigate the basic features of pedestrian flow, and the other is an evacuation [4-6], which is an essential theme since it is strongly connected with people's life in an emergency situation.

A queue is one of the typical phenomenon, which is observed everywhere in the cities, for instance, at a supermarket, a bank, a concert hall, and so on. Thus, research on a queuing system is one of an important topic in pedestrian dynamics; however, there are not as many researches on it as those of one dimensional flow and evacuation since queuing dynamics is usually studied by using queuing theory [7].

Although the queuing theory is very successful and has been used to study queuing systems, it does not consider the effect of walking distance in the system, which is a very important factor to study pedestrian queuing systems. Thus, we combine the queuing theory and the cellular automaton to obtain a realistic model which includes the effect of walking distance. Two kinds of advanced queuing systems are studied in this paper. In the following sections, the recipes for improving efficiency of queuing systems with plural service windows (in Sec. 2) and in amusement parks (in Sec. 3) are proposed.

## Queuing System with Plural Service Windows

There are mainly two kinds of queuing systems with plural service windows, which are a *parallel-type* queuing system (Parallel), i.e., queues for each service

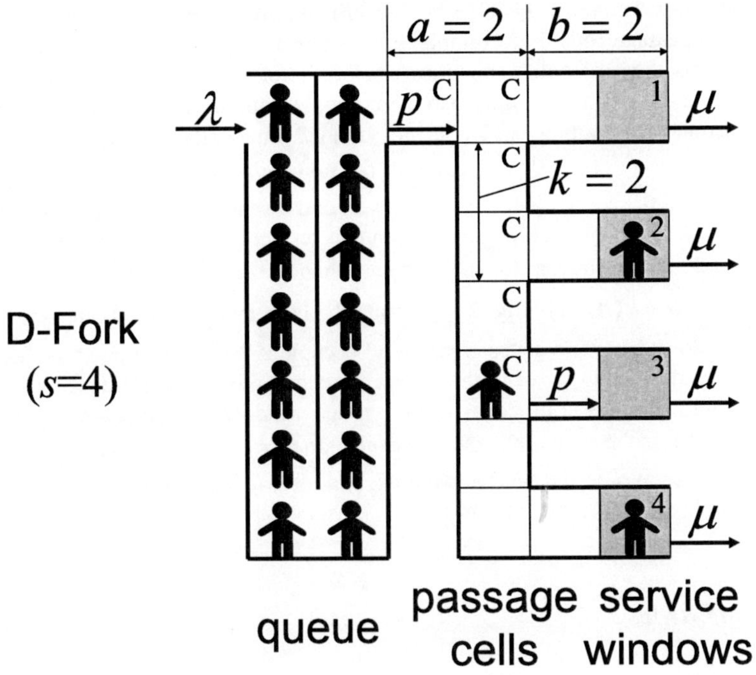


**Fig. 2. Schematic view of D-Fork ($s = 4$)**

windows, and a *fork-type* queuing system (Fork), which collects people into a single queue. According to the queuing theory, the waiting time in Fork is always shorter than that in Parallel (Fig. 1 (a)). However, Fork considered in the normal queuing theory (N-Fork) (Fig. 1 (b)) does not reflect the effect of the walking distances from the head of the queue to the service windows. The effect of the distance may be possible to neglect in a small queuing system; however, it should not be ignored in a large one such as an immigration inspection floor in an international airport. Thus, we have introduced the effect of walking-distance to the queuing theory as follows.

## *Distance introduced Fork-type Queuing System: D-Fork*

In Parallel, people wait just behind the former person, so that there is no delay in walking. While, in N-Fork, people take some time to walk from the head of the queue to the service windows; therefore, we consider D-Fork as in Fig. 2 by representing the walking distance using cellular automaton. The gray cells are window cells, and the numbers described in them are window numbers. The white cells are passage cells. Note that the letter "C" described in the passage cells represents the common passage cells. For example, both persons who are going to the window 3 and 4 pass them. People sometimes cannot go forward in the common passage cells since there is a possibility that other people stand in front of them. The place that people are waiting, which is not divided into cells, is a queue. $s \in \mathbf{N}$, $\lambda \in [0,\infty)$, and $\mu \in [0,\infty)$ represent the number of service windows, the arrival rate, and the service rate, respectively. *a* and *b* represent the length of the passage, and *k* is the interval length between two service windows. The distance from the head of the queue to the service window $n \in [1,s]$, is described as $d_n = a + b + k(n-1)$. Fig. 2 represents the case *s*=4, *a*=2, *b*=2, and *k*=2. Service windows have two states: vacant and occupied. When a person at the head of the queue decides to move to the vacant service window *n*, it changes into occupied state. The person proceeds to the service window by one cell with the rate $p \in [0,\infty)$ as the asymmetric simple exclusion process [8]. A service starts when the person arrives at the service window, and after it finishes the state of the service window changes into vacant state.

## *Update Rules*

The simulation of D-Fork consists of the following five steps per unit time step.

1. If there is at least one vacant service window and one person in the queue, and the first cell of the passage is vacant, then the person decide to proceed to a vacant service window which is the nearest to the head of the queue, and the state of the service window becomes occupied.
2. Add one person to the queue with the probability $\lambda\Delta t$, where $\Delta t$ is the length of the unit time step.
3. Proceed each person in the passage cells to his/her service windows with the probability $p\Delta t$ if there is not other person at their proceeding cell.
4. Remove people at the service windows and change their states into vacant state with the probability $\mu\Delta t$.
5. If 1. takes into practice, proceed the person at the head of the queue to the first passage cell with the probability $p\Delta t$.

### *Stationary Equations*

We define the sum of the walking time and the service time at service window $n$ as a throughput time $\tau_n$ and its reciprocal as a throughput rate $\mu_n$. Here, we calculate the mean throughput rate $\hat{\mu}_n$ when $n$ service windows are occupied, and obtain stationary equations of D-Fork. We suppose that all passage cells are vacant by mean field approximation. Then, the mean value of the throughput time $E(\tau_n)$ is described as follows.

$$E(\tau_n) = \frac{1}{\mu} + \frac{a+b+k(n-1)}{p}. \tag{1}$$

The throughput rate $\mu_n$ is obtained as

$$\mu_n = \frac{1}{E(\tau_n)} = \frac{1}{\frac{1}{\mu} + \frac{a+b}{p} + \frac{k(n-1)}{p}} = \frac{\mu}{1+\alpha+2\beta(n-1)}, \tag{2}$$

where

$$\alpha = \frac{\mu(a+b)}{p}, \qquad \beta = \frac{k\mu}{2p}. \tag{3}$$

In the case $2\beta(n-1)/(1+\alpha) \ll 1$, we calculate the mean throughput rate $\hat{\mu}_n$ as

$$\hat{\mu}_n = \frac{1}{n}\sum_{l}^{n} \mu_l \approx \frac{\mu}{1+\alpha+\beta(n-1)}. \tag{4}$$

By using (4) the stationary equations are described as follows:

$$\begin{aligned} &\lambda P_0 = \hat{\mu}_1 P_1 \\ &\lambda P_{n-1} + (n+1)\hat{\mu}_{n+1} P_{n+1} = (\lambda + n\hat{\mu}_n) P_n \qquad (1 \le n \le s-1) \\ &\lambda P_{n-1} + s\hat{\mu}_s P_{n+1} = (\lambda + s\hat{\mu}_s) P_n \qquad (n \ge s). \end{aligned} \tag{5}$$

We obtain the mean waiting time $W_q$ by solving (5) analytically. In the case $\alpha=\beta=0$, we have the stationary equations of M/M/s [7] from (5), thus $\alpha$ and $\beta$ represent the effect of walking time.

In our simulation the distribution of the throughput time is gamma distribution. We approximate it as exponential distribution in this calculation, however, when $\beta$

is small the results from the exponential distribution approximated well to those from gamma distribution.

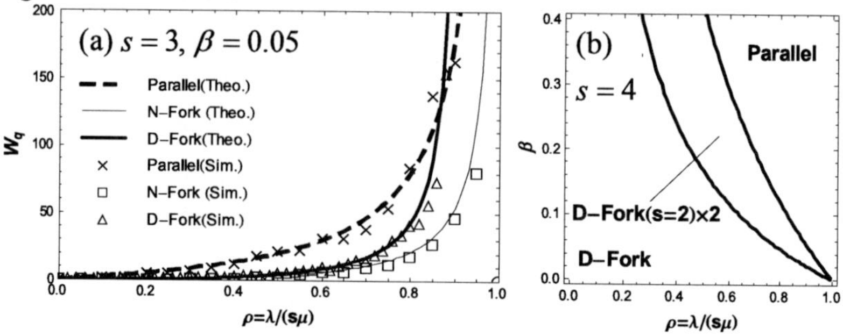


**Fig. 3. (a) Comparison of $W_q$ of Parallel, N-Fork, and D-Fork in the case $\alpha$=0, $\beta$=0.05, $\mu$=0.05. (b) Queuing system which makes the mean waiting time $W_q$ minimum in the case $s$=4.**

## *Comparison between Parallel Queue and Fork Queue*

We compare the mean waiting time $W_q$ of Parallel (Fig. 1 (a)), N-Fork (Fig. 1 (b)), and D-Fork (Fig. 2 (a)). Figure 3 (a) shows $W_q$ against the utilization $\rho = \lambda/(s\mu)$. The results of analysis agree with those of the simulation very well. We see that $W_q$ of N-Fork is smaller than that of Parallel and D-Fork in $0 \leq \rho < 1$. There is a possibility that more than one person is waiting in one queue and no one is in the other queue in Parallel $(s \geq 2)$, however there is no vacant service window in N-Fork when people are waiting in the system. This is the reason why $W_q$ of N-Fork is always smaller than that of Parallel. Since N-Fork does not take into account of the effect of the walking distances, i.e., $\alpha = \beta = 0$, it is obvious that $W_q$ of N-Fork is smaller than that of D-Fork. The N-Fork is the most efficient of the three; however, it is an ideal system and does not exist in reality. By focusing on the curves of Parallel and D-Fork, we can clearly observe the crossing of them. This means that when the utilization $\rho$ is small, i.e., there are not sufficiently many people in the system; we should form D-Fork to decrease the waiting time. On the contrary, when the utilization $\rho$ is large, i.e., there are many people in the system, we should form Parallel. When $\beta$ become large, the crossing point moves to the left. The strong effect of the walking distances extends the suitable region for Parallel. This agrees with our intuition, since D-Fork is influenced by the distances but Parallel does not. The reversal phenomenon of $W_q$ is obtained for the first time by introducing the effect of distance.

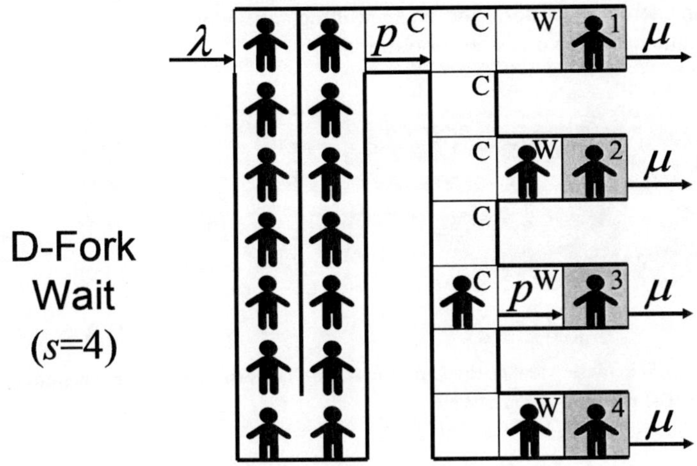


**Fig. 4.** **Schematic view of D-Fork-Wait. People can wait at the cells, which are described as "W".**

Figure 3 (b) shows the type of queuing system, which minimize $W_q$ against $\rho$ and $\beta$ in the case $s$=4. This figure is useful for designing queuing systems. The curves divide the plane into three regions. In the lower left region $W_q$ of D-Fork ($s$=4) is the smallest, and in the upper right region $W_q$ of Parallel is the smallest. Surprisingly, $W_q$ of D-Fork (s=2) $\times$ 2 is the smallest in the middle region. This indicates that the choice of the type of queuing systems is not only Parallel and D-Fork, but also a combination of them. According to (3), $\beta$ represents the ratio of walking time and service time. Therefore, D-Fork is suitable when service time is much longer than walking time. The value of $\beta$ is small in most D-Fork in reality, however, in large queuing system such as an immigration inspection floor in an international airport, we should divide the large D-Fork into the several small D-Forks to decrease the effect of the walking distance.

## *Keep One Person Waiting at the Windows: D-Fork-Wait*

The walking distance in D-Fork is essentially problematic, since it delays the start of services, i.e., people have to walk the passage before they start to receive the service. Thus, we propose to keep one person waiting at the service windows. We call a queuing system which this method is applied to as D-Fork-Wait (Fig. 4). Since people are waiting just next to the service windows, they can receive service instantaneously when their former people leave there. The delay in walking is almost removed by this method, i.e., the effect of walking distance does not need to be considered. Therefore, the mean waiting time is approximately calculated by the mathematical expression for N-Fork.

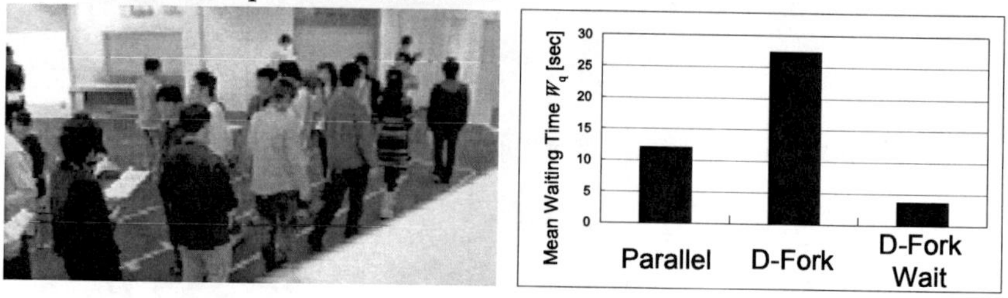


(a) (b)

**Fig. 5. (a) Schematic view of the experiment. (b) The experimental mean waiting time for Parallel, D-Fork, and D-Fork-Wait.**

## *Experiments*

We have performed the experiments to examine the two results in the former section:

1. There is a case that $W_q$ in Parallel becomes smaller than that in D-Fork.
2. We can decrease $W_q$ by keeping one person waiting at each service window.

We have constructed the queuing system, whose parameters are $s$=4, $a$=1[m], $b$-=0.5 [m], $k$=3 [m], $\lambda$=188/600 [persons/sec], and $\mu$=1/8 [persons/sec]. Figure 5 (a) is the snap shot of the experiment. Participants of the experiments enter the system and line up in the queue when the staff says to do so. They proceed to the windows and receive service. After that they wait at the starting position until the staff let him/her enter the system again. We put 188 people in 600 [sec] in one experiment. Note that the arrival was random while the service was deterministic for simplicity. According to the Pollaczek-Khintchine formula [7], $W_q$ becomes small when the service is deterministic. Since this effect acts on the all kinds of the queuing systems in the same way, the results are not critically influenced by deterministic service. Therefore, we can examine the result of the theoretical analysis and simulations by these experiments. Figure 6 (b) shows the result of the experiments. We see that $W_q$ in Parallel is smaller than that in D-Fork. This result verifies our theoretical analysis and simulation by using the walking-distance introduced queuing theory. The reversal of $W_q$ between Parallel and D-Fork is observed experimentally. We also find that $W_q$ becomes dramatically small in D-Fork-Wait. This new result indicates that the method "Keep one person waiting" is an effective way to shorten the waiting time empirically.

(a) (b)

**Fig. 6. (a) Schematic view of the experiment. (b) The experimental mean traveling time for the normal start and the moving start.**

## Queuing System in Amusement Parks

We study queuing systems observed in amusement parks, bus stops, and concert halls in this section. In such queuing systems, plural pedestrians begin to move when a bus arrives or a gate of the hall opens.

Two starting methods, which are the *normal start* and the *moving start*, are compared. In the normal start case, people start to move when their former pedestrian moves, therefore, the last pedestrian has to wait some time from the movement of the first pedestrian. In the moving start case, people keep walking in the queue until the gate opens. The last pedestrian can move some distance before the information of the first pedestrian's movement is transmitted. Thus, the total travel time is expected to become smaller when we adapt the moving start.

We have performed the experiment of the normal start and the moving start as in Fig. 6 (a). First, twenty pedestrians are stand in the circuit. The density in the circuit is approximately 1 [person/m]. In the normal start case, the first pedestrian starts to move and go out the circuit when our staff claps his hands. Then the next pedestrian follows the first one and so on. Finally, the last pedestrian gets out of the circuit. In the moving start case, all pedestrians go around the circuit once and then start to go out. We measure the total travelling time, which is a time between the first pedestrian's departure and the last pedestrian's departure from the circuit. The results are described in Fig. 6 (b). We see that the total traveling time in the moving start case is smaller than that of the normal start case since pedestrians need not to wait their former pedestrian's start in the moving start case. The normal start in the various density conditions are studied with cellular automata simulation in Ref. [9].

## Conclusion

We have studied the methods for improving the efficiency of the two kinds of queuing systems. In a queuing system with plural service windows, the mean waiting time decreases when we form the suitable system according to the degree of congestion and keep one person waiting at the service windows. In a queuing system at an amusement park, people can get on a roller coaster quickly if they keep walking when they are waiting in the queue.

**Acknowledgments** We thank Dai Nippon Printing Co. and Kozo Keikaku Engineering Inc. in Japan for the assistance of the experiments, which are described in Sec. 2 and Sec. 3, respectively. This work is supported by JSPS and JST.

## References

1. Helbing, D.: Traffic and related self-driven many-particle systems. Rev. Mod. Phys. **73**, 1067–1141 (2001)
2. Nishinari, K., Kirchner, A., Namazi, A., Schadschneider, A.: Extended floor field CA model for evacuation dynamics. IEICE Trans. Inf. Syst. **E87-D**, 726-732 (2004)
3. Seyfried, A., Steffen, B., Klingsch, W., Boltes M.: The fundamental diagram of pedestrian movement revisited. J. Stat. Mech. P10002 (2005)
4. Yanagisawa, D., Kimura, A., Tomoeda, A., Nishi, R., Suma, Y., Ohtsuka, K., Nishinari, K.: Introduction of frictional and turning function for pedestrian outflow with an obstacle. Phys. Rev. E. **80**, 036110 (2009)
5. Kretz, T., Grunebohm, A., Schreckenberg., M.: Experimental study of pedestrian flow through a bottleneck. J. Stat. Mech. P10014 (2006)
6. Seyfried. A., Rupprecht, T., Passon, O., Steffen, B., Klingsch, W., Boltes., M.: New insights into pedestrian flow through bottlenecks. Transportation Science. **43,** 395–406 (2009)
7. Bolch, G., Greiner, S., de Meer, H., Trivedi, K.: Queuing Networks and Markov Chains. A Wiley-Interscience Publication, U.S.A. (1998)
8. Blythe, R.A., Evans, M.R.: Nonequilibrium steady states of matrix product form: A solver's guide. J. Phys. A: Math. Gen. **40,** R333 (2007)
9. Tomoeda, A., Yanagisawa, D., Nishinari, K.: Sonic Speed on Pedestrian Dynamics: Relation between Sonic Speed and Density (in this proceedings)

# A Stochastic Evacuation Model for Fire Life Safety Assessment in Transportation System

**K. Kang**

Department of Mechanical Engineering, Columbia University
New York, NY 10027 USA
Corresponding author: kk444@columbia.edu

**Abstract** This paper presents a stochastic model for emergency evacuation simulation such as for a subway station. The model, MCEVAC was developed using Monte Carlo and a macro-simulation technique; the model includes a number of random variables, such as occupant load, initial occupant load, and pre-evacuation time. A random event of a train fire including fire location, fire growth rate, and the effects of fire smoke, can be used to determine the accessibility of egress paths. As model output, a probability representation of the evacuation time can be obtained by taking into account all or selected random variables. For validation, the calculated exiting time shows agreement with that from a previous station evacuation study. For a hypothetical train fire, the predicted evacuation time consists of likely consequences of the required safe evacuation time in the form of a statistical distribution. While still under development, the model is expected to be used for performance-based fire safety design of transit stations and tunnels.

## Introduction

As lessons learnt from incidents such as the Daegu city subway station fire [1] and the fire in the escalator passage of the London Kings Cross station, safe evacuation from an underground station during a fire emergency poses significant challenges to both transit operators and designers. Subway stations are unique as the smoke from a fire would most likely coincide with the path and direction of evacuation and so would adversely impact the availability of egress elements such as stairways. An appropriate station egress design would then need to take into account the blockage of egress paths due to fire smoke, in addition to other requirements such as capacity, way finding and circulation control. The study on end-loading and centre-loading types of stations using the computer evacuation model, EMBER [2] is among the earliest to address the issue of fire smoke and its implications on mass evacuation in underground stations. The effects of fire smoke are emphasized recently [3] based on NFPA 130 [12]. The results of that study are further discussed with code requirements and are compared with a micro-simulation model [4, 5].

R.D. Peacock et al. (eds.), *Pedestrian and Evacuation Dynamics*,
DOI 10.1007/978-1-4419-9725-8_28, 

In addition to fire smoke, human behavior factors have long been recognized, i.e., [6, 7]. For example, the delay time to start the evacuation is considered important and in situations may even be longer than the exiting movement time. The delay includes the time intervals for occupants to receive, recognize and react to fire cues, and could involve activities such as to investigate. This initial delay is referred to as "pre-evacuation time" in this paper. In a station evacuation exercise conducted in London Underground [10], the pre-evacuation time was shown to be as long as 9 min when alarm bell was the only warning signal, whereas verbal announcement via the public address system significantly reduced the pre-evacuation time. In the latter case, it was observed only 15 s after the voice communication, passengers start to move.

Consideration of pre-evacuation time in practice, however, has been mostly based on qualitative interpretation or by using a safety factor. This is largely due to the sparse data in the literature, and the limited quantitative estimate of pre-evacuation time [7]. In general, a scenario-based prescriptive approach remains probably the most widely used [3]. While reasonable assumptions can be made for most parameters in the scenario-based approach, such as occupant load, the uncertainties of these parameters on evacuation time would be largely limited to the selected scenario(s) and the engineer's judgment. As noted by many [9], there is a need for means to integrate human behavior factors in evacuation modeling, rather than using prescriptive approach. Based on the previous studies on transit station evacuation [3-5], this paper discusses the development of a stochastic simulation model and demonstrates its use for fire safety assessment in underground transport.

## MCEVAC Model Description

MCEVAC is a network model based on macro-simulation. The model has three basic *geometry definitions*: "Element", "Interface" and "Link". Element is defined as a walking surface with a finite area. The properties of each element include effective width, area, travel distance and occupant load. Interface is used to connect Elements that are at the same level, for example, when a large area is divided into small Elements. Interface can also be used to represent various pedestrian flow constraints, such as a door or fare turnstile. Elements at different levels, such as between platform and mezzanine, is connected by a Link. Each link has an effective width, travel distance and flow capacity. Both Interface and Link are open by default, but can be closed in a simulation.

The distinction of MCEVAC from micro-simulation models and others is that instead of tracking each individual [3], an Eulerian or *control volume* approach is used for each Element. Within each Element, continuity equation of pedestrian flow is to be satisfied, that is:

$$\frac{\partial \rho}{\partial t} = -\rho \frac{\partial u_i}{\partial x_i} + \sum S_{in} - \sum S_{ex} \qquad [1]$$

Where,

$\rho$ is the occupant density within element (people/m²);
$u$ is the pedestrian flow speed (m/s);
$x$ is the pedestrian flow direction;
$S_{in}$ is the net inflow from link and connected elements;
$S_{ex}$ is the net outflow to link and connected elements; and
$i$ is the subscript for each flow direction with element.

For each flow direction within the element, a first-in-first-out (FIFO) queue is used to model the spatial characteristics of pedestrian movement. The Interface direction is determined by a number of rules: primarily, it is based on the shortest distance to an exit Link, and other rules include comparison of travel time with queuing time, occupant load and occupant density. The interface direction can be adjusted during a simulation. FIFO is also used for the Link, which is essentially a single queue.

The model can consider *occupant characteristics* including walking speed, occupant load and pre-evacuation time. The speed-density correlation [11] is implemented and in general form, it is,

$$u_d = f(\rho) \qquad [2]$$

Eqs 2 and 1 are solved for density (occupant load) and walking speed. Note that the walking speed is adjusted constantly during a simulation, while the occupant load and the pre-evacuation time are specified as initial conditions for each element. The occupant load and the pre-evacuation time are based on parametric probability density distribution functions; the walking speed is "perturbed" using a standard normal function, that is,

$$U = u_d \times P(f) \qquad [3]$$

Where $P(f)$ is the standard normal function between zero and one.

## Station Description

Fig. 1 depicts the configuration of the subway station, which has a center platform with two trains. Each train has ten cars, and each car is approximately 50 ft (15 m) long. The platform is 14 ft (4.3 m) wide and is 550 ft (168 m) long. Along the platform, there are eight stairways numbered as P1 to P8 from south to north. The spacing varies from 30 ft (9 m) to 75 ft (23 m). Four stairways S1 to S4 are located on the mezzanine. In addition, emergency exit stairways are at each end of the platform, each having two sets of stairs leading to the street. The effective width of the stairs is 4 ft (1.2 m), except that S1 is 15 ft (7.5 m) wide and S2 to S4 are 7.5 ft (2.3 m) wide. The longest travel distance is 160 ft (49 m) from the south end of the platform to P1, or 80 ft (24 m) when taking into account the emergency exit stairways. On the mezzanine, the longest travel distance is 200 ft (61 m) from P1 to S1. Fig. 1 also shows a train fire near P4 [3].

## MCEVAC Model Setup

The occupant load was estimated to be 1,980 people including the entraining load and the link load [3]. The number consists of 660 people from each train and 660 on the platform. This initial load is Level of Service (LoS) B on platform and LoS E/F on each train [11]. For each element, the occupant load is treated as a random variable using a normal distribution with a standard deviation of 30%.

A log-normal distribution is used for the pre-evacuation time [8]. From the observations [10], an average of 40 s with a standard deviation of 20 s is assumed for occupants on the platform; for those on the trains, an average of 30 s with 10 s standard deviation is assumed. On a horizontal walking surface, the maximum and average speed is 235 fpm (1.2 m/s); and it is 70 fpm (0.36 m/s) on stairways.

The smoke exhaust rate of 350 kcfm (165 $m^3/s$) is located at the mezzanine ceiling. Tenability analysis was conducted previously using CFD for a train fire scenario [3]. The results showed that at 80 s, smoke is mostly between P2 and P5 on the mezzanine and likely comes through P4 that is closest to the fire. As fire develops, at about 240 s, smoke reaches stairways P1 and P7 on the mezzanine.

The estimated time of tenability [3] may be considered as a correlation, using a polynomial curve fit as a function of the distance from the fire (Fig. 2). When fire growth rate varies, Zukoski's correlation of smoke plume mass flow rate can be used to scale the curve for different fire heat release rate, i.e., $m \propto Q^{1/3}$. A uniform probability distribution is assumed for the fire location, i.e., a fire could start from any train car, whereas a log-normal distribution is assumed for fire growth, with an average of 12 $W/s^2$ and a standard deviation of 6 $W/s^2$ (Fig. 4).

## Results and Discussion

When the fire event option is turned off and the pre-evacuation time is set to zero, the model predicts an exiting movement time of 5.8 min with standard deviation of 15 s. 5.8-min is within 10% of the 5.3-min calculated using NFPA 130 [3]. This may be explained by "local shortest travel distance", which means that occupants would unlikely to automatically balance the queuing between the emergency exit and other platform stairways. As occupant load is not balanced between all the exits/stairways at the same level, there will be a situation when the queue at a particular exit is much longer than at other exits. As illustrated in Fig. 3, the difference in egress utilization between exits S2/S3 and S1/S4 is probably because the platform stairways are clustered towards S2/S3. It may also be because of the initial occupant load that has a higher density near the platform center. For S2 and S3, the flat curve section implies that queuing has exceeded the maximum allowed density for the respective element area. Such "unevenness" in egress utilization reflects the fact that occupants prefer the closest and familiar routes in emergency evacuation [6]. Note also that the average walking speed, 235 fpm (1.2 m/s) is higher than the 200 fpm (1.0 m/s) used previously [3].

## *Predicted station emergency evacuation*

With all the uncertainties taken into account and including a probabilistic fire event, Fig. 4 shows the results for the occupant load, the pre-evacuation time, the fire growth rate and the calculated distribution of station evacuation time. The left vertical axis indicates the frequency fraction for the block chart obtained from histogram, and the right axis is the cumulative frequency fraction distribution with indications at 25%, 50%, 75% and 100%.

The occupant load and the pre-evacuation time are "aggregated" probability distributions from what has been specified for each platform element and train car. These results are consistent with the input as normal and log-normal distributions, for example, the average occupant load is 1,980 with standard deviation of 82. At 95% confidence interval, the range of the occupant load would 1,980±164, or with approx 8% variations from the average. The pre-evacuation time is similar to a log-normal distribution, and shows that 50% of the occupant starts to evacuate within 60 s and 90% within approx 100 s.

For the fire growth rate, the average of 12 $W/s^2$ represents medium growth rate. This distribution indicates that most of the fire growth rate would be medium or slow, whereas some would be medium-fast. Fire growth rate more than fast (44 $W/s^2$) is excluded from the simulation.

Fig. 4d shows the predicted probability distribution of the station evacuation time. Note that the histogram shows a "multi-modal" behavior, that is, there are multiple peaks over the entire range up to 600 s, namely, at 350 s, 440 s, and 540 s. The results suggest that there is a 50% chance for the station to be evacuated within 360 s or 6 min per NFPA 130 [12]; while for the other 50%, the evacuation time would be between 360 and 600 s. In practice, this information is useful. It may be concluded that for the results of this case study, appropriate fire safety measures should ensure tenability for at least 9 min,, which gives 90% chance to evacuate, or 10 min, which gives 100% of the chance to evacuate the station, regardless of the fire location and the fire growth rate. In this case, the emergency ventilation performance would have to be re-evaluated, which is left as future work.

## *Sensitivity analysis*

The number of iterations or samples is important for meaningful statistical correlation. This has been tested from 1,000 to 1,000,000 and the results indicate that a minimum 100,000 iterations in general should provide sufficient sampling of the specified probability distribution, beyond which the predicted evacuation time would have little change. The cumulative frequency distribution converges rather quickly, whereas slight differences can still be noted for occupant load and pre-evacuation time when the number of iteration is over 100,000. The latter is probably because the frequency fractions are based on discrete histogram and are ensembles from that of each element. For all the results presented in this paper, a minimum 200,000 iterations is used.

It should be noted that as the model devleopment is still on-going, other sensitivity tests, especially validation and verification efforts, would still be needed. Further development of MCEVAC includes faster and more efficient alogrithms such as for data storage, more "intelligent" control of occupants movement and decision-making, and a windows-based graphical user interface. The results presented in this paper, however, have demonstrated that a full understanding with regard to the implications and uncertainties of the egress parameters would be important in engineering an effective fire life safety solution for mass transit stations.

## Conclusions

MCEVAC is presented in this paper which is a stochastic evacuation model using Monte Carlo simulation. One of the objectives of MCEVAC development is to assess the uncertainties associated with a wide range of parameters for evacuation analysis, including occupant load, occupant walking speed, fire location, fire growth rate and pre-evacuation time. For a underground railway station studied previously using a prescriptive approach, a probabilistic evacuation analysis is demonstrated using MCEVAC. Taking into account the uncertainties of all the parameters, the model results have shown that the required station evacuation time would be 50% likely within 360 s, but could be up to 600 s or 10 min. The model predicted probability distribution indicates a "multi-modal" behaviour of the evcaution time, that is, an evacuation time of 350 s, 440 s, or 540 s is more likely than others, with 350 s being of the highest likelihood. This is useful in understanding the level of fire life safety that needs to be provided in practice; and in this case, it may be concluded that fire safety measures, such as emergency ventilation, would have to ensure a tenable environment for evacuation within a minimum of 10 min.

## References

1. Chien, S-W., Chen, W-L., Shen T-S., Cheng, C-C., Lee, D-C., Hsuse Y-L, Chen, T-K.: A study on the people evacuation safety for the underground MRT station. In: Proc. of the 6th Asia-Oceania Symp. on Fire Science and Technology, March 17-20, Daegu, Korea (2004)
2. Berlin, G.N., Hathaway, W.: An analysis of emergency transit station evacuation using EMBER. Presented at the Annual Conference on Fire Research, National Bureau of Standards, Gaithersburg, MD (1984)
3. Kang, K.: Application of code approach for emergency evacuation in a rail station. Fire Tech. Vol. 43, No. 4, pp. 331-346 (2007)
4. Kang, K., Till, R.C.: Underground rail station evacuation: review of current code requirements and a case study on emergency evacuation. Presented at the World Safety Conference & Exhibition of National Fire Protection Association (NFPA), Boston, MA (June 2007)
5. Kang, K.: Application of NFPA 130 for emergency evacuation in an underground station. ASHRAE Transactions. Vol. 112, Part 2, pp. 266-274 (2006)
6. Bryan, J.L.: Section 3 Hazard Calculations, Chapter 12 Behavioral Response to Fire and Smoke. In: DiNenno et al. (eds.), The Handbook of Fire Protection Engineering (3rd Ed.). Society of Fire Protection Engineers (SFPE) and NFPA, Quincy (2002)
7. Proulx, G.: Section 3 Hazard Calculations, Chapter 13 Movement of People: The Evacuation Timing. In: DiNenno et al. (eds.), The Handbook of Fire Protection Engineering (3rd Ed.). SFPE and NFPA, Quincy (2002)
8. Fahy, R.F.: Calculation Methods for Egress Prediction. In: Cote et al. (eds.), The Fire Protection Handbook (20th Edition), Section 4, Chapter 2. NFPA, Quincy (2008)
9. Meacham, B.J.: Integrating human factors issues into engineered fire safety design. Fire Mater. 23, 273-279 (1999)
10. Proulx, G., Sime, J.D.: To prevent "panic" in an underground emergency: why not tell people the truth?. In: Proceedings of the 3rd Intl Symp. for Fire Safety Science. Elsevier, London (1991)
11. Fruin, J.J.: Pedestrian Planning and Design, Revised ed., Elevator World Inc., Mobile, AL (1987)
12. NFPA 130 Standard for Fixed Guideway Transit and Passenger Rail System (2007 Ed.). Quincy (2007)

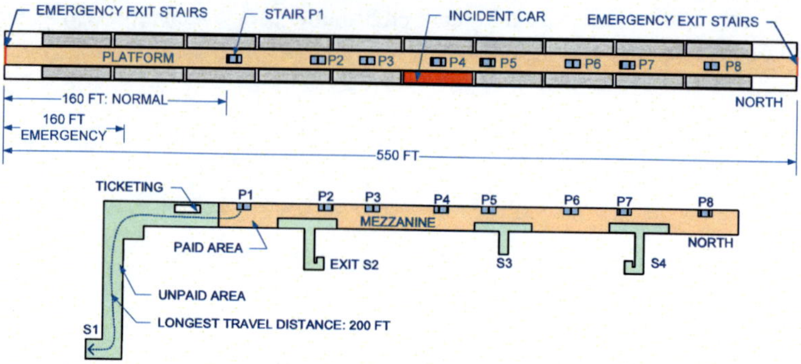


**Fig. 1. Description of subway station**

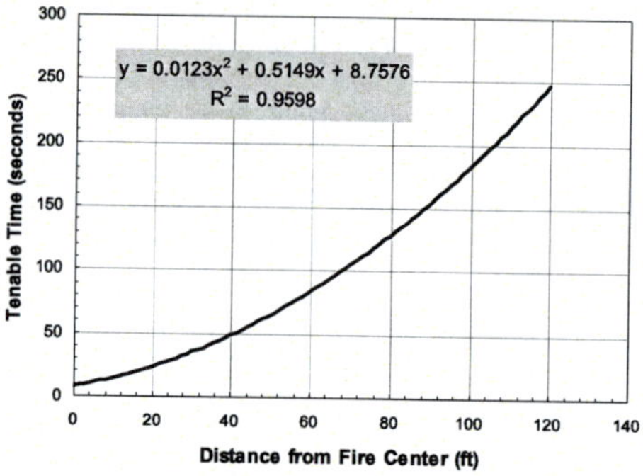


**Fig. 2. Estimate of time of tenability/availability for a train fire scenario**

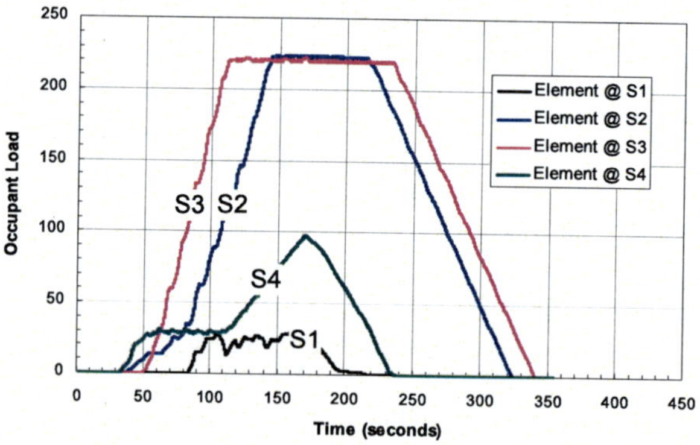


**Fig. 3. Predicted occupant load on Mezzanine near Exit S1 to S4**

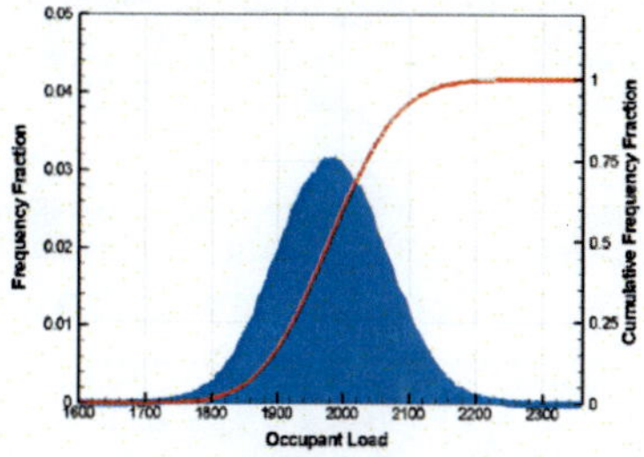


(a) Occupant load

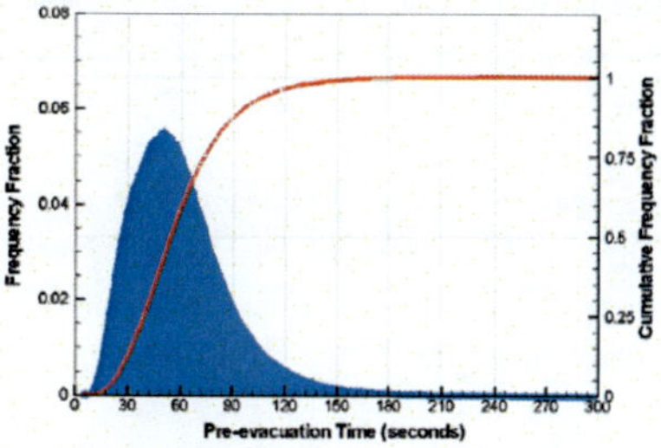


(b) Pre-evacuation time

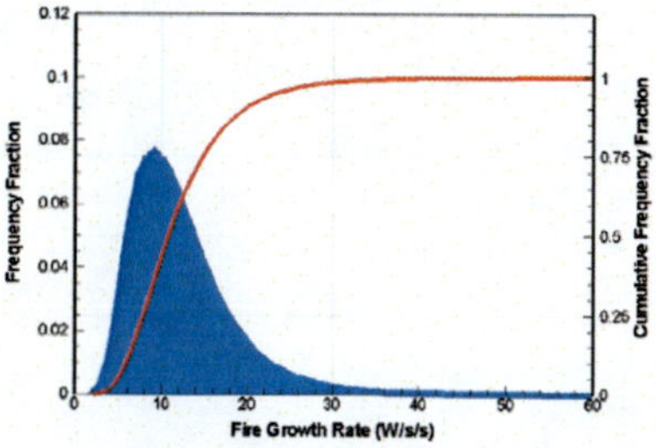


(c) Fire growth rate

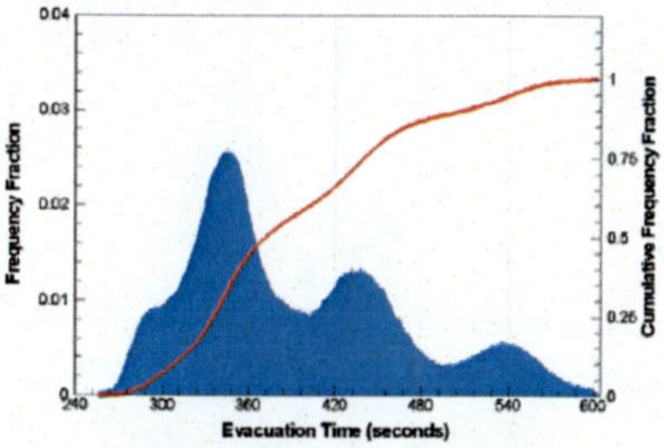


(d) Predicted evacuation time

**Fig. 4. Model predicted distributions of occupant load, pre-evacuation time, fire growth rate and evacuation time**

# Cellular Automata Evacuation Model Considering Information Transfer in Building with Obstacles

**L. Yang, K. Zhu, and S. Liu**

State Key Laboratory of Fire Science, USTC
Hefei, Anhui 230027 China
Corresponding author: yanglz@ustc.edu.cn

**Abstract** The buildings are usually divided into two categories in light of whether obstacles exist: the large space with no obstacles (C-type buildings) and the complex buildings with lots of obstacles (L-type buildings). In this paper, considering the aisle region attraction factor, we proposed a revised model which is suitable to simulate the evacuation process in the L-type buildings (such as classroom, theater, and stadium bleacher) based on our original cellular automata occupant evacuation model. Furthermore, the revised model is able to implement the function that transfers real time information of the occupant density at exit area to evacuees. At last, a case study of simulation evacuation process in a theatre was proposed.

## Introduction

An emergency occupant evacuation is mainly affected by either individual characteristics factors, including gender, age, education, experience, physical condition and so on, or environmental factors which are twofold, that is, dynamic environment information (i.e. fire spreading, evacuation instructions, and occupant density distribution) and static structural characteristics of the evacuation region. The buildings are usually divided into two categories in light of whether obstacles exist: the large space with no obstacles and the complex buildings with lots of obstacles. The former is usually called C-type buildings because pedestrians can directly move towards the target exits, such as main hall, and squares; in contrast, for the latter buildings, pedestrians have to change their movement direction to sheer obstacles, and the route is like the letter L, therefore this type of structure is L-type buildings, e.g. classroom, theater, stadium bleacher etc.

For evacuation in the C-type buildings, researchers pay considerable attention on the pedestrian distribution characteristics at exit bottleneck[1-3] and the mechanism of phase transition in the large-scale pedestrian flow[4-7]. Generally in the L-type buildings, the mass population evacuation is to a large extent affected by the obstacles. Unfortunately, only little research has been done with respect to

R.D. Peacock et al. (eds.), *Pedestrian and Evacuation Dynamics*,
DOI 10.1007/978-1-4419-9725-8_29, 

the effect of obstacles[8-11]. Consequently, it is significant to investigate how apperceive and avoid the obstacles, how to arrange the obstacles, how to design the L-type buildings evacuation routes etc., especially in emergency.

It is crucial to grasp the dynamic real time information accurately for effective evacuation. In order to simulate evacuation process more realistically, a sophisticated model could declare the current environmental information to pedestrians in a particular area, such as information about the change of occupant density near the exit, and the spread of fire smoke, to provide pedestrians a basis for decision-making, or to guide people to choose a more secure and efficient evacuation route.

In this paper, we proposed a revised model which is suitable to simulate the evacuation process in the L-type buildings based on our original cellular automata occupant evacuation model. Furthermore, the revised model is able to transfer the information of the occupant density near the exits to evacuating pedestrians. At last, we investigate an evacuation process in a large theater using this new model.

## Model

Cellular automaton [12] is a kind of multi-dimensional systems with finite states whose space and time are both discrete. It is widely used in traffic flow[13-16], occupant evacuation [17-20] and other complex non-linear sciences[21, 22]. In a cellular automata model, individual changes its state according to the states of its neighbors at the last time step, and the entire system updates by communicating and cooperating between individual and individual or between individuals and environment at each discrete time step. When simulated occupant evacuation process, a pedestrian chooses a destination cell for next time step in accordance with the current sates of his or her several neighboring cells.

### *Space segment and valuation*

Building space is first divided into a series of square grids, and each grid is called a cell whose size is typically 0.5 m × 0.5m. At any time step, a cell can be occupied by one person at most. Each cell has a weight value, which we call total attraction value (TAV), representing each cell's attraction degree to pedestrians. In addition, pedestrians choose route based on the TAV. It means that they will select the cell which has large TAV with a larger probability as the target cell at next time step. There are two main aspects affecting the TAV of a cell: the static location information of building structure and the dynamic environmental information in evacuation process.

It is intuitive that the cells which belong to exits or much vacant space are more attractive and corresponding larger TAV. In contrast, the cells occupied by obstacles are smaller TAV since the obstacles will impede pedestrian movement. In addition, people are usually accustomed to follow the footprints of the crowd, namely, following behavior. Our previous model [23-25] has considered lots of factors, such as the exit location, the movement direction of the pedestrian crowd, the repulsion force between people and people or between people and obstacles. In this revised model, considering characteristic of L-type buildings, we introduce aisle area factors, which is considered in similar way as the exits attraction factor because both are static position information, the calculation of TAV can be obtained by the following equation:

$$N_{ij} = \exp[k_s \times (S_1 + S_2) + k_r \times R + k_d \times D]$$

where, $S_1$ is the attraction of exit position; $S_2$ is the attraction of aisle region; $R$ is the repulsive force between people and people or between people and obstacles; $D$ is a factor which represents the influence of people around, driven by a psychology of following others; $k_s$, $k_r$, $k_d$ are the impact factors of parameters mentioned above, respectively. They are used to determine which parameter plays the dominant role.

The reminder of this section will discuss how to identify each factor:

Two main factors should be considered when calculate the $S_1$ value of a cell, namely, the distance between cell and exits and pedestrians' familiarity with the exits. In general, $S_1$ is larger if the cell is closer to the exit or pedestrians are more familiar with the exit. $S_1$ is identified in two steps: firstly, calculate the distance between the cell $(i, j)$ and the exit $k$, then choose the minimum as the distance from the cell $(i, j)$ to the exits; the second step is to choose the largest value among distance from all cells to exits as a benchmark, and establish the relative distance value from cell $(i, j)$ to exits. The following type is in details:

$$S_1 = \max\{\min[\alpha_k \cdot \sqrt{(i-e_{1k})^2 + (j-e_{2k})^2}]\} - \min(\alpha_k \cdot \sqrt{(i-e_{1k})^2 + (j-e_{2k})^2})$$

Where, $(e_{1k}, e_{2k})$ is the coordinates of the exit $k$; $\alpha_k$ is the degree of familiarity with the exit $k$.

The establishment of $S_2$ is shown as the below following: if the cell is in aisle region, then assign; otherwise not be considered. To simplify matter, we assume $S_2$ is an integer,

$$S_2 = \begin{cases} \lambda & \text{if the cell } (i,j) \text{ is aisle region} \\ 0 & \text{else} \end{cases}$$

The factors $S_1$ and $S_2$ are both static information and cannot be changed during evacuation. The value of each cell is the same for different pedestrians, therefore they can be identified before evacuation. In contrast, the factors $R$ and $D$ are dynamic information, thus their values of cell $(i, j)$ are different for different pedestrians at different time steps. Pedestrians choose movement direction according to local rules. Each cell has 8 neighboring cells at most (see Fig. 1). The values of $R$ and $D$ of the cell $(i, j)$ and its neighboring cells update at each time step. The calculation process is as follows:

$$R = \begin{cases} \beta & \textit{if the cell } (i,j) \textit{ is occupied} \\ 0 & \textit{else} \end{cases}$$

Where, $\beta < 0$, represents the repulsive force between people and people or between people and obstacles;

$$D = \begin{cases} \delta & \textit{if the cell } (i,j) \textit{ orientats } \text{crowd} \\ 0 & \textit{else} \end{cases}$$

where, $\delta > 0$, represents the extent of following behavior of pedestrians. By counting all individuals' movement directions within a specific scope of cell $(i, j)$, it could choose the direction which occurs most frequently as the crowd movement direction.

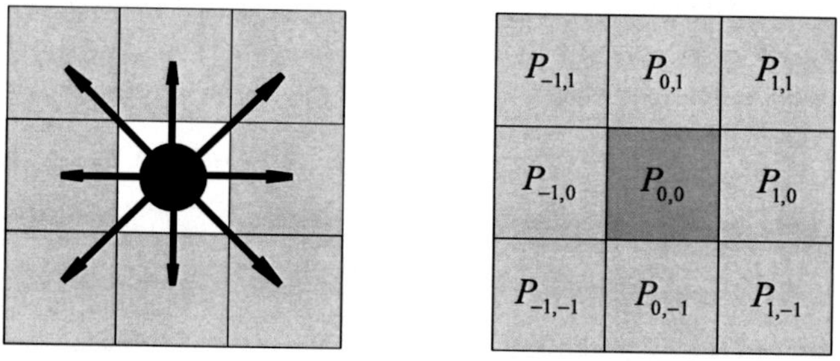


**a) Moore neighbor** **b) Transition probability**

**Fig. 1. Moore neighbor and Transition probability: one pedestrian has 8 possible movement directions at most; the sum of $P_{ij}$ is 1, namely, $\sum_j \sum_i P_{i,j} = 1$**

So far, we have given the calculation process of several considered factors in revised model, the impact coefficient $k_s$, $k_r$, $k_d \in [0, 1]$, and if a coefficient is 0, it represents that we do not consider this factor; similarly, if it is 1, it means that this factor plays a dominant role. In this proposed model, it can be introduced different factors for different situations and for different investigation focus, thus it is an extended model. For example, we may introduce the factor of fire repulsion in fire situation; we can also introduce the impact of the attraction between family members considering the close relationship among a family.

### ***Transition probability***

Generally speaking, pedestrians choose preference direction as target for next time step according to combined effects of their neighboring cells' TAV values. Transition probability (see Fig. 1) is given by following type:

$$P_{ij} = \frac{N_{ij}}{\sum N_{ij}} \cdot (1 - n_{ij}) \cdot w_{ij}$$

Where, $P_{ij}$ is the probability for pedestrians transfer to the cell $(i, j)$; $n_{ij}$ is the free coefficient for cell $(i, j)$, if cell $(i, j)$ has been occupied by another pedestrian at current time step, then $n_{ij} = 1$, otherwise $n_{ij} = 0$; in addition if cell $(i, j)$ is occupied by obstacles, then $w_{ij} = 0$, or else $w_{ij} = 1$.

### ***Occupant distribution***

Occupant distribution is identified before evacuation simulation, and it mainly includes two aspects: namely, occupant density and occupant initial position. In China, Code of Design on Building Fire Protection and Prevention (GB50016-2006) provides that: the size of evacuation occupants in the video theatre should be calculated by 1.0 persons/$m^2$ multiplying the building area; and in other entertainment concourses, it should be determined by 0.5 persons/$m^2$ multiplying the building area. Furthermore, occupant initial position is either randomly or determined distribution. Considering a maximal capacity of L-type buildings, the maximal occupant density determined distribution is chosen, namely, each seat subjects to one people. In following study, we will select this type occupant distribution, if no special instructions are supported.

Until now, we have analyzed several steps of evacuation simulation process, and the whole simulation process is schematically in Fig. 2:

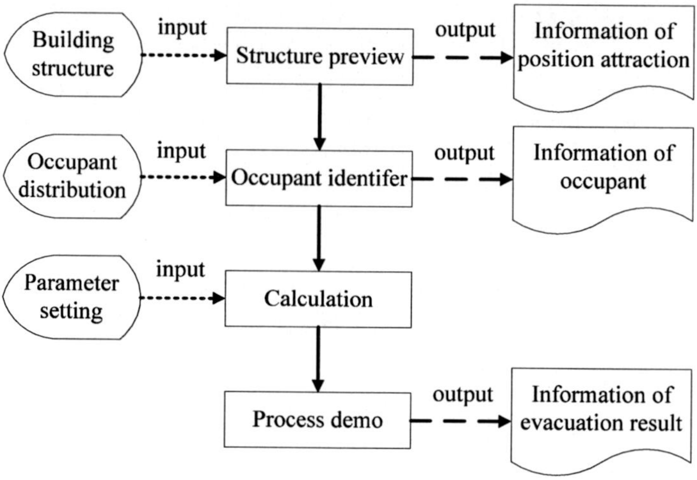


**Fig. 2. Schematic drawing of evacuation simulation process**

## Information transfer

In emergency situation, it is significant to select reasonably a safe escape route for evacuees. Moreover, it is necessary that pedestrians access to real time information when they choose escape route. Thus, in real world, pedestrians often observe rapidly the situation around them to obtain current useful information before they take next action. For example, at an intersection, pedestrians would make a simple comparison about the situation in a visible region (the occupant density in visible region, the aggravating degree of environment, etc.) and then select a branch (turn left or turn right) as the next evacuation direction. In order to simulate evacuation process realistically, it is necessary to introduce information transfer function in an excellent evacuation model, in which pedestrians make decisions according to achieved environment information.

Take a video theater as an example, shown as Fig. 3. There are two exits (exit A and exit B) in front of the theater, 108 seats are arranged in 6 rows and 18 columns, and divided into three parts by four longitudinal passages. There is also a horizontal passage leading to the exits. A rectangular area is defined as the exit area at each exit, and two “message release” (MESSAGE RELEASE 1 and MESSAGE RELEASE 2) are also set at the crossing of longitudinal passages and horizontal passage (see Fig. 3), which transmit occupant density of exit area and update at each time step. Pedestrians decide to turn left or turn right based on the distance to each exit and the information of occupant density.

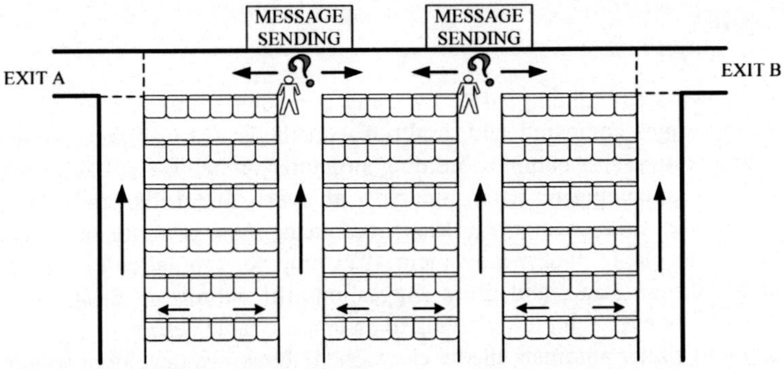


**Fig. 3. Schematic illustration of simulation setup**

The pedestrians who enter the horizontal passage from the passages close to the wall will evacuate directly out through exit A or exit B. However, the pedestrians entering from the middle longitudinal passages maybe have an alternative evacuation direction. Consequently, evacuees make decision by light of occupant density which received from "MASSAGE RELEASE" and the distance to each exit. The specific flow chart of the comparison process is shown in Fig. 4, where, $D_1$ and $D_2$ are the distance to exit A and exit B, respectively; $\rho_1$ is occupant density of exit area A, similarly, $\rho_2$ is that of exit area B.

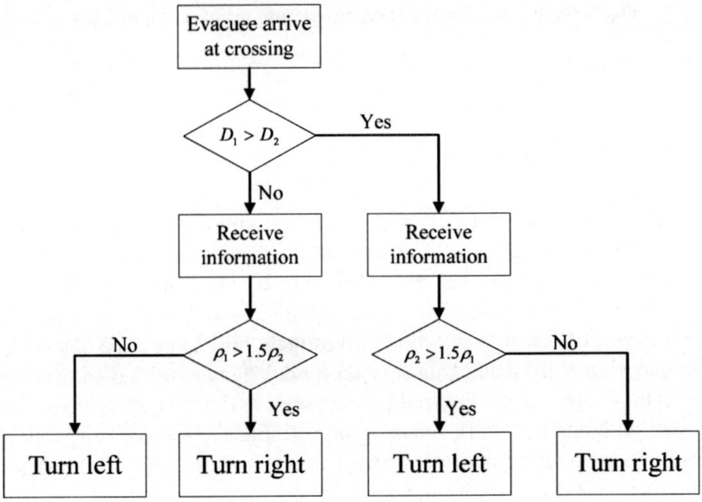


**Fig. 4. the specific flow chart of the comparison process**

## Case study

It is wide to assess building safety by using computer simulation in view of its unique advantages compared with traditional methods: (1) Computer simulation can be applied to assess complex building structures, which is not traditional specification design competent; (2) Compared with evacuation drills, computer simulation is shorter time-consuming, less investment, more security, and will not cause accidents due to poor management; (3) Computer simulation can be applied at structure design stage, and gives suggestions for optimizing design to avoid money-waste, or even rebuilding owing to an unreasonable design.

Based on cellular automata theory mentioned above, we develop a software to simulate an evacuation in a theater. The theater's dimension is the same as that talked in section transfer information (see Fig. 3). In addition, Fig. 5 is snapshots of simulation evacuation process in the theater. Generally, this model is able to reproduce the evacuation process in L-type buildings.

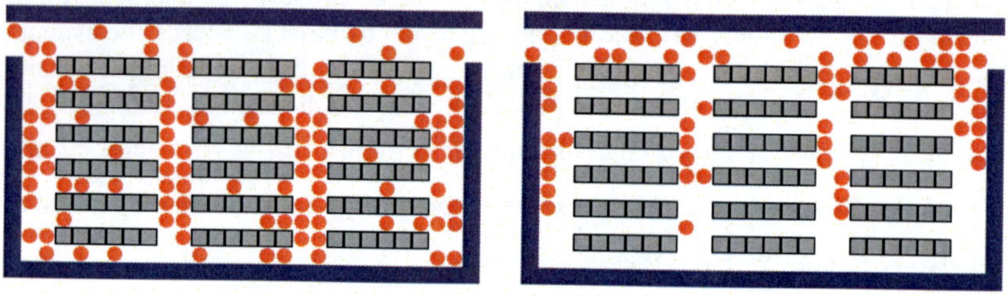

**Fig. 5.** **Snapshots of simulation evacuation process in a theater**

## Conclusion and future work

We proposed a revised model which is suitable to simulate evacuation process in L-type buildings based on our original cellular automata occupant evacuation model. Furthermore, the revised model is able to implement the function that transfers real time information of the occupant density at exit area to evacuees.

The next phase of research will be to investigate how to arrange obstacles in L-type buildings rationally using this revised model. It is expected to provide suggestions for the design of the internal layout (seats and aisles) of theaters. In addition, we will continue our work focusing on refining the revised model perfectly, for example, to introduce fire effect factor for evacuation in fire. Furthermore, we will analyze the effect of making-decisions by comparison rules, such as occupant density ($\rho_1 > 1.5\rho_2$ or $\rho_2 > 1.5\rho_1$, see Fig. 4). In this paper, an experiment value, 1.5 is chosen.

**Acknowledgments** This research was supported by National Natural Science Foundation of China (No. 90924014) Specialized Research Fund for the Doctoral Program of Higher Education (No. 200803580007). The authors deeply appreciate the supports.

## References

1. Helbing, D., Farkas, I., Vicsek, T.: Simulating dynamical features of escape panic. Nature. **407**, 487-490 (2000)
2. Song, W.G., Yu, Y.F., Wang, B.H., Fan, W.C.: Evacuation behaviors at exit in CA model with force essentials: A comparison with social force model. Physica A. **371**, 658-666 (2006)
3. Kirchner, A., Nishinari, K., Schadschneider, A.: Friction effects and clogging in a cellular automaton model for pedestrian dynamics. Phys Rev E. **67**, 056122-1-056122-10 (2003)
4. Nagai, R., Nagatani, T.: Jamming transition in counter flow of slender particles on square lattice. Physica A. **366**, 503-512 (2006)
5. Li, J., Yang, L.H., Zhao, D.L.: Simulation of bi-direction pedestrian movement in corridor. Physica A. **354**, 619-628 (2005)
6. Yang, L.Z., Li, J., Liu, S.B.: Simulation of pedestrian counter-flow with right-moving preference. Physica A. **387**, 3281-3289 (2008)
7. Nakayama, A., Sugiyama, Y., Hasebe, K.: Instability of pedestrian flow in two-dimensional optimal velocity model. In: Waldou, N., Garttermann, P., Knoflacher, H., Schreckenberg, M. (eds.), Proceedings of the International Conference on Pedestrian and Evacuation Dynamics, pp. 321-332. Springer-Verlag Berlin, Heidelberg (2007)
8. Varas, A. *et al.*: Cellular automaton model for evacuation process with obstacles. Physica A. **382**, 631-642 (2007)
9. Zhang, J., Song, W.G., Xu, X.: Experiment and multi-grid modeling of evacuation from a classroom. Physica A. **387**, 5901-5909 (2008)
10. Helbing, D., Isobe, M., Nagatani, T., Takimoto, K.: Lattice gas simulation of experimentally studied evacuation dynamics. Phys Rev E. **67**, 067101-067104 (2003)
11. Liu, S.B., Yang, L.Z., Fang, T.Y., Li, J.: Evacuation from a classroom considering the occupant density around exits. Physica A. **388**, 1921-1928 (2009)
12. Neumann, J.V., Burks, A.W.: Theory of Self-reproduction Automata. University of Illinois Press, Urbana (1966)
13. Ebersbach, A., Schneider, J.J.: Two-lane traffic with places of obstruction to traffic. Int J Mod Phys C. **15**, 535-544 (2004)
14. Nagel, K., Wolf, D.E., Wagner, P.,Simon, P.: Two-lane traffic rules for cellular automata: A systematic approach. Phys Rev E. **58**, 1425-1437 (1998)

15.Rickert, M., Nagel, K., Schreckenberg, M., Latour, A.: Two lane traffic simulations using cellular automata. Physica A. **231**, 534-550 (1996)
16.Tian, J.F. *et al.*: Synchronized traffic flow simulating with cellular automata model.Physica A. **388**, 4827-4837 (2009)
17.Yuan, W.F., Tan, K.H.: An evacuation model using cellular automata, Physica A. **384**, 549-566 (2007)
18.Burstedde, C., Klauck, K., Schadschneider, A., Zittartz, J.: Simulation of pedestrian dynamics using a two-dimensional cellular automaton. Physica A. **295**, 507-525 (2001)
19.Kirchner, A., Schadschneider, A.: Simulation of evacuation processes using a bionics-inspired cellular automaton model for pedestrian dynamics. Physica A. **312**, 260-276 (2002)
20.Schultz, M., Lehmann, S., Fricke, H.: A discrete microscopic model for pedestrian dynamics to manage emergency situations in airport terminals, In: Waldou, N., Garttermann, P., Knoflacher, H., Schreckenberg, M. (Eds.), Proceedings of the International Conference on Pedestrian and Evacuation Dynamics, pp. 369-375. Springer-Verlag Berlin, Heidelberg (2007)
21.Malamud, B.D., Turcotte, D.L.: Cellular-automata models applied to natural hazards. Comput Sci Eng. **2**, 42-51 (2000)
22.Ohgai, A., Gohnai, Y., Watanabe, K.: Cellular automata modeling of fire spread in built-up areas - A tool to aid community-based planning for disaster mitigation, Comput Environ Urban. **31**, 441-460 (2007)
23.Yang, L.Z., Fang, W.F., Fan, W.C. Modeling occupant evacuation using cellular automata - Effect of human behavior and building characteristics on evacuation, J Fire Sci. **21**, 227-240 (2003)
24.Yang, L.Z., Fang, W.F., Huang, R., Deng, Z.H.: Occupant evacuation model based on cellular automata in fire, Chinese Sci Bull. **47**, 1484-1488 (2002)
25.Yang, L.Z., Li, J., Zhao, D.L., Fang, W.F., Fan, W.C.: A microcosmic discrete occupant evacuation model based on individual characteristics, Sci China Ser E. **47**, 608-615 (2004)

# Quickest Cluster Flow Problems

**H. W. Hamacher, K. Leiner, and S. Ruzika**

Technical University of Kaiserslautern
67663 Kaiserslautern
Corresponding author: leiner@mathematik.uni-kl.de

**Abstract** Macroscopic models based on dynamic network flow theory are successfully applied to obtain lower bounds on real evacuation times [1]. The goal of our research is to tighten this lower bound and to make this macroscopic approach more realistic by taking into account clustering of evacuees - a sociological phenomenon observed in evacuation scenarios. A cluster of flow units in the network flow model represents families or cliques which tend not to move independently but as groups [2]. This fact is not covered by macroscopic approaches based on classical network flow theory. In this article, we take clustering into account and thus improve existing macroscopic network flow models. We focus on two different sizes of groups traversing the network, modeled as single flow units and cluster flow units the latter of which occupy $d$ times as much capacity as single flow units. In this novel approach, we are given fixed amounts of single flow units and cluster flow units and minimize the time at which the last (single or cluster) unit reaches the target. We present an algorithm that gives a *2*-approximation for general networks and is optimal for the subclass of series-parallel networks.

## Introduction

In the context of evacuation modeling, dynamic network flow theory can be used to provide lower bounds on evacuation times and optimal evacuation routes [1]. An optimal dynamic network flow can be computed efficiently and yields a best achievable evacuation plan, provided all evacuees follow the routes corresponding to this flow. Hence it provides provable lower bounds for egress times. This information is certainly invaluable, especially if complemented by results from microscopic models [3, 4]. Macroscopic approaches are often criticized for their lack of representation of sociological components among evacuees during egress movement. As a consequence, while still being correct, the lower bounds of these pure macroscopic approaches without sociological components may be improved.

This article presents first steps toward incorporating social components into network flow models while still keeping above mentioned desirable properties of macroscopic approaches such as efficient computation times and lower bound guarantees. The goal of our work is to tighten lower bounds for real evacuation times by extending existing dynamic network flow models: certain interdependen-

R.D. Peacock et al. (eds.), *Pedestrian and Evacuation Dynamics*,
DOI 10.1007/978-1-4419-9725-8_30, © Springer Science+Business Media, LLC 2011

cies between evacuees, i.e., flow units, are taken into account. Our dynamic network flow model incorporates some types of group behavior of evacuees, which is assumed to be maintained throughout any kind of emergency [2].

In classical dynamic network flow theory, flow units are considered to be independent users of equal size sharing capacities of the network, whereas in the modified model presented in this paper two different kinds of flow units are considered: Those (smaller ones) representing single pedestrians not committed to any group during egress, and those (larger ones) representing groups, or more specifically, individuals in strong affiliation to some group. In this context, we consider the *quickest cluster flow problem*: Given a network, a cluster (group) size and a number of cluster flow units (groups) as well as an amount of single flow units (single pedestrians), both kinds of flow units are to be routed from a specified source to a target in shortest possible time while sharing the network capacities. Subsequently, we specify that each cluster flow unit occupies the $d$-fold capacity of a single flow unit.

To the best of our knowledge this model has not yet been considered, neither in the area of dynamic network flow theory nor in its application in evacuation planning. In the literature on static network flows there exists, however, a related problem, the *unsplittable flow problem* which was first introduced by Kleinberg [5]. The research in this area goes into the direction of minimizing the amount that the capacities have to be enlarged to allow routing of all flow from the source to the target. This is, however, not desired in modeling evacuation, since we do not want to enlarge pathways, but aim for the best schedule to evacuate pedestrians in the given topology of the evacuation region.

## Preliminaries on Dynamic Network Flow Theory

A *(discrete-time) dynamic network* $N = (G,c,\tau)$ is a graph $G = (V,E,T)$, where $V$ is the set of nodes, $E$ is the set of directed edges and $T \in \mathbb{N}$ is the time horizon of interest. This time horizon is discretized into the set $\mathcal{T} = \{0,...,T\}$. Two designated nodes $s, t \in V$ model source and target, respectively, and we intend to route flow from node $s$ to node $t$ using the edges contained in the set $E$ of our network. For each edge $e \in E$, we are given two attributes, the capacity $c_e$ of e, i.e., the maximum number of flow units that can enter $e$ at any time $\theta \in \{0,...,T\}$, and the transit time $\tau_e$. The latter is the number of time steps that a unit of flow needs to traverse edge $e$, i.e., for $e = (i, j)$ a flow unit starting at time $\theta$ at node $i$ will reach node $j$ at time $\theta + \tau_e$. If the capacity $c_e$ changes depending on the time, we denote the capacity at time $\theta \in \{0,...,T\}$ by $c_e(\theta)$. A path $P$ is defined as a sequence of nodes such that two consecutive nodes are connected by an edge $e \in E$. The length $\tau(P)$ of a path $P$ is the sum over the transit times of all edges contained in $P$. Additionally, we define the capacity $c(P)$ of path $P$ as the minimum of the capacities of all edges contained in $P$.

A *feasible flow f* in a dynamic network is a mapping of the edge set and the time steps to the nonnegative real numbers ($f: E \times \mathcal{T} \rightarrow \mathbb{R}^+$) such that all flow units have reached the target $t$ in the network after $T$ time periods and the flow entering any edge does not exceed its capacity at any time in the considered time horizon.

The *maximal dynamic flow problem* for a given time horizon $T$ asks for a feasible dynamic flow that routes the maximal number $b$ of flow units from the source $s$ to the target $t$ such that any flow has reached $t$ at or before time $T$. The value $b$ is then called the maximal dynamic flow value. Ford and Fulkerson [6] showed that this problem can be solved efficiently by the technique of *temporally repeated flows,* where paths are computed in a static network from a path decomposition of a minimum cost flow (cf. Ahuja et al. [7]) with transit times interpreted as costs. Then the dynamic flow is constructed from these paths and their lengths by repeating these flows $T\text{-}\tau(\text{P})+1$ times starting at time step 0. Each of these paths carries as much flow units as the corresponding path $P$ does in the static network.

A particular maximum dynamic flow problem is the *earliest arrival flow problem,* where we not only want to maximize the amount of flow that has reached the target at time $T$, but also the total amount at any intermediate time step $\theta \in \{0, ..., T\text{-}1\}$.

Another well-studied dynamic network flow problem in the context of evacuation modeling is the *quickest flow problem (QFP)*: Given a dynamic network and a number $b$ of flow units, find the minimal time $T$ needed to route all flow units from the source $s$ to the target $t$. The QFP models the fastest possible evacuation of a specified number of evacuees. The resulting minimal time $T$ therefore gives a lower bound on the minimal evacuation time. Burkard et al. [8] propose several polynomial time algorithms for this dynamic network flow problem. One approach is based on binary search, where iteratively the feasibility of a dynamic flow for varying time horizons $T'$ are tested, until the smallest of these values is delivered as optimal solution of the quickest flow problem. Note that in this approach any polynomial time algorithm solving the maximum dynamic flow problem can be used to compute a quickest flow problem in polynomial time.

The quickest flow problem, as well as many other dynamic network flow problems, can be solved by a flow computation in a special static network, the so-called *time-expanded network (TEN)* first introduced by Ford and Fulkerson [6]. In a time-expanded network, each node $v$ of the original network is copied $T+1$ times resulting in $v(0), \ldots, v(T)$. Each of these copies represents this node at the corresponding time step. Each edge is also multiplied such that edge $e = (i,j)$ with transit time $\tau_e$ connects node $i(\theta)$ with node $j(\theta+\tau_e)$ for $\theta \in \{0, \ldots, T\text{-}\tau_e\}$. Additionally, we add a *supersource* (*SS*) that is connected by an edge with infinite capacity and transit time zero to every copy $s(\theta)$ of the source of the original network. Analogously, a *supertarget* (*ST*) is added. Every dynamic network corresponds in a unique way to a time-expanded network and vice versa. An example is depicted in Fig. 1. As a consequence, many algorithms for static network flow problems can be used to solve dynamic flow problems like the quickest flow problem. The drawback of this general approach is the size of the time-expanded network which is only pseudo-polynomial in the input of the instance, since it depends on the considered time horizon $T$. This suggests avoiding computations in the time expanded network.

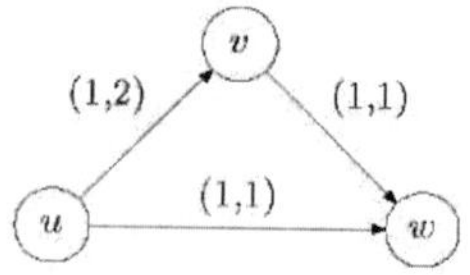


**Fig. 1a. Dynamic network $N$. The edge-labels represent capacities and transit times ($c_e$,$\tau_e$).**

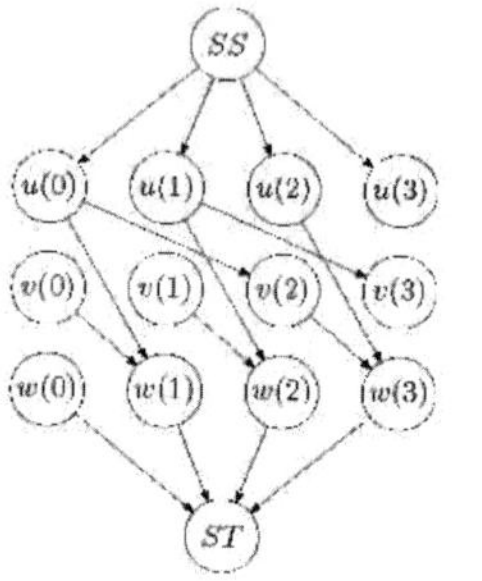


**Fig. 1b. Corresponding time-expanded network $N^T$ for $T$=3.**

## Quickest Cluster Flow Problems

We introduce the *quickest cluster flow problem (QCFP),* a problem closely related to the quickest flow problem. An instance $C = (N,d,b_d,b_1)$ of this problem is given by a dynamic network $N = (G,\tau,c)$ with constant capacities on the edges, a cluster size $d$, a number of cluster flow units $b_d$ and a number of single flow units $b_1$. We aim to route all flow units from the source $s$ to the target $t$, where a single flow unit takes one unit of the edge-capacity when entering it and a cluster flow unit takes $d$ units of the capacity. A feasible solution to the quickest cluster flow problem is given by a set of $s$-$t$-paths $\{P_1,\ldots,P_{b_1},P_{b_1+1},\ldots,P_{b_1+b_d}\}$ as well as the corresponding starting times $\{\theta_1,\ldots,\theta_{b_1},\theta_{b_1+1},\ldots,\theta_{b_1+b_d}\}$. The goal is to find the "best" $s$-$t$-paths in the sense that the time when the last (single or cluster) flow unit reaches the target is minimized without violating capacity constraints on any edge at any point in time. This time is called the *makespan* of the instance. A mathematical formulation of this problem is given in Fig. 2:

$$\min\ T$$

$$\text{s.t.} \sum_{\theta=0}^{T}\left(\sum_{e\in\delta^+(v)} f_e^k(\theta) - \sum_{e\in\delta^-(v)} f_e^k(\theta-\tau_e)\right) = \begin{cases} b_k & \text{for } v = s,\\ -\ b_k & \text{for } v = t,\\ 0 & \text{else,}\end{cases}$$

$$\forall\, v \in V,\ k \in \{1, d\},$$

$$0 \le f_e^1(\theta) + df_e^d(\theta) \le c_e\ \forall\, e \in E,\ \theta \in \{0,\ldots,T\},$$

$$f_e^k(\theta) \text{ integer for } k \in \{1,d\},\ \theta \in \{0,\ldots,T\}.$$

**Fig. 2. Mathematical formulation of the quickest cluster flow problem**

In the formulation given in Fig. 2, $\delta^+(v)$ denotes the set of all edges emanating from node $v$. Analogously, $\delta^-(v)$ is the set of all edges entering it.

We assume that in any considered cluster flow instance there is at least one $s$-$t$-path with capacity greater than or equal to $d$ as otherwise the instance is infeasible.

Note that holding over of flow at any intermediate node in the network does not improve the makespan: Assume that $T^*$ is the makespan of an optimal solution for a cluster flow instance and let $q$ be the size of a flow unit that waits at an intermediate node $v \in V$, i.e., it stays at node $v$ from time step $\theta_1$ to $\theta_2$. We choose $v$ as the last node of the path in which this holdover of flow occurs and assume that $\theta_2$ is the earliest possible time when it can leave node $v$. Then in the time-expanded network of the dynamic network it occupies $q$ units of capacity of a path from the supersource $SS$ to $v(\theta_1)$ and $q$ units of capacity of a path from $v(\theta_2)$ via $t(T^*)$ to the supertarget $ST$. Since the sum over all capacities of edges leaving $v(\theta_2)$ is equal to the sum of all capacities leaving $v(\theta_1)$, and flow conservation holds for all intermediate nodes, there must be $q$ units of free capacity on paths from $SS$ to $v(\theta_2)$. The amount of flow entering $v(\theta_1)$ is equal to the amount of flow leaving it and it is not possible to reroute the flow on the paths from $v(\theta_1)$ to $SS$ such that there is a single path that has more than $q$ units of remaining capacity. Then there is a path $P$ from $SS$ to $v(\theta_2)$ such that the flow on all paths from $SS$ to $v(\theta_2)$ can be rerouted giving $P$ remaining capacity $q$. Hence there is a feasible cluster flow solution of the same makespan $T^*$ where one flow unit less has to wait at some intermediate node. Applying the above argument iteratively yields our claim that holdover cannot improve the makespan of an optimal solution.

## *Lexicographical Solution Scheme for QCFP*

For a given $s$-$t$-network $N$ let $N_d = (G,\tau,c^d)$ be the dynamic network that originates from $N$ by substituting the capacities $c_e$ by rounded fractions

$$c_e^d = \left\lfloor \frac{c_e}{d} \right\rfloor.$$

The lexicographical solution scheme for the quickest cluster flow problem can be outlined as follows: In a first step we compute a quickest flow problem with $b_d$

many $d$-clusters in the network $N_d$. Due to the construction of $N_d$ we can apply any standard quickest flow algorithm. This yields the makespan $T_d^L$. The amount of flow that is routed on any edge at any point in the discretized time interval (i.e., $d$ times the amount of clusters) is stored in $f$. Next, we update the capacities with respect to $f$ and define the network $N_1$ as indicated in line 3 of the algorithm shown in Fig. 3. Note that $N_1$ has time-dependent capacities, i.e., the quickest flow problem with $b_1$ units of flow in line 4 of the algorithm cannot be computed by a standard quickest flow algorithm. However, this quickest flow problem can be solved in $O(nmb_1T;^{-2})$ by the algorithm developed by Tjandra [9], where $n$ is the number of nodes, $m$ is the number of edges and $T;^{-}$ is an upper bound on the minimal time horizon. The flow values in $f$ are then updated by the total amount of flow that is routed on the respective edge at each time step. The paths and starting times can then be reconstructed by a path decomposition of $f$ (cf. Ahuja et al. [7]).

**Algorithm 1**: Lexicographical Solution Scheme for QCFP

**Input**: Cluster Flow Instance

**Output**: Feasible Cluster Flow Solution $f$ with makespan $T^L$

1 Compute quickest flow $f^d$ with value $b_d$ in $N_d \rightarrow$ makespan $T_d^L$;
2 Set $f_e(\theta) = df_e^d(\theta)$ for any edge $e \in E$ and $\theta \in \{0, 1, \ldots, T_d^L\}$;
3 Set $c_e^1(\theta) = c_e(\theta) - f_e(\theta)$ for $\theta \in \{0, \ldots, T_d^L\} \rightarrow N_1$;
4 Compute quickest flow $f^1$ with value $b_1$ in $N_1 \rightarrow$ makespan $T_1^L$;
5 Set $T^L = \max\{T_d^L, T_1^L\}$;
6 Set $f_e(\theta) = f_e(\theta) + f_e^1(\theta)$ for any edge $e \in E^d$; and $\theta \in \{0, 1, \ldots, T^L\}$;
7 Reconstruct the paths and starting times from $f^d$ and $f^1$;

**Fig. 3. Pseudocode of the lexicographical solution scheme for the quickest cluster flow problem**

Since the cluster flow units cannot be routed faster than in Algorithm 1, the following result is an obvious consequence of the construction of the flow $f$:

**Lemma 1:** For a given cluster flow instance let $f$ be the feasible solution computed by Algorithm 1. If $T^L = T_d^L$, then $T^L$ is the optimal makespan for the cluster flow instance.

**Example 1:** Fig. 4 gives an example of a network for which Algorithm 1 fails to find a solution with optimal makespan for $d=2$, $b_d=2$ and $b_1=5$. A quickest flow for the *2*-clusters can send both cluster units via path *u-v-w-x* starting at times *0* and *1*, resulting in $T_2^L = 5$. Then the single flow units cannot be routed such that all of them have reached node $t$ before time *7*, hence we have an overall makespan $T^L=7$. A solution with a strictly better makespan consists of routing the *2*-clusters on path *u-w-x* starting at times *0* and *1*, giving $T_2=6$. Then it is possible to route all the single flow units such that they also reach the target $t$ before or exactly at time *6*, hence this solution has a better makespan of $T$=*6*.

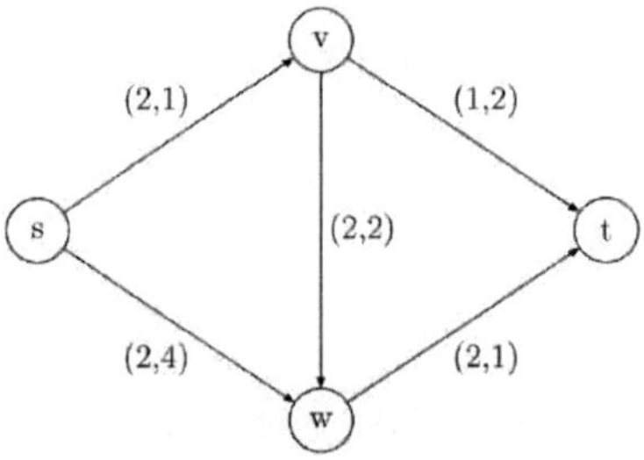


**Fig. 4. Special network. The edge-labels represent capacities and transit times ($c_e,\tau_e$)**

In the following, we will show that Algorithm 1 finds, however, an optimal solution for a subclass of dynamic *s-t*-networks, so-called series-parallel graphs. An *s-t*-graph $G$ is called *(two-terminal) series-parallel (SP)* if it consists either of the nodes $s$ and $t$ connected by a single edge $(s,t)$ or it can be obtained from two series-parallel graphs $G_1$ and $G_2$ with sources $s_1$ and $s_2$ and targets $t_1$ and $t_2$ by one of the following operations:

- *Serial composition:* Identify $s_2$ with $t_1$. Then the source $s$ of $G$ is node $s_1$ and the target $t$ is $t_2$.
- *Parallel composition:* Identify $s_1$ with $s_2$ and $t_1$ with $t_2$. Then the source $s$ of $G$ is the common node $s_1 = s_2$ and the target is $t_1 = t_2$.

Additionally, series-parallel graphs can be characterized as follows: A graph $G$ is called series-parallel if and only if it does not contain a subgraph which is homeomorphic[1] to the graph shown in Fig. 4 [10]. A *series-parallel dynamic network* $N = (G,c,\tau)$ is a dynamic network with $G$ being series-parallel.

The algorithm proposed in [11] computes an earliest arrival flow for a given time horizon on series-parallel networks in polynomial time. Since any earliest arrival flow is in particular a maximum dynamic flow, this algorithm can be combined with binary search on the time horizon as shown in [8] to compute a quickest flow for a given amount of flow units in polynomial time.

---

[1] [10]: A graph $G$ contains a subgraph *homeomorphic* to a graph $H$ if $H$ can be obtained from $G$ by a sequence of the following operations:

- Removal of an edge
- Replacement of two edges $(u,v)$ and $(v,w)$ by $(u,w)$ when these edges are the only ones that are incident with $v$

**Theorem 1:** If the quickest flow problem in line 1 of Algorithm 1 is computed by the algorithm of [11], then it finds an optimal solution for the quickest cluster flow problem in series-parallel dynamic networks.

A proof of Theorem 1 can be found in the full version of the paper.

It is crucial for this result that the single flow units have size 1, since then the capacity along paths cannot be wasted. Example 2 illustrates this fact. This example also shows that the approach of Algorithm 1 is not suitable for more than two different cluster sizes.

**Example 2:** Fig. 5 shows a series-parallel network for which the approach of Algorithm 1 fails for two different cluster sizes none of which is equal to one (non-unit cluster flows). For $d_1=3$, $b_{d_1}=2$, $d_2=2$, $b_{d_2}=2$ a solution found by an analogue procedure to Algorithm 1 routes the *3*-clusters first using the upper edge and achieves by routing the *2*-clusters with respect to the remaining capacities a makespan of at least *3*. A better solution can, however, be obtained by routing one of the *3*-clusters via the lower edge and both *2*-clusters on the upper one resulting in an overall makespan of *2*.

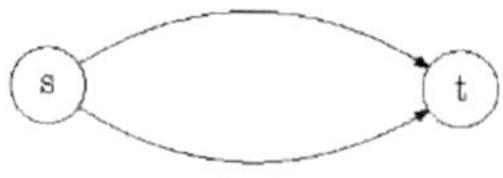


**Fig. 5. Counterexample for cluster flows with non-unit clusters. The edge-labels represent capacities and transit times ($c_e$,$\tau_e$)**

## 2-Approximation for QCFP

In this section we examine upper and lower bounds on the optimal makespan of the QCFP and derive a polynomial time 2-approximation algorithm for quickest cluster flows. This implies that the makespan of the solution obtained by Algorithm 1 is at most twice the value of the optimal makespan.

Let $C = (N,d,b_1,b_d)$ be a cluster flow instance. For the network $N = (G,c,\tau)$ let $T_1$ be the time horizon of an independent quickest flow problem with $b_1$ single flow units and $T_d$ the time horizon of a independent quickest flow problem with $b_d$ cluster flow units. As explained in the previous section, the latter problem can also be computed as a quickest flow problem.

**Theorem 2:** For $\underline{T} = max\ \{T_1, T_d\}$, $\overline{T} = \underline{T} + min\ \{T_1, T_d\}$, and $T^*$, the optimal makespan for the quickest cluster flow problem, we get the following results.

1. $\underline{T} \leq T^* \leq \overline{T}$;
2. A 2-approximation for $T^*$ can be computed in polynomial time.

**Proof:**

1. $\underline{T} \leq T^*$, since $\underline{T}$ is the shortest possible time needed to route either the single or the cluster flow units independently.
   $T^* \leq \overline{T}$, since $\overline{T}$ is obtained by starting the routing of the second class of flow units after finishing the routing of the first class.
2. Follows since $T_l$ and $T_d$ can be computed as quickest flow problems in polynomial time [8] and from part 1 with $\underline{T} \leq T^* \leq T_l + T_d \leq 2\underline{T} \leq 2T^*$.

*QED*

Notice that the 2-approximation bound $\underline{T} \leq T^* \leq 2\underline{T}$ may be tight as can be seen in the following example:

**Example 3:** Fig. 6 gives a simple network for which the approximation bound of $2$ is tight. For $d=3$, $b_d=2$ and $b_l=6$ we have $\underline{T} = max\ \{T_l, T_d\} = 2$. For the optimal makespan it holds that $T^* = 4 = \overline{T} = \underline{T} + min\ \{T_l, T_d\} = 2\underline{T}$.

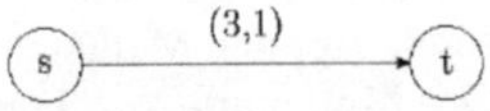


**Fig. 6. Simple network where the approximation bound of 2 is tight. The edge-labels represent capacities and transit times ($c_e, \tau_e$)**

**Acknowledgments** This paper is supported in part by the Federal Ministry for Education and Research (Bundesministerium für Bildung und Forschung, BMBF), Project REPKA, under FKZ 13N9961 (TU KL). We thank Heike Sperber, Technical University of Kaiserslautern, for several fruitful discussions on flows in series-parallel networks.

## References

1. Hamacher, H.W., Tjandra, S.A.: Mathematical Modelling of Evacuation Problems – A State of the Art. Pedestrian and Evacuation Dynamics. 227-266 (2002)
2. Mawson, A.R.: Understanding Mass Panic and Other Collective Responses to Threat and Disaster. Psychiatry: Interpersonal and Biological Processes. 68(2), 95-113 (2005)
3. Hamacher, H.W., Heller, S., Klein, W., Köster, G., Ruzika, S.: A Sandwich Approach for Evacuation Time Bounds. To appear in conference proceedings PED (2010)

4. Kneidl, A., Thiemann, M., Borrmann, A., Ruzika, S., Hamacher, H.W., Köster, G., Rank, E.: Bidirectional Coupling of Macroscopic and Microscopic Approaches for Pedestrian Behavior Prediction. To appear in conference proceedings PED (2010)
5. Kleinberg, J.M.: Single-Source Unsplittable Flow. Annual Symposium on Foundations of Computer Science. 37, 68-77 (1996)
6. Ford, L., Fulkerson, D.: Flows in Networks. Princeton University Press. Princeton, New Jersey (1962)
7. Ahuja, R.K., Magnanti, T.L., Orlin, J.B.: Network Flows – Theory, Algorithms and Applications. Prentice Hall, New Jersey (1993)
8. Burkard, R.E., Dlaska, K., Klinz, B.: The Quickest Flow Problem. Mathematical Methods of Operations Research. 37(1), 31-58 (1993)
9. Tjandra, S.A.: Dynamic Network Optimization with Application to the Evacuation Problem. Ph.D. Thesis, University of Kaiserslautern (2003)
10. Valdes, J., Tarjan, R.E., Lawler, E.L.: The recognition of series parallel digraphs. Proceedings of the Eleventh Annual ACM Symposium on Theory of Computing. ACM (1979)
11. Ruzika, S., Sperber, H., Steiner, M.: Earliest Arrival Flows on Series-Parallel Graphs. Report in Wirtschaftsmathematik 122. TU Kaiserslautern (2009)
12. Bein, W.W., Brucker, P., Tamir, A.: Minimum-Cost Flow Algorithms for Series-Parallel Networks. Discrete Applied Mathematics. 10, 117-124 (1985)
13. Klinz, B., Woeginger, G.J.: Minimum-Cost Dynamic Flows – The Series-Parallel Case. Networks. 43(3), 153-162 (2004)

# Virtual Reality Simulation of Architectural Clues' Effects on Human Behavior and Decision Making in Fire Emergency Evacuation

**Sh. F. Abu-Safieh**

Applied Science University
Amman, Shafa Badran 11931 Jordan
Corresponding author: shaden74@yahoo.com

**Abstract** Disasters analyses have brought the interest in improving response to emergencies. In many cases there is a need to understand how people within a built environment react in a fire building emergency. Architectural clues play a significant part in the decision making and time taken to evacuate due to an emergency. In this study, virtual reality simulated experiments have been constructed to study the human ability to stay oriented while moving through spaces, evaluating alternatives and making decisions. Two variable models were constructed and implemented in a simulated built environment, in which 100 subjects were tested as in evacuation experiment. Data analysis shows how these variables affect human behavior on each critical decision points. Results showed that some architectural clues such as: windows and colors are important factors in the process of determining and following a route in fire emergency evacuation.

## Introduction

In built environments we need to understand human behavior in fire emergency situations. These studies involve the interaction between psychologists, sociologist, engineers, and others professionals.

Most of the evacuation models concern on how people think and act before the start of their evacuation. In fires, there are major factors affecting human wayfinding actions, spatial behavior, and decision making. Such as: persons' age, education, experience, culture, mental and physical capabilities, social situations and architectural clues (Spearpoint, M J., 2008). In building fire cases, people are put in stressful unfamiliar situation, where they have to take rapid decision making using the information that is available. This information collection and decision making may sometimes indicate panic. And this will happen when people are under extreme life threatening conditions especially when they are going through evacuation process (Fahy, 2009). At panic situations, most of the people do not use the emergency direction signs; as a result, they look for other environment's

R.D. Peacock et al. (eds.), *Pedestrian and Evacuation Dynamics*,
DOI 10.1007/978-1-4419-9725-8_31, 

clues to escape (Malek, 2006). Achieving an efficient, clear and comfortable circulation system is an important aspect in large complex building.

Many researchers studied the pre-movement behavior model to represent the rational human behavior pattern in fires or emergencies, and to integrate the major variables that affect human perception, cognition process and behavior reactions for people in fire emergency cases. Some were interested in reading supporting information of an emergency task, trying to speed up the time for decision making in emergency situations. While Fire safety engineers try to realize the full value mathematical models that help in human behavior prediction during evacuations. This helps them in designing and implements safety measurements which reduce and control the impact of fire (Chen, 2008).

And to understand the wayfinding process in a complex building, various virtual escape games have been developed. These studies help in explain people's wayfinding behavior in unfamiliar buildings. The aim of this research is to find architectural clues which will help people choosing correct decisions in their decision-making process. We focused on virtual reality simulation In order to predict some spatial clues which people aware of, while performing a wayfinding task.

## Background

To understand the human response to clues and decision-making in wayfinding, important concepts and scientific backgrounds for this research have been introduced below.

### *Spatial cognition*

Spatial Cognition is concerned with the acquisition, organization, utilization, and revision of knowledge about spatial environments. This will enable humans to manage basic and high level cognitive tasks in everyday life. Many disciplines work together to understand spatial cognition in humans and in technical systems.

According to Raubal, cognition is acquisition, storage and retrieval, manipulation, and use by humans (Raubal, 2001). Cognitive is part of the mind: brain and nerve system which is part of social and physical world. Location, size, distance, direction, separation and connection are spatial properties. Spatial orientation is one of the cognitive skills and it is the ability to orient oneself in space.

## *Wayfinding*

Wayfinding term has been introduced in the late 70s; it replaced the term "spatial orientation." Wayfinding reflects the approach to studying people's movements and their relationship to space. It is "the process used to orient and navigate. The overall goal of wayfinding is to accurately relocate from one place to another in a large-scale space" (Gluck, 1990). In other words, way finding is "the ability to find a way to a particular location in an expedient manner and to recognize the destination when reached" (Peponis, et. al., 1990). Even more importantly, this approach opens up new ways to design for people's spatial behavior. It is a very important aspect of daily life, especially in emergency situations. In large building complexes, wayfinding shape the setting, affect the choice of the circulation system, and contribute to the design of the interior.

Signs are not the most important means of providing wayfinding information in a built environment settings. People during their movements have to collect circulation information and clues in order to find their way. They have to find out where to enter a building and where to exit it. They have to recognize destination and to identify the horizontal and vertical circulation systems. In other words, they have to understand circulation systems. This will help in their decision-making process.

## *Evacuation process*

Evacuation process contains several overlapping stages: awareness stage, pre-evacuation stage, evacuation stage, and post-evacuation (Spearpoint M J., 2008).

*Awareness stage,* in this stage people become aware of fire by seeing fire or smoke or by smelling unfamiliar smell or by hearing an alarm system.

*Pre-evacuation stage,* or the pre-movement stage. In this stage people evaluate the information available and decide their actions. This will last for few seconds or hours. The behaviors in this stage are complicated and a subject of continuous research.

*Evacuation stage,* after decision has been made, evacuation process will start. If the building is unfamiliar to people they will look for exit route by using their wayfinding strategies. Their decisions depend on individual's perceived level of threat and other behavioral aspects. Choosing the escape rout will depend on the exit route signage: lighted signs, reflective signs, floor lightning, and other means. At the same time, well designed built environment with clear clues will help in this stage, and in correct decision making.

## *Virtual reality*

Virtual is "something that can be done or seen using a computer and therefore without going anywhere or talking to anyone" (Cambridge University Press, 2006). Virtual Reality (VR) has been described as "a magical window onto other worlds, from molecules to minds," (Rheingold, 1991). It helps us in learning how people visualize and interact with objects, in order to enhance their navigational awareness efficiently. In this technique, we can provide a controlled environment in wayfinding researches. VE is a logical technology as an evaluation tool in architecture (Bertol, 1997). VE will allow humans to actively interact with the environment in various ways. De Kort, Ijsselsteijn, Kooijman, and Schuurmans (2003), confirmed that VE could be a prospective tool for environmental behavioral research. According to Riecke, van Veen, and Bülthoff (2002) VE offers experimental research conditions that are easy to define, control, and duplicate. Moreover, VE is able to create experimental research conditions that would be difficult to create in the real environment (Péruch, Gaunet, Thinus-Blanc, & Loomis, 2000). New techniques have been developed by researchers as a Disaster Management Simulator to study the evacuation behavior and the effect of the building design on that evacuation behavior, in particular on way finding (Kobes, 2009) . In this research, virtual environments were used to test some architectural variables and their effect on people's decisions making.

# Methods

Most of the wayfinding problem refers to lack of signage system, or the panic which occurs within people during evacuation in fire emergency cases. And the hardest decision making process while navigate through built environments occurs at the intersection points.

The two experiments in this research focus on environmental clues that perceived by wayfinder while they are trying to solve a wayfinding task in a crisis situation and their actions based on such information. Virtual animated complex buildings were designed to investigate if color (warm / light) and transparency (windows) affect the spatial orientation and decision making of humans during emergency cases. Two different environments were designed to measure this effect. Both of them are consisting of ten T intersections (figure 1).

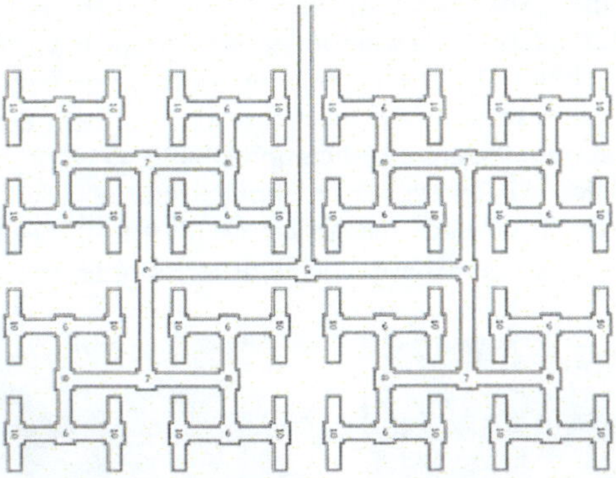

**Fig. 1. Floor plan of the animated experiment (ten intersections)**

***Experiment 1*:** In this experiment, we intended to test the transparency factor's effect. With different orders two kinds of corridors were distributed at each intersection: the first one has no windows at all while the second one has windows (80% voids) (figure 2). At each intersection, the subjects were asked to choose one of the two corridors they are facing, while navigating to escape from the virtual animated complex building.

**Fig. 2. Experiment one - transparency variable**

***Experiment 2*:** In this experiment, we intended to test the color factor's effect. Two corridors at each intersection were found. One of them was colored with light color (white, green, blue) and the other with warm color (red, orange, yellow) (figure 3). The subjects were asked to choose one of them during their spatial orientation to escape from the animated complex building.
100 students (53 males, 47 females) from applied science university were participated in this research. Their ages ranged between 19 to 22 years old. The same subjects went through the two experiments in different order at different days.

**Fig. 3. Experiment two - color variable**

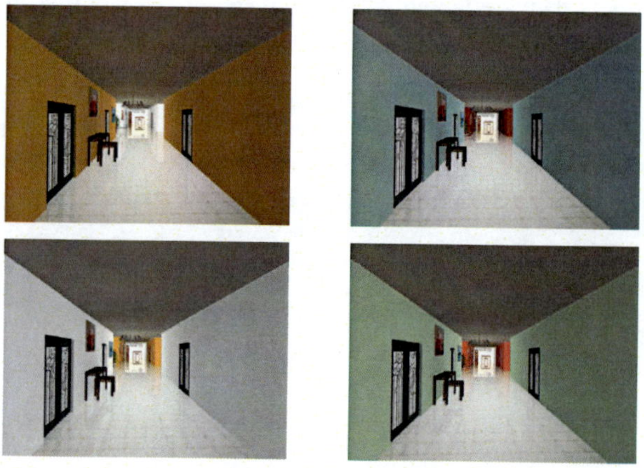

**Fig. 4. Experiment two - choosing different light and warm colors**

## Results

This paper presents a model of the wayfinding process in a fire building emergency. The model integrated elements of people's perception and cognition, focusing on how people make sense and take of their wayfinding environment and take decisions at intersection points. The wayfinder gains knowledge about the building through visual perception of architectural clues at decision points. One would decide which path to choose with the use of these clues at decision points in order to find his/her way to escape from the building.

In experiment 1, for each subject, the number of closed corridors and transparency corridors which he/she chose were counted. Same procedure was done for experiment 2, where number of light colored corridors and warm colored corridors were counted. The frequency test was applied to the number of transparent and closed corridors for all subjects. Figure 5 shows the relation between the frequencies for each kind of corridors. In figure 6, percentage of each corridors choices, opened and closed, were calculated.

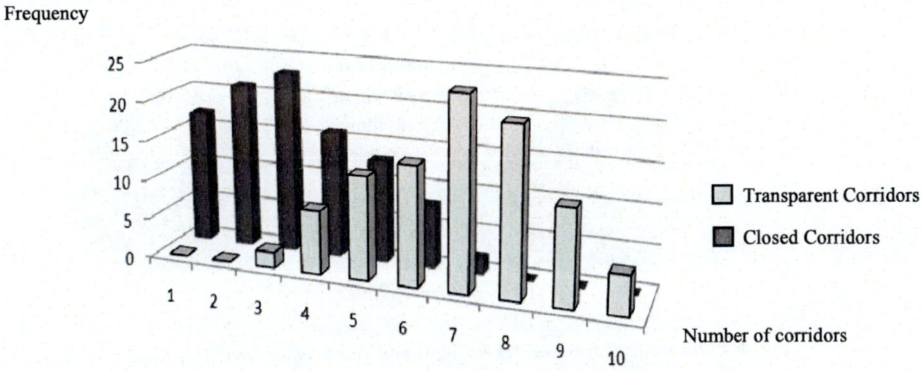


**Fig. 5. Frequencies for opened and closed corridors for experiment one**

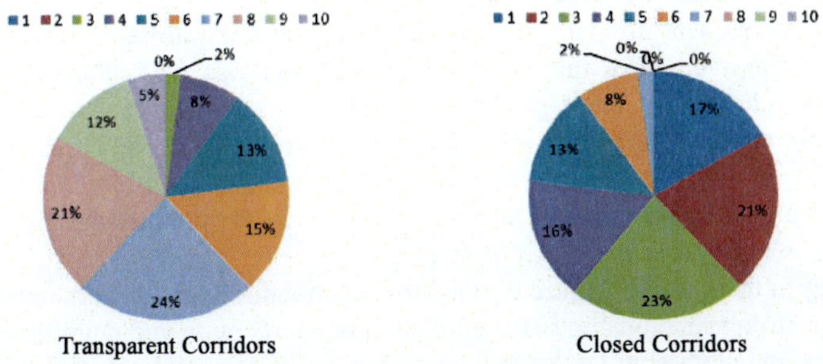


**Fig. 6. Percentage of frequencies for opened and closed corridors for experiment one**

The frequency test was applied to the number of cool colored corridors and the warm colored corridors for all subjects. Figure 7 shows the relationship between the frequencies for each kind of corridors. In figure 8, percentage of each corridor choices, cool and warm, were calculated.

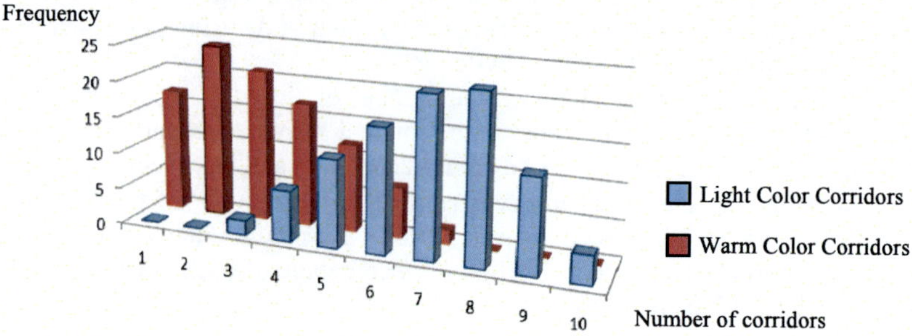


**Fig.7. Frequencies for light and warm color corridors for experiment two**

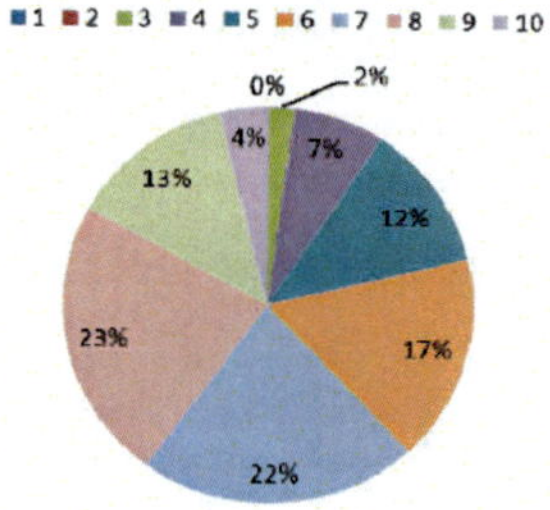


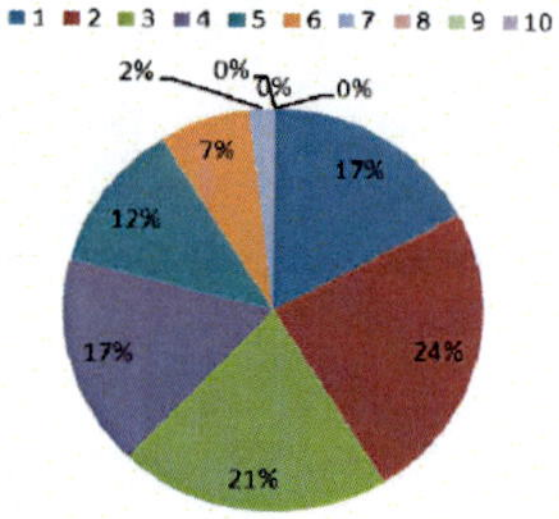


**Fig. 8. Percentage of frequencies for light and warm color corridors for experiment two**

The average of the chosen transparent corridors was 6.87. Based on this value, the percentage of subjects who chose more than 7 transparent corridors is 62%. This was with a standard variation value of 1.6. These results show that the majority of people are moving towards open transparent corridors during their evacuation process. The average of the chosen light corridors was 6.91. Therefor, and based on this value, the percentage of subjects who chose more than 7 light colored corridors equal to 64%. The standard variation value was 1.6. These results show that the majority of people are moving towards light colored corridors through their process of navigation to escape from complex building during emergency cases.

According to the statistics, there was a correlation value of 0.73. This indicates that there is a strong relationship between the human perception and cognition of the environmental clues (light color and transparency), which affect decision making at intersection points in spatial orientation during fire emergency cases. Figure 9 shows the relation between light color and transparent corridors frequencies.

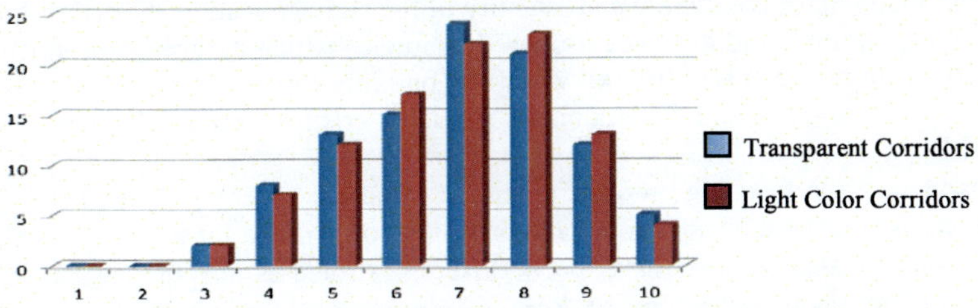


**Fig. 9. Frequencies for transparency and light color corridors**

Table 1 shows the final results of the frequency test, which were found in the two experiments conducted in this paper.

**Table 1. Frequency test results for the two experiments**

| Number of corridors | Transparent corridors | Light color corridors |
|---|---|---|
| 1 | 0 | 0 |
| 2 | 0 | 0 |
| 3 | 2 | 2 |
| 4 | 8 | 7 |
| 5 | 13 | 12 |
| 6 | 15 | 17 |
| 7 | 24 | 22 |
| 8 | 21 | 23 |
| 9 | 12 | 13 |
| 10 | 5 | 4 |
| Mean | 6.87 | 6.91 |
| Percentage more than 7 | 62% | 64 |
| STDEV | 1.68 | 1.64 |

## Discussion and Conclusions

Virtual environment were used as a prospective tool for wayfinding research because of its ability to control the different variables of the environment. VE also provides similar movement experience as the experiments in the real environment (Raubal, 2001). In this research, by using virtual animated experiments, some architectural clues were tested to improve evacuation plans with lower probability of failure, enabling faster and more efficient evacuation of a building in the event of emergency escape. Hajibabai, et. al found that the better placement of the cues and optimum planning of the quality and quantity of the signage lead to shorter evacuation time from the building. They proposed strategies which describe how spatial cognition of human can find the exit ways in a building fire emergency by the use of the building signage and the landmarks in such a crisis situation (Hajibabai, et. al., 2007).

This research is focusing on other elements of people's perception and cognition, such as color and transparency, and on how people make sense of their way-finding environment. There have been different strategies for the evacuation process in fire emergencies. One of them is helping people to use environmental clues in way finding, to enhance correct decisions for escaping from the building. The significance of this paper is the architectural elements in the built environment that helps people in choosing the right path to escape. People need information about the destination to be presented at every decision point (Leila 2006). When we use the transparency in corridors which coming next to each decision making points, we will enhance the ability to collect more information about our environment and our final destination. Moreover it will help people in reducing the panic situation which happened usually in emergency cases.

Many researches investigate Color psychology which refers to the effect of color on human behavior and feelings. They found that color plays a significant role in behavior and performance. For example, "Red turns up on top for associations for a number of different things: angry, aggressive, strong, courageous, frustrated and lustful," according to Stephen E. Palmer, a professor of psychology and cognitive science at the University of California, Berkeley. Orange the color of energy and warmth while green is the color of balance, harmony, caring, tenacious self-reliance, and healing. Blue the color of calmness, concentration, healing, and relaxation. Using light colors like blue, green and white for corridors, will gives people kind of balance, concentration and relaxing at panic situations occurring during fire emergencies evacuation.

Some limitations were noted and could be used as factors for establishing future studies. The experiment was conducted with students most of them are architects and civil engineers. The limited diversity of these participants raised an issue of the external validity of the study, which was the problem of generalizing the Results, because these students might have better skills in understanding building layouts. Additional experiments should be conducted with different types of participants to confirm the results of this study. In addition, individual differences can be an interesting are to study according to the research foundlings.

## References

1. Chen, P., Nguyen, T.: Computer-Aided Visual Communication for Wayfinding in Emergency in Indoor environment. The 25$^{th}$ International Symposium on

Automation and Robotics In Construction, June 26-29, Vilnius, Lithuania (2008)
2. Fahy, R.F., Proulx, G.: 'Panic' and Human Behavior In Fire. 4th International Symposium on Human Behaviour in Fire, July 13, Robinson College, Cambridge, UK, pp. 387-398 (2009)
3. Gluck, M.: Making Sense of Human Wayfinding: A Review of Cognitive and Linguistic Knowledge for Personal Navigation with a New Research Direction. Myke Gluck School of Information Studies, Syracuse University, Syracuse (1990)
4. Hajibabai, L., Delavar, M., Malek, M.R., Frank, A.U.: Spatial Cognition and Wayfinding Strategy During building Fire. Dynamic in Spatial Interactions. ICSC (2006)
5. Hajibabai, L., Delavar, M., Malek, M.R., Frank, A.U.: Agent-Based Simulation of Spacial Cognition and Wayfinding in Building Fire Emergency Evacuation. Geomatics Solution for Disaster Management, pp. 255-270. Springer, Berlin, Heidelberg (2007)
6. Kobes, M., Oberijé, N., Groenewegen, K., Morsche, T.: Serious Gaming for Behavioural Assessment and Research in Case of Emergency: An Evaluation of Experiments in Virtual Reality. Netherlands Institute for Safety, Arnhem, The Netherlands VU University Amsterdam, The Netherlands (2009)
7. Passini, R.: Wayfinding in Architecture, Second Edition. Van Nostrand Reinhold, NY (1992)
8. Peponis, J., Zimring, C., Choi, Y.K.: Finding the Building in Wayfinding. Environment and Behavior, 22-5, pp. 555-590 (1990)
9. Péruch, P., Gaunet, F., Thinus-Blanc, C., Loomis, J.: Understanding and learning virtual spaces. In: Kitchin, R.M., Freundschuh, S. (eds.), Cognitive mapping: Past, present and future, pp. 108-124. Routledge, London (2000)
10. Proulx, A.: The Use of Virtual Environment as a method of wayfinding research in architecture (2006)
11. Raubal, M.: agent based simulation of human wayfinding: a perceptual model for unfamiliar buildings. PHD thesis, Vienna University of Technology, Astria (2001)
12. Riecke, B.E., von der Heyde, M., Bülthoff, H.H.: Spatial updating experiments in Virtual Reality: What makes the world turn around in our head?. Tübingen Perception Conference (TWK), p. 162. Tübingen, Germany (2002)
13. Satalich, G.A.: Navigation and wayfinding in Virtual Reality. Finding the Proper Tools and Cues to enhance Navigational Awareness. Washington (1995)
14. Spearpoint, M.J.: Fire emergencies and people. New Zealand Science Teacher. No. 117, pp. 14-16 (2008)

# A Multi-Grid Model for Evacuation Coupling With the Effects of Fire Products

**Z. Fang, W. Song, and H. Wu**

SKLFS, University of Science and Technology of China.
Hefei, Anhui 230026 P.R.China
Corresponding author: wgsong@ustc.edu.cn

**Abstract** The effects of fire products on pedestrians are introduced into a multi-grid evacuation model, in which the space is discretized into smaller grids with the size of 0.1 m × 0.1 m and each pedestrian occupies 5×5 grid sites. The effects of fire products on pedestrians consist of two parts: the desired movement direction and the step frequency. The data of fire products obtained from the simulation results of a well-founded computational fluid dynamic (CFD) program, the Fire Dynamics Simulator (FDS) developed by the National Institute of Standards and Technology (NIST). With the multi-grid model, we investigated the walk routes of pedestrians in fires, and the evacuation time in scenarios with different fire intensities, pre-movement times or door widths.

## Introduction

By now, computer models have been widely used to predict the movement of people during egress and then evaluate whether an evacuation process is efficient, for instance, social force model [1, 2], lattice-gas model [3, 4] and floor field model[5, 6]. These models adopt reasonable movement algorithm and mode, thus they can reproduce typical pedestrian behavior and capture microscopic and macroscopic characteristics of pedestrian traffic during egress. In order to study the evacuation process in more detail, researchers have also paid more attention to finer discrete model, in which a pedestrian occupies more lattice sites. Kirchner [7] has studied the discretization effects in cellular automata models for pedestrian dynamics by reducing the cell size. Song and Xu [8] quantified the force concept of the social force model and built the multi-grid model. They also investigated the effects of finer discretization on pedestrian dynamics using multi-grid model and found that the evacuation time is associated with the grid size and the length of the time step [9]. Weng et al. [10] investigated crowd flow going outside a hall the same as that in Ref.[4] with a small-grid analysis model. The utilization of finer grids is helpful to simulate the collective phenomena such as jam, block and free movement, and at the same time remains relative high simulation efficiency as well.

R.D. Peacock et al. (eds.), *Pedestrian and Evacuation Dynamics*,
DOI 10.1007/978-1-4419-9725-8_32, 

It has been recognized that exposure to toxic smoke products is one of the hazards confronting people in fires [11-13]. Both smoke density and irritation appear to effect the walking speed [11]. Poor visibility, irritation of the eyes, heat or a combination of factors may cause evacuees to stop or turn back [12]. The physiological effects of exposure to toxic smoke and heat in fires result in varying degrees of incapacitation which may also lead to death or permanent injury[13]. So it is important and necessary to introduce the effect of fire products on the occupants into the evacuation model. As well-founded computational fluid dynamic (CFD) program, the Fire Dynamics Simulator (FDS) [14]and Consolidated Model of Fire and Smoke Transport (CFAST)[15], developed by National Institute of Standards and Technology (NIST), have been selected as the section of fire spread simulation in some evacuation models. Using the export data form FDS, effect of the fire products like temperature, CO concentration and visibility on the occupants is considered in some agent based evacuation models [16-19] and in the evacuation simulation of a large stadium[20]. There is a software link between buildingEXODUS evacuation model and the CFAST zone model [21], which allows CFAST history files to be automatically passed to the buildingEXODUS model, thereby enabling the buildingEXODUS and CFAST models to interact in a relatively straight forward manner.

As mentioned above, the model considering not only the evacuation process but also the effect of fire products on this process has attracted more and more attention. However, it should be noticed that the fire simulator like FDS or CFAST reproduces a given building structure exactly, but most of the presented discrete evacuation models could not do for the reason that the size of their basic grid is comparable to the personal size and then the size of exits, obstacles and channels has to be integral multiples of personal size. Thus, there would be unavoidable error in the process of data transfer between fire simulator and egress model. To decrease this error, in this study, we introduced the effect of fire products on people into a multi-grid model, in which the space is discretized into small grids with the size of 0.1 m × 0.1 m. Meanwhile the FDS is used to predict the generation and spread of fire hazard.

## Model

Generally speaking, fire products include heat, soot and toxic gases etc. It is hard to consider the effect of all the fire products in the evacuation model, as a result we only focus on the effect of smoke temperature and visibility on the human behavior, including the selecting of desired movement direction and the movement speed. We adopt the data of temperature and visibility of a plane at the height of 1.5m, because most pedestrians can feel directly the change of temperature and visibility of this height.

## *Desired movement direction*

The desired movement direction in this paper denotes the direction pedestrians prefer moving to with a higher probability. It is recognized that the direction leading to the exits directly is the first choice to move to when pedestrians escape. In order to reproduce this fact, the static floor field is introduced in the floor field model [5, 6] to describe the shortest distance to the exits for each grid and then to calculate the transition probability to the four around directions. Besides, we introduce the static floor field into the multi-grid model [22] to decide the desired movement direction. However, when fire happens, movement routes of pedestrians not only point to the direction with shortest distance to the exits but are influenced by the fire products. As specified in the SFPE handbook, poor visibility and heat, the two factors we focus on, can cause evacuees to stop or turn back [12]. So we adopt a modified static floor field $S$ in this paper to present the effect of fire products on the desired movement direction.

The modified static floor field $S$ is calculated using a repetitive function of $S^n_{i,j} = S_{i,j} + 1 + S_{add}$. Where $S_{i,j}$ represents the potential distance to the exits of each grid $(i, j)$, $S^n_{i,j}$ is for each neighboring grid in forward, backward, left and right of the grid $(i, j)$ , and $S_{add}$ represents the effect of smoke temperature and visibility. It is obvious that if $S_{add}$ is 0, the action of $S_{i,j}$ is just same as in Ref. [5, 6, 22]. In this present, the value of $S_{add}$ depends on the temperature and visibility of corresponding neighboring grid. Moreover, we assume that the value of $S_{add}$ is a modified hyperbolic tangent function of the value of temperature and visibility as shown in Fig.1.

It is known that the basic hyperbolic tangent function is $y=\tanh(x)$ with the range of $-1<y<1$, moreover absolute value of y has been more than 0.96 when $|x|\geq 2$, although its domain is $-\infty<x<+\infty$, so $x=\pm 2$ can be regarded as turning points of $y=\tanh(x)$ that value of y changes obviously as x increases if $|x|\leq 2$ but be almost changeless otherwise. There are also turning points in the function of fire products' effect on movement direction, for instance, when the extinction coefficient reach 0.151/*m*, occupants began to feel uneasy[11], while smoke density of 1.53 to 2.26/*m* is indicated as severity limit to what normal people will endure[12]. As a result, we assume that effect of smoke density on the desired movement direction is significant when extinction coefficient is between 0.151 to 1.53/*m* but slight otherwise. To agree with the two turning points of basic hyperbolic tangent function, a transition is carried out and then the extinction coefficient part of $S_{add}$ is as follows:

$$S^1_{add} = \tanh(2.9C - 2.438) + 1 \tag{1}$$

where $C$ represents the value of extinction coefficient. Similarly, temperature of $37\,^oC$, which is the normal body temperature, and $100\,^oC$ are assumed as the two turning points for the temperature part of $S_{add}$.

$$S_{add}^{2} = \tanh(0.063T - 4.349) + 1 \tag{2}$$

At last, $S_{add}$ is summation of the two parts:

$$S_{add} = S_{add}^{1} + S_{add}^{2} \tag{3}$$

In addition, as fire spread, the values of temperature and extinction coefficient are changeable every moment. To promote the computational efficiency, we update the temperature and visibility field in the model every 4 seconds instead of each time step. After updating, the static floor field is recalculate, and then the desired movement direction of each pedestrian is determined by the minimal average value of $S_{i,j}$ in the four neighboring areas, just like in Ref. [22]. Thus, pedestrians in our model will avoid smoke of higher temperature and lower visibility when egress in fire.

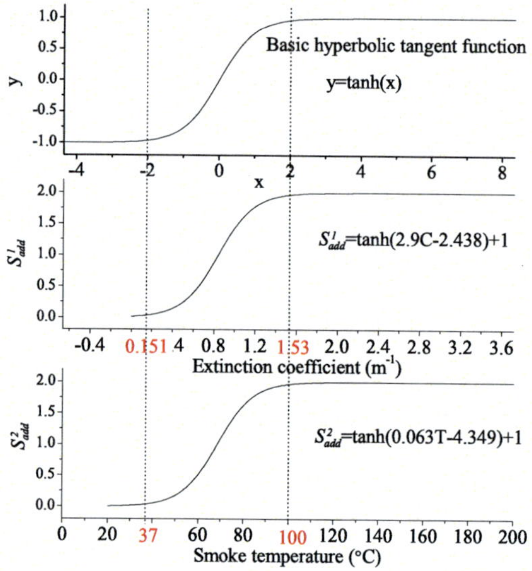


**Fig. 1. Hyperbolic tangent relationship between $S_{add}$ with temperature and visibility**

## *Movement speed*

Previous study carried out by Song et al.[23]and Hoogendoorn et al. [24] indicated that the step frequency is about 2 Hz in normal situation. However, pedestrians would heighten their step frequency when egress under emergency, especially in fire. We analyzed the video about people evacuation process through a hall when an earthquake happened in Panzhihua, China at October 30, 2008 and found that the average step frequency is 3.8 Hz. In addition, it is indicated in Ref. [11] that the walking speed will decrease rapidly as the smoke density increases, and the sharp drop of speed value will happen when the extinction coefficient reaches around 0.5 $m^{-1}$. Therefore, we assume the step frequency of pedestrian decreases exponentially from 3.8 as the extinction coefficient increases and then keeps a constant value of 0.8 Hz when the extinction coefficient reaches 0.5 $m^{-1}$, as shown in Fig. 2. The step frequency of 0.8 Hz is corresponding to a desired speed of 0.56 m/s in this paper for the reason that the step length is 0.7 m i.e. 7 grids' length. Thus, the actual speed when the extinction coefficient reaches $0.5m^{-1}$ in this model will be no more than 0.56 m/s, which fits the theory in Ref. [12] that "ultimate speed in heavy smoke conditions " are within 0.2 to 0.5 m/s.

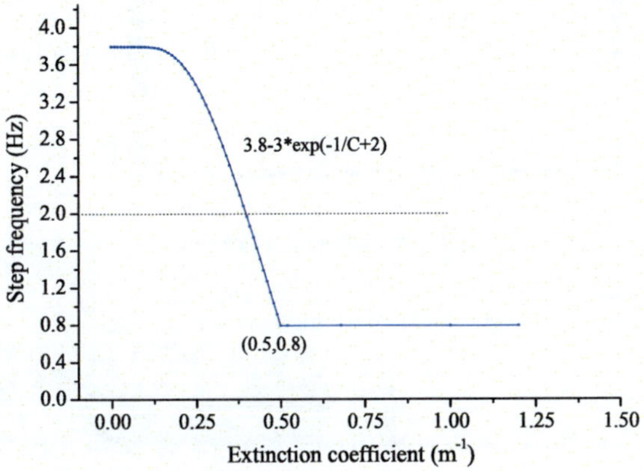


**Fig. 2. Step frequency in fire smoke**

## Simulation and result

We carried out a series of computer simulation based the model described above that 400 pedestrians leave a hall with length×width of 50 m × 50 m and one exit. The width of the exit is D and there is a fire source with length×width of 2 m × 2 m in front of the exit, as shown in Fig. 3. We first investigate the difference of evacuation route between normal condition and fire condition. In order to view the evacuation process directly and conveniently, we import the result of pedestrian movement into a tool for visualizing fire dynamics simulation data, the Smokeview [25]. Fig. 4 shows snapshots from SmokeView of the evacuation process under the two different conditions. In normal condition, the fire source is treated as an obstruction, so pedestrians move towards exit and bypass the obstruction in the way. By contrast, when fire happens, pedestrians will avoid moving towards the fire source, as a result of the effect of temperature and visibility on the selecting of desired movement direction.

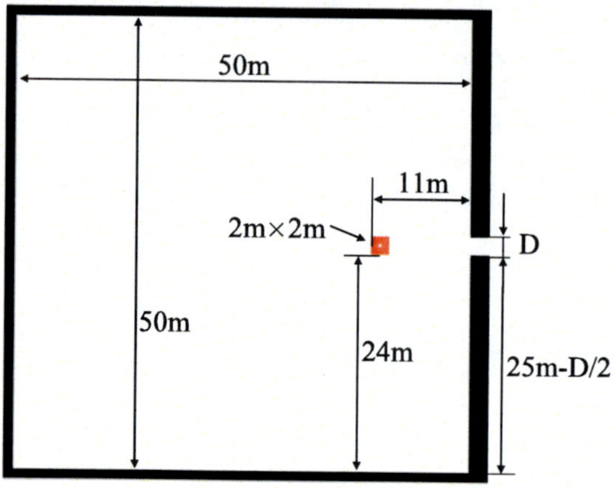


**Fig. 3. The hall with a fire source**

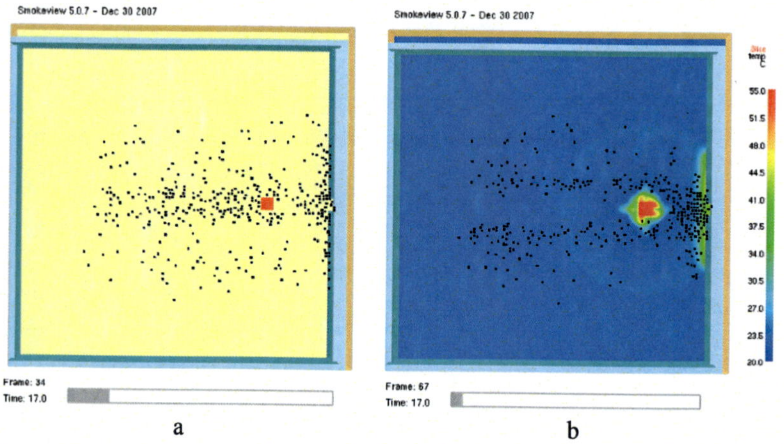


**Fig. 4. Snapshots of evacuation process under normal condition (a) and fire condition (b)**

To further study the effect of Heat Release Rate Per Unit Area (HRRPUA), door width (D) and pre-movement time ($T_p$) on the evacuation process, we simulated the evacuation process in different scenarios and repeated each scenario 10 times. Using temperature and visibility data when HRRPUA is set as 100, 200,..., 2000 $KW/m^2$ in FDS respectively, we simulated the egress process under different fire intensities. It is shown in FIG. 5 that evacuation time increases rapidly as

HRRPUA increases for the reason that the worse and worse visibility results in slower and slower walking speed. However, if visibility is bad enough the effect of visibility on walking speed wouldn't be more serious, so there is little difference among the evacuation times when HRRPUA is more than about 1000 KW/m$^2$. Furthermore, it is shown is Fig. 5 that evacuation time is less than the normal condition when fire is little, which reflects the fact that little fire will make egress process faster but great fire will reduce evacuation efficiency greatly.

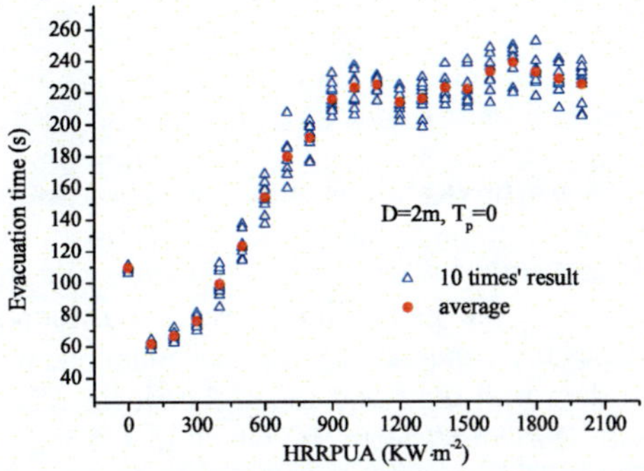


**Fig. 5. Evacuation time against Heat Release Rate Per Unit Area**

Pre-movement time is the time from the ignition of a fire to beginning to escape of pedestrians, and we use movement time to denote the time from beginning to finishing of pedestrians' escape process, so the evacuation time in this paper is their summation. In traditional models, the relationship between pre-movement time and movement time is always ignored. However, it's obvious that when fire happens the longer the pre-movement time is, the more difficult to escape it is for pedestrians due to accumulative fire products. Since the effect of fire products on pedestrians is introduced, our model can present the relationship between pre-movement time and movement time. It is shown in FIG. 6 that the movement time first increases rapidly as the pre-movement increases, because as fire spread, visibility become worse and worse. Furthermore, once the extinction coefficient exceeds 0.5 m$^{-1}$ as smoke accumulates pedestrians will always walk with their ultimate speed, so at last movement time keeps almost changeless. As a result, the pre-movement time affects not only the whole evacuation time directly but also the movement time seriously, so it is necessary to reduce the pre-movement time by promoting fire detection devices, fire alarm equipments, human response and so on.

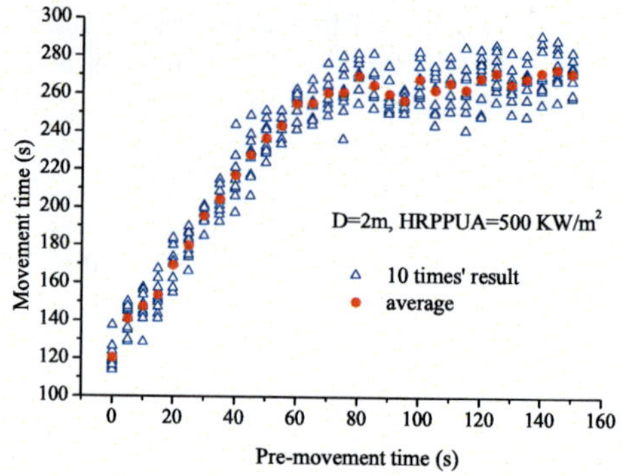


**Fig. 6. Relationship between pre-movement time and movement time**

Figure 7 gives a comparison between normal situation and fire situation of HRRPUA=1000 KW/m$^2$ when the door width increases. It is found that the evacuation time in fire decreases more quickly than normal situation. We further carried out linear fit based on the result plots and found the slopes of the two different conditions are -50.6 s/m, -17 s/m respectively, which means that one more meter width of the exit would result in a reduction in the evacuation time of 50.6 s in fire condition while 17 s in normal condition, so the door width plays an important role in egress process, especially when fire happens.

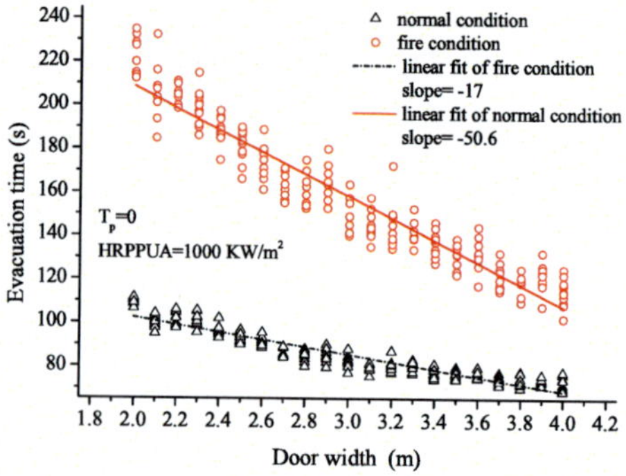


**Fig. 7. Evacuation time against door width**

## Summary

We have introduced the effects of fire products on pedestrians into a multi-grid model for evacuation. The data of fire products comes from the Fire Dynamics Simulator (FDS) developed by National Institute of Standards and Technology (NIST), and the result of pedestrian movement could be imported into the SmokeView in order to view the evacuation process directly and conveniently. Two typical fire products, smoke heat and visibility, are emphasized in our model, and their effects on pedestrians consist of two parts: on the desired movement direction and the step frequency. We have studied the difference of evacuation routes between normal condition and fire condition and repeated the reasonable and actual phenomenon that pedestrians will avoid moving towards the fire source when fire happens. We have also investigated the effect of Heat Release Rate Per Unit Area, door width and pre-movement time on the evacuation process using our model. The results reflect the fact that little fire will make egress process faster but great fire will reduce evacuation efficiency significantly, the movement time increases rapidly as the pre-movement increases, and the door width plays a more important role for occupants escaping from fire.

**Acknowledgments** The study is supported by China National Natural Science Foundation (No. 50678164), Program for New Century Excellent Talents in University (NCET-08-0518)

## References

1. Helbing, D., Farkas, I., Vicsek, T.: Simulating dynamical features of escape panic. Nature. **407**, 487-490 (2000)
2. Helbing, D., Molnar, P.: Social Force Model for Pedestrian Dynamics. Physical Review E. **51**, 4282-4286 (1995)
3. Muramatsu, M., Irie, T., Nagatani T.: Jamming transition in pedestrian counter flow. Physica A. **267**, 487-498 (1999)
4. Tajima, Y., Nagatani, T.: Scaling behavior of crowd flow outside a hall. Physica A. 292, 545-554 (2001)
5. Kirchner, A., Schadschneider, A.: Simulation of evacuation processes using a bionics-inspired cellular automaton model for pedestrian dynamics. Physica A. **312**, 260-276 (2002)
6. Burstedde, C., Klauck, K., Schadschneider, A., Zittartz, J.: Simulation of pedestrian dynamics using a two-dimensional cellular automaton. Physica A. **295**, 507-525 (2001)
7. Kirchner A., Klupfel, H., Nishinari, K., Schadschneider, A., Schreckenberg, M.: Discretization effects and the influence of walking speed in cellular automata models for pedestrian dynamics. Journal of Statistical Mechanics-Theory and Experiment (2004)

8. Song, W.G., Xu, X., Wang, B.H., Ni, S.J.: Simulation of evacuation processes using a multi-grid model for pedestrian dynamics. Physica A. **363**, 492-500 (2006)
9. Xu, X., Song, W.G., Zheng, H.Y.: Discretization effect in a multi-grid egress model. Physica A. **387**, 5567-5574 (2008)
10. Weng, W.G., Pan, L.L., Shen, S.F., Yuan, H.Y.: Small-grid analysis of discrete model for evacuation from a hall. Physica A. **374**, 821-826 (2007)
11. Jin, T.: Visibility and Human Behavior in Fire Smoke. In: DiNenno, P.J., Walton, W.D. (Eds.), The SFPE Handbook of Fire Protection Engineering (Third ed.), pp. 2-42-2-53. Society of Fire Protection Engineers, Bethesda, MD (2002)
12. Jensen, G.: Wayfinding in Heavy Smoke: Decisive Factors and Safety Products Findings Related to Full Scale Tests. IGP AS (2002)
13. Purser, D.: Toxicity Assessment of Combustion Products. In: DiNenno, P.J., Walton, W.D. (Eds.), The SFPE Handbook of Fire Protection Engineering (Third ed.), pp. 2-83-2-171. Society of Fire Protection Engineers, Bethesda (2002)
14. McGrattan, K., Klein, B., Hostikka, S., Floyd, J.: Fire Dynamics Simulator (Version 5), User's Guide. NIST Special Publication, **1019** (2004)
15. Peacock, R., Jones, W., Reneke, P., Forney, G.: CFAST–Consolidated Model of Fire Growth and Smoke Transport (Version 6) User's Guide. NIST Special Publication, **1041** (2005)
16. Tang, F., Ren, A.: Agent-Based Evacuation Model Incorporating Fire Scene and Building Geometry. Tsinghua Science & Technology. **13**, 708-714 (2008)
17. Shi, J., Ren, A., Chen, C.: Agent-based evacuation model of large public buildings under fire conditions. Automation in Construction. **18**, 338-347 (2009)
18. Korhonen, T., Hostikka, S., Heliovaara, S., Ehtamo, H.: FDS+ Evac: Modelling Social Interactions in Fire Evacuation. In: Proceedings of the 7th International Conference on Performance-Based Codes and Fire Safety Design Methods. Bethesda (2008)
19. Korhonen, T., Hostikka, S., Heliovaara, S., Ehtamo, H., Matikainen, K.: Integration of an agent based evacuation simulation and the state-of-the-art fire simulation. In: Proceedings of the 7th Asia-Oceania Symposium on Fire Science & Technology. Hong Kong (2007)
20. Zhu, W., Ding, H., Wang, T.: The application of fire spread and evacuation simulation technology in large stadium. Stochastic Environmental Research and Risk Assessment. **23**, 433-439 (2009)
21. Gwynne, S., Galea, E.R., Lawrence, P.J., Filippidis, L.: Modelling occupant interaction with fire conditions using the buildingEXODUS evacuation model. Fire Safety Journal. **36**, 327-357 (2001)

22. Fang, Z.M., Song, W.G., Zhang, J., Wu, H.: Experiment and modeling of exit-selecting behaviors during a building evacuation. Physica A. **389**, 815-824 (2010)
23. Liu, X., Song, W.G., Zhang, J.: Extraction and quantitative analysis of microscopic evacuation characteristics based on digital image processing. Physica A. **388**, 2717-2726 (2009)
24. Hoogendoorn, S.P., Daamen, W.: Pedestrian behavior at bottlenecks. Transportation Science. **39**, 147-159 (2005)
25. Forney, G., McGrattan, K.: User's Guide for Smokeview Version 5-A Tool for Visualizing Fire Dynamics Simulation Data. NIST Special Publication (2004)

# Large-scale Modeling

# Runtime Optimization of Force Based Models within the Hermes Project

A. Seyfried[1], M. Chraibi[1], U. Kemloh[1], J. Mehlich[1] and A. Schadschneider[2]

[1]Jülich Supercomputing Centre, Forschungszentrum Jülich GmbH,
52425 Jülich, Germany
[2]Institut für Theoretische Physik, Universität zu Köln,
50937 Köln, Germany
Corresponding author: a.seyfried@fz-juelich.de

**Abstract** The aim of the Hermes project is the development of an evacuation assistant to support security services in case of emergency in complex buildings and thus to improve safety at mass events. One goal of the project is to build models for pedestrian dynamics specifically designed for forecasting the emergency egress of large crowds faster than real-time using methods applied in high performance computing. We give an overview of the project and the modeling approaches used focusing on the runtime optimization and parallelization concepts.

## Introduction

Multifunctional building structures in combination with a wide range of large-scale public events present new challenges for the quality of security concepts. Prescriptive construction and planning regulations ensure in general that in case of an emergency everyone present is able to quickly leave the danger zone by specifying e.g. minimal width and maximal length of escape routes. In the event of loss of rescue routes due to fire or other risks, however, dangerously high crowd densities and bottlenecks can occur. To prevent such critical situations optimal crowd management needs accurate and actual information about the current status. Usually in complex buildings the decision makers miss information e.g. the number of people in the danger zone, how the loss of escape routes influences the evacuation time or where dangerous congestions with long waiting times will occur in the course of the evacuation. The evacuation assistant, developed within the Hermes project and outlined in this contribution will close this gap and support the decision makers to rate the actual danger, to decide a successful evacuations strategy and to optimally employ the security staff. The ESPRIT arena in Düsseldorf (Germany) provides a venue for testing the evacuation assistant. The example of this multifunctional arena with a capacity of 60,000 spectators will show how crowds of people at big events can be guided – also considering the current risk situation. A test system of the assistant will be installed in 2011.

R.D. Peacock et al. (eds.), *Pedestrian and Evacuation Dynamics*,
DOI 10.1007/978-1-4419-9725-8_33, 

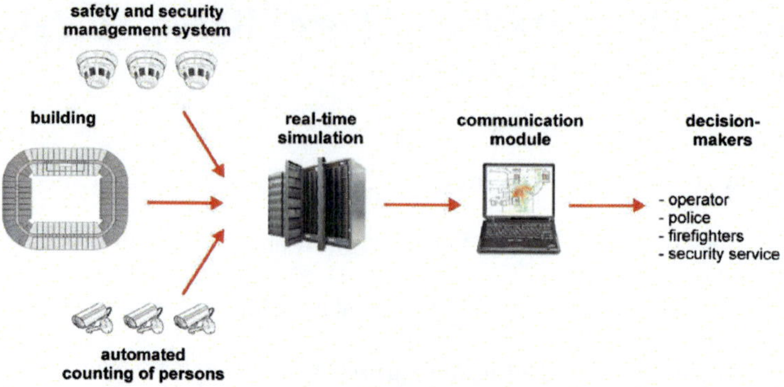


**Fig. 1. Schematic diagram of the evacuation assistant**

Fig. 1 shows the layout of the evacuation assistant. Using automated counting of persons at entrances and doors the present dispersal of people in the building is delivered to the decision makers and the simulation core. The safety and security management system provides information about escape routes blocked due to smoke, locked doors or other dangers. Using the actual data about the distribution of people and the availability of rescue routes, a parallel computer will generate faster than real-time simulations to predict the movement of all people during the next 15 minutes and updates it at one-minute intervals. The simulation results include e.g. potential dangerous congestion areas or evacuation times. Moreover a macroscopic network model will calculate the optimal distribution of occupants on the available routes. A communication module will provide this information to the emergency teams on site. Various universities, industrial partners and users cooperate in this project. For an overview we refer to [1].

## Modeling and Simulation

Different modeling strategies are combined to obtain reliable predictions from the simulations and achieve an optimal performance. One approach uses cellular automata (CA) the other is a spatially continuous forced-based model. Both approaches have advantages and drawbacks and we refer to [2] for a deeper discussion. A crucial point in the application of these models to security sensitive tasks within the Hermes project is their quantitative verification and calibration. Regarding the reliability of simulation results basing on these models we are still at the beginning, see e.g. [3]. This lack is due to the contradictory experimental data base for model testing [2] and an open discussion about the principles of validation and calibration, see section 2.1 in [4]. In the following we introduce the basics of the modeling approaches performed for the Hermes project.

## *Cellular Automata*

In CA space, time and state variables are discrete. One of their attractive properties is that they allow for an intuitive definition of the dynamics in terms of simple rules. In CA models of pedestrian dynamics the space is discretized into small cells which can be occupied by at most one pedestrian (exclusion principle). The cell size corresponds to the space requirement of a person in a dense crowd. Time is also discrete and the pedestrians move synchronously in each timestep which corresponds to a few tenths of a second in real-time. Most models represent pedestrians by particles without any internal degrees of freedom which move to one of the neighbouring cells with transition probabilities determined by three factors: (1) the desired direction of motion (e.g. given by origin and destination), (2) interactions with other pedestrians, and (3) interactions with the infrastructure (walls, doors, etc.). In the simplest models, the latter two factors are only taken into account through an exclusion principle, i.e. occupied or wall cells are not accessible.

In the Hermes project the *floor field model* [5,6] is used which is based on a more sophisticated representation of these interactions that allows the capture of longer-ranged interactions. This leads to more realistic results, especially for collective effects, self-organization phenomena and in evacuation scenarios [5,7]. For a detailed definition of the model, we refer e.g. to [5,6,8]. Here it is essential that the transition probabilities do not only depend on the current local configuration of pedestrians in the neighbourhood, but also takes into account their dynamics. This is achieved through the introduction of *floor fields* that allow the model to be kept simple, but nevertheless yielding realistic results.

Cellular automata are through their discreteness very efficient in simulations. On the other hand, the reduced spatial resolution determined by the cells is often unsatisfactory in applications, e.g. if the infrastructure needs to be represented in more detail. Therefore a reduction of the cell size is necessary [8].

## *Generalized Centrifugal Force Model*

The Generalized Centrifugal Force Model is a spatially continuous force-based model. The acceleration $\vec{a}_i$ of a pedestrian $i$ results from the superposition of forces acting on her/him at a certain moment. In analogy to Newtonian dynamics the equations of motion of pedestrian $i$, is written:

$$\vec{F}_i = \vec{F}_i^{drv} + \sum_{j \neq i}^{N} \vec{F}_{ij}^{rep} + \sum_{B} \vec{F}_{iB}^{rep} = m_i \vec{a}_i. \qquad (1)$$

Where $\vec{F}_i^{drv}$ is the driving force of pedestrian $i$ to move towards a desired destination, $N$ is the number of interacting pedestrians, $\vec{F}_{ij}^{rep}$ the repulsive force emerging from pedestrian $j$ to $i$ and $\vec{F}_{iB}^{rep}$ the force between the pedestrian $i$ and walls or other stationary obstacles. The driving force and repulsive forces are defined by:

$$\vec{F}_i^{drv} = m_i \frac{\vec{V}_i^0 - \vec{V}_i}{\tau} \text{ and } \vec{F}_{ij}^{rep} = -m_i K_{ij} \frac{(\nu \| \vec{V}_i^0 \| + V_{ij})^2}{\| \vec{R}_{ij} \| - \frac{1}{2}(D_i(\| \vec{V}_i \|) + D_j(\| \vec{V}_j \|))} \vec{e}_{ij}. \quad (2)$$

For a detailed definition of the model we refer to [10]. While the repulsive force is decreasing with increasing distance $R$ between pedestrians, only the influence of adjacent pedestrians is taken into account. Two pedestrians influence each other if their distance is smaller than a certain cut-off radius $R_c$. To guarantee robust numerical integration a Hermite-interpolation of the repulsive force is implemented. The interpolation guarantees that the norm of the repulsive force decreases smoothly to zero at $R_c$. This manipulation is important for the runtime optimization by means of neighbor list methods discussed in the next section.

The initial value problem (1) was solved using an Euler scheme with fixed-step size $\Delta t = 0.01$ s. The desired speeds of pedestrians are Gaussian distributed with mean $\mu = 1.34$ m/s and standard deviation $\sigma = 0.26$ m/s. Other parameter specifications and numerical details can be found in [10]. For model verification the fundamental diagram in corridors of a length of 20 m and different widths is measured. The shape of the resulting velocity density relation is in good agreement with the empirical data [11-14], see Fig. 2 (left). Furthermore, the flow of pedestrians through a bottleneck as described in [15] was simulated and compared experimental data [15,16], see Fig. 2 (right).

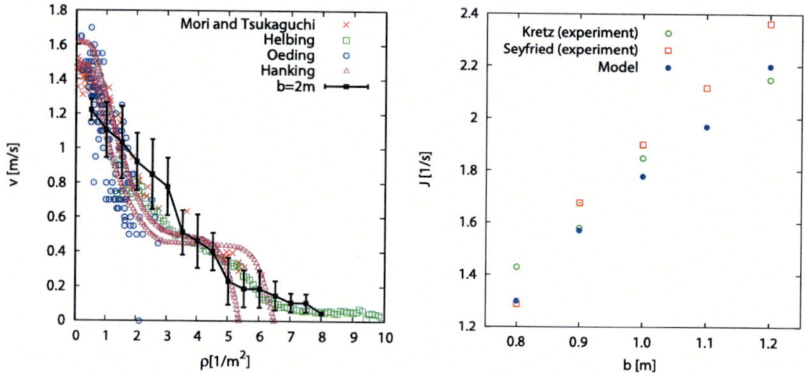


**Fig. 2. Fundamental diagram in wide corridor (20m x 2m) in comparison to empirical data (left). Flow through bottleneck for different widths (right)**

After an adequate calibration of the free parameters the model describes quantitatively well the dynamics of pedestrians in bottlenecks and wide corridors. Nevertheless, the shape of the volume exclusion of each simulated pedestrian makes it difficult to describe the dynamic with one set of parameters in one and in two dimensional spaces.

## Runtime Optimization

Even if previous experience shows that current CA models need less computational effort than actual spatially continuous models, general comparative statements about the runtime of implementations are difficult to make. The runtime depends on the spatial and temporal resolution of the model, the situation modeled, the implementation and the underlying hardware as well as other parameters. A reliable comparative statement would need at least a definition of a scenario together with the phenomena the model has to reproduce and specific implementations of these models on the same hardware. For CA models we expect that the runtime will increase if models with smaller cell sizes are used, see section Modeling and Simulation. Runtime problems in forced based models are due to the small time steps needed to enable an accurate integration and due to the non local interactions between pedestrians, see Eq. (1). A straightforward implementation (brute force method) of the repulsive forces requires the calculation of $N$-1 × $N$ terms, leading to a complexity of $O(N^2)$. In the next sections we focus on neighbor list approaches to reduce the runtime of force-based models and the appropriate parallelization strategy on a special hardware.

### *Neighbor list methods*

We tested two neighbor list concepts which are widely used in molecular dynamics simulations with short-range forces. One is the Linked-Cell algorithm [17,18] and the second is the Verlet-List algorithm [18,19].

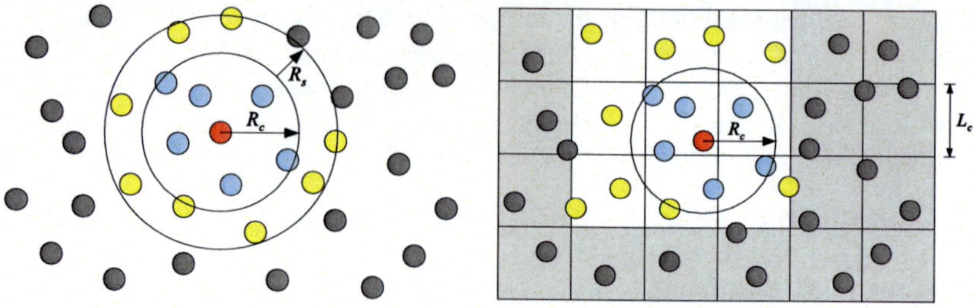


**Fig. 3. Sketch of the Verlet-List method (left) and the Linked-Cell method (right)**

The concept of the Verlet-List algorithm is sketched in Fig. 3 (left). For each person the neighbors in the region with radius $R_c$+$R_s$ are saved in a list. $R_c$ gives the cutoff-radius of the interaction force while $R_s$ is the skin-radius determining the size of the reservoir. In each time step only interactions with pedestrians in the list have to be calculated. The list must be updated in time intervals given by $R_s$ and the maximal speed of the pedestrians. This leads to a total complexity of $O(N^2)$ with a small coefficient. A drawback of the Verlet-List algorithm is the amount of memory needed for the neighbor lists. In case of the Linked-Cell algorithm the space is decomposed in cells of size $R_c \times R_c$ as illustrated in Fig. 3 (right). Only the interactions with pedestrians in adjacent cells have to be considered, reducing the complexity to $O(N)$. In each time step of the simulation the pedestrians are reassigned to their particular cells. Opposed to the Verlet-List algorithm the memory is of order $O(N)$. The indirect addressing which is caused by managing the cells and the corresponding pedestrians in two lists might be a drawback because of possible cache misses and thus a loss of performance.

Aside from further parameters, the efficiency of the methods mentioned above depends on the local distribution of the pedestrians and the range of the repulsive force. For example the runtime of a simulation using the Linked-Cell method will not differ from the brute force method when all pedestrians are located in one cell. Thus we analyzed the dependency of the performance gain on the degree of congestion, the system size, and the range of the interaction force. For this purpose we introduce the speedup $S=t_{BF}/t_{NL}$ given by the runtime of the brute force method $t_{BF}$ divided by the runtime of the respective neighbor-list method $t_{NL}$.

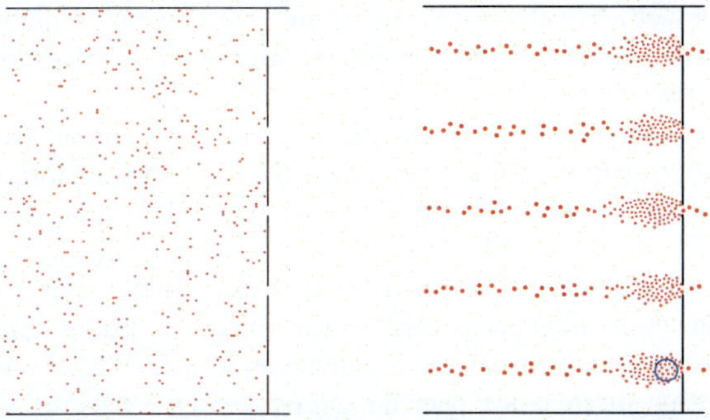

**Fig. 4. Examples for a homogeneous (left) and an inhomogeneous (right) distribution of pedestrians. The speedup reachable using the Linked-Cell or Verlet-List method depends on the degree of congestion. The blue circle (right) shows the cutoff-radius $R_c$ of the force.**

We investigated the performance gain using two different test scenarios: a homogeneous and an inhomogeneous distribution of pedestrians see Fig. 4. The speedup was determined by measuring the corresponding runtime in relation to an increasing number of pedestrians. For the tests a simulation time of 100 s and a room size of 35 m x 50 m were chosen. The time interval $\Delta t = 0.01$ s for integrating the equations of motion (1) results in $10^4$ iteration steps. Runtime measurements were performed using the operating system openSUSE 10.2 and an Intel Pentium D Proc. with 3.40 GHz, a L2-Cache of 2048 KB and 2048 MB RAM.

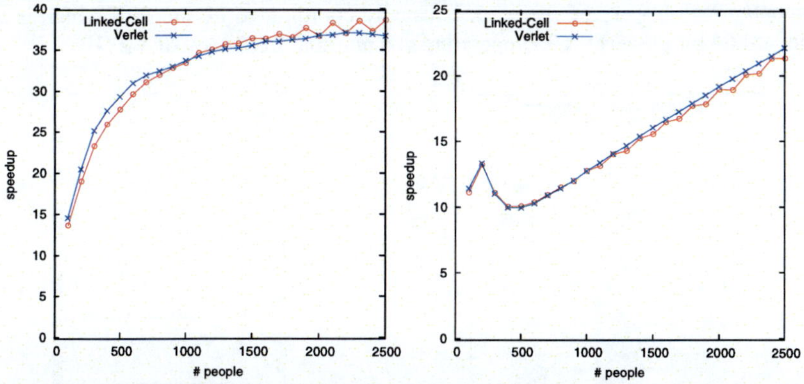


**Fig. 5. Speedup of the homogeneous (left) and inhomogeneous (right) distribution of pedestrians. For 2500 pedestrians a speedup of more than 35 could be reached. The speedup for the inhomogeneous distribution is smaller than for the homogeneous. For the latter case with constant density of 3 [$m^{-2}$] the speedup reaches 80 for 10,000 pedestrians.**

In the inhomogeneous scenario the speedup scales differently for an increasing number of pedestrians as shown in Fig. 5 (right). Looking at the developing of the speedup the local maximum for a simulation with 200 persons after a strong increase and followed by a decrease until about 500 persons is quite remarkable. This can be explained as follows: For few pedestrians no congestions occur and the speedup increases due to fewer calculations of interaction forces. The congestion areas grow with an increasing number of pedestrians and as soon as the size of the congestion areas is in the order of $R_C$ most interaction forces are located within this area, thus reducing the performance gain obtained by the neighbor list methods, see Fig. 4 (right). When the majority of pedestrians are located in the congestion areas we obtain a distribution of a quasi-constant density and the speedup scales as in the homogeneous case. Since for an increasing number of pedestrians the inhomogeneous distribution converges to a homogeneous distribution in the limit case the speedup is expected to converge to a constant.

## *Parallelization*

For parallelization the Cell Broadband Engine Architecture (CBEA) [20] is used. First introduced in the PlayStation 3 from Sony Computer Entertainment, the CBEA has a wide range of applications in the high performance computing domain including large crowd simulations [21]. The neighbor-lists methods presented above are implemented on an IBM BladeCenter®. Each blade features two CBEAs (see Fig. 6, left), each of which has 9 elements: 1 Power Processor Element (PPE) and 8 Synergistic Processing Elements (SPE) all connected to a high bandwidth bus, the Element Interconnect Bus (EIB). The SPEs give the CBEA its computational power and the work is usually coordinated by the PPE which is a standard 64 bit PowerPC CPU operating at a nominal frequency of 3.2 GHz.

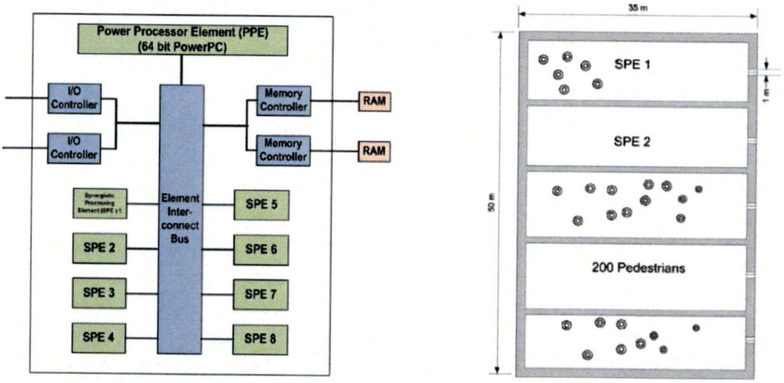


**Fig. 6. Left: Cell Broadband Engine Architecture featuring 1 Power Processor Element (PPE) and 8 Synergistic Processing Elements (SPE). Right: Simplified model of the Esprit arena divided in 5 areas, each with 200 pedestrians.**

The parallelization principle is simple and straightforward. The idea is to perform one simulation per blade. The simulation area is ideally divided into rooms, which are allocated to the SPEs at the beginning of the simulation by the PPE. The outputs (trajectories and velocities) of each simulation step are then collected by the PPE and written to a file. A maximum of 800 pedestrians could be simulated on one SPE due to memory restrictions. With a total of 16 SPEs per blade 12800 pedestrians can be updated in one simulation step.

**Table 1. Speedup (1600 pedestrians, evacuation time ~130 seconds)**

| Number of SPEs | brute force (s) | Linked-Cell (s) | Speedup |
|---|---|---|---|
| Laptop* | 1509.72 | 180.84 | 8,4 |
| 2 | 1205.55 | 158.07 | 7,6 |
| 4 | 323.04 | 80.48 | 4,0 |
| 8 | 91.68 | 40.69 | 2,3 |

* 2.50 GHz, 4GB RAM, Dual-Core. The CBEA operates at 3.2 GHz.

In the tests, only brute force and Linked-Cell are considered. The Verlet-Lists presented earlier have very high memory consumption, which is a big issue for the SPE local store memory of 256 KB. The first simulation is performed with 1600 pedestrians distributed in 8 areas which are represented as separated rooms (see Fig. 6, right). The simulation is performed on a standard laptop and on one blade of a Cell BladeCenter using only one CBEA processor computing with 2, 4 and 8 SPEs. The results are presented in Table 1. Considering the results achieved on the laptop and the results achieved by the CBEA (8 SPE's) an overall speedup of round 38 is reached. The total evacuation time is 130 seconds, the simulation is thus on the CBEA by a factor of 3.25 faster than real-time.

## Outlook

To improve the modeling we follow several ideas. For CA we will realize smaller cell sizes. For spatially continuous models we will modify the forces to realize elliptical volume exclusion. Moreover several routing concepts will be incorporated. For model calibration and validation more test cases will be used oriented on the special geometries given by the test venue. The runtime will be further optimized using SPE Single Instruction Multiple Data (SIMD) instructions to perform up to 4 operations in a single clock together with a better memory management to fit more pedestrians on one SPE. An efficient workload distribution has to be implemented together with inter-connected rooms. The introduction of transitions (doors) lead to data exchange between the SPEs which can be efficiently done using Direct Memory Access (DMA).

**Acknowledgment:** This work has been performed within the program "Research for Civil Security" in the field "Protecting and Saving Human Life" funded by the German Government, Federal Ministry of Education and Research (BMBF).

## References

1. Hermes – Investigation of an evacuation assistant for use in emergencies during large-scale public events. Juliche Supercomputing Centre (JSC). http://www.fz-juelich.de/jsc/hermes.
2. Schadschneider, A., Klingsch, W., Klüpfel, H., Kretz, T., Rogsch, C., Seyfried, A.: Evacuation dynamics: Empirical results, modeling and applications. In: Meyers, R.A. (ed.), Encyclopedia of Complexity and System Science, pp. 3142. Springer (2009)
3. Rogsch, C., Seyfried, A., Klingsch, W.: Comparative Investigations of the Dynamical Simulation of Foot Traffic Flow. In: Waldau, N., Gattermann, P., Knoflacher, H., Schreckenberg, M. (ed.), Pedestrian and Evacuation Dynamics 2005, p.357-362. Springer, Berlin (2006)
4. Schadschneider, A., Seyfried, A.: Empirical Results for Pedestrian Dynamics and their Implications for Cellular Automata Models. In: Timmermans, H. (ed.), Pedestrian Behavior: Data Collection and Applications, p. 27-43. Emerald Group Publishing Limited (2009)
5. Burstedde, C., Klauck, K., Schadschneider, A., Zittartz, J.: Simulation of pedestrian dynamics using a two-dimensional cellular automaton. Physica A. **295**, 507-525 (2001)
6. Schadschneider, A.: Cellular automaton approach to pedestrian dynamics-Theory. In: Schreckenberg, M., Sharma, S.D. (Eds.), Pedestrian and Evacuation Dynamics, pp. 75. Springer (2002)
7. Kirchner, A., Klüpfel, H., Nishinari, K., Schadschneider, A., Schreckenberg, M.: Simulation of competitive egress behavior: Comparison with aircraft evacuation data. Physica A. **324**, pp. 689 (2003)
8. Schadschneider, A., et al.: Runtime Optimization of Force Based Models within the Hermes Project (2010)
9. Kirchner, A., Klüpfel, H., Nishinari, K., Schadschneider, A., Schreckenberg, M.: Discretization effects and the influence of walking speed in cellular automata models for pedestrian dynamics. J. Stat. Mech. P10011 (2004)
10. Chraibi, M., Seyfried, A., Schadschneider, A., Mackens, M.: Quantitative verification of a force based model for pedestrian dynamics. http://arxiv.org/abs/0912.4044. To appear in Traffic and Granular Flow 2009 (2010)
11. Mori, M., Tsukaguchi, H.: A new method for evaluation of level of service in pedestrian facilities. Transp. Res. **21A(3)**, 223–234 (1987)
12. Helbing, D., Johansson, A., Al-Abideen, H.Z.: Dynamics of Crowd Disasters: An Empirical Study. Phys. Rev. E. **75**, 046109 (2007)

13. Oeding, D.: Verkehrsbelastung und Dimensionierung von Gehwegen und anderen Anlagen des Fußgängerverkehrs. Forschungsbericht 22, Technische Hochschule Braunschweig (1963)
14. Hankin, B.D., Wright, R.A.: Passenger Flow in Subways. Operational Research Quarterly. **9**, 81–88 (1958)
15. Seyfried, A., Rupprecht, T., Passon, O., Steffen, B., Klingsch, W., Boltes, M.: New insights into pedestrian flow through bottlenecks. Transportation Science. **43**, 395-406 (2009)
16. Kretz, T., Grünebohm, A., Schreckenberg, M.: Experimental study of pedestrian flow through a bottleneck. J. Stat. Mech. **10,** P10014 (2006)
17. Allen, M.P., Tildesley, D.J.: Computer Simulation of Liquids. Oxford Science Publications, Oxford (1987)
18. Sutmann, G., Stegailov, V.: Optimization of neighbor list techniques in liquid matter simulations. Journal of Molecular Liquids, Complex Liquids - Fundamental Properties to Industrial Applications. Volume 125, Issues 2-3, **15,** 197-203 (2006)
19. Verlet, L.: Computer experiments on classical fluids, I. Thermodynamical properties of Lennard-Jones molecules. Phys. Rev. E. **159**, 98 (1967)
20. Cell Broadband Engine Architecture. Version 1.02. https://www-01.ibm.com/chips/techlib/techlib.nsf/techdocs/1AEEE1270EA2776387257060006E61BA/$file/CBEA_v1.02_11Oct2007_pub.pdf. Assessed 11 October 2007
21. Reynolds, C.: Big fast crowds on PS3. Proc. 2006 ACM SIGGRAPH symposium on Videogames, p. 113-121. Boston (2006)

# A Dynamic Simulation on Crowd Congestion in Large-Scale Terminal Station Complex in an Official Announcement Advisory Information

**Q. Cui [1], M. Ichikawa[2], T. Kaneda[1], and H. Deguchi[2]**

[1] Graduate School of Engineering, Nagoya Institute of Technology
Gokiso-cho, Showa-ku, Nagoya 466-8555, Japan
[2] Interdisciplinary Graduate School of Science and Engineering, Tokyo Institute of Technology
Nagatsuta-cho, Midori-ku, Yokohama 226-8503, Japan
Corresponding author: qinglin.cui@gmail.com

**Abstract** On the assumption of advisory information concerning an imminent Tokai earthquake being officially announced, as a case example we developed a spatial-spot type agent-based simulation model for the Nagoya Station area, where several terminal stations are concentrated; in the model, agents played people on their way home, and such factors as the routes selected by agents and the spatial restrictions, e.g. passages, were taken into consideration. Basic on SOARS(Spot Oriented Agent Role Simulator)platform, we conducted a large-scale crowd simulation with 160,000 agents and analysis the change of space density in one hour to compare to the estimates given by Nagoya City, we analysis the result and also refer to this kind of project for implementing much higher functions.

## Introduction

At present the Japanese Government has designated the possibility of a Tokai earthquake as the only example of a predictable earthquake. In 2004, the Meteorological Agency newly added "advisory information" assuming a predictable case. Advisory information will first be officially announced, followed later by a public "warning" .the municipality of Nagoya expects all people, commuters and the like to return to their homes immediately. It is reasonable to make the area around a large-scale terminal station, such as Nagoya Station, due to the massive numbers of people all returning home simultaneously; such a situation requires the application of measures to prevent crowd accidents [1].

We first do the research with a maximum scale of 6,000 people, employing cell-spatial-type agent-based simulation [2] [3] [4] and the process of such an increase and measures to prevent accidents is possible. But it is difficult to deal with 100,000 pedestrians. Earlier research employs a spatial-spot type agent-based simulation method, as a kind of network-type agent-based simulation [5], and con-

R.D. Peacock et al. (eds.), *Pedestrian and Evacuation Dynamics*,
DOI 10.1007/978-1-4419-9725-8_34, © Springer Science+Business Media, LLC 2011

ducts a large-scale crowd simulation [6]. This time we main to analyze the congestion of complicated flows around terminal stations.

Therefore, beside the model, this research uses a spatial model for the Nagoya Station area where five railway lines converge, the research measures density at each space throughout the simulation. The research demonstrates a process in which the spatial distribution of crowd density is changing, and examines the density distribution and the tendency of changes according to different cases, e.g. flow coefficients or origin-destination data, and the application of measures to prevent accidents.

## A framework of the simulation

### *Characteristics of the model employed in the research*

To handle a large-scale crowd simulation, the research introduced two characteristics into the model. Firstly the entire space was divided into spots but nor cells which connect with a link. Secondly, the pedestrian selects their own walking route and this changes the density and affects other pedestrians.

### *Introducing SOARS*

In addition, the simulation conducted in the research employed SOARS (Spot Oriented Agent Role Simulator) as a platform that has strong advantages in spot-agent representation and enables us to run large-scale agent simulation [7].

SOARS, Spot Oriented Agent Role Simulator, is an empirical method for simulating social interactions in which social scientist can observe the dynamics of human related activities, which is being developed and released for academic and educational use by Deguchi Lab., Tokyo Institute of Technology. All the components of SOARS are fundamentally in public domain software. Being different from popular computer languages, such as C, Fortran and Basic, SOARS does not necessitate any mathematical programming skills for writing a simulation model. And also you do not have to be trained much about this simulation language like Stella simulator (which is not micro-oriented but macro-oriented). What SOARS only requires them to represent a simulation model are to select items of elements of social systems, to describe them by some natural words and to write, drag and drop them on a virtual workbench, that is, a artificial world in a Web browser.

SOARS follows the agent based modeling (ABM) approach for representing a simulation model like Repast and MASON. With SOARS, you can simulate easily this kind of the social phenomenon emerged among the activities of human. As we explained above, any experience and knowledge about step-by-step programming languages is not needed to operate SOARS. In SOARS, there are a lot of convenient rules readily provided or to be highly combined these rules, you can create a social simulation model and can represent the social phenomena. All you have to do to construct a social simulation model are to provide agents with how to conduct in particular conditions, who are the decision-making entities, and the environment, where agents act, and action rules for these agents and the environment. Besides, SOARS also has a log analyzing function and a large-scale execution support. SOARS is an all-in-one simulation developing environment for not only modeling, simulating, but also analyzing.

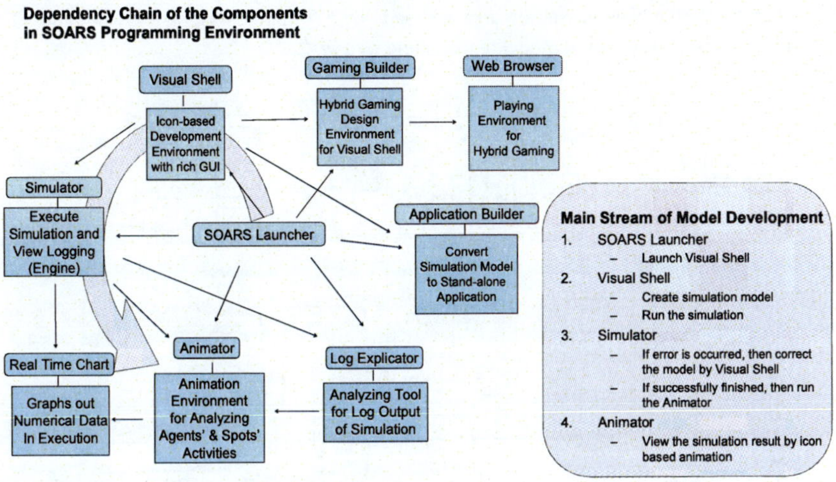


**Fig. 1. Programming environment of SOARS**

## Trial production of the system

### *Details of the model*

This section describes details of the model from two points of view – for each of the spots and pedestrians – and then, explains the computing process by spots and the behavior rules of pedestrians.

## Spatial spots

The model used for the research consists of pedestrian agents and a spatial model. The spatial model is classified into two types of spatial spots: internal spot S, inside the station, and external spot O, outside the station, and spot T, the inside of a train, and spot P, platform. Spot has length $L_m$ and width $W_m$; this is approximated by a rectangular space with an area of $L_m * W_m$.

## Spatial model of Nagoya Station area

For the research, the Nagoya Station area are made by 5 T-spot, 5 P-spot, 9 O-spot and 30 S-spot which shows the spatial model representing by Fig. 2.

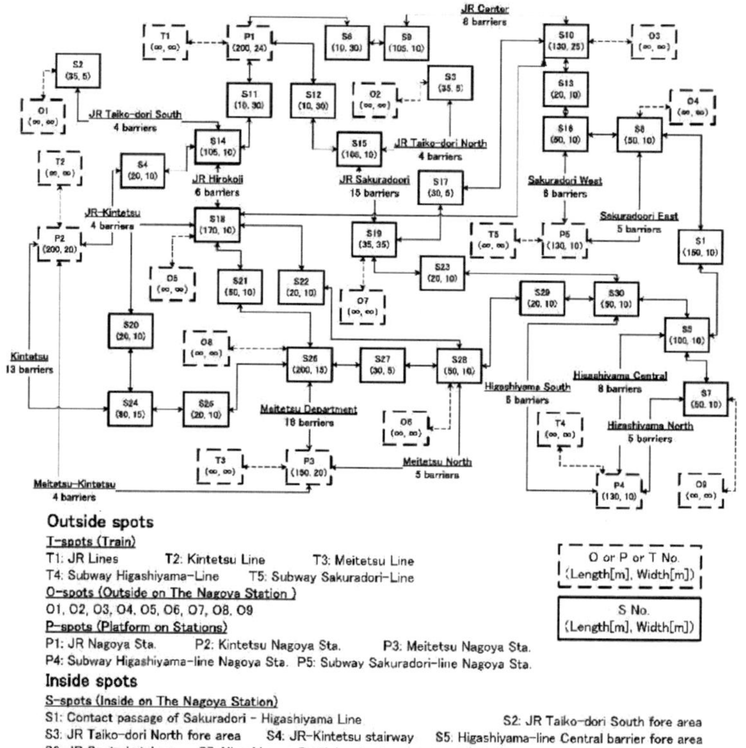


**Fig. 2.** Nagoya station area spatial model expressed as spatial spots

**Pedestrian agents**

Spots O and T have random variables $w_1$ and $w_2$ which belong to the agent $u \in C$. $w_1$ is a random variable for determining a starting point, and $w_2$ is a random variable for determining the destination of an agent who departs from that starting point.

A pedestrian agent selects the shortest route without the change of density from their starting point to the ending point.

## *Assessment of the shortest distance*

Concerning the calculation of a distance, firstly it is assumed that there is a node in the center point of a crossover spot expressed as a rectangle. As a method to search for the shortest route, Dijkstra's method. (label-setting method) is used. The found shortest route is set as the basic walking route of an agent.

## *Rules concerning movement of pedestrian agents*

This section describes the method to calculate assessment indicators, which affect walking and route selection, and the overall process.

**Rules for movement within a spot**

When a pedestrian agent enters in spot $S_m$, the agent reads an Lm value of spot $S_m$. Movement distance for each step ΔL is given by Equation (1).

One of the most influential to affect the speed is considered to be density. A straight line model measured by an actual measurement experiment for two-way flows, and an estimate equation of walking speed V is given by Equation (2).

$$\left.\begin{aligned} \Delta L &= V \times T \quad \text{---} \quad (1) \\ V &= a \times \rho + V_0 \quad \text{---} \quad (2) \end{aligned}\right\} \Rightarrow \Delta L = (a \times \rho + V_0) \times T \quad \text{---} \quad (3)$$

Here, T: Time/step (second/step)
V: Walking speed (m/second)
$V_0$: Standard walking speed
ΔL: Movement distance/step (m/step)
$\rho$: density (people/m²)
a: Parameter

ΔL is added to L, the distance which an agent has already moved within a spot. When $L \geq L_m$, movement within the spot ends. Set values for $V_0$ and a differ according to one-way or two-way flow. In addition, in two-way flow the set values differ according to the ratio of pedestrians moving in the opposite direction. (Table 1)

**Table 1. Walking speed used for the model [8]**

| Crowd flow | The number of people rate (α) | *Flow of people | a | $V_0$ |
|---|---|---|---|---|
| One-way flow | - | All | -0.28 | 1.48 |
| Two-way flow | 1 time | All | -0.275 | 1. 605 |
| | 2 times | More | -0.285 | 1.675 |
| | | Less | -0.39 | 1.958 |

*Flow of people:
All: The flow of all people, More: The fbw of more people, Less: The fbw of less people

In the research one-way flow and single and double pedestrian ratios for two-way flow are implemented. The ratio of number of pedestrians is rounded off, and less than one is regarded as a one-way flow, and two or more is regarded as a two-way flow.

## Rules for restricting movement between spots

A stationary state is caused by pedestrians stopping or reducing walking speed, or by a location with a width narrowing. In the research, we only consider the stationary states caused by reduction of spot width are taken into account.

When spots $S_m$ and $S_m$' are adjacent to each other, and their width $W_m$ and $W_m$' are compared, the in-flow upper limit for the side with the smaller width is applied. In this case, if $W_m \leq W_m$', the in-flow upper limit Q is expressed as below:

$$Q = k \times \gamma \times W_m \times T \quad \text{--- (4)}$$

Here, Q: In-flow upper limit, k: Flow coefficient (= 1.5 people/m·sec), γ: The effective rate of the width, and T: Time (sec)/step. Agents more than Q stay and remain in the same spot. The setting of the passage and the ticket barrier below

**Table 2. Expression of the ticket barrier**

| The Passage type | The effective rate of the width (γ) | Method of counting number of streets |
|---|---|---|
| Ticket barrier none | 1 | Each of the two directions |
| Ticket barrier | 0.5 | Two directions total |

**Rules for in-flow/out-flow of pedestrians**

In-flow and out-flow of pedestrians is controlled by spot O outside the station and spot T on each platform and platform P.

An outside-the-station spot calculates and controls a maximum number of pedestrians that flow in or out per unit of time, based on the width of the spot for in-flow or out-flow, and a flow coefficient.

Platform spot, considered the train carry ability , each up-line and down-line controls the largest number of people per time who flow out.

A train spot on each platform allows a set number of pedestrians to flow in with 360 steps and averagely flow out.

## *Rules for route selection by pedestrian agents*

When an agent arrives at a crossover spot, they need to determine their next route. But if the shortest route can not be chosen, people can choose the rest but the shorter line. However, the condition for being a possible route choice is any route that enables the agent to reach their destination

For route selection, two kinds of criteria are provided: the shortest route$\beta_1$ and density assessment$\beta_2$.

When the density of the next spot located on the shortest route is less than $\beta_1$ (people/m$^2$), the agent can now take the shortest route.

When the density of the next spot located on the shortest route is $\beta_1$ (people/m$^2$) or more, the agent moves to an alternative route spot with a density of less than$\beta_1$ (people/m$^2$). If it is the agent takes the shortest route. If density is $\beta_2$ (people/m$^2$) or more, the agent remains in the present spot.

## Simulation analysis

This chapter describes the simulation analysis. It was assumed that advisory information is officially announced from 2:00 p.m. onward. The Nagoya City Local Government has given estimates of the number of people likely to converge on the station; for the simulation three cases were set: 0%, 50% and 100% of the estimated numbers. In addition, for each case 4 patterns of detour criteria density were provided: 0, 1, 2, or 3 (people/m$^2$).

The simulation was conducted to confirm the agent behavior needed to analyze stationary states, and spatial bottlenecks; the time scale was set as 10 seconds/step with a total of 360 steps.

## *Parameter setting and classification of cases*

Parameters were set as shown in Table 3. OD data matrix and the number of agents were set as shown in Table 3 (train side), and Table 4 (outside). The number of agents was set as 158200

**Table 3. Estimated OD matrix used for the simulation (Train) [9]**

| | T-line | T-line | K-line | C-line | H-line | H-line | S-line | S-line | M-line | M-line | K-line | Outside | Total (people) |
|---|---|---|---|---|---|---|---|---|---|---|---|---|---|
| | up train | down train | up train | up train | up train | | down train | | up train | down train | up train | | |
| T-Line up train | - | 0 (0%) | 285 (6%) | 999 (22%) | 1070 (24%) | 272 (6%) | 285 (6%) | 0 (0%) | 188 (4%) | 83 (2%) | 574 (13%) | 772 (17%) | 4529 |
| T-Line down train | 0 (0%) | - | 630 (9%) | 31 (1%) | 355 (12%) | 212 (7%) | 71 (2%) | 0 (0%) | 0 (0%) | 430 (15%) | 681 (24%) | 448 (16%) | 2858 |
| K-Line down train | 144 (12%) | 225 (19%) | - | 152 (13%) | 153 (13%) | 0 (0%) | 33 (3%) | 0 (0%) | 87 (7%) | 227 (19%) | 0 (0%) | 173 (14%) | 1194 |
| C-Line down train | 48 (%) | 3059 (39%) | 1077 (14%) | - | 0 (0%) | 187 (2%) | 0 (0%) | 0 (0%) | 57 (1%) | 702 (9%) | 2299 (29%) | 365 (5%) | 7793 |
| H-Line up train | 607 (9%) | 920 (14%) | 65 (1%) | 500 (7%) | - | 0 (0%) | 661 (11%) | 0 (0%) | 1566 (25%) | 1501 (23%) | 335 (5%) | 300 (7%) | 6455 |
| H-Line down train | 1177 (10%) | 2410 (20%) | 296 (2%) | 54 (0%) | 0 (0%) | - | 18 (0%) | 13 (0%) | 1392 (12%) | 3306 (27%) | 2800 (23%) | 718 (7%) | 12183 |
| S-Line up train | 1211 (18%) | 1343 (19%) | 138 (1%) | 929 (13%) | 852 (12%) | 114 (1%) | - | 0 (0%) | 348 (4%) | 1673 (24%) | 444 (6%) | 0 (0%) | 7052 |
| S-Line down train | 1038 (10%) | 2048 (20%) | 378 (4%) | 123 (1%) | 99 (1%) | 241 (2%) | 0 (0%) | - | 1130 (12%) | 1919 (20%) | 2686 (25%) | 490 (5%) | 10153 |
| M-Lineup train | 460 (4%) | 504 (5%) | 564 (5%) | 687 (7%) | 1865 (18%) | 560 (5%) | 183 (2%) | 60 (0%) | - | 0 (0%) | 3766 (35%) | 1949 (18%) | 8070 |
| M-Linedown train | 0 (0%) | 197 (3%) | 176 (3%) | 133 (3%) | 687 (12%) | 444 (6%) | 136 (3%) | 0 (0%) | 0 (0%) | - | 3119 (47%) | 1562 (23%) | 6454 |
| K-Line up train | 358 (5%) | 831 (12%) | 0 (0%) | 911 (13%) | 1195 (17%) | 73 (1%) | 391 (5%) | 0 (0%) | 1193 (17%) | 1685 (24%) | - | 503 (7%) | 7140 |
| Total | 5043 | 11537 | 3609 | 4519 | 6276 | 2103 | 1778 | 73 | 5961 | 11526 | 16704 | 7280 | 73881 |

* T-Line: Tokaido Line; K-Line: Kansai Line; C-Line: Chuo Line; H Line: Higashiyam a-Line; S-Line: Sakuradoori-line; M-line: M ei-tetsu-Line; K-line : K intetsu-line

**Table 4. Estimated OD matrix used for the simulation (Outside) [9]**

| | T-Line | T-Line | K-Line | C-Line | H-Line | H-Line | S-Line | S-Line | M-Line | M-Line | K-Line | Total (people) |
|---|---|---|---|---|---|---|---|---|---|---|---|---|
| | Up train | Down train | Up train | Up train | Up train | Down train | Up train | Down train | Up train | Down train | Up train | |
| 01 | 7% | 11% | 2% | 9% | 12% | 2% | 3% | 0% | 20% | 23% | 11% | 6392 |
| 02 | 7% | 11% | 2% | 9% | 12% | 2% | 3% | 0% | 20% | 23% | 11% | 6392 |
| 03 | 7% | 11% | 2% | 9% | 12% | 2% | 3% | 0% | 20% | 23% | 11% | 15523 |
| 04 | 7% | 11% | 2% | 9% | 12% | 2% | 3% | 0% | 20% | 23% | 11% | 12784 |
| 05 | 7% | 11% | 2% | 9% | 12% | 2% | 3% | 0% | 20% | 23% | 11% | 6392 |
| 06 | 7% | 11% | 2% | 9% | 12% | 2% | 3% | 0% | 20% | 23% | 11% | 6392 |
| 07 | 7% | 11% | 2% | 9% | 12% | 2% | 3% | 0% | 20% | 23% | 11% | 21915 |
| 08 | 7% | 11% | 2% | 9% | 12% | 2% | 3% | 0% | 20% | 23% | 11% | 9131 |
| 09 | 7% | 11% | 2% | 9% | 12% | 2% | 3% | 0% | 20% | 23% | 11% | 6392 |
| Total | 6517 | 9930 | 1622 | 8479 | 10979 | 2260 | 2595 | 38 | 18337 | 20937 | 9617 | 91311 |

*T-Line: Tokaido Line; K-Line: Kansai Line; C-Line: Chuo Line; H-Line: Higashiyam a-Line; S-Line: Sakuradoori-Line; M-Line: Mei-tetsu-Line; K-Line: K intetsu-line

**Table 5. Cases in the simulation**

| Case No. | | The number of people (People) | | The density strandard value (People/m²) | |
|---|---|---|---|---|---|
| | | Train side | In-flow | β 1 | β 1 |
| Case 1 | 1 | 73881 | 0 | 3 | 0 |
| | 2 | 73881 | 0 | 3 | 1 |
| | 3 | 73881 | 0 | 3 | 2 |
| | 4 | 73881 | 0 | 3 | 3 |
| Case 2 | 1 | 73881 | 45656 | 3 | 0 |
| | 2 | 73881 | 45656 | 3 | 1 |
| | 3 | 73881 | 45656 | 3 | 2 |
| | 4 | 73881 | 45656 | 3 | 3 |
| Case 3 | 1 | 73881 | 91311 | 3 | 0 |
| | 2 | 73881 | 91311 | 3 | 1 |
| | 3 | 73881 | 91311 | 3 | 2 |
| | 4 | 73881 | 91311 | 3 | 3 |

It was assumed that advisory information of a Tokay earthquake is officially announced at 2:00 p.m. on a weekday and the simulation experiment focused on the congestion state in the station space between 2:00 to 3:00 p.m. Table 5 gives classifications for each case.

This section describes the classification of cases. Assuming that all passengers traveling by train flow into the station, the number of passengers was set at 73,881, the number estimated by Nagoya City. The number of pedestrians arriving from outside the station to take a train was set at 3 cases – 0%, 50% and 100% of the number estimated by Nagoya City; the number of in-flow pedestrians from outside was 0 for Case 1; 45,656 for Case 2; and 91,311 for Case 3. Moreover, according to the differences of $\beta_2$, which is a density decision criteria for taking a detour, 4 patterns were set for each case. $\beta_2$ (people/m²) is a decision criteria for an agent to take a detour. The agent takes a detour under the following conditions: the density of the front spot – the shortest route – is $\beta_1$ (people/m²) or more, and any possible detour route has a density of $\beta_2$ (people/m²) or less. When $\beta_2$ is 0 (people/m²), no detour is necessary. The upper limit of $\beta_1$ was set as 3 (people/m²).

## *Results analysis*

This section illustrates the measurement data used for analysis, and the data is compared with congestion analysis using estimates given by Nagoya City. The performance of the model and analysis results are then described.

**Measurement data**

Fig. 3 illustrates the results of the measured densities (Case 2-2, Case 3-4).

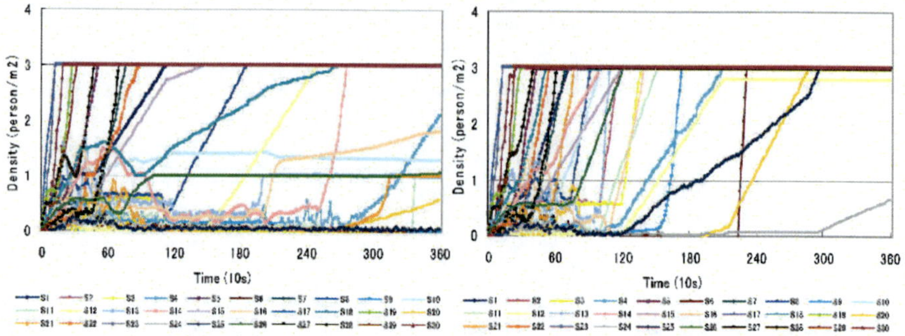


(Case 2-2) (Case 3-4)

**Fig. 3. Illustration of the density measurement results**

**Examination for crowd management**

When the measurement results of each case are examined, the points found by simulation analysis are summarized as follows:

1) In each case, the highest number of agents processed, and the lowest number in a stationary state were found when no detours were taken. For both $\beta_1$ and $\beta_2$, in the case of 3people/m², the number of agents who stayed inside increased and the number of agents processed showed the greatest decrease.

2) When the number of in-flow agents from outside increased, locations with increasing density also tended to increase.

3) When the same patterns in different cases were compared, it was found that as the number of in-flow agents from outside increased, processing capacity tended to decline.

Concerning processing capacity, when there was no in-flow from outside, the number of agents processed in the simulation model was about 80% of the estimates of Nagoya City; when the time required for movement and the impact of congestion are taken into consideration, this result is reasonable. However, along with an increase of in-flow from outside, congested spaces triggered a decrease in walking speeds and consequent stationary phenomenon, and it was found that the processing capacity dropped to about 40%. Particularly, in the conditions of Cases 2-4 and 3-4 where a detour was allowed to the upper limit, it was confirmed the processing capacity decreased to 30% (Fig. 4.).

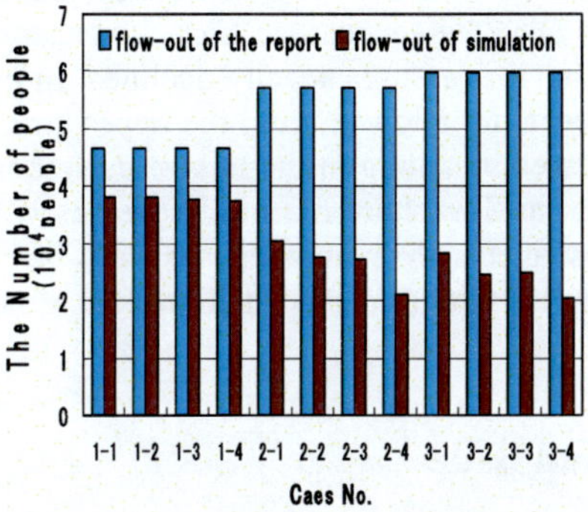


**Fig. 4. Comparison with the estimates given by Nagoya City. (processing capacity in the initial one hour)**

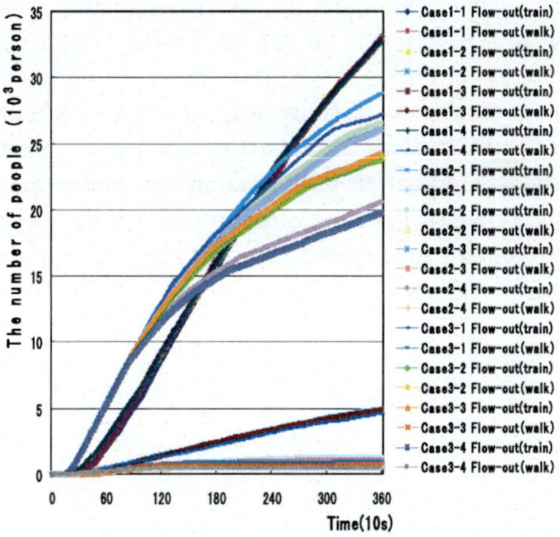


**Fig. 5. Graph: number of out-flow agents (train/walking)**

The changes to the number of agents processed were examined using the graph for the number of out-flow agents (Fig. 5.). Case 1 – no in-flow from outside – recorded the largest accumulated number of agents processed. In Cases 2 and 3, up to about 120 steps, more out-flow agents than Case 1 were found, but gradually due to increasing congestion, processing capacity declined. In particular, in Cases 2-4 and 3-4, with detour behavior allowed up to the upper limit, congestion in the inside area further increased, resulting in the largest decline in processing capacity. The case of pedestrian out-flow showed a similar tendency; however, compared to the number of in-flow agents their number was very low, and therefore in the second half, processing capacity remained at almost the same level.

## Research result and issue

On the assumption of Tokai earthquake advisory information being officially announced, this research created a model of the behavior of pedestrian agents, some of whom were eventually forced to remain in a stationary state, and developed the model to analyze an agent-based simulation at the terminal station. Through the simulation experiment the following was found: (1) due to congestion caused by agents getting off trains, increasing density spread in several areas; as in-flow agents from outside increased, the sections in front of the ticket barriers and internal passages experienced an increase of density, causing a further decline of processing capacity; (2) as a result of (1), when the simulation results for the number of agents processed during the initial one hour are compared to the estimates released by Nagoya City, they show a figure of 30 to 80% of the upper limit of the estimate; and (3) even though agents who flowed in from outside were able to take a detour in response to congestion, the number of stationary agents actually increased, and taking a detour had little effect on the increasing density; therefore it is suggested to implement measures to restrict the total number of pedestrians who flow in from outside the station.

## References

1. Accident Investigation Report on the Firework Festival at the 32nd Akashi City Summer Festival, pp. 84-114 (January 2002)
2. Kaneda, T.: Pedestrian Flow Agent-Based Simulation. Measurement and Control. Vol. 43, No. 12, pp. 944-949 (December 2004)
3. Kaneda, T., Suzuki, T.: A simulation analysis for pedestrian flow management. In: Terano, T., Kita, H., Arai, K., Degchi, H. (Eds.) Agent-Based Simulation from Modeling Methodologies to Real-World Applications, pp.220-229. Springer (2005)
4. Cui, Q., Taniguchi, H., Kaneda, T.: Crowd Simulation of Terminal Station Transfer Passage when Tokai Earthquake Advisory Information is An-

nounced Officially. Journal of Social Safety Science. No.10, pp.153-159 (November 2008) (in Japanese)
5. Fujioka, M., Ishibashi, K., Tsukagoshi, I., Kaji, H.: "A Research of the Multi-Agent Simulation Model for Evacuees Escaping from Tsunami Disaster." Paper presented at Sixth inter University Seminar on Asian Megacities in Taipei, Taiwan, Proceedings of the sixth inter University seminar on Asian Megacities (2001) (In Print)
6. Cui, Q., Ichikawa, M., Kaneda, T., Deguchi, H.: Large Scale Crowd Simulation of Terminal Station Area when Tokai Earthquake Advisory Information is Announced Officially. The Sixth International Workshop on Agent-Based Approaches in Economic and Social Complex Systems, pp. 45-53 (2009) (CD-ROM)
7. Deguchi, H., Kanatani, Y., Kaneda, T., Koyama, Y., Ichikawa, M., Tanuma, H.: "Social Simulation Design for Pandemic Protection." Proceeding of the First World Congress on Social Simulation. Vol.1, pp. 21-28 (2006)
8. Report on comprehensive technology development project about city disaster prevention measures technique by Ministry of Construction. Ministry of Construction (1982)
9. "Survey and Research for Measures to Support Pedestrians Remaining in front of Nagoya Station," pp.80-123. CTI Engineering Co., Ltd (2002)

# Evacuation Assistance for a Sports Arena Using a Macroscopic Network Model

**A. Schomborg[1], K. Nökel[2] and A. Seyfried[3]**

[1,2] PTV Planung Transport und Verkehr AG
[1] 40472 Düsseldorf / [2] 76131 Karlsruhe, Germany
[3] Forschungszentrum Jülich GmbH (JSC)
52425 Jülich, Germany
Corresponding author: Andreas.Schomborg@ptv.de

**Abstract** The macroscopic network model for pedestrians described in this paper is part of the German research project Hermes. The purpose of Hermes is the development of an evacuation assistant. Key outputs of the network model are travel (evacuation) times, identification of bottlenecks, pointing out alternative escape routes and/or optimization of escape route usage. The model uses Dynamic User Equilibrium (DUE) to analyse the network. This way it is able to show pedestrian flow with transient congestion effects, leading to time-varying route choice during an evacuation. The results can be calculated before or during an emergency case. The calculation time is much shorter and the results can be shown quicker than with a common microsimulation movie or animation. But arguably a microsimulation can be more detailed and for absolute information like evacuation time more conventional.

## Introduction

We report on a macroscopic network model used as part of an electronic online evacuation assistant for a 50.000 seat indoor sports arena. The heart of this work is a network model of the ESPRIT-Arena in Düsseldorf, Germany. The network model was built with the transportation planning software tool VISUM and consists of about 1.780 nodes, 4.260 links and 550 zones. This modeling is performed as part of the research project Hermes (funded by the German Ministry of Education and Science BMBF) [1]. The Hermes project aims to improve safety for people in large multifunctional buildings and also at big events by exploring the effectiveness of an evacuation assistant.

The main aim of the Hermes project is the development of the assistant for the decision-makers (operators, security services, police and fire fighters). To this end it is planned to install an online microscopic simulation tool that is able to simulate the evacuation scenario with up to 66.000 people, much faster than real time, to predict possible critical situations during the course of evacuation.

R.D. Peacock et al. (eds.), *Pedestrian and Evacuation Dynamics*,
DOI 10.1007/978-1-4419-9725-8_35, © Springer Science+Business Media, LLC 2011

The evacuation assistant system will be tested in the ESPRIT arena in Düsseldorf in 2011 [1]. The macroscopic network model is intended as a complementary system in two aspects. First, microsimulating so many pedestrians in only a fraction of real time is a very ambitious goal and takes current computer systems to the limit. The macroscopic model has the big advantage of needing much less computer power. Secondly, while a macroscopic model can not deliver in dynamic situations as detailed results as a microscopic model and a number of situations cannot be simulated with it at all, it can optimize the distribution of spectators over available routes. In this respect the model provides more guidance to the decision-maker than a microsimulation, which only calculates the performance of an externally given evacuation strategy.

## Construction of the network

The model includes all five floors of the arena (Fig. 1) and consists of nodes, links, connectors and zones. Each corridor, mouth, stair or door is modeled as a link.

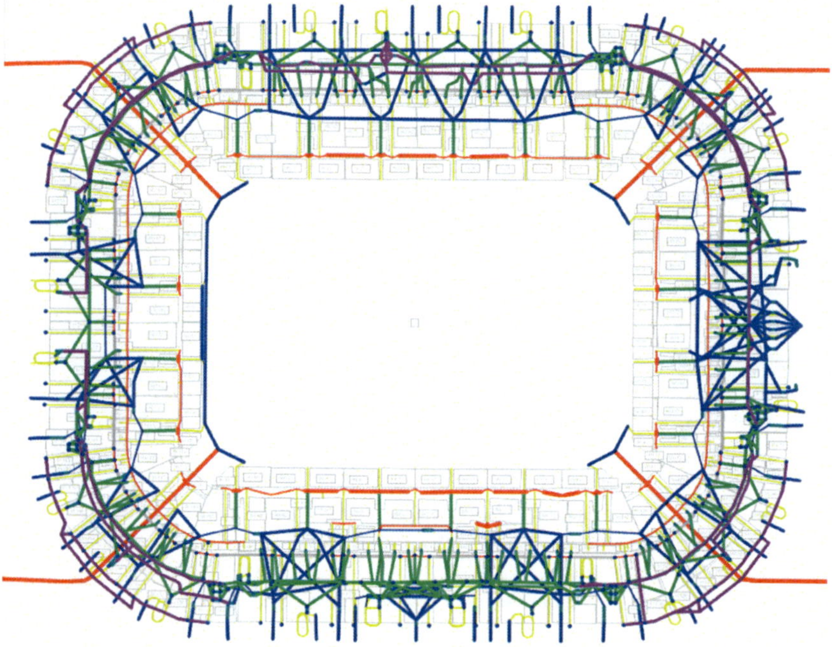

**Fig. 1. Network Model of the Esprit Arena Level 0 – 5**

The links are connected by the nodes. The nodes are in turn linked by the connectors to zones. Each of the 500 zones represents a part of the grandstands and the number of spectators exiting from any such area. The basic assumption is that when an emergency occurs the event is in progress and the spectators are seated. The rest of the people being in a restroom or in a VIP bar will be ignored in the model. Network links belong to various link types, depending on the facility they represent, as is shown in table 1 for different types of doors.

**Table 1. Link Types - doors**

| Name | Number | Width in meter | Max. spec. Flow in Pers/m*s [1] | Capacity in Pers./hour |
|---|---|---|---|---|
| External 2 wing door | 40 | 2,30 | 1,3 | 10.764 |
| External 2 wing door | 41 | 2,35 | 1,3 | 10.998 |
| Intermediate door asymmetric 2 wings | 42 | 1,60 | 1,3 | 7.488 |
| External door 1 wing | 43 | 1,10 | 1,3 | 5.148 |
| VIP Area door | 44 | 1,00 | 1,3 | 4.680 |
| Intermediate door 2 wings | 45 | 2,10 | 1,3 | 9.8298 |
| Sluice door | 46 | 2,55 | 1,3 | 11.934 |
| VIP main entrance | 47 | 5,25 | 1,3 | 24.570 |
| Fire safety gate | 48 | 7,50 | 1,3 | 35.100 |

Other network elements like corridors, stairs or mouths are divided into link types in the same way.

$$\text{Capacity/h: Width * Max Spec. Flow * 1h} = \text{m*1/(m*s)*3600s}$$

The resulting capacity per hour is a parameter for the model as is the velocity. Downstairs each link has a velocity of 2,52 km/h [3] and within one level it is 5,4 km/h [4][2]. These speeds are average values for free flow conditions. If the links are crowded the velocity decreases. The three most important attributes are the capacity, the velocity and the length of the links. The nodes connect the links but do not restraint capacity.

## Assignment - Dynamic User Equilibrium (DUE)

By Dynamic User Equilibrium we mean a distribution of spectators to available routes where nobody can reduce his/her time to exit the stadium by unilaterally switching to a different route [6].

---

[1] RiMEA [2]

[2] The velocities are average values for people with different ages.

The analysis of road network traffic performed through static assignment models reaches equilibrium between demand and supply under the assumption of constant flow and capacity within a certain period. Therefore static assignment models cannot reproduce dynamic phenomena like the formation and dispersion of queues due to the temporary over-saturation of corridor sections, and spillback, i.e. queue propagation towards upstream links [5]. The following specific features of the DUE are useful for our application:

- Simulation of networks with transient congestion effects, leading to route choice varying during assignment period,
- Simulation of incident effects and incident management,
- Simulation of evacuation plans, in particular when the maximum evacuation time is required.

DUE uses the starting route choice and the impedance calculation out of a static assignment (all or nothing assignment). The assignment period is divided into intervals and for each time interval we assume that the demand is known, i.e. the number of spectators in each zone wishing to exit during this particular time interval. After that the routes become loaded for each interval according to their specific demand. In contrast to the static case route search considers not only the spatial, but also the temporal dimension in the determination of the link flows.
For queue spillover modeling, the interaction among the flows on adjacent arcs is propagated in terms of time-varying link exit capacities. The approach is then to model the spillover phenomenon as a hypercritical flow state, which propagates backwards from the final section of a link, until it reaches its initial section. At that point it reduces the exit capacities of the upstream links and eventually influences their flow states [5].

## Fundamental Diagram

A link ($\alpha$) is a connection with two bottlenecks located at the beginning and at the end. Flow, velocities and densities are constant over the link. The flow along the link is determined on the basis of the Simplified Theory of Kinematic Waves (STKW), assuming the concave parabolic-trapezoidal fundamental diagram depicted in figure 2. Pedestrian flow ($q_\alpha$) at a given section (x) of the link ($\alpha$) and period ($\tau$) is a function of the pedestrian density $k_\alpha(x,\tau)$ at the same section and period. The link ($\alpha$) is then characterized by.
$L_\alpha$ = length of link $\alpha$
$Q_\alpha$ = capacity of the initial bottleneck and of the link
$S_\alpha$ = capacity of the final bottleneck associated to link $\alpha$
$V_\alpha$ = maximum speed allowed on link $\alpha$, called free flow speed
$KJ_\alpha$ = maximum density on link $\alpha$, called jam density

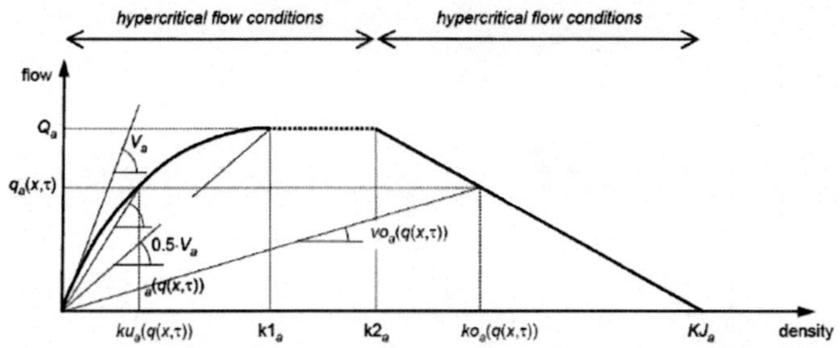


**Fig. 2. The parabolic-trapezoidal fundamental diagram [5]**

Based on the fundamental diagram, it is possible to identify two families of flow states:

- Hypocritical flow conditions, corresponding to uncongested or slightly congested traffic. Under these conditions, pedestrian flow increases or stays constant if density increases.
- Hypercritical flow conditions, corresponding to heavily congested flow, where queues and "stop and go" phenomena occur. Under these conditions pedestrian-flow decreases if density increases.

## Basic scenario

The basic scenario is defined as an evacuation during a sports event. The arena is sold out and about 99% of the spectators are on their seats. Only an insignificant number of people remain in the access areas, restaurants or restrooms. This way the evacuating demand is about 30.000 people from the upper level and 21.000 from the lower level grandstands. Another assumption of the basic situation is the 100% availability of all emergency exits. Moreover the basic route choice is designed as a lowest resistance (fastest) route to get out of the arena.

As basic parameters for the people uninfluenced walking speed specifications from RiMEA [2] are used.

The next figure shows a basic evacuation route from the upper level. It begins with the stairway inside the arena between the grandstands. Then there is a short crosswalk leading into the mouth. Afterwards the people reach the promenade, pass through the door and finally reach the staircase. The five indicated network elements are basis of the following link evaluations shown in the next two diagrams.

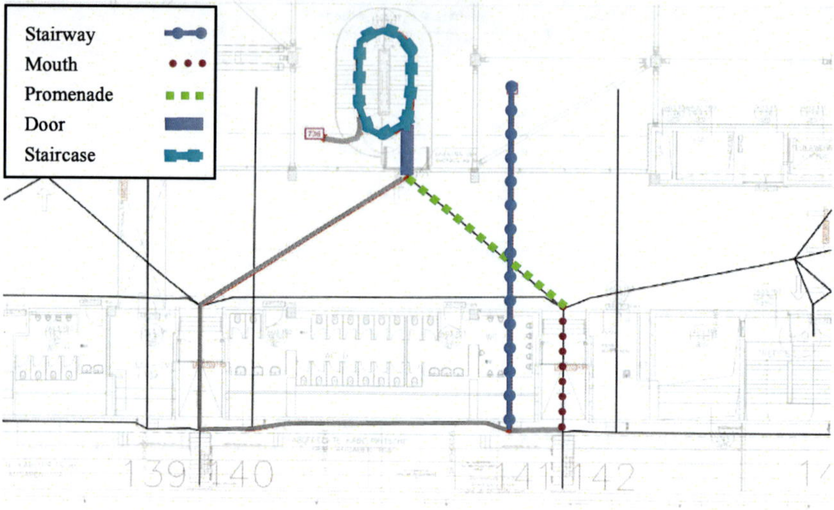


**Fig. 3. Evacuation way (example) Level 3**

Figure 4 shows the link flow aggregated for every 60 seconds on the five network elements for the basic scenario. The evacuation begins at 17:30 o'clock. After 16 Minutes the last people are passing the mouth and only one minute later the last people left the staircase.

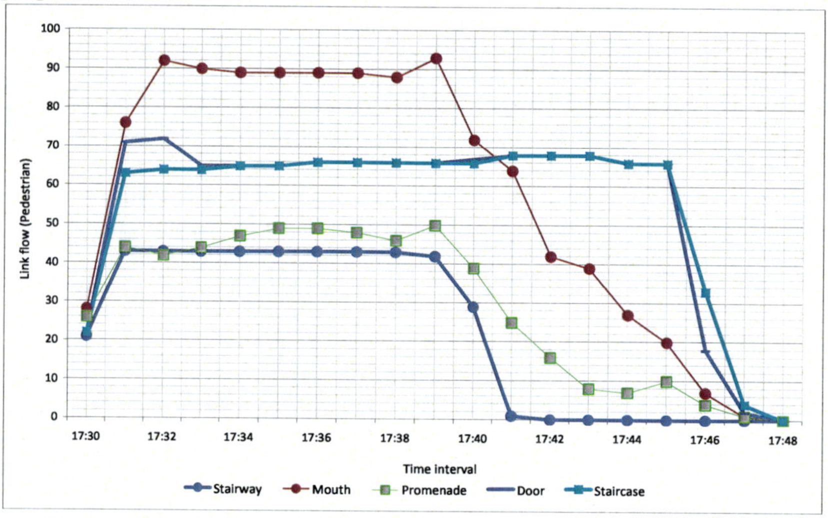


**Fig. 4. Link flow – basic scenario**

Figure 5 illustrates the walking speed on the same five network elements.

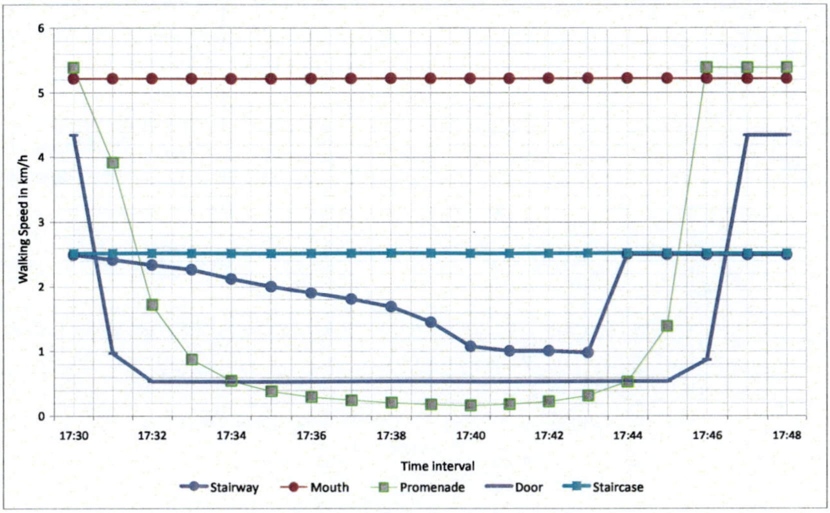


**Fig. 5. Walking speed – basic scenario**

The biggest speed drop occurs on the promenade and at the doors leading to the outside stairs. The speed falls under 0,2 km/h indicating congestions and spillback. After reaching the outdoor stairs they are able to walk at their desired speed. The same applies to the grandstand and the mouths. As soon as the people are inside the mouth they are able to walk at their desired speed. They are already losing most of the time and speed getting to the long stairway in the grandstands. The total evacuation time for the upper level is about 18 minutes. For the lower level it takes about 16 minutes.

## Evacuation Scenario

The evacuation scenario relies on the same basic conditions as the basic scenario except one mouth is closed during the evacuation. The following figure shows the influence of the missing mouth. The mouth breakdown leads to a detour for the spectators sitting in that area. They have to divert over a longer crossway to one of the neighbouring mouths to get on the promenade and afterwards out of the arena, as example shown in Figure 6. The six indicated network elements are the basis of the following link evaluations shown in the next diagram (Fig. 7) 20 minutes after the evacuation begins the people left the stairway, 2 minutes later they left the crossway and after 3 more minutes the last people left the mouth, the promenade and the staircase. In summary after 25 minutes the evacuation is finished. Compared with the basic scenario it took about seven minutes longer.

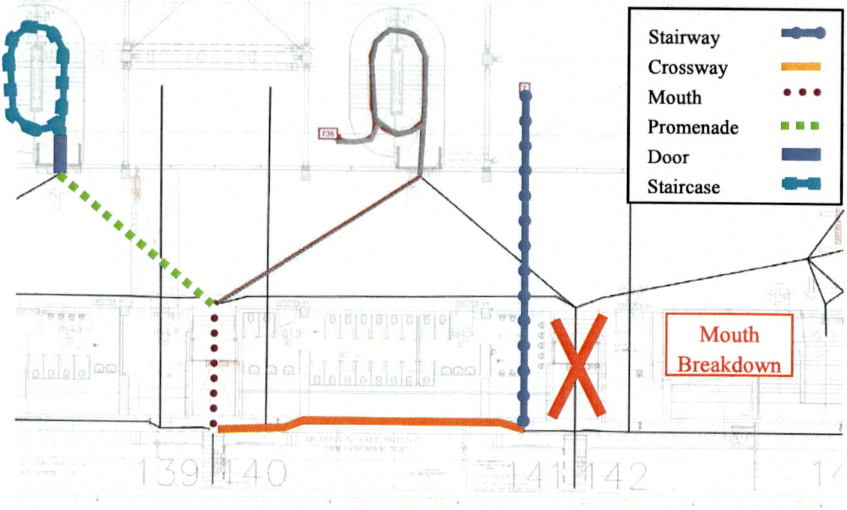


**Fig. 6. Evacuation way (example) – evacuation scenario**

The comparison between the walking speeds shows that the desired speeds on the promenade and at the door are in the evacuation scenario as well as in the basic one already after about 15 minutes back on their default level. This implies the people lose the time inside the arena at the grandstands, mainly to get on the stairway and walk the crossway.

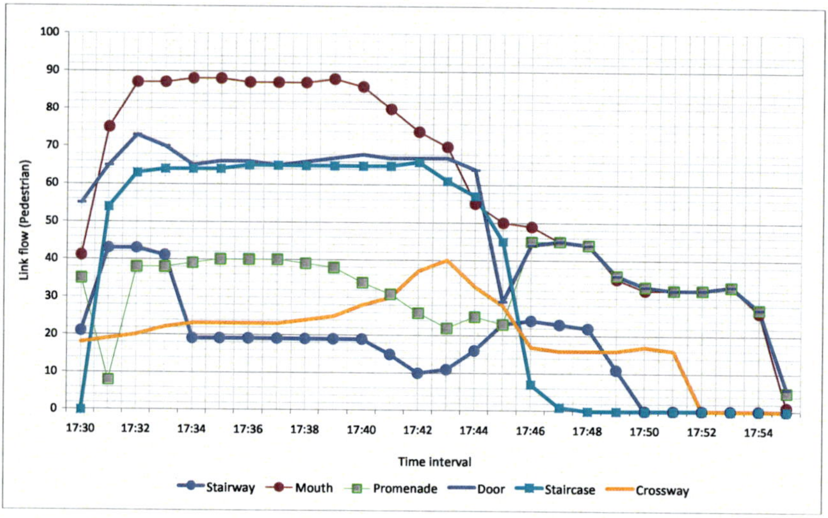


**Fig. 7. Link flow – evacuation scenario**

The following figure 8 shows the link flow for the evacuation scenario and the difference of the link flow between the two scenarios. Clearly, most of the affected people use the two bordering mouths to get on the promenade and the two bordering staircases are used by more people in the emergency scenario. There are at least three critical points where the spectators have to be advised to find the fastest way to get out. About 1.500 are using the mouth at point one, almost a third of them should be directed to point 2 to use the inside staircase to get on level 0. The mouth at point 3 is used by about 1.100 hundred people, almost a third of them should walk parallel to the grandstands to use the next outside staircase and none of them should walk against the direction of the emergency door towards the inside staircase at point 2. The third point where the spectators need advice is at point 4. At this point the people have the opportunity to go upstairs on level 5 towards point 5 and use an extra emergency exit to the next outside staircase behind point 5. In the basic scenario about 400 use the stairway, half of them should be directed upstairs.

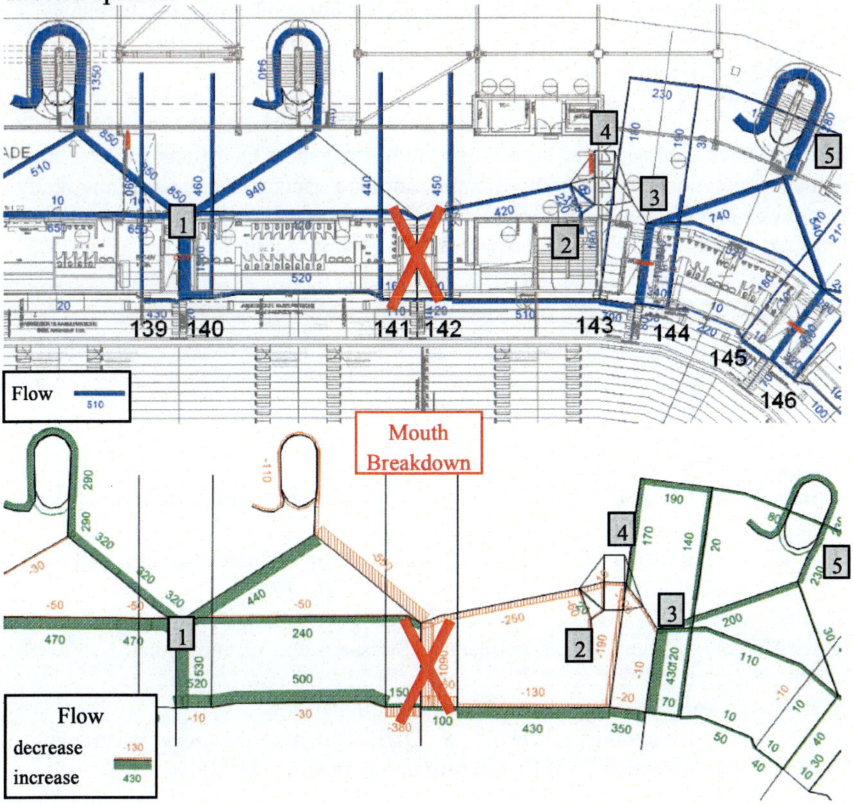


**Fig. 8. Link flow evacuation scenario (above) - difference evacuation minus basic scenario (below)**

## Conclusion

At the current state of work of the research project Hermes this network model is still a stand alone system. It is supposed to become fully integrated into the evacuation assistant. The goal is to have interfaces with the online pedestrian count system as well as with the fire protection system. In case of an emergency the model should be able to semi-automatically run assignments with online information of spectator numbers and missing evacuation corridors.

A benefit of the model is that these calculations can be done in advance of an event to optimize route guidance system [7] as well as in the beginning of an evacuation, to find the best distribution of evacuees on the remaining routes, if one or many corridors are unavailable due to smoke or structural damage.

The key for success of this model and even for the whole research is to produce output that is useful in the decision-making process for the stakeholders from the fire service, the police, the security personal and the arena staff. The results need to be very clear, easy to interpret and readily available even for non-experts because of the special emergency situation when to decide.

For a detailed description a microsimulation is needed. The network model results are beneficial for relative comparison. E.g. the differences between the basic and the evacuation scenario both online and in advance. Questions such as, how many minutes longer, how many people more are using this corridor, how many percent of people should use this way or that way or which evacuation route has most capacities available?, can answered by a network model.

## References

1. Julich Supercomputing Centre (JSC). http://www.fz-juelich.de/jsc/Hermes/. Last accessed 23 January 2010.
2. Inititors of the RiMEA Project: Richtlinie für mikroskopische Entfluchtungsanalysen, Version: 2.0.2 (2007)
3. Fruin, J.: Pedestrian planning and design. Metropolitan Association of Urban Designers and Environmental Planners, New York (1971)
4. Weidmann, U.: Transporttechnik der Fußgänger. In: Schriftenreihe des Institut für Verkehrsplanung, Transporttechnik, Strassen- und Eisenbahnbau Nr. 90, S. 35-46. Zürich (1992)
5. PTV AG, VISUM 11.0, Basics, Karlsruhe (2009)
6. Gentile, G., Meschini, L., Wekeck, S.: DUE - Die neue dynamische Umlegung in VISUM. Karlsruhe PTV Vision user group meeting (2008)
7. Geoerg, P.: Optimierung eines statischen Evakuierungskonzeptes mit Hilfe dynamischer Netzwerkmodellierung, Köln 2009 (unpublished)

# Validation of a Potential-based Evacuation Model of City Residents in Post-earthquake Fire

**T.Nishino, S.Tsuburaya, K.Himoto and T.Tanaka**

Disaster Prevention Research Unit, Kyoto University
Gokasyo, Uji, Kyoto 611-0011, Japan
Corresponding author: tomoaki.nishino@t02.mbox.media.kyoto-u.ac.jp

**Abstract** Evacuation of a large number of residents is conceivable in case of urban fires following a large earthquake in a city. It is essential to implement effective evacuation measures for ensuring residents' safety in the regional disaster prevention plan. We have been developing a simulation model based on a potential method for the evacuation behaviors of city residents in a post-earthquake fire. The model has been validated by comparing the evacuation behaviors of Tokyo City residents in the Kanto Earthquake Fire, where the distribution of fatalities calculated by this model was qualitatively similar to that reported by the survey of that time. In this paper, the evacuation behaviors of Tokyo City residents in the Kanto Earthquake Fire were simulated for validating the prediction function of this model in terms of the traveling trajectory of an evacuee.

## Introduction

In Japan, fire is the second largest causes of disaster that causes numerous deaths and major earthquakes take place frequently, e.g. the Kanto Earthquake in 1923 and Kobe Earthquake in 1995. Numerous fatalities were caused by fire in the Kanto Earthquake, because institutions for fire prevention and fire control were not developed satisfactorily and numbers of wooden buildings were densely built. After that event, several approaches for fire prevention were taken, but buildings are not fully non-combustible. In the case of large earthquakes in which multiple fires may break out simultaneously, the spread of such fires may overwhelm the ability of firefighters and damage large areas. It was recognized again in the Kobe Earthquake that such a risk remains in a modern city.

Evacuation of a large number of residents is conceivable in case of urban fires following a large earthquake in a city, which is predicted to take place in near future in Japan. It is essential to implement effective evacuation measures for ensuring residents' safety in the regional disaster prevention plan. We have been developing a simulation model based on a potential method for the evacuation behaviors of city residents in a post-earthquake fire. The potential theory means that the direction of evacuee's traveling is determined according to the gradient of the potential field where several factors governing the evacuation behaviors distri-

R.D. Peacock et al. (eds.), *Pedestrian and Evacuation Dynamics*,
DOI 10.1007/978-1-4419-9725-8_36, 

bute changing with time. The model has been validated by comparing the evacuation behaviors of Tokyo City residents in the Kanto Earthquake Fire, where the distribution of fatalities calculated by this model was qualitatively similar to that reported by the survey of that time [1].

In this paper, for the additional validation of the prediction function of this model, the evacuation behaviors of Tokyo City residents in the Kanto Earthquake Fire were simulated, where the traveling trajectories of evacuees were discussed:

## Potential-based Model for the Evacuation Behaviors of City Residents in Post-earthquake Fire

The conceptual diagram of this model is shown in **Fig.1**, which is based on a potential method [1]. This model consists of the following models:

A) A hazard evaluation model of spreading fire (the danger potential model)
B) A safety evaluation model for evacuees (the safety potential model)
C) A traveling model of residents

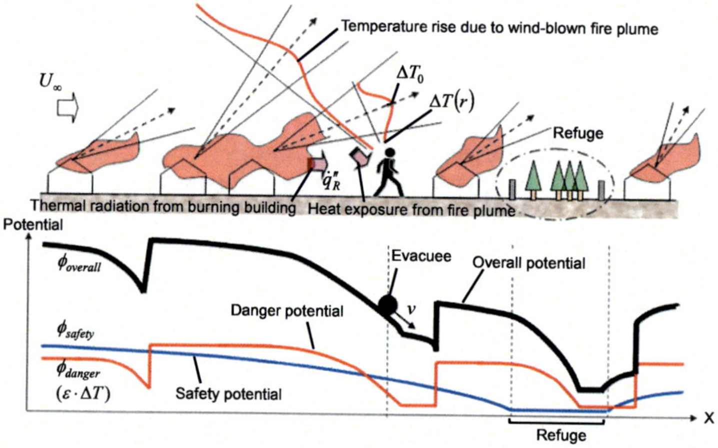


**Fig.1. Conceptual diagram of a potential-based evacuation model in post-earthquake fire**

### *A) The Danger Potential Model*

In the danger potential model, temperature rise due to fire plumes and thermal radiation from burning areas are evaluated to generate the danger potential $\Phi_{danger}$. For calculating the temperature rise, an urban area is divided into meshes and burning state of each mesh is estimated from the survey data of the fire. Temperature rise $\Delta T$ of each downwind mesh due to fire plumes and thermal radiation from burning areas is calculated and is converted proportionally to the danger potential $\Phi_{danger}$ as follows:

$$\Phi_{danger} = \varepsilon \cdot \Delta T \tag{1}$$

where, $\varepsilon$ is the conversion factor.

### B) The Safety Potential Model

In the safety potential model, safety of an urban area recognized by residents is evaluated as the safety potential $\Phi_{safety}$, where an urban area is modeled as an evacuation route network consisting of nodes and links. The safety potential $\Phi_{safety}$ is defined as the hypothetical psychology field, which means the level of psychological inclination to be attracted to a space recognized to be safe. The lower the value of the safety potential, the higher the psychological safety. The safety potential $\Phi_{safety}$ at each node is evaluated by solving the following governing equation of psychological safety:

$$\sum_j \frac{\Phi_{safety,i} - \Phi_{safety,j}}{L_{ij}} B_{ij} = 0 \tag{2}$$

where, $i$ is the identification mark of a node, $j$ is the connecting node with a node $i$, $L$ is the distance between nodes $i$ and $j$, and $B$ is the width of a link connecting nodes $i$ and $j$.

### C) The Traveling Model of Residents

In the traveling model of residents, evacuees travel on the evacuation route network toward the direction of descending the potential selecting a link stochastically at a node according to the gradient of the overall potential $\Phi_{overall}$. The overall potential $\Phi_{overall}$ corresponds to the integration of the danger potential $\Phi_{danger}$ and the safety potential $\Phi_{safety}$ as follows:

$$\Phi_{overall} = \Phi_{danger} + \Phi_{safety} \tag{3}$$

Potential $\Phi_{overall}$ means overall risk which an arbitrary point in urban area has.

The probability that a link consisting of nodes $i$ and $j$ is selected is given as follows:

$$P_{ij} = \begin{cases} 0 & (\Delta\Phi_{ij} \le 0) \\ \Delta\Phi_{ij} \Big/ \sum_{k=1}^{N} \Delta\Phi_{ik} & (\Delta\Phi_{ij} > 0) \end{cases} \tag{4}$$

where,

$$\Delta\Phi_{ij} = \frac{\Phi_{overall,i} - \Phi_{overall,j}}{L_{ij}} \times v_{ij} \tag{5}$$

where, $v_{ij}$ is the travel speed of an evacuee. As an evacuee travels from a high potential point to a low potential point, one of links where $\Delta\Phi_{ij} > 0$ are fulfilled is selected by an evacuee.

The travel speed $v$ of evacuees is given as follows by population density $\rho$ of a link [2]:

$$v = \begin{cases} 1.3 & (\rho < 0.64) \\ 1.48 - 0.28\rho & (0.64 \le \rho < 5.3) \\ 0.1 & (5.3 \le \rho) \end{cases} \tag{6}$$

where, the travel speed in free ambulation is 1.3 m/s as that of standard pedestrians in normal situation [3]. On the other hand, it is not assumed that in high-density crowd walking, pedestrians cannot travel at all. Hence, in this model, the latest travel speed is 0.1 m/s.

During an evacuation, an evacuee is exposed to heat due to fire plumes and thermal radiation, and physically influence due to heat is accumulated. In the traveling model of residents, the start and end of an evacuation are governed by the amount of exposure to heat $E(t)$, which is calculated as follows by using the temperature rise $\Delta T$ of a mesh calculated in the danger potential model:

$$E(t) = \int_{t_0}^{t} \{T_\infty + \Delta T - T_0\}^n dt \tag{5}$$

where $T_\infty$ is the temperature of ambient air, $T_0$ is critical temperature when physiological reaction arises, $t_0$ is the time when $T_\infty + \Delta T$ exceeds $T_0$ for the first time, and $C$ and $n$ are constants governed by the hazardous property of a hot gas [3].

An evacuee starts to evacuate as $E(t)$ exceeds some criteria, and dies as $E(t)$ exceeds the fatal dose. An evacuation is completed as reaching the destination refuge before the amount of exposure to heat $E(t)$ exceeds the fatal dose.Simulation of the Evacuation Behaviors in the Great Kanto Earthquake Fire

The Kanto Earthquake occurred at 11:58 AM on September 1st in 1923. Cities such as Tokyo City, Yokohama City and so on were damaged. Over 100,000 buildings collapsed and numbers of fires broke out at multiple locations. An outline of the effects of the Kanto Earthquake is shown in Table.1 [4]. The number of fatalities over the whole ruined area was 105,385 and in Tokyo City it was 68,660. The number of fatalities caused by fire in Tokyo City was 65,902.

**Table 1. Outline of the great Kanto Earthquake**

| | |
|---|---|
| The Date of Fire Breakout in Tokyo City | 11:58 AM, 1st, Sep, 1923 |
| The Data of Fire Extinguishment in Tokyo City | 10:00 AM, 3rd, Sep, 1923 |
| Number of Burnt Buildings | 293,387 |
| Number of Fatalities | 105,385 in the Ruined Area<br>68,660 in Tokyo City (Fire: 65,902 Collapse: 2,758) |
| Burnt Area in Tokyo City | 34,664,251 $m^2$ |
| Number of Afflicted People in Tokyo City | 1,356,595 |

## *Numerical Conditions*

In Fig.2, the burnt area in Tokyo City in the earthquake fire is shown, and the area of computation corresponds to the area enclosed by a dotted line. The Calculation time was 43hrs from the outbreak and the calculation time step was 1min. Fire

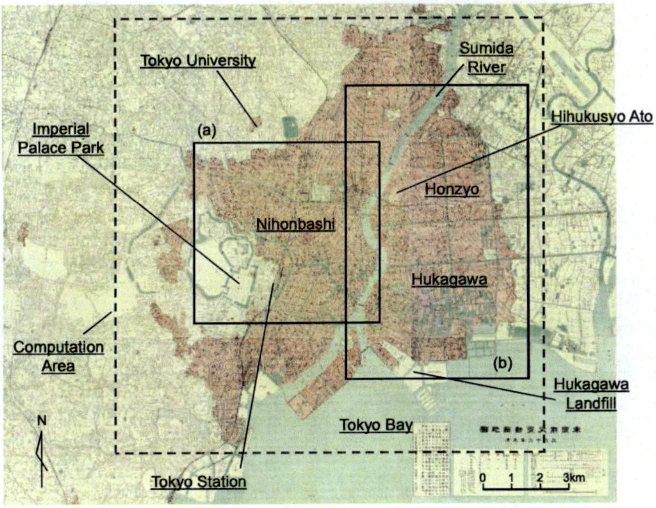


**Fig. 2. Burnt area in Tokyo City and computation area of simulation**

spread condition for computation was obtained from the after event field survey data of that time. The computation number of residents was 1,355,906 and initial distribution of residents was set according to the demographic statistics. The evacuation route network was reconstructed from the city map of that time, and the representative refuges were set, and their maximum capacity was set to the number of evacuated people reported.

## *Results and Discussion*

This model can trace the evacuation route of each evacuee. For example, the traveling trajectories of two residents calculated by this model are shown in Fig.3 along with the burning areas at the illustrated time. The areas shown in Fig.3 correspond to the areas enclosed by solid lines in Fig.2.

In Fig.3.(a), the evacuee did not start evacuation until the fire, which had originated at Nihonbashi area, approached close to him at 16:03, September 1st. The wind direction around this time was SSW, so that the evacuee traveled in the southern direction as getting across wind-blown fire plume generated from burning area. The evacuee passed the route again which he had passed once. The eva

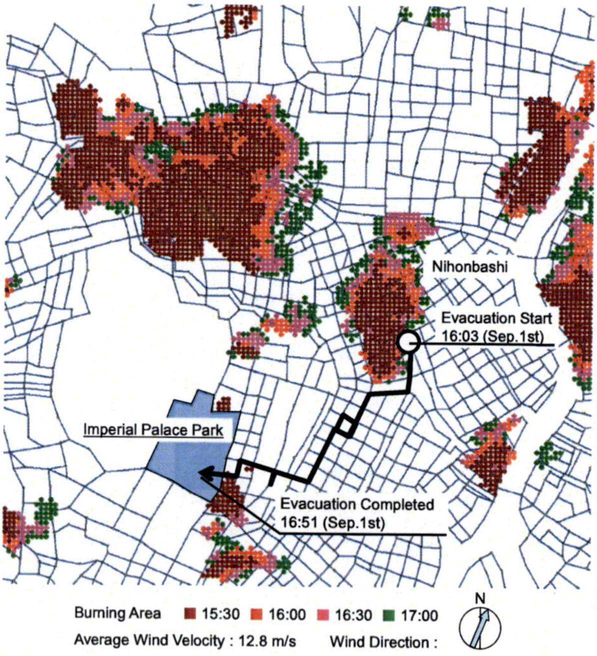


**Fig. 3a. Traveling trajectory of a resident at Nihonbashi area by the model**

cuee reached the refuge called Imperial Palace Park at 16:51, September 1st, and the travelling distance was calculated as about 3,700m. This traveling distance is considered to be within the range of escaping ability except that the evacuee is a child or an elderly, so the evacuee was considered to enable to finish evacuating without feeling physiological difficulty, and the amount of exposure to heat of the evacuee by that time was insignificant. The average traveling speed was calculated as about 1.2m/s, so the population density of evacuation routes which the evacuee has passed was thought to be insignificant, and the decrease of ambulatory ability due to the interference from others was thought to be insignificant.

In Fig.3.(b), the evacuee started evacuation at 13:44, September 1st as the fire,

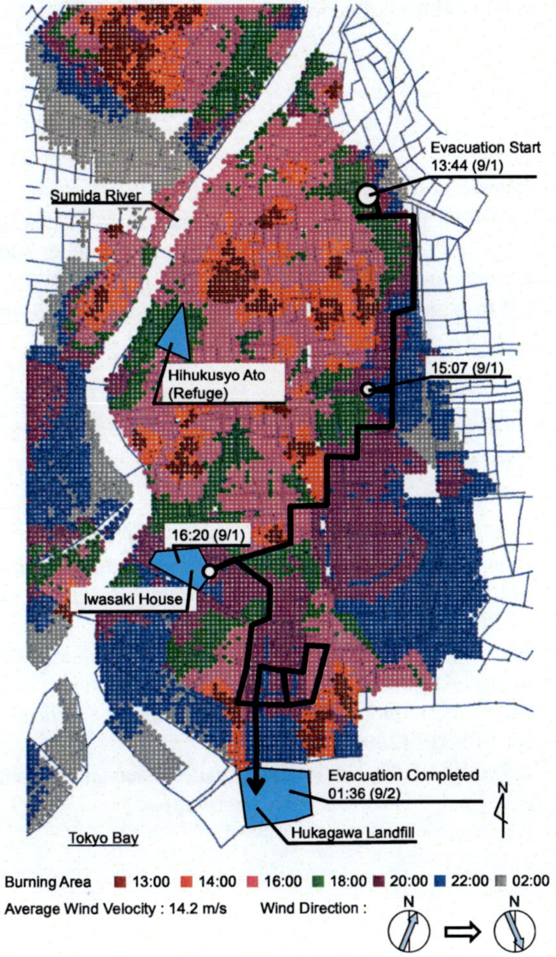


**Fig. 3b. Traveling trajectory of a resident at Honzyo area by the model**

which had originated at Honzyo area, was far from him. The evacuee, who was downwind from the fire, was exposed to heat of wind-blown fire plumes, so it is considered that he started evacuation much earlier than the fire approached close to him. After that, the evacuee traveled in the southwestern direction as getting around fires being exposed to temperature rise due to fire plumes. Once, the evacuee reached the refuge called Iwasaki House at 16:20, September $1^{st}$, but he continued evacuation because the refuge was filled with other evacuees. After that, as the area exposed to wind-blown fire plumes was changing quickly due to the change of wind direction, the evacuee wandered around at Hukagawa district for a long time, and reached the refuge called Hukagawa Landfill at 01:36, September $2^{nd}$. The travelling distance was calculated as about 18km, which was considered to be over the escaping ability of people.

## Conclusions

The evacuation behaviors of Tokyo City residents in the Great Kanto Earthquake Fire were simulated for validating the prediction functions of this model in terms of the traveling trajectory of an evacuee. The obtained results showed that this model could reasonably simulate the evacuation that an evacuee traveled avoiding the fire hazard due to wind-blown fire plumes and thermal radiation, and psychologically-inclining to a space recognized to be the safest.

## References

1. Nishino, T., Tsuburaya, S., Himoto, K., Tanaka, T.: A Study on the Estimation of the Evacuation Behaviors of Tokyo City Residents in the Kanto Earthquake Fire. Proceedings of 9th Symposium. International Association for Fire Safety Science. 453-464 (2008)
2. Okada, M., Yoshida, K., Kashiwahara, S., Tsuzi, M.: Human Dynamics of a Building and a City. Kashima (1977) (In Japanese)
3. Tanaka, T.: An Introduction to Building Fire Safety Science. The Building Center of Japan. 315-317 (2001) (In Japanese)
4. Nakamura, S.: Record of the Kanto Earthquake. Disaster Preventive Research Institute in Japan, pp. 81-134 (1925) (In Japanese)

# Transport Modeling

# Assistance of Evacuation Planning with High-Speed Network Model-based Pedestrian Simulator

**T. Yamashita, S. Soeda, and I. Noda**

Information Technology Research Institute (ITRI)
National Institute of Advanced Industrial Science and Technology (AIST)
2-41-6 Aomi, Koto-ku, Tokyo 135-0064, Japan
Corresponding author: tomohisa.yamashita@aist.go.jp

**Abstract** As described in this paper, we analyzed the influence of the time necessary to begin coping behaviors on the damage caused by chemical terrorism. To calculate the damage of a chemical attack in a major rail station, our network model-based pedestrian simulator was applied with systems designed to predict hazards of indoor gas diffusion. Our network model is designed to conduct simulations much faster, taking less than few minutes for simulation with ten thousands of evacuators. Results of our analyses were used for the instruction of rail station managers in a tabletop exercise held by the Kitakyushu City Fire and Disaster Management Department.

## Introduction

Terrorist acts using chemical and biological agents and radioactive (CBR) materials have typically been nonselective attacks on crowds in urban areas. These hazardous materials might be sprinkled, vaporized, or spread by an explosion. First responders to these accidents, such as fire protection and police agencies of municipalities must prepare practical plans of coping behaviors against CBR terrorism, but they typically have insufficient experience and knowledge of CBR terrorism. Therefore, useful tools supporting planning and preparation are necessary to estimate and illustrate the damage done by CBR attacks. To meet those needs, the Ministry of Education, Culture, Sports, Science and Technology of Japan (MEXT) ordered a research consortium of Tokyo University, Advanced Industrial Science and Technology (AIST), Mitsubishi Heavy Industries (MHI), and Advancesoft to develop a new evacuation planning assistance system for use before CBR attacks.

With our evacuation planning assist system, a user can estimate the damage caused by CBR terrorism. These disasters, which are most likely to occur in urban areas, have numerous characteristics that differ from those of natural disasters. These disasters are caused intentionally, which means that responsible persons must prepare for the worst. There are still few cases in which CBR terrorism has

R.D. Peacock et al. (eds.), *Pedestrian and Evacuation Dynamics*,
DOI 10.1007/978-1-4419-9725-8_37, 

actually been conducted, which means that we still know little about what damage might result from such terrorism.

As described in this paper, we have built a network model-based pedestrian simulator as a part of the evacuation planning assist system. Compared to previous grid-based and continuous space-based models that required hours to conduct simulations with fewer than thousands of evacuees, our network model is designed to conduct simulations much more rapidly, taking less than few minutes for simulations with tens of thousands of evacuees. Our pedestrian simulator is designed to function with hazard prediction systems of outdoor and indoor gas diffusion, which calculate how rapidly and at what concentrations harmful gases spread. Using data provided by hazard prediction systems, our pedestrian simulator is useful to estimate how much damage will be incurred under various evacuation scenarios. These results are expected to be useful to produce and evaluate evacuation plans as countermeasures against CBR terrorism.

As described in this paper, we explain our evacuation planning assist system, and share an example of practical use of our system. We dealt with coping behaviors against chemical attacks at a major rail station at the request of the Fire and Disaster Management Department of Kitakyushu City. In our simulation, we showed a relation between the time necessary to begin coping behaviors of the managers and the damage to passengers. Our analysis was used for enlightening the managers of the rail station in a tabletop exercise held by the Fire and Disaster Management Department of Kitakyushu City.

## Evacuation Planning Assist System

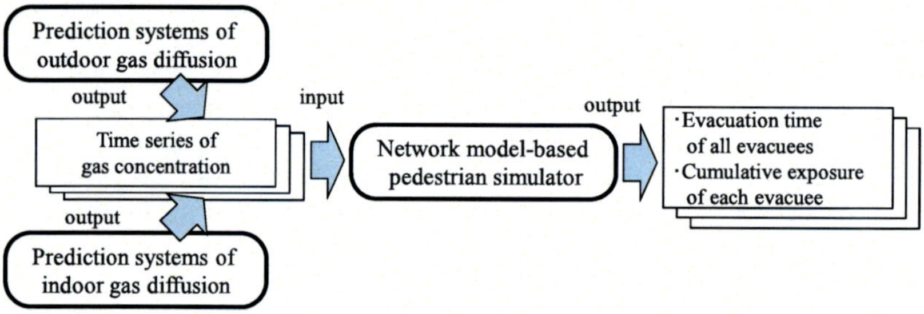


**Fig. 1.** **Outline of dataflow of the evacuation planning assist system**

Our evacuation planning assist system consists of three components: a pedestrian simulator constructed by AIST, a prediction system of outdoor gas diffusion by MHI, and a prediction system of indoor gas diffusion by Advancesoft. First, the prediction systems of indoor and outdoor gas diffusion calculate the concentration of hazardous gases. The output of these systems is a time series of gas concentration in designated areas. Then, the pedestrian simulator calculates the evacuation time of all evacuees and the cumulative exposure of each evacuee using a time series of gas concentration. An dataflow of our system is presented in Fig. 1.

## *Pedestrian simulator*

Pedestrian simulators of various kinds have been developed for various purposes. Pan roughly classified them into three categories: fluid and particle systems, matrix-based systems, and emergent systems [5]. However, all of these systems are two-dimensional systems, which allow pedestrians to move around two-dimensionally. Unlike other pedestrian simulators, our simulator simplifies traffic lines by representing it with a graph model: a model with links and nodes (Fig. 2). The paths where the pedestrians move around are represented as links.

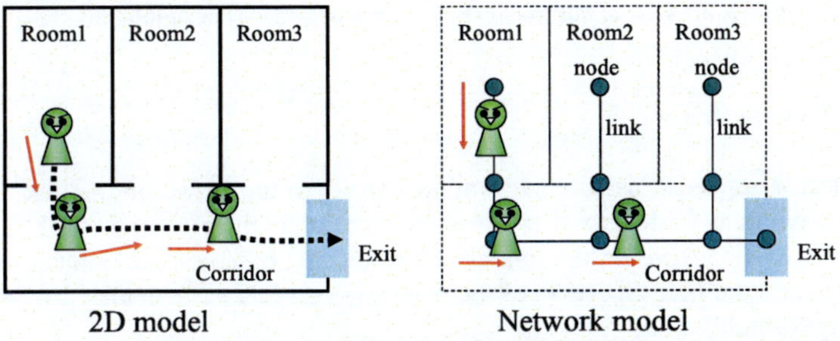


**Fig. 2. 2D model and network model**

These links are connected at nodes. Because the pedestrians were able to move only along the links, our model is more one-dimensional than two-dimensional. This approach has often been used in traffic simulators [1], but not for pedestrian simulators. We chose a network based-model for our simulator because we need a high-speed simulator to examine the evacuation behaviors of many evacuees on the macroscopic side. A network based-model is unsuitable for simulating many pedestrians evacuating a large space precisely, but it might be useful to reveal bottlenecks and evacuation times quickly and thereby enable the comparison of many evacuation plans. The appearance of a network-based pedestrian simulator is shown in Fig. 3.

### Pseudo-lane model

The speed of a pedestrian is calculated using a model we call "pseudo-lane model". We assume that each path is made up of several pseudo-lanes, based on the width of the path. The speed of each pedestrian is calculated from the free flow speed (the speed of the pedestrian without any other pedestrians), and from the pedestrians immediately front of the pedestrian in the same pseudo-lane.

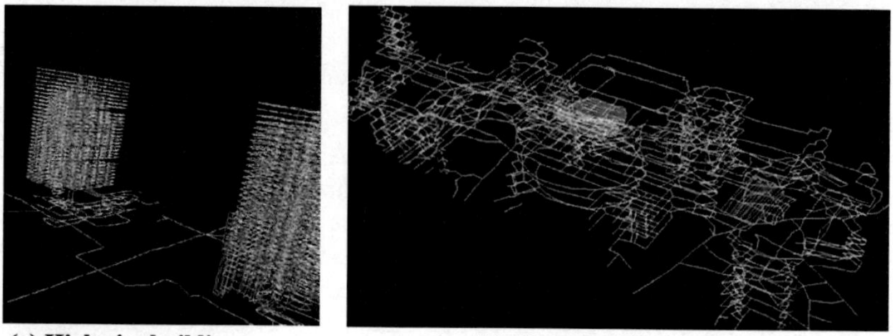

**(a) High-rise building** **(b) Commercial complex**

**Fig. 3. 3D view of our network-based model pedestrian simulator**

### Speed of a pedestrian

The pedestrian speed is calculated from the density of the crowd on the link. Each link has a width and a length; they are used to calculate the area of the link. Then, the density of the crowd on the link can be calculated based on the number of the pedestrians on the link. Speed $V_i$ of the pedestrian on link $i$ is calculated from the following formula,

$$V_i = \begin{cases} V_f, d_i < 1 \\ d^{-0.7945} V_f, 1 \leq d_i \leq 4 \\ 0, d_i > 4, \end{cases}$$

where $V_f$ represents free flow speed of the pedestrians, which is the speed of the pedestrian when not in a crowd, and $d_i$ represents the density of the pedestrians on link $i$. The exception to this formula is the pedestrian at the head of a crowd on the link. For this pedestrian, $V_f$ is used, irrespective of how crowded the link is. When the density of a link exceeds four pedestrians/m$^2$, no pedestrian on the link can move except for the one who is at the head of the crowd.

### Confluence

Confluences - where two or more paths meet together - slow the pedestrian speed. To elucidate the slow-down by a confluence, we used a simple model of limiting the number of the pedestrians who can enter a link. The maximum number of the pedestrians entering link $l_{out}$ shown in Fig. 4 is determined from the link width. When pedestrians exist on lout already, the number of the pedestrians who can enter $l_{out}$ is decreased at some ratio. When there are more than two links where the pedestrians are trying to enter $l_{out}$, this number is divided among the links depending on the number of the pedestrians trying to enter $l_{out}$. In addition, in this case, the number of pedestrians who can enter $l_{out}$ is also reduced.

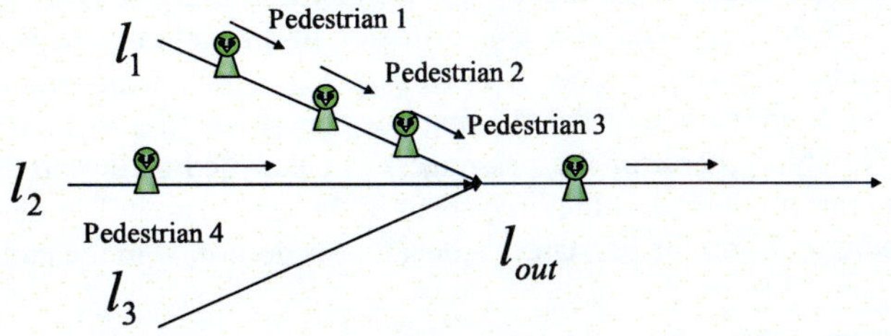


**Fig. 4. Modeling confluence**

## *Prediction systems of gas diffusion*

Recently, more subways, shopping malls, and high-rise buildings have large-scaled and intricate passages. Accordingly, the casualties of CBR attacks or fires at such locales might mount disastrously. For prevention or reduction of these disasters, a hazard prediction system of indoor gas diffusion called "Enhanced Virtual Environment Simulator for Aimed and Yielded Fatal Accidents" (EVE SAYFA) has been developed to facilitate preparation for such events and to evaluate safety of different plans and preparations [2, 7].

A hazard prediction system has been developed for CBR attacks in urban areas using the mesoscale meteorological model, RAMS and its dispersion model HYPACT. In fact, RAMS is equipped with an optional scheme to simulate airflow around buildings based on the volume fraction of the buildings within each grid cell. The HYbrid PArticle and Concentration Transport (HYPACT) code is an atmospheric diffusion code that can be coupled to RAMS. This code is based on a Lagrangian particle model that satisfies mass conservation in a complex airflow and which can adopt the finite difference method at large distances downwind to reduce the computational time.

## Simulations

With our evacuation planning assist system, we investigated a simulated chemical attack on a major rail station because of a request from the Fire and Disaster Management Department of Kitakyushu City. In our simulation, for enlightening the managers of the rail station, we showed a relation between the time necessary to begin coping behaviors of the managers and the damage to passengers.

***Simulation setting***

In our simulations, a chemical attack with chloropicrin is set to take place in the station yard of a conventional line. Gas diffusion in the station yard is calculated using the systems to predict indoor gas diffusion. The movements and damage to about 9000 passengers are calculated using our pedestrian simulator. The amount of exposure to chloropicrin of the passengers is calculated as the product of the chloropicrin concentration and the time spent in the area. The influence of chloropicrin exposure [6, 8] on the passenger's behavior is described in the table. 1.

**Table 1. Influence of chloropicrin exposure**

| Amount of exposure (mg · min/m3) | 1 | 200 | 1,000 | 2,000 | 20,000 |
|---|---|---|---|---|---|
| Influence in behavior | Pain in the eye & throat | Nausea & headache | Breathing trouble | Lethal dose 50% | Lethal dose 100% |
| Implementation in evacuation simulation | Decrease in speed (-40%) | Decrease in speed (-90%) | Stop | Stop | Stop |
| Damage level | Mild | Moderate | Severe | Severe | Severe |

This rail station is a complex facility. It has facilities of four kinds: a conventional line, a new bullet train line, a monorail, and a hotel. Each facility has a manager. We assume the following 10 coping behaviors of the managers and the times required to begin these coping behaviors described in the table 2. The times required to begin these coping behaviors influence the damage because the damage to passengers increases the beginning of these coping behaviors are delayed. The sequence of the coping behaviors is portrayed in Fig. 5.

**Table 2. Coping behaviors of the managers and the times required for them**

| Manager | Coping behavior | | Time required to begin(min) | |
|---|---|---|---|---|
| | | | quick | slow |
| Conventional line | 1 | Detecting chemical attack | 5 | 10 |
| | 2 | Reporting to the fire department and the other managers | 3 | 6 |
| | 3 | Shutting down the trains | 3 | 6 |
| | 4 | Ordering an evacuation to the passengers | 3 | 6 |
| New bullet train line | 5 | Ordering an evacuation to the passengers | 3 | 6 |
| | 6 | Shutting down the trains | 3 | 6 |
| Monorail station | 7 | Ordering an evacuation to the passengers | 3 | 6 |
| | 8 | Shutting down the trains | 3 | 6 |
| Hotel | 9 | Ordering an evacuation to the passengers | 3 | 6 |
| Fire and Disaster Management Department | 10 | Rescuing insured passengers | 20 | 30 |

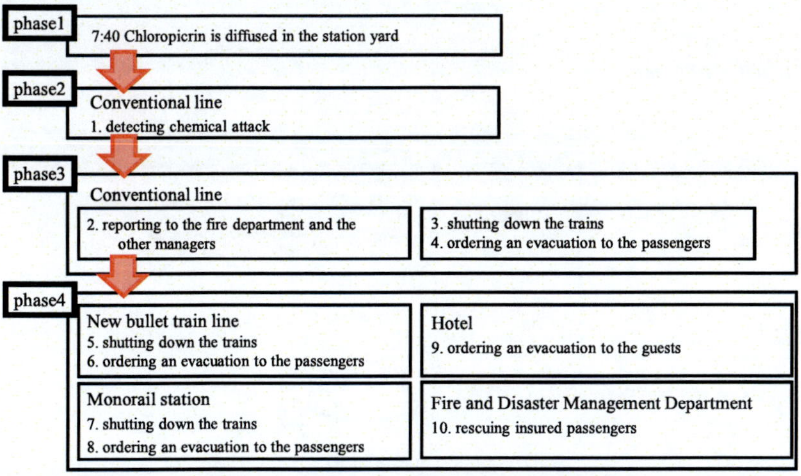


**Fig. 5. Sequence of the coping behaviors**

For example, the manager of the conventional line has four coping behaviors: (1) detecting the chemical attack, (2) reporting to the fire station and the other managers of the new bullet train line, monorail, and the hotel, (3) shutting down conventional trains, and (4) ordering an evacuation of the passengers of the conventional line. Each coping behavior is set to have the time of two kinds. For example, if coping behavior 4 is begun quickly, the time necessary to begin is 3 min. Otherwise (begun slowly), the time required is 6 min.

The evacuation scenarios are 1024 because the number of combinations of the times required to begin 10 coping behaviors is $2^{10}$ (=1024). We calculate damage to approximately 9000 passengers in 1024 evacuation scenarios. In our pedestrian simulation, each passenger walks around normally, from and to outside areas of the station and from and to platforms, until the attack is detected and an alarm is given. After ordering an evacuation of the passengers, the passengers evacuate using a route directed by the station staff.

To assign a sequential serial number to each scenario according to whether 10 coping behaviors are begun quickly or slowly, we use a ten-digit number in the binary system. For example, if coping behavior 1 (detecting chemical attack) is begun quickly, then the first bit of the ten-digit number is set as 0. This scenario is represented as 0000000000 in the binary system if all coping behaviors are begun quickly. Then, the 10-digit number in the binary system is transferred to a serial number in the decimal system. The scenario represented as 0000000000 in the binary system is assigned serial number 0 in the decimal system; the scenario represented as 1111111111 is assigned serial number 1023.

## *Simulation result*

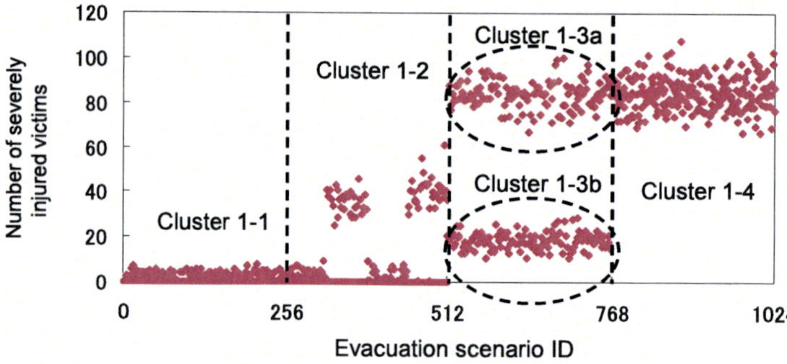


**Fig. 6. Number of severely injured victims**

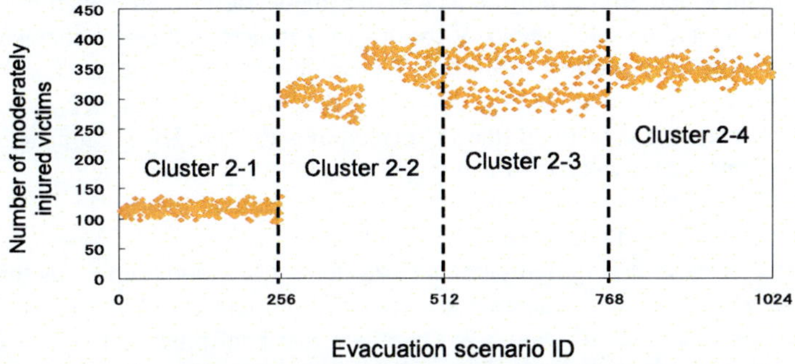


**Fig. 7. Number of moderately injured victims**

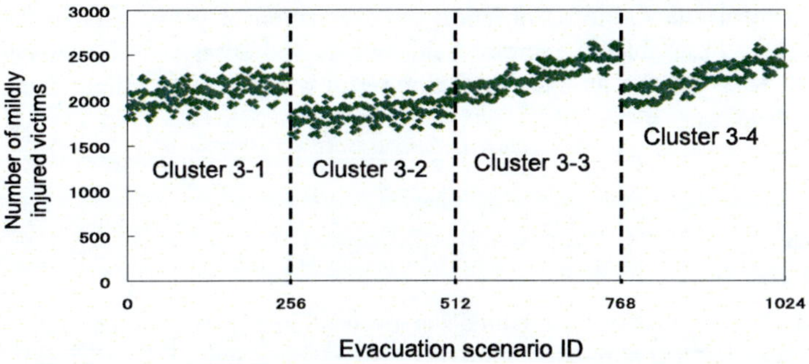


**Fig. 8. Number of mildly injured victims**

The results of our simulation are shown in Figs. 6-8. In Fig. 6, the graph shows the number of the severely injured victims in 1024 scenarios. Based on the number of severely injured victims, five characteristic clusters can be inferred. In each cluster, the scenarios have the same tendencies of coping behaviors 1 and 4. In scenarios of cluster 1-1, both (1) detecting chemical attack and (4) ordering evacuation of passengers of the conventional line are begun quickly. In scenarios of cluster 1-2, (1) detection is begun quickly, and (4) ordering an evacuation is begun slowly. In scenarios of cluster 1-3a and 1-3b, (1) detection is begun slowly, and (4) ordering an evacuation is begun quickly. The difference between cluster 1-3a and 1-3b is coping behavior 10. In the scenarios in cluster 1-3a, (10) rescuing injured passengers by the Fire and Disaster Management Department is begun slowly. On the other hand, in scenarios in cluster 1-3b, (10) rescuing is begun quickly. In scenarios of clusters 1-4, both (1) detection and (4) ordering an evacuation are begun slowly. Therefore, it is confirmed that both (1) detecting, (4) ordering an evacuation, and (10) rescuing injured passengers are more important to decrease the number of severely injured victims.

In Fig. 7, the graph shows the number of the moderately injured victims in 1024 scenarios. Based on the number of the severely injured victims, 4 characteristic clusters exist. The victims in cluster 2-1 are fewer than in other clusters. There is no great difference among the number of the victims of clusters 2-1, 2-2, and 2-3. Therefore, it is confirmed that (1) detection and (4) ordering an evacuation are more important to decrease the number of moderately injured victims.

In Fig. 8, the graph shows the number of the mildly injured victims in each of 1024 scenarios. Based on the number of the severely injured victims, four clusters can be inferred. The mildly injured victims in cluster 3-1 are more numerous than in cluster 3-2. However, the number of all victims in clusters 1-1, 2-1, and 3-1 is equal to that in cluster 1-2, 2-2, and 3-2. Therefore, it is confirmed that (1) detection is extremely important to decrease the damage to passengers. Through comparison of the damage incurred in each of the 1024 scenarios, we confirm that the most effective coping behaviors for decreasing the damage to passengers are i) detection of chemical attack and ii) ordering the evacuation of passengers of the conventional line. In a tabletop exercise held by Fire and Disaster Management Department of Kitakyushu City, our simulation result was shared to enlighten the rail station managers.

## Conclusion

As described in this paper, we explained our evacuation planning assist system consisting of a pedestrian simulator, a prediction system of indoor gas diffusion, and a prediction system of outdoor gas diffusion. A chemical attack at a major rail station in Japan was taken as an example of the practical use of our system because of a request from the Fire and Disaster Management Department of Kitakyushu City. In our simulation, we showed the relation between the time necessary to begin coping behaviors by managers and the damage to passengers. Results of comparisons of the damage incurred in 1024 scenarios confirmed that, to decrease damage to the passengers, it is important to begin coping behaviors quickly by i) detecting the chemical attack and ii) ordering an evacuation of passengers.

**Acknowledgments** This work was supported as a national project for urban safety commenced from 2007 by the Ministry of Education, Culture, Sports, Science and Technology of Japan (MEXT).

## References

1. Burmeister, B., Haddadi, A., Matylis, G.: Application of Multi-Agent Systems in Traffic and Transportation. Software Engineering, Vol. 144,

Issue 1, pp. 51-60 (2002)

2. Nara, M., Kato, S., Huang, H., Zhu, S.W.: Numerical analysis of fire on disasters with "EVE SAYFA", a fire simulator, and its validation ? LES for thermal plume. Summaries of Technical Papers of 2006 Annual Meeting of Architectural Institute of Japan, pp. 317-318 (2006) (In Japanese)
3. Ohba, R., Kouchi, A., Hara, T.: Hazard Projection System of Intentional Attack in Urban Area. 11th Annual George Mason University, Conference on "Atmospheric Transport and Dispersion Modeling" (2006)
4. Ohba, R., Yamashita, T., Ukai, O., Kato, S.: Development of Hazard Prediction System for Intentional Attacks in Urban Areas. Seventh International Symposium on New Technologies for Urban Safety of Mega Cities in Asia (USMCA2008), pp. 687-696 (2008)
5. Pan, X.: Computational Modeling of Human and Social Behaviours for Emergency Egress Analysis. Ph.D. Dissertation, Department of Civil and Environmental Engineering, Stanford University (2008)
6. Chloropicrin. Wikipedia. http:// en.wikipedia.org/wiki/Chloropicrin.
7. EYE SAYFA: Simulation System for Urban Safety and Environment. http://venus.iis.u-tokyo.ac.jp/english/Research/field-e/evesayfa/index.html
8. FACTSHEET: CHLOROPICRIN. Workcover NSW. http://www.workcover.nsw.gov.au/formspublications/publications/Documents/chloropicrin_fact_sheet_1371.pdf

# An Evacuation Model for High Speed Trains

**J.A. Capote, D. Alvear, O. Abreu, M. Lázaro and A. Cuesta**

GIDAI Group, Fire Safety – Research and Technology. University of Cantabria (Spain)
Santander, 39005, Spain
Corresponding author: capotej@unican.es

**Abstract** In this paper we present a stochastic evacuation model specifically for high speed passenger trains. The proposed model is an object-oriented model in which passengers are represented using a cellular automata method and the train space by a fine network of 0.5 m x 0.5 m cells. The model is based on Monte Carlo methods in order to simulate the probability and effects of passengers' actions and decisions during the evacuation process. The datasets used as default by the model are taken from video recordings of evacuation drills and virtual experiments conducted at the University of Cantabria. However, the flexibility of the model allows the user to modify this data. The results of this model are then compared with other validated evacuation models. The proposed model has a simple user interface and the results are given in real-time. This model could be a useful tool for evacuation management during real emergencies. The advantages of using a stochastic approach for modelling passengers' behaviour in relation to a deterministic approach are discussed.

## Introduction

In the last few decades, many egress models have been developed to improve human safety in environments such as buildings, stadiums, aircrafts, ships etc. Some of these have been applied for evacuation analysis in passenger trains [1, 2], however, no such evacuation models have been developed specifically for this particular scenario. Compared to buildings, high speed passenger trains have the following special evacuation characteristics:

1. Train vehicles can be in motion. In some circumstances, it may be necessary to move passengers from one coach to another before the train stops. Therefore, the RSET calculations should take into account the time it takes for the train to stop as well as the crew and passengers pre-evacuation activities inside the train.
2. High speed trains are multi-enclosure environments where passengers are seated in separate coaches.

R.D. Peacock et al. (eds.), *Pedestrian and Evacuation Dynamics*,
DOI 10.1007/978-1-4419-9725-8_38, 

3. There may only be a few crew members for the entire train, therefore public address systems are usually applied to issue the emergency instructions.
4. Trains have simple geometries. Most of the passengers will be familiar with the geometry of each coach.
5. Trains have limited escape routes. There are no more than two possible directions to reach an exit.
6. Trains have narrow spaces. The aisles, passageways and exits are no wider than 0.90m, and the exit steps are narrower and steeper than building stairways.

In fact, people's actions and decisions may have a relevant influence on the evacuation process [3]. This is particularly important inside trains, where the space is limited, especially in comparison to most buildings. As soon as the train passengers have been warned about the emergency, their actions before and during evacuation movements, such as preparing for evacuation, gathering information, waiting for others etc. may cause a block in the aisle and interrupt continuous movement.

In order to reproduce any actions the occupants might perform during the evacuation process, most egress models currently assume a pre-evacuation time as the time in which individuals will wait in their initial position before beginning evacuation movements. Other models assign a sequence of behavioural actions (a "behavioural itinerary") to the occupants in order to simulate an interruption in the evacuation process. However, this "behavioural itinerary" is assigned by the user before the simulation begins rather than being predicted by the model [4].

Inside trains, a major problem is that the actions performed during the evacuation are unknown prior to an event. The behaviour of passengers is assumed to be complex and varies amongst individuals. Deterministic models are not able to analyse all the possible behaviours and their effects on the evacuation process. A probabilistic approach can be an alternative to solve the problem [5]. This paper presents a stochastic evacuation model for passenger trains. By using Monte Carlo methods, the proposed model has the capability to input random characteristics and probabilities of passengers performing actions, such as time to prepare ($T_1$) and time to pick up baggage ($T_2$), in order to simulate interruptions in the evacuation process. A probability of occurrence is assigned to these behavioural actions by assigning minimum and maximum values. This process allows for variations in the outcomes by repeating thousands of simulations.

The paper begins by describing the features of the stochastic model. The second part of the paper describes the data collection to determine distributions of the input variables in the model. The third part of the paper outlines a comparison between the developed stochastic model and other validated evacuation models.

## Description of the Model

The purpose of the model is to capture and process stochastic variations in evacuation times by using Monte Carlo methods in order to simulate the random characteristics, decisions and actions of passengers in trains. In this first version, the model focuses on specific scenarios: three types of passenger trains (AVE S 102, Alvia S 130 and TALGO VI) and two evacuation procedures – the first is an evacuation through one exit from two passenger coaches to the platform, and the second is an evacuation from two passenger coaches to an adjacent coach within the train (see Fig. 1 and 2).

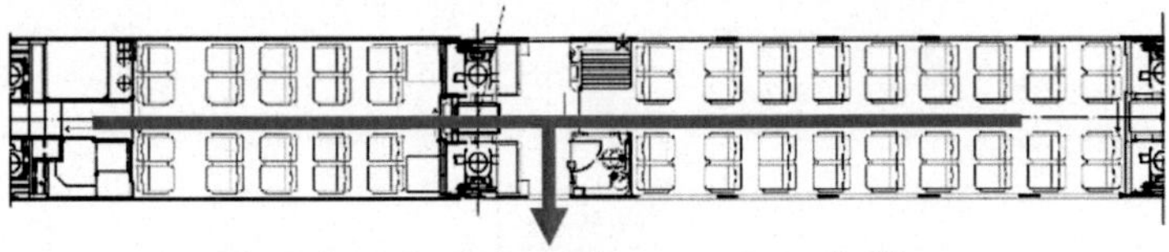

**Fig. 1. Sample of evacuation scenario to platform.**

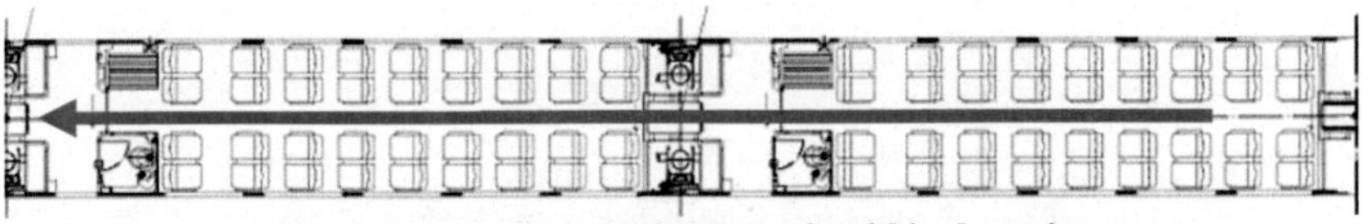

**Fig. 2. Sample of evacuation scenario within the train.**

According to the review of evacuation models by Kuligowsky [6], the model is a cellular automata (CA) in which passengers move throughout the train from one cell space to another (with a cell size of 0.5 x 0.5m). We established a numerical code for way-finding where each cell has its own information. Passengers find their way towards the exit by following the information of neighbouring cells. A passenger will not move to another occupied grid cell and will wait until the next cell is empty. If more than one passenger is waiting for the same cell (i.e. in merging flows at exit doors) and they have the same characteristics (i.e. walking speed), the model will resolve the conflict randomly to decide which passenger moves first.

As shown in Fig. 3 the numbered cells 32, 33 and 44 are assigned as potential locations where behavioral actions may occur that could interrupt evacuation movements. Cells 32 correspond to the time to prepare ($T_1$), cells 33 correspond to the time to pick up baggage ($T_2$) and cell 44 corresponds to a personal time it takes to negotiate the exit steps. The algorithm for movement of passengers is described in Fig. 4.

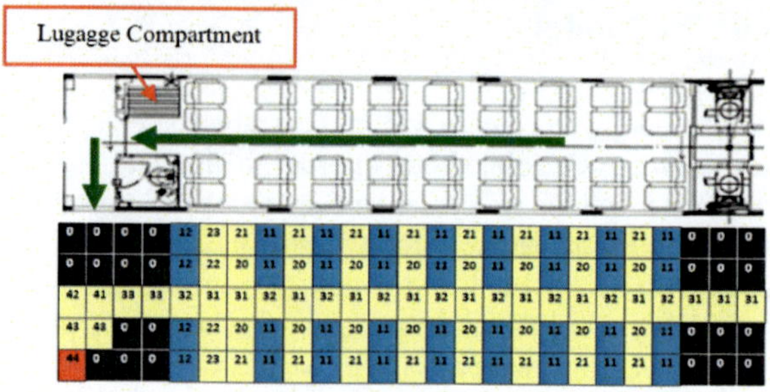


**Fig. 3. The fine network system of a passenger coach.**

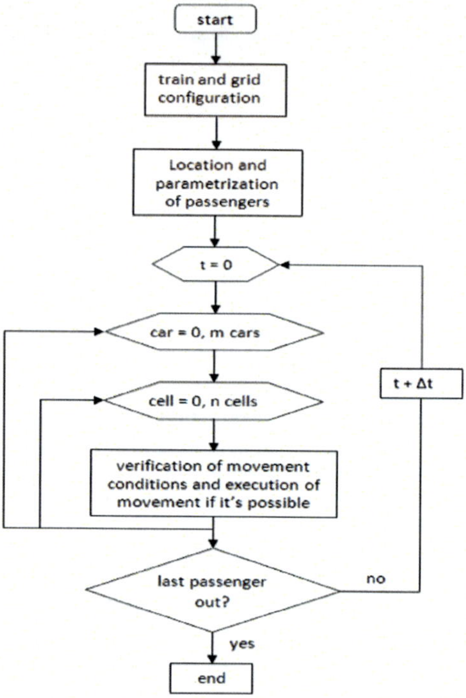


**Fig. 4. Movement algorithm.**

This is a microscopic approach which incorporates the probability of passengers performing actions in addition to moving towards the exits. Monte Carlo methods are used to assign stochastic parameters to each individual. The stochastic parameters are as follows:

- Unimpeded walking speed: $W_s$ (m/s).
- Personal response time. $T_{pr}$ (s).
- Time to prepare: $T_1$ (s) – the time spent by a passenger blocking the aisle.
- Probability of occurrence for $T_1$.
- Time to pick up baggage $T_2$ (s) – the time spent by a passenger blocking the aisle in front of a luggage compartment.
- Probability of occurrence for $T_2$.
- Personal exit flow $T_3$ (s) – the time spent by each passenger to negotiate the exit steps.

The model permits to statistically treat the sample of total evacuation times and fit it to a known distribution (if possible). Otherwise, a density estimation is given. The main output parameter is a percentile of egress times (0.90, 0.95 and 0.99). The model also provides other statistical characteristics: mean, variance, maximum and minimum values. The resulting model is quick and easy to set-up and the results of thousands of simulations can be obtained in real-time.

## Dataset

The datasets used are taken from video recordings of passenger behaviour in evacuation drills and virtual experiments conducted at University of Cantabria. Firstly we classified and measured the actions and behaviour of passengers that may have the greatest impact during the evacuation process. This was achieved by analysing video recordings of three evacuation drills in trains (see Fig. 5), conducted by RENFE Operadora (the Spanish Railroad Administration).

**Fig. 5. Passengers' behaviour during an evacuation drill, and an experiment participant of performing $T_1$.**

Additionally, a dataset of unimpeded walking speeds of passengers in the aisle was obtained inside a high speed train (Alvia S 130) during 8 journeys in normal conditions. In order to complement and increase the dataset, simple experiments were conducted at University of Cantabria. The times of 22 participants (12 males and 10 females) performing discrete actions were measured.

Each participant performed the following variety of actions:

- ($T_{pr}$) Personal response time: 1) Reading, 2) Listening to music in earphones and 3) Using a laptop.
- ($T_1$) Time to prepare: 1) Putting on jacket and 2) Collecting hand bag from overhead baggage rack (see Fig.5).
- ($T_2$) Time at luggage compartment: 1) Collecting a large suitcase and 2) Collecting a small suitcase.
- ($T_3$) Personal time to negotiate exit steps: 1) Normally, 2) Carrying a large suitcase and 3) Carrying a small suitcase.

In assessing whether a given distribution is suited to datasets, the following tests and their underlying measures of fit are used [7-9]:

- D' Agostino's K-square normality test (for samples greater than 25).
- The Anderson-Darling normality test (for samples smaller than 25).
- Hypothetical log-normal test applying the normality test.
- The Anderson-Darling uniformity test (modified normality test).

Samples of data obtained from video recordings of drills and experiments were combined according to the Mann-Whitney non-parametric test for assessing whether two samples come from the same distribution.The histograms for these combined datasets for $T_{pr}$ (personal response time) and $W_s$ (unimpeded walking speed) are shown in Fig. 6. For probabilities $P_{T1}$ (time to prepare) and $P_{T2}$ (time at luggage compartment), a Bernoulli trial was performed [7] with a significance level of $\alpha = 0.05$.

The default values used in the model are the input distributions and probabilities as obtained by data collection, as shown in Table 1. However, the inherent flexibility of the model allows the user to change the passenger input parameters.

**Table 1. Input variables for proposed model.**

| | $T_{pr}$ (s) | $P_{T1}$ | $T_1$ (s) | $P_{T2}$ | $T_2$ (s) | $T_3$ (s) | $W_s$ (m/s) |
|---|---|---|---|---|---|---|---|
| **PDF** | LN | - | U | - | LN | LN | N |
| **m** | 11.92 | - | 12.04 | - | 4.38 | 2.27 | 0.99 |
| **$\sigma$** | 16.25 | - | 7.98 | - | 2.15 | 1.26 | 0.27 |
| **pmin** | - | 0.70 | - | 0.076 | - | - | - |
| **pmax** | - | 0.93 | - | 0.31 | - | - | - |
| **a** | - | - | 1.48 | - | - | - | - |
| **b** | - | - | 26.06 | - | - | - | - |

**PDF** = Probability Density Function; LN (log-normal); U (uniform); N (normal).
**m** = mean; = standard deviation; **pmin** = probability minimum value; **pmax** = probability maximum value; **a** = minimun value; **b** = maximum value;

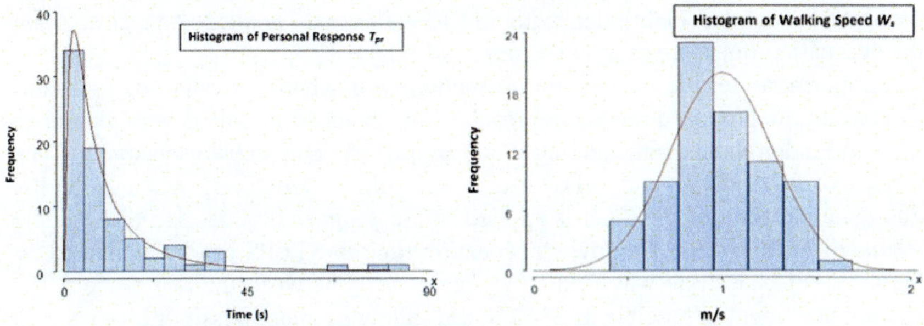


**Fig. 6. Histograms of personal response and walking speed.**

# Comparison with other Models

The current version of the program is still under development. However, for the first stage of the validation process, the model is compared with the following validated evacuation models: STEPS [10], PathFinder [11] and FDS+Evac [12]. The comparison is made for a single-exit scenario with two coaches and 50 passengers in a high speed train (Ave S 102).

The comparison is divided into two tests. In Test 1, no behaviour is performed in order to check that the simulation of passenger movement is working satisfactorily. No pre-movement is considered and a fixed unimpeded walking speed of 1m/s is assigned to all of the passengers. In this case, the simulation results are dependent on the input flow rate through an exit 0.83m wide.

**Table 2. Evacuation times for Test 1. "Proposed model" is the model of the current study.**

| Model (flow inputs) | Time (s) |
|---|---|
| STEPS (NFPA door/gate) | 69 |
| PathFinder (SFPE) | 74.90 |
| Pathfinder (steering) | 53 |
| FDS+Evac | 68 |
| Proposed model ("no interaction") | 55.9 |
| Proposed model (SFPE) | 74.90 |
| Proposed model (NFPA door/gate) | 79.9 |

Results from Test 1 show a correct functioning of the model and basic components of movement are adequate. As shown in Table 2, the results vary for the different flow inputs considered. Of particular note are the evacuation times of the proposed model and Pathfinder, with an SFPE flow rate applied. As can be seen, the evacuation times are exactly the same (74.90s).

Furthermore, for the case of "no interaction" (i.e. when any value for $T_3$ is considered) in the proposed model, the result of the evacuation time is very similar to the PathFinder results with steering mode applied (53 and 55.9s respectively). The results of FDS+Evac and STEPS are very close. However, when a NFPA door/gate capacity of 0.789p/s is applied to the proposed model and STEPS, the results differ by 10.9s. The lower evacuation time in STEPS can be explained by the different pedestrian way-finding strategies used between the two models. Diagonal movement is possible in STEPS, thus allowing more possibilities to access the aisle. However, diagonal movement is not possible in the proposed model. There is only one possible way to access the aisle. Ongoing research being performed by the authors, however, suggests that the latter method may be more applicable to high speed passenger trains.

In Test 2, a behaviour comparison is performed in order to check the effect of passenger-passenger interactions simulated by the proposed model and their possible impact on predicted evacuation times. A total of 100 runs were performed with each model in order to estimate the probable maximum egress time as well as the average egress time. All pre-evacuation activities are defined in STEPS and FDS+Evac by only one parameter. However, in the proposed model these activities are defined by a set of parameters. In order to solve this problem, a program was developed using Microsoft Visual Studio 2008. NET Framework 3.5 SP1. The program simulates 1000 random values for $T_{pr}$, $T_1$ and $T_2$ taking into account the probabilities of occurrence, $P_{T1}$ and $P_{T2}$. The program then sums all of the values and fits them to a distribution. The results fit to a log-normal distribution with a mean of 53.18s and a standard deviation of 47.08s. This pre-movement time, as well as the walking speed used as default in the proposed model were also used for the STEPS and FDS+Evac simulations.

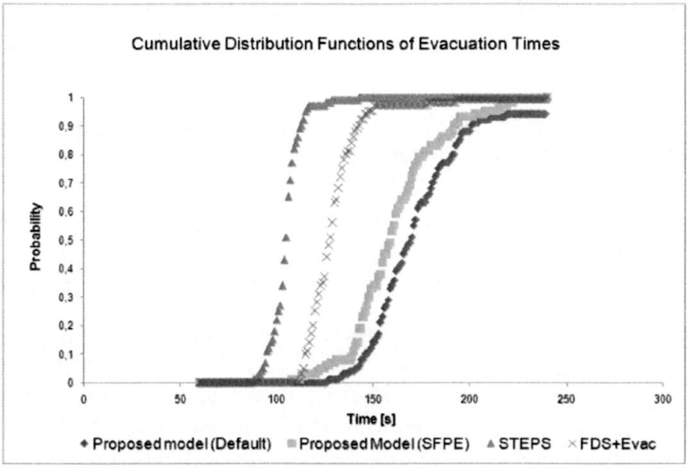


**Fig. 7. Cumulative distributions of probabilities of evacuation times.**

Fig. 7 shows the probability distribution functions (CDF) of evacuation times. The predicted evacuation times vary among each model. It can be seen that results from STEPS and FDS+Evac have lower evacuation times in comparison with the results of the proposed model. Predicted evacuation times of STEPS have a low variability with a mean of 105.25s and a standard deviation of 7.69s. The predicted evacuation times of FDS+Evac are higher, with a mean of 130.40s and a standard deviation of 17.52s. Furthermore, the predicted FDS+Evac evacuation times have a higher variability than STEPS. The pre-evacuation times in these models have no impact in the evacuation movement of passengers within the aisle. In the proposed evacuation model, the evacuation times are strongly dependent on the activities of individuals whose actions interrupt the continuous movement of other passengers in the aisle. This phenomenon, which cannot be well represented in STEPS and FDS+Evac, produces longer evacuation times with a mean of 177.74s when applying the default input values and a mean of 162.47s when applying the SFPE flow rate through the exit.

The results of the proposed model show a wide range of possible evacuation times. This is expected for a stochastic model because of the intrinsic uncertainty in such complex systems. In this case, the proposed model simulates the worse case where the probability of each passenger blocking the aisle ($T_1$) is between 0.7 and 0.9. A similar percentage of people performing these actions have been observed in video recordings of evacuation drills. Test 2 suggests that currently, evacuation models – particularly those primarily designed for buildings – can make simplifications about the behaviour of passengers and are likely to produce inaccurate results in passenger trains.

## Conclusions

A stochastic evacuation model for passenger trains has been presented in this paper. The proposed model is an alternative approach to solve the problem of predicting additional behaviours of passengers during the evacuation process, and their impact on egress calculations. The model is based on a stochastic formulation as opposed to the traditional deterministic approach. Actions and behaviours of passengers have been classified and measured in order to generate a dataset for the proposed model. As a first stage of the validation process, the proposed model has been compared with other current models.

The results of two tests have been presented. Test 1 showed that the basic components of the proposed model work adequately. In Test 2, a behaviour comparison was performed in order to check the effect of passenger-passenger interactions simulated by the proposed model and their possible impact on evacuation times.

Current egress models allow the user to input distributions of the walking speeds and pre-movement times of passengers. However, the average predicted evacuation times of the proposed model suggest that more parameters related to human behaviour in trains should be considered. The current version of the proposed model has limitations that need to be addressed. Future research will include:

- Simulation of evacuations in a wider range of scenarios (railway tunnels, evacuation to track level etc.) and different evacuation procedures.
- More data collection in order to increase the statistical samples and quantify more behavioural parameters.
- Further validation of the model.

## References

1. Kagendal, P., Nilson, D.: Fire Safety on intercity and interregional multiple unit trains. Department of Fire Safety Engineering, Lund University, Sweden (2002)
2. Daamen, W., Hoogendoom, S.: Assessment methodology for safety of passengers in railway tunnels. Proceedings of Pedestrian and Evacuation Dynamics 2008. Wuppertal, Germany (2008) (In press)
3. Capote, J.A. et al.: Evacuation modelling of fire scenarios in passenger trains. 4th Conference on Pedestrian and Evacuation Dynamics 2008. Wuppertal, Germany (2008) (In press)
4. Kuligowski, E.D.: Modeling Human Behavior during Building Fires. Technical Note 1619, Natl. Inst. Stand. Technol., Gaithersburg (2008)
5. Averill, J.D, Reneke, P., Peacock, R.D.: Required Safe Egress Time: Data and Modeling. 7th Conference on Performance-Based and Fire Safety Design Methods. Auckland, New Zealand (2008)

6. Kuligowski, E.D., Peacock, R.D.: Review of Building Evacuation Models. Technical Note 1471, Natl. Inst. Stand. Technol., Gaithersburg (2005)
7. NIST/SEMATECH e-Handbook of Statistical Methods. http://www.itl.nist.gov/div898/handbook (2009)
8. D'Agostino, R.B., Belanger, A.: A Suggestion for Using Powerful and Informative Tests of Normality. The American Statistician. Vol. 44, No. 4: 316-321 (Nov.1990)
9. Rahman, M., Pearson, L.M., Heien, H.C.: A Modified Anderson-Darling Test for Uniformity. Bull. Malays. Math. Sci. Soc. (2) 29(1): 11-16 (2006)
10. MacDonald, M.: STEPS Simulation of Transient and Pedestrian movements User Manual. Unpublished, available with egress model from Mott MacDonald.
11. Thunderhead Engineering. Pathfinder Technical Reference. May 2009 ed. Thunderhead Engineering Consultants, Inc., Manhattan (2009)
12. Korhonen, T., Hostikka, S.: Fire Dynamic Simulator: Technical Reference and User's Guide. VTT Technical Research Centre of Finland (October 2009)

# A Comparison of Grid-based and Continuous Space Pedestrian Modelling Software: Analysis of Two UK Train Stations

**C. J. E. Castle, N. P. Waterson, E. Pellissier, and S. Le Bail**

Mott MacDonald Ltd.
8-10 Sydenham Road, Croydon, CR0 2EE, UK
Corresponding author: christian.castle@mottmac.com

**Abstract** The present paper outlines different approaches to pedestrian modelling, classifying them according to their movement and collision avoidance algorithms. An interesting question, not yet fully explored, is whether the simulation output is markedly different between the two main agent-based approaches. The paper explores this question by considering the operational performance of two rail stations in the UK. A software tool with a grid-based approach to pedestrian movement (STEPS 4.0), and a tool with a continuous approach (Legion Studio 2006) have been used to evaluate the gate-line clearance and platform circulation of passengers with respect to proposed station upgrades and future passenger demand. The paper concludes by summarizing the advantages of each approach, in the light of the findings from these studies.

## Introduction

Pedestrian modelling software tools can be distinguished by their modelling approach (e.g. movement and collision avoidance algorithms), and more generally by attributes such as software availability, developer background, features / functionality, validation and technical support. Detailed frameworks of criteria for assessing software tools have been developed (see e.g. [1, 2]), and a number of software reviews have also been conducted [1-5].

Of these criteria, most notable are the approaches used to simulating occupant movement and collision avoidance, which depend in particular on the spatial resolution employed. The space in which pedestrian movement is simulated can be represented with various levels of discretization and similarly the pedestrians moving through it can be represented on a scale from a homogeneous ensemble to a set of discrete individuals.

R.D. Peacock et al. (eds.), *Pedestrian and Evacuation Dynamics*,
DOI 10.1007/978-1-4419-9725-8_39, 

Figure 1 illustrates a taxonomy of software tools based on spatial discretization. At the coarsest level of discretization, the space is simply divided into a number of large blocks, representing for example whole walkways, small rooms or parts of larger spaces which can be occupied by many people. At this level of resolution the pedestrian movement is typically treated as a continuum flow based on observed characteristics such as flow-density relations, as in the Pedroute model [6]. Though models of this type lack the detail provided by finer-grain models, they have the advantage of fast set up and computation times.

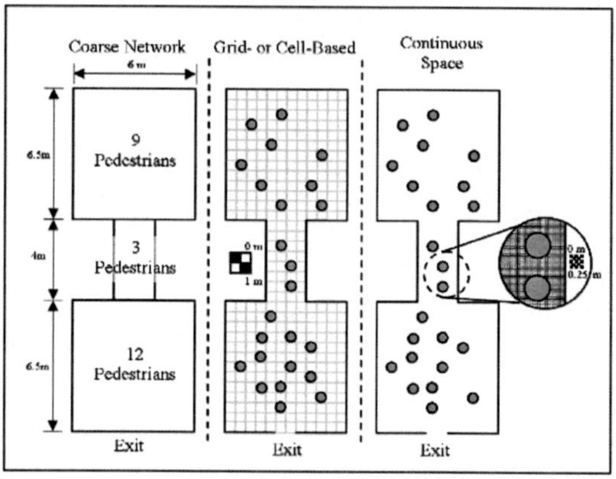


**Fig. 1. Illustration of the three different types of enclosure representation**

A second alternative (Figure 1, centre) is to tesselate the floor area with a grid of cells of a scale similar to the space occupied by a single person. The cells are typically, though not necessarily, square with area of the order of $0.25m^2$. More sophisticated tools can tessellate discrete sections of a building individually, permitting irregular cell shapes and different cell sizes in different areas, as required. Examples of this type of model are given in [7,8,9]. Typically pedestrians are represented as discrete individuals ("agents") and the grid is used both to define the movement of people through the space and to resolve collisions. Movement is typically defined using some form of potential table, as reviewed in [10]. The latter is achieved by requiring that each cell is occupied by only one person at a time. The main advantages of this approach over coarse scale network models are that the internal geometry of a structure can be represented in detail and each pedestrian can be represented as an individual with different characteristics.

The third alternative (Figure 1, right) discretizes the floor space with cells of a size less than that occupied by a single person, potentially allowing even more detailed resolution of geometric details. In this case the grid is typically used to define the movement of people through space though not their mutual interaction. The interaction between people is handled separately without reference to the grid, hence the term "continuous space". Published examples of continuous space approaches include Helbing's "social forces" model [11] and that proposed by Thompson [12]. The continuous space approach has the potential to achieve finer geometric resolution and a more realistic representation of the details of pedestrian interaction than the grid-based approach described above. However, it should be noted that the full potential benefit is not necessarily realized by the currently available generation of models. The main disadvantage of this approach is the significantly increased computational cost which can limit its applicability to large-scale cases. This can also mean that, for practical reasons, the results of studies are based only on a single simulation rather than the ensemble average of multiple simulations as should be the case for all stochastic models.

Broadly, the above taxonomy of models traces the evolution of pedestrian modelling from aggregate to individual-level. With respect to the advantages and disadvantages to each modelling approach, an interesting question not yet fully explored, is whether the simulation output is significantly different between these types of model. The purpose of the present study is to explore this question, with respect to the grid-based and continuous space approaches.

## Research study

In this context, an evaluation of passenger circulation within two UK train stations has been undertaken using two agent-based pedestrian modelling software tools (STEPS and Legion). A key distinction between these tools lies in the movement and collision algorithms, with the first using a grid-based approach and the second being continuous space, based on the classification given above. The two tools are first briefly reviewed, before presenting the simulation output for each station.

### *STEPS software tool*

STEPS is a microsimulation tool developed for the analysis of pedestrian movement under both normal and emergency conditions [8,13]. It employs an agent-based approach to simulate the movement of discrete individuals through three-dimensional space with each person defined using a set of unique attributes and personal objectives. The geometry of a building can be imported directly in either 2D or 3D CAD formats. From this CAD geometry the space in which people can move is defined by tessellating with a quadrilateral grid. STEPS provides more flexibility that a traditional cell-based model since it can tessellate irregular shaped building geometries including curved surfaces and allows different cell sizes in different sections of the model.

## *Legion software tool*

Legion is also an agent-based microsimulation tool which can simulate the movement of people under normal and emergency conditions [14]. The walking speed distribution is usually selected from a library and the objective/target of each pedestrian during their journey can be defined. Legion can import the geometry of a building directly into the software in 2D CAD format and the model building process is essentially 2D in nature, though 3D output representations can be created in a post-processing step, if desired. As noted above, pedestrian interaction is handled using a continuous-space approach.

## *Simulation output*

The simulation output presented here represents the typical required output defined by industry standards such as [15]. Of these, two types of output that may require explanation are *cumulative mean* and *cumulative high* density maps which are used to represent the density experienced by people in the model. The Legion software tool identifies the number of people within a circle of radius 1.5m (i.e. 7.07m$^2$), to determine the density. STEPS adopts a very similar approach, calculating the density using the cells that fall within 2.5m of each person (i.e. 6.25m$^2$). It should be noted that the default size of the footprint that Legion plots density is 0.6m in radius, using a grid of cells 0.1m in dimension. STEPS plots density in the cells adjacent to each person, though a smoothing algorithm is used to plot the output maps.

*Cumulative mean density* is the sum of density experienced by every person to have occupied that location divided by the number of people who have been located at that cell. *Cumulative high density* is exactly the same except that only density above a defined threshold is recorded (e.g. 0.72 p/m$^2$). Density output maps are often presented and assessed in relation to the Fruin Levels of Service (LOS): walkway, stair and queuing. In addition, output maps are often presented for discrete 15 minute periods (e.g. because flow is sustained but density varies through time), or a specific duration in relation to an event (e.g. a train arrival), as opposed to the entire simulation.

## *Case 1: Urban commuter station*

The first station considered serves a large town in the south of England and experiences significant commuter traffic. In order to cope with the anticipated increase in demand, the station infrastructure must be upgraded to increase capacity. Specifically, platforms will be added and the current layout of the station will be altered to improve passenger circulation, see Figure 2. Pedestrian modelling of passenger circulation under normal conditions was commissioned to inform the design of the proposed station and evaluate different layout options.

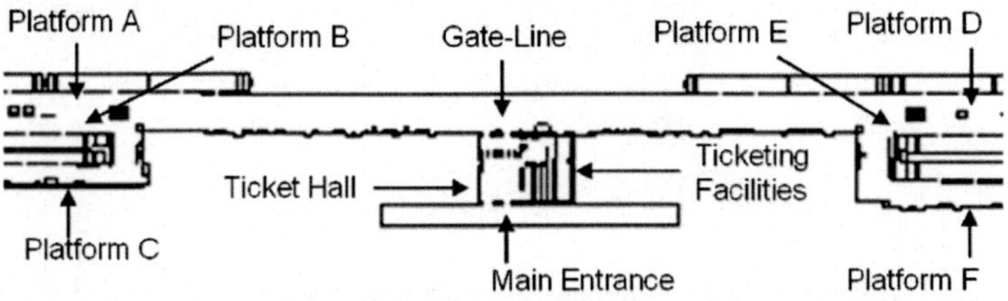


**Fig. 2. Current ticket hall and platform layout of station 1**

The details of the modelling assumptions and configuration are beyond the scope of this paper. The Legion model, built with Legion Studio 2006 (EP5), was undertaken according to industry best practice [15], using observed passenger demand. A detailed survey was conducted of the station to determine the passenger demand and movement, supplemented two years later by gate-line counts and passenger boarding/alighting data provided by the rail operator. The latter were used to uplift the original survey data, thereby providing current demand data for the model. The available passenger data were also used to validate the simulation output of the base case demand year (e.g. flow-rates and queuing length at pivotal locations within the station). Subsequent to this study, a model of the station was also built using the STEPS software tool (version 4.0). The same basic assumptions and model configuration settings were applied, and the calibration and validation process replicated.

The objective of the initial study was to explore the circulation of passengers in relation to AM and PM demand, multiple demand years and several design options. The following discussion presents the modelling output of passenger circulation during the 2012 AM peak period under normal conditions. Specifically, the following output from both software tools is presented (refer to Figures 3-5):

- Output maps of cumulative mean density (Walkway LOS): 08:00-08:15
- Output maps of cumulative mean density (Queuing LOS): 08:00-08:15
- Gate-line queue length: 08:56 (busiest egress period)
- Snapshot of gate-line queue: 08:56 (busiest egress period)

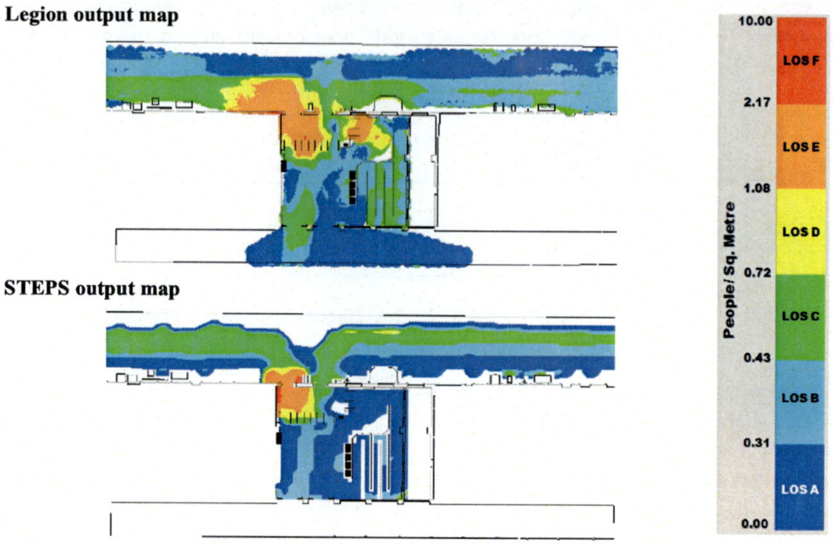


**Fig. 3.** **Legion (top) and STEPS (bottom) cumulative mean density (Walkway LOS) output maps for 2012 AM peak 15 minutes (08:00-08:15)**

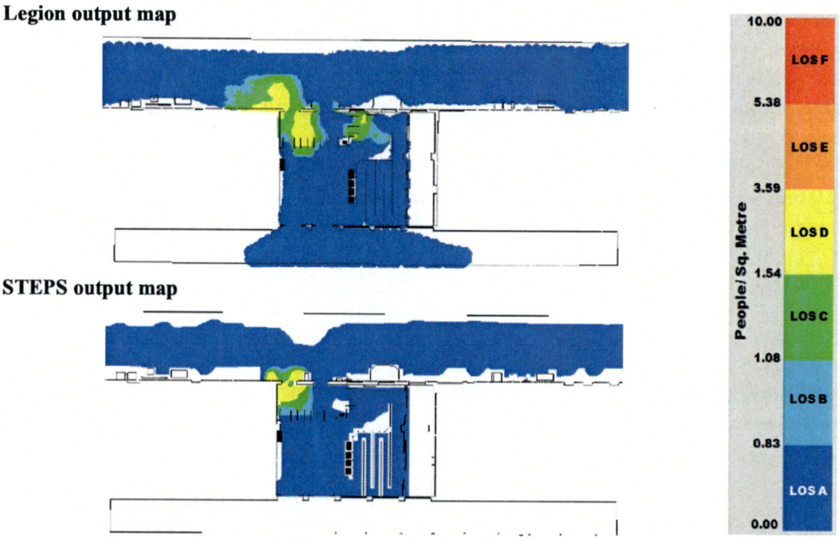


**Fig. 4.** **Legion (top) and STEPS (bottom) cumulative mean density (Queuing LOS) output maps for 2012 AM peak 15 minutes (08:00-08:15).**

**Legion screen shot and maximum queue length**

168 people

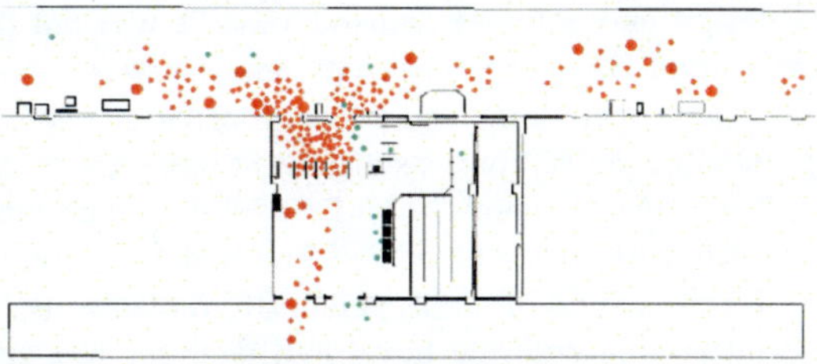

**STEPS screen shot and maximum queue length**

146 people

**Fig. 5. Legion (top) and STEPS (bottom) output of gate-line queue length and a screenshot of passenger circulation at the busiest egress period during the 2012 AM peak (08:56)**

## *Case 2: Suburban interchange station*

Case 2 is a busy suburban interchange station. Upgrades to the station are planned to accommodate anticipated future passenger demand. These include improved ticket hall facilities, increased platform lengths to permit longer trains with greater capacity, and a secondary access route between platforms to decrease interchange journey times. Figure 6 presents the proposed station design, indicating the secondary access route and the increased platform lengths to the north end of platforms A and B. Pedestrian modelling of passenger circulation under normal conditions was carried out to evaluate the proposed design in relation to the opening year and a subsequent future demand year during the AM peak period.

Again, the detailed modelling assumptions and configuration are beyond the scope of this paper. The Legion model was built with Legion Studio 2006 (EP5), according to industry best practice [15] using real-world passenger demand and observations. Unlike Case 1, a specific passenger survey was beyond the remit of this study. However, the rail operating company supplied current peak period passenger demand and the projected proportional increase for expected future growth. These numbers were used to determine alighting and boarding numbers per train as well as interchange at the station. Given that the model of the station was of a proposed design and a future demand year, it was not possible to validate the simulation output. However, sensitivity analysis was conducted to assess the full impact of various operational and demand scenarios. On completion of this study, a partial model of the station was also built using the STEPS software tool (version 4.0). The same assumptions and model configuration settings were applied. However, for practical reasons of time and budget, sections of the station that did not affect the circulation of passengers to/from platforms A and B were not simulated, and interaction with the ticket hall facilities was simplified, as this would have no impact on the circulation of passengers on platforms A and B.

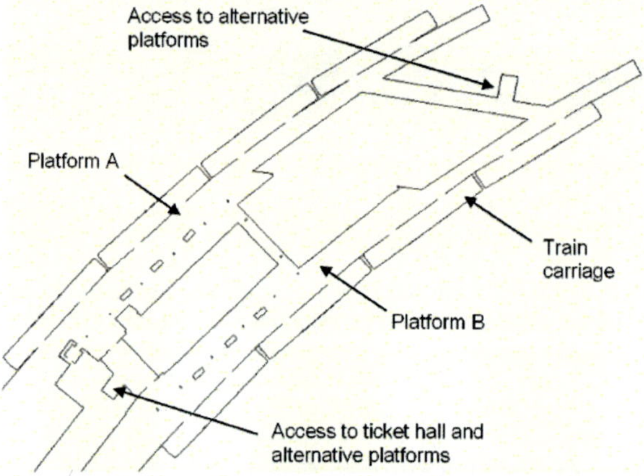


**Fig. 6. Proposed platform layout of station 2**

The primary objective of this pedestrian modelling study was to evaluate the operational performance of the extended platform areas, and the secondary access, particularly with respect to passenger density. Output in relation to passenger circulation during the 2026 AM peak period under normal conditions was captured and is presented below for both software tools (see Figures 7 and 8).

- Output maps of cumulative mean density (Walkway LOS): busiest train
- Output maps of cumulative mean density (Queuing LOS): busiest train

For the purpose of this discussion, output relating to only the busiest train arrival during the AM peak is presented. The busiest train correlates with the highest number of boarding and alighting passengers at platform B during the AM peak.

**Legion output map**

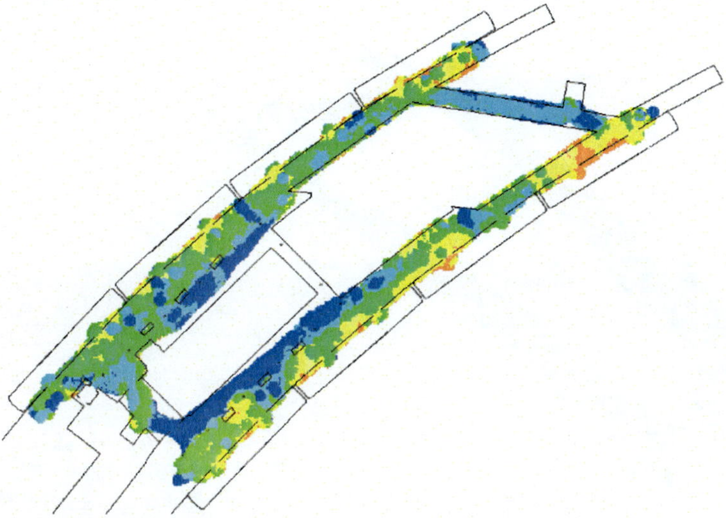

**STEPS output map**

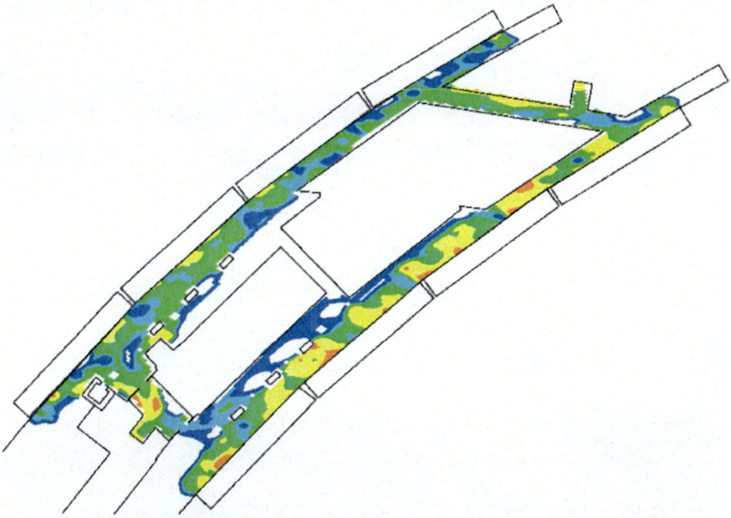

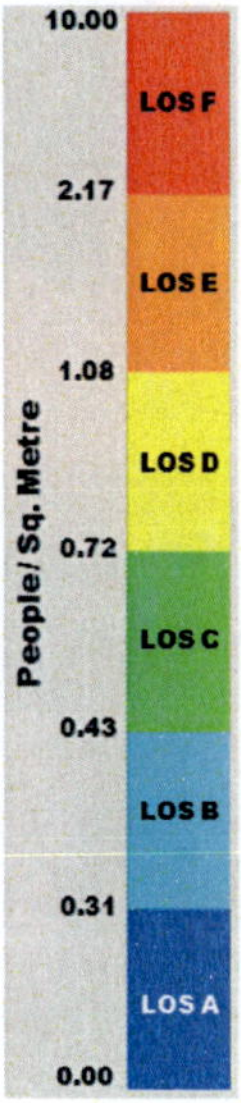


**Fig. 7.** **Legion (top) and STEPS (bottom) cumulative mean density (Walkway LOS) output maps for the busiest 2026 AM train service (08:34:30-08:40:30)**

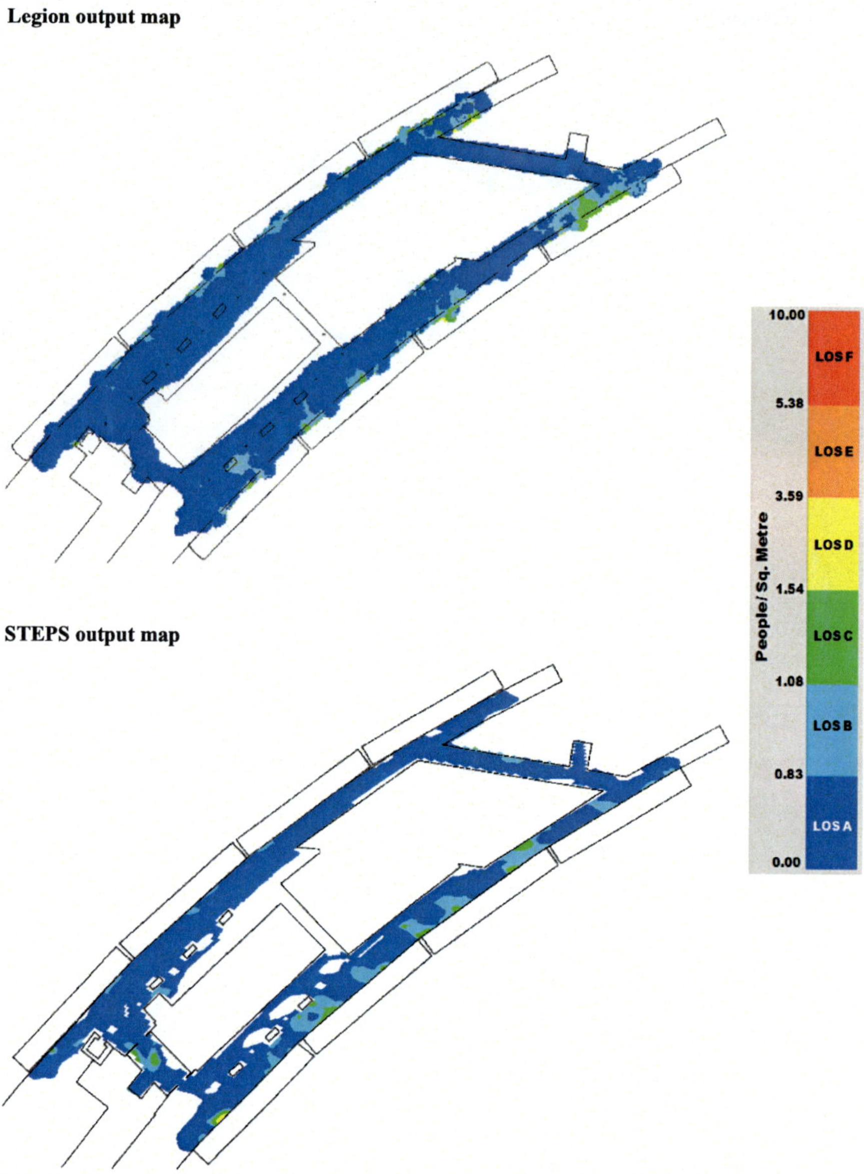


**Fig. 8. Legion (top) and STEPS (bottom) cumulative mean density (Queuing LOS) output maps for the busiest 2026 AM train service (08:34:30-08:40:30)**

## Comparison of simulation results

The simulation output of the two case studies is very comparable between the two software tools. For reasons of brevity, the results presented here are only for a single demand year for each case. However, output for additional demand years has been generated and is also very comparable.

The peak 15 minute (08:00-08:15) cumulative mean density output maps for Station 1 (Figure 3 and 4), illustrate that both software tools predict sustained queuing at the gate-line, which results in elevated density levels within the ticket hall. In addition, both software tools simulate the formation of a large queue for the gate-line during the busiest egress time (08:56) of the 2012 AM peak (Figure 5). Finally, in relation to Station 1, both STEPS and Legion indicate that 'arched' queuing will form around the doorways between the platforms and the ticket hall. STEPS calculates a maximum queue length of 146 people at this busy period, while Legion predicts a maximum queue length of 168 passengers. With respect to Station 2, the cumulative mean density output maps for the busiest train service during the 2026 AM peak period illustrate that both tools predict very similarly Levels of Service on the platforms (Figure 7 and 8). In particular, elevated Walkway and Queuing LOS are observable on platform B.

Though the results are generally similar, the output of the two applications is clearly not identical. Every attempt has been made to ensure consistency between the configurations of the two tools with respect to the boarding/alighting passenger numbers, the passenger arrival distribution, the walking speed distribution, the proportional split of passenger loading between train carriages, platform waiting zones, exits/entrances, etc. Nevertheless, there are multiple reasons why there are differences between the results, and some of which are considered here.

Both tools are fundamentally stochastic with respect to the movement and behavioural considerations of the simulated people. For this reason, the output could never be identical, even for two runs of the same model. For instance, while the waiting zones defined were virtually identical in their size and position relative to each platform, the specific location chosen by each passenger to wait in a zone was completely random. Consequently, the density observed in relation to these waiting zones could vary significantly. To this end, both simulations were run multiple times in an attempt to identify simulations with approximately the same distribution of waiting boarders, to enable a fair comparison of the output.

As noted in the simulation output section above, the size of the personal space used to calculate each passenger's density, the footprint size that each software tool uses to plot each passenger's density, the resolution of the grid cells used to plot the output density, and the algorithm used to average/interpolate density between passenger-density footprints are not the same. As a consequence, it is inevitable that the density maps will not be identical. Analysis of the effect that each of these parameters has on the simulation output is beyond the scope of this paper, and indeed some of these settings are not definable by the user.

In the light of the subtle but important differences between these, and indeed any, software tools, it is imperative that the model building, analysis and interpretation are undertaken by a trained and experienced practitioner. This research has highlighted to the authors that minor differences between two 'identical' models can have a considerable effect on the simulation, and consequently significant impact on the interpretation of the resulting output.

The overview of the different modelling approaches presented above, indicated that continuous space modelling seems to be the logical end point of disaggregate modelling of pedestrian dynamics. However, the computational expense / run-time is a potential reason why very few software tools to date have adopted the continuous space approach, and why the vast majority of tools employ a grid or cell-based approach. In an attempt to explore this hypothesis, the run-time of both software tools was recorded while running the simulations of the two case studies.

It should be noted that the STEPS software uses a single interface in which the model is built, the simulation run and the output analyzed. Moreover, the simulation and output generation can be undertaken simultaneously. In contrast, Legion Studio is comprised of three applications. Firstly a user must build and compile a model in one application and then import the compiled model into a separate application which will run it and generate a simulation file. Finally, this simulation file is imported into a further application for analysis and output generation. Ten simulations of each model were run using both tools. The time to compile the model, run the peak 15 minutes of the simulation, and generate the output was recorded, and the average is presented in Table 1. In order to make a fair comparison, the time required to export and import the output between each Legion application has not been included within these times, and both tools were configured to generate the same output.

**Table 1. Run time of STEPS 4.0 and Legion Studio 2006 in relation to two case studies**

| | Time (mm:ss) | | | |
|---|---|---|---|---|
| | Case 1 | | Case 2 | |
| | STEPS 4.0 | Legion Studio 2006 | STEPS 4.0 | Legion Studio 2006 |
| Compile model | | 00:17 | | 00:28 |
| Run simulation | 02:13 | 06:24 | 03:38 | 05:52 |
| Generate output | | 01:25 | | 04:41 |
| **Total** | **02:13** | **08:06** | **03:38** | **11:01** |

Whilst this comparison is clearly only applicable to these two software tools, and indeed these two cases, is does provide some indication as to the additional run-time required to undertake modelling using a continuous space, as opposed to a grid-based, approach. Other software tools that employ similar modelling approaches may show significant variation in their run-times.

## Conclusions

The purpose of this paper has been to review the grid-based and continuous-space approaches to pedestrian modelling, in order to evaluate the influence of these approaches on the output generated. Approaches to pedestrian modelling are generally distinguishable by the spatial resolution of the movement and collision avoidance algorithm employed, and have been classified here as: 1) coarse-scale network or flow, 2) grid / cell-based and 3) continuous space.

Output derived from two leading pedestrian modelling software tools has been presented for two case studies of rail stations in the UK. Specifically, a software tool with a grid-based approach (STEPS 4.0), and a tool with a continuous-space approach (Legion Studio 2006) have been used to evaluate each station's operational performance with respect to considerations such as gate-line clearance and passenger density. The output from the two tools proved to be very similar, despite the quite different approaches to representing movement and collision avoidance.

Differences in the output of the two tools have been appraised and some explanations suggested for the minor differences observed. In addition, the comparative run-times have been examined, and it was demonstrated the continuous space approach took considerably longer (based on the performance of the two tools considered here). The present study also highlights the importance of using trained and competent practitioners for pedestrian modelling studies.

## References

1. Nelson, H.E., Mowrer, F.E.: Emergency Movement. In: DiNenno, P.J. (eds.) The SFPE Handbook of Fire Protection Engineering (Third ed.), pp. 3- 367-3-380. Society of Fire Protection Engineers, Quincy (2002)
2. Kuligowski, E.D., Peacock, R.D.: A Review of Building Evacuation Models, Technical Note 1471. Natl. Inst. Stand. Technol., Gaithersburg (2005)
3. Gwynne, S., Galea, E.R., Owen, M., Lawrence, P.J., Filippidis, L.: A Review of the Methodologies Used in the Computer Simulation of Evacuation from the Built Environment. Building and Environment. **34**, 741-749 (1999)
4. Hoogendoorn, S.P.: Microscopic Simulation of Pedestrian Flows. Transportation Research Board Annual Meeting. National Academy Press, Washington DC (2003)
5. Castle, C.J.E., Longley, P.A.: Emergency Pedestrian Evacuation Analysis: A Review and Interpretation of Software Applications. In: Sui, D.Z., Cutter, S.L. (Eds.), Geospatial Technologies and Homeland Security: Research Frontiers and Challenges. Springer (2008)

6. Buckman, L.T., Leather, J.A.: Modelling Station Congestion the Pedroute Way. Traffic Engineering and Control. **35**:6, 373-377 (1994)
7. Galea, E.R., Perez Galparsoro, J.M.: A Computer Based Simulation Model for the Prediction of Evacuation from Mass Transport Vehicles. Fire Safety Journal. **22**, 341-366 (1994)
8. Hoffmann, N.A., Henson, D.A.: Simulating Transient Evacuation and Pedestrian Movements in Stations. International Conference on Mass Transit Management. Kuala Lumpur, Malaysia (1997)
9. Tobias, K., Schreckenberg, M.: F.A.S.T. - Floor Field and Agent-based Simulation Tool. International Symposium of Transport Simulation 2006 (ISTS06), 4 - 6th September. Lausanne, Switzerland (2006)
10. Kretz, T., Bönisch, C., Vortisch, P.: Comparison of Various Methods for the Calculation of the Distance Potential Field. In: Klingsch, W.W.F., Rogsch, C., Schadschneider, A., Schreckenberg, M. (eds.), Pedestrian and Evacuation Dynamics 2008. Springer (2008)
11. Helbing, D., Molnár, P.: Social Force Model for Pedestrian Dynamics. Physical Review E. **51**, 4282-4286 (1995)
12. Thompson, P.A.: Developing New Techniques for Modelling Crowd Movement. PhD Thesis, University of Edinburgh (1994)
13. MacDonald, M.: STEPS Software, http://www.steps.mottmac.com (2010)
14. Legion, Legion Software, http://www.legion.com (2010)
15. LUL, Best Practice Guide - Station Modelling With Legion. London Underground Limited, UK (2009)

# Modeling Methods

# Implementing a Hybrid Space Discretisation within an Agent Based Evacuation Model

**N. Chooramun, P.J. Lawrence and E.R.Galea**

Fire Safety Engineering Group
University of Greenwich, London SE10 9LS, UK
Corresponding author: E.R.Galea@gre.ac.uk

**Abstract** Egress models typically use one of three methods to represent the physical space in which the agents move, namely: coarse network, fine network or continuous. In this work, we present a novel approach to represent space, which we call the 'Hybrid Spatial Discretisation' (HSD), in which all three spatial representations can be utilised to represent the physical space of the geometry within a single integrated software tool. The aim of the HSD approach is to encompass the benefits of the three spatial representation methods and maximise computational efficiency while providing an optimal environment to represent the movement and interaction of agents.

## Introduction

Within all evacuation and pedestrian dynamics models, the physical space in which the agents move and interact is discretised in some way [1,2]. Models typically use one of three basic approaches to represent space [1], a continuous representation of space e.g. SIMULEX [3], a fine network of nodes e.g. buildingEXODUS V4.07 [4] or a coarse network of nodes e.g. PEDROUTE [5]. Each approach has its benefits and limitations, the continuous approach allows for an accurate representation of the building space and the movement and interaction of individual agents but suffers from relative poor computational performance, the coarse nodal approach allows for very rapid computation but suffers from an inability to accurately represent the interaction of individual agents with each other and with the structure. The fine nodal approach represents a compromise between the two extremes providing an ability to represent the interaction of agents while providing good computational performance.

In this paper, we present a novel approach to represent space, which we call the 'Hybrid Spatial Discretisation' or HSD. In the HSD approach all three spatial representations can be utilised to represent the physical space of the geometry within a single integrated software tool. The aim of the HSD approach is to encompass the benefits of the three spatial representation methods and maximise computational efficiency while providing an optimal environment to represent the

R.D. Peacock et al. (eds.), *Pedestrian and Evacuation Dynamics*,
DOI 10.1007/978-1-4419-9725-8_40, © Springer Science+Business Media, LLC 2011

movement and interaction of agents. Using the HSD approach, the fine nodal approach is used to map the majority of the geometry, providing reasonable speed and an ability to model agent interaction. In parts of the geometry where greater precision is required to model detailed interaction between agents, the continuous approach is used and in regions where knowledge of detailed agent interaction is not required the coarse nodal approach can be used, providing improvements in speed and computational efficiency. This approach is particularly useful in modelling very large complex spaces and urban environments.

In this paper, we examine where the various spatial representations are appropriate for use and we briefly describe the key algorithms developed for the bEX-H implementation, in particular those required to model the continuous and coarse nodal regions. In addition, we provide a demonstration example of how the technique can be applied and discuss related accuracy and performance issues. In the work presented here, the HSD approach is implemented within the buildingEXODUS V4.07 software and is identified as the buildingEXODUS-Hybrid prototype or bEX-H. The buildingEXODUS (bEX) model has been frequently described in other publications [4, 6] and will therefore not be described here.

## Software Architecture

The bEX-H makes use of the core architecture of the buildingEXODUS software. The buildingEXODUS software has been modified to allow plug-in modules to be included into the core software using a component oriented engineering approach. This architecture provides a platform whereby new functionalities can be independently developed and incorporated into the model as required. The coarse network and continuous region are examples of two components which have been developed as plug-in modules for the bEX-H model.

## Continuous Region Component

When using a continuous approach for the discretisation of space, it is possible to take into consideration a larger number of agent attributes allowing for a wider range of agent behaviours to be modelled. In this section, we describe the planning approach of the agents and some of their advanced behaviours. bEX is based on a multi-agent system whereby each agent is modelled as an autonomous agent which exhibits some forms of adaptive behaviour. In other words, the agent has the ability to navigate in a life-like manner and react to stimuli in its environment. The agent will at some point in its trajectory encounter obstacles. The obstacles can be static such as walls or tables or dynamic, for instance, other agents navigating in the same environment. In this respect, the agent is able to detect these obstructions and react to them accordingly. Some of the additional attributes of the continuous agents are as shown in Table 1 below.

**Table 1. Additional agent attributes within bEX-H**

| Attributes | Description | Quantity |
|---|---|---|
| 2-D Position | The location of the person in continuous space | Vector |
| Velocity | Rate of change of displacement | Vector |
| Acceleration | Rate of change of velocity | Vector |
| Max Acceleration | Maximum acceleration of a continuous person | Scalar |
| Orientation | The possible movement directions (headings) | N basis vectors |
| Body Frame Width | The width of the person excluding the size of the shoulders | Scalar |
| Body Shoulder Width | The width of the person's shoulders | Scalar |

The agent navigates around its environment using two levels of navigation comprising of local and global strategies each influencing different aspects of the individual's movements. Local navigation relates to low level reactive behaviours which are required for collision avoidance. Whereas the global strategy relates to navigation and high level decision making processes for example an agent deciding which route to adopt from their current location to their target.

The path of an agent within continuous space from a start location to an end location can be described as a continuous map [7]. However, the complexity of this map can increase significantly in large geometries for example, buildings with multiple internal rooms and floors. Moreover, the presence of static obstacles within the enclosure makes the path planning process of the agents even more complex. In order to reduce the complexity of the continuous map, the continuous region in bEX-H uses a Navigational Graph. This is a network of waypoints and path segments which is automatically generated in the pre-processing phase. Each waypoint is assigned a potential value which represents the shortest visible arc distance from the external door. Illustrated in Figure 1 is a geometry with multiple compartments and its corresponding navigational graph. However, unlike other roadmap methods such as the visibility graph [7] where the links (path segments) are connected to each and every visible vertex, in the navigational graph, the waypoints are generated only at locations where the internal angles are concave.

The behaviour of the agents is modelled as simple components which can serve as building blocks for simulating more complex behavioural routines. Moreover, this approach allows behaviours to be implemented and tested individually, followed by an incremental integration. Examples of behaviours in bEX-H include:

- Seek - used to steer a person towards a specific goal, which takes into account the agents speed and turning rate,
- Wall Avoidance – ability to detect collisions which could happen in the future given the current trajectory,
- Agent Avoidance and Lane Formation - enables the agents to maintain a desired interpersonal space from each other which is proportional to the agents velocity and the body width of neighbouring agents.

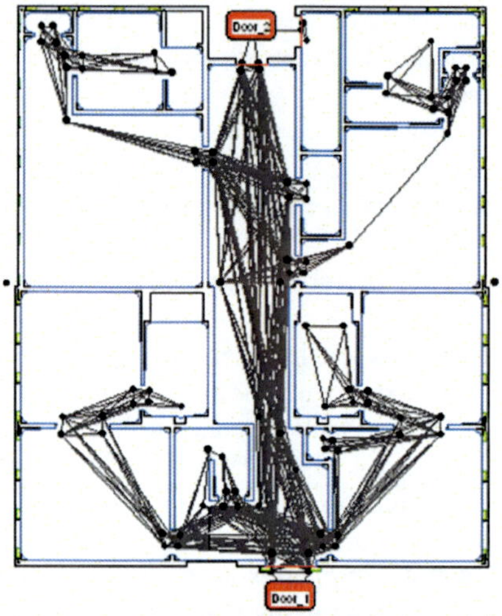

**Fig. 1. The Underlying navigational graph**

The Continuous Region component includes both an adaptive and prescriptive setting for agent travel speeds. In the adaptive setting, the agent considers each neighbouring agent within its perception box (see Agent Avoidance) in turn, and computes a repulsive force. This force is normalised and scaled to be inversely proportional to the squared of the inter-person distance so that distant neighbours have a negligible effect on the assessing agent. The sum of the vectors has a decelerating effect on the agents thereby reducing the travel speed of the agent. In other words, the agents adapt their travel speeds according to population density in their surroundings. In the prescriptive setting, the travel speed of each agent is dictated by a speed density relationship such as, $S = k - akD$ [8], where S is the speed, D is the population density and a and k are constants. As the population density within a room or compartment may be non-uniform, bEX-H uses a localised density calculation approach.

## Coarse Network Component

In this section we describe the coarse network approach for representation of physical space, and its implementation within bEX-H. Using this approach, the available physical space can be segmented into partitions whereby each partition can represent a section of the geometry such as a room or corridor. Each partition is a node which is connected via arcs or links to represent doorways or other forms of connectivity in the structure. Each node has a limit on the number of agents it can contain (maximum capacity) while the arcs have a maximum flow capacity, that is, the maximum number of agents that can traverse the arc per time period. There are seven types of coarse node implemented within bEX-H, these are; compartment (general region in which agents exit or enter), intersection (complex region in which flows from different directions merge), interchange (larger version of intersection), gates (narrow metered passage such as turnstiles), stairs, escalators and travelators. The agents within coarse networks transit from one segment to another while the physical movement within the segment itself is not represented. The flow rate through a coarse node in bEX-H is a function of travel speed, travel distance and population density in the region. The travel distance is defined as the distance the agent has to travel, within the node, from their entry point to their exit point. This is implemented using a "Flow to Density Equation" also known as the F-D Model **[5]**. The rate of flow, in a coarse node, is also subject to two other limitations, the maximum capacity of the coarse node and the connecting arcs. When an agent enters a coarse node, their path, travel distance and speed is fixed based on the Flow to Density Equation. The only dynamic events which can affect their dwell time in the node are the capacity of the arcs out and congestion in adjacent nodes.

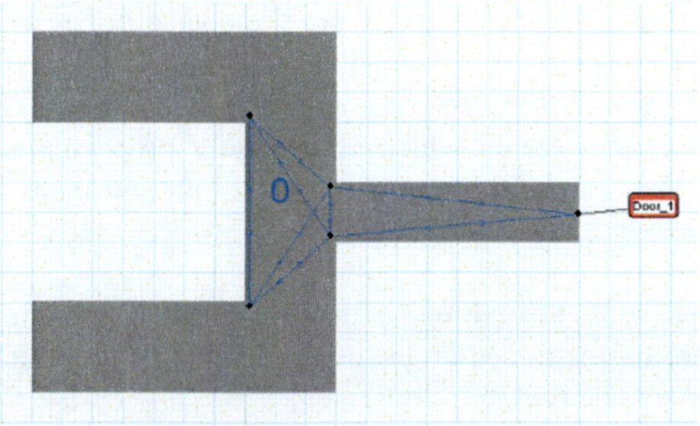


**Fig. 2. Coarse node implementation of non-convex geometry in bEX-H**

The nature of the coarse network approach restricts the modelling of a region with simple geometrical shapes such as rectangles. However, these simple shapes might not be sufficient to accurately model all the complex segments within an enclosure. Unlike other coarse node models, the coarse node implementation in bEX-H allows the creation of non-convex regions as shown in Figure 2. The key data structure behind a coarse node is a mesh which is based on the same principle as the Navigational Graph.

The coarse node model in bEX-H features a behavioural mechanism which enables agents entering a coarse node to adjust their paths depending on the evolving conditions of population density within the coarse node. This feature is implemented through a load balancing algorithm. This algorithm dynamically adjusts the potential value of the waypoints to account for the number of agents heading towards them.

The HSD approach involves the mixing of macroscopic (coarse node) and microscopic (continuous and fine) modelling methodologies which in itself presents several challenges. In the continuous and fine node models, the agents are modelled from an individual perspective whereby their movements, exact locations and behaviours can be tracked. However, in the coarse network approach, the agents are modelled from a global perspective whereby the population is treated as a homogenous ensemble, such that the locations and physical space occupancy of the agents are not represented. bEX-H incorporates some behavioural and movement mechanisms to facilitate the transition of agents across the transition regions. The current implementation of bEX-H features the representation of all six possible interface transition regions namely: Coarse Node ↔ Fine Node; Fine Node ↔ Continuous and Coarse Node ↔ Continuous. The mechanisms for some of the key transitions are briefly described below.

**Coarse Node/Continuous Region Transition:** This transition represents the two extremes on the scale of granularity. When an agent traverses from the coarse node to the continuous region, its starting location in the continuous region is set equal to the coordinates of the waypoints from which it emerges. This may result in a narrow stream of agents emerging from the coarse node, which may appear unrealistic. bEX-H incorporates a behaviour called Separation which is invoked temporarily by the agents upon entering the continuous region. This generates a repulsive force between agents which is inversely proportional to their inter-person distance. This allows the agents to spread out and make a more realistic use of the available continuous space.

**Coarse Node/Fine Node Transition:** When agents traverse from coarse nodes to fine nodes, they are re-positioned on the available fine nodes which are connected by arcs to the coarse node region.

**Fine Node/Continuous Region Transition:** The transition of agents from the fine nodes to the continuous regions is based on the same approach as for the Coarse Node/Continuous Region.

## Demonstration Case using HSD approach

In this section we demonstrate the application of the HSD approach to the complex geometry depicted in Figure 3a. In this example a multi-compartment geometry is modelled using all three spatial representations. The building population initially occupies every compartment, thus the scenario investigated demonstrates all six possible interface transition regions: Coarse Node ↔ Fine Node; Fine Node ↔ Continuous and Coarse Node ↔ Continuous. The simulations were performed using a PC with Intel E8600 Core 2 Duo CPU, 3.3 GHz and 8GB of RAM with a 512 MB GeForce GTS 250 graphics card. The experimental set up in bEX-H was as follows:

- 300 agents in total with average density of 2 persons/$m^2$ in each compartment.
- Free-Flow conditions were imposed on the exits i.e. no flow rate limitations were imposed.
- Fast Walk (Unimpeded walking speed) : 1.5 m/s
- Response Time: 0 s
- Both external exits available and all internal exits are 1.0 m wide.

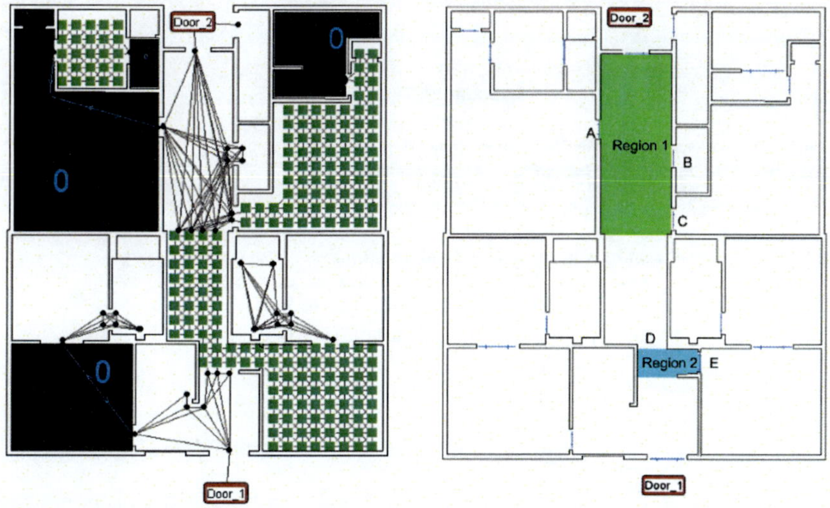


**Fig. 3. Complex geometry used in demonstration example**

**(a) Geometry modelled using bEX-H involving 40% Continuous (white regions), 30% Coarse (dark regions) and 30% Fine node regions (grid of nodes).**

**(b) Location of Intersection Nodes (shaded regions 1 and 2) in the geometry. All the other areas modelled using Compartment Nodes.**

In order to demonstrate the differences and similarities between the various approaches, the demonstration case is repeated using; all Coarse Nodes, all Fine Nodes and all Continuous Regions. Ten simulation runs were conducted for each case. In the all Coarse Nodes case, the geometry was modelled by using a combination of compartment and intersection blocks as shown in Figure 3b.

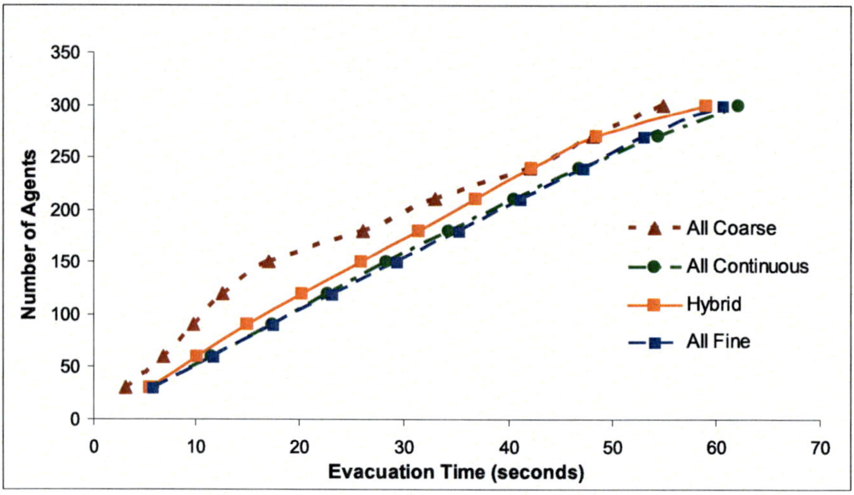


**Fig. 4. Time taken for a specific percentage of people to evacuate using All-Fine, All-Continuous, All-Coarse and Hybrid spatial representations**

**Table 2. Summary of results averaged over 10 simulations**

| Spatial Representation | Average Total Evacuation Time (sec) |
|---|---|
| All-Fine | 60.6 |
| Continuous | 62.2 |
| All-Coarse | 54.9 |
| Hybrid | 59.1 |

The evacuation time curves for each case are shown in Figure 4. As can be seen, the evacuation curves for the All-Coarse, All-Fine and Continuous cases are similar, with the All-Coarse case consistently underestimating the egress times throughout the evacuation. The All-Fine and Continuous simulations produce virtually identical evacuation histories up to the final part of the evacuation. During the last 10% of the evacuation, the exit flow rate in the Continuous model tails off at a slightly greater rate than in the All-Fine model. This produces a slightly longer total evacuation time for the Continuous model as shown in Table 2. The average total evacuation time for the Continuous model is some 2.6%

slower than the All-Fine model while the average total evacuation time for the All-Coarse model is some 9.4% faster than the All-Fine model.

As can be seen the Hybrid simulation curve falls between the curves for the All-Coarse simulation and the All-Continuous simulation (see Figure 4). The average total evacuation time falls between the two extremes produced by the All-Coarse and the All-Continuous models and is marginally (2.5%) smaller than that produced by the All-Fine model (see Table 2).

More effort is required to determine the impact of using different combinations of the three discretisation approaches on both the accuracy of predictions and the speed of performance. These factors are also expected to be influenced by the type of discretisation that is used to represent specific regions of the geometry and the size and location of the simulated population. However, two primary applications are anticipated for bEX-H, the first are large complex structures such as airport terminals and high-rise buildings or systems of structures such as long tunnels with complex interchanges. In these types of applications, the majority of the structure would primarily consist of a combination of Fine and Coarse nodes. The Fine node structure would be used throughout the bulk of the structure where detailed analysis was required while the Coarse nodes would be used in the far field to represent parts of the structure which are not central to the analysis. The Continuous approach would only be utilised in special areas to represent regions such as pinch points or exits.

The second type of application involves urban environments such as a town or city. In applications involving urban scale geometries, the bulk of the geometry would be represented using the Coarse node approach with key areas such as assembly points or interchanges represented using the Fine node approach. It is unlikely that the Continuous approach would be utilised in such large scale applications.

In addition to the performance enhancements offered by the hybrid version of buildingEXODUS, the software can also be run in parallel using multiple computers [9]. This capability will also be expanded to include the hybrid version of the software.

## Conclusions

In this paper we have presented a novel approach, known as the HSD to represent the discretisation of space in circulation and evacuation models. The HSD approach allows enclosures to be modelled using a mixture of the three basic techniques for space discretisation, coarse networks, fine networks and continuous. In the example presented, in which 30% of the domain was represented by the coarse discretisation, 30% by the fine discretisation and 40% by the continuous discretisation, the HSD approach was shown to produce results of similar accuracy to that produced by the All-Fine and All-Continuous approach. While further testing is required, the HSD approach appears to provide flexibility in defining the mix of approaches used in discretising the circulation space within the geometry. Further work is also required to optimise the numerical efficiency of the various plug-in components so that numerical efficiencies offered by using a mix of discretisation schemes can be fully exploited. In addition, the HSD approach will be extended so that it works efficiently within the parallel computing environment currently offered by buildingEXODUS.

## References


1. Gwynne, S., Galea, E.R., Lawrence, P.J., Owen, M., Filippidis, L.: A Review of the Methodologies used in the Computer Simulation of Evacuation from the Built Environment. Building and Environment. Vol. 34, pp. 741-749 (1999)
2. Kuligowski, E.D., Peacock, R.D.: A Review of Building Evacuation Models. Technical Note 1471. National Institute of Standards and Technology (NIST), U.S Department of Commerce (2005)
3. Thompson, P., Marchant, E.: A Computer Model for the Evacuation of Large Building Populations. Fire Safety Journal. **24**, pp. 131-148 (1995)
4. Galea, E.R., Galparsoro, J.M.P.: A Computer Based Simulation Model for the Prediction of Evacuation from Mass Transport Vehicles. Fire Safety Journal. Vol. 22, pp. 341 - 366 (1994)
5. Buckmann, L.T., Leather, J.: Modelling station congestion the PEDROUTE way. Traffic Engineering and Control. 35, pp. 373-377 (1994)
6. Gwynne, S., Galea, E.R., Lawrence, P.J., Filippidis, L.: Modelling Occupant Interaction with Fire Conditions Using the buildingEXODUS Evacuation Model. Fire Safety Journal. 36, pp. 327-357 (2001)
7. Doyle, A.B.: Algorithms and Computational Techniques for Robot Path Planning. Thesis (PhD), University of Wales, UK (1995)
8. Nelson, H.E., Mowrer, F.W.: Emergency Movement. In: DiNenno, P.J, Walton, W.D. (Eds.), The SFPE Handbook of Fire Protection Engineering, Third ed., pp. 3-367-3-380. SFPE, Bethesda, MD (2002)
9. Grandison, A., Muthu, Y., Lawrence, P., Galea, E.R.: Simulating the Evacuation of Very Large Populations in Large Domains using a Parallel Implementation of the buildingEXODUS Evacuation Model. Proceedings of the 11th International Fire Science & Engineering Conference, Interflam 2007, 3-5th September 2007, Royal Holloway College, University of London, UK, Volume 1, pp. 259-270, ISBN 978 0 9541216-8-6 (2007)

# Bidirectional Coupling of Macroscopic and Microscopic Approaches for Pedestrian Behavior Prediction

**A. Kneidl, M.Thiemann, A. Borrmann, S. Ruzika, H.W. Hamacher, G. Köster, and E. Rank**

Technische Universität München, 80290 Munich, Germany
Technische Universität Kaiserslautern, 67663 Kaiserslautern, Germany
Siemens AG, CT PP2, 80200 Munich, Germany
Corresponding author: kneidl@bv.tum.de

**Abstract** We combine a macroscopic and a microscopic model of pedestrian dynamics with a bidirectional coupling technique to obtain realistic predictions for evacuation times. While the macroscopic model is derived from dynamic network flow theory, the microscopic model is based on a cellular automaton. Output from each model is fed into the other, thus establishing a control cycle. As a result, the gap between the evacuation times computed by both models is narrowed down: the microscopic approach benefits from route optimization resulting in lower evacuation times. The network flow approach is enriched by including data of microscopic pedestrian behavior, thus reducing the underestimation of evacuation times.

## Introduction

Modeling pedestrian dynamics to predict pedestrian behavior for both standard and panic situations has been examined using various approaches, as described in [1]. One goal of modeling pedestrian crowds is to find lower bounds for the evacuation time of a given scenario, e.g. buildings, regions, etc. Network flow-based approaches are capable of yielding this information [2]. However, since some aspects of pedestrian behavior - such as interaction - are not taken into account, the estimated times will probably never be reached in real, nevertheless they can serve as a lower bound for evacuation times.

Our goal is to find a more realistic lower bound. To this end, we propose to combine two different approaches. First, we use a macroscopic model based on quickest flows in dynamic networks to compute optimal routing strategies [3]. Then we apply a microscopic model to capture pedestrian behavior and derive a heuristic upper bound. The model is based on a cellular automaton [4-7] using a potential field to describe forces (according to [1]). The two estimates enfold the

R.D. Peacock et al. (eds.), *Pedestrian and Evacuation Dynamics*,
DOI 10.1007/978-1-4419-9725-8_41, © Springer Science+Business Media, LLC 2011

true evacuation time like a sandwich [8]. We couple the two approaches by means of a control cycle feeding output from one model into the other, and vice versa.

This article is organized as follows: to begin with, we describe the macroscopic and microscopic models. Then we specify the setup of the bidirectional coupling relating to the two models. Results and an outlook on further research conclude the article.

## Description and Setup of the Macroscopic Model (Optimization)

The scenario (building, region, etc.) is modeled using a discrete-time dynamic network $G = (N, A, T)$, where $N$ is a set of nodes, $A$ is a set of directed arcs, and $T$ is a finite time horizon discretized into the set $\{0,...,T\}$. The node set $N$ subsumes a *source* $s\mu N$ and a *target* $t\mu N$. Each arc $(i,j)\mu A$ has an associated time-dependent *capacity* $u_{ij}(\theta)\mu ]_0^+$ and a time-dependent *travel time* $\tau_{ij}(\theta)\mu ]_0^+$ for all time steps $\theta=0,...,T$. Here, $u_{ij}(\theta)$ limits the number of flow units that can enter arc $(i,j)$ at time $\theta$. We assume that the node capacity is zero for all nodes and all time steps, i.e. no waiting at nodes is permitted. The travel time $\tau_{ij}(\theta)$ defines the time needed to traverse arc $(i,j)$ for flow departing from node $i$ at time $\theta$, i.e. the flow will arrive at node $j$ at time $\theta+\tau_{ij}(\theta)$.

A *flow* is a function $x : A \times \{0,...,T\} \rightarrow ]_0^+$ which assigns a non-negative value to each arc for all time steps and which is subject to flow conservation and capacity constraints. For a more detailed introduction on network flows we refer readers to the book of Ahuja et al. [9].

The goal of the *quickest flow problem* (see [10]) is to find a feasible flow $x$ which sends a given number of flow units $U\mu ]_0^+$ from $s$ to $t$ in the shortest time $T_U \leq T$. With the setting given above, the problem is called *discrete-time quickest flow problem with time-dependent attributes.* We refer to [11] for mathematical details.

### *Network Setup for Realizing the Coupling*

To model pedestrian movements using dynamic network flows, we represent corridors, walkways, streets etc. in a given scenario as arcs in the network. Every arc $(i,j)\mu A$ has a corresponding fixed *width* $w_{ij}[m]$ and *length* $l_{ij}[m]$. In the coupling setup, we predefine the maximum possible rate of flow per unit width $M_{ij}[peds/ms]$ for every arc $(i,j)\mu A$, henceforth called the *specific flow rate* of arc $(i,j)$. We fix the length of the basic time unit for the network parameters as $z=1s$. Based on this data we compute the *capacity* as $u_{ij} = \lfloor M_{ij} \cdot w_{ij} \cdot 1/z \rfloor$. Note that the capacity is constant over time. Moreover, an *average velocity* $v_{ij}(\theta)$ for every arc

$(i,j)\mu A$ and $\theta=\{0,\dots,T\}$ with corresponding *travel time* $\tau_{ij}(\theta)=\lfloor l_{ij}\cdot v_{ij}(\theta)\cdot z\rfloor$ is assumed to be known.

## Description and Setup of the Microscopic Model (Simulation)

Our microscopic model is based on a cellular automaton [4-7]. The whole area of interest is discretized by hexagonal cells, each of which can accommodate an average European male [12]. At each time step, each cell can be occupied either by a pedestrian, an obstacle, a source or a target. Pedestrians move according to specific behavior rules from sources to targets. The movement of a pedestrian is influenced by different forces, namely the repellent forces of obstacles and other pedestrians and the attraction of the targets. All forces are represented by a common potential field. At each time step, each person moves to an accessible neighboring cell with minimum potential field value. Once the target has been reached, the person vanishes from the model. Each pedestrian is "born" with a certain desired walking speed – the so-called free flow velocity [1,12]. Depending on the local density, i.e. the number of pedestrians in the surrounding cells, pedestrians are forced to slow down. The code is calibrated in such a way as to reproduce Weidmann's fundamental diagram [12]. For a more detailed description of the microscopic model, please refer to [13].

### *Extensions of the Model for Realizing the Coupling*

For the coupling, we define a graph on top of the cellular automaton that is automatically derived from the underlying topography – this is done by finding orientation points on the bisector of each convex obstacle corner (see Fig. 1). These orientation points refer to graph nodes and each point is subsequently connected to all orientation points in sight by means of an arc. In addition, they are connected to the source and the target in the same manner. This graph, including the arc parameters width and length, is used in the macroscopic setup to construct the network. In the microscopic model we use this graph to replace the target and obstacle function of our potential.

In the simulation, a pedestrian traverses along arcs of the graph leading to his target. The macroscopic model yields a distribution rate for each node, according to which the pedestrians choose their next intermediate target.

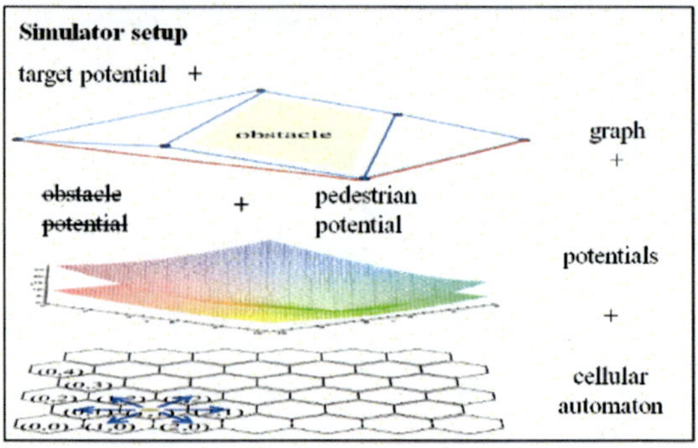


**Fig. 1. Microscopic setup**

## Control Cycle Setup and Constraints for Realizing the Coupling

We define shared (fixed) parameters and variable parameters, which are adapted in each control cycle. A cycle consists of one optimization run followed by one simulation run.

**Table 1. Shared and variable parameters**

| Shared (fixed) parameters | Variable parameters |
|---|---|
| Scenario including network derived with arc width and length | Distribution ratios for each arc and each time step |
| Number of pedestrians | Source flow quantity for each time step |
| Time step size | Average velocity on each arc for each time step |
| Specific flow rate for all arcs | |

**Scenario including the network derived** - We choose different scenarios to test our method. Each scenario consists of one source and one final target plus some intermediate targets and the derived network.

**Number of pedestrians** - The number of pedestrians has to be large enough to observe interaction between the two models.

**Time step size** - The time step size describes the common interval size of the parameter exchange between the two models. In each time step the values are averaged and adapted by both models, respectively.

**Specific flow rate (SFR) for each arc** - The SFR on an arc corresponds to the maximum number of pedestrians who can move through unit width in one second along that arc. We determine the SFR for each arc within the simulation in a pre-processing phase.

**Source flow quantity (SFQ) for each time step** - The source flow quantity is a result of the quickest flow calculated in the optimization network. For each time step, the amount of flow leaving the source in the network defines the number of pedestrians to be generated in the corresponding time step of the simulation run. Interaction between pedestrians may prevent the creation of all required pedestrians in a single time step. In this case, they are generated in the subsequent time step.

The number of effectively generated pedestrians is fed back to the optimization to serve as a reference. In our tests we consider two cases: one with feedback and adaptation of the flow quantities by the optimization, and one without. The adaptation of the flow quantity within the optimization works as follows: if the optimal flow quantity is not achieved in the simulation in a single time step, then the overall capacity of the source (i.e. the total amount of flow that can be sent from the source) in this time step is reduced to the smaller value from the simulation.

**Time-dependent distribution ratios for each arc** - The second output of the optimization is the time-dependent flow distribution ratio on each arc incident to some node. The distribution ratios are calculated as an average value at each time step. The pedestrians are distributed according to these ratios at the corresponding orientation points during the simulation.

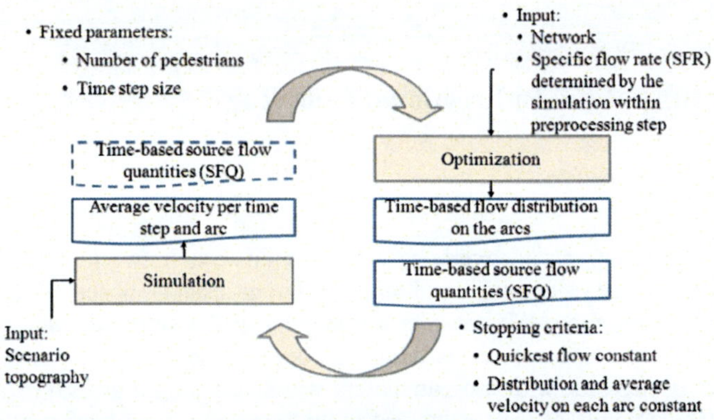


**Fig. 2. Coupling setup**

**Average velocity on each arc for each time step** - The simulation returns the average velocities for each arc and time step.

In coupling cycle $i\mu\{1,\ldots,\textit{number of coupling cycles}\}$, these velocities are read in by the optimization in the following manner:

$$v^i_{regulation} = \alpha v^i + (1-\alpha) v^{i-1}_{regulation}.$$

Here, $v^0_{regulation}$ is the velocity used in the initial dynamic network of cycle zero. The parameter $\alpha \in [0,1]$ refers to the predefined weight of the new average velocities.

Fig. 2 gives an overview of all parameters and the way they are exchanged within one coupling cycle.

**The Control Cycle**

Before the actual control cycle starts, the SFR is derived on each arc of the network by means of a pre-processing phase, as described above. The initial dynamic network is currently established with the arc parameters (width and length) and, for simplicity, an average walking speed of 1.34 m/s for all pedestrians, as in [12]. The quickest flow is computed in this network. The corresponding time-dependent flow distribution ratios on each arc and the time-dependent flow quantity of the source are returned as input parameters for the simulation. The simulation sends the pedestrians from the source towards the target according to these two variable parameters. We get time-dependent average walking speeds on each arc as a result. The time-dependent travel times of the arcs in the dynamic network are adjusted on the basis of these average velocities. The quickest flow is computed in the modified network, the source quantities and flow distributions are updated and, once again, returned to the simulation. This cycle is repeated for a fixed number of times or until a stopping criterion is satisfied.

## The Results of the Bidirectional Coupling

### *Choice of Scenario*

We tested the coupling in several scenarios. In this article, we present results for one representative scenario, combining different effects observed during the testing of the coupling. We consider a triangular walkway in combination with a bottleneck. The topography used in the simulation and the corresponding optimization network with arc parameters width and length are given in Fig. 3. There are two points of special interest in the scenario – the junction at *a* and the bottleneck *b*.

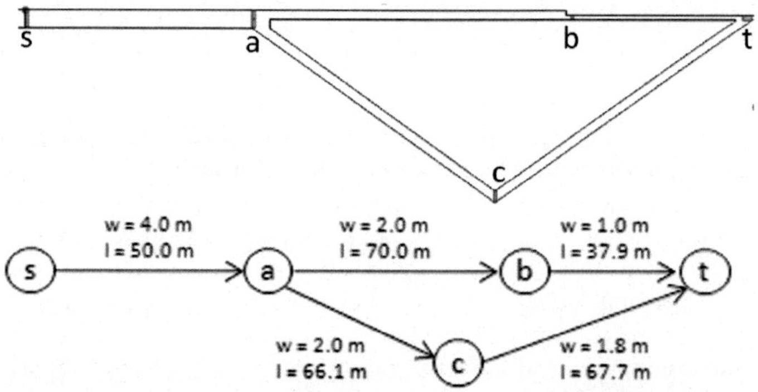


**Fig. 3. Topography of example scenario (a) and corresponding network (b)**

## *Parameter Variation*

We investigate the scenario with various parameter configurations.

The first varying parameter is the source flow quantity (SFQ) feedback of the simulation. Secondly, we adjusted the time step size - we contemplate the results for 5 and 10 seconds.

**Table 2. List of tested configurations**

| Configuration | SFQ feedback | Time step size (s) |
|---|---|---|
| Config_a | no | 10 |
| Config_b | yes | 10 |
| Config_c | no | 5 |
| Config_d | yes | 5 |

**Table 3. Parameter overview**

| Parameter | Value |
|---|---|
| No. of pedestrians | 1,000 |
| α | 0.3 |
| Initially assumed velocity | 1.34 m/s |
| Number of coupling cycles | 25 |
| SFR | Computed in preprocessing phase |

The different configurations are summarized in Table 2. . Table  shows all other parameters which are left unchanged for all configurations.

## *Results*

The results for the different configurations are summarized in
Fig. 4. The curves show the time of the quickest flow obtained by network flow optimization compared to the egress time computed by the simulation. For the simulation we define as egress time the moment when 99% of the pedestrians have reached the target. In the simulation, pedestrians may be diverted from their original path onto the secondary path in a dense crowd. Not yet having implemented personal strategy changes, they must resume their original path after it has been cleared. Thus they become extreme latecomers. Without the 99% rule, they would distort the results. In the following, we refer to the egress time as simulation time. Simulation times are plotted for each cycle starting with the values of "cycle zero" showing the output of both models without coupling. We ran 25 cycles to get a representative statement.

Fig. 4 shows a steady increase of the quickest flow in all four configurations during the first few cycles of the coupling. In the remaining cycles it oscillates around a steady state. The simulation time, on the other hand, decreases from its initial value to approach a steady state from above. Again, we observe the typical oscillations of a control cycle. In this case, the oscillations are caused by the difficulty of controlling the pedestrians in front of the junction *a*. A slightly different distribution at node *a* may result in slightly more or less congestion at the junction and therefore in an increased or reduced total egress time, respectively.

The steady states enclose a small corridor. The only exception is Config_d, where the time step size of 5 seconds is too small to avoid disturbing fluctuations in the gliding average values for the feedback circle. We observe an approximation of both simulation times with a certain overshooting of the lower estimate. The results suggest that, with a sufficiently small time step, it is possible to achieve the convergence of the two approaches to a common steady state. In practice, this is restricted by the fact that meaningful average values can only be obtained with a sufficiently large time step size. But even without convergence, we achieve a substantial improvement of the evacuation time estimates for both approaches.

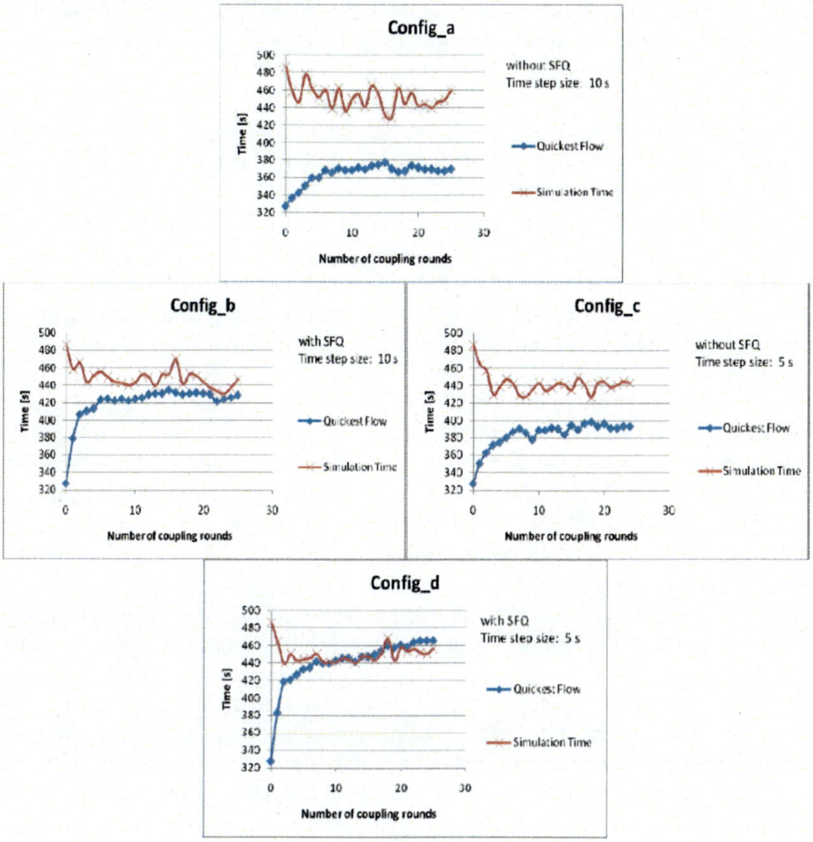


**Fig. 4. Coupling results for scenario of Fig. 3 comparing the quickest flow and simulation time for Config_a (top left), Config_b (top right), Config_c (bottom left), Config_d (bottom right)**

The incremental value of the quickest flow results from increasing travel times on the arcs of the network induced by the simulation feedback: the simulation captures pedestrian interaction which slows down the crowd. It returns a realistic average velocity below the optimal average velocity to the flow model.

The decreasing value of the simulation is caused by both the adapted source flow quantity (SFQ) in the simulation and the adapted distribution ratio at the node *a*:

In the stand-alone simulation run, all pedestrians take the shortest path from *a* via *b* to *t*. The adaptation of the distribution ratio from the optimal flow ensures that, for certain time steps, a given percentage of the pedestrians deviates to the longer route from *a* via *c* to *t*. This results in less congestion on the original path, especially in front of the bottleneck *b*, and the pedestrians move faster. After a few coupling cycles, the last individuals in the pedestrian streams on the two paths from *a* to *t* reach the target almost simultaneously. This is exactly what one expects from an optimal flow in a network.

The effect of the SFQ is as follows: the capacities on the arcs of the network are bounded by the maximum possible specific flow rate (SFR) in the simulation. Since every feasible flow in the dynamic network maintains the SFR on every arc, congestion is averted. By adapting the value for the SFQ in each time step and sending the pedestrians along paths corresponding to the optimal flow, the simulation similarly maintains the SFR on every arc for all time steps, thus preventing congestion (especially in front of the bottlenecks). This in turn reduces the total time needed for all pedestrians to reach the target.

Comparing Config_a and Config_b in

Fig. 4 shows the positive influence of the SFQ feedback. For a time step size of ten seconds, near-convergence of the steady states of the two approaches is reached with the SFQ feedback (Config_b). The SFQ feedback reduces network capacities whenever the simulation is unable to actually generate the number of pedestrians suggested by the flow model. Otherwise the two models tend to decouple and converge to their separate steady states.

Config_c and Config_d in

Fig. 4 show the same results for a time step size of five seconds with the overshooting due to fluctuations in the average velocity values. We describe one cause of the overshooting: Very small average velocities derived from a short period of observing congestion induce smaller source flow quantities on the optimization side in the next cycles. While these velocities rise again in the next coupling cycle (because no congestion occurs), the source quantities will not increase due to our definition of how the SFQ values are handled. This leads to a capacity limit that is too strict, hence the value of the quickest flow increases. To summarize, both parameters influence each other very sensitively and can therefore induce a chain reaction causing an overshooting of the lower steady state.

The bidirectional coupling leads to an improvement in the evacuation time estimate for both models. Using parameter variation we are able to adjust the interaction sensitivity of the two models during the control cycle. However, parameters have to be adapted with caution. For the given scenario, the optimization curve approximates the simulation curve best with the adapted SFQ feedback.

## Outlook

We have presented a control cycle to combine a macroscopic and a microscopic approach for modeling pedestrian behavior. The coupling leads to a modification of the dynamic network so that the computed quickest flow is almost reproducible by the simulation. At the same time, the total egress time of the pedestrians in the simulation is lowered due to the source flow quantity and flow distribution given by the optimization.

For the given example, the egress times computed by both models approach steady states. The steady states enclose a small corridor. This effect could also be observed for other scenarios. The results of the simple scenario suggest that, with the right choice of parameters, the gap can almost be closed.

We propose to continue our work by testing the control cycle for bigger scenarios with more junctions and bottlenecks to answer questions such as: Will the results remain as smooth as in the small examples? Will it be possible to close the gap between the two models in a more complex setup? Also, in the experiments presented here, all pedestrians share the same free flow velocity. This is unrealistic. In a follow-up step, we will assign individual free flow velocities through a Gauss distribution, as described in [12] and observe the effects on the control cycle.

**Acknowledgments** This research is partly supported by the Federal Ministry for Education and Research (Bundesministerium für Bildung und Forschung, BMBF), Project REPKA, under FKZ 13N9961 (TU Kaiserslautern), 13N9964 (Siemens)

## References

1. Schadschneider, A., Klingsch, W., Kluepfel, H., Kretz, T., Rogsch, C., Seyfried, A.: Evacuation Dynamics: Empirical results, Modeling and Applications. In: Meyers, R.A. (Ed.), Encyclopedia of Complexity and System Science Vol. 3, pp. 3142-3176. Springer, Berlin Heidelberg (2009)
2. Chalmet, L.G., Francis, R.L., Saunders, P.B.: Network Models for Building Evacuation. Management Science. **28**, 86-105 (1982)
3. Hamacher, H.W., Tjandra, S.A.: Mathematical Modelling of Evacuation Problems: A State of the Art. In: Schreckenberger, M., Sharma, S.D. (Eds.), Pedestrian and Evacuation Dynamics, pp. 227-266. Springer, Berlin Heidelberg (2002)
4. Emmerich, H., Rank, E.: An Improved Cellular Automaton Model for Traffic Flow Simulation. Physica A. **234**, pp. 676-686 (1997)
5. Burstedde, C., Klauck, K., Schadschneider, A., Zittartz, J.: Simulation of Pedestrian Dynamics Using a 2-dimensional Cellular Automaton. Physica A. **295**, 507-525 (2001)
6. Kinkeldey, C.: Fussgängersimulation auf der Basis zellulärer Automaten. Studienarbeit im Fach Bauinformatik, Universität Hannover (2003)
7. TraffGo Handbuch: www.traffgoht.com/de/pedestrians/downloads/index.html

8. Hamacher, H.W., Heller, S., Klein, W., Köster, G., Ruzika, S.: A Sandwich Approach for Evacuation Time Bounds. To appear in conference proceedings (PED 2010)
9. Ahuja, R.K., Magnanti, T.L., Orlin, J.B.: Network Flows: Theory, Algorithms, and Applications. Prentice Hall, Englewood Cliffs, New Jersey (1993)
10. Burkard, R., Dlaska, K., Klinz, B.: The Quickest Flow Problem. ZOR Methods and Models of Operations Research. **37**, 31-58 (1993)
11. Tjandra, S.A.: Dynamic Network Optimization with Application to the Evacuation Problem. PhD thesis, Universität Kaiserslautern. Shaker, Aachen (2003)
12. Weidmann U.: Transporttechnik der Fussgänger. Schriftenreihe des IVT 90. Zürich (1992)
13. Klein, W., Köster, G., Meister, A.: Towards the Calibration of Pedestrian Stream Models. Lecture Notes in Computer Science: PPAM 2009. Springer, Berlin Heidelberg (2010)

# Emergency Evacuation Modeling: A Novel Approach to Layout Designs and Evacuation Procedures

**R. A. Kady**[1] **and A. Tolk**[2]

[1]Department of Industrial Engineering
Alhosn University, P.O.Box 38772, Abu Dhabi, UAE
Corresponding author: r.kady@alhosnu.ae

[2]Department of Engineering Management and Systems Engineering
Old Dominion University, Norfolk, VA 23529 USA

**Abstract** The development of evacuation models in the last three decades has mainly contributed to the assessment of occupant safety and evacuation procedures in a variety of building designs, under a range of environmental conditions. The effectiveness of such evaluation relies mainly on the models ability to reflect the detailed interactions between the occupant, building design, and environment. The purpose of this study is to present emergency evacuation modeling as a novel approach to layout designs and evacuation procedures. The approach is based on the development of a novel evacuation model that adjusts its outputs to evaluate a range of layout designs. The proposed evacuation model relies on the application of evolutionary computation techniques to assess the means of egress by evolving the location and number of exits needed to ensure occupants safety. The performance of the algorithms varies by occupant behavior. The study suggests that the algorithms have the potential to be implemented in more complex design problems. The study further suggests the need to validate the configurations found by the algorithms by conducting actual evacuation drills.

## Introduction

The ongoing trend of advancing knowledge in building designs and structures has raised major concerns for occupant safety. Innovative methods and approaches are needed to understand and assess these designs to assure occupant safety and verify building compliance with standards and guidelines. Traditionally, prescriptive codes have been applied to building designs to establish occupant safety without the need to demonstrate the level of safety achieved, or the effectiveness of evacuation procedures [1]. A potential alternative to these challenges and obstacles lies in computer-based evacuation models. The development of evacuation models in the last three decades has mainly contributed to the assessment of occupant safety

R.D. Peacock et al. (eds.), *Pedestrian and Evacuation Dynamics*,
DOI 10.1007/978-1-4419-9725-8_42, © Springer Science+Business Media, LLC 2011

and evacuation procedures in a variety of building designs, under a range of environmental conditions. The effectiveness of such evaluation relies mainly on the models' ability to reflect the detailed interactions between the occupant, building design, and environment.

According to the *Life Safety Code®* [2], the components of a building design and structure must provide protection to the occupants of a building in order to reach safety. The *Code* uses the term *means of egress* to reflect the compliance of those components with standard and guidelines. The *Code* defines means of egress as "a continuous and unobstructed way of travel from any point in a building or structure to a public way consisting of three separate and distinct parts: (1) the exit access, (2) the exit, and (3) the exit discharge." In addition, the geometry of a building, the location of exits, and the number of exits influence the means of egress for all those occupying a building. The purpose of this study was to evaluate the application of evolutionary computation technique, namely Genetic Algorithm (GA) in building designs to assess the means of egress for occupants by evolving the location and number of exits required to minimize total evacuation time.

In order to present emergency evacuation modeling as a novel approach to layout designs and evacuation procedures, an evacuation model needs to be developed. Furthermore, a layout must be selected that is representative in terms of the area, number of occupants, exit width, and occupant load factor to be in compliance with the *Life Safety Code®* [2]. For validation purposes, the performance (results) of the evacuation simulation model should be compared to the results of other evacuation models. Finally, several evolutionary computation techniques must be implemented to investigate the optimal location and number of exits that minimized the overall evacuation time. Figure 1 illustrates the approach followed in this study.

## The Model Development

In order to evaluate the application of evolutionary computation techniques in building designs and evacuation procedures, an evacuation model is needed to simulate occupant movement and behavior. Helbing and Molnár [3] suggested that occupant motion can be realistically described using a mathematical model named the *social forces model*. The main effects that determine the motion of an occupant are reaching a certain destination at certain period of time, which requires desired direction and velocity, and a repulsive effect which is the influence of an occupant on others or that provided by a boundary.

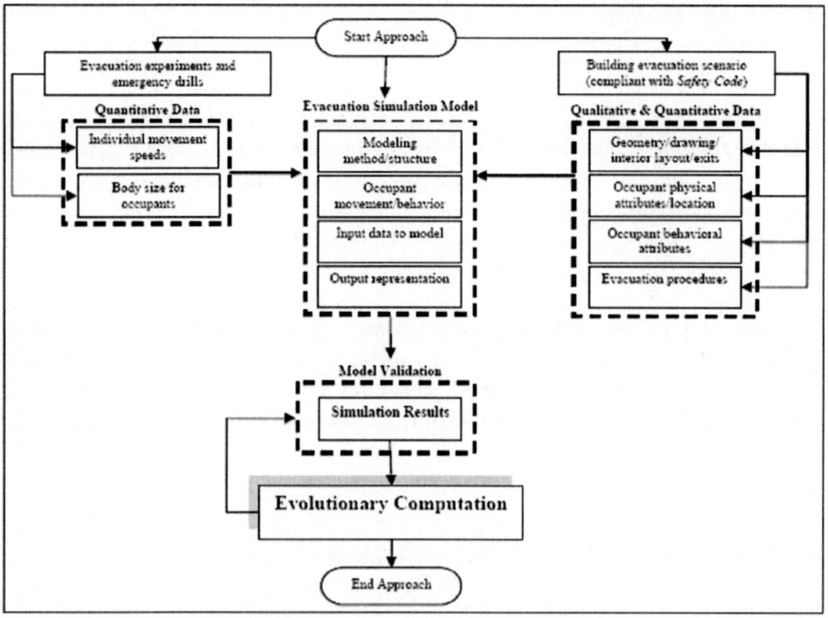


**Fig. 1. The novel approach to layout designs and evacuation procedures**

The evacuation model in the study was based on a simplified framework of the *social forces model*, namely the *artificial potential field* approach [4]. In the model, an exit location creates an attractive force for an occupant, while obstacles/barriers and other occupants act as repulsive forces. The resultant force acting on an occupant forms a potential field, whose gradient drives the occupants at every time-step of the simulation. In order to calculate the movement of an occupant, each force is inversely proportional to the square of the distance between the occupant and the source of the force. In order to determine occupant change in location during a time step, the gradient vector is normalized and multiplied by the occupant speed.

The repulsive forces acting on an occupant from obstacles/barriers and other occupants diminish when the distance to the occupant exceeds a pre-determined minimum value. The minimum distance applied here is more conservative when compared to that applied in other studies [5, 6]. The model also accounts for trampled and/or crushed occupants during the evacuation process. At each simulation time-step, the model checks each occupant for overlapping by other occupants or obstacles by a distance of the occupant radius. Therefore, dead occupants do not obstruct the movement of the other ones since they no longer act as repulsive forces.

## *The Layout Design*

The researchers suggest a layout design for this study that consists of a fixed number of obstacles (furniture) and occupants. The area of the room, number of occupants and occupant load factor are all in compliance with the *Life Safety Code®* [2]. The possible exit locations are determined by calculating how many a predetermined exit-width doors would fit along each wall without separation. If a wall does not allow an integral number of doors, the set of doors are centered along the wall with the excess space placed in each corner. The doors are then numbered starting with the far left door along the north wall and continuing clockwise. In this way, the presence or absence of the exits is represented by a one-dimensional Boolean array. Although this approach is limited to a set of fixed door locations, it uses a fixed-length encoding scheme to represent a variable number of doors. In all, a suggested layout should have the potential for a specific number of exit door locations.

## *Occupant Size and Shape Representation*

The size and shape of individuals in an evacuation model influence occupant movement, density, and response to surroundings. Many evacuation models rely on movement techniques to model the dynamic of spatial systems during the evacuation process, where a discrete environment is updated in steps according to global rules. As a result, occupant shape is a critical element of that spatial system representation and dynamic environment. Meanwhile, the size of individuals assesses the method of movement throughout a building and simulates the presence of occupants and building enclosure such as walls, rooms, exits, corridors, stairs, and obstacles. The majority of the models obtain their occupant shape and size from anthropometric studies [7, 8], or people movement research [9-13].

The scarcity of civilian anthropometric data led researchers to seek alternative anthropometric measurements to represent occupant shape and size in evacuation models. Predtechenskii and Milinskii [9] and Ando et al. [14] observed a projected area of people based on the *average* dimensions (measurements) of a person's width and breadth obtained at the shoulder and chest levels, respectively, while Fruin [10] applied shoulder breadth and body depth measurements obtained from U.S. Army human factors design recommendations.

## *Model Verification and Validation*

One of the most challenging tasks facing evacuation model developers is the verification and validation of their models. According to Banks et al. [15], verification relates to the correct structure of a model by comparing the computer representation to the conceptual model, whereas validation attempts to confirm that the model is a true representation of a real system. With respect to evacuation models, verification can be conducted by testing the performance and functionality of various modules of the computer model. This includes checking the code to examine the performance of model components, and testing a series of model capabilities to ensure the accurate representation of the model functionality [16]. In this study, the evacuation simulation model has been verified by inspecting its code routinely and assessing its ability to perform scenarios with expected outcomes. The majority of evacuation modeling literature has concentrated on model validation, assuming verification is an integral part of the development phase of the model. Evacuation models have been validated against building code requirements [17], fire drills and movement experiments [18-24], literature on past evacuation trials [25-28], or other evacuation models [29-30].

## Evolutionary Computations (ECs)

Evolutionary computation is the discipline devoted to the design, development, and analysis of problem solvers based on natural selection [31]. Evolutionary computation techniques have been applied to a range of design, scheduling, and optimization problems [32]. Figure 2 illustrates the basic structure of an EC. A set (population) of candidate solutions (individuals) for the optimization problem is randomly initialized and evaluated with respect to an objective function. The evaluation function assigns candidate solutions fitness values corresponding to how well the solutions optimize the problem. After the initial population is evaluated, a subset of the population is chosen to become parents for the next generation, allowing the selected parents to create offspring through procreation operators such as crossover and mutation. The procreation operators modify and combine the genetic composition of the parents to create offspring. A subset of the offspring is evaluated and selected for inclusion in the next generation of the population. This process is repeated until some stopping criterion is reached; the discovery of an optimal solution or exceeding a maximum number of iterations. Examples of evolutionary computation techniques that can be implemented include:

- Estimation of Distribution Algorithms: EDAs attempt to leverage the statistical properties of the fitness landscape in order to create children.
- Genetic Algorithms: GAs simulate natural evolution using binary strings (chromosomes), which represent candidate solutions for the problem of interest.
- Particle Swarm Optimization Algorithms: PSOs simulate the movement and behavior of bird flocks and insect swarms to solve optimization problems

```
Procedure EC {
   t = 0;
   Initialize P(t);
   Evaluate P(t);
   While (Not Done) {
      Parents(t) = SelectParents(P(t));
      Offspring(t) = Procreate(Parents(t));
      Evaluate(Offspring(t));
      P(t+1)= SelectSurvivors(P(t),Offspring(t));
      t = t + 1;
   }
}
```

**Fig. 2. Pseudocode structure of an evolutionary computation**

## *Evaluation (Fitness) Function*

The assessment of the fitness of a candidate solution with respect to evaluating a set of exit locations required several criteria. Total evacuation time is the most important one. It is defined as the time needed for all occupants to leave the place of occupancy to reach safety. Another element to take into account is the number of occupants who are crushed to death in their attempt to evacuate. The penalty for a crushed occupant, in the fitness function, is significantly higher than that for an occupant who is unable to escape by the end of the allotted simulation time. This is because a crushed occupant has no chance of escaping regardless of time. Since the EC algorithms would likely generate solutions where each occupant has a door nearby, which is noncompliant with the *Life Safety Code*® [2] guidelines, the third and final element of the fitness function is the number of exit locations.

In order to calculate the fitness value of a given configuration of exit locations, the simulation evacuation model should be set at a maximum simulation time ($T_{max}$).

At any simulation run, when all occupants safely evacuate the place of occupancy in a time less than the allotted maximum evacuation time ($t < T_{max}$), the fitness value for a configuration of exits is calculated according to Equation 1, where $n$ represents the number of exit doors in a configuration. The fitness function favors configurations with $n \leq 2$ due to the high penalty score against configurations with three or more exits. If some occupants are unable to evacuate due to trampling to death or running out of time, the fitness score is computed via Equation 2, where $a$ represents the number of evacuees who are still alive but could not escape in the time allowed, and $d$ represents the number of evacuees who are trampled to death. Thus, the fitness function penalizes for long evacuation times, the number of occupants who are crushed, and configurations with three or more exits.

$$t + 10{,}000\,(\max(n,2) - 2) \tag{1}$$

$$T_{\max} \times a + 1.5T_{\max} \times d + 10{,}000\,(\max(n,2) - 2) \tag{2}$$

## *Encoding Scheme and EC Setup*

Each chromosome in the population is encoded as an array of N binary values, where N denotes the maximum number of exits that may be located along the perimeter of the banquet hall. The binary values represent whether a door should be placed at that location of the room or not (0 for no door, 1 for door). An elitist would be used to search for the best layout configuration. This indicates that the best individual for each generation would be allowed to survive. A pre-determined population size would be generated in a binary format. A binary tournament is then used to select parents. This implies that two individuals are randomly chosen from the population, their fitness values are compared, and the individual with the lower fitness value becomes a parent. Each pair of parents then undergoes a uniform crossover, at the rate of 100%, to generate offspring. No modifications are made on the offspring (mutation rate was set at 0%). For a given layout configuration, each EC will be executed for an independent number of runs with a pre-determined maximum number of function evaluations per run.

## Conclusions

The purpose of this study was to present emergency evacuation modeling as a novel approach to layout designs and evacuation procedures. The approach was based on the development of a novel evacuation model that adjusts its outputs to evaluate a range of layout designs. The proposed evacuation model relied on the application of evolutionary computation techniques to assess the means of egress by evolving the location and number of exits needed to ensure occupants safety. The researchers predicted that the performance of the algorithms would vary by occupant behavior. The suggested algorithms have the potential to be implemented in more complex designs. However, since such implementation is solely dependent on the outcome of the evacuation model used, a continuous development of evacuation models with accurate representation of occupant performance and behavior characteristics is highly needed, especially in deteriorating environmental conditions. Another future work is the validation of the design suggestions found by the algorithms in real evacuation scenarios.

## References


1. Kady, R., Gwynne, S., Davis, J.: The Incorporation and Validation of Empirical Data into the building EXODUS Model. Safety Journal. **47**, 97-104 (2009)
2. Coté, R., Harrington, G.E.: Life Safety Code Handbook®. National Fire Protection Association. Quincy (2006)
3. Helbing, D., Molnár, D.: Social Force Model for Pedestrian Dynamics. Physical Review E. **51**, 5, 4282-4286 (1995)
4. Khatib, O.: Real-time Obstacle Avoidance for Manipulators and Mobile Robots. In: Proceedings of the IEEE International Conference on Robotics and Automation, 500-505. St. Louis (1985)
5. Helbing, F., Farkas, I., Vicsek, T.: Simulating Dynamical Features of Escape Panic. Nature. **407**, 487-490 (2000)
6. Kirkland, J.A., Maciejewski, A.A.: A Simulation of Attempts to Influence Crowd Dynamics. In: Proceedings of the IEEE International Conference Systems, Man, and Cybernetics, 4328-4333. Washington, DC (2003)
7. Abraham, S., Johnson, C.L., Najjar, M.F.: Weight and height of adults 18-74 years of age, United States, 1971-1974 (Vital and Health Statistics, Series 11, No. 211). US Department of Health Education and Welfare. Hyattsville (1979)
8. Pheasant, S., Haslegrave, C.M.: Bodyspace: Anthropometry, Ergonomics, and the Design of Work (3rd Edition). Taylor & Francis, Inc., London (2006)
9. Predtechenskii, V.M., Milinskii, A.I.: Planning for Foot Traffic in Buildings. Amerind Publishing Co. Pvt. Ltd, New Delhi (1978)
10. Fruin, J.J.: Pedestrian Planning and Design (Revised Edition). Elevator World, Inc., Mobile (1987)
11. Eubanks, J., Hill, P.: Pedestrian Accident Reconstruction and Litigation (2nd Edition). Lawyers & Judges Publishing Co., Tucson (1998)
12. Sutherland, D.H., Kaufman, K.R., Moitoza, J.R.: Kinematics of Normal Human Walking. In: Rose, J., Gamble, J.G. (Eds.), Human Walking, pp. 23-44. Williams and Wilkinsore. Baltimore (1994)
13. Pauls, J.: Building Evacuation: Research Methods and Case Studies. In Canter, D. (Ed.), Fires and Human Behavior, pp. 251-275. John Wiley & Sons, Ltd. London, UK.
14. Ando, K., Ota, H., Oki, T.: Forecasting the Flow of People. Railway Research Review. **45**, 8-14 (1988)
15. Banks, J., Carson II, J.S., Nelson, B.L., Nicol, D.M.: Discrete-Event System Simulation (4th Edition). Pearson Education, Inc. Upper Saddle River (2005)
16. Galea, E., Gwynne, S.: Principles and Practice of Evacuation Modeling. Center for Numerical Modeling and Process Analysis. The University of Greenwich, London, Paper 04/IM/112 (2004)

17. Matt McDonald Group, Ltd.: STEPS Software. Available online http://www.mottmac.com/skillsandservices/software/stepssoftware. Last accessed 28 July 2006
18. Schneider, V.: Application of the Individual-based Evacuation Model ASERI in Designing Safety Concepts. In: Shields, T.J. (ed.), Proceedings of the Second International Symposium on Human Behavior in Fire, Boston, MA, 41- 51 (2001)
19. Fraser-Mitchell, J.N.: An Object Oriented Simulation (CRISP II) for Fire Risk Assessment. In: Kashiwaga, T. (Ed.), Proceedings of the 4th International Symposium of Fire Safety Science, Ontario, Canada, 793-804 (1994)
20. Kendik, E.: Methods of Design for Means of Egress: Towards a Quantitative Comparison of National Code Requirements. In: Grant, C., Pagni, P. (Eds.), Proceedings of the First International Symposium on Fire Safety Science, Natl. Inst. Stand. Technol., Gaithersburg, MD, 497-511 (1985)
21. Ketchell, N., Cole, S., Webber, D.M., Marriott, C.A., Stephens, P.J., Brearley, I.R., Fraser, J., Dohney, J., Smart, J.: The EGRESS Code of Human Movement and Behaviour in Emergency Evacuation. In: Smith, R.A., Dickie, J.F. (Eds.), Proceedings of the International Conference on Engineering for Crowd Safety, UK, 361-370 (1993)
22. Fahy, R.: Exit 89-An Evacuation Model for High-rise Buildings: Model Description and Example Applications. In: Kashiwagi, T. (Ed.), Proceedings of the 4th International Symposium on Fire Safety Science, Ontario, Canada, 657-688 (1994)
23. Takahashi, K., Tanaka, T., Kose, S.: An Evacuation Model for Use in Fire Safety Design of Buildings. In: Wakamatsu,T., Hasemi,Y., Sekizawa, A., Seeger, P., Grant, C., Pagni, P. (Eds.), Proceedings of the 2nd International Symposium on Fire Safety Science, Tokyo, Japan, 551-560 (1989)
24. Shestopal, V.O., Grubits, S.J.: Evacuation Model for Merging Traffic Flows in Multi-room and Multi-storey Buildings. In: Kashiwagi, T. (Ed.), Proceedings of the Fourth International Symposium on Fire Safety Science, Ontario, Canada, 625-632 (1994)
25. Proulx, G.: Occupant Response During a Residential High-Rise Fire. Fire and Materials. **23**, 317-323 (1999)
26. Galea, E., Galparsoro, J.M.P.: A Computer-based Simulation Model for the Prediction of Evacuation from Mass-transport Vehicle. Fire Safety Journal. **22**, 341-366 (1994)
27. Bensilum, M., Purser, D.: GridFlow: An Object-oriented Building Evacuation Model Combining Pre-movement and Movement Behaviours for Performance-based Design. In: Evans, D. (Ed.), Proceeding of the 7th International Symposium on Fire Safety Science, Worcester Polytechnic Institute, MA, 941-952 (2002)
28. Thompson, P.A., Marchant, E.W.: A Computer Model for the Evacuation of Large Building Populations. Fire Safety Journal. **24**, 131-148 (1995)
29. Heskestad, A.W., Meland, Ø.J.: Determination of Evacuation Times as a Function of Occupant and Building Characteristics and Performance of Evacua-

tion Measures. In: Shields, T.J. (Ed.), Proceedings of the 1st International Symposium on Human Behaviour in Fire, Belfast, 673-680 (1998)
30. Lo, S.M., Fang, Z., Lin, P., Zhi, G.S.: An Evacuation Model: the SGEM Package. Fire Safety Journal. **39**, 169-190 (2004)
31. Spears, W.M., De Jong, K.A., Bäck, T., Fogel, D.B., De Garis, H.: An Overview of Evolutionary Computation. In: Brazdil, P.B. (Ed.), Proceedings of the European Conference on Machine Learning (ECML-93),Vienna, Austria, 442-459 (1993)
32. Back, T., Hammel, U., Schwefel, H.P.: Evolutionary Computation: Comments on the History and Current State. IEEE Transactions on Evolutionary Computation. **1(1)**, 3-17 (1997)

# The Use of Fine – Coarse Network Model for Simulating Building Evacuation with Information System

**M. Cisek and M. Kapalka**

The Main School of Fire Service SGSP
Warsaw, 52/54 Slowackiego Street, Poland
Military University of Technology WAT
Warsaw, 2 Gen. Kaliskiego Street, Poland
Corresponding author: mcisek@sgsp.edu.pl, mkapalka@wat.edu.pl

**Abstract** In this article, the optimization problem of evacuation of a building with many escape routes with consideration of dynamic changes in people – building- threat interactions is described. The model of system conveying information of the best way of evacuation with taking into account the people who will not adapt to the recommendations is proposed. On the base of chosen strategy of evacuation in every place of the building where direction of evacuation can be changed, the proposed system will give full information about the evacuation possibilities and recommends one of the routs. In the agent model, the individual characteristics affecting the decision-making process are been proposed. Self developed microscopic model of human behavior was used to simulate the evacuation problems. The representation of model environment uses a combination of two approaches: coarse network model and fine network model simultaneously. During the simulation, the main factor influencing the path of movement of individuals is the information supplied by the shared information system of the best way. In addition to a certain probability can decide to change the route of the evacuation.

## Introduction

Studies of the people behavior during emergency situations in buildings which are conducted over the past few years provide a better understanding of the mechanisms that influence people who are evacuating. In addition, these studies allow for a proper adjustment of the geometry of the buildings, the size of escape routes and evacuation strategies for a changing people characteristics. However, the authors believe there is a need not only to adapt to the characteristics of the occupants but also to influence the decisions taken by them. Especially in buildings with complicated configuration of evacuation routes , when there are alternative directions of escape, and the distances to travel for evacuees can be long, there informing or in-

R.D. Peacock et al. (eds.), *Pedestrian and Evacuation Dynamics*,
DOI 10.1007/978-1-4419-9725-8_43, 

structing system may be necessary. This article presents the description of fine-course network evacuation modeling software and the possibility of using this type of software to implement the proposed model for the assessment of evacuation routes. Combining these two models enables simulations of evacuation in buildings with information system.

## Fine – Coarse Network Model

For implementing building information system self developed evacuation model was used [5]. Its main features include:

- the representation of space through a network of cells and a graph,
- microscopic approach for the representation of people,
- traffic modeling using graph algorithms and potential fields method [3,4],
- behavior modeling using rule based systems or more complex intelligence models [6].

The figure 1. shows the basic elements of the model:

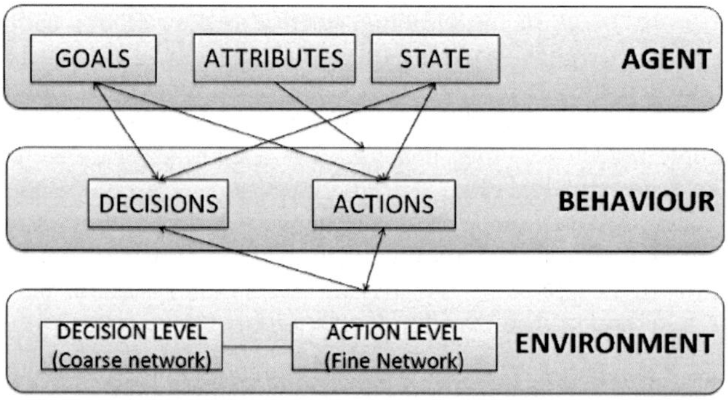


**Fig. 1. Basic elements of the model**

In the presented model the space is defined on two levels: decision level and operating level. The decision level is represented by network $<N,E,F_D>$ in which every node $n \in N$ responds to logical subarea of model space (e.g.: one compartment, corridor, passage, fragment of large compartment). If there is physical possibility of passing from one to another subarea then the nodes representing them are connected by an edge $e \in E = \{(n_1, n_2) \in N \times N: n_1 \neq n_2\}$. For nodes and edges the set of functions $F_d$ was defined, from which most important are:

- attractiveness – a function describing node's attractiveness of the node as a number of people or medium people congestion in the node,
- distance – a function describing edge's medium distance between nodes.

The figure 2. shows an example of the model space and the model on the decision level:

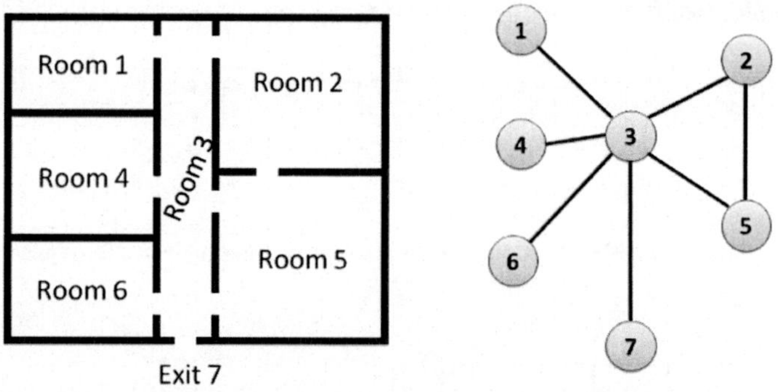


**Fig. 2. Model space and the model on the decision level**

At the operating level of the model, the space is represented by a network of cells $< C, D, F_A >$. The space is divided into tfhe adjacent square cells. The size of each cell is the same and it is 0,3 m x 0,3 m. In similar models, which are using a network of cells as a representation of space, the size of the side of the cell is from 0,3 m [1] to 0,5m (e.g. Exodus [8]).
In the model vector $< i , j >$ represents cell $c$ located in i-th row and j-th column. Additionally a set of possible directions of movement of person between cells was defined $D = \{< d_x, d_y > \in \{-1,0,1\} \times \{-1,0,1\}\}$. The vector $< d_x , d_y >$ described as the direction of movement of person located in cell $< i,j >$ means that the person is going to make a move to cell $< x + d_x, y + d_y >$. The set of functions $F_A$ was defined for cells. The most important functions are:

- $f^{CSTATE}$ – state – a function describing the content of the cell; in basic approach 3 possible states of cell are considered: 0 – unavailable, 1 – free, 2 – person;
- potential – a function determining for each cell the distance from it to the selected cells;
- attractiveness –a function affecting the probability of cell being occupied by a person.

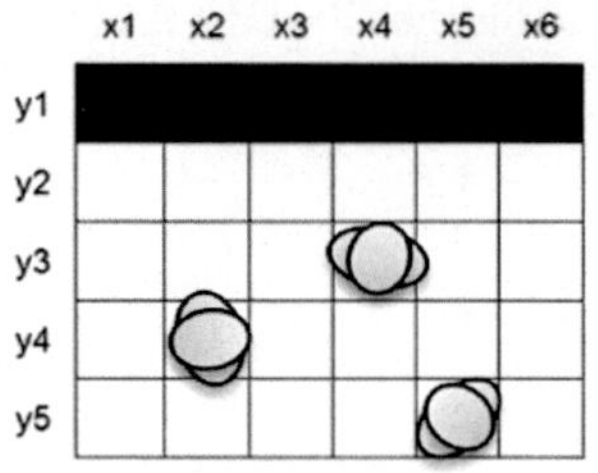


$$f^{CSTATE}(x2, y4) = 2$$

$$f^{CSTATE}(x3, y3) = 1$$

$$f^{CSTATE}(x5, y1) = 0$$

**Fig. 3. Fragment of the model, different states of cells are representing: persons, free space and unavailable space**

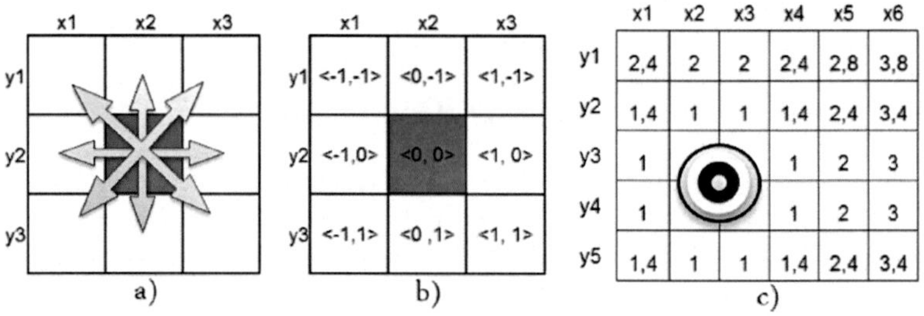


**Fig. 4. The assumptions of movement model: a) 8 possible directions of movement, b) the vectors describing the direction, c) the potentials**

The space representations in the model are used on both levels simultaneously. The connection was achieved by assigning the cells of the model at the operating level to the corresponding nodes on the decision level.

In the developed model, each person has its own:

- characteristics (average and maximum speed, maximum energy level, obedience, knowledge of the surroundings),
- state (position, speed, energy),
- the objectives to be achieved (to leave the building, to escape the room).

In the implemented simulator the person is represented by an agent who makes decisions and performs actions with consideration of a set of characteristics, states, the objectives to be achieved and the state of the environment.

The movement of agents in the model is executed in two stages. In the first stage the agent decides where is it going to move (considering the objective it wants to achieve). In this stage the route between the nodes is set with use of the space model on decision level. In basic approach graph – network algorithms are used for setting the shortest route (e.g.: Dijkstra, $A^*$). During the route setting the congestion in node may have influence on cost of movement between the nodes. Additionally it is possible to consider the probability of the agents' familiarity of surroundings for setting the incomplete and approximate routes. In the second stage, the agents' route is set on the operating level. In this stage, the field potential method was used for setting the direction of movement to any node in accordance with rule: for every node $n \in N$ potential values are determined only for cells assigned to the nodes adjacent to node n.

Apart from potentials also the cell's attractiveness has the influence on the final direction of movement of agent. In addition it was assumed that the "best" direction of movement is chosen with certain probability. In the model, the agent will move not only to find itself in a cell with the lowest (highest) potential, but also will select the most attractive cells. In the model, the attractiveness of the cell is determined in accordance with assumptions:

- the most attractive cell doesn't change the direction of movement,
- the cells located near walls and around other agents are less attractive,
- the cells located in the route of movement of other agent are less attractive.

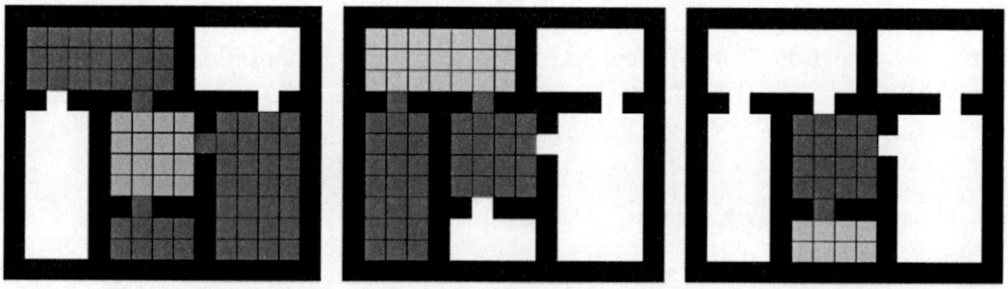

**Fig. 5. Potential field method – target nodes (light grey color), nodes for which potentials describing the distance from target nodes are determined (dark grey color)**

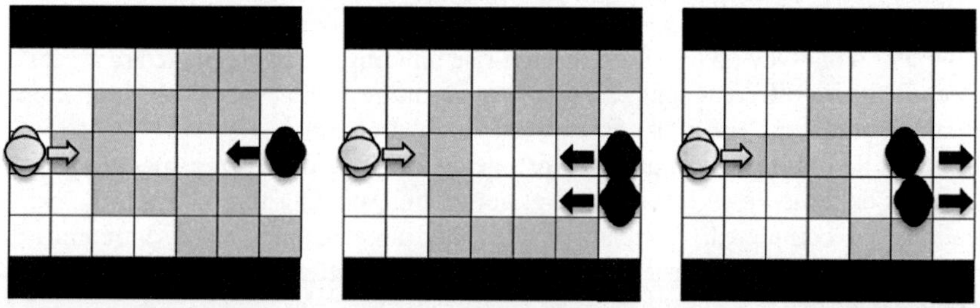

**Fig. 6. The most attractive cells (light grey color) for the agent (grey color) in different situations**

This approach of movement modeling allows for the consideration of dynamic changes in the model environment without the necessity of re-designation of the potentials. Changes in model environment like fire or door blockage influence only change in movement between the nodes.

One of the most important elements of described model is the module responsible for agent behavior. Behavior modeling is executed by constructing decision making network on basis which the agent's behavior is determined. An example of simplified decision making model for evacuees can be described by rules:

- if the agent knows the route to exit – it will choose the shortest route,
- if the agent doesn't know the route to exit – it will follow other agents,
- if the agent stands for certain time in congestion – it will again choose the shortest route without taking into consideration the crowded node,
- if the agent is located near the place of hazard – it will find the route without the dangerous nodes.

Of course, there can be much more rules, what on one hand, can improve the appropriateness of behavior of agents but on the other hand affects the computational complexity.
In addition to the above rules for the purpose of carrying out experiments, taking into account of the system informing the agents there was one more rule added: if the agent is located in the information node – it will with a certain probability change the direction of movement according to the one proposed by the information system.

## Route Assessment Model

The primary objective of the assessment model of the escape route is the speed of the analysis, so as to produce at any chosen time of evacuation simulation fast and reliable calculations. The final result of this calculations is the optimal direction of evacuation at places where the choice is possible. An examples of such place is the intersection of corridors, from which there is possibility to go outside of the building or a hallway on any floor of high-rise building, on which there are at least two exits to two different stair cases. In certain times of the evacuation simulation the simulation stops and route assessment calculation are carried out. These moments can be established in advance or in case of situation when a specific parameter has exceeded a certain level (e.g. congestion, waiting time). When the calculations are completed, the result is presented as a recommended direction of evacuation in areas where such a choice can be made. Result of the analysis may be represented as:

- Recommending the direction of movement,
- No recommendations,
- Disallowing one of the directions of movement,
- Information about distance to the exit,
- Information about assessed time needed to reach the exit.

Based on the given information, people in a given area may or may not decide to evacuate in the recommended direction. The probability that the person's decision will be accordant with the direction proposed by the information system is a matter of individual characteristics of the occupant. In addition, it is possible to consider many types of evacuation routes, including elevators, escalators and travelators. The change of the recommended direction of movement is possible after certain period of time.

Before the beginning of simulation and after a certain periods of time of evacuation simulation the required data from simulating model is imported to route assessment model. These are:

$N = \{1,2,3,\ldots\}$ - set of nodes,

$U = \{(i,j) \in N \times N\}$ – set of edges,

$L_{i,j}$ – maximum lenght in node to next one,

$W_{i,j}$ – width of the passage from node $i$ to $j$ $(i,j) \in U$,

$P_i^{t}$ – number of people in node $i$ in time $l$ $t \in N$,

$V_i$ – speed of movement [m/s],

$FR$ – unit flow rate [occ/(m*s)].

According to the coarse network in the simulating model, the model of the building space is divided to subareas, which correspond to nodes $N$. Four types of nodes are distinguished: $I$ – informative (with inflows and outflows), $C$ – corridor (with inflows and outflows), $S$ – spaces and compartments (only with outflows) and $D$ – the exit doors (only with inflows). Figure 7. represents an example of building floor where the information system was implemented.

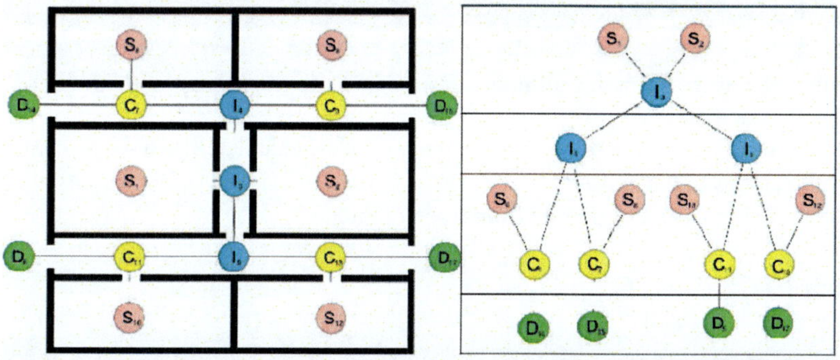

**Fig. 7. Coarse network model for complex one story building: divided to nodes (left), Layer graph for calculation (right)**

The analysis start from the ranking of informative nodes $I$ in the order depending on the distance to the nodes $D$. As the first the nodes $I$ nearest to the nodes $D$ are analyzed. Then the calculations for further nodes $I$ are being made with taking into account the decisions taken by the previous nodes $I$. During the analysis of each node $I$ all routes $R$ to exits $D$ are evaluated with consideration of the all inflows from nodes $S$ and all the flows through nodes $C$. Calculations are performed in time steps with consideration of events $E$. Those events are moments in time when the first or the last person passes from one node to another.

$E^{f}_{i,j}$ - the time of the first person coming from the node $i$ to $j$, measured from the pause time of evacuation simulation,

$E^{l}_{i,j}$ - the time of last person coming from the node $i$ to $j$,

where:

$E^{f}_{i,j} = 0 \; for \; P_i^{t_0} > 0$

$E^{f}_{i,j} = E^{f}_{h,i} + T_{i,j} \; for \; P_i^{t_u} = 0 \; and \; P_h^{t_u} > 0$

$E^{l}_{i,j} = T_{i,j} \; for \; P_i^{t} > 0 \; and \; P_h^{t} = 0$

## Calculations

1. Calculate the maximum number of people in node, can leave the node without congestion at the door.

$$P_i^{con} = \frac{L_{i,j} \cdot W_{i,j} \cdot FR}{V_i}$$

2. Calculate the minimum time for evacuating people from node $S_i$.

$$T_{i,j}(t) = \begin{cases} T^{1}_{i,j}(t) = \dfrac{L_{i,j}}{V_i} \; for \; P_i(t) < P_i^{con} \\ T^{2}_{i,j}(t) = \dfrac{P_i}{W_{i,j} \cdot FR} \; for \; P_i(t) \geq P_i^{con} \end{cases} \quad (i,j) \in U$$

3. Calculate the event times for all nodes from exit $D_k$ to the nearest node not calculated before node $I_i$.

4. By analysis of minimum flows $F^{t}_{i,j}$ between nodes, calculate the time of evacuating all persons to the exit $D_k$.

$F^{t}_{i,j}$ - flow rate of passing of people from node $i$ to node $j$.

$$F^{t}_{i,j} = \min\left( F^{con}_{i,j}, \sum_{h \in N^{+}(i)} F^{t}_{h,i} \right)$$

$N^{+}(i)$ - set of inflow nodes

$F^{t_0}_{i,j} = \min\left( F^{con}_{i,j}, \frac{P_i(t_0) \cdot V_i}{L_{i,j}} \right)$ for nodes $S_i$

$F^{con}_{i,j} = W_{i,j} \cdot FR$ - maximum flow rate of passing of people from node $i$ to node $j$.

In every time step with consideration of event times $E$, calculate the number of people in every node $N$.

5. The total duration of the evacuation for the calculated route $R_{i,k}$ is determined by the time after which all evacuating people come to the node $D_k$.

$R_{i,k}$ – route – sequence of nodes from informative node $I_i$ exit node $D_k$,
To make the decision of evacuating direction in node $I_i$ it is necessary to determine a route $R_{i,k^*}$, for which:

$$T_{R_{i,k^*}} = \min_{k\in D_i} T_{R_{i,k}}$$

, where $D_I$ – set of all possible exits $D$ for node $I_i$, $T_{R_{i,k}}$ - evacuation time for all persons on route $R_{i,k}$, and $T_{R_{i,k}} - t^* \rightarrow \forall_{i \in R_{i,k}} P_i(t^*) = 0$.

6. After calculating the evacuation time for the informative node $I_i$ in every direction, the decision of the best route $R_{i,k^*}(I_i)$ is made.
7. The node $I_i$ of next level is calculated with taking into account the decision of previous informative nodes. The calculation can't be made for those parts of routs which were previously evaluated as worse choice. The calculations are conducted to the last informative node according to the examples of layer graph on figures 4 and 5.

After completing the calculations for the decisions of directions for every node $I_i$, all the decisions are exported to simulation model (decision level). The subareas assigned to informative nodes provide the information to all the agents passing through them. The change in the decision can be made after next pause in simulation and after recalculating all the routes with new data from the simulating model.

During the work on combining the fine – coarse network simulating model and the route assessing model two types of buildings were tested: complex one story building and multistory building with two staircases. Complex one story large area building model is represented on figure 7. and the multistory building is represented on figure 8.

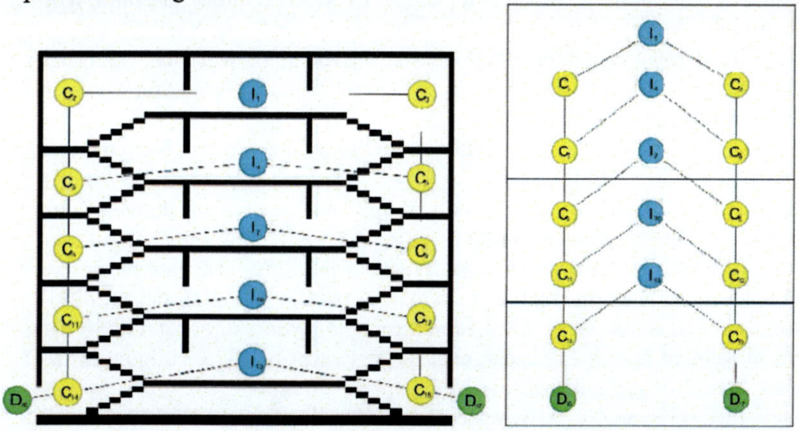

**Fig. 8. Coarse network model for multistory building: divided to nodes (left), Layer graph for calculation (right)**

## Conclusions

Described existing model with representation on two levels and with implemented model for the assessment of evacuation routes can be used for testing of the real information systems in buildings. The model allows for accurate tracking of agents' location and theirs behavior. The possibility for dynamic changes in the model on the decision level (coarse network) without any additional changes to potentials in cells' network allows to quickly simulate such situations as: decision changes in information system, blocking and unblocking the routes, cutting off access to exit by the fire, etc. A simulation of the information system allows evaluation of usefulness of the system with taking into account: different ratio of obedience of people, the optimum time to re-designate the information, different places where should be the points of information.

**Acknowledgments** The work described in this paper is part of BeSeCu Project (Behavior, Security and Culture) Human behaviour in crisis situations: A cross-cultural investigation in order to tailor security-related communication. The BeSeCu project (n°: 218324) is funded by the European Commission within the Seventh Framework Programme, Priority 10: Security.

## References

1. Fruin, J.J.: Pedestrain Planning and Design, (Revised Edition). Elevator World, Inc ,Mobile, AL (1987)
2. Still, G.K.: Crowd Dynamics. Ph.D. thesis, University of Warwick, UK (2000)
3. Schreckenberg, M., Sharma, S. (eds.): Pedestrian and Evacuation Dynamics. Springer, Berlin (2002)
4. Schreckenber, M.:"A Cellular Automaton Model for Crowd Movement and Egress Simulation" (2003)
5. Kapałka, M.: "Modelowanie i symulacja przemieszczania się tłumów." Praca magisterska. WAT, Warszawa (2006)
6. Dzida, D.: "Model i implementacja sztucznej inteligencji agenta na potrzeby symulacji życia w aglomeracji", praca magisterska. WAT, Warszawa (2009)
7. Kuligowski, E.D.: Review of 28 Egress models. Workshop on building occupant movement during fire emergencies. In: Peacock, R.D., Kuligowski, E.D. (eds.), Proceedings, Gaithersburg, M.D., pp. 68–90 (2005)
8. Galea, E.R., Gwynne, S., Lawrence, P.J., Filippidis, L., Blackshields, D.: buildingEXODUS V4.0 User Guide and Technical Manual. Fire Safety Engineering Group, University of Greenwich, UK (2004)
9. Pauls, J.: Movement of People. In: DiNenno, P.J., Beyler, C.L., Custer, R.L.P., Walton, W.D., Watts, J.M., Drysdale, D., Hall, J.R., (eds.), The SFPE Hand-

book of Fire Protection Engineering (Second ed.), pp. 3-263-3-285. Society of Fire Protection Engineers, Bethesda (1995)

10. Veeraswamy, A., Lawrence, P.J., Galea, E.R.: Implementation of Cognitive Mapping, Spatial Representation and Wayfinding Behaviours of People Within Evacuation Modelling Tools. Human Behavior in Fire Symposium, pp. 501 – 512 (2009)

# Simulation of Pedestrian Flow outside a Single-exit Room in Mean-field Approximation Model

**J. Zhang[1,2] and W. Song[2]**

[1]University of Wuppertal, Institute for Building Material Technology and Fire Safety Science, Pauluskirchstr. 7, D-42285 Wuppertal, Germany
[2]University of Science and Technology of China, Hefei, 230026, P.R.China
Corresponding author: cjjz@mail.ustc.edu.cn

**Abstract** In this study, a framework model, based on lattice gas model and Mean-Field Approximation model, is proposed to analyze pedestrian evacuation uncertainty. The model is focused on the probability that each grid is occupied by a pedestrian, but not the specific pedestrians. By calculating the evolution of occupancy probabilities, more information about the simulation uncertainty can be gotten. The model is presented in terms of a series of nonlinear equations and complete probability formula. In each time step, the probabilities that pedestrians exist on each site, as well as the transition probabilities to the neighboring sites, are calculated and updated using random sequential update rule. The pedestrian flow going outside a single-exit room is investigated numerically. The cumulative distribution and probability density distribution of the total evacuation time can be obtained by a single simulation using the model. In this case, the uncertainty and reliability of the simulation results can be easily analyzed and improve the calculation efficiency. In addition, the time dependent of mean flow rate and the effect of the width of exit on the total evacuation time are studied. The framework model can be extended using other Cellular Automaton model with different rules to analyze their uncertainty.

## Introduction

Along with the development of society and economy, there appear more and more emergency incidents such as fire, explosion, terrorist attack etc. and thus it attracts many attentions of researchers to study the public safety in recent years. Meanwhile, the large crowd gathering events also appear frequently because of the appearing of the huge constructions. For example, in the concert, cinema, and other public places. In this case, it is important to guarantee the safety of the human beings and evacuation efficiently in case there is some emergency.

Evacuation models, as one of the important methods to study pedestrian dynamics, developed rapidly and were broadly used not only in the designing of the construction but the evacuation management. Generally, these models can be classified into two categories. One is macro method that pedestrians are regarded as

R.D. Peacock et al. (eds.), *Pedestrian and Evacuation Dynamics*,
DOI 10.1007/978-1-4419-9725-8_44, 

liquid and thus hydro mechanical findings were applied directly, the other is micro-method which considers the human behavior feature and individual differences. Microcosmic model includes continuous models, such as the social force model [1-4], and discrete models. Most of these models can reproduce some typical egress phenomena and behaviors during the evacuation, for example, lane formations, oscillations at bottleneck [3, 4], the "fast is slow" effect and clogging at exit doors [1]. The social force model is based on a system of coupled differential equations which have to be solved using, e.g., a molecular dynamics approach as in the study of granular matter. The results of continuous model are accurate. However, it is not suitable to simulate the evacuation of huge crowd due to the high computational complexity and time consuming. As for discrete models, there are lattice gas models (*LG*) [5-7], cellular automata models (*CA*) [8-10] and Mean-field Approximation models [6, 11]. The discrete models are popular for the properties of simple rule and high operation speed etc..

Accurate predictions of the time required to egress a building under various conditions are crucial to decisions regarding the acceptability of a building design. However, there exist some great differences on evacuation time among various simulations because of the model randomness, the uncertainty of the number and spatial distribution of pedestrians in the building etc. In this case, it usually needs to run many simulations and calculate the mean value to eliminate errors, which leads to low calculation efficiency, especially for the continuous models. Therefore, it is necessary to use a new method to analyze the uncertainty of an egress model efficiently and make the evaluation results more credible.

In this paper, we try to propose a model based on classical lattice gas model [7] and mean-field approximation to analyze the evacuation uncertainty. The model is focused on the probability that each grid is occupied by a pedestrian, but not the specific pedestrians. The pedestrian flow going outside a single-exit room is investigated numerically. The cumulative distribution, probability density distribution of the total evacuation time, and the time independent of mean flow rate are investigated.

## Model

We consider the pedestrian flow going outside a single exit room. In the model, the space is divided into square lattice of $0{:}5m \times 0{:}5m$. That is to say, the grid is either empty or occupied by a pedestrian and a grid is unlikely to be occupied by more than one pedestrian each time step. For simple, we suppose that the pedestrians in the model only move towards 3 directions without backward. We define the initial probability that a pedestrian exists on the site $(i, j)$ at time $t$ as $P(i, j; t)$. As shown in Fig. 1, there exist 8 situations (from case1 to case8) according to the state of the site $(i, j)$ and its 3 neighboring sites. The circles indicate that the site was occupied by a pedestrian, and arrow indicates that this site is empty and the neighboring pedestrian is allowed to enter. The probabilities, $P_{casei}(i, j; t)$, of each situation appearing are defined as follows:

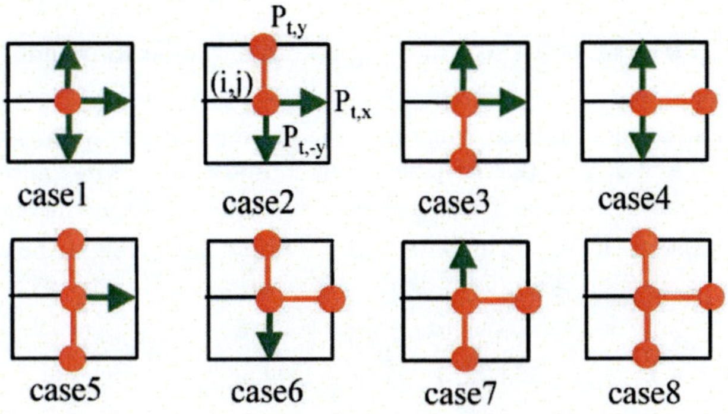


**Fig. 1. All possible configurations of a walker on the square lattice. The full circle indicates the site occupied by a pedestrian walker. The arrow indicates that the site is empty and a pedestrian can move into.**

$$
\begin{aligned}
P_{case1}(i, j; t) &= (1-P(i, j+1; t))\times(1-P(i-1, j; t))\times(1-P(i+1, j; t));\\
P_{case2}(i, j; t) &= (1-P(i, j+1; t))\times P(i-1, j; t)\times(1-P(i+1, j; t));\\
P_{case3}(i, j; t) &= (1-P(i, j+1; t))\times(1-P(i-1, j; t))\times P(i+1, j; t);\\
P_{case4}(i, j; t) &= P(i, j+1; t)\times(1-P(i-1, j; t))\times(1-P(i+1, j; t));\\
P_{case5}(i, j; t) &= (1-P(i, j+1; t))\times P(i-1, j; t)\times P(i+1, j; t);\\
P_{case6}(i, j; t) &= P(i, j+1; t)\times P(i-1, j; t)\times(1-P(i+1, j; t));\\
P_{case7}(i, j; t) &= P(i, j+1; t)\times(1-P(i-1, j; t))\times P(i+1, j; t);\\
P_{case8}(i, j; t) &= P(i, j+1; t)\times P(i-1, j; t)\times P(i+1, j; t);
\end{aligned}
$$

We then define $\Delta P_{t,x}(i, j; t)$ , $\Delta P_{t,y}(i, j; t)$ , $\Delta P_{t,-y}(i, j; t)$ as the transfer probability from site $(i, j)$ to the three neighboring sites at time t. According to the total probability formula, the probabilities for each case are as follows:

$$
\begin{aligned}
&\Delta P_{t,x}(i, j; t+1)|\text{casei}= P(i, j; t)\times \mathrm{p}_{t,x}|\text{casei} + (1-P(i, j; t)\times 0;\\
&\Delta P_{t,y}(i, j; t+1)|\text{casei}= P(i, j; t)\times \mathrm{p}_{t,y}|\text{casei} + (1-P(i-1, j; t)\times 0;\\
&\Delta P_{t,-y}(i, j; t+1)|\text{casei}= P(i, j; t)\times \mathrm{p}_{t,-y}|\text{casei} + (1-P(i + 1, j; t)\times 0;
\end{aligned}
$$

In these three equations above, the first terms represent the probability that the site $(i, j)$ is occupied by an occupant at time $t$ and the occupant transfer to the 3 directions at time $t$+1 according to the different situation separately. The second terms represent the transition probabilities when the site $(i, j)$ is not occupied and thus it equal zero. Where the determinations of moving probability $p_{t,x}$, $p_{t,y}$, $p_{t,-y}$ for each pedestrian are as following according to the surrounding situation and the spatial place.

Each pedestrian is supposed moving in the preferential direction with no back step. The preferential direction of pedestrians is toward the exit. The room is divided into two parts, the upper and nether area, by the center of the exit. For the pedestrians existing in upper area, $p_{t,x}$, $p_{t,y}$, $p_{t,-y}$ are given as below corresponding to each cases in Fig. 1:

for case1: $p_{t,x} = D_x + (1-D)/3;\ p_{t,y} = (1-D)/3;\ p_{t,-y} = D_y + (1-D)/3;$

for case2: $p_{t,x} = D_x + (1-D)/2;\ p_{t,y} = 0;\ p_{t,-y} = D_y + (1-D)/2;$

for case3: $p_{t,x} = D + (1-D)/2;\ p_{t,y} = (1-D)/2;\ p_{t,-y} = 0;$

for case4: $p_{t,x} = p_{t,y} = 0;\ p_{t,-y} = 1;$

for case5: $p_{t,x} = 1;\ p_{t,y} = 0;\ p_{t,-y} = 1;$

for case6: if $D_x < D_y$, $p_{t,x} = p_{t,y} = 0;\ p_{t,-y} = 1;$

Otherwise, $p_{t,x} = p_{t,y} = p_{t,-y} = 0;$

for case7: if $D_x < D_y$, $p_{t,x} = 0;\ p_{t,y} = 1;\ p_{t,-y} = 0;$

Otherwise, $p_{t,x} = p_{t,y} = p_{t,-y} = 0;$

for case8: $p_{t,x} = p_{t,y} = p_{t,-y} = 0;$

Similarly, for the pedestrians existing in the nether area, $p_{t,x}$, $p_{t,y}$, $p_{t,-y}$ are given by replacing $p_{t,-y}$ with $p_{t,y}$.

Where, $D$ ($D \in [0,1]$) is the drift pointing to the exit, which is the same as Ref.[7]. $D_x = D \cdot |x - x_0| / (|x - x_0| + |y - y_0|)$ is the x component of drift. $D_y = D \cdot |y - y_0| / (|x - x_0| + |y - y_0|)$ is the y component of drift. $(x, y)$ is the coordinate of the walker, and $(x_0, y_0)$ is the coordinate of the exit.

Then the probability $P(i, j; t+1)$ that the site $(i, j)$ is occupied by an occupant at time $t+1$ can be described by the following:

$$
\begin{aligned}
\mathrm{P(i, j; t{+}1)} = & \sum_{case1}^{case8} [P_{casei}(i-1, j; t) \times \Delta P_{t,-y}(i-1, j; t)|_{casei}] + \\
& \sum_{case1}^{case8} [P_{casei}(i+1, j; t) \times \Delta P_{t,-y}(i+1, j; t)|_{casei}] + \\
& \sum_{case1}^{case8} [P_{casei}(i, j-1; t) \times \Delta P_{t,-y}(i, j-1; t)|_{casei}] + \\
& \{P(i, j; t) - \sum_{case1}^{case8} [P_{casei}(i, j; t) \times (\Delta P_{t,-y}(i, j; t)|_{casei} + \\
& \quad \Delta P_{t,y}(i, j; t)|_{casei} + \Delta P_{t,x}(i, j; t)|_{casei})] \}
\end{aligned}
$$

The first three terms on the right-hand side represents the inflow probability of a pedestrian moving to the site $(i, j)$ from site $(i-1, j)$ (site $(i+1, j)$, $(i, j-1)$). The last term represents the probability that the pedestrian might stay still on the site $(i, j)$ without moving because of the occupied neighboring sites. On each time step, all lattices are updated only once using random Shuffled Sequential update and the probabilities that each site is occupied by an occupant can begotten.

## Simulation and Results

We carry out a computer simulation. The room is represented by the square of 100 × 100 sites with a single exit 10 sites at the middle of the right wall. Initially, the uniform distribution is considered and thus the initial probabilities for all the sites are distributed uniformly between [$P_0$–0.025; $P_0$+0.025] (here we set $P_0$ as 0.1 for example). The drift D = 0.9 represents that the pedestrians nearly move towards the exit without hesitate. Each lattice is updated in order following above rules. After all the lattices in the room are updated, one time step is completed. The above procedure is repeated. When the sum of the probability of all the 10000 lattices is less than 0.1, we consider that all of the pedestrians are evacuated and the procedure finished.

Fig. 2 is some snapshots of the simulation results obtained by the above procedure. The different values of the probability that there exists a pedestrian in a site are presented in different colors. Thus, the probability evolution for each site can be observed intuitively. The pattern of the evacuation can be observed. At the beginning, the entire pedestrians prefer to move toward the exit, and thus the probability in front of the exit increase gradually. Then, the clogging occurs due to the narrow exit and thus the arch formed near the exit. Most of the pedestrians gather in front of the exit, where the probabilities for most of the site become close to 1. In the end, the clogging decays and the probabilities near the exit start to decrease. In some extent, the local probability here can be thought as the local density. Thus, the evolution of the probability above equals to the density evolution during an evacuation.

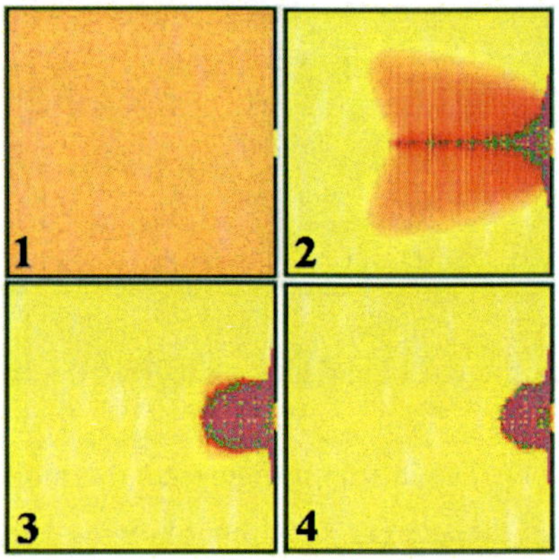


**Fig. 2. Some snapshots of the simulation results from a single exit room. The initial density is 0.1 and the width of the room is 10 sites.**

As the probabilities that there exist a pedestrian are calculated for every site each time step, the cumulative distribution and probability density distribution of the total evacuation time can be obtained easily. The probability that all the pedestrians have evacuated for a certain time (the total evacuation time) is defined as the product $\prod(1-P(i, j; t)$ of probabilities $(1-P(i, j; t)$ for all sites. Thus, the probability density function can be obtained (As shown in Fig. 3). In this case, the uncertainty and reliability of the simulation results can be easily analyzed and improve the calculate efficiency.

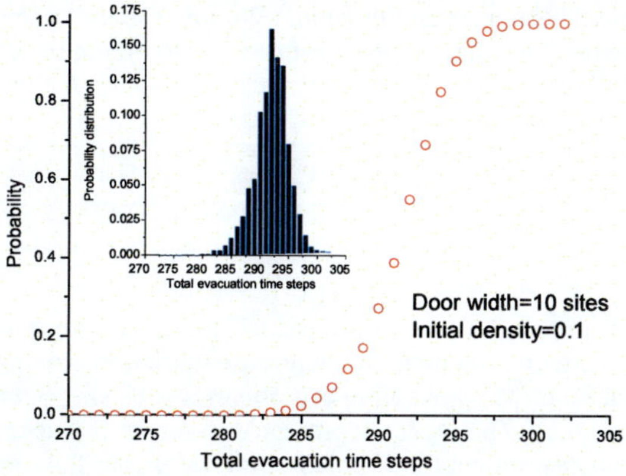


**Fig. 3. The cumulative distribution and probability density distribution of the total evacuation time.**

Fig. 4 shows the time dependent of the flow rate, which can also be calculated and analyzed using the model easily, at the exit for different door width. As mentioned above, the probability can be thought as the local density. Thus, the mean flow rate can be defined as the sum of the probabilities transferred outside the room per unit time. According to the Ref.[7], we also The current is obtained by averaging over 50 time steps. It is shown that the simulation results have the same trend with Tijama's common model.

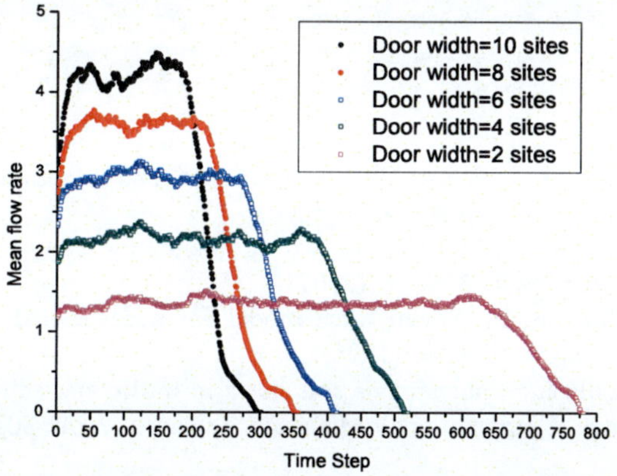


**Fig. 4. The plot of mean flow rate against evacuation time for different door width.**

Here, we will not analyze the quantitative simulation results of our model, because most of the transfer rules have been studied by others in common models and the relative results have been analyzed previous. We just expect to develop a framework model or method that can analyze the uncertainty refer to the evacuation easily and efficiently. Our model can be extended using other Cellular Automaton model with different rules to analyze their uncertainty.

## Summary

In this study, we proposed a framework model, based on lattice gas model and Mean-field Approximation method, to analyze pedestrian evacuation uncertainty. The model is focused on the probability evolution for each site, but not the specific pedestrian movement. The model is presented in terms of a series of nonlinear equations and complete probability formula. The pedestrian flow going outside a single-exit room is investigated numerically. The cumulative distribution and probability density distribution of the total evacuation time are obtained. The time dependent of mean flow rate are derived. The effect of the width of the exit on the total evacuation time is studied. What's more, our framework model can be extended using other Cellular Automaton model with different rules to analyze their uncertainty. This work is important to approach the problems related to the uncertainty analysis of the common egress model and their simulation results.

**Acknowledgments** The study is supported by China National Natural Science Foundation (No. 50678164), Program for New Century Excellent Talents in University (NCET-08-0518), National Science and Technology Pillar Program (No.2006BAK06B00).

## References

1. Helbing, D., Farkas, I., Vicsek, T.: Simulating dynamical features of escape panic. Nature. Vol. 407, pp. 487-490 (2000)
2. Yu, W., Johansson, A.: Modeling crowd turbulence by many-particle simulations. Physical Review E. Vol. 76 (2007)
3. Helbing, D., Molnar, P.: Social force model for pedestrian dynamics, Phys. Rev. E. Vol. 51, pp. 4282-4286 (1995)
4. Helbing, D.: Collective phenomena and states in traffic and self-driven many-particle systems. Computational Materials Sciences. Vol. 30, pp. 180-187 (2004)
5. Tajima, Y., Takimoto, K., Nagatani, T.: Scaling of pedestrian channel flow with a bottleneck. Physica A. Vol. 294, pp. 257-268 (2001)
6. Nagatani, T.: Dynamical transition and scaling in a mean-field model of pedestrian flow at a bottle-neck. Physica A. Vol. 300, pp. 558-566 (2001)
7. Tajima, Y., Nagatani, T.: Scaling behavior of crowd flow outside a hall. Physica A. Vol. 292, pp. 545-554 (2001)

8. Song, W., Xu, X., Wang, B.H., Ni, S.: Simulation of evacuation processes using a multi-grid model for pedestrian dynamics. Physica A. Vol. 363, pp. 492-500 (2006)
9. Yang, L., Li, J., Liu, S.: Simulation of pedestrian counter-flow with right-moving preference. Physica A: Statistical Mechanics and its Application. Vol. 387, No. 13, pp. 3281-3289 (May 2008)
10. Kirchner, A., Schadschneider, A.: Simulation of evacuation processes using a bionics-inspired cellular automaton model for pedestrian dynamics. Physica A. Vol. 312, pp. 260-276 (2002)
11. Nagatani, T.: Freezing transition in the mean-field approximation model of pedestrian counter flow. Physica A. Vol. 388, pp. 4973-4978 (2009)

# A Sandwich Approach for Evacuation Time Bounds

**H.W. Hamacher, S. Heller, W. Klein, G. Köster and S. Ruzika**

TU Kaiserslautern
67663 Kaiserslautern

Siemens AG
80333 München

Corresponding author: hamacher@mathematik.uni-kl.de

**Abstract** In this article, we propose a novel modeling approach – the sandwich approach – to deal with evacuation time bounds (ETB) - in which lower and upper bounds for the evacuation time are calculated. A provable lower bound is achieved by computing a quickest flow, using a dynamic network flow model, an upper bound is obtained via simulation using a cellular automaton model. Coherence between the macroscopic network flow and the microscopic simulation model will be discussed. In order to validate our theoretical results, we report on our practical experiences with the Betzenberg, the region containing the Fritz-Walter soccer stadium in Kaiserslautern, Germany.

## Model Overview

An Evacuation Time Bound (ETB) for the evacuation of objects like ships, airplanes, buildings, regions, etc. is defined as the maximum number of time units until the last of the evacuees must be able to leave the evacuation object. They are required by law for certain classes of evacuation objects or are part of agreements between partners concerned with evacuation issues. However the proof of adherence to ETB is usually obtained by evacuation exercises under appropriate conditions. Only scenarios in which the case of danger has not occurred yet are studied. The exercises are very expensive and must be limited in number. Therefore it is desirable to substitute them by applying suitable mathematical models and software tools.

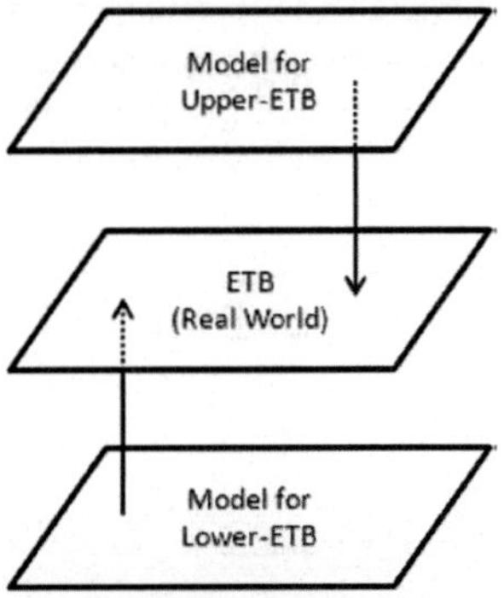


**Fig. 1: Visualization of the Sandwich-Approach**

In this article the sandwich approach – a new approach to deal with ETBs – is introduced. This method yields provable lower bounds for the evacuation time using a dynamic network flow model and at the same time substitutes the need of implementing expensive exercises by using a cellular automaton as a simulation tool. In both models the same parameters are used. For this reason a minimal

R.D. Peacock et al. (eds.), *Pedestrian and Evacuation Dynamics*,
DOI 10.1007/978-1-4419-9725-8_45, © Springer Science+Business Media, LLC 2011

evacuation time and a heuristic estimate of the time needed for the evacuation are achieved.

## *The Dynamic Network Flow Model*

To obtain lower bounds for the evacuation time, a network flow model is used. In this model streets are modeled as edges, intersections as vertices of a graph $G=(V,E)$ (see Figure 4). People moving in the network are modeled as flow units. Every edge $e=(i,j) \in E$ and every vertex $v \in V$ has a set of parameters. The parameters of the edges are capacity $u(i,j)$ and travel time $\tau(i,j)$, the parameter of the vertices is balance#$b(i)$. All these parameters are positive integer numbers. The capacity restricts the number of flow units which can enter the edge at every time step and the travel time expresses how long it takes a flow unit to traverse the edge. The travel times are constant, i.e., they are independent of the time and the load of an edge. Positive balances at a vertex represent supply stored there, negative ones demand. Vertices with balance equal to zero are called traversal-vertices and have neither demand nor supply. For a more detailed description of the dynamic network flow model see e.g. [1].

The optimization algorithm calculates the minimal time needed to implement the balances in each vertex – i.e. sending flow from all vertices with supply to those with demand – while satisfying the capacity- and flow conservation-constraints. This problem is called quickest flow problem (see [2]) in a dynamic network. For a mathematical formulation of the problem see Figure 2.

The goal is to send $v$ units of flow from a source $s$ to a sink $t$ while satisfying the flow conservation constraints – given by the second equation – and the capacity constraints – given by the third equation. The flow conservation constraint ensures that all flow units which enter a vertex other than the source or sink will leave it again. The capacity constraint ensures that in each time step at most $u(i,j)$ flow units enter an edge and no negative flow is sent.

$$
\begin{aligned}
&\min T\\
&s.t. \sum_{\vartheta=0}^{T}\left(\sum_{(s,i)\in E} x_{si}(\vartheta) - \sum_{(i,s)\in E} x_{is}(\vartheta-\tau_{is})\right) \geq v\\
&\sum_{\vartheta=0}^{T}\left(\sum_{(t,i)\in E} x_{ti}(\vartheta) - \sum_{(i,t)\in E} x_{it}(\vartheta-\tau_{it})\right) =\\
&\sum_{\vartheta=0}^{T}\left(\sum_{(s,i)\in E} x_{si}(\vartheta) - \sum_{(i,s)\in E} x_{is}(\vartheta-\tau_{is})\right)\\
&\sum_{(i,j)\in E} x_{ij}(\vartheta) - \sum_{(j,i)\in E} x_{ji}(\vartheta-\tau_{ji}) = 0 \qquad \forall i \in V, 0 \leq \vartheta \leq T\\
&0 \leq x_{ij}(\vartheta) \leq u_{ij} \qquad \forall (i,j) \in E, 0 \leq \vartheta \leq T
\end{aligned}
$$

**Fig. 2. Mathematical formulation of the quickest flow problem**

The quickest flow algorithm is only applicable to graphs with a single source and single sink. In the studied scenario there are nine sources and five sinks (see Fig. 4), thus it is necessary to introduce a supersource and a supersink. Since it is desirable to assimilate the optimization model and the simulation model, we use the time-expanded network (for a definition see e.g. [1]) to compute the quickest flow and adapt the sources such that they are similar to the ones used in the simulation model. This is realized as follows (see Fig. 3): For each source $s^i$ a vertex $\tilde{s}^i$ is introduced as well as additional edges $(\tilde{s}^i, s^i)$ with capacity $b(s^i)$ and travel-time zero, edges $(s^i, s^i_9)$ with capacity equal to the number of people *f* leaving the source $s^i$ in each time-step. The edges $(s, \tilde{s}^i)$ connecting the supersource with the additional vertices $\tilde{s}^i$ do not have any capacity restrictions and travel-time equal to zero.

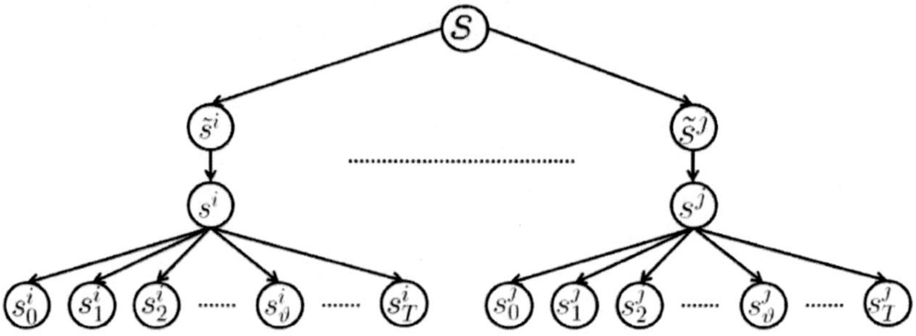


**Figure 3. Supersource**

Using this modification of the time-expanded network, two main properties of the sources are true for the simulation and the optimization model: Each source has an a priori given amount of people – represented by the supply – starting from there, and in each time-step only $f$ people can start at the source. The frequency $f$ is given by capacity restrictions due to the size of the exits or empirical observations and can be time-dependent. Although this model has the advantage of making the parameters of both models more comparable, its obvious disadvantage is that the quickest flow calculation has to be done in the larger time-expanded network (see [3]) and hence worsens the running time of the algorithms. At the same time using the time-expanded network allows time-dependent parameters like capacities or travel times with very little extra effort.

In contrast to the simulation model the optimization problem yields an optimal solution, meaning the minimal time in which all evacuees coming from the sources can reach one of the sinks. With the given parameter setting there is no possibility to accomplish this task any faster than indicated by the optimal solution of the optimization problem. Hence the model provides a lower bound for the evacuation time (lower-ETB).

## *The Simulation Model*

We choose a cellular automaton model that is capable to simulate very large crowds in a fraction of real time. Additionally this model allows incorporating interaction between entities in a very simple and intuitive way.

In a cellular automaton model, the whole area of interest is covered by cells: in our case by, hexagonal cells. At each time step each cell has a certain state: it is either empty or occupied by a person or an obstacle like a piece of a wall. We choose a cell size to accommodate an average-sized European male. Persons either are located when the simulation run is initialized or appear through sources strategically placed at entrances. They disappear in the model as soon as they reach their targets, i.e. one of the exits. The 'automaton' is the core of the model: a set of rules, according to which the states of the cells are updated over time. To this end we employ ideas from electrodynamics. Pedestrians are treated as electrons which are attracted by positive charges or targets, and repelled by negative charges such as other pedestrians or obstacles. The forces between pedestrians, targets and obstacles are calculated via a potential field, using the properties of conservative force fields from physics in which the force can be expressed as the gradient of a suitable scalar function: the potential. This model resembles other cellular automaton models based on potentials as, for example described in [4, 3 and 5] or in the web-published handbook of the TraffGo tool [6]. For a detailed description of our model we refer to [7].

In contrast to the flow calculations described above, the simulation can model the movement of each individual and their interaction with other members of the same crowd. Each person is equipped with an individual speed which it tries to achieve – and indeed does achieve, provided the path is free: the free flow velocity [8]. The distribution of the speeds follows the suggestions in [9]: a normal distribution about some average free flow velocity. Some persons wish to go faster, if given the chance, others are slower by habit. Different velocities are enabled by allowing a person to move forward multiple cells in each simulation step. Furthermore, we increase the reliability of our results by calibrating our code according to socio-cultural behavior: Pedestrians slow down as the pedestrian density increases. This dependency of pedestrian velocities – or alternatively flow – on crowd density is quantitatively described by so-called *fundamental diagrams*. Fundamental diagrams can also be considered as a behavioral model that aggregates a multitude of socio-cultural and even scenario-dependent parameters [10]. Ultimately, the differences in – say gender, nationality, fitness – are modeled by the way people walk as individuals and surrounded by a crowd. In this paper we use the fundamental diagram published by Weidmann [9] and the automatic calibration described in [11] to calibrate human behavior in a crowd.

## Simulation and Optimization Parameters

Optimization and simulation both seek to predict evacuation times in emergency situations around the stadium. We consider the same area: The optimization algorithm constructs a graph from the physical paths represented as a series of edges with appropriate capacities and travel times using a constant walking speed of 1.34 m/s along each edge. The simulation covers the physical paths with cells through which the pedestrians pass at speeds governed by individual characteristics and crowd interaction. The free flow velocity is Gaussian distributed with mean 1.34 m/s. Both models start with 40,000 people in the stadium. In the simulation as well as in the optimization model people leave the stadium with a frequency $f$ of two to five persons per second. Sinks in the graph and targets in the simulation coincide. They are placed where safety has been reached. In the case of the optimization they have unlimited capacity, in the simulation the capacity depends on local parameters such as the number and frequency of buses leaving a parking lot. The influence on the simulation outcome is not significant in this context.

Figure 4 shows the studied area for both models. On the left hand side the graph representation can be seen. The width of the edges represents their capacities. On the right hand side the map used for simulation is shown. It is very close to the real topography, with the exception of the exact width and form of some bottlenecks. The white areas are free paths; the darker an area, the less attractive it is to walk on. Pedestrians cannot step onto black areas representing solid obstacles such as houses. Some areas represent stairs, where people slow down and congestions form (see Fig. 4). Congestions also occur in front of bottlenecks. From eye observation it is also evident, that the free flow velocity and, probably also, the fundamental diagram depend on time. Right before the end of the game we mostly observed very fast walking persons who avoided getting caught in the crowd. Unfortunately, we still lack precise measured data. The simulations are therefore based on Weidmann's fundamental diagram with an average free flow velocity of 1.34 m/s. We suspect that the present choice of input parameters leads to an overestimation of the evacuation time.

**Figure 4. Observed Area:** Graph used in the optimization model and map used for the simulation model, both of the same area surrounding the soccer-stadium in Kaiserslautern

When planning an event or designing an emergency plan, there are several parameters one would like to vary for a better understanding of the situation and which have to be determined and adapted for each scenario. Examples of such parameter are the following: The distribution of the persons to the sources and the sinks, i.e.

the balances, or the frequency in which they can leave the sources and enter the sinks.

## *Refinement of the Network*

To better understand the interrelation between the network model and the simulation model we refine the network, such that its structure resembles a cellular automaton. Each edge with travel time $\tau(i,j) > 1$ is replaced by $\tau(i,j)$ edges each with travel time equal to one. Therefore it is necessary to introduce dummy-vertices between those edges (see Figure 5). All edges have travel time and capacity one, except for the edges with end vertex *i* or *j* which have travel time zero.

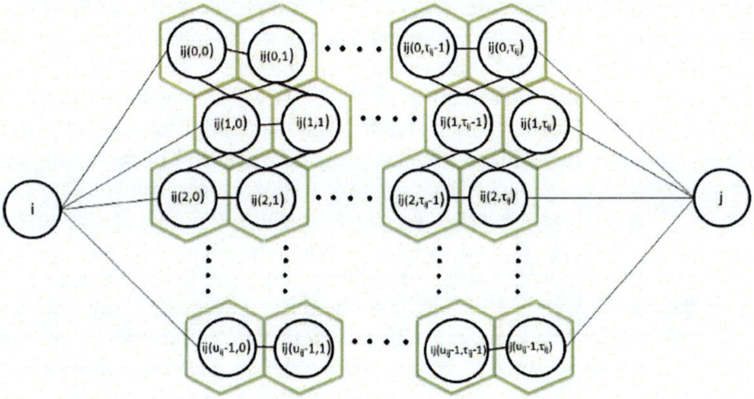


**Fig. 5. Refined Graph** for one edge *(i,j)* with capacity *u(i,j)* and travel time $\tau(i,j)$

The properties of this refined network correspond to the cellular automaton model with constant travel times of value 1.34 m/s. Obviously, each path in the refined graph corresponds to a path in the cellular automaton. One can show that the maximum flow algorithm and the minimum cost flow algorithm with travel times interpreted as costs yield the same results in the refined graph and the original graph. This result shows the interrelation between the optimization and simulation model and provides a basis for the combination of the two approaches.

In the simulation model the respective paths chosen by the evacuees play an important role, since the interaction among the people is an essential factor. In the optimization model the only factor distinguishing the different paths from each other is the parameter travel time. This explains why the optimization results in the original and the refined graph do not differ from each other. This is also true for obstacles. While an obstacle in the graph is modeled via a reduction of the capacity, an obstacle in the simulation model is realized by deleting the corresponding cells. In the refined graph it is modeled by deleting the corresponding vertices and the adjacent edges. This modification reduces the number of paths between two (or more) vertices and thus the capacity of paths from a source to a sink. This reduction of the capacity relates exactly to the diminishing of the capacity in the graph model.

## Comparison of the Results for the Betzenberg Area

In the optimization we observe the utilization of each edge, the number of persons arriving at each sink and the overall evacuation time: when the last person has reached safety. In the simulation, we observe densities on the physicals paths instead of the edge utilization. The other observation parameters are the same.

As parameter-setting we choose a supply of 40,000 people in the stadium and a frequency $f$ of two to five people per second leaving the sources. The distribution of the people to the sources and sinks can be seen in Table 6. The optimum calculated with the optimization algorithm is 1860 sec. Given the same setting in the simulation model 2091 seconds are needed. In the simulation model 90% of all evacuees have reached their final destination after 30 minutes. The remaining persons are extremely slow walkers with an average walking speed below 0.3 m/s despite the fact that crowds are relatively loose towards the end. After another 5 minutes even those have reached safety. In a true evacuation scenario it is reasonable to assume, that people will not dawdle and one may assume a biased distribution of the free flow velocity. With that, the time to evacuate – say – 90% of the people will be closer to the lower bound from the optimization algorithm. However, extremely slow walkers cannot be completely neglected because fans tend to get inebriated during the game. Also, there is quite a number of elderly and handicapped fans.

We can conclude that an evacuation time of less than 31 minutes is definitely not achievable, and that more than 35 minutes will most likely not be necessary. The real evacuation time, computed without an expensive evacuation exercise, is contained in the time window of 31 to 35 minutes. Any improvement in the lower and upper bounds of 31 and 35, respectively, will close this time window further.

**Table 6. Distribution of the people to the sources and sinks**

| Source | # Persons | Sink | # Persons |
|---|---|---|---|
| Horst-Eckel-Tor | 16990 | Station | 9125 |
| South 4 | 2820 | East 1 | 3804 |
| South 3 | 2440 | East 2 | 2650 |
| South 2 | 2350 | P+R Ost | 8927 |
| South 1 | 2060 | P+R Uni + Opel | 11178 |
| Wemer-Kohlmeyer-Tor | 6730 | South | 3560 |
| Wemer-Liebrich-Tor | 3170 | South-west | 1756 |
| Ottmar-Walter-Tor | 2970 | | |
| North | 470 | | |

## Conclusion and Outlook

As far as the authors know there exist only few approaches incorporating simulation as well as optimization models in one large model. There are several methods for optimizing certain parameters of simulations (see e.g. [12]). The sandwich-method introduced in this article combines the two approaches to get a provable lower bound for the evacuation time using optimization and an almost-realistic time using simulation. The lower bound states if it is theoretically possible to achieve a given bound. Using the simulation to get a realistic evacuation time supersedes partly the need of accomplishing expensive evacuation exercises to verify the time-bound estimates. The simulation model can also be used to get virtual experience and to test different ways of directing people to targets. The results encourage us to further pursue the sandwich approach and substantiate our estimates with more accurate and realistic input data.

A large ETB-gap can indicate an unthrifty distribution of pedestrians to paths. In this case it is possible to reduce the gap by guiding evacuees to use alternative evacuation routes or by reducing/augmenting the number of people leaving the sources per time-unit. Therefore the results of the optimization algorithm can be utilized and the parameter change can be tested by the simulation model before employing it in a real-world scenario.

There are several open topics for future research. As mentioned before the time-expanded network can easily be adapted for the use of time-dependent parameters. Nevertheless it is desirable not to work with the time-expanded version of the network. Miller-Hooks and Patterson ([13]) introduced an algorithm to calculate a quickest flow in a network with more than one source and sink and time-dependent parameters without considering the time-expanded version of the network and thus achieved a better running-time. This concept can also be adapted for our problem with one restriction: The algorithm requires that a holdover-edge exists at each vertex, having unlimited capacity, which is not a realistic assumption for the observed scenario. An adaption of their algorithm such that the time-expansion is not required and the use of time-dependent parameters is possible, is a topic for further research. Furthermore topics like modeling of different groups of people ([14]) or coupling of both models ([15]), as well as an integration of location planning (e.g. for the placing of booths) leave room for further studies.

**Acknowledgments** This paper is supported in part by the Federal Ministry for Education and Research (Bundesministerium für Bildung und Forschung, BMBF), Project REPKA, under FKZ 13N9961 (TU Kaiserslautern), 13N9964 (Siemens)

## References

1. Ford, L., Fulkerson, D.: Flows in networks (1962)
2. Burkard, R.E., Dlaska, K., Klinz, B.: The quickest flow problem. Mathematical Methods of Operations Research. **37**, 31-58. Springer (1993)
3. Hamacher, H.W., Tjandra, S.A.: Mathematical Modelling of Evacuation Problem: A State of the Art. In: Schreckenber, M., Sharma, S.D. (eds.) Pedestrian and Evacuation Dynamics. Springer, Verlag (2002)
4. Burstedde, C., Klauck, K., Schadschneider, A., Zittartz, J.: Simulation of pedestrian dynamics using a 2-dimensional cellular automaton. Physica (Amsterdam) 295A, 507 (2001)
5. Kinkeldey, C.: Fußgängersimulation auf der Basis zellularer Automaten. Studienarbeit im Fach Bauinformatik, Universität Hannover (2003)
6. TraffGo HT. http://www.traffgo-ht.com/de/index.html
7. Klein, W., Köster, G., Meister, A.: Towards the Calibration of Pedestrian Stream Model. Workshop on Complex Collective Systems, PPAM 2009, Wroclav 2009. To appear in Springer LNCS series (2010)
8. Nagel, K., Schreckenberg, M.: A cellular automaton model for freeway traffic. J. Phys. I France. **2**, 2221–2229 (1992)
9. Weidmann, U.: Transporttechnik für Fußgänger. Schriftenreihe des IVT, 90 (1992)
10. Chattaraj, U., Seyfried, A., Chakroborty, P.: Comparison of pedestrian fundamental diagram across cultures. arXiv: 0903.0149 [physics.soc-ph] (2009)
11. Davidich, M., Köster, G.: Towards Automatic and Robust Adjustment of Human Behavioral Parameters in a Pedestrian Stream Model to Measured Data. To appear in PED conference proceedings (2010)
12. Carson, Y., Maria, A.: Simulation optimization: methods and applications. Proceedings of the 29$^{th}$ conference on Winter simulation, 118-123 (1997)
13. Miller-Hooks, E., Patterson, S.: On solving quickest time problems in time-dependent, dynamic networks. Journal of Mathematical Modelling and Algorithms, Springer. **3**(1), 39-71 (2004)

14. Hamacher, H.W., Leiner, K., Ruzika, S.: Quickest Cluster Flow Problems. To appear in conference proceedings (PED 2010)
15. Kneidl, A., Thiemann, M., Borrmann, A., Ruzika, S., Hamacher, H.W., Köster, G., Rank, E.: Bidirectional Coupling of Macroscopic and Microscopic Approaches for Pedestrian Behavior Prediction. To appear in PED conference proceedings (2010)

# A Knowledge-based Approach to Crowd Classification

**S. Bandini, L. Manenti, S. Manzoni and F. Sartori**

Department of Informatics, System and Communications
University of Milano – Bicocca
Viale Sarca 336, 20126, Milano, Italy
{bandini, manenti, manzoni, sartori}@disco.unimib.it

**Abstract** This paper illustrates a formal tool for knowledge representation and management in the crowding area, in order to improve on the sharing of knowledge, data and information produced by different models and simulation tools. The presented approach exploits knowledge-based methods for acquisition and representation phases.

After a short discussion about crowd research area, we will present the methodology we have used to develop the tool. It is composed by an ontology and a set of fuzzy rules, which provided crowd classification according to a sociological theory previously formalized.

At last, in order to show how using the tool, a case study on a particular crowd scenario is proposed.

## Introduction

Crowding phenomena are an interesting topic traditionally studied in the context of human sciences (sociology, anthropology, and so on), but also scientific disciplines, such as Mathematics, Physic and Computer Science [1], have recently proposed models and tools to satisfactory describe human behaviors and interactions within a crowd. In particular, in the Computer Science community, several authors have worldwide proposed modeling and software tools aiming at collecting data and knowledge about events and situations where a large gathering of people share spatial areas. The proposed models and simulation tools can be very different and can be very difficult to use and sharing their results. Taking into account only one of the most evident modeling aspects, the proposed models and tools can refer to many different scales e.g. buildings [2], urban areas, public stations[3], airports, exit doors[4], elevator halls [5].

These applications are based on different approaches and data, knowledge and information used and produced can hardly be shared outside the group that has developed them.

Besides this high risk for researchers and practitioners working in pedestrian and crowd dynamics community of loss of data, information and shared seman-

R.D. Peacock et al. (eds.), *Pedestrian and Evacuation Dynamics*,
DOI 10.1007/978-1-4419-9725-8_46, 

tics, in the Computer Science field, several interesting initiatives oriented to data sharing for model benchmarking and validation are also recently appearing (e.g. tracking technology [6]).

This paper illustrates a formal method for knowledge representation that has been designed with the aim of providing a sounding computational framework for the development of knowledge bases and related knowledge exploitation functionalities. This work is part of an ongoing research project aiming at the development of decision support crowd management where techniques and methods from knowledge-based area are used [7].

The general methodological approach we have exploited is based, in fact, on principles from knowledge engineering (i.e. domain knowledge acquisition and modeling) and knowledge representation (i.e. domain modeling with sounding formal tools) .

The main feature of the proposed approach is the explicit representation of crowds domain according to a formal and sounding model which provides domain semantics to knowledge users. According to this reference domain model, software tools for knowledge extraction, sharing and exploitation can coherently access the framework and, possibly, cooperate.

A proposal for a tool for knowledge sharing within the heterogeneous, multidisciplinary community of pedestrian and crowd dynamics is presented, developed based on the integration of ontological languages (i.e. OWL[1]) and fuzzy logic theory [8]. The tool proposes a crowd classification based on features that could be measured on any crowd.

We have used the following general methodology in order to develop the tool: starting from a theory (first step), it has been analyzed to obtain a formal representation (second step) that allows to identify the principal concepts and create a conceptual model (third step). This conceptual model supports the phase of creation of a software tool (fourth step).

The paper is organized as follows: after a short discussion about the reference crowd theory adopted as background domain knowledge, we will present the developed ontology to model it and integrate it with experiential knowledge acquired by knowledge engineering with domain experts (i.e. pop-rock concert music organizers and managers). Finally concluding remarks and future works will be briefly point out.

## Crowd Knowledge Base Development

As previously anticipated, the first step to accomplish has been choosing a reference theory for crowd analysis: among available alternatives, we referred to Elias Canetti [9] contribution.

---

[1] http://www.w3.org/TR/owl-features/

The result of crowd knowledge base development phase has been an ontology in which several crowd phenomenologies can be described according to a set of (quantitatively or qualitatively observable) features according to the Canetti's work. On top of this ontology we have developed a first prototype of a software tool for the classification of crowds.

## *Crowds Types and Features*

Six types of crowds can be identified (see Figure 1): *Open-Closed* crowds, *Stagnating-Rhythmic* crowds and *Slow-Quick* crowds.

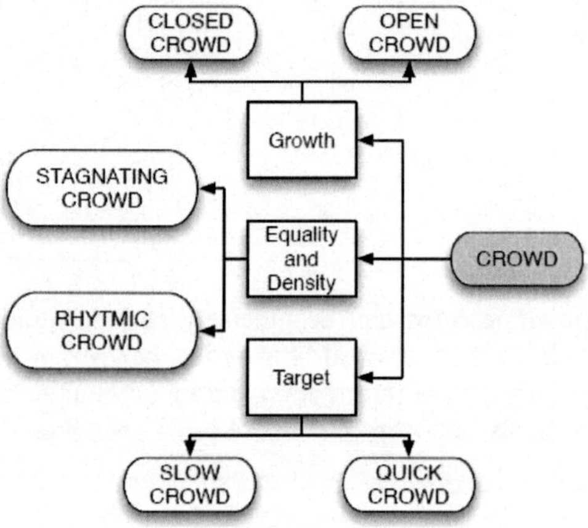


**Fig. 1. Canetti types of crowd.**

Crowd features taken into account for domain ontology development are:

- *physical boundaries,* which can be *present* if the crowd is inside a building, *absent* if the crowd is situated in an open space or *not influent* if this feature is not important to characterize the kind of crowd;
- *psychological boundaries,* which can be *present* if there are some social/psychological limits to take part in the crowd, *absent* if there are no limits to take part in the crowd or *not influent*;
- *movement,* which can be *present* if the individuals move according to external solicitations, *absent* or *not influent;*
- *density,* which can be *high, medium, low* or *not influent* and it refers to the fuzzy interpretation of the ratio between people number and space dimensions;
- *attitude to grow,* which can be *high, medium, low* or *not influent* and it refers to the increase of crowd population;

- *lifespan* which can be *long*, *medium*, *short* or *not influent* and it refers to the crowd disappearance time;
- *destination* which can be *near*, *far* or *not influent,* and described as the time that crowd spends to reach its goal.

Table 1 shows the relationships between crowd features and crowd types. It can be read as a set of rules that allows to identify the type of crowd through its observable features (second step).

**Table 1. Relations between types of crowd and observable features.**

| Features | Open | Closed | Stagnating | Rhythmic | Slow | Quick |
|---|---|---|---|---|---|---|
| Ph. boundaries | absent | present | - | - | - | - |
| Ps. boundaries | absent | present | - | present | - | - |
| Movement | - | - | absent | present | - | - |
| Density | - | - | high/medium | low | high/medium | low |
| Growth | high | medium/low | - | low | high | medium/low |
| Lifespan | - | - | - | medium/short | long | short |
| Destination | - | - | - | near | far | near |

In some cases, crowd features can be precisely and quantitatively collected (e.g. the presence or absence of physical boundaries, pedestrian movement, local densities). Otherwise, only partial information and mostly qualitative descriptions characterized by uncertainty and imprecision are often available after data collection.

## *From Features to Ontological Concepts*

After the theory formalization, following the methodology we have introduced, the third step consists in the identification of main concepts in order to build an ontological model and a domain representation. In fact, these concepts overlap with the main elements derived from theory analysis: crowd is the topic argument, with its six specializations. The other concepts are related to the observable features identified from theory formalization.

Due to the domain complexity, features management can be difficult: while it is quite simple to evaluate the presence or absence of physical and psychological boundaries and movement, since they can only assume boolean values, the evaluation of others is more complicated due to their level of uncertainty that makes difficult to deal with them. For this reason the ontology exploited fuzzy rule sets theory [10] to describe the values of uncertain features (i.e. *density*, *growth*, *lifes-*

*pan* and *destination*) and to create a relationship between observable values and the set of linguistic quantifiers adopted (i.e. *low, medium, high*).

A *fuzzy set* is a couple <U, f> where U is a set called *universe of definition of fuzzy set* and $f: U \rightarrow [0, 1]$ is a function that returns the membership degree of a specific element to a given fuzzy set. The function *f* is called *membership function* and every fuzzy set is completely defined by its own membership function.

Starting from this definition, the application of fuzzy logic to concepts is obtained by specification of fuzzy sets and by creation of membership functions in order to set the membership degree of a value to the fuzzy set. We have created a fuzzy set for each sub-concept, defined from the uncertain features and overlapped with linguistic quantifiers. According to fuzzy set theory, we have described the sub-concepts of density and growth by means of *trapezoidal functions*, and the sub-concepts of lifespan and destination by means of *bell-shaped functions*. In the following, we describe the parametric functions we have used to create density and growth functions:

- *LowFunction*($n$, $m$) with $m>n$. The parameter $n$ represents the $x$-coordinate value where the function assumes the true value (equal to 1) and $m$ represents $x$-coordinate value where the function assumes the false value (equal to 0);

$$y = \begin{cases} 1 & \text{if} \quad 0 < x < n \\ \dfrac{m-x}{m-n} & \text{if} \quad n \le x \le m \\ 0 & \text{if} \quad x > m \end{cases}$$

- *HighFunction*($w$, $z$) with $z > w$. The parameter $w$ represents the $x$-coordinate value where the function assumes the false value (equal to 0) and $z$ represents $x$-coordinate value where the function assumes the true value (equal to 1);

$$y = \begin{cases} 0 & \text{if} \quad x < w \\ \dfrac{x-w}{z-w} & \text{if} \quad w \le x \le z \\ 1 & \text{if} \quad x > z \end{cases}$$

- *MediumFunction(s, t)* with $t > s$ and $k=(s+t)/2$. The parameters $s$ and $t$ represent the $x$-coordinate values where the function assumes the false value (equal to 0) and $k$ represents the $x$-coordinate value where the function assumes the true value (equal to 1);

– if $(s+t) \bmod 2 = 0$

$$y = \begin{cases} 0 & \text{if} \quad x < s \\ \dfrac{x-s}{k-s} & \text{if} \quad s \leq x \leq k \\ \dfrac{t-x}{t-k} & \text{if} \quad k < x \leq m \\ 0 & \text{if} \quad x > t \end{cases}$$

– if $(s+t) \bmod 2 = 1$

$$y = \begin{cases} 0 & \text{if} \quad x < s \\ \dfrac{x-s}{k-s} & \text{if} \quad s \leq x < k \\ 1 & \text{if} \quad k \leq x \leq k+1 \\ \dfrac{t-x}{t-(k+1)} & \text{if} \quad k+1 < x \leq m \\ 0 & \text{if} \quad x > t \end{cases}$$

Lifespan and destination sub-concepts are described using bell-shaped functions. In particular, we have decided to used the Beta Curve function:

$$B(x,\gamma,\beta) = \frac{1}{1+\left(\dfrac{x-\gamma}{\beta}\right)^2}$$

The Beta Curve depends on two parameters, $\gamma$ and $\beta$, which are related as follows: while $\gamma$ defines the point that assumes value 1 in the *y*-axis, $\beta$ settles the size of the bell-shaped curve and it is also used to fix the largeness of the fuzzy set related to $\gamma$.

All these functions depend on some parameters: in order to establish parameters values, it has been necessary to consider the crowd scenario to be studied and collect experimental data and information related to density, growth, lifespan and destination. In the following we will present a little discussion about the concert scenario, showing how to obtain parameters value and how to build appropriate membership functions.

In the next section we will show the use of membership degrees in order to define rules in the ontology to classify crowds.

## *Ontology Implementation*

After the identification of concepts, we have focused on ontology implementation (fourth step). We have decided to use Protégé[2], the standard de-facto editor for ontologies, which supports the OWL language. In order to develop our ontology it has been necessary to consider the following information (Figure 2): concepts, elements and relationships; membership functions (and their fuzzy values); rules based on fuzzy values.

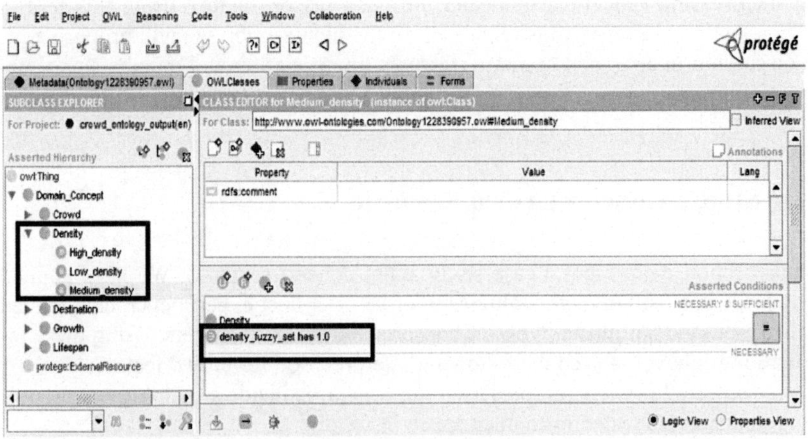


**Fig. 2. The Protégé Interface: concepts and fuzzy values are emphasized on the left and on the right respectively.**

As previously introduced, the main concepts are crowd, density, growth, lifespan and destination: each concept is specialized by its sub-concepts. There is an unambiguous relationship between crowd and features concepts. Moreover, for each concept a set of elements is defined: they are related to observation values (on the studied crowd), parameters values and membership degrees (to be calculated). In order to compute membership degrees, we have implemented an external program written in the Java[3] language, using the OWL API[4] library, that allows setting them in the ontology.

Then a reasoner allows classifying input instances on the basis of the rules showed in Table 1. In order to obtain the classification, we have used Pellet[5], an Open Source OWL DL Reasoner.

---

[2] http://protege.stanford.edu/

[3] http://java.sun.com/

[4] http://owlapi.sourceforge.net/

[5] http://clarkparsia.com/pellet/

Given a specific crowd to be studied , the ontology can be used following three steps:

1. to collect data about density, growth, lifespan and destination and verify the presence of movement, physical and psychological boundaries;
2. to put the observation values (true, false or obserable values) in the ontology (input step);
3. to start the reasoner that produces crowd classification (output step).

The data must be collected according to features interpretation used to establish membership function parameters. For this reason, before using this tool, a study about crowd scenario is requested. In the next section we will propose a little discussion in a concert scenario, features interpretation and membership functions definition.

## Case Study: Concert Crowd Scenario

In this section we present a case study about concert scenario. We have applied knowledge acquisition methodologies in order to collect experimental data related to different kinds of music, types of concerts and types of singers. Using these information we have inferred the following interpretation for crowd features:

- *physical boundaries*: present if the concert is inside a building or absent if concert is located in an open space;
- *psychological boundaries*: present if you must pay to take part in the concert or absent if the concert is for free;
- *movement*: present if people dance and are very involved by music;
- *density* is described as the ratio between people number and the available space;
- *growth* is the number of people who add to the crowd in a given period of time;
- *lifespan* is described as the concert duration;
- *destination* is described as songs average duration.

Starting from this interpretation, we have created different data sets. For each set we have calculated the minimum, medium and maximum value. We have used these values in order to establish membership functions parameters (Table 2).

**Table 2. Parameters values for density and growth concepts**

| | *m* | *n* | *t* | *s* | *w* | *z* |
|---|---|---|---|---|---|---|
| Density | 3 | 2 | 5 | 2 | 4 | 6 |
| Growth | 10 | 5 | 18 | 8 | 15 | 20 |
| | $\gamma min$ | $\gamma medium$ | $\gamma max$ | $\beta min$ | $\beta medium$ | $\beta max$ |
| Lifespan | 1.5 | 5 | 18 | 0.5 | 2 | 6 |
| Destination | 4.3 | - | 2 | 2 | - | 50 |

The parameters evaluation allows to define and represent membership functions: in Figure 3 the membership functions related to the concert scenario are shown. On the left the trapezoidal functions of growth sub-concepts are presented whereas on the right the bell-shaped functions of lifespan sub-concepts are depicted.

After the study of crowd scenario and the definition of membership functions, the tool can be finally executed, inputting the parameters and values derived from the given crowd and obtaining its classification as output.

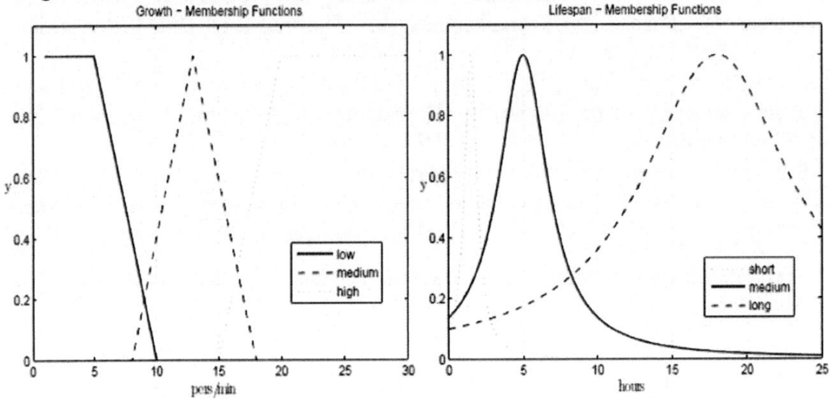


**Fig. 3. Membership functions related to concert scenario.**

## Concluding Remarks and Future Works

In this paper we have presented a knowledge-based tool for the management of knowledge and data in the crowding domain in order to provide a method to share information derived and used from different tools and models. The tool is composed by an ontology and a set of fuzzy rules that allows to classify a crowd on the basis of its own features.

Future works are related to test ontology on different crowd scenarios, using observations collected during events and situations by means of knowledge-based methodologies. At this moment the ontology has been tested on values collected during a concert of an Italian artist, Lorenzo Cherubini alias Jovanotti. The ontology results are significant and encouraging, agreeing with theory formalization [11,12].

## References

1. Timmermans, H. (ed.): Pedestrian Behavior Models, Data Collection and Applications. Emerald Group Publishing Limited (2009)
2. Teknomo, K., Gerilla, P.G.: Pedestrian Static Trajectory Analysis of a Hypermarke. Proceedings of the 8th Int. conference of the Eastern Asia Society for Transportation Studies, November 16-18. Surabaya, Indonesia (2009)
3. Schadschneider, A.: Cellular Automaton Approach to Pedestrian Dynamics – Theory. In: Schreckenberg, M., Sharma, S.D. (Eds.), Pedestrian and Evacuation Dynamics, p. 75. Springer (2001)
4. Morishita, S., Shiraishi, T.: Evaluation of Billboards Based on Pedestrian Flow in the Concourse of the Station. In: El Yacoubi, S., Chopard, B., Bandini, S., Cellular Automata. LNCS 4173, pp. 716-719. Springer (2006)
5. Susi, T.: Passenger Behaviour in Elevator Simulation. In: A. Lustig (ed.) Proc. of Int. Congress on Vertical Transportation Technologies. IAEE publisher (2004)
6. PETS (Performance Evaluation of Tracking and Surveillance) workshop series, http://www.cvg.cs.rdg.ac.uk/
7. Hayes-Roth, F., Jacobstein, N.: The State of Knowledge-Based Systems. Commun. ACM. 37(3), pp. 27-39 (1994)
8. Zadeh, L.A.: Fuzzy Logic = Computing with Words. IEEE Transactions on Fuzzy Systems. Vol. 4, No. 2, pp.103-111 (1996)
9. Canetti, E.: Crowds and Power. The Noonday Press/Farrar, Straus and Giroux (1984)
10. Calegari, S., Ciucci, D.: Fuzzy Ontology, Fuzzy Description Logics and Fuzzy-OWL. In: Masulli, F., Mitra, S., Pasi, G. (eds.), Applications of Fuzzy Sets Theory, Lecture Notes in Computer Science 4578, pp. 118-126. Springer (2007)
11. Bandini, S., Manzoni, S., Sartori, F.: A Fuzzy Ontology for the Classification of Crowds at Concerts. In: Kop, C., Sicilia, M.A., Sartori, F., Dubois, E., Johannesson, P. (eds.): Proceedings of ONTOSE 2009, CEUR Workshop Proceedings.Vol. 460, pp 13-24 (2009)

12. Magnolo, E., Manenti, L., Manzoni, S., Sartori, F.: Towards a MAS Model for Crowd Simulation at Pop-Rock Concerts Exploiting Ontologies and Fuzzy Logic. In: Proc. of 10th workshop from Objects to Agents, WOA 2009, 9-10 July, Parma (2009)

# Towards Realistic Modeling of Crowd Compressibility

**J. Wąs, W. Myśliwiec, and R. Lubaś**

Institute of Automatics
AGH University of Science and Technology
Al. Mickiewicza 30, 30-059 Kraków, Poland
jarek@agh.edu.pl

**Abstract** The article presents a new approach to crowd compressibility modeling in pedestrian evacuation. The model is based on Social Distances Model and implements a compressibility coefficient. The main purpose was to study specific flow of pedestrians through bottlenecks. Real data from two experiments were used to validate received results. Differences in compressibility parameters significantly influence pedestrian behavior and simulation scenarios. Higher values of compressibility coefficient lead to increased densities and flow of pedestrian stream.

## Introduction

Accurate prediction of crowd dynamics makes its possible to organize pedestrian traffic in a safe and effective way. Changeability and non-homogenity of the density parameter enables the observation of local fields of higher or lower density . It is influenced by a few factors, such as: occurrence of subgroups and familiarities [2] and occurrence of density waves [1] (start-stop wave or forward-backward compression waves).

The aim of the article is to propose an evacuation model of crowd compressibility as well as compare simulation outcomes with experimental results.

## Representation

The model proposes an elliptic representation of a pedestrian allocated on square lattice (the center of a cell coincides with the center of a pedestrian) just as the Social Distances Model [3]. Another assumption is that social zones are asymmetric (fig. 1):

R.D. Peacock et al. (eds.), *Pedestrian and Evacuation Dynamics*,
DOI 10.1007/978-1-4419-9725-8_47, © Springer Science+Business Media, LLC 2011

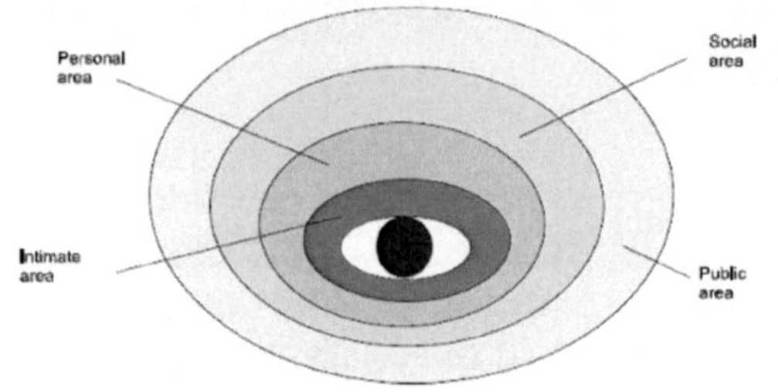


**Fig. 1. The proposed interpretation of proxemics theory – asymmetrical and elliptical social distances. The upper part marks the front of a person**

Parameter ε in the presented model describes pedestrian allocation in space, with the use of allowed and forbidden configurations. Thanks to this parameter, crowd compressibility can be taken into consideration.

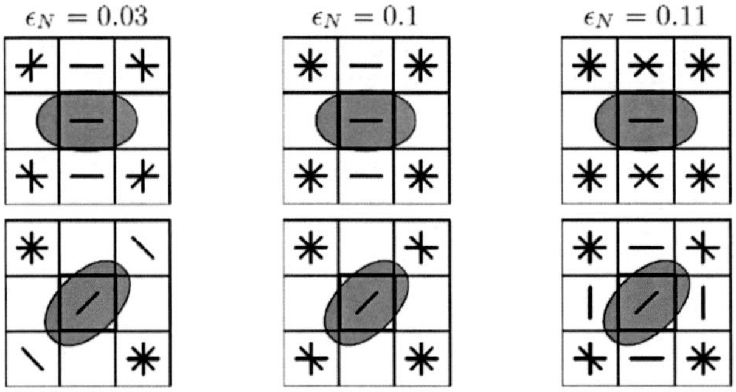


**Fig. 2. Allowed neighborhood configurations for different tolerance parameters**

## Model Description

Taking into account discretization errors, additional measures have been applied to reduce the negative influence on correspondence in reality. One of them is the elliptical model of a pedestrian; the second one is the parameter which specifies the allowed degree of crowd compressibility.

Algorithm treats the shape of a pedestrian as an ellipsis: 45 cm by 27 cm (WHO data), according to Social Distances Model [3]. Realization of the compressibility measure is carried out through a matrix of values, similarly to the potential layer. There are two ways of coefficient ε solving: constant value depending on the distance to exit and two-criterion which is based on both: potential value and local density.

Constant value ε which equals zero models calmly walking pedestrians. In this method all cells of the matrix of ε coefficients have the same value. During the algorithm iteration, after the selection of movement path and after solution of any possible conflicts, the elimination of forbidden configurations takes place. When $\varepsilon = 0.0$, only two configurations are possible (fig.3).

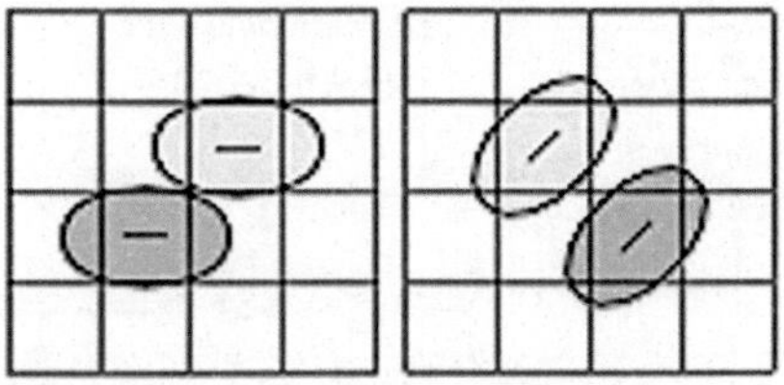

**Fig. 3. Possible configurations while** $\varepsilon = 0.0$

The situation is similar to a calm walk, when average velocity and local densities are characterized by low values.

Two-criterion method takes into account pedestrian distance to the exit changing appropriately the ε parameter. Changes are fluent and happen within the range of 0.0 to 0.2. At the same time, in each algorithm's iteration local densities on the area of the whole room are measured. Sub-areas of one square meter are used for the measurement that also takes into account elliptical pedestrian shapes.

Depending on the local density value, the appropriate ε coefficient within the range of 0.0 to 0.2 is applied. Both those methods complement each other while in use.

Three profiles of pedestrian's behavior have been created based on the application of parameter series, such as: average pedestrian velocity, panic factor, or the ways of calculating crowd compressibility. They reflect the following situations: normal conditions, non-competitive evacuation, and the situation of competitive evacuation.

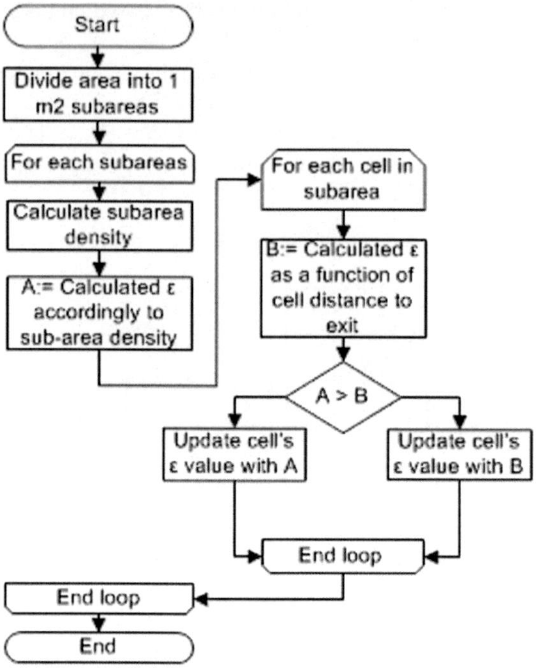


**Fig. 4. Flowchart for ε coefficient calculation**

## Results and Their Validation

Described approaches have been confronted with real data from the two experiments [4, 5].

On the basis of the experiment [4], the room has been modeled together with initial placement of N pedestrians, where N = 31. Due to a cell's size of 25 cm as used in the algorithm, some discretization errors could not have been avoided. The purpose of this simulation was to validate its results with real data and to analyze the influence of ε parameter on the time of evacuation for each pedestrian. There were three simulations carried out, each with different and specific settings. Each of them was conducted for a different scenario: normal conditions, controlled evacuation, and the situation of competitive evacuation.

Each pedestrian's evacuation time was recorded during the simulation. Results are presented in the comparative diagram (fig. 6).

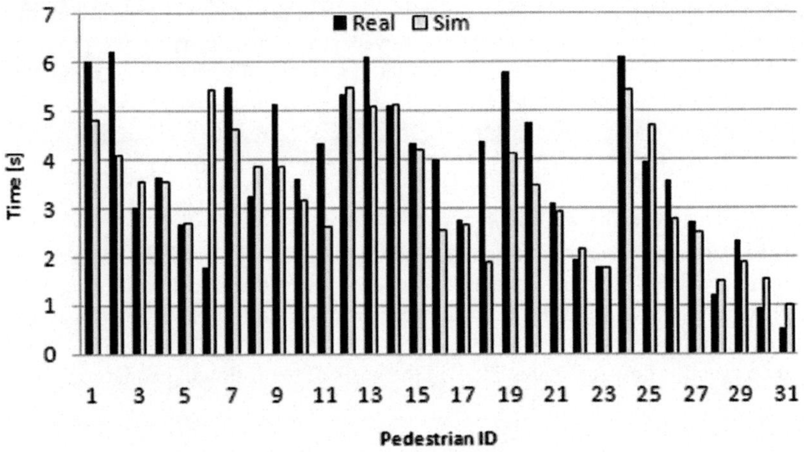


**Fig. 5. Comparison of experiments [4] vs. simulation – controlled evacuation**

The differences in ε value determine pedestrian behavior, consequently also evacuation scenarios. This is demonstrated by received times of evacuation.
For further analysis, a series of simulations have been conducted, which varied only in the way of obtaining ε coefficient. Results are presented in figure 7.

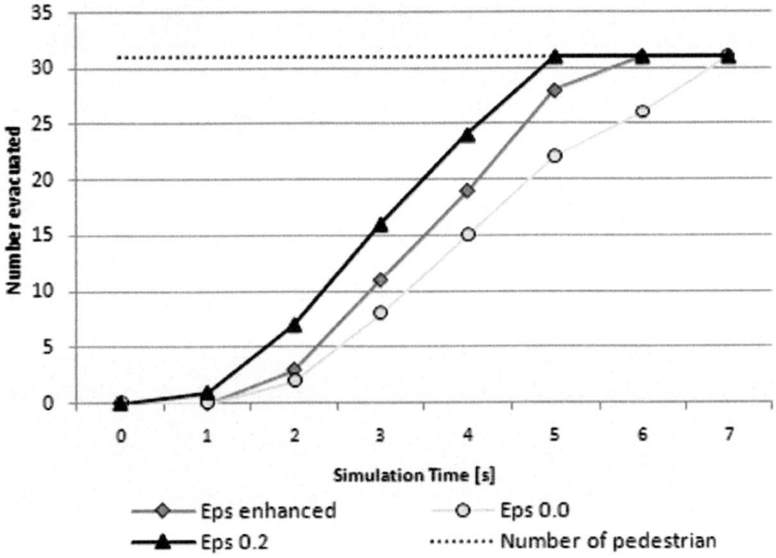


**Fig. 6. Comparison of different ε coefficient values and its influence on evacuation time**

The diagram above shows the diversity of evacuation efficiency depending on the selection of ε value. For ε = 0.0, time and evacuation characteristics are the least favorable. Whereas, the use of constant value ε = 0.2 (and higher) yields shortest evacuation times. The curve for $\varepsilon_{enhanced}$ located between ε = 0.0 and ε = 0.2 confirms the positive influence on accuracy of reality correspondence of pedestrian behavior in the simulation. It is also worth mentioning that constant value ε >= 0.2 leads to excessive local densities build-up, which does not exist in real data in such types of evacuation.

Further simulation series were conducted on the basis of Seyfried's experiment assumptions [5]. Selected cell size of 25 cm enforced the application of three bottleneck widths: 75, 100, and 125 cm. Initial position and density of pedestrians have been maintained.

Obtained specific flow $J_s$ is calculated on the basis of the following equation:

$$J = J_s b\,,\; J_s = \rho v$$

Where ρ is the average density and *v* is the average velocity of pedestrian stream in the bottleneck. Obtained data is shown in the table [tab.1] – the case of normal condition, therefore used ε = 0.0.

**Table 1. The specific flow – $J_s$ [$(ms)^{-1}$] for different bottleneck widths – b [cm]**

| b [cm] | N = 60 | N = 40 | N = 20 |
|---|---|---|---|
| 75 | 0.92 | 1.06 | 1.04 |
| 100 | 0.72 | 0.84 | 0.94 |
| 125 | 0.67 | 0.68 | 0.81 |

For each and every bottleneck width, the increase in flow together with the decrease of the number of pedestrians participating in the simulation is visible. Similar trends are visible in the results of the experiment [5]. Received results let us validate the functioning of the presented model with the aid of a fundamental diagram.

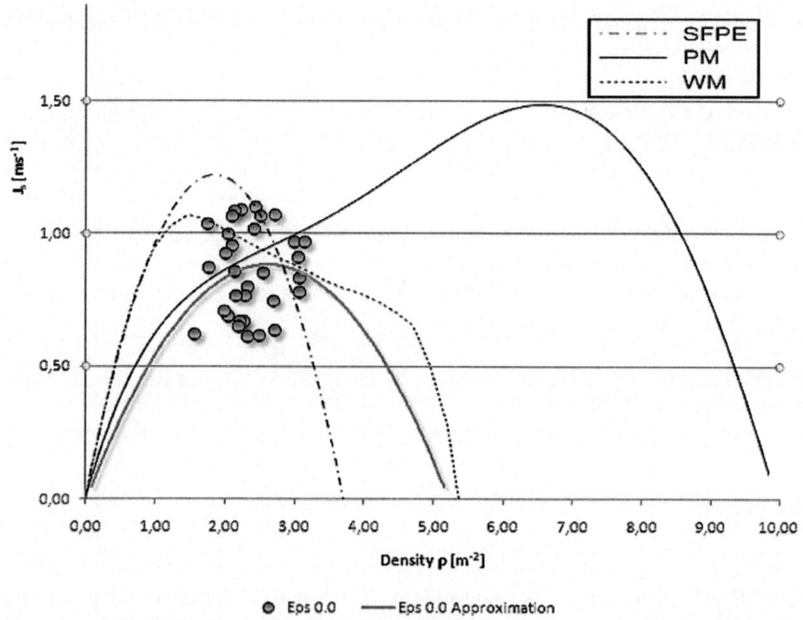


**Fig. 7. Simulation data of flow and associated density in the bottleneck in comparison with fundamental diagrams for pedestrian movement through openings and doors according to the SFPE Handbook (SFPE) and the Guidelines of Predtechenskii and Milinski (PM) and Weidmann (WM) [6],[7],[8]**

## Conclusions

The parameter of compressibility proves to be significant factor in describing various evacuation situations such as: normal situation, non-competitive evacuation, competitive evacuation, or panic (experiments from [4]). Growing pedestrian nervousness in these situations is accompanied by increasing crowd density in the neighborhood of bottlenecks (exits). At the initial stage of pedestrian mobilization, growing compressibility exerts positive influence on evacuation time. However, with the situation becoming competitive, the possibility of individual motion control for each pedestrian can be partially (or sometimes completely) lost. In such a case, the generated pressure can endanger pedestrians' health or life.

To implement models of compressibility for different situations, authors have leant on Social Distance model [3]. During evacuation, pedestrian social zones are smaller than in normal situations. Pedestrians are oriented towards a realized target.

Different values of compressibility parameter were taken into account. Simulations have confirmed that the growth of ε parameter increases evacuation effectiveness. Another tested solution was dynamically changed ε, when value of the parameter was increased towards gradient of potential field. Gained results show that for different values of e parameter we can receive different statistics of evacuation. For example: the result of evacuation effectiveness for $\varepsilon_{dyn}$ was better than for $\varepsilon = 0$ and worse than for $\varepsilon = 0.2$ (fig. 3).

The presented approach needs further validation and it can be helpful for different evacuation scenario modeling.

**Acknowledgments** This research is financed by the Polish Ministry of Science and Higher Education, Project no: N N516 228735

## References

1. Helbing, D., Johansson, A., Al-Abidem, H.: Crowd Turbulence: the physics of crowd disasters. Conference on Nonlinear Mechanics, 967 669 (2007)
2. Singh, A., Arter, R., Louise, D., Langston, P., Lester, E., Drury, J.: Modelling subgroup behavior in crowd dynamics DEM simulation. Applied Mathematical Modelling (2009). doi:10.1016/j.apm.2009.03.020
3. Wąs, J., Gudowski, B., Matuszyk, P.J.: Social Distances Model of Pedestrian Dynamics. International Conference on Cellular Automata For Research & Industry, Lecture Notes in Computer Science. Vol. 4173. Perpignan, France (2006)
4. Wąs, J.: Experiments on evacuation dynamics for different classes of situations. Conference on Pedestrian and Evacuation Dynamics. Wuppertal (2008) (in print)
5. Seyfried, A., Passon, O., Steffen, B., Boltes, M., Rupprecht, T., Klingsch, W.: New Insights into Pedestrian Flow Through Bottlenecks. Transportation Science. **43**, 395-406 (2009)
6. Predtechenskii, V.M., Milinskii, A.I.: Planning for Foot Traffic Flow in Buildings. Amerind Publishing, New Dehli, India (1978)
7. Hoogendoorn, S.P, Daamen, W.: Pedestrian Behavior at Bottlenecks. Transportation Science. Vol. 39, No. 2, pp.147-159 (May 2005)
8. Nelson, H.E., Mowrer, F.W., DiNenno, P.J.: "Emergency movement." SFPE Handbook of Fire Protection Engineering, p. 367. National Fire Protection Association, Quincy (2002)

# Modeling Calibration / Validation

# Towards Automatic and Robust Adjustment of Human Behavioral Parameters in a Pedestrian Stream Model to Measured Data

**M. Davidich and G. Koester**

Siemens AG, CT PP2, Virtual Design
Otto-Hahn-Ring, 6, 81739 Munich, Germany
Corresponding author: maria.davidich@siemens.com

**Abstract** People die or get injured at mass events when the crowd gets out of control. Urbanization and the increasing popularity of mass events, from soccer games to religious celebrations, enforce this trend. Thus, there is a strong need to find a better means to control crowd behavior. Here, simulation of pedestrian streams can be very helpful: Simulations allow to run through a number of scenarios in a critical situation and thereby to investigate adequate measures to improve security. In order to make realistic, reliable predictions, a model must be able to reproduce quantitatively the data, known from experiments. Therefore, automatic and fast calibration methods are needed that can easily adapt model parameters to different scenarios. Also, the model must be robust: Small changes in the crucial input parameters must not lead to large changes in the simulation outcome. In this paper we represent two methods to automize the calibration of pedestrian simulations. We then introduce a concept of robustness to compare the two methods. In particular, we propose a quantitative estimation of parameter quality and a method of parameter selection based on robustness criteria.

## Introduction

Larger and larger pedestrian crowds can be observed daily: in so-called critical infrastructures such as subways and railway stations, at airports, in shopping malls and high rise buildings, and at different mass events. This phenomenon entails problems of comfort and safety that become more and more pressing.. Therefore, adequate and well-organized crowd management is necessary. The aim of crowd management is to avoid crowd disasters in which people are injured or even killed by crushing, trampling or suffocation. Today, crowd management is usually done by careful planning. Emergency plans are based on previous experience. However, experience does not cover all possible future scenarios. Here mathematical models and simulation tools can provide virtual experience, where real experience is missing. Simulations allow to run through a number of scenarios and to observe the outcome. Decisions for a larger number of emergencies can be based on this.

R.D. Peacock et al. (eds.), *Pedestrian and Evacuation Dynamics*,
DOI 10.1007/978-1-4419-9725-8_48, 

However, to make a crowd simulations an effective tool for, say security staff, especially during the crisis, it must be several times faster than real time even in a complex scenario and reproduce crowd phenomena and empirical observations not only qualitatively but also quantitatively. This high simulation speed can be achieved, even on off-the-shelf hardware, with a cellular automaton model. For quantitative reproduction of observations, the simulation parameters must be adjusted to the measured data, that is, they must be calibrated.

Recent progress in calibrating pedestrian stream models has encouraged the authors to reconsider this problem. According to [1-4] calibration can be achieved in principal. However, up to now, automatic algorithms have been discussed for social force models, only. In this paper we present two approaches to automize calibration for a *cellular automaton* model and introduce a new quantitative criterion to measure the robustness of each calibration. It turns out, that the optimal calibration parameter sets derived by the two methods do not only differ in how well the subsequent simulations capture the measured phenomenon, but also in how sensitive the simulation model reacts to small changes in the parameter sets. This leads us to the concept of *sensitivity* or, equivalently, *robustness* of a calibration as an additional decision criterion for the optimal choice of parameters.

Our paper is organized as follows. The next section will shortly introduce our cellular automaton model. Then the automatic calibration and a method to select a robust parameter set are described and results are presented. A brief outlook on future potential concludes the paper.

## A glance at the model

We choose a cellular automaton (CA) model for two reasons. First, we are interested in large-scale pedestrian simulation in a fraction of real time. CA models have proven to show faster-than-real-time speed [5-7] even for large systems. Furthermore, cellular automata provide an intuitive representation of interactions between entities that can be incorporated in a very simple way.

In a cellular automaton model, the whole area of interest is covered by cells: In our case by, hexagonal cells. Each cell, at each time step, is either empty or occupied by a person, a piece of a wall or an obstacle. We choose a cell size to accommodate an average sized European male. Persons appear through so-called sources and disappear when they reach their targets, namely, entrances and exits. The core of the model is in the 'automaton', that is, a set of rules, according to which the states of the cells are updated in time. Here, we borrow ideas from electrodynamics. Pedestrians are treated as electrons, which are attracted by positive charges, that is, targets, and repelled by negative charges, such as other pedestrians or obstacles. The forces between pedestrians, targets and obstacles are calculated via a potential field, using the properties of conservative force fields from physics in which the force can be expressed as the gradient of a suitable scalar function: the potential. This model resembles other cellular automaton models based on potentials as, for example described in [7-14].

This type of a cellular automaton has been widely used for pedestrian movement modeling, effectively capturing collective behavior. However, not much attention has been paid to qualitative validation, which is vital to make the model useful for applications. In particular, calibration of the model is of the utmost importance. Therefore we will focus our model description on those parameters that the calibration algorithm needs.

We choose the walking speed for calibration. Its impact on crowd movement is immediate and obvious. It is, furthermore, directly accessible through experiments and measurements. Each person has an individual speed that the person tries to achieve – and in deed does achieve when the path is free: the free flow velocity [5, 14]. The distribution of the speeds follows the suggestions in [15]. That is, we assume a normal distribution about some medium desired speed. Different velocities are made possible by allowing a person to move forward multiple cells per simulation step. However, there is another very important phenomenon that must be captured: It is known that pedestrians slow down as the pedestrian density increases. This dependency of pedestrian velocities – or alternatively flow – on crowd density is quantitatively described through so-called *fundamental diagrams*. Fundamental diagrams can also be considered as a behavioral model that aggregates a multitude of socio-cultural and even scenario dependent parameters [16]. Ultimately, the differences in – say gender, nationality, fitness – find their expression in the way people walk as individuals and surrounded by a crowd: Obviously, fundamental diagrams differ considerably. For our calibration, we choose the most widely used diagram, that is, the one described by Weidman. However, taking the Weidman diagram as a reference, one notices, that in most cases, pedestrians move too fast in the cell automata-based model and are "short sighted". They do not decelerate before they literally "bump" into a dense crowd. Therefore, velocities have to be calibrated according to the fundamental diagram relevant to the situation.

## Calibration

The main challenge is to decelerate pedestrians according to a fundamental diagram while preserving individual differences in speed. We propose the following two criteria to individually adapt each pedestrian's speed: The number of persons in the field of vision (local density) and the person's individual speed. Note that the discrete nature of the model leads to a finite number of possible densities and speeds, and thus, to a finite number of adjustment parameters, which we call *deceleration classes*. Each person's velocity is then reduced individually. This preserves the individual differences in speed. With a well calibrated set of deceleration classes a given fundamental diagram can be faithfully reproduced. For a detailed description of the approach please refer to [3].

Due to the discrete nature of the model, the parameter adjustment can be reformulated as a multidimensional optimization problem on discrete sets. In general, one may apply any number of different optimization methods to solve this problem: evolution strategies, extremal optimization, memetic algorithms, reactive search optimization, simulated annealing, tabu search and many more. We choose two algorithms, a genetic and threshold accepting algorithm, since they are resource efficient and easily adapted to our problem. Both speed up previous manual off-line calibration by, roughly, a factor of fifty.

We also propose a scalar measure, *fitness*, to quantitatively estimate the parameter quality. It indicates how well the behavior obtained from simulation runs with a particular parameter set approximates the measured fundamental diagram:

$$Fitness = 1/\sum(S_i - R_i)^2,$$

where $\boldsymbol{S_i}$ is a value obtained from simulation runs, and $\boldsymbol{R_i}$ is a value from the reference curve, the fundamental diagram, for the same argument. That is, $\boldsymbol{R_i}$ is the *i*-th sample density. The values of $\boldsymbol{S_i}$ and $\boldsymbol{R_i}$, and with that of the fitness, depend on scaling, e.g. the unit of measurement (cm, m, ...).

Both of optimization algorithms, genetic and threshold accepting, deliver a set of well adjusted parameter values, capturing Weidmann's fundamental diagram with a good fitness of 2,36 and 2,66 respectively (Fig.1, 2). We successfully applied the proposed algorithms to calibrate the model according to further measured data, such as "London subway", "rush hour" [17].

While fitness may be essential for a good selection of parameter values, it is by no means sufficient: Real life applications have to deal with ever changing situations. So another crucial question is: How will changes in a fit parameter choice affect the results? Will the impact on fitness be significant? Obviously, from two parameter sets with similar fitness the one is preferable that experiences the smaller effects when slightly altered. This brings us to the question of robustness of parameter values.

## Robustness: a general procedure to judge parameter robustness

In order to explore the robustness, one needs to investigate the response of the system, in particular, the changes in fitness, when parameter values are disturbed. More precisely, we aim to estimate the sensitivity, or equivalently the robustness, for a given set of parameters, or deceleration classes, $(p_1, ..., p_n)$ with fitness $f_p$. For this we disturb the parameter set in the following way: Each value, $p_i$ is changed to the next bigger one $p_i^+$ and then to the next smaller one $p_i^-$. Note that our deceleration classes are discrete, so that choosing the next bigger and the next smaller value makes sense. For each $p_i$, we then determine the fitness $f_{p,i}^+$ of the parameter set $(p_1,...p_{i-1}, p_i^+, p_{i+1}...,p_n)$, where $p_i$ has been augmented to the next discrete bigger value $p_i^+$ and all other parameters values $p_j$, $j \neq i$, have been left undisturbed. The same is repeated for $p_i^-$.

Then for all $i$, varying from $1$ to $n$, the *mean local fitness* of the parameter set locally disturbed in $p_i$ is calculated as follows:

$$f_{p,i}^{local} = (f_{p,i}^+ + f_{p,i}^-)/2.$$

Now we propose a measure of local sensitivity in $p_i$:

$$s_{p,i}^{local} = |(f_p - f_{p,i}^{local})|/f_p,$$

where $f_p$ is the fitness of the undisturbed parameter set $p$.

A measure of local robustness in $p_i$ is given by the inverse of the local sensitivity:

$$r_{p,i}^{local} = 1/s_{p,i}^{local}$$

The next step is to summarize information on the response of the system if any one of the parameter values has been changed. An intuitive way is to determine an overall disturbed fitness $f_p^{disturbed}$:

$$f_p^{disturbed} = (f_{p,1}^{local} + ... + f_{p,n}^{local})/n.$$

Based on this, we propose a sensitivity measure

$$s_p = |(f_p - f_p^{disturbed})|/f_p$$

where $f_p$ is the fitness of the undisturbed parameter set. We divide by $f_p$ to make different choices of parameter sets $p=(p_1,...,p_n)$ and $q=(q_1,...,q_n)$ more comparable. The corresponding measure of overall robustness is:

$$r_p = 1/s_p$$

The concept can be enlarged by varying the parameters $p_i$, $(i=1,..,n)$ by not only one value (augment to the next bigger or decrease to next smaller) but by 2 or more values simultaneously. It can be generalized by disturbing more than one parameter at a time. This makes sense to capture interdependencies between parameters. Generally speaking, one disturbs $k$-tuples taken from $p=(p_1,\ldots,p_n)$, with $k=1,\ldots,n$. Altogether, for each $k$, there are $n!/(k!(n-k)!)$ possible $k$-tuples. The disturbed fitness, for each choice of $k$, is then calculated by going through all possible selections and calculating the mean.

$$F_p^{k,\ disturbed} = (f_1^{\ local} + f_2^{\ local} + \ldots f_{k!\ (n-k)!/n!})\ k!\ (n-k)!\ /\ n!$$

For every particular choice of *k-tuples* of disturbed calibration parameters, one can calculate the corresponding sensitivity:

$$S^k_p = (f_p - F_p^{k,\ disturbed})/f_p$$

Finally, we introduce a more general overall disturbed fitness of a parameter set:

$$F_p^{disturbed} = (f_p + F_p^{1,\ disturbed} + \ldots + F_p^{n,\ disturbed})/(n+1).$$

This is a mean value of the fitness, when all possible $k$-tuples of calibration parameters, from 1-tuples to $n$-tuples, are disturbed. The fitness $f_p$ of the undisturbed parameter set $p$ is included to give a relation to the initial precision of the model.

The measures proposed here are immediately useful to compare parameters sets of similar fitness and to detect very sensitive and very robust parameter choices among parameter sets of similar fitness. In addition, we suggest the use of thresholds to group parameter sets in clusters of similar fitness and/or robustness, to make practical comparison possible and to exclude inacceptable parameter choices. E.g. parameter sets with fitness below a certain threshold can be excluded. E.g. among parameter sets with similar fitness, the more robust can be chosen and among parameter sets with similar robustness, the fittest can be chosen.

Based on the measures introduced above, we suggest the following automatic selection method. Note, that if one needs to select one parameter set among several sets, it is enough to compare pairs. Therefore, without loss of generality, we describe our selection method comparing two parameter sets $(p_1,\ldots,p_n)$ and $(q_1,\ldots,q_n)$. The selection criterion is as follows:

1. Selection based on fitness only: If the initial fitness $f_p$ and all local fitness values, $F_p^{k,\ disturbed}$ for all $k=1,..n$, are higher than for parameter set $q$, that is, $f_p > f_q$ and $F_p^{k,\ disturbed} > F_q^{k,\ disturbed}$, then select parameter set $p$. If the inverse is true for $q$, select $q$.
2. Else, selection based on sensitivity:
   i. Selection on sensitivity only: Calculate the sensitivity $S^k_p$ and $S^k_q$ for each choice of $k$. If for all $k$, $S^k_p < S^k_q$, select $p$. If the inverse is true for $q$, select $q$.

ii. Else, selection based on the generalized overall fitness: Compare $F_p^{overall}$ and $F_q^{overall}$ and select the parameter set with the bigger fitness.

If the number of parameters *($p_1$... $p_n$)* is large (*n* large) it may not be economical to go through all possible disturbances. Then we suggest restricting the selection method to disturbances of one at a time or pairs of parameters.

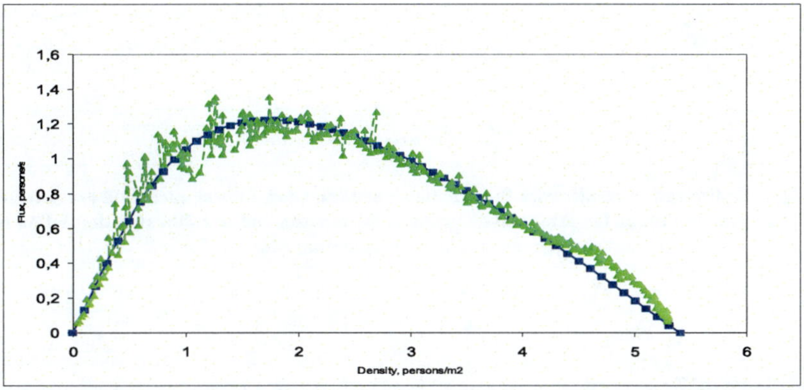


**Fig. 1. Optimization results with the genetic algorithm. Dark blue squares and green triangles correspond to the fundamental diagram by Weidmann and to calibrated simulation results respectively. The corresponding fitness is 2,35.**

## *Application to the pedestrian stream simulation*

Let us now apply the proposed procedure to compare the parameter values obtained for the pedestrian stream model from the genetic and the threshold accepting algorithms. The initial fitness of the simulations calibrated with the threshold accepting algorithm is a little higher, than the one for the genetic algorithm ($f_{threshold\_accepting} > f_{genetic}$). Figures 1 and 2 suggest that both methods yield a good calibration. However, when we disturb the deceleration classes, we clearly obtain better fitness values for the genetic algorithm (Fig. 3). Therefore, the sensitivity analysis is the next step. Fig. 4 shows the results: The parameters obtained from the genetic algorithm prove to be more robust.

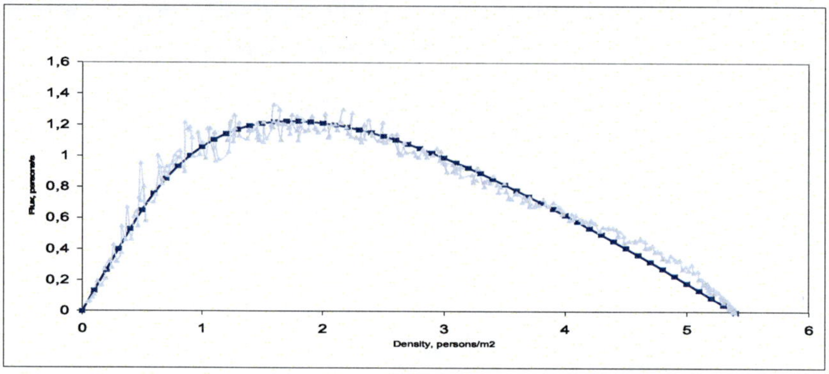


**Fig. 2. Optimization results with the genetic algorithm. Dark blue squares and green triangles correspond to the fundamental diagram by Weidmann and to calibrated simulation results respectively. The corresponding fitness is 2,66.**

## Conclusions

For practical applications of pedestrian stream simulations, reliable quantitative predictions are required. In order to fulfill these requirements, a model for pedestrian simulation must be calibrated and robust. In this paper, we propose two methods to automatically calibrate pedestrian stream simulations and a method to select robust parameter sets. The two fast automatic calibration methods rely on a genetic and a threshold accepting algorithm. The walking velocity of the pedestrians is then adjusted according to the optimal values found in the automatic calibration so that a given fundamental diagram is faithfully reproduced.

To achieve robustness we introduce a method to first measure and then assess the sensitivity of our calibration. We are thus able to select robust parameter sets so that small deviations from the optimal parameter set will not lead to disproportionally large changes in the simulation results. The method systematically disturbs parameter sets, then measures and compares the sensitivity of the system's reaction. The approach is fast and flexible. It allows easy automatic adaptations of the model to social-cultural behavior and changing circumstances. It can be applied to a wide class of models. Future developments will focus on analysing real time data and will demand fast model adaptations to situational changes, e.g in the walking behavior. Thus, automatic and robust calibration, as suggested here, is an important step towards real time application of pedestrian stream simulations.

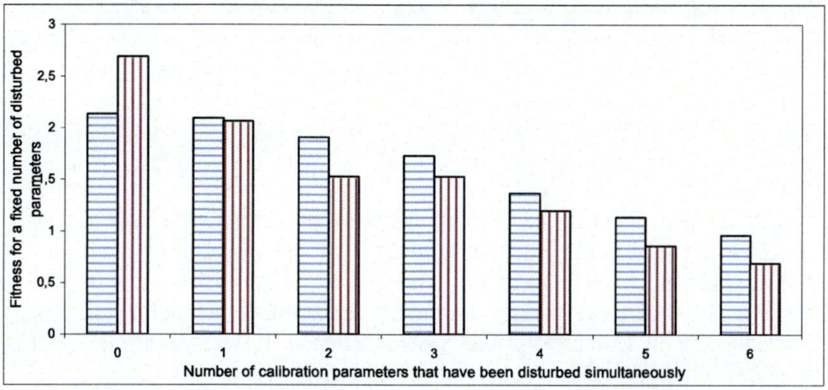


**Fig. 3.** $F_p^{k\ disturbed}$ **for up to 6 deceleration classes disturbed simultaneously. Horizontal/vertical lines correspond to optimal parameter sets from a genetic /threshold accepting algorithm respectively.**

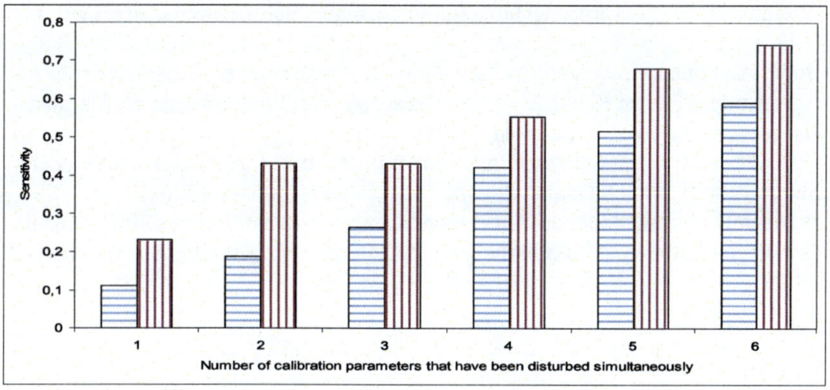


**Fig. 4. Comparison of the sensitivity $S^k_p$ of optimal parameter sets obtained by two different optimization algorithms genetic (horizontal lines) and threshold accepting algorithm (vertical lines).**

## References

1. Klein, W., Koester, G., Meister, A.: Towards the Calibration of Pedestrian Stream Models, Lecture Notes in Computer Science: PPAM 2009. Springer, Verlag (2010)

2. Hoecker, M., Milbracht, P.: Genetic Algorithms as a Means of Adjusting Pedestrian Dynamics Models. Proceedings of the First International Conference on Soft Computing Technology in Civil, Structural and Environmental Engineering (CSC). Funchal, Madeira (2009)
3. Johansson, A., Helbing, D., Shukla, P.K.: Specification of the Social Force Pedestrian Model by Evolutionary Adjustment to Video Tracking Data. Advances in Complex Systems. Advances in Complex Systems. **10(4)**, 271–288 (2007)
4. Nagel, K., Schreckenberg, M.: A Cellular Automation Model for Free-way traffic. J. Phys. I France. 2, 2221-22229 (1992)
5. Schadschneider, A., Klingsch, W., Kluepfel, H., Kretz, T., Rogsch, C., Seyfried, A.: Evacuation Dynamics: Empirical results, Modeling and Applications. Encyclopedia of Complexity and System Science, Robert A. Meyers (Ed.), ISBN 978-0-387-75888-6, Vol. 3, pp. 3142. Springer (2009)
6. Hamacher, H.W., Tjandra, S.A.: Mathematical Modelling of Evacuation Problem: A State of the Art. Springer, Verlag (2002)
7. Burstedde, C., Klauck, K., Schadschneider, A., Tittartz, J.: Simulation of Pedestrian Dynamics Using a 2-dimensional Cellular Automation. Physica A. **295** (4), 507-525 (2001)
8. Weidmann, U.: Transporttechnik für Fussgänger. Schriftenreihe des IVT, 90 (1992)
9. John, A., Schadschneider, A., Chowdhury, D., Nishinari, K.: Characteristics of Ant-inspired Traffic Flow: Applying the social insect metaphor to traffic models. Swarm Intelligence. **2,** 25-41 (2008)
10. Kinkeldey, C.: Fussgängersimulation auf der Basis zellularer Automaten: Studienarbeit im Fach Bauinformatik. Universität Hannover (2003)
11. Rogsch,C.: Vergleichende Untersuchungen zur dynamischen Simulationen von Personströmen. Diplomarbeit an der Bergischen Universität Wuppertal (2005)
12. TraffGo.Handbuch http://www.trafgoht.com/de/pedestrians/downloads/index.html.

# Comparing Pedestrian Movement Simulation Models for a Crossing Area Based on Real World Data

**D. Bauer**

Austrian Institute of Technology,
Department Mobility
Giefingg. 2, A-1210 Vienna, Austria
Corresponding author: Dietmar.Bauer@ait.ac.at

**Abstract** In this paper, two different pedestrian movement simulation models (a model of the social force type and a queuing network model) are compared with respect to their capability to predict individual walking times in a crossing area. Both models are calibrated using a trajectory data set and their relative performance on the estimation data set as well as on a separate validation data set is discussed. The social force type model is found to better predict the walking times as well as space usage in both in- and out-of sample comparison.

## Introduction

Public infrastructures such as junctions of mass public transport need to be carefully designed in order to avoid overcrowding and congestion and thus to ensure the safety and convenience of the passengers. Consequently the design process needs tools in order to assess the typical usage patterns given the demand for the infrastructure. Pedestrian movement simulation models (PMSMs) constitute a main component in this respect as they allow predicting space usage as well as delays due to congestion in the infrastructure.

During the past decade a multitude of different PMSMs have been proposed. Many of these models describe the movement of individuals such as the cellular automata models [1], social force models [2], models based on utility maximizing behaviour [3] or mesoscopic models such as the model proposed in [4], to cite just a few of the many suggested specifications.

This multitude of modeling paradigms and model specifications is to date not countered by sufficient knowledge on the relative merits of the various approaches. This holds true both in terms of computational burden involved as well as with respect to predictive accuracy. While there are a number of papers justifying the extension of previously proposed models by hinting at certain deficiencies [5-9] there seems to be not much literature on a comparison of different modeling paradigms. The only citation in this respect appears to be [10]. This lack of quantita-

R.D. Peacock et al. (eds.), *Pedestrian and Evacuation Dynamics*,
DOI 10.1007/978-1-4419-9725-8_49, © Springer Science+Business Media, LLC 2011

tive comparisons is complemented with prejudices such as cellular automata being numerically much faster than social force based models which are typically seen to be more accurate.

This paper adds to the knowledge on the relative merits of two modeling paradigms. To this end a simple but relevant real world scenario is chosen and a corresponding real world trajectory data set is obtained. The scenario subsequently is modeled using a social force type model and the mesoscopic model of [4]. Hereby part of the data set is used for calibration and a part is retained for model validation. These two models will be compared with respect to their accuracy in replicating (on the estimation data set) and predicting respectively (on the validation data set) the distribution of travel times needed to cross the study area as well as a quantitative measure of space usage.

The paper is organized as follows: In the next section the scenario is described and the data collection process detailed. Afterwards the two models used and their respective calibration is discussed. Subsequently the results for the two models are compared. Finally some conclusions are given.

## Scenario and Data Collection

In this paper a comparatively simple but relevant situation is investigated: **Fig. 1** (a) shows a screenshot of the rectangular area considered which is located inside the upper level of the main hall of the Westbahnhof in Vienna. The area measures 10m square and lies in front of a set of escalators and stairs leading down to the lower level. Four main traffic streams are observed crossing the area: Passengers coming from the tracks continue to the lower level which provides access to local trams and taxis. At the left side of the image the access to the underground station is to be found while the right end of the image leads to snack stands and another taxi stand. There are no fixed obstacles interfering with movement inside the area.

Visual inspection of the video shows that while most passengers (559 out of 569 observed) only cross the area, some (the remaining 10) stop to orientate themselves or wait for connecting trains or friends. The scene contains periods of low densities in which passengers can cross the area without changing direction or speed in order to avoid collisions as well as periods of considerably higher densities where changes in speed and direction occur frequently.

Fifteen minutes of video coverage have been obtained on Sunday March 30$^{th}$, 2008 (the Sunday following Easter Sunday, a highly frequented day) starting at 18:17. The video sequences have been manually annotated using the system VIPER (viper-toolkit.sourceforge.net/) which allows tracking of persons by fitting a rectangle over the outline of the body. The middle point of the lower end of the rectangle is used as the position of the person reducing the impact of body sway. Tracking then is obtained by moving the rectangle from frame-to-frame in order to follow the passenger. As a result for each individual a sequence of (time stamp, position) pairs is obtained constituting the raw data set.

(a) (b)

**Fig. 1. Observed region (a) and trajectories (b)**

The resulting movement trajectories for all persons crossing the study area is shown in **Fig. 1** (b). A total of 569 trajectories have been observed. The trajectories subsequently have been converted to real world coordinates, smoothed coordinatewise using smoothing splines (using the MATLAB function `csaps` with smoothing parameter 0.99) and resampled (sampling frequency 1/18 seconds) in order to provide the data set used in the analysis.

The fifteen minutes of observations are partitioned into two sets: The first 10 minutes are used for the calibration of the models while the last 5 minutes are kept as a validation sample.

## Social force type model

The social force model proposed in [2] is a so called microscopic PMSM which describes the movement of every individual as the result of goal directed behaviour. Pedestrians react to the geometrical layout of the infrastructure and the position and movement of the other pedestrians present. In general terms the social force paradigm explains the directed acceleration $a_i(t)$ of a pedestrian $i$ at time $t$ (i.e. the change of velocity) as a function of the deviation of the velocity $v_i(t)$ from some desired velocity $v_i^d(t)$, the distance to obstacles (given as the difference between the current position $p_i(t)$ and the closest point $w_o(t)$ on the obstacle or wall for all obstacles or walls o) and the relation to other pedestrians (characterized using the vector $x_j(t)$):

$$a_i(t) = f(v_i^d(t) - v_i(t)) + \sum_o f_o(p_i(t) - w_o(t)) + \sum_{j,j \neq i} f_{rep}(x_i(t), x_j(t)) \quad (1)$$

Here the first sum ranges over all obstacles and the second over all pedestrians. Since the acceleration is the first derivative of velocity which itself is the first derivation of the position, these equations define a system of ordinary differential equations. Together with the imposition of boundary values specifying where and when pedestrians enter the simulated region the numerical solution to the system of differential equations provides the simulated trajectories.

Reference [2] provides specifications for the functions $f$, $f_o$ and $f_{rep}$, which have been adapted based on real world data in a number of subsequent papers (i.a. [7-9,5,11]). Also the values of the parameters contained in these formulas have been challenged [5]. Typically the desired velocity $v_i^d(t)$ points into the direction of the desired exit point and the corresponding speed equals some preferred value (sometimes called free speed) which hence varies between individuals. Additionally in (1) terms modeling the attraction to persons or places and random fluctuations have been suggested but are not typically used (see e.g. [2]).

In this paper the system of differential equations is solved using a forward Euler scheme with constant time step translating the differential equations into difference equations. Somewhat arbitrarily a constant time step of 1/18 second is used. Different time steps in the interval [1/36,1/6] have been tested and the differences are minor. Calibration has been performed using the first 600 seconds of trajectory data where the objective is to minimize the mean absolute deviation (MAD) of simulated travel times from measured travel times. A total of 6 parameters are contained in the model (i.e. fit to the specific purpose, see below). From a mathematical viewpoint the calibration amounts to the minimization of a nonlinear function which is implicitly given by the simulation. Minimization is performed using the MATLAB function fminsearch implementing a Nelder-Mead algorithm. Calibration resulted in the following specification of the various terms in (1):

- $v_i^d(t)$: For each passenger the points where the observed area is entered and left is extracted from the movement trajectory. The desired velocity is always directed towards the exit point. For the desired speed the resampled trajectory is used in order to calculate the distribution of the instantaneous speeds (i.e. the speed corresponding to the distance travelled within each time step in the trajectory). Desired speed then is defined as the q% percentile of the speed distribution. For the particular data at hand q was estimated to equal q=52.8.
- $f(x)=g(\|x\|)\frac{x}{\alpha}$: Calibration of the model on a different data set (which will be discussed elsewhere) led to the specification of the nonparametric function g(x). A graph of the function g is provided in **Fig. 2** (a). The function implies that small deviations are not corrected much. $\alpha$=0.013 led to the best results.
- Since there are no obstacles the second term is not included.
- The function $f_{rep}$ has the following form:

$$f_{rep}(p_i, p_j, v_i, v_j) = f_{rep,1}(p_j - p_i)(\|v_i - v_j\| + 0.1) + f_{rep,2}\left(\varphi_{ij}, \frac{r_{ij}}{\|v_i - v_j\|}\right)$$
$$f_{rep,1}(x) = \frac{x}{\|x\|}(\theta_1(\theta_2 - \|x\|) + \theta_3((\theta_2 - \|x\|)^2)I(\|x\| \le \theta_2), \tag{2}$$

Here $p_i$ and $v_i$ denote the current position and velocity respectively of pedestrian $i$, $d_{ij}$ the distance between pedestrian i and j. Further $r_{ij}$ denotes the displacement of the two pedestrians, if their position is forecasted (assuming constant velocity) for $\theta_4$ seconds. $\varphi_{ij}$ denotes the angle between the forecasted displacement and the relative velocity. $f_{rep,2}$ is only nonzero, if $r_{ij}$<4 and if the angle between the forecasted displacement and the current velocity of pedestrian $i$ is smaller than 90° such that the forecasts predict that pedestrian $i$ will "walk through" pedestrian $j$.
The first term of the repulsive force here tends to separate pedestrians that are currently too close. It is proportional to the relative velocity and a quadratic function of the distance. The parameter estimates are $\theta_1$=2.44, $\theta_2$=1.62 and $\theta_3$=0.80 such that the force acts within approx. 1.6 m.
The second term implements forward looking behaviour where $\theta_4$ is estimated to equal approx. 2 seconds.

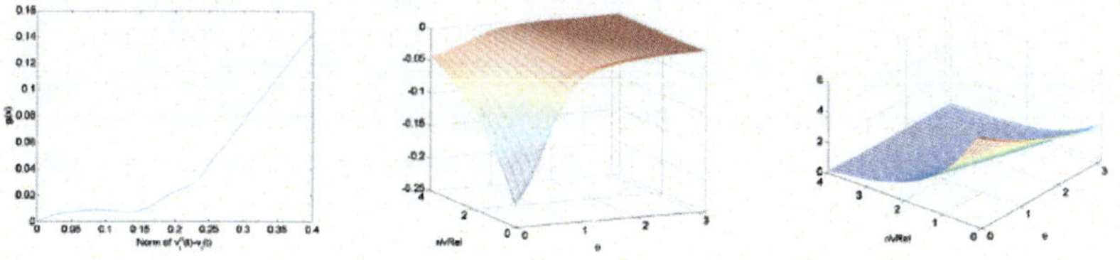

(a) (b) (c)

**Fig. 2. (a) Graph of $g(x)$. (b) First component of $f_{rep,2}$. (c) Second component of $f_{rep,2}$**

The form of the nonparametric function $f_{rep,2}$ has been obtained on a different data set (due to space limitations details will be presented elsewhere). It is composed of a component in the direction of the relative velocity (**Fig. 2** (b)) and one perpendicular to it (see **Fig. 2** (c)). If a collision is predicted the largest acceleration occurs in the direction perpendicular to the relative velocity such that for opposing flows the pedestrian takes a sidestep. Acceleration decreases with increasing ratio of expected distance to relative speed and hence with increasing distance or decreasing relative speed respectively. Additionally there is a component of smaller size reducing relative speed.

- Boundary conditions: Every pedestrian enters the simulated area at the point and the time according to the real data. The initial velocity equals the desired velocity.

- The speed of every pedestrian is restricted to be smaller than the desired speed.

## Mesoscopic model

The social force model is compared to a mesoscopic model which is a slight adaptation of the model described in [4]. In the current context the mesoscopic model acts on a grid of square cells of size 1m representing the walkable space. A maximum of 7 persons per cell (that is per $1\text{m}^2$) is allowed for.

The movement of persons is modeled as a queuing network in discrete time (time step equals 1/6 seconds; a time step of 1/18 seconds analogous to the social force type model did not lead to an improvement). Within each cell persons move according to a speed given as

$$s_i(t) = s_i^d s(d_i(t)/C), s(u) = 1 - F_\beta(u;\theta_1,\theta_2). \tag{3}$$

Here $F_\beta(u, \theta_1, \theta_2)$ denotes the cumulative distribution function of the beta distribution with parameters $\theta_1$ and $\theta_2$, $C=7$ denotes the maximal capacity of the cells and $s_i^d$ denotes the maximal speed of individual $i$ estimated using the instantaneous speeds analogous to above where now the 60% percentile (calibrated value) is used.

Each cell is considered to possess a diameter representing the length to be walked in order to cross the cell. If an individual reaches the end of the cell, she selects the next cell to enter. When a person enters a new cell the distance walked too far (i.e. distance walked minus the diameter of the last cell) is transferred to the new cell. The selection trades off the decrease of distance to the exit achieved by transiting to the new cell with the current density prevailing there. To this end for each of the nine cells (i.e. the current cell and its 8 neighbouring cells in the grid) the distance $n_{i,j}$ of the center point to the exit point (defined as in the social force model above) is calculated ($i=-1,0,1$, $j=-1,0,1$ in a local coordinate system such that the current cell is indexed as (0,0)). This is compared to the current density $d_{i,j}$ in these cells. The next candidate cell then is selected as the cell maximizing

$$C_{i,j} = \frac{1}{1+\exp(\lambda_1(n_{i,j} - n_{0,0}) + \lambda_2)}\left(1 - F_\beta(d_{i,j}(t)/C;\phi_1,\phi_2)\right). \tag{4}$$

Provided $\lambda_1>0$ the first factor is large if the gain $(n_{0,0}-n_{i,j})$ in distance to the exit is large. The second factor equals 1 if the cell candidate is empty and equals 0 if it is full, thus preventing persons to enter already full cells.

For this type of models the sequence of the time update is crucial in terms of collision avoidance. Collisions occur if – according to the selection criterion given above – more persons want to enter a cell than space is available there. In such cases decisions need to be taken which person is allowed to enter. Different collision avoidance algorithms have been discussed in the literature. The most common methods are sequential updates, where persons are moved one after the other in a fixed order and random shuffled updates where the order of persons for the time update is drawn randomly in every time step. It is simple to see that the sequential update depends on the ordering of persons in the internal representation of the simulation which is clearly unwanted. The randomly shuffled sequential update does not depend on the internal representation but also leads to unrealistic behaviour in certain situations. As a simple example consider persons walking in single file movement in a small corridor.

This example motivates the collision avoidance procedure used in this paper. To this end persons are sorted according to their remaining distance to be travelled inside the simulated region in ascending order (i.e. persons closer to the exit move first). If a move to the preferred cell is not possible, the persons stop (and do not consider alternatives at this stage).
The individuals enter the simulation in the cell corresponding to their entry point in the social force type model and exit in the cell containing their exit point.

On total there are seven parameters to calibrate for the mesoscopic model. Calibration again is performed in order to minimize the MAD in travel time estimates using the Nelder-Mead algorithm. Since the decisions taken in the algorithm are discrete the parameters of the model are not identified and hence several sets of parameters provide the same behaviour.

Due to the calibrated values of $\lambda_1$=0.43 and $\lambda_2$=0.17 the transformation of the density is practically linear for values occurring in the simulation. The same applies to the transformation of the density (see the red broken line in **Fig. 3** below). The relation between density and velocity is estimated as expected; see the blue solid line in **Fig. 3** below.

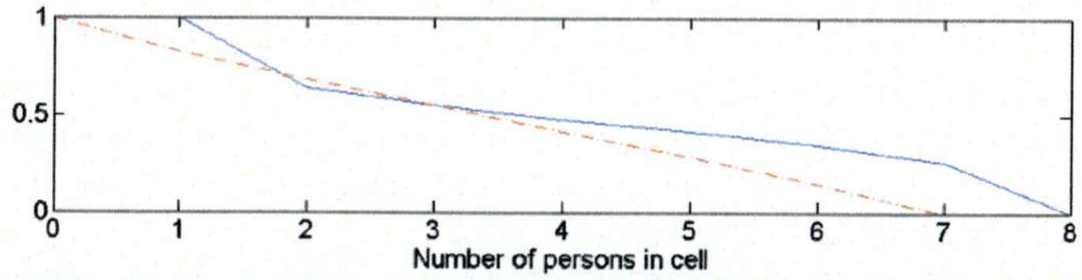


**Fig. 3. Functions specifying the mesoscopic model. Blue solid line: Velocity-density relation (s(u), see (3)). Red broken line: transformation of density for tradeoff (see (4)).**

## Model comparison

The calibrated models are compared on the estimation and validation data set with respect to the following three measures of accuracy:

- **Replication of travel times**. The accuracy measure here is the sum of absolute errors between the measured and simulated travel times (denoted as MADTT in the following), as well as the sum of absolute deviation of the travel time histograms with bin size 1 second (ADH).
- **Replication of the overall usage of space**. Here the accuracy is measured using the MAD of the two dimensional histogram on a regular grid of cellsize 1m over the full simulation period (denoted as MADSU in the following).

Twelve persons travelling more than 25 seconds are assumed to be waiting inside the area and hence are taken out of the comparison. On both datasets the information used in the simulation comprises the entry and exit points as well as free flow speed for all persons. This provides an optimistic setting since in real applications much less information will be present for the simulation. At best distributions of entry and exit points as well as speeds will be available. In this sense the results can be seen as a lower bound on achievable accuracy. The investigation of more realistic situations is left for future research.

**Table 1: Comparison of the accuracy of the various models**

| Measure of accuracy | In sample | | | Out of sample | | |
|---|---|---|---|---|---|---|
| | MADTT | MADSU | ADH | MADTT | MADSU | ADH |
| Social force type model | **0.74** | **0.27** | **0.16** | 0.90 | **0.34** | **0.15** |
| Mesoscopic model | 1.45 | 0.44 | 0.27 | 1.62 | 0.42 | 0.35 |
| Simple model (median speed) | 0.82 | - | 0.20 | **0.88** | - | 0.16 |
| Simple model (free speed) | 3.15 | - | 0.90 | 3.22 | - | 0.62 |

In order to provide a reference frame the results are compared to the results of a simple model predicting travel time as the distance between entry and exit point times speed taken to be the median (50% percentile) and the 85% percentile (sometimes taken to be free flow conditions) respectively. The median here is seen as a tough test since for straight paths and symmetric distribution for instantaneous speed the corresponding predicted travel time equals the actual time. For this approach no space usage is calculated.

The main results are provided in **Table 1** above and can be summarized as follows: As expected the social force method performs better than the mesoscopic approach for all accuracy measures on the estimation and the validation sample. The MAD in travel times equals less than one second both in the estimation and the validation data set for the social force type model. For the mesoscopic model the MAD equals 1.45 and 1.6 seconds respectively on the estimation and validation part. This has to be compared with an average travel time of approximately 9 seconds.

The simple model using the median performs slightly worse than the social force type model while using only the free speed and thus ignoring effects of density clearly produces the worst predictions. Qualitatively this ranking holds for all measures provided both on the estimation and on the validation data set. The ranking is also stable for a range of time step selections for the simulation models as well as smoothing parameters in the preprocessing step. Overall the differences between the social force type model and the simple model using the median are small in terms of the estimation of the mean travel time. The social force model is always (marginally) superior in terms of the estimation of the whole travel time distribution and contrary to the simple model is able to provide predictions also for space usage.

## Conclusion

The main result of this paper is to quantify for a particular real world scenario the gap in the accuracy between a more detailed social force type model and a simpler queuing network model, both of which have been calibrated on the situation analysed. The results show that indeed the social force type model is more accurate in out-of-sample prediction showing approximately an error of 10% compared to approximately 19% for the queuing network model. Also the error in space usage of the queuing network model is about a factor 1.2 as large as the corresponding one of the social force type model. Nevertheless the queuing network model performed far better than a simple model using free speed distributions and hence neglecting speed density relations.

There are many open questions that deserve a more profound analysis. First additional models such as cellular automata or the original social force model should be included in the comparison.

Second the models need to be validated on more extensive data sets. This also includes the specification of the various functions involved in the definition of the social force type model. While the selection of influence factors (i.e. the variables included in the specification) of the specific model used in this paper is motivated by prior work as well as heuristic arguments the specific functional form is only motivated empirically. More extensive research is needed.

This also relates to the important question of transferability. Validation in this paper is only performed with respect to a specific scenario of a crossing region. It is unclear whether the same model can be used in order to model situations of counterflow or unidirectional flow in a corridor or other more general settings. This is not only a deficiency of this paper but a general shortcoming in the literature.

Finally the scenario did not include any obstacles or walls. Clearly for the most interesting applications of PMSMs space restrictions and the resulting high densities play a major role. Thus the analysis needs to be extended to scenarios including walls and obstacles.

Despite these obvious shortcomings the paper is seen as a first step towards a better understanding of the strengths and weaknesses of the various modeling paradigms.

**Acknowledgments** The data sets discussed in this paper have been obtained at the Westbahnhof in Vienna, Austria, with the permission of the Austrian Federal Railways. The data has been manually annotated by Fenni Kang, which is gratefully acknowledged. The work leading to this paper has been financed in part by the project PACE-AOM funded by the Austrian Ministry for Traffic, Innovation and Technology (BMVIT) which is gratefully acknowledged.

## References

1. Burstedde, C., Klauck, K., Schadschneider, A., Zittartz, J.: "Simulation of Pedestrian Dynamics Using a Two-Dimensional Cellular Automaton." Physica A. Vol. 295, pp. 507–525 (2001)
2. Helbing, D., Molnar, P.: "Social Force Model for Pedestrian Dynamics." Physical Review E. Vol. 51, pp. 4282–4286 (1995)
3. Hoogendoorn, S., Bovy, P.H.L.: "Simulation of Pedestrian Flows by Optimal Control and Differential Games." Optimal Control Applications and Methods. Vol. 24, pp. 153-172 (2003)
4. Teknomo, K., Bauer, D., Matyus, T.: "Pedestrian Route Choice Self-Organisation." Proceedings of the ISTS08 Conference (2008)
5. Lakoba, T., Kaup, D., Finkelstein, N.: "Modifications of the Helbing-Molnar- Farkas-Vicsek Social Force Model for Pedestrian Evolution." SIMULATION. Vol. 81, pp. 339-352 (2005)
6. Johansson, A., Helbing, D.: "Analysis of empirical trajectory data of pedestrians." Proceedings of the PED conference. Wuppertal, Germany (2008)
7. Johansson, A., Helbing, D., Shukla, P.: "Specification of the Social Force Pedestrian Model by Evolutionary Adjustment to Video Tracking Data." Advances in Complex Systems. Vol. 10, pp. 271-288 (2007)
8. Johansson, A.: "Data-Driven Modeling of Pedestrian Crowds." PhD thesis, TU Dresden (2009)
9. Steffen, B.: "A simple but useful modification of the social force model." Proceedings of the 4th PED conference. Wuppertal, Germany (2008)
10. Steffen, B., Seyfried, A.: "Modeling of pedestrian movement around 90 and 180 degree bends." Proceedings of the Workshop on Fire Protection and Life Safety in Buildings and Transportation Systems (2009)
11. Chraibi, M., Seyfried, A., Schadschneider, A., Mackens, W.: "Quantitative Verification of a Force-based Model for Pedestrian Dynamics." Proceedings of Traffic and Granular Flow (2009)

# Towards a Calibration of the Floor Field Cellular Automaton

**A. Schadschneider, C. Eilhardt, S. Nowak, and R. Will**

Institut für Theoretische Physik, Universität zu Köln
509337 Köln, Germany
Corresponding author: as@thp.uni-koeln.de

**Abstract** We discuss several aspects related to the validation and calibration of cellular automata based models of pedestrian dynamics. Empirical fundamental diagrams obtained in large-scale experiments are compared with simulations of the floor field model. Although this kind of macroscopic calibration gives insights into the relevant interactions that govern the collective behaviour of pedestrians, microscopic validation based on the trajectories is expected to be more reliable. As a simple scenario we consider trajectories of the motion of individual pedestrians around corners. It is found that besides interactions with walls inertia effects play an important role for the correct reproduction of the empirical trajectories.

## Introduction

Although a large number of models for the simulation of pedestrian dynamics has been proposed in recent years [1,2] only a few have been tested quantitatively by comparing with empirical data. Instead the qualitative reproduction of collective phenomena like lane formation is often used as criterion to judge the validity of a model. One reason for this is that empirical data often is either not available or unreliable. Even for simple scenarios like fundamental diagrams of corridors or flow through bottlenecks the situation is still controversial [1,3]. Therefore only a few attempts to calibrate models of pedestrian dynamics properly have been made. However, this appears to be absolutely essential for any application, especially if quantitative predictions like evacuation times are required.

Here we discuss several aspects of the calibration of cellular automata (CA) models of pedestrian dynamics. CA are discrete in space, time and state variables and therefore ideally suited for large-scale computer simulations. Even large crowds can be simulated faster than real time with modest computational effort. We focus on the floor field model [4,5] which has become the standard CA approach for the description of pedestrian dynamics and evacuation processes. As a microscopic rule-based model e.g. psychological effects can be incorporated in an intuitive way. This flexibility has been used to develop numerous variants of the model which have been applied to study different situations. On the other hand, so

R.D. Peacock et al. (eds.), *Pedestrian and Evacuation Dynamics*,
DOI 10.1007/978-1-4419-9725-8_50, 

far the validation and calibration of the model has not been pursued sufficiently. Emphasis of earlier work was the qualitative reproduction of collective phenomena like lane formation or flow oscillations at bottlenecks. Although this might be sufficient to draw qualitative conclusions, a more detailed microscopic comparison with empirical data is highly desirable.

Recent large-scale experiments [6,7] performed within the Hermes project [8] have provided the data for the calibration. The trajectories of all persons can be generated automatically from video footage with an accuracy of a few centimeter [9,10]. In a first step data fundamental diagrams are derived which allow a straightforward comparison with computer simulations. Going beyond this quantitative, but macroscopic approach, we consider trajectories of individual pedestrians for the motion around a 90° bend. We also discuss general aspects of CA modelling which are relevant for calibration, e.g. the influence of the spatial discretization on the qualitative and quantitative predictions.

## General Remarks on Validation and Calibration

Currently there is no real consensus about the principles which should be used for validation and calibration [3,11,12]. Various criteria have been proposed which often appear to be biased to overstate the performance of the preferred model. In principle, different forms of validation and calibration can be distinguished [3]. *Qualitative validation* is most common because it can easily be done. Here certain collective phenomena like lane formation or the formation of jams are reproduced qualitatively. In contrast, *quantitative validation* requires testing whether these phenomena are reproduced in a quantitative way, e.g. with realistic jam densities. Here one can further distinguish between macroscopic and microscopic validation. *Macroscopic validation* uses macroscopic observables averaged over time or space. In contrast, *microscopic validation* tests individual properties like velocities and their distribution at a certain density or properties of single trajectories.

Experimental data of pedestrian flow are often connected with inhomogeneities in space and time, finite size effects and non-equilibrium conditions. For quantitative macroscopic validation it is therefore important that the same system sizes as well as measuring methods are used for comparison of experimental data with simulation results. An ideal validation procedure would guarantee that the model works in very general settings, not just in the scenarios tested. Currently it is unclear how this could be achieved. Therefore one should try to design a collection of tests which cover a variety of scenarios. The most detailed information is contained in the microscopic trajectories of individual pedestrians. As a first step, however, macroscopic trajectories can be considered, like the formation of lanes in counterflow and in narrow bottlenecks.

## *A Related System: Vehicular Traffic*

For vehicular traffic the situation concerning empirical data is much better than for pedestrian dynamics, both in quantity and quality. Therefore quantitative microscopic calibration is standard although far from trivial.

Data collection in highway traffic is simpler than in pedestrian dynamics for two main reasons: 1) the motion is (quasi-) one-dimensional and 2) automatic detection of vehicles is no problem. The latter is usually done using inductive loops which allows gathering *single-vehicle data* like individual velocities and flows in real time. The determination of the density from such local measurements, however, suffers from the same problems which occur in pedestrian dynamics [1].

Besides single-vehicle data so-called *car-following data* are used. These are collected using vehicles equipped with detectors that measure the distance to the preceding car (headway) and the current velocity. Since the motion is quasi-one-dimensional this gives the trajectories of individual vehicles.

A typical calibration procedure [13] relates the parameters of the model to characteristic features of the single-vehicle data, e.g. free flow or jam velocities, maximal flow, maxima of headway distributions, outflow from a jam etc. Usually not all of these characteristics are necessary to determine the parameters so that the others can be used for a consistency check.

## Cellular Automata Models

In cellular automata (CA) space, time and state variables are discrete. Space is divided into small cells that can be occupied by at most one pedestrian (exclusion principle). The cell size corresponds to the space requirement of a person in a dense crowd, e.g. a maximal density of 6 persons/m$^2$ yields a cell size of 40*40 cm$^2$. Time is also discrete and pedestrians move synchronously in each timestep which can be identified with the reaction time of a pedestrian. Thus one timestep corresponds to a few tenths of a second in real time.

One of the attractive properties of CA models is that they allow for an intuitive definition of the dynamics in terms of simple (stochastic) rules. Most models represent pedestrians by particles without any internal degrees of freedom although extensions to multi-agent systems [14] are easily possible. Particles move to one of the neighbouring cells with transition probabilities $p_{ij}$ determined by three factors: (1) the desired direction of motion (given by origin and destination), (2) interactions with other pedestrians, and (3) interactions with the infrastructure (walls, doors, etc.). In the simplest models, the latter two factors are only taken into account through an exclusion principle, i.e. occupied or wall cells are not accessible. More sophisticated approaches like the floor field model try to model these interactions in more detail which leads to more realistic results, especially for collective effects and self-organization phenomena. Fig.1 illustrates the definition of the transition probabilities $p_{ij}$ for a particle representing a pedestrian located at (0,0) to one of the neighbor cells (including the current position).

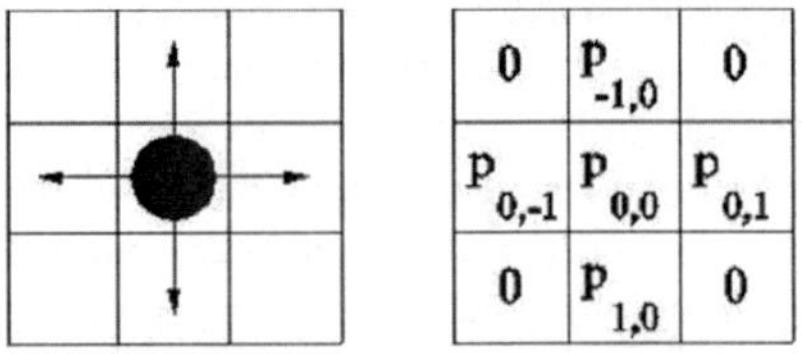


**Fig. 1. Definition of the transition probabilities $p_{ij}$.**

One of the most popular models used in pedestrian dynamics is the floor field model developed in [4,5]. Here the transition probabilities are mainly determined by the preferred walking direction of the pedestrians. This information is encoded in the *matrix of preference*. Its matrix elements $M_{ij}$ are directly related to observable quantities, namely the average velocity and its fluctuations [4].

Two discrete *floor fields* $D_{ij}$ and $S_{ij}$ modify the basic transition probabilities given by $M_{ij}$ in such a way that a movement in the direction of higher fields is preferred. The *dynamic floor field D* represents a virtual trace left by moving pedestrians. Similar to the process of chemotaxis responsible for the formation of ant trails, this trace has its own dynamics (diffusion and decay).

The *static floor field S* reflects the surrounding infrastructure and does not change in time. It has become common to avoid using the matrix of preference explicitly. Instead the information about the preferred direction of motion is directly encoded in the static floor field. Typically it describes the shortest distance to the destination (e.g. an exit) and is calculated for each lattice site using some distance metric. The field value increases in the direction of the destination.

Then for each pedestrian the transition probabilities $p_{ij}$ for a move to a neighbour cell $(i,j)$ (see Fig. 1) are given by

$$p_{ij} = N \exp(k_S S_{ij}) \exp(k_D D_{ij})(1 - n_{ij}) \zeta_{ij} \qquad (1)$$

where we have not included the matrix of preference $M_{ij}$ since it will not be used in the examples discussed here. The relative influence of the fields $D$ and $S$ is controlled by sensitivity parameters $k_S$ and $k_D$. The occupation number is $n_{ij} = 0$ for an empty and $n_{ij} = 1$ for an occupied cell where the occupation number of the cell currently occupied by the considered particle is taken to be $n_{00} = 0$. The obstacle number is $\xi_{ij} = 0$ for forbidden cells, e.g. walls, and $\xi_{ij} = 1$ otherwise, and the factor $N$ ensures the normalization $\sum_{ij} p_{ij} = 1$ of the probabilities. A more detailed description of the update rules can be found in [4,5].

The floor field model has successfully been validated qualitatively, in some cases even quantitatively. Its realism is further improved by various modifications, e.g.

1. Transitions beyond nearest neighbors [15]: In each timestep, agents can now move $v = 0,1,\ldots,v_{max}$ cells. The basic model corresponds to $v_{max} = 1$.
2. Friction effects [16] are related to the treatment of conflicts, i.e. situations where several particles choose the same destination cell in one timestep.
3. Spatial resolution [15]: Smaller cells are used to represent the infrastructure in more detail. For 20*20 cm$^2$ cells each pedestrian occupies 2*2 cells.
4. Inertia [4,17,18,19] suppresses abrupt changes in the direction of motion.
5. Politeness factor [18]: Suppresses transitions into high density areas.

## *Validation and Calibration of CA*

The validation and calibration of CA models brings about special problems, typically related to the discreteness of the approach. The discreteness in time is actually an advantage of CA since it provides a natural timescale for calibration. It is of the order of 1/10 sec, which is small enough for most applications (even laboratory experiments). For models which are continuous in time (e.g. social-force type models) no such timescale exist. The discretization used in numerical investigations of the dynamics is not related to any real timescale like reaction times and should be taken as small as possible to achieve a sufficient numerical accuracy.

The discreteness in space is a problem in applications if normal cell sizes are used. These are not sufficient to represent the infrastructure in detail. Also microscopic validation with trajectories becomes difficult if the spatial resolution is too low. In computer simulations the discreteness leads to a 'quantization' of measured values, e.g. all times are multiples of the timescale. This makes the comparison with empirical data more difficult. For a more general discussion of this point in the context of CA models for vehicular traffic we refer to [13].

## Fundamental Diagram

The most natural form of calibration is based on the fundamental diagram. We use empirical data from two experiments [6,7] with up to 70 persons obtained for single-file motion (no overtaking) along a circular corridor of length 17.3m and 26m, respectively. All measurements have been made in sections of length 2m and 4m.

Due to the single-file character of the motion we use a one-dimensional variant of the floor field model for the simulations. The corridor consists of $L$ cells $j = 1,\ldots,L$ with periodic boundary conditions. The static floor field has a constant gradient taken to be 1. The transition probabilities for a particle at site j are then given by $p_{j+d} = N\exp(dk_s)\exp(k_D D_{j+d})(1-n_{j+d})$ with $d = +1,-1,0$ corresponding to a forward, backward step or no movement, respectively.

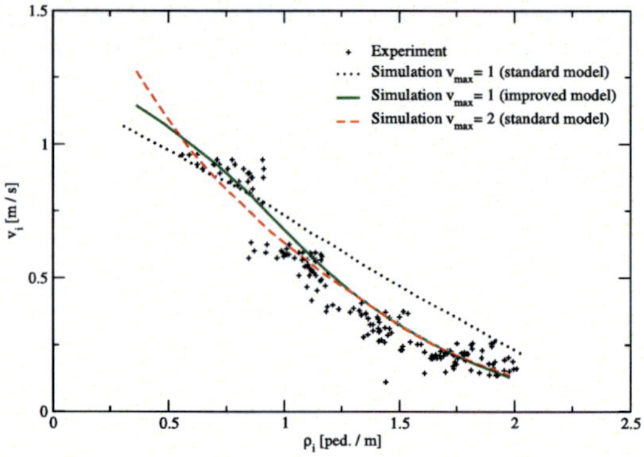


**Fig. 2. Comparison of empirical and simulated fundamental diagrams for three model variants. In the simulations the global density definition has been used.**

Fig. 2 shows a comparison of the empirical fundamental diagram with simulations of three different variants of the floor field model. In contrast to the experiments, the global density $\rho = N/L$ is used where $N$ is the total number of particles. As expected [15], the standard model with $v_{\max} = 1$ does not give a good agreement with the empirical data. The agreement is much better if a politeness factor [18] is included or the $v_{\max} = 2$ variant [15] of the model is used.

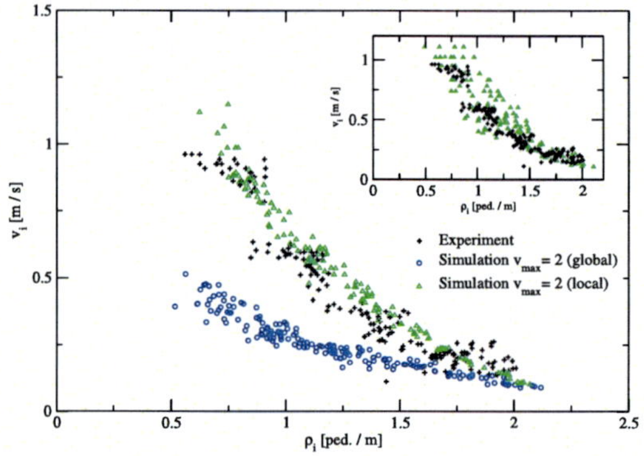


**Fig. 3. Comparison of empirical and simulated fundamental diagrams for the model with $v_{\max} = 2$. The blue curve (○) uses the optimal parameter set for the global fundamental diagram whereas the green curve (Δ) has been optimized for the local measurement. To reduce fluctuations all sizes have been enhanced by a factor of 10. The inset shows simulations for the system size used in the experiment.**

Fig. 3 shows the comparison for simulations where densities are measured locally by emulating the method used for the empirical densities [20]. It is found that the parameter set which has been optimized for the global measurements does not give a good agreement with the actual data. This shows the strong influence of the measurement method [7] and that one should avoid calibrating models with locally measured data by using simulations based on global measurements. To reduce the influence of fluctuations the simulated system has been increased by a factor of 10 compared to the experimental setup, i.e. it corresponds to a length of 260m. Using the original system size of 26m and especially the short measurement section of 4m the fluctuations become much larger (inset of Fig. 3), but the agreement is still good.

## Trajectories

The most reliable form of calibration is based on trajectories which represent the precise position at any given time. However, such microscopic calibration requires more sophisticated fitting procedures [12] than calibration using macroscopic quantities. As example we discuss the motion around a 90° corner. This is also relevant for practical purposes, e.g. the modeling of motion on a staircase.

An inappropriate choice of the floor field leads to unrealistic behavior in the pedestrian motion around bends, e.g. the creation of strong bottlenecks [21]. To get a better understanding of the origin of these problems we consider here as a first step the individual motion. This will also provide a first example for the calibration using trajectories (Fig. 4). The empirical data used as basis for the calibration have been obtained in series of experiments within the Hermes project.

**Fig. 4. Snapshot of the experiment with some measured trajectories superimposed (left) and static floor field used in the simulations (right). Stronger fields correspond to darker colors. The direction of motion is from left to right bottom.**

The static floor field has been determined using a combination of a Manhattan and an Euclidian metric with intermediate reference points [22]. It takes the interaction of pedestrians with walls explicitly into account by modifying the basic static field to $S_{ij}^{(W)} = S_{ij} - \max(0,(\eta+1-d_{ij})k_W/\eta)$ where $d_{ij}$ is the distance to the next wall, $k_w$ the interaction strength and $\eta$ the range. The resulting floor field is shown in Fig. 4. Since only individual motion is relevant, no dynamic field is needed.

Furthermore inertia has been included which suppresses abrupt changes of the direction of motion by introducing an additional factor $p_{ij}^{(I)} = \exp(k_I T_{ij})$ in the transition probabilities (1). The interaction constant $k_I$ controls the strength of the inertia effects. In the simplest case the matrix elements $T_{ij}$ are zero except for the direction $(i,j)$ of the previous timestep for which $T_{ij} = 1$, i.e. the additional factor enhances transitions into the same direction as in the previous timestep. This can be generalized to an enhancement factor that depends on the turning angle [19,22].

Fig. 5 shows a comparison of empirical and simulation results. A good agreement for the average trajectories as well as the standard deviations is found if a combination of wall interactions and inertia effects is used. Further improvements can be made especially through modifications of the wall interaction near the corner.

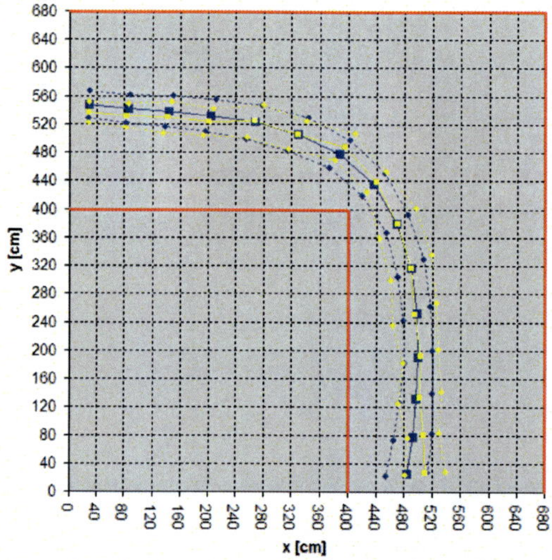


**Fig. 5. Comparison of empirical (yellow) and simulated (blue) trajectories. Thick lines indicate walls, the thin broken lines the standard deviation σ of the trajectories. The direction of motion is from left to right bottom.**

## Conclusions and Outlook

We have discussed several aspects of the quantitative calibration of cellular automata models of pedestrian dynamics. Despite problems due to the discrete nature of this model class we have demonstrated that in principle both macroscopic and microscopic calibration is possible. For macroscopic calibration we have used the fundamental diagram obtained in large-scale experiments whereas microscopic calibration was done using trajectory data for the motion round a bend. This has shown the relevance of the interactions with walls and inertia effects.

In a next step more complex situations should be considered, e.g. fundamental diagrams for two-dimensional scenarios. For the microscopic calibration, trajectories obtained for the *collective* motion can be used to gain further insights into the relevant interactions which govern the behavior of large crowds.

**Acknowledgments** This work was performed within the project Hermes funded by the Federal Ministry of Education and Research (BMBF) Program on "Research for Civil Security - Protecting and Saving Human Life". We thank Maik Boltes and Armin Seyfried for providing the empirical data.

## References

1. Schadschneider, A., Klingsch, W., Klüpfel, H., Kretz, T., Rogsch, C., Seyfried, A.: Evacuation dynamics: Empirical results, modeling and application. In: Meyers, R.A. (ed.), Encyclopedia of Complexity and System Science, p. 3142. Springer (2009)
2. Timmermans, H. (ed.): Pedestrian Behavior: Models, Data Collection and Applications. Emerald (2009)
3. Schadschneider, A., Seyfried, A.: Empirical results for pedestrian dynamics and their implications for cellular automata models. In [2], p. 27.
4. Burstedde, C., Klauck, K., Schadschneider, A., Zittartz, J.: Simulation of pedestrian dynamics using a two-dimensional cellular automaton. Physica A. 295, 507–525 (2001)
5. Schadschneider, A.: Cellular automaton approach to pedestrian dynamics - Theory. In: Schreckenberg, M., Sharma, S.D. (eds.), Pedestrian and Evacuation Dynamics, p.75. Springer (2002)
6. Seyfried, A., Steffen, B., Klingsch, W., Boltes, M.: The fundamental diagram of pedestrian movement revisited. J. Stat. Mech. P10002 (2005)
7. Seyfried, A., Boltes, M., Kähler, J., Klingsch, W., Portz, A., Schadschneider, A., Steffen, B.,Winkens, A.: Enhanced empirical data for the fundamental diagram and the flow through bottlenecks. In: Klingsch, W.W.F, Rogsch, C., Schadschneider, A., Schreckenberg, M. (eds.), Pedestrian and Evacuation Dynamics 2008, p. 145. Springer (2010)

8. Seyfried, A., Chraibi, M., Kemloh, U., Mehlich, J., Schadschneider, A.: Run-time optimization of forced based models within the Hermes Project. http://www.fz-juelich.de/jsc/hermes
9. Boltes, M., Seyfried, A., Steffen, B., Schadschneider, A.: Automatic extraction of pedestrian trajectories from video recordings. In: Klingsch, W.W.F, Rogsch, C., Schadschneider, A., Schreckenberg, M. (eds.), Pedestrian and Evacuation Dynamics 2008, p. 43. Springer (2010)
10. Boltes, M., Seyfried, A., Steffen, B., Schadschneider, A.: Using stereo recordings to extract pedestrian trajectories automatically in space (2010)
11. Klüpfel, H.: Crowd dynamics phenomena, methodology, and simulation. In [2], p.215.
12. Hoogendoorn, S.P., Daamen, W.: A novel calibration approach of microscopic pedestrian models. In [2], p. 195.
13. Knospe, W., Santen, L., Schadschneider, A., Schreckenberg, M.: Empirical test for cellular automaton models of traffic flow. Phys. Rev. E70, 016115 (2004)
14. Bandini, S., Manzoni, S., Vizzari, G.: Agent based modeling and simulation. In: Meyers, R.A. (Ed.), Encyclopedia of Complexity and System Science, p. 184. Springer (2009)
15. Kirchner, A., Klüpfel, H., Nishinari, K., Schadschneider, A., Schreckenberg, M.: Discretization effects and the influence of walking speed in cellular automata models for pedestrian dynamics. J. Stat. Mech. P10011 (2004)
16. Kirchner, A., Nishinari, K., Schadschneider, A.: Friction effects and clogging in a cellular automaton model for pedestrian dynamics. Phys. Rev. E 67, 056122 (2003)
17. Nishinari, K., Kirchner, A., Namazi, A., Schadschneider, A.: Extended floor field CA model for evacuation dynamics. IEICE Trans. Inf. & Syst. E87-D, 726 (2004)
18. Kretz, T., Schreckenberg, M.: The F.A.S.T.-model. Lect. Notes Comp. Sci. 4173, 712 (2006)
19. Yanagisawa, D., Kimura, A., Tomoeda, A., Nishi, R., Suma, Y., Ohtsuka, K., Nishinari, K.: Introduction of frictional and turning function for pedestrian outflow with an obstacle. Phys. Rev. E 80, 036110 (2009)
20. Portz, A., Seyfried, A.: Analyzing stop-and-go waves by experiment and modeling, these proceedings
21. Steffen, B., Seyfried, A.: Modeling of pedestrian movement around 90 and 180 degree bends. http://arxiv.org/abs/0912.0610 (2009)
22. Will, P.R.: Simulation von Fußgängerströmen. Schriftliche Hausarbeit im Rahmen der Ersten Staatsprüfung. Universität zu Köln (2009)

# To See Behind the Curtain – A Methodical Approach to Identify Calculation Methods of Closed-Source Evacuation Software Tools

**C. Rogsch and W. Klingsch**

University of Wuppertal
Institute for Building Material Technology and Fire Safety Science
Pauluskirchstr. 7
D-42285 Wuppertal
Corresponding author: christian@rogsch.de

**Abstract** Commercial software-tools for evacuation calculation are mostly closed-source (also called "proprietary") software-tools. Contrary to open-source software tools users of closed-source software-tools have to trust the technical manual of the software to understand how the software works. For users it is important to know how the model they use works, because without this knowledge results calculated by the software-tool can be interpreted in a wrong way. We will present a methodology how to identify basics of evacuation modelling and how to interpret and understand calculated results in a better way. This methodology should give users a "better feeling" in a very short time for the model they use for evacuation analysis and calculation.

## Introduction

Modelling of pedestrians and/or evacuation dynamics can be done by various approaches [1]. Commercial software-tools normally use one approach, but due to commercial reasons the description of these models is sometimes "very inexact", otherwise other companies would be able to reprogram the model and use it in their software. Based on this fact, users of closed-source software-tools are not able to understand the basics of the model they use for daily work. Sometimes this kind of software is called "black box" [2]. In the following we want to describe different scenarios which should give a possible view inside the models to allow users to understand the model they use in a better way. It should be mentioned that this is only a very basic approach and does not fit to every model, but it should be a starting point for ongoing work.

We will show how users can test based on simple scenarios the implementation of

- pedestrian movement (acceleration, deceleration, overtaking)

R.D. Peacock et al. (eds.), *Pedestrian and Evacuation Dynamics*,
DOI 10.1007/978-1-4419-9725-8_51, © Springer Science+Business Media, LLC 2011

- potential field (von-Neumann vs. Moore-neighbourhood)
- boundary conditions (exit implementation)
- exit selection
- path selection (quickest vs. shortest path)
- speed-density relation (fundamental diagram)
- update schemes.

For this basic and very simple tests the user has to ensure that all additional features of the model are disabled. This is important, because sometimes additional features may override the basics of a model in some situations.

## Pedestrian Movement

One of the basics of a model is to understand how pedestrians accelerate or decelerate. Especially for high crowded scenarios this is an important point which can have a large influence on calculated evacuation times. Furthermore it is important to know when (and how) faster people overtake slower pedestrians. This will be discussed in the following.

### *Acceleration and Deceleration of Pedestrians*

To show how acceleration and deceleration is implemented a simple and narrow floor has to be modelled (see Figure 1). Place one pedestrian with a selected free walking velocity and no response time at the end of the floor and measure the distance to a selected point (ca. 2m away from starting point). Based on speed and distance calculate the time the pedestrian needs to reach that point.

$$\text{time} = \frac{\text{distance}}{\text{velocity}}$$

If the pedestrian reaches the chosen point in the calculated time, there is no implementation of an acceleration to reach the free walking speed. If the pedestrian needs more time as calculated, then an acceleration is implemented.

To show deceleration, model a simple and narrow floor (length ca. 10 m). In front of the exit place one pedestrian who does not move (e. g. use a high response time). Another pedestrian has to be placed at the end of the floor with a selected free walking velocity and no response time. Then start the simulation. Measure the travel distance “walking pedestrian” to “blocking pedestrian” and measure the time until the walking pedestrian stops. Then restart the simulation without the blocking pedestrian and get the time until the pedestrian reaches the position where he stops in the previous simulation (the pedestrian will walk to the exit). Compare both times. If both times are identical, no deceleration is implemented, if the pedestrian needs more time in the simulation with the blocking pedestrian than in the simulation without the blocking pedestrian, a deceleration is implemented.

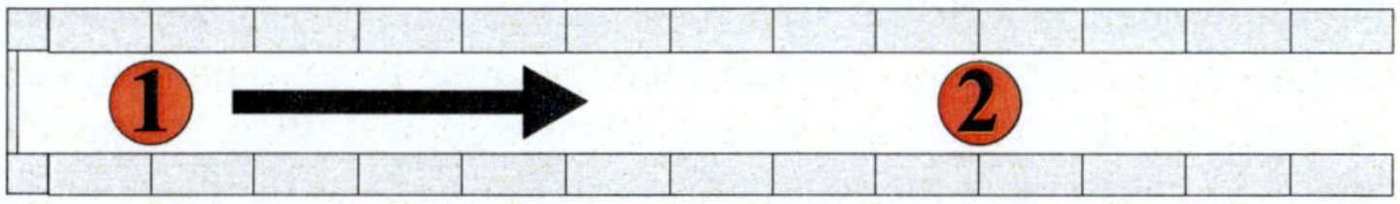


**Fig. 1. Acceleration and deceleration. For acceleration measurement place only pedestrian 1 (red circle), for deceleration place pedestrian 1 and 2 in the simulation. Lines in walls are for distance measurement used.**

## *Overtaking*

Model a simple floor (length ca. 30 m) where overtaking is possible (width ca. 2 m). Place two pedestrians in the floor on the same side with a distance between the two pedestrians (see Figure 2). The pedestrian placed closer to the exit should have no or half of the free walking velocity then the other pedestrian. Both pedestrians have to start at the same time (no response time). If the faster pedestrian overtakes the slower or not walking pedestrian, overtaking is implemented, if not, no overtaking is implemented. To get the distance when overtaking starts, the user has to measure the distance between both pedestrians, when the faster pedestrian changes his forward movement into a forward/sideward movement.

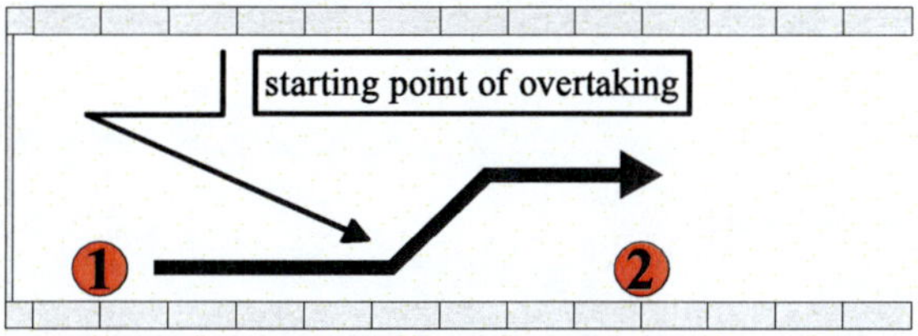


**Fig. 2. Schematic draw for testing overtaking. Pedestrian 1 has no response time, pedestrian 2 has a long response time. Lines in walls are for distance measurement used.**

## Potential Field

Methods, which are mostly for calculation of a potential field used, are the von-Neumann- and the Moore-Neighbourhood. More methods can be found in [3]. In scenarios with large space areas (e. g. large fair halls) the movement of pedestrians will be represented in different ways by using the two different approaches. To find out which neighbourhood is implemented, a simple large room with one exit has to be modelled. The room should have 10 m width and 20 m length, and the exit (width 1 m) has to be in the centre of one of the 10 m walls. Place on pedestrian in one corner of the room, which is on the opposite side of the wall with the exit. Run the simulation. If the pedestrian moves straight to the exit, a Moore-neighbourhood is implemented, if he walks in a 45° angle to the middle of the room and then straight forward to the exit, a von-Neumann-neighbourhood is implemented (see Figure 3). If the pedestrian walks along the wall to the exit, a fluid dynamics based potential field is implemented.

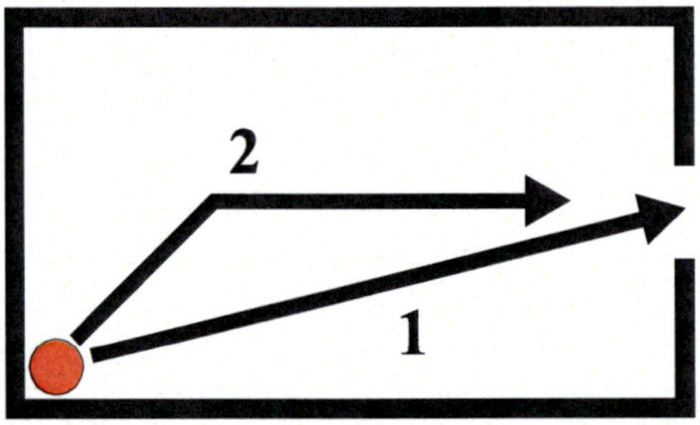


**Fig. 3. Implementation of Moore- or von-Neumann-neighbourhood potential field. If the pedestrian walks way (1), a Moore-neighbourhood potential field is implemented, if he walks way (2), a von-Neumann-neighbourhood potential field is implemented.**

## Boundary Conditions

The implementation of boundary conditions is a very important thing in evacuation calculations, because if a pedestrian has reached the boundary (= exit) he is safe. To reach a boundary or an exit normally means (in the case of evacuation simulation) that the pedestrian is taken out of the simulation, because he has reached the "safe region". But to take out a pedestrian of a simulation means, that this pedestrian does not have any influence of following pedestrians. Based on the implementation of e. g. speed-density relations, like the inter-person-distance [4] this could have a large influence of calculated results.

That's why it is important to implement exits in a correct way. Correct means, that it has no influence on the evacuation time, where the exit is placed in a simulation to define a safe area. To show if boundaries (or exits) are correctly implemented, a simple two-room scenario has to be modeled (see Figure 4). The two rooms should be 5 m x 8 m, and connected at the 5 m wall. Between the two rooms a 1 m door has to be placed, and at one of the rooms a 2 m door has to placed in the centre of the 5 m wall at the opposite of the 1 m door. The room with only the 1 m door has to be filled with a large number of pedestrians (e. g. 160 pedestrians without response time). Walls should have a thickness of 0.5 m. Two simulation runs have to be made. In one simulation, the exit of the simulation has to be placed directly after the 1 m door, in the other simulation the exit has to be placed directly after the 2 m door. As evacuation time it must be measured when the last pedestrian has left the room (crossed the 1 m door), which was filled with pedestrians. This measurement must be done at both simulations. If both evacuation times are the same, boundary conditions are correctly implemented, otherwise the implementation is incorrect.

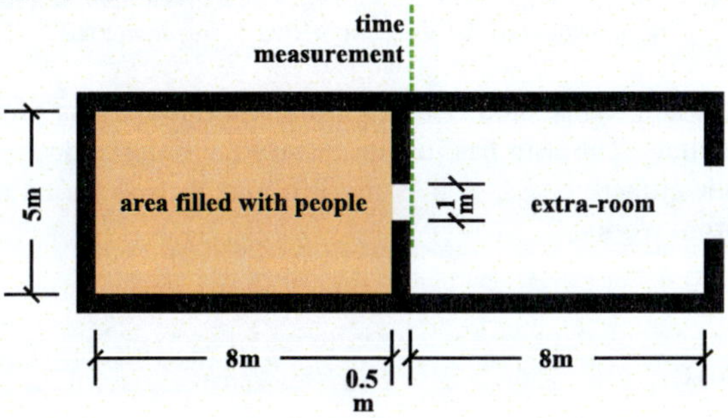


**Fig. 4. Testing correct implementation of boundary conditions**

## Exit Selection

The implementation of exit selection can be done in different ways. In this section we want to show how the selection of visible exits is implemented, we do not consider any smoke movement. It should be tested, if the exit selection is based on a "shortest distance" algorithm or if any other algorithm is implemented. This case is a modification based on the German RiMEA-Richtlinie [5].

To show if both exits are in the same way used, a simple room with two exits has to be modelled. The room should be 10 m x 20 m, 1 exit should be located at each 20 m wall at one end and they should not have the same difference to the 10 m wall (see Figure 5). At the opposite side of the room some pedestrians should be placed in row in the middle of the room (the orthogonal distance to the 20 m walls must be the same). Pedestrians should start sequentially (this means the first one should have no response time, the last one should have the longest response time), thus they do not try to overtake. If all pedestrians walk through the same exit, the exit selection is based on a "shortest distance" algorithm, otherwise a different algorithm is implemented.

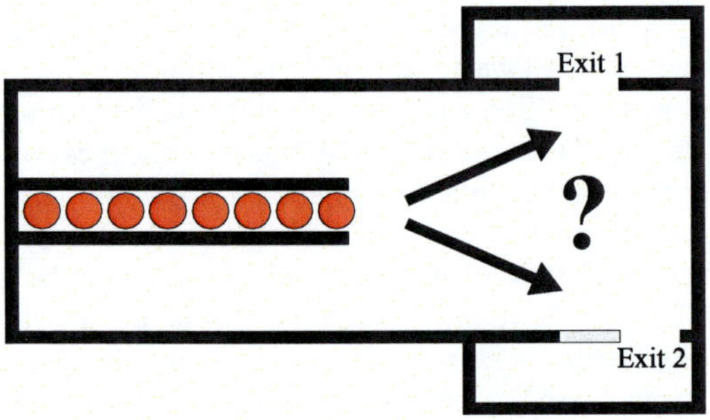


**Fig. 5. Testing "exit selection algorithm". If all pedestrians (red dots) use the same exit (Exit 1), a shortest distance algorithm is implemented.**

This scenario can easily modified e. g. to test attractiveness of exits or the influence of exit signs. The user has just to modify the most remote exit with e. g. a sign or a higher attractiveness, if it is more frequent used (in an expected way), then the algorithm works.

## Path Selection

The selection of a walking path (quickest path vs. shortest path) has to be divided in two different scenarios:

- Path selection in only one floor
- Path selection using multiple floors

Both test-cases are based on the work done by Kretz [5, 6]. First we discuss the path selection in one floor, based on congestions at a door, which is a classical "quickest vs. shortest path" scenario. Then we discuss how it is possible to test how stairs are implemented in the path selection algorithm.

### *Quickest vs. Shortest Path*

To show if a quickest or a shortest path algorithm is implemented, a room with two corridors has to be modelled (see Figure 6). Each corridor must have a corner, thus the exit at the end of the floor is not visible. Pedestrians have to be placed inside the room; it is necessary that a congestion appears at the opening between room and floor. If pedestrians only walk through one corridor, a shortest path algorithm is implemented, if both corridors are used after congestions appear, a quickest path algorithm is implemented. It is important that both corridors have the same exit to ensure that no "exit selection algorithm" is used.

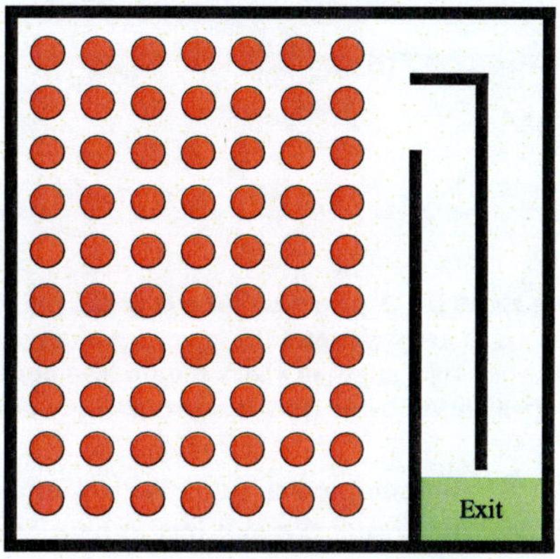


**Fig. 6. Quickest vs. shortest path. If only the left corridor is used, a shortest path algorithm is implemented. If a congestion appears at the beginning of the left corridor and pedestrians use the right corridor, a quickest path algorithm should be implemented.**

## *Path Selection Using Multiple Floors (Stair Implementation)*

The implementation of stairs can be done in different ways. One possibility is to compute a potential field using stairs as a "normal" part of the geometry, which means that the calculation does not stop at any stair. Another possibility is to stop the calculation at the end of each stair and then start at each level a new calculation beginning with the same value at each staircase. Both types of stair implementation will calculate different ways to an exit.

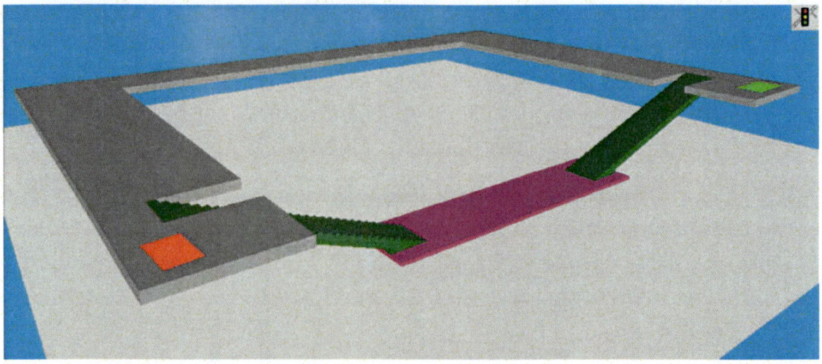

**Fig. 7. Path selection using stairs [5]. A pedestrian starts at the red area to reach the green area (exit). One possibility is to use stairs and the lower (or changed to upper) level or to walk in the same level to the exit. Length of stairs should also be changed in the way to have different distances (way in the same level is shorter/longer/equal than using stairs)**

To show which case is implemented a floor with a square layout has to be modelled (see Figure 7). At one corner the exit has to be placed, at the next corner the user has to place some pedestrians. The shortest way between pedestrians and exit has to be the way using an upper (or lower) level with two stairs. The longer way has to be in the same level. If pedestrians only use one level and no stairs, then stairs are calculated as separate elements in the building, if they use the stairs and the upper or lower level, stairs are calculated as a "normal" part of the building like the corridor.

## Speed-Density Relation

Speed-density relation means, that a higher crowded mass of people moves slower to an exit than a lower crowded mass. This basic behaviour of pedestrians is also shown in different fundamental diagrams [1]. Due to different measurements of densities [7] we only can provide two simple test-cases to show if this feature is implemented or not, or if it works only under certain conditions.

To show if this kind of behaviour is implemented, two corridors have to be modelled:

- a narrow corridor, where no overtaking is possible
- a two meter wide corridor, where overtaking is possible

Each corridor should have a length of ca. 50m to show this effect in a clear way. In both corridors pedestrians should be placed (response time = 0 seconds) uniformly. With a higher number of pedestrians in each type of corridor the total evacuation time (time when the last pedestrian leaves the corridor) should rise. If this does only happen in the wide corridor, this indicates a problem with the implemented update scheme, which is discussed in the following section.

## Update Schemes

Update schemes are the basics of evacuation modelling, because they determine in which order the calculation of new positions for each pedestrian is done [8, 9]. The user should also mention, that it is possible to "reproduce" different kind of human behaviour with different kind of update schemes [9].

Based on the type of the model, it is easier to identify update schemes at cellular automata software tools because of using cells for movement; in continuous models it is more difficult to see some effects, which are clearly shown in a cellular automata model. The method we present in this case works well for cellular automata software-tools, but the principles are the same for continuous models, too.

To find out, if the software has an ordered sequential front-to-back, random-shuffled, or parallel/back-to-front update implemented, the user has to model a narrow corridor, where no overtaking is possible. The corridor should be filled with the maximum number of pedestrians with no response time. Then the simulation should be started. If a clear start wave can be seen, a parallel or ordered sequential back-to-front update is implemented. If the last pedestrian starts walking when the first pedestrian starts walking, an ordered sequential front-to-back update is implemented. If something "between start-wave and no start-wave" can be seen, a random shuffled update is implemented. A start-wave is the normal human behaviour [10], thus a parallel update should be implemented in each software-tool.

## Conclusion

In this paper we presented some simple test-cases which should help the users to identify calculation methods implemented in software-tools they use for evacuation calculation, in detail we have discussed test-cases for the implementation of

- pedestrian movement (acceleration, deceleration, overtaking)
- potential field (von-Neumann vs. Moore-neighbourhood)
- boundary conditions (exit implementation)
- exit selection
- path selection (quickest vs. shortest path)
- speed-density relation (fundamental diagram)
- update schemes.

With this knowledge the user should be able to understand calculated results in a better way, and he/she should be able to understand how different methods can influence results of calculations for whole buildings. Users should also be able to understand why changes in the building layout sometimes have no influence on the calculation of evacuation times.

**Acknowledgments** The authors thank Hubert Klüpfel, Tobias Kretz, Andreas Schadschneider and Armin Seyfried for useful discussions. Furthermore we would like to thank different software companies for providing their evacuation software free of charge or as student version.

## References

1. Schadschneider, A., Klingsch, W., Kluepfel, H., Kretz, T., Rogsch, C., Seyfried, A.: Evacuation Dynamics: Empirical Results, Modeling and Applications. In: Encyclopedia of Complexity and System Science. Springer
2. Paulsen, T., Schneider, V.: Qualitative Analysis of Evacuation Data Technical Note WP320. In: Paulsen, T. et al. (eds.), Evaluation of Simulation Models of Evacuation from Complex Spaces, SINTEF Report
3. Kretz, T., Bönisch, C., Vortisch, P.: Comparison of Various Methods for the Calculation of the Distance Potential Field. In: Klingsch, W., Rogsch, C., Schadschneider, A., Schreckenberg, M. (eds.), Pedestrian and Evacuation Dynamics. Springer (2008)
4. Thompson, P.: Simulex: simulated people have needs too. In: Workshop on Building Occupant Movement During Fire Emergencies. National Institute of Standards and Technology (2004)
5. RiMEA-Richtlinie für Mikroskopische Entfluchtungs-Analysen, www.rimea.de
6. Kretz, T.: The use of dynamic distance potential fields for pedestrian flow around corners. ICEM 09, http://arxiv.org/abs/0906.2667
7. Steffen, B., Seyfried, A.: Methods for measuring pedestrian density, flow, speed and direction with minimal scatter. Physica A (Article in Press). doi:10.1016/j.physa.2009.12.015
8. Rogsch, C., Schadschneider, A., Seyfried, A.: How to Select the "Right One" – Update Schemes for Pedestrian Movement in Simulation and Reality. To Appear in: Traffic and Granular Flow 2009.
9. Rogsch, C., Schadschneider, A., Seyfried, A.: Simulation of human movement by cellular automata using different update schemes. In: 4th International Symposium on Human Behaviour in Fire (Conference Proceedings)
10. Rogsch, C.: Start Waves and Pedestrian Movement – An Experimental Study. In: Klingsch, W., Rogsch, C., Schadschneider, A., Schreckenberg, M. (eds.), Pedestrian and Evacuation Dynamics. Springer (2008)

# Analyzing Stop-and-Go Waves by Experiment and Modeling

**A. Portz and A. Seyfried**

Jülich Supercomputing Centre, Forschungszentrum Jülich GmbH
52425 Jülich, Germany
Corresponding author: a.portz@fz-juelich.de

**Abstract** The main topic of this paper is the analysis and modeling of stop-and-go waves, observable in experiments of single lane movement with pedestrians. The velocity density relation using measurements on a 'microscopic' scale shows the coexistence of two phases at one density. These data are used to calibrate and verify a spatially continuous model. Several criteria are chosen that a model has to satisfy: firstly we investigated the fundamental diagram (velocity versus density) using different measurement methods. Furthermore the trajectories are compared to the occurrence of stop-and-go waves qualitatively. Finally we checked the distribution of the velocities at fixed density against the experimental one. The adaptive velocity model introduced satisfies these criteria well.

## Introduction

To enhance crowd management or to improve escape route systems computer simulations are used to describe the dynamics of pedestrians [1-4]. Here one main aim is to analyze when and where long lasting jams or congestions occur and how they could be prevented. But tests to show whether the models used for simulation are able to describe e.g. formation of jams or the degree of congestion are still in their infancy and suffer from a poor experimental data base. In particular specifications of the density, where jamming starts, are inconsistent and range from 3.8 to 10 [Pers. per m²] for planar corridors [2]. Moreover there is no consensus in the community as to which criteria have to be chosen for model verification [5]. To improve this situation we present experimental results about stop-and-go waves in single file movement. Even in this simple case pedestrians interact in many ways and not all factors, which have an effect on their dynamics, are known. One proven factor is the cultural differences occurring in the fundamental diagram [6]. Along with a new modeling approach we introduce quantitative criteria to calibrate and test models regarding characteristics of stop-and-go waves. For modeling pedestrian motion, we start with the simplest case: single lane movement. If

R.D. Peacock et al. (eds.), *Pedestrian and Evacuation Dynamics*,
DOI 10.1007/978-1-4419-9725-8_52, 

the model is able to describe the dynamics of pedestrians quantitatively and qualitatively for that simplified case, it is a good candidate for adaption to two-dimensional situations. In the past we followed different modeling approaches continuous in space and validated them with empirical data [7]. Only one model satisfied our criterion, reproduction of the fundamental diagram. This model, with adaptive velocity, is now discussed in detail regarding stop-and-go waves. At first we systematically analyze the experimental data to get significant criteria for stop-and-go waves. This is done in the first section. After this the adaptive velocity model is introduced and validated. To avoid deviations due to different measurement methods [8] we used the same methods for the comparison of experimental data with model results.

## Experimental Data

We obtained empirical data from two experiments of single lane movement we had performed in previous years. There a corridor with circular guiding was built to realize periodic boundary conditions. The density was varied by increasing the number of the test persons in the corridor. The experiment was performed on two occasions using different corridor lengths and test persons.

The first experiment was conducted at the auditorium Rotunde of the Jülich Supercomputing Centre (JSC), Forschungszentrum Jülich [9]. The group of pedestrians was composed of students and staff of JSC. The length of the measured section was $l = 2$ [m] and the whole corridor was 17.3 [m]. We executed runs with N=15, 20, 25, 30 and 34 pedestrians in the passageway.

The second experiment was performed in 2006 in the wardroom of Bergische Kaserne Düsseldorf [8]. The test group was of soldiers. Here the length of the system was about 26 [m], with a $l = 4$ [m] measurement section. We were able to do several runs with up to 70 pedestrians.

Detailed information about the experimental set-up and analysis, is given in [8, 9]. In the first experiment the measurement of velocity and density was done manually. For the second experiment we used advanced cameras and the tool PeTrack for automatic extraction of pedestrian trajectories from video recording [10].

### *Fundamental Diagram*

Firstly we compare the data of both experiments to test whether system's length and composition of the test group lead to differences in the results. For this, the velocity $v_i$ is calculated using the entrance and exit times $t_i^{in}$ and $t_i^{out}$ of pedestrian $i$ into and out of the measurement section $l$,

$$v_i = \frac{l[m]}{(t_i^{out} - t_i^{in})[s]}.$$

To avoid discrete values of the density leading to a large scatter, we define the density by

$$\rho(t) = \frac{\sum_{i=1}^{N} \Theta_i(t)}{l[m]},$$

where $\Theta_i(t)$ gives the fraction of the space between pedestrian $i$ and $i+1$, which can be found inside the measurement area, see [8]. $\rho_i$ is the mean value of all $\rho(t)$ for $t \in [t_i^{in}, t_i^{out}]$

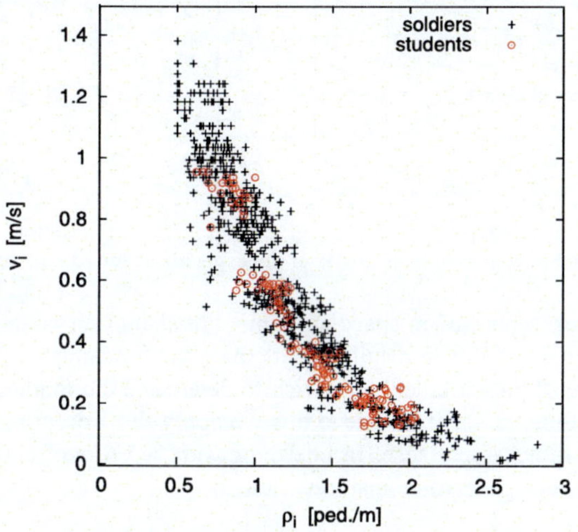


**Fig. 1. Fundamental diagrams for both experiments (soldiers and students) for the single file movement**

The results are shown in Fig. 1. The fundamental diagram of the students lies on the same curve as the one of the soldiers. This indicates that the data gained by the experiments are representative and reproducible, so we can treat data sets equally for the next step.

## *Stop-and-Go Waves*

In Fig. 2 the *x*-component of trajectories of the pedestrians are plotted against time. For the extraction of the trajectories, the pedestrians' heads were marked in advance and tracked. Backward movement in the trajectories leading to negative velocities is caused by head movement of the pedestrians during a standstill. Irregularities in the trajectories increase with increasing density. In both plots the stop-and-go waves occur throughout the experiments with the wave propagating opposite to the movement direction. Stopping is first observed during the runs with 45 pedestrians, at 70 pedestrians they can hardly move forward. Stop-and-go phases occur simultaneously. Some pedestrians stop, but elsewhere pedestrians could walk slowly.

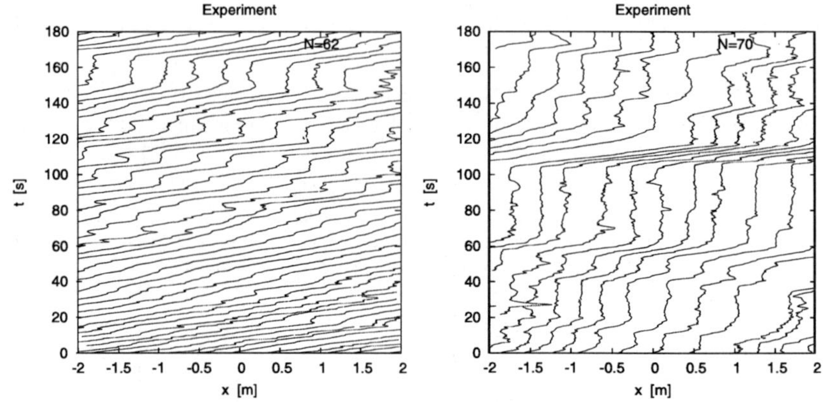


**Fig. 2. Stop-and-go waves in trajectories of the experiment for 62 and 70 pedestrians**

These two phases, separated in space and time, are also observable at a fixed density. Before we discuss this in detail, we calculate the fundamental diagram with a different method of measurement. We wish to determine the fundamental diagram on the scale of individuals. To achieve this we apply the Voronoi density method [11] to the one dimensional case. In one dimension a Voronoi cell is defined by the center $z_i$ between pedestrian position $x_i$ and $x_{i+1}$

$$z_i = \frac{x_{i+1} + x_i}{2}, \quad L_i = z_{i+1} - z_i \text{ and } \rho_i(x) = \begin{cases} \frac{1}{L_i}, & x \in [z_i, z_{i+1}[ \\ 0, & \text{otherwise} \end{cases}.$$

The fundamental diagram using the Voronoi method is shown in the left part of Fig. 3. Regular stops occur at densities higher than 1.5 pedestrians per meter. On the right side of Fig. 3 the distribution of the velocities for fixed densities from 1.8 to 2.6 pedestrians per meter are shown. There is a continuous change from a single peak near $v$ = 0.15, to two peaks, to a single peak near $v$ = 0. The right peak represents the walking pedestrians, whereas the left peak represents the stopping pedestrians. For a density between 2.0 and 2.2 pedestrians per meter both peaks coexist. Several causes could be responsible for the double peak structure of the velocity distribution. One is, that different pedestrians react to the situation differently and have different personal space requirements. Some pedestrians prefer a larger personal space than other. Also differences in the reaction time could be responsible. These differences in the perception and personal space requirements could be combined with differing step phases and stopping stances.

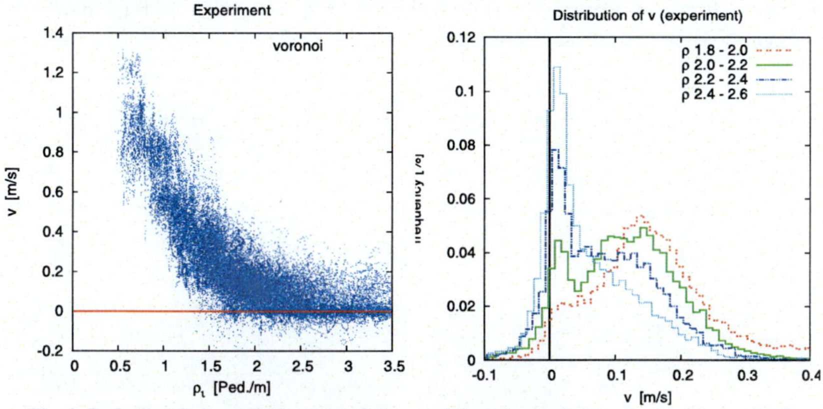


**Fig. 3. Left: Fundamental diagram of the empirical data measured with Voronoi cells Right: Distribution of *v* at fixed densities**

## Modeling

### *Adaptive Velocity Model*

In this section we introduce the adaptive velocity model, which is based on an event driven approach. A pedestrian can be in different states and a change between these states is called event. The calculation of the velocity of each pedestrian is straightforward and depends on these states. The model was derived from force based models, where the dynamics of pedestrians are given by the following system of coupled differential equations

$$m_i \frac{dv_i}{dt} = F_i \quad \text{with} \quad F_i = F_i^{drv} + F_i^{rep} \quad \text{and} \quad \frac{dx_i}{dt} = v_i\,,$$

where $F_i$ is the force acting on pedestrian $i$. The mass is denoted by $m_i$, the velocity by $v_i$ and the current position by $x_i$. $F_i$ is split into a repulsive force $F_i^{rep}$ and a driving force $F_i^{drv}$. The dynamic is regulated by the interrelation between driving and repulsive forces. In our approach the role of repulsive forces are replaced by events. The driving force is defined as

$$F_i^{drv} = \frac{v_i^0 - v_i}{\tau_i}\,,$$

where $v_i^0$ is the desired velocity of a pedestrian and $\tau$ their reaction time. By solving the differential equation

$$\frac{dv_i}{dt} = F_i^{drv} \quad \Rightarrow \quad v_i(t) = v_i^0 + \exp\left(-\frac{t}{\tau_i}\right),$$

the velocity function is obtained. This is shown in Fig. 4 together with the parameters governing the pedestrians' movement.

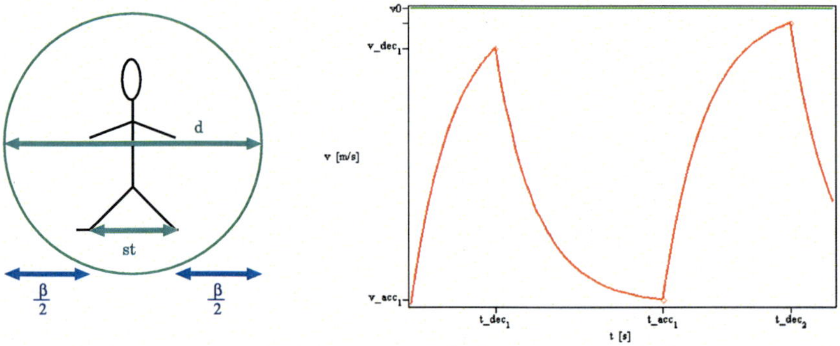


**Fig. 4. Left: connection between the required space *d*, the step length *st* and the safety distance *β***
**Right: the adaptive velocity with acceleration until $t_{dec1}$ than deceleration until $t_{acc1}$, again acceleration until $t_{dec2}$ and so on**

In this model pedestrians are treated as hard bodies with a diameter $d_i$ [12]. The diameter depends linearly on the current velocity and is equal to the step length *st* in addition to the safety distance $\beta$

$$d_i(t) = e + f\, v_i(t) = st_i(t) + \beta_i(t). \tag{1}$$

Based on [13] the step length is a linear function of the current velocity with following parameters:

$$st_i(t) = 0.235[m] + 0.302[s] v_i(t). \tag{2}$$

*e* and *f* can be specified through empirical data. Here *e* is the required space for a stationary pedestrian and *f* describes the additional space a moving pedestrian requires. For *e* = 0.36 [m] and *f* = 1.06 [s] the last equations (1) and (2) can be summarized to

$$\beta_i(t) = d_i(t) - st_i(t) = 0.125[m] + 0.758[s] v_i(t).$$

A pedestrian accelerates to his/her desired velocity $v_i^0$ until the distance to the pedestrian in front ($\Delta x_{i,i+1}$) is smaller than the safety distance. From this time on, he/she decelerates until the distance is larger than the safety distance and so on. Via $\Delta x_{i,i+1}$, $d_i$ and $\beta_i$ events can be defined: deceleration (3) and acceleration (4). To implicitly include a reaction time and to ensure good computational performance for high densities, no events are explicitly calculated. Instead in each time step of $\Delta t = 0.05$ seconds, it is checked whether an event has taken place and $t_{dec}$, $t_{acc}$ or $t_{coll}$ are set to *t* accordingly. The discreteness of the time step could lead to configurations where overlapping occurs. To guarantee a minimal volume exclusion, case (5) is included, in which the pedestrians are too close to each other and have to stop. To avoid artifacts related to the update procedure a recursive update procedure is necessary: Each person is advanced one time step according to equations (3-5). If after this step a pedestrian is in a different state because of the new distance to the pedestrian in front, the velocity is set according to this state. Then the state of the next following person is reexamined. If the state is still valid the update is completed. Otherwise, the velocity is again calculated and so on.

$$t = t_{dec}, \quad \text{if:} \quad \Delta x_{i,i+1} - 0.5 * \left(d_i(t) + d_{i+1}(t)\right) \leq 0 \tag{3}$$

$$t = t_{acc}, \quad \text{if:} \quad \Delta x_{i,i+1} - 0.5 * \left(d_i(t) + d_{i+1}(t)\right) > 0 \tag{4}$$

$$t = t_{coll}, \quad \text{if:} \quad \Delta x_{i,i+1} - 0.5 * \left(d_i(t) + d_{i+1}(t)\right) \leq -\beta_i(t)/2 \tag{5}$$

To incorporate the dispersion in the characteristics and behavior of pedestrians we choose the following parameters from a normal distribution, $\tau_i \sim N(1.0, 0.1)$, $e_i \sim N(0.36, 0.1)$ and $f_i \sim N(1.06, 0.5)$.

## *Fundamental Diagram*

The fundamental diagram of our modeled and empirical data is displayed in Fig. 5. The adaptive velocity model yields the right relation between velocity and density.

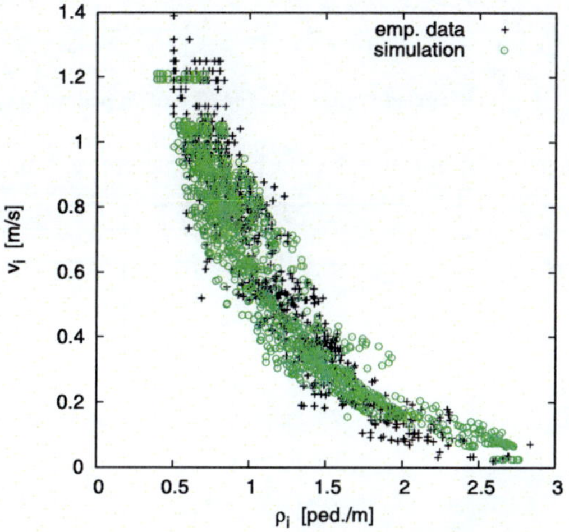


**Fig. 5. Validation of the modeled fundamental diagram with the empirical data**

We also tested different time steps $\Delta t$ and $\tau_i = 1.0$, $e_i = 0.36$, $f_i = 1.06$ for all pedestrians. For $\Delta t$ from 0.001 to 0.1[s] the included reaction time has no significant influence on the shape of the fundamental diagram. To avoid interpenetrations of pedestrians we choose $\Delta t = 0.05$[s] for further simulations. Variation of the personal parameters $\tau_i$, $e_i$ and $f_i$ affects the scatter of the fundamental diagram.

## *Stop-and-Go Waves*

For a qualitative comparison we plot the modeled trajectories in Fig. 6. Stop-and go waves occur as in the experiment, see Fig. 2. The pattern as well as the change of the pattern from $N = 62$ to $N = 70$ are in good agreement with the experimental results. Even the start and stop phases match qualitatively. However the phases of stop and go traffic appear more regular in the modeled trajectories.

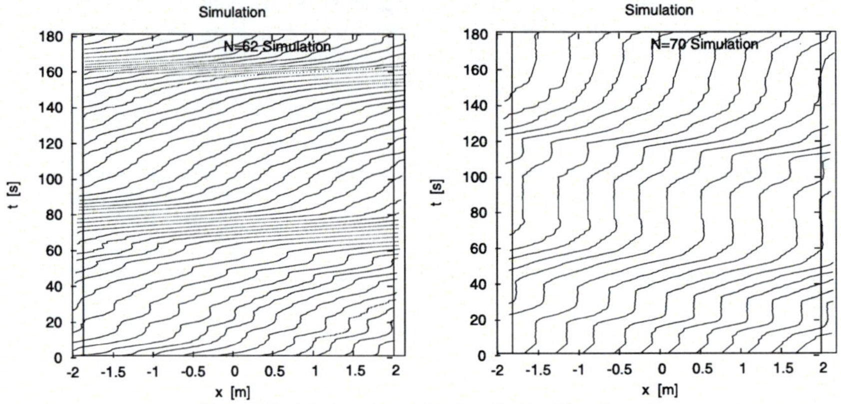


**Fig. 6. Modeled trajectories for 62 and 70 pedestrians**

Furthermore we examine the double peak structure, exhibited by the experimental data, see Fig. 3. This acts as a quantitative criterion.

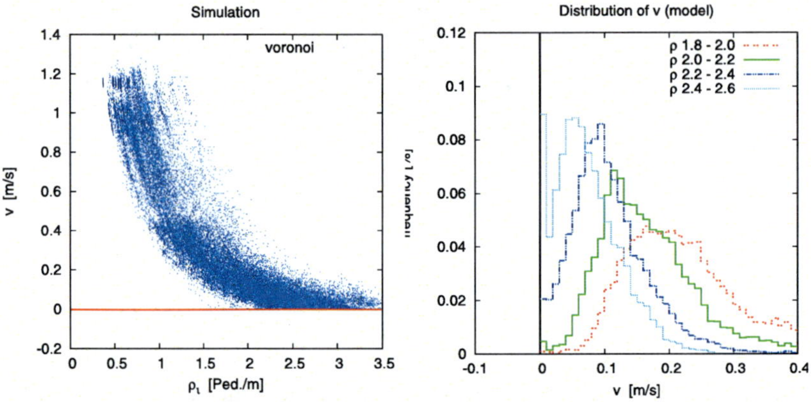


**Fig. 7. Left: Fundamental diagram of the modeled data measured with Voronoi cells Right: Distribution of *v* at fixed densities**

There is also a double peak in the distribution of the modeled velocities, see Fig. 7. The velocity distribution is similar but not identical to the empirical one. The double peak structure given by the height and position of the peaks differ from experimental results. These differences could be caused by the irregularities of the experimental data. This is also recognizable in the comparison of the trajectories in Fig. 6 and Fig. 2. To avoid these irregularities a step-detection and a cleaning of the empirical trajectories is necessary. We will consider this in future work.

## Conclusion and Perspectives

Choosing an adequate distribution of individual parameters, the adaptive velocity model is able to generate the fundamental diagram and to create stop-and-go waves without unrealistic phenomena, like overlapping or interpenetrating pedestrians. Even the change of the characteristics of stop and go waves for varying densities is in good agreement with the experiments. For a more detailed analysis of stop-and-go waves we need to clean the experimental data, so that the irregularities caused by the self dynamic of pedestrians' head vanish. In the future we will try to get a deeper insight into to occurrence of stop-and-go waves with other criteria like average stopping time. Furthermore we plan to include steering of pedestrians. For the validation of these models more test cases, like flow characteristics at bottlenecks and junctions will be used..

**Acknowledgments** This study was supported by the German Government's high-tech strategy, the Federal Ministry of Education and Research (BMBF). Program on "Research for Civil Security - Protecting and Saving Human Life". Execution of experiments was supported by the German Research Foundation (DFG) KL 1873/1-1 and SE 1789/1-1.

## References

1. Hermes – Investigation of an evacuation assistant for use in emergencies during large-scale public events. Julich Supercomputing Centre (JSC). http://www.fz-juelich.de/jsc/hermes
2. Schadschneider, A., Klingsch, W., Kluepfel, H., Kretz, T., Rogsch, C., Seyfried, A.: Evacuation Dynamics: Empirical Results, Modeling and Applications, In: Meyers, R.A (Ed.), Encyclopedia of Complexity and System Science, 3142-3176. Springer (2009)
3. Klingsch, W.W.F., Rogsch, C., Schadschneider, A., Schreckenberg, M. (Eds.): Pedestrian and Evacuation Dynamics 2008, Springer (2010)
4. Waldau, N., Gattermann, P., Knoflacher, H., Schreckenberg, M. (Eds.): Pedestrian and Evacuation Dynamics 2005. Springer (2006)
5. Schadschneider, A., Seyfried, A.: Empirical Results for Pedestrian Dynamics and their Implications for Cellular AutomataModels. In: Timmermans, H. (Ed.)

Pedestrian Behavior: Data Collection and Applications, p. 27-43. Emerald Group Publishing Limited (2009)
6. Chattaraj, U., Seyfried, A., Chakroborty, P.: Comparison of pedestrian fundamental diagram across cultures. Advances in Complex Systems. 12, 393-405 (2009)
7. Portz, A., Seyfried, A.: Modeling Stop-and-Go Waves in Pedestrian Dynamics. In: PPAM 2009, Eighth international conference on parallel processing and applied mathematics. Wroclaw, Poland (2009) **arXiv: 1001.3283**
8. Seyfried, A., Boltes M., Kähler, J., Klingsch, W., Portz, A., Schadschneider, A., Steffen, B., Winkens, A.: Enhanced empirical data for the fundamental diagram and the flow through bottlenecks. In: Klingsch, W.W.F, Rogsch, C., Schadschneider, A., Schreckenberg, M. (Eds.), Pedestrian and Evacuation Dynamics 2008, 145-156. Springer (2010)
9. Seyfried, A., Steffen, B., Kingsch, W., Boltes, M.: The fundamental diagram of pedestrian movement revisited. J. Stat. Mech., P10002 (2005)
10. Boltes, M., Seyfried, A., Steffen, B., Schadschneider, A.: Automatic Extraction of Pedestrian Trajectories from Video Recordings. In: Klingsch, W.W.F., Rogsch, C., Schadschneider, A., Schreckenberg, M. (Eds.), Pedestrian and Evacuation Dynamics 2008, 43-54. Springer (2010)
11. Steffen, B., Seyfried, A.: Methods for measuring pedestrian density, flow, speed and direction with minimal scatter. Physica A (2010). doi:10.1016/j.physa.2009.12.015
12. Chraibi, M., Seyfried, A.: Pedestrian Dynamics With Event-driven Simulation. In: Klingsch, W.W.F, Rogsch, C., Schadschneider, A., Schreckenberg, M. (Eds.), Pedestrian and Evacuation Dynamics 2008, 713-718. Springer (2010)
13. Weidmann, U.: Transporttechnik der Fussgänger. Institut für Verkehrsplanung,Transporttechnik, Strassen- und Eisenbahnbau. Schriftenreihe des IVT, Nr. 90. ETH Zürich (1993)

# Experimental Study of Crowd Flow Passing through Simple-shaped Room and Validation for an Evacuation Simulator

**T. Sano[1], Y. Yoshida[2], N. Takeichi[2], T. Kimura[3], and Y. Minegishi[2]**

[1]Faculty of Human Sciences, Waseda University, Mikajima 2-579-15, Tokorozawa, Saitama, 359-1192, JAPAN
[2]Takenaka Corporation
[3]A & A Co., Ltd.
Corresponding author: sano-t@waseda.jp

**Abstract** This paper describes the characteristics of crowd flow passing through simple-shaped rooms and to validate an evacuation simulation model called "SimTread". It presents experiments on crowd flow with 43 subjects and intends to quantify the aspects of crowd flow in simple rooms. Recently, for estimating evacuation of buildings on fire, several computer simulation models have been developed and applied. However, human evacuation data-sets for validating simulation are scarce. The results of this study show that pedestrian flow rate at the opening changes depending on the density of the space connected. Flow rate rises if the opening is connected to larger space, which is less dense and, in result, increases speed of pedestrians. For validating an evacuation simulator, evacuation data from actual building fires are too complex for proving the equivalence. Therefore, we carried out the experiments of crowed flow passing through simple-shaped rooms, and compared experimental data with our simulation results. There was a good agreement between the result of experiments and simulations. The differences were less than 10%.

## Introduction

A computer evacuation model can provide the knowledge about the movement of people through building and provide an evaluation of building plan. But reality of evacuation modeling accounts for a degree of human behaviours. There is a severe lack of data for use in predicting crowd movement in buildings, and for the data that do not exist. In this paper, the measured data of actual crowd flow are reported, and utilizing these, validation of a computer simulator named SimTread is discussed.

Performance-based codes, provided by both Building Standard Law and Fire Service Law, have opened the way for advanced evaluation to safety to fire evacu-

R.D. Peacock et al. (eds.), *Pedestrian and Evacuation Dynamics*,
DOI 10.1007/978-1-4419-9725-8_53, 

ation in Japan. In validating evacuation safety, above government-made methods are widely utilized, while computer simulation is also useful. However, when we utilize a computer simulation, basic pedestrian data are not gathered enough, so we have to collect, analyze and accumulate these data to make the simulator realistic.

Multi-agent models, simulate pedestrian flow by defining mechanism of individual agent, are world widely studied, developed and utilized today[1]. In these models, crowd flow is formed by interactions of individual agent and obstacles such as a wall. We can produce various character of crowd flow by changing parameters of each agent. A calculation is processed in every delta-T with a regard to the location of agents those are defined by coordinates, however, major indexes used in the validation of the model are evacuation time, flow rate and density. In this paper, measured data of crowd flow are reported, and utilizing these, validation of a computer simulator named SimTread is discussed.

## Methodology

A five-step process has been used to carry out the study, as follows;

Step 1: Experiments of crowd flow passing through the simple-shaped room
Step 2: analysis of the crowd flow at the opening in each experiment
Step 3: simulation of crowd flow in the same condition of experiments
Step 4: Calculations of flow rate at the opening in each simulation
Step 5: Comparison of the result between the experiments and the simulation

### *Procedure*

The experiments of crowd flow passing through a simple-shaped room were carried out. There were two test subject groups. Each group had 43 people. All subjects were college students, and the average age was 20 years old. The experiments were performed using a large laboratory room in a building. Experiment settings, such as the wall with opening, the corridor and the small room were made of cardboard boxes (Figure 1). There was no door at the entrance and the exit.

In total 18 tests were conducted, and every test was carried out twice by different subject group. The entrance and corridor width, room size and shape connected to the opening were changed and the subjects passed through settings (Table 1). The test subjects walked through the experiment settings. Before the experiments, we indicated them, "Please walk safety and a little rapidly as you are in the evacuation drill of the building." Three video cameras were mounted overhead on the ceiling to record the time passing through the opening. Then, flow rate through the entrance and exit and a walking speed was calculated.

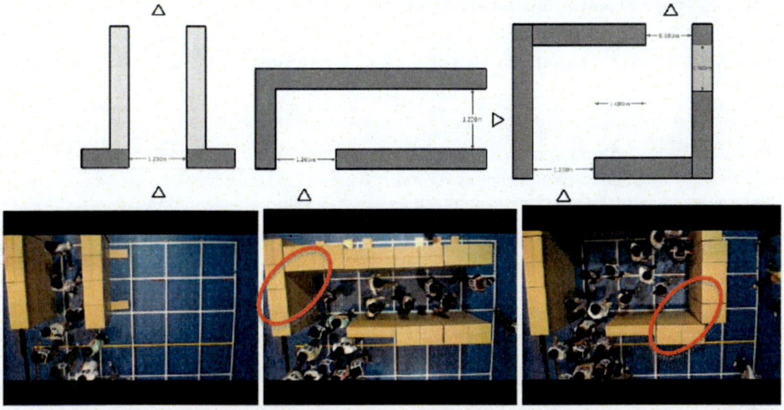

Fig. 1. Experiment settings (I shape, L shape, Small room). I shape Corridor (exp.02,03) L shape Corridor(exp.04-10) Small Room(exp.11-13).

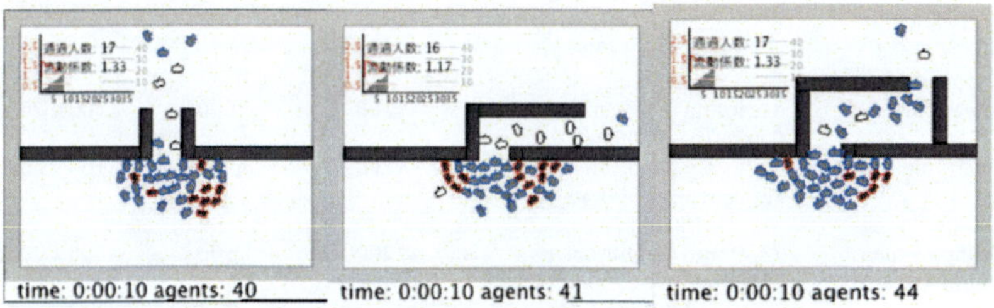


Fig. 2. Image of evacuation simulation (sim) result

## *Configurations for Evacuation Experiments*

The experiments were categorized into four groups to investigate the characteristics of crowd flow according to the room shape and size. (Table 1). The category for analysis is as follows;

Category I: I shape corridor, with or without wall.
Category II: L shape corridor, changing the exit (corridor) widths.
Category III: L shape corridor, changing the entrance widths.
Category IV: Small room, changing the room size.
Category V: Opening, changing the opening widths

**Table 1. Type of crowd flow experiments**

| Category of Experiment | Type of Experiment | Entrance Width | Exit Width | Size of the room |
|---|---|---|---|---|
| Category I: | 01. Opening: Entrance. 1200mm-Exit.1200mm | 1200 | 1200 | - |
| I Corridor | 02. Corridor (I): Entrance. 900mm-Exit. 900mm with wall | 900 | 900 | - |
| (with or without Wall) | 03. Corridor (I): Entrance. 1200mm-Exit. 1200mm with wall | 1200 | 1200 | - |
| Category II: | 04. Corridor (L): Entrance. 1200mm- Exit. 1200mm | 1200 | 1200 | - |
| L Corridor | 05. Corridor (L): Entrance. 1200mm- Exit. 1500mm | 1200 | 1500 | - |
| (Changing the Widths of the Corridor) | 06. Corridor (L): Entrance. 1200mm-Exit. 1800mm | 1200 | 1800 | - |
| | 07. Corridor (L): Entrance. 1200mm-Exit. 2100mm | 1200 | 2100 | - |
| Category III: | 08. Corridor (L): Entrance. 1200mm-Exit. 1200mm | 1200 | 1200 | - |
| L Corridor | 09. Corridor (L): Entrance. 1500mm-Exit. 1200mm | 1500 | 1200 | - |
| (Changing the Widths of the Entrance) | 10. Corridor (L): Entrance. 1800mm-Exit.1200mm d.2000mm | 1800 | 1200 | - |
| Category IV: | 11. Room (S): Entrance. 1200mm-Exit.900mm d.0mm | 1200 | 900 | 2100 x 2000 |
| Small Room | 12. Room (S): Entrance. 1200mm-Exit.900mm d. 1000mm | 1200 | 900 | 3100 x 2000 |
| (Changing the Room Size) | 13. Room (S): Entrance. 1200mm-Exit.900mm d.2000mm | 1200 | 900 | 4100 x 2000 |
| Category V: | 14. Opening: Entrance. 800mm-Exit.800mm | 800 | 800 | - |
| Opening | 15. Opening: Entrance.900mm-Exit.900mm | 900 | 900 | - |
| (Changing the Widths of the Exit) | 16. Opening: Entrance. 1000mm-Exit. 1000mm | 1000 | 1000 | - |
| | 17.Opening: Entrance. 1100mm-Exit.1100mm | 1100 | 1100 | - |
| | 18.Opening: Entrance. 1200mm-Exit. 1200mm | 1200 | 1200 | - |

## Results of Crowd Flow Experiments

### *Analysis of Crowd Flow at the opening in various condition of experiment*

Flow rate at the opening is stable despite the entrance width (2.17 ~ 2.26 persons/m/s) (Category V: exp.14-18). And the results for various configurations of experiments are summarized in Figure 3 and Figure 4. Figure 3 shows the average flow rate in two experiments at the entrance and the exit. Figure 4 shows the average walking speed in two experiments.

**Flow rate:** Figure 3 shows flow rate at the entrance and the exit. Flow rate at the entrance almost exceeded 1.5 person/m/second, which is used as a standard flow rate value for evacuation calculation in Japanese Building Law.

- Category I: Flow rate at the opening without walls (exp. 01) was larger than that of the opening with walls (exp. 03). Subjects walked and extended transversely after passing the opening without wall. And the density descended there.
- Category II: Flow rate at the entrance increased, as the exit (corridor) width is wider (exp. 04 – 07).
- Category III: Flow rate at the exit was stable despite the entrance width (exp. 08 – 10). The left end of the corridor becomes a bottleneck that decides the total flow rate.
- Category IV: Flow rate at the entrance increased, as the room size is larger (exp. 11 – 13).

Flow rate at the opening varies according to the situation of the space connected. Flow rate rises if the space connected to the opening is wide.

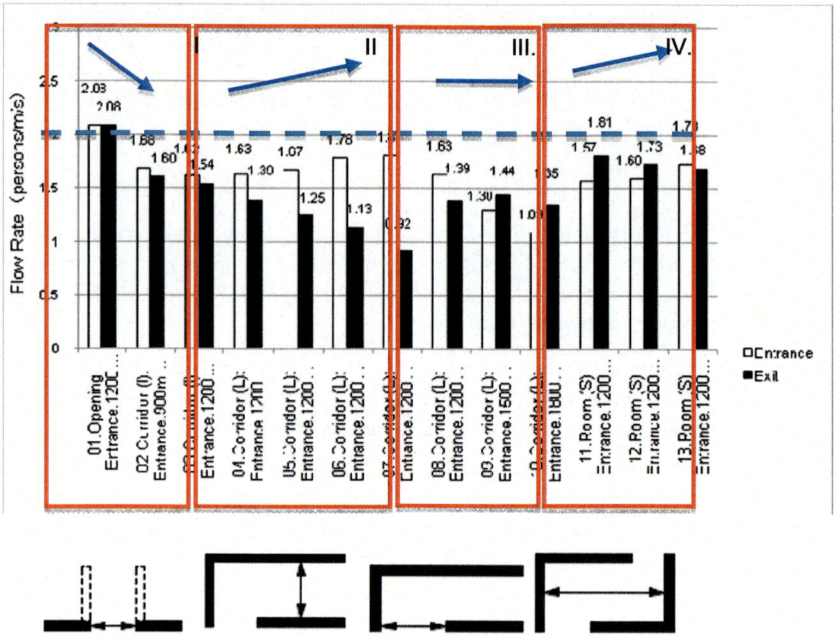


**Fig. 3. Flow rate**

**Walking speed:** Figure 4 shows walking speed in the corridor and the room.

- Category I: walking speed passing through the corridor increased as the opening becomes wider (exp. 02, exp. 03).
- Category II: walking speed in the corridor increased, as the exit (corridor) width is wider (exp. 04 – 07).
- Category III: walking speed in the corridor was stable despite the entrance width (exp. 08 – 10). Because the left end of the corridor becomes a bottleneck and the situation was similar at the right side.
- Category IV: Walking speed increased as the room size becomes larger (exp. 11 – 13).

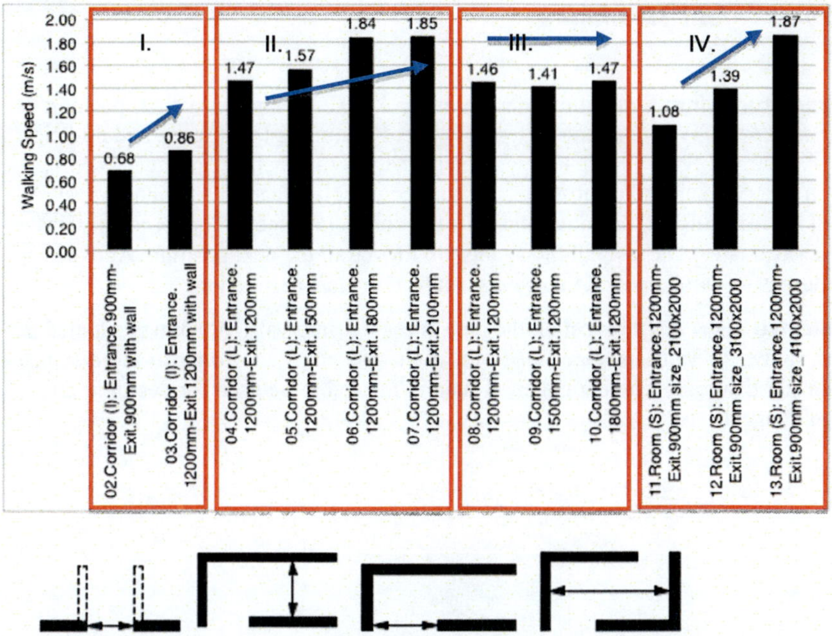


**Fig. 4. Walking speed**

# Pedestrian Simulator "SimTread"

## *Elements of the Model*

SimTread is a type of multi-agent model, where the characteristic of a crowd flow is made from movement of individual agents to which the same behavioral rule is given.

Model elements of SimTread are as follows;

- Agent, a walker: position, direction, and maximum speed,
- Space, building plan: obstacles such as walls and furniture, and
- Destination: target to which agents walk toward.

Operation starts with drawing of a plan, then arrangement of agents and destinations follows next; those are prepared on a CAD drawing software. Figure 5 illustrates it using a simple plan.

### *Process of the simulation*

**Time processing:** Simulation runs on every $\Delta t$, 0.2 seconds. At first, position at the next $\Delta t$ is given temporarily to every agent according to its speed and direction at the current time. Next follows determination of whether any agent conflicts with other agents and/or obstacles. If any conflict is found, the temporal position will be recalculated, repeating until all the conflicts dispel. This process will be described later. The last, all the agents are moved to the next position. Above procedure is repeated until all the agents arrive at the last destination.

**Potential map:** A grid is set on the plan at regular intervals; 100 mm is adopted in our study. A value is apportioned on each grid point according to the distance from destinations, where a potential map is formed to be used for deciding direction of agents.

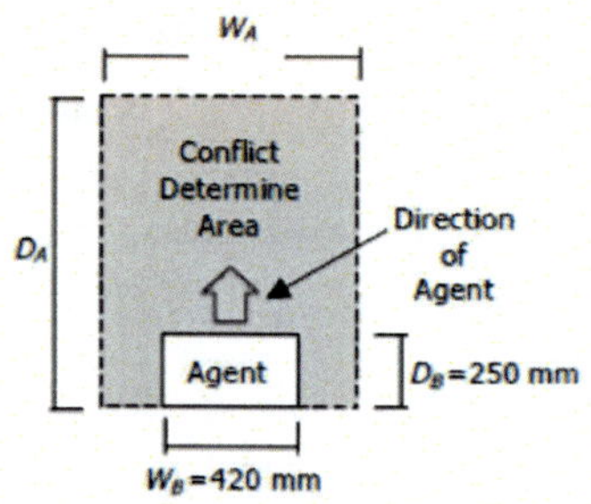


**Fig. 5. An example of making a model on a CAD**

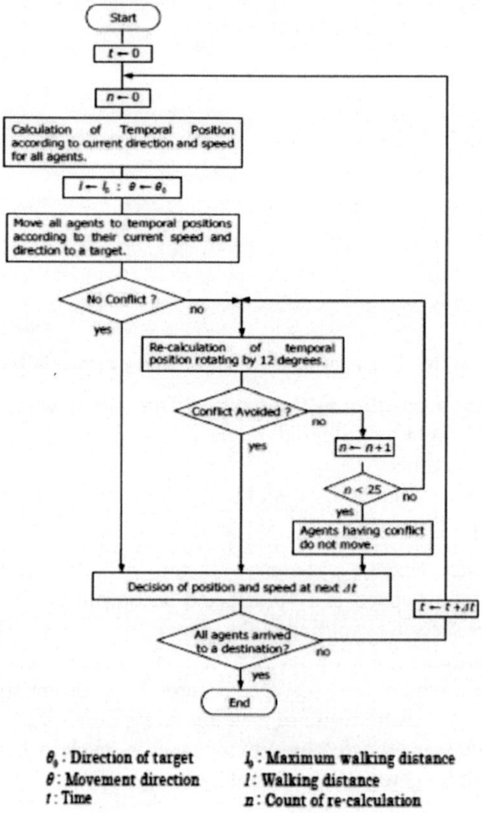


**Fig. 6. Flow diagram of SimTread**

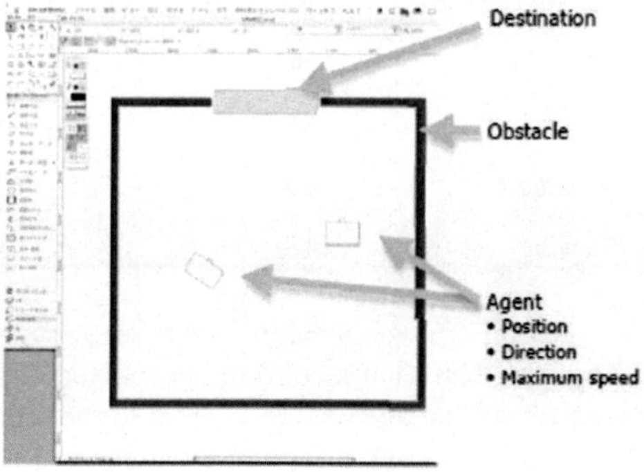


**Fig. 7. Configuration of agent and conflict determine area**

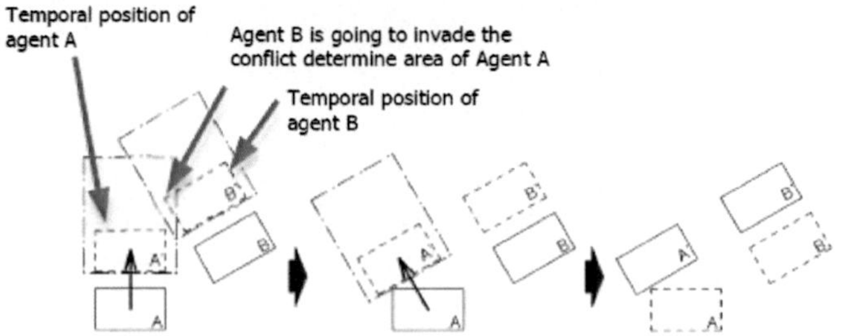


**Fig. 8. Deciding process of next position by recalculation**

**Calculation of agent's position at the next $\Delta t$:** Direction toward next $\Delta t$ is calculated from a slope formed by potential values of five grid points, the closest point and neighboring four points.

Temporal position is decided by both the direction obtained from the slope and walk distance in a $\Delta t$ (Figure. 8).

**Parameters of agents:** Size of an agent is fixed to width 420mm, depth 250 mm in SimTread. Maximum speed can be selected according to character of human flow; 1.3 meters per second is applied in this paper.

A "Conflict Determine Area" is set up around an agent to examine having conflicts or not, shape of which has two types according to current speed of the agent (Figure7 and Table 2). As the result of a simulation is controlled by these parameters, the actual value was adjusted by previous studies such as data by various investigations and results by regal calculation method in Japan.

**Table 2. Specifications of Conflict Determine Area**

| Object of Conflict | Current Speed | $W_A$ [mm] | $D_A$ [mm] |
|---|---|---|---|
| With Agents | Maximum | 560 | 810 |
| | Slowing | 420 | 120 |
| With Obstacles | Maximum | 440 | 600 |
| | Slowing | 420 | 210 |

**Mechanism of avoiding conflict:** That a conflict is expected or not is determined by as follows; at first, as for all the agents, move them to the temporary positions, then check if other agents and/or obstacles are in the area. If anything is found in any area, it means that conflict will happen in the next $\Delta t$, then the position will be recalculated as far as conflicts dispelled.

**Recalculation of the temporary position:** In case a conflict is expected, the agent changes the direction to rotate 12 degrees increments until it can be avoided. If, however, all else fails speed of the agent will be reduced to 70 %, which will be repeated until the conflict is avoided. When above all processes failed, the agent keeps the previous position (Figure 8). After recalculation to all the conflicting agents, all the agents will be moved to new position and will be recalculated to check conflict again. This repetition is limited to 25 times in single $\Delta t$. Positions of all agents at next $\Delta t$ are not decided one after another, but they are decided when all the conflicts in the plan disappear. Therefore, values of calculated positions do not depend on the order of calculations.

### *Output of the simulation*

SimTread outputs next two results;

- Movies, in which you can see agents moving with elapse of time and number of agents finished evacuation. Besides, statistical graphs that illustrate the number of agents and flow rate at designated spot can be expressed on this frame as well.
- Log data of position, direction and walk distance of all the agents with every $\Delta t$ as a text file. Also, flow rates of simple moving average in five $\Delta t$ are provided. These data are used for various analysis of crowd flow.

## Comparison between Experiments and Simulation.

Comparisons between experiments and simulation in various configurations are summarized in Figure 9. Flow rate at entrance and exit for each configuration are shown with results of experiment (Exp-) and simulation (Sim-) in pair. The numerical values of flow rate in the simulations are shown in Figure 9. And flow rate data in the experiments are identical to those in Figure 3.

Figure 10 indicates the ratio of flow rate in simulation to that in experiments. The ratios mostly ranged from 90% to 110%. This represents that the result of experiment and simulation has a good agreement.

Through the detailed investigation of the movies, we found similarity in crowd behavior between experiment and simulation, such as the area occupied by pedestrians (Figure 1 and 2).

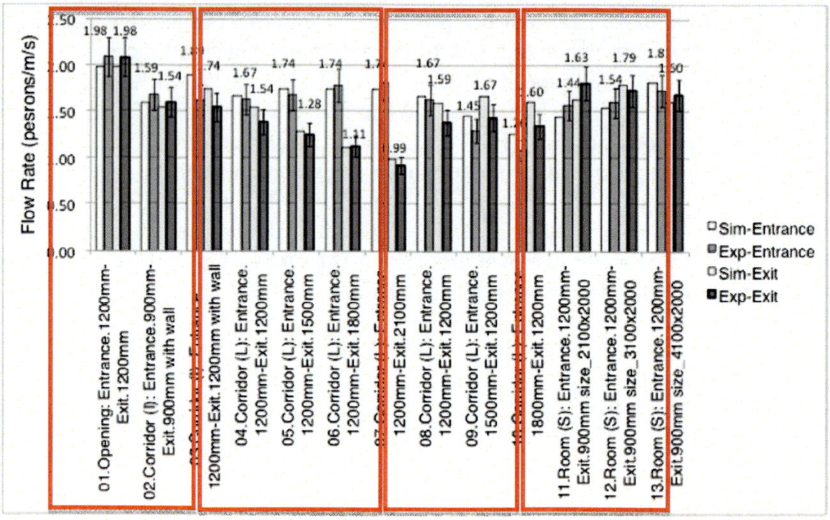


**Fig. 9. Comparison between Experiment and Simulation (Flow rate)**

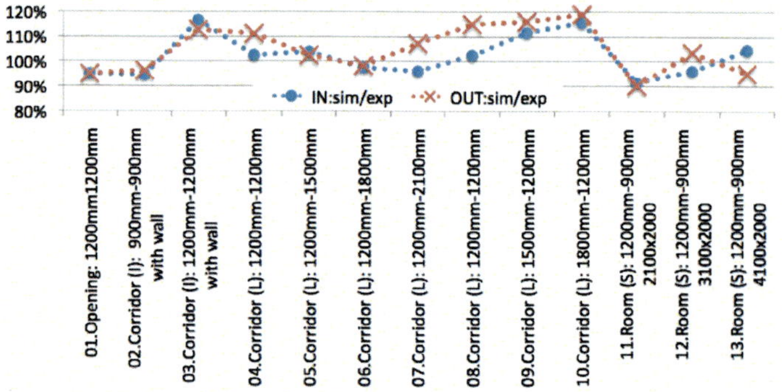


**Fig. 10. Ratios of flow rate in Experiments to Simulation.**

## Conclusion

This study shows the characteristics of crowd flow passing through a simple-shaped model. Flow rate at the entrance and the exit of experiments under various configurations were analyzed.

The results show that flow rate at openings changes depending on the density of the space connected. Flow rate rises if the opening is connected to larger space which is less dense and, in result, increases speed of pedestrians. Flow rate of the openings were compared in experiments and simulations with equivalent configurations. And there was a good agreement between the result of experiments and simulations. The differences were less than 10%.

**Acknowledgments** These experiments in the paper are funded by KAKENHI, Grant-in-Aid for Scientific Research (20560593). The author would like to thank the Building Research Institute for offering the large laboratory for the observation of evacuation experiments. Thanks are also due to Dr. Ken NUNOTA and Dr. Ichiro Hagiwara from Building Research Institute for advice of the experiments.

## References

1. Kady, R.A., Gwynne, S., Davis. J.: A Review of the Sources of Occupant Performance Data Used in Building Evacuation Models. In: 4th International Symposium on Human Behaviour in Fire 2009. Cambridge, UK.

# Calculating and Verifying the Staircase-length for Evacuation Analysis

**Hyun-seung, Hwang.[1] Jun-ho, Choi.[2] and Won-hwa, Hong.[3]**

[123] School of Architecture, Kyungpook National University.
#1370 Sankyuk-Dong, Buk-gu, 702-701, Daegu, South Korea.
[2]Corresponding author: kissist@knu.ac.kr

**Abstract** Recently, evacuation analysis, which is one of the performance-based evacuation design methods, has been widely used to estimate the egress capacity of buildings and analyze evacuees' characteristics. But almost all the models do not consider accurate evacuation scenarios because of the limitation of simulations, especially when setting up the staircase-length in building geometry mode. Therefore, this study selected a subject of a high-rise building with 351 invited participants, and conducted an experiment to investigate the evacuation from the 30th floor to the ground level and then classified overall 5,265 movement cases to calculate the participants' evacuation route. After the trial experiment, the paper calculated the precise average distance of participants' evacuation route in staircases considering each zone's length and dimension and derived the numerical equations for inputting the parameter of staircase-length in evacuation models. And the verifications were also conducted by comparing simulations with the results of the experiment.

## Introduction

Evacuation simulation models, which try to mimic a real evacuation scenario as precisely as possible, have been generally used as an alternative to the full-scaled investigation into the occurrence and characteristics of fire, smoke flow and occupants' decisions. However, many evacuation simulators have the delicate problem when the user operates the simulation especially for high-rise buildings. One of the problems is that most simulators recognize the staircase as rectilinear only in the same level of floor plan; therefore, users must insert the true length of the staircase as measured by the slope length and consider the shape of the staircase-length. And staircases with landings should also be inputted as a total length represented by the sum of the sloping parts and the landings. But there are no accurate standards or examples which compute the staircase-length in this situation currently and standard research is not concerned with the pattern of evacuees' evacuation routes. Hence, this paper selected as a subject a high-rise building and executed a trial evacuation experiment. All procedures of the experiment were recorded on CCTVs while the 351 invited participants were descending the stair-

R.D. Peacock et al. (eds.), *Pedestrian and Evacuation Dynamics*,
DOI 10.1007/978-1-4419-9725-8_54, © Springer Science+Business Media, LLC 2011

case from the 30th floor to the ground level and the participants' movements in the stair landings were classified in all 5,265 cases that were divided among A, B and C zone-passed participants to derive the numerical equations of participants' passed distance and dimension through stair-landings. Finally, derived numerical equations about the staircase-length were verified by the simulation program and compared with actual results of the experiment.

## Geometric confusions of staircase-length

When evacuation simulator users run the occupants' evacuation in high-rise structures, they should connect each floor using the staircases and set up to make links with each other. Users could decide the staircase-width and staircase-length when connecting the staircases. At that time, the staircase-width can be inserted the correct size which is showed from the floor plan, but most users might have difficulties in setting up staircase-length in programs. Because users must insert the true length of the staircase measured as the slope length and staircases with landings have a total length represented by the sum of the sloping parts and the landing. The evacuation simulators mostly recognize the staircases following the same level of the floor plan as in *Fig.1*.

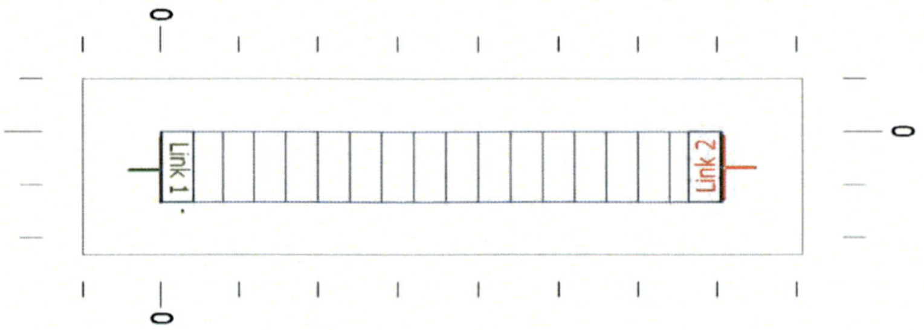


**Fig. 1. Representative staircase-image in Simulex**

The confusions occurred in this procedure when trying to determine the length of the staircase. Usually, some users insert the staircase-length as a decided average distance of staircase-length; others insert the staircase-length as a decided rectilinear staircase-length. There haven't been any standards or examples of calculating the staircase-length so that the guideline must be set up through the analysis of occupants' escape route on the stair landings in order to input the accurate staircase-length. The escaping routes of evacuees through staircases in high-rise buildings are also changed individually by the various factors such as overcrowding, congestion and empirical personal habits inside stairwells. However, simulation models could recognize the staircase-length only and there aren't parameters of evacuees' flexible evacuation routes. Hence, movement patterns of many participants should be analyzed through the repetitive experiments and derived as a numerical equation for simulating staircase-length.

## Procedure of the trial experiment

351 participants' evacuation route was recorded in the staircases from the 30th floor to the ground level of a high-rise building in South Korea for calculating the precise staircase-length in an evacuation model. The evacuation scenario was set up for placing the participants who did not recognize any fire situations in the pre-assigned locations before the occurrence of a fire alarm and they were descending to the final exit through the staircases after the fire alarm was operated. All of the participants' movements in staircases were recorded by CCTVs installed above the landings vertically. The paper analyzed how the participants decided their escape routes and which zone of the stair landings was passed by the participants the most frequently after the trial experiment. As the paper described before, users should input the precise staircase-length with the detailed slope angle and length because most simulators can not recognize stair landings and various shape of high-rise buildings' staircases. The size of the detailed staircase is the same as shown in *Fig.2* below.

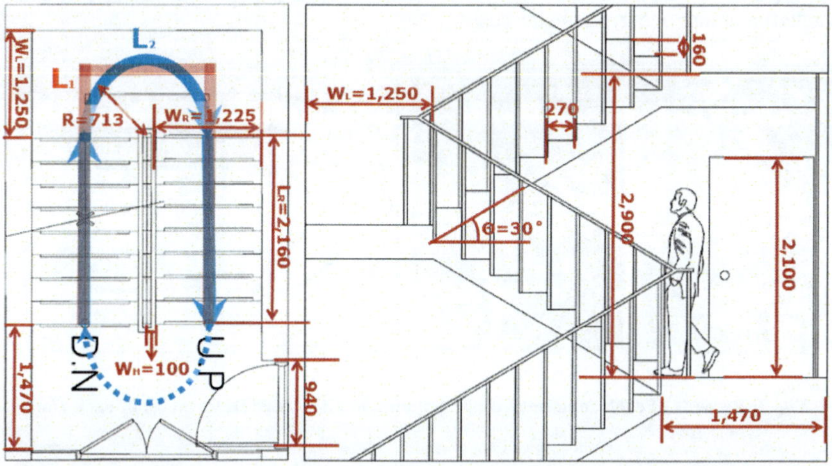


**Fig. 2. Plan & sectional size of the staircase (Including the length of $L_1$ and $L_2$)**

Almost all the users who were taking part in the simulation programs assumed that the occupants moved following the trace of $L_1$ or $L_2$ when they passed through stair landings. The equation is the ordinary method of most users for calculating the staircase-length mathematically;

$$L_1 = \frac{L_R}{\cos\theta} \times 2 + 2W_H + W_R + W_L \qquad (1)$$

$$L_2 = \frac{L_R}{\cos\theta} \times 2 + \pi R \qquad (2)$$

where; $L_1$ = Distance of participants' escape route following the trace of $L_1$

$L_2$ = Distance of participants' escape route following the trace of $L_2$

R = Radius of the center route on the stair landings; $\pi \fallingdotseq 3.14$

According to the equations, $L_1$ and $L_2$ are represented as *7,641mm* and *7,205mm* respectively. Although many users applied the staircase-length results as calculated, the paper classified as single-passed participants and simultaneous-passed participants at first for investigating the density of evacuees on stair landings and also classified three cases divided into A, B and C zone like *Fig.3*. Because analyzing each case of zone-passed participant data can allow deduction of the average distance of evacuees' route in staircases. There was no overcrowding when the single-passed participants through the stair landings, while the simultaneous-passed participants through the stair landings signified that congestion or evacuation delay occurred in staircases.

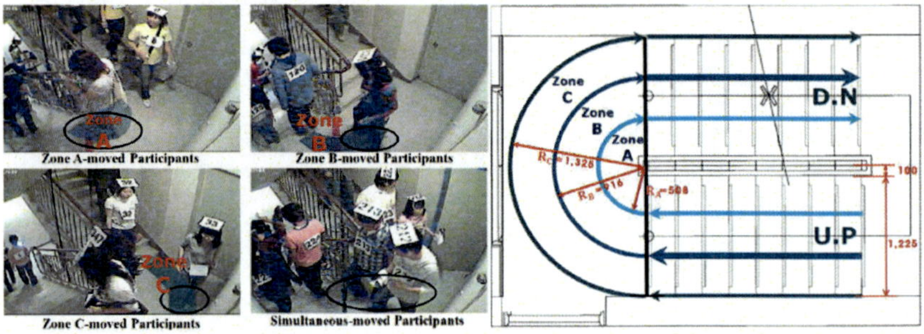


**Fig. 3. Records of each zone-passed participants in CCTVs & Description of each zone**

## Results determination and verification

### *Results of trial experiment in the staircase*

In order to determine the staircase-length in the parameter of evacuation simulators through analyzing the participants' escape route in stair landings using the CCTVs' records, the results are presented as shown in *Table 1*. Simultaneous-passed participants' cases were regarded as two cases because two participants passed through the stair-landings simultaneously.

**Table 1. Indicating which group each one represents**

| | Zone A-passed Cases | Zone B-passed Cases | Zone C-passed Cases | Simultaneous-passed Cases | Total |
|---|---|---|---|---|---|
| Number of cases | 1,057 | 1,118 | 1,588 | 751(x2) | 5,265 |

Zone C-passed cases were the most frequent and zone A-passed cases were the least frequent in the investigation results. The ratio of single-passed cases and simultaneous-passed cases are as shown in *Fig.5* below.

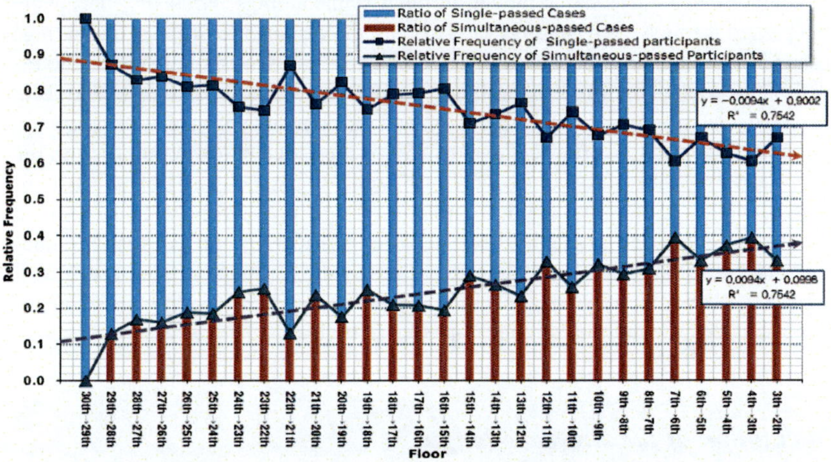


**Fig. 4. Relative frequency of passed participants through the stair landings of each floor**

The relative frequency of simultaneous-passed participants increased and the relative frequency of single-passed participants gradually decreased while the participants descend from the 30th floor to the ground, because the staircases were overcrowded with the participants who descended from each departed floor to the final exit on the ground and the fluctuation and congestion also occurred inside the staircase as shown in *Fig. 4* above.

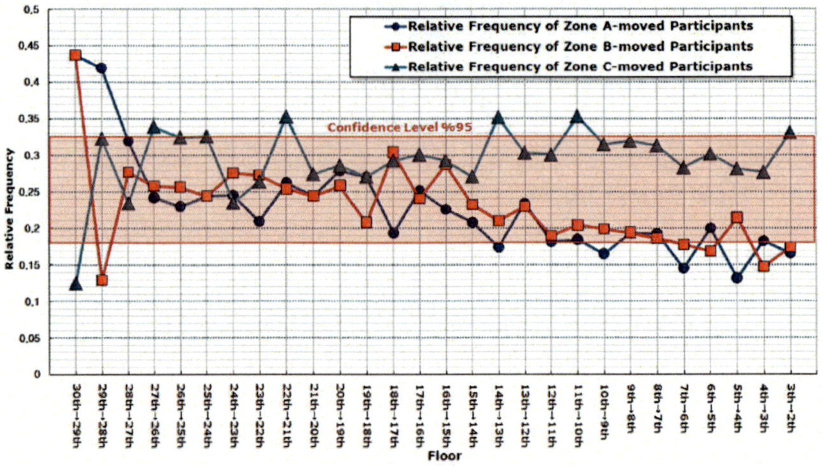


**Fig. 5. Relative frequency of each zone-passed participant through the stair landings**

Fig. 5 illustrates which zone is the most frequently used by participants when the single-passed participants moved through stair landings. Basically, occupants who passed through stair landings rarely break the center-line of the stair landings because the radius of occupants' physical characteristics was smaller than the radius of the center-line in stair landings. So if there are no obstacles blocking in staircase, relative frequency of zone C-passed participants only appeared sparsely. But the graphs in the confidence level of 95% represented evidence contrary to the general theory. The relative frequency of zone C-passed participants also increased, while on the other hand, the relative frequency of zone A and B-passed participants decreased gradually because of overcrowding and congestion situations that occurred inside the staircases. In descending evacuation, the centrifugal force of participants' movement occurred by the descending velocity is also one of the reasons that the number of zone C-passed participants increased.

## *Calculation of the staircase-length*

The precise staircase-length in simulation analysis was calculated by using the two method-equations through analyzing the represented experiment data. The equation for calculating the staircase-length is used the average escape distance and dimensions mathematically;

≫The first method for calculating the participants' average escape distance using the center-radius of each zone in the stair landings

$$L_3 = \frac{\pi R_A^{'} \cdot \sum_{f=2}^{30} M_A + \pi R_B^{'} \cdot \sum_{f=2}^{30} M_B + \pi R_C^{'} \cdot \sum_{f=2}^{30} M_C}{\sum_{f=2}^{30} (M_A + M_B + M_C)} + \frac{L_R}{\cos\theta} \times 2 \qquad (3)$$

where: $L_3$ = Staircase-length using distance of the half of width in each zone when participants passed through the stair landing;

$R_A^{'} = 0.5(R_A - W_H) + W_H$; $R_B^{'} = 0.5(R_B - R_A) + R_A$; $R_C^{'} = 0.5(R_C - R_B) + R_B$

$M_A$ = Zone A-passed cases of each floor;

$M_B$ = Zone B-passed cases of each floor;

$M_C$ = Zone C-passed cases of each floor

f = Floor; $\pi \fallingdotseq 3.14$; $\cos 30° \fallingdotseq 0.87$

As a result, $L_3$ represented the staircase-length equaled *7,384mm*.

≫The second method for calculating the participants' average escaping distance using each zone's dimension passed by participants in stair landings

$$S = \frac{0.5\pi R_A^{''} \cdot \sum_{f=2}^{30} M_A + 0.5\pi R_B^{''} \cdot \sum_{f=2}^{30} M_B + 0.5\pi R_C^{''} \cdot \sum_{f=2}^{30} M_C}{\sum_{f=2}^{30} (M_A + M_B + M_C)} \qquad (4)$$

$$L_4 = \pi \sqrt{\frac{2S}{\pi}} + \frac{L_R}{\cos\theta} \times 2 \qquad (5)$$

where; S = The average dimension of passed participants on stair landings

$L_4$ = Staircase-length using the each zone's dimension passed by participants in the stair landing

$R_A^{''} = R_A^2$; $R_B^{''} = R_B^2 - R_A^2$; $R_C^{''} = R_C^2 - R_B^2$; f=Floor; $\pi \fallingdotseq 3.14$; $\cos 30° \fallingdotseq 0.87$

As a result, $L_4$ represented the staircase-length equaled *7,607mm* using each zone's dimension following above equation.

## *Evaluation of the results*

The results are as below through the trial experiment for calculating staircases-length in high-rise buildings.

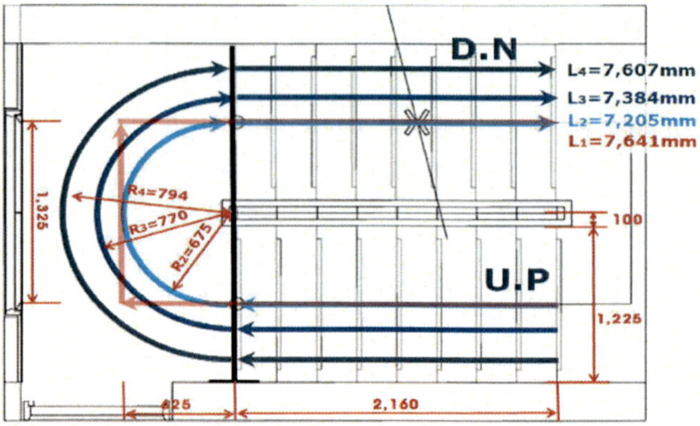


**Fig. 6. Experiment results of the calculating the staircase-length**

The numerical value of $L_3$ and $L_4$ analyzed the escape movement of the participants in a high-rise building experiment is bigger than the numerical value of $L_2$ inputted by almost all the simulation users. The numerical value of $L_1$ is the biggest value of all methods which calculates the staircase-length. However, there were no participants who moved through $L_1$ in the recorded CCTVs because ordinary occupants tend to evacuate through stair landings searching for the optimum angles. And the reasons that the numerical value of $L_3$ and $L_4$ is bigger than $L_2$'s value were occurrences of overcrowding, delay and congestion phenomenon inside the staircases and it was impossible for participants to make use of the most efficient escape route and were gradually shoved out of the center-line of the staircases' escaping route due to the centrifugal force. Finally, the value of $L_3$ and $L_4$ were different by as much as 2.5%~5.6% from the value of $L_2$. When inputting the staircase-length of the high-rise building in evacuation models, if one of the staircase-length is different by approximately 5% from the real length of the staircases, differences of the final evacuation time will be even more and more expanded. Thus, inputting the staircase-length must be executed precisely during the simulation of the evacuation models.

### *Verification of the staircase-length*

Simulation results inputting parameter of staircase-length equations used by most users and derived equations by this paper in Simulex are as shown in *Fig.7* below.

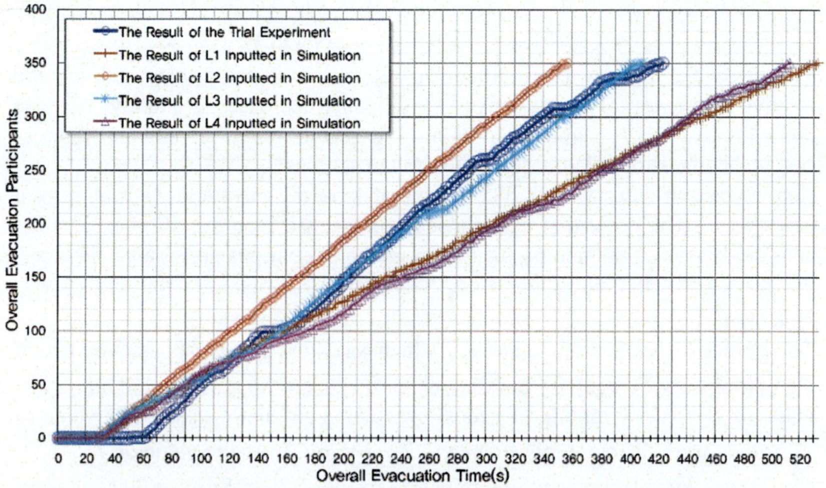


**Fig. 7. Comparison with the results of trial experiment and simulation**

The overall evacuation time of the trial experiment is *422s* and simulations results of $L_1$, $L_2$, $L_3$ and $L_4$ were represented as *530s, 308s, 406s* and *510s* respectively; therefore, difference of results reach about 72% at the most and the fluctuation phenomenon can be shown if the height of a building is taller and taller. As $L_3$ result approaches the closest to the result of trial experiment, therefore, the $L_3$ length can increase the reliability of simulation results.

## Conclusion

This paper analyzed the participants' escape routes in stair landings and operated the procedure for revising the confusing problem of how to calculate the precise staircase-length reflecting the slope and stair landings in the evacuation analysis. For that, this study executed a full-scaled experiment, classified as single-moved participants or simultaneous-moved participants and A, B and C zone-passed participants on the stair landings. It was observed which zone is the most frequent passed by participants when the single-passed participants passed through stair landings. The verification was also executed through the comparisons of simulation and trial experiment results.

1. The cases of passed participants through the stair landings of each floor appeared 5,265 cases except 351 cases in which the participants didn't reach the staircases.
2. Most users who performed the simulators assumed that the occupants moved following the trace of $L_1$ or $L_2$ when they passed through the staircase and stair landings. As a result, $L_1$ represented the staircases-length equaled *7,641mm* and $L_2$ was *7,205mm* reflecting the slope of staircases.
3. The relative frequency of simultaneous-moved participants increased and the relative frequency of single-moved participants decreased gradually, because the staircases were overcrowded with the descending participants and fluctuation and congestions also occurred inside the staircase.
4. Zone C-passed cases were 1,588 cases that were the most and Zone-A passed cases were 1,057 cases that were the least on the stair landings.
5. This study was calculated by using two method-equations considered the average passed distance and dimension. As a result, $L_3$ was *7,384mm* and $L_4$ was *7,607mm*.
6. The numerical values of $L_3$ and $L_4$, which analyzed the escape movement of the participants in high-rise building experiments are bigger than the numerical value of $L_2$ usually inputted by most evacuation simulator users. The value of $L_3$ and $L_4$ were different by as much as 2.5%~5.6% from the value of $L_2$.
7. The overall evacuation time of simulation result of $L_3$ in Simulex has the closest value with the result of a trial experiment, so it can be regarded as that $L_3$ staircase-length has the highest reliability of all.

**Acknowledgments** The research presented in this paper was supported by Ministry of Education, Science and Technology through the 'Global Expert Training Center for Disaster Prevention' of the Second Stage of Brain Korea 21 and 'Development of Smoke Control System of a High-rise Building', as a part of the flagship research project for development of the next generation fire safety technology funded by the National Emergency Management Agency.

## References

1. Paul, J.L.: Calculating Evacuation Times for Tall Buildings. Fire Safety Journal. Vol. **12**, 1987, pp 213-236.
2. Ando, K., Ota, H., Oki, T.: Forecasting the Flow of People. Railway Research Review. **45**, 8, pp. 8-14 (1988)
3. Daly, P.N.: Pedestrian Speed/Flow Relationships for Underground Stations. Traffic Engineering and Control. **32**, 2, pp. 76-78 (February 1991)

4. Thompson, P.A., Marchant, E.W.: A Computer Model for the Evacuation of Large Building Populations. Fire Safety Journal. **24**, pp. 131-148 (1995)
5. Thompson, P.A., Marchant, E.W.: Testing and Application of the Computer Model 'Simulex'. Fire Safety Journal. **24**, pp. 149-166 (1995)
6. Galea, E.R.: A general approach to validating evacuation models with an application to EXODUS. Journal of Fire Science. **16**, 414-436 (1998)
7. Ashe, B., Shields, T.J.: Analysis and Modelling of the Unannounced Evacuation of a Large Retail Store. Fire and Materials. **23**, pp. 333-336 (1999)
8. Ko, S.Y., Spearpoint, M.: Comparison of Evacuation Times Using Simulex and EvacuatioNZ Base on Trial Evacuations. Fire Engineering Research Report, University of Canterbury (2003)
9. Thomson, P.A.: "Simulex: simulated people have needs too." NIST Workshop on Building Occupant Movement (2004)
10. Galea, E.R., Gwynne, S., Lawrence, P.J., Filippidis, L., Blackshields, D.: User Guide and Technical Manual for building EXODUS. University of Greenwich (2004)
11. Kuligowski, E.D., Peacock, R.D.: Review of building evacuation models. NIST Technical Note 1471, Building and Fire Research Laboratory (2005)
12. Choi, J., Hwang, H., Hong, W., Choi, Y.: Analysis on Occupants' Escape Speed and Reason of Bottle-neck Occurrence through the Trial Evacuation Experiment at a High-rise Apartment Housing. Human Behaviour in Fire 2009: 4th International Symposium (2009)
13. Peacock, R.D., Averill, J.D., Kuligowski, E.D.: Stairwell evacuation from buildings: What we know we don't know. NIST Technical Note 1624, Building and Fire Research Laboratory (2009)

# Vertical Egress

# Addressing the Needs of People using Elevators for Emergency Evacuation

**R. W. Bukowski, P.E., FSFPE**

Rolf Jensen & Associates
Laurel, Maryland 20708 USA

**Abstract** US model codes and building regulations are recognizing the provision of protected elevators for occupant self-evacuation after more than two decades of training people that elevators are unsafe in fires. This reversal will require that people can readily identify those elevators that are safe to use and be provided with information and reassurances during use that the system is functioning safely. Lengthy discussions on the interactions between the systems and users have resulted in requirements for visual, audible, and voice messaging systems and operational protocols designed to provide reliable, real-time information needed by users to make informed decisions. The discussions have further identified the need for public education to provide for effective use by infrequent visitors to buildings equipped with these systems.

The paper will discuss the approaches being developed to address these needs by a consortium of public and private organizations including the American Society of Mechanical Engineers (ASME), National Institute of Standards and Technology (NIST), National Elevator Industry Inc. (NEII), disability advocacy groups, and the fire alarm and model building code developers. There is an expectation that since a building's elevators are used daily by the occupants, by keeping the system used in emergencies as close as possible to normal use, the provision of additional information on status and safety will represent sufficient reassurance to users.

One of the outstanding issues identified in the discussions is the need for testing and verification that the approaches will be effective. Due to the difficulties inherent in human testing, there is a need for the inclusion of these features into observational research being conducted through required evacuation drills. Since elevator use is not prohibited now for non-fire emergency egress, this may provide the opportunity to test public response to the approaches being contemplated. The paper will suggest ideas for such research being included in planned studies and as a part of building commissioning.

R.D. Peacock et al. (eds.), *Pedestrian and Evacuation Dynamics*,
DOI 10.1007/978-1-4419-9725-8_55, 

## Background

Passage of the National Construction Safety Team Act (*PL 107-231*) in October of 2002 marked the beginning of the National Institute of Standards and Technology's (NIST) federal investigation of the collapse of the World Trade Center Towers on September 11, 2001. Early concern about the long times required for fire service access and for occupant egress from the Towers led NIST to approach the American Society for Mechanical Engineers (ASME) committee responsible for the elevator code [1] to discuss the feasibility of elevators that could be used safely during fire incidents. That group decided to organize a workshop to gage the opinion of the fire service and elevator industry on this subject. The workshop was held in March, 2004.

The workshop concluded that, while there were technology and human factors issues, these could be resolved and the benefits of fire safe elevators were worth the effort. The technology and operational issues have been discussed in a series of papers by this author [2,3,4,5,6,7,8] the human factors issues are the primary focus of this paper. Following the workshop, two task groups of the A17.1 Emergency Operations Committee were formed to address the issues and to develop requirements for the model building codes and the Elevator Code. Following many hours of work by these task groups provisions were adopted in the 2009 editions of both US model building codes [9,10] and provisions for ASME A17.1 are being written.

## Elevator Safety

The elevator industry is extremely safety conscious and (statistically) elevators are the safest mode of transportation. These high levels of safety are achieved by a philosophy that many safety controls are provided and any hint of failure causes the system to stop. Occupant entrapment is considered acceptable because the occupants are safe and a system is in place by which responsible parties can be notified (by the in-car phone) and immediately respond to get occupants out safely.

It is recognized that there are two circumstances when entrapped occupants might be at risk; during earthquakes and fires. Elevators in seismic zones incorporate sensors and a special earthquake operating mode where sensor activation by a lateral acceleration exceeding 0.15g causes the elevator to stop and then move slowly to the nearest landing where the passengers are let out and the elevator is automatically taken out of service. In fires, smoke detectors located on every floor near the landing doors and in the machine room initiate a recall (called Firefighters Emergency Operation, FEO) of the cars to a designated level (usually the main lobby) any time smoke directly threatens the elevator system. In this condition, elevators cannot be used by occupants but can be used under manual control by the fire service, by use of a special key.

Two decades of experience with elevators equipped with FEO show that the system is safe in fires because most occupants use the stairs and the elevators are taken out of service before a fire can compromise their safety. Other system safety provisions incorporated in this system configuration include a fully sprinklered building, elimination of the shunt trip that disconnects power before sprinkler water is released onto the elevator brakes, standby power, protected wiring, and an operational protocol that is coordinated with firefighting practice so that the egress operation runs smoothly.

## Human Factors

A building's elevators are the everyday means of ingress and egress for nearly all occupants. Thus the need for special training is minimized where the use in emergencies is as similar as possible to normal. It can be expected that there will be some natural apprehension in a fire evacuation, but this can be minimized by providing information that can be used by occupants to make informed decisions. Concerns by the fire service can be reduced by providing the means to monitor the system in real time and to make necessary adjustments if problems occur. These are the observations that shaped the human factors provisions in the code development process. The following includes references to code provisions in brackets. The required operational protocol considers fire service operations in high rise incidents and the needs of building owners and managers to limit unnecessary disruptions. Thus (in a building provided with both fire service access[1] and occupant egress elevators) on any alarm to which the fire department is responding, the designated fire service elevator(s) are recalled to the main level to await fire department use on their arrival. The remaining elevators are used to evacuate the fire zone consisting of the fire floor and two floors above and below, so that occupants potentially in harm's way are safe and the floors on which the fire department will establish their forward command and stage firefighters and equipment, are clear. Once the fire zone is evacuated the remaining elevators are recalled temporarily to limit occupant movement into and within the building. The voice communication system is used by the fire department to provide instructions and information to remaining occupants during their evaluation.[IBC 2009 3008.5]

If the situation warrants full building evacuation that decision is made by the Incident Commander and is initiated manually from the fire command center. The occupant elevators begin to unload the building from the top down to first remove people with the longest distance to travel. Occupants wait for elevators in the pro-

[1] Fire service access elevators are required in new buildings >120 ft (36.6 m) [IBC 2009, 3007; NFPA 5000-2009, 54.12] and follow similar requirements. See references 2, 3, and 4.

tected lobby on every floor and are provided continuous information on the status of the evacuation. Indicator lights (green light and associated text "Elevators in evacuation service" or red light "Elevators out of service, use the stairs") and a display showing estimated time to elevator arrival (to the nearest minute) provides reassurance and information needed to decide to wait or take the stairs [IBC 2009 3008.12; NFPA 72-2010, 21.6.2]. Direct access to a stair from the lobby (without the need to go back onto the floor) permits occupants to take the stairs to the street or to reenter the elevator lobby at another floor if they choose. [IBC 2009 3008.11.1]. Two-way communication to the fire command center permits speaking directly to the fire department with special information such as an injured person needing immediate attention. [IBC 2009 3008.13]

The importance of the required real time information and communication systems to reassuring occupants and facilitating an orderly and effective evacuation cannot be overstressed. In their paper on human behavior in fire, Fahy and Proulx [11] state,

> *"It has been stressed repeatedly by human behavior experts that what is needed for occupants to make timely decisions is information. By providing information, people can refine their situation awareness, making them more competent at weighing their options before engaging in proper actions. During a fire emergency, information can take many forms. There is the information that should be provided prior to an event through education and training. If occupants cannot be trained, as in shopping malls or airport terminals, it is essential that staff in place has received training and are fully aware of their role as leaders in case of an emergency. At the time of the event, information that is provided by signage, announcements and staff should provide the timely information to support occupants' decision-making. After the event, debriefing the occupants regarding what happened, what went well or less well should take place so that occupants can understand the situation and be better prepared for a future event. Information is the key to a successful building evacuation during an emergency."*

## Real Time Monitoring

The fire department will be managing the incident from the fire command center from which specific conditions that relate to the safety of the evacuation can be monitored. Heat detectors capable of reporting actual temperatures are required in the lobbies on every floor and in the elevator machine room [IBC 2009 3008.14; NFPA 72-2010 21.5.1]. Activation of any of these heat detectors or any lobby smoke detector is displayed in the fire command center; smoke detector activation also initiates elevator recall (FEO).

Other functions monitored in the fire command center include the status of both main and emergency power to the elevators, elevator controllers, and the machine room ventilation necessary to keep the system running safely. The status of all elevators including position within the hoistway, direction of travel, position of landing doors, and whether the cars are occupied is also displayed in fire command. The availability of all this information in real time allows the fire department to react quickly to any condition that threatens the safe operation of the system and to take appropriate action.[NFPA 72-2010, 21.5 and 21.6, IBC 2009, 3007.5 and 3008.14]

Concern has been expressed by some that the amount of information that is being monitored in the fire command center will overwhelm the fire service staff. Another new technology implemented in the 2007 edition of the National Fire Alarm Code [12] was the Standard Fire Service Interface, also known as the SB30 panel (for the NEMA standard that applies).

The SB30 panel was developed jointly by the National Electrical Manufacturers Association (NEMA), NIST, and the fire alarm industry to address fire service complaints that differences between fire alarm panels makes it difficult for them to operate and obtain information from these systems. Since NIST was interested in significantly increasing the range of information available during an incident, the development of a display standardized across the industry was initiated. This panel not only has a common interface so that different company's products are similar, but it also facilitates the provision of a broad range of information from many building systems that can be useful in incident management. Both NIST and alarm manufacturers held focus groups and extensively tested prototype displays to ensure the systems were both intuitive and useful.

## People with Disabilities

The disability community has long complained that the Americans with Disabilities Act has facilitated their constituents' access to most buildings but has not adequately addressed emergency egress. The provision of occupant egress elevators clearly rectifies that shortcoming and is of significant interest to regulators in many countries as part of their disability regulations.

Egress elevators eliminate the need for special equipment (such as evacuation chairs) and special arrangements (such as "buddy systems") while meeting the disability community's primary objective of being able to self-evacuate with everyone else. By utilizing all the elevators normally provided in a building there is no need to give special priority to people with disabilities since the entire building population can evacuate any building of any height in less than one hour [13]. Assistive equipment such as powered wheelchairs and service animals need not be left behind since they do not materially impact the speed of evacuation. The ability to self-evacuate by elevator is available at any time of day, eliminating concerns about people with disabilities working alone, outside normal hours. Since the elevators are used daily, it is not necessary to egress by an unfamiliar route, which can be problematic for the blind. Since everyone self-evacuates, there is no need for the fire service to assign resources to evacuation assistance.

## Training

In recent years, US building and fire regulators have taken a stronger position on requiring larger buildings to have approved emergency plans and to conduct emergency drills once or twice a year [IFC 401.2, 404; UFC 10.9]. These now include both fire evacuations and response drills for non-fire emergencies that may involve relocation to safe areas within the building. In most cases occupants are required to at least enter the stairway and many require egress to the street where this does not involve descending so many floors that there is significant risk of fatigue or injury. Where the primary means of egress would be by elevator, it is likely that all occupants would be required to evacuate to the designated assembly point, resulting in a more realistic experience.

Emergency plans and drills are effective tools for training occupants of buildings with limited transient populations like offices and residential buildings but for others like hotels it is up to employees to lead the occupants in an efficient evacuation. Once again, since elevators are the normal means of egress, training is less important if the emergency use is similar to normal use. As fire safe elevators are introduced into the built environment it is important that the public is able to recognize elevators that incorporate the features that make them safe to use in emergencies. There is a critical need for a graphic marking that is intuitive and easily recognized by both trained individuals and the general public. One possibility that combines the international graphic for an emergency exit (the running figure) with the international graphic for an elevator (a box with up and down arrows above) is shown in Fig. 1.

**Fig. 1. Possible graphic for evacuation elevator**

Coupled with training is the increasingly common requirement for fire wardens in tall buildings. Fire wardens were introduced in New York City as part of Local Law 33/1978. Since then other major cities such as San Francisco, Chicago, and (recently) Los Angeles have begun to require fire wardens in high rise buildings. Fire wardens provide a trained authority figure to guide evacuations and can respond to conditions that might require the standard procedure to be modified (e.g., blockage of an egress route by debris or smoke). Fire wardens also enhance accountability in that they keep track of people in their care and ensure that there are no stragglers.

Recently, Australian authorities have been struggling with a problem of how to get residents to volunteer to be fire wardens in residential high rise [14]. Most people do not want the responsibility and are uncomfortable being the last one out in a real emergency. This has led to a debate in Australia about going to a "defend-in-place" strategy for residential buildings as is the case in England and Wales. There the codes permit single stair buildings since occupants are not expected to evacuate in case of a fire. Even in office occupancies there will be cases where some wardens take their responsibilities seriously and some may abandon their duties in an emergency. Examples of both were seen in the World Trade Center evacuation on September 11, 2001 [15].

A recent survey of people who live or work in high rise buildings found that 84% of respondents knew that they were not to use elevators for egress during a fire [16], demonstrating the effectiveness of the message posted in every elevator lobby (Fig. 2). However, other indications exist that people would like to use the elevators but trust the experts' advice that this is not safe (along with a fundamental fear of entrapment). These signs are intentionally misleading since the elevators are not taken out of service during a fire unless a smoke detector in an elevator lobby (or in the machine room) activates. People should be equally trusting if the experts now say that these new elevators are safe to use in fires, especially when faced with the need to descend many flights of stairs and a desire to evacuate quickly.

## System Evaluation

The 2009 editions of both US model building codes do not require evacuation elevators but rather permit them with no explicit credit against the required number, arrangement or capacity of egress stairs. Before evacuation elevators are mandated, the system as proposed needs to be evaluated for,

- reliability of operation during fires,
- functionality and usability by trained and untrained occupants, and

- the conditions under which some occupants may continue to use stairs.

**Fig. 2: Sign currently required by building codes in every elevator lobby**

There are a number of evacuation elevator systems and arrangements that are being provided in very tall buildings or where assembly occupancies on upper floors would require increased stair capacity in the entire building. In addition, standard elevators are permitted to be used for evacuation in non-fire emergencies and it is increasingly common for buildings to hold non-fire emergency drills that may involve evacuation or relocation to a safe location within the building.

While modern human-research protocols make it difficult to conduct evacuation research, it is possible to gather data by observing drills that would be conducted anyway to comply with regulations or policies. While these are generally announced in advance (occupants know that a drill will be conducted on a certain day but not the time), these can be the source of useful data on functionality, usability, and continued stair use. By comparing data from fire drills (using only stairs) and non-fire drills (using stairs and elevators) it is possible to answer many of the questions and to provide the data needed for the development of evacuation simulation software that includes elevators.

With funding from the General Services Administration (GSA), NIST has been collecting observational data in evacuation drills in various government and non-government buildings, none of which have yet included elevator use [17]. Data are collected by the video cameras installed in stairways and near egress doors to record the event. Individuals are identified by clothing and tracked past different cameras to obtain movement speed data and to document behavior such as deference at merging flows. While care must be taken with respect to privacy issues, the data obtained is useful and reasonably inexpensive to collect. The most difficult and costly issue is the time required for data collection from the frame-by-frame analysis of many hours of video.

The availability of this data can be significantly increased by collecting data in the commissioning of new or extensively remodeled buildings, particularly where the buildings include evacuation by elevators. Commissioning could include an evacuation drill to collect baseline data on the efficiency of the egress system used in accordance with the building's emergency plan. The company conducting the commissioning could collect and analyze the data, and provide a report to the owner that includes total evacuation time. This would also identify in advance of any real emergency, problems with the emergency plan so that these can be addressed. The greatest obstacle to this will be arranging for a full occupant load to participate in the drill, since most buildings are occupied in stages following their completion.

## Conclusions

The ability to use a building's elevators as a primary means of egress and access in fires and other emergencies presents numerous advantages to building designers, owners, occupants and first responders and especially to people with disabilities. Because people have been trained for the past two decades that elevators are not safe to use during fires, the new arrangements must provide assurances during emergency use that the system is operating safely. Use of the system during emergencies should be as close as possible to normal use with any differences supported by good human factors design. Intuitive signage and communications must be provided to reassure users and to ensure that the public can tell systems safe for emergency use from those that should not be used. Regular training through the conduct of drills is needed to reinforce written emergency plans.

If properly operated, occupant egress and fire service access elevators can provide for timely egress and efficient emergency operations for the tallest buildings and can permit disabled occupants to self-evacuate with everyone else.

## References

1. Safety Code for Elevators and Escalators, ASME A17.1. American Society for Mechanical Engineers, New York (2007)

2. Bukowski, Protected Elevators for Egress and Access During Fires in Tall Buildings, Strategies for Performance in the Aftermath of the World Trade Center. CIB-CTBUH Conference on Tall Buildings. Proceedings. Task Group on Tall Buildings: CIB TG50. CIB Publication No. 290. October 20-23, Kuala Lumpur, Malaysia, Shafii, F.; Bukowski, R.; Klemencic, R. (eds.), pp. 187-192 (2003)
3. Kuligowski, E.D., Bukowski, R.W.: Design of Occupant Egress Systems for Tall Buildings, Use of Elevators in Fires and Other Emergencies Workshop. Proceedings. Co-Sponsored by American Society of Mechanical Engineers (ASME International); National Institute of Standards and Technology (NIST); International Code Council (ICC); National Fire Protection Association (NFPA); U.S. Access Board and International Assn of Fire Fighters (IAFF). Mar 2-4, Atlanta, GA, pp 1-12 (2004)
4. Bukowski, R.W.: Protected Elevators for Egress and Access During Fires in Tall Buildings, Workshop on Building Occupant Movement During Fire Emergencies. Proceedings. Session 2.4. June 10-11, 2004, Gaithersburg, MD, Peacock, R. D., Kuligowski, E.D. (eds.), pp. 14-21, NIST SP1032 (January 2005)
5. Bukowski, R.W.: Protected Elevators and the Disabled. J Fire Protection Engineering. 42, 44-46, 48-49 (Fall 2005)
6. Bukowski, R.W., Fleming, R.P., Tubbs, J., Marrion, C., Dirksen, J., Duke, C., Prince, D., Richardson, L.F., Beste, D., Stanlaske, D.: Elevator Controls. NFPA Journal. Vol. 100, No 2, 42-57 (March/April 2006)
7. Bukowski, R.W.: Emergency Egress from Ultra-Tall Buildings, Tall & Green, Typology for a Sustainable Urban Future. CTBUH, Dubai UAE (March 2008)
8. Bukowski, .W.: Status of the Use of Elevators in Fires, Emerging Trends (online). SFPE, Bethesda (2008)
9. International Building Code. International Code Council, Country Club Hills (2009)
10. NFPA 5000: Building Construction and Safety Code. National Fire Protection Association, Quincy (2009)
11. Fahy, R.F., Proulx, G.: Panic and Human Behavior in Fire, Proc of the 4$^{th}$ International Symposium on Human Behavior in Fire, July 13, Robinson College, Cambridge, UK, pp. 387-398 (2009)
12. National Fire Alarm Code, NFPA 72 2007 edition. National Fire Protection Association, Quincy, MA (2007)
13. Bukowski, R.W.: Emergency Egress Strategies for Buildings. International Interflam Conference, 11th Proceedings, Interflam 2007, September 3-5, London, England, pp. 159-168 (2007)
14. Barber, D.: High-rise Apartment Fires, Evacuate or Not?. *Facility Management*. http://www.fmmagazine.com.au/high-rise-apartment-fires-%E2%80%93-evacuate-or-not/ (Feb/Mar 2009)

15. Averill, J.A., Mileti, D.S., Peacock, R.D., Kuligowski, E.D., Groner, N., Proulx, G., Reneke, P.A., Nelson, H.E.: Federal Building and Fire Safety Investigation of the World Trade Center Disaster, Occupant Behavior, Egress and Emergency Communications. NIST NCSTAR 1-7. National Institute of Standards and Technology, Gaithersburg (2005)
16. Zmud, M., Public Perceptions of High-rise Building Safety and Emergency Evacuation Procedures Research Project. Fire Protection Research Foundation, Quincy http://www.nfpa.org/assets/files//PDF/Research/NuStats_Final_HighRise.pdf
17. Peacock, R.D., Averill, J.A., Kuligowski, E.D.: Stairway Evacuation From Buildings: What We Know and What We Don't Know. NIST Technical Note 1624, p. 16. National Institute of Standards and Technology, Gaithersburg, MD (January 2009)

# Stairs or Lifts? - A Study of Human Factors Associated With Lift/Elevator Usage during Evacuations Using an Online Survey

**M.J Kinsey, E.R Galea, and P.J Lawrence**

Fire Safety Engineering Group (FSEG),
University of Greenwich, London, UK
Corresponding author: m.j.kinsey@gre.ac.uk

**Abstract** This paper presents an overview of human factors data collected via an online survey related to the use of lifts (elevators) and stairs during both circulation and evacuation scenarios. Survey participants were presented with a series of hypothetical situations and asked how they would behave. The survey was split into two broad sections, the first dealing with normal circulation usage of lifts/stairs and the second dealing with evacuation usage of lifts/stairs. Detailed demographic information about each participant was also collected. In total some 468 people from 23 countries completed the survey. An overview of the survey and initial results are presented in this paper.

## Introduction

How will people behave when given the option to use lifts during emergency evacuation situations within high-rise buildings? In countries such as the UK, Australia, Malaysia, China and USA, lifts are either being used or being considered for use as part of building evacuation systems. In past ad-hoc egress situations lifts have been used to good effect to assist in the rapid evacuation of high-rise buildings [1]. In such cases lifts were not intended to form part of the evacuation system but were used by residents for rapid egress. Computer modelling also suggests that if used correctly, the combined use of lifts and stairs can speed up full building evacuation process by as much as 50% compared to the use of stairs alone [2]. However, in these modelling examples, due to lack of human factors data, ideal "compliant" occupant behaviour was assumed. This meant that all the agents that were designated to use the lifts waited to use the lifts for as long as required. However, how many people would actually consider using a lift rather than the stairs? How long would people wait for a lift? Some evidence suggests that when faced with large queues occupants will not be prepared to wait for lifts [3]. Under what conditions will people wait for the lift? Would people in different countries behave differently? Answers to these questions are essential if engineers

R.D. Peacock et al. (eds.), *Pedestrian and Evacuation Dynamics*,
DOI 10.1007/978-1-4419-9725-8_56, 

are to realistically model building evacuation using lifts and design reliable evacuation systems in which both stairs and lifts are used.

While several studies have postulated human response to the use of lifts during evacuations [4, 5], certain studies have interviewed survivors who used lifts during real evacuations/drill [1, 3, 6] and some studies have conducted surveys [6], there is still a lack of understanding regarding the key factors which influence human behaviour relating to lift/stair selection during evacuations. Indeed whilst past studies have provided insight into such behaviour, most have been narrow in their focus resulting in questionable general applicability, for example, focusing on narrow population age groups (e.g. students, elderly); involving populations with little or no experience of high-rise buildings; drawn from potentially biased populations (e.g. businesses involved in fire engineering), or from a very narrow cultural diversity. Further to this, very little publically accessible data pertaining to human factors associated with normal lift usage is available. Use of lifts for evacuations may be related to experiences and expectations drawn from normal lift usage and so an understanding of human factors associated with normal lift usage is considered important. To address the above issues and attempt to gain a better understanding of human factors associated with lift/stair use during circulation and evacuation scenarios, an online survey (http://fseg.gre.ac.uk/elevator) was developed, asking participants how they would behave with regards to lift/stair usage within a series of hypothetical situations. The use of a publically accessible online survey was intended to reach as wide an international audience as possible coming from a broad variety of different cultural backgrounds.

## Survey Description

The survey was made available in two languages English and Chinese. The later was selected as it enabled a specific cultural group, other than English only speakers, to respond to the survey. In addition, in 2009 China possessed six of the world's ten tallest completed buildings and cities such as Shanghai and Beijing have a large number of high-rise residential and office buildings. The survey is split into three parts, the first addresses circulation issues, the second evacuation issues while the third part concerns requests participant demographic information. The survey requires approximately 20 minutes to complete. The first part of the survey explores the influence of travel distance, queues and groups on exit/stair choice. Here participants are requested to state the maximum number of floors they would consider walking on the stairs in a variety of situations. Each situation explored the influence of direction (travelling up/down), familiarity (being familiar/unfamiliar), trip purpose (being in a leisure/business activity) and time pressure (having/not having time pressure). The second part of the survey focused specifically on evacuation usage and informed participants that it was safe to use a lift during the hypothetical evacuation. Participants were then asked a series of questions related to whether they would consider using a lift and, if so, a variety of questions as to some of the influences effecting this selection and the amount of time they would wait for a lift.

## Participant Characteristics and Demographics

In total 468 participants either fully or partially completed the survey, of which 424 provided complete main demographic information. Of all participants 60.6% (269) were male and 39.4% (175) were female. Of all participants who provided age data (N=444), the average age was 35.0 years: 44.6% between 18-30 years, 26.6% between 31-40 years, 15.3% between 41-50 years, 9.7% between 51-60 years and 3.8% were over 60 years. Considering participants who provided their occupation (N=449): 18.9% were students, 7.6% were from the fire safety/protection profession and 1.6% came from the lift industry. The remaining 71.9% of participant occupations were either classified as coming from other professions or non-specific (e.g. office worker, staff, assistant etc). Of all the participants, 63.5% confirmed that their place of work/study possessed lifts with these buildings varying from 2 to 78 floors with an average of 10.1 floors, with over half (54.9%) of those buildings being over 5 floors in height. Approximately 15.6% of all participants had at least one lift in their place of residence, varying from 3 to 35 floors with an average of 10.8 floors in height, with approximately three quarters (75.3%) of those buildings being greater than 5 floors in height. Whilst overall participants came from some 23 different countries, six countries made up approximately 88.9% of all participants: UK (30.8%), China (25.9%), US (12.8%), Germany (11.1%), Japan (5.6%), Australia (2.8%). Using the WHO (World Health Organisation) classification of body mass indexing (BMI), of the participants who provided plausible height/weight information (N=445), 6.7% were classed as underweight, 56.4% were normal weight, 24.7% were overweight, 11.0% were obese and just 1.1% were classed as being morbidly obese.

## Results - Circulation and Evacuation Usage

Each section within the survey is based around a hypothetical scenario. The core part of the scenario description, unless stated otherwise, is identical for each question and consists of the following information:

- You are familiar with the layout of the building.
- The lifts/stairs are located in the same area.
- You are not carrying or wearing anything to restrict your movement.

- A lift is not currently on your floor and you do not know how long you will have to wait for a lift to return.

## *Circulation Usage*

The first part of the survey, addressing circulation behaviour, explored issues to do with vertical travel distance, queue length in the lift waiting area and group behaviour. Three specific variations of the core scenario were presented to the participants. Additional situational information relating to the nature of these various scenarios is presented in Table 1. Given these specific situations, participants were asked what is the maximum number of floors they would consider travelling on the stairs before electing to use a lift. Participant responses either stated that they always consider using the stairs, never consider using the stairs (always use the lift), or sometimes consider using the stairs (specifying a finite number of floors they would walk on the stairs). Answers to the various questions were further categorised according to: building familiarity, whether or not travel was time critical and whether or not the travel was for leisure or business. While these factors have varying influences upon the responses, due to space limitations these various categories have been collapsed into direction of travel and trip purpose with the average results presented in Table 2.

**Table 1. Additional situation information provided for each section**

| Base Case | Queues | Groups |
|---|---|---|
| You are alone in a lift waiting area on your floor. | There are a number of people in the lift waiting area on your floor. | You are travelling with a group of 2-4 people.<br>The people in the group are all of similar physical ability and fitness to yourself.<br>The lift waiting area on your floor is empty. |

In the base case, 87.8% of the participants would always or sometimes consider using the stairs to travel down and 84.2% to travel up. This is rather a high percentage of people who would consider using the stairs, with slightly more participants prepared to travel down the stairs compared to up. On average participants were prepared to walk 2.0 floors further down than up, 6.7 floors down and 4.7 floors up.

When faced with a queue in the lift waiting area, slightly more participants would always or sometimes consider using the stairs compared to the base case, with 89.4% of participants always or sometimes consider using the stairs to travel down (compared with 87.8%) and 87.3% to travel up (compared with 84.2%). This highlights a slight decrease in attractiveness of the lift due to congestion in the waiting area. When faced with a queue, participants were prepared to walk slightly further up/down (mean 5.0/7.0 floors) compared to the base case (mean 4.7/6.7 floors).

When travelling in a small group, slightly fewer participants would consider using the stairs compared to the base case, with 81.0% of participants always or sometimes considering using the stairs to travel down (compared with 87.8%) and 76.4% to travel up (compared with 84.2%). This highlights a decrease in attractiveness of the stair when travelling in groups compared to the queue scenario where an increase was observed. On average participants were prepared to walk 5.3 floors down (median 4.0) and 4.2 floors up (median 3.0). This represents a 20.9% (1.4) and 10.6% (0.5) decrease in the number of floors participants would consider walking on the stairs in the down and up direction respectively compared to the base case. When travelling in groups there is a considerable reduction in the distance people are prepared to travel on stairs.

**Table 2. Overall Combined Average Results Irrespective of Time Pressure or Familiarity for the Base, Queue and Groups cases**

| | | Base Case | Queues | Groups |
|---|---|---|---|---|
| **Up** | **Always use lift** | 15.8% [592] | 12.7% [474] | 23.5% [875] |
| | **Always consider using Stairs** | 3.7% [138] | 4.5% [169] | 4.3% [161] |
| | **Sometimes consider using Stairs** | 80.5% [3008] | 82.8% [3091] | 72.1% [2682] |
| | **Median Stair Travel (floors)** | 3.8 | 4.0 | 3.0 |
| | **Mean Stair Travel (floors)** | 4.7 | 5.0 | 4.2 |
| | **Total Frequency** | 3738 | 3734 | 3718 |
| **Down** | **Always use lift** | 12.2% [450] | 10.6% [392] | 19.0% [701] |
| | **Always consider using Stairs** | 5.6% [208] | 7.6% [281] | 5.0% [184] |
| | **Sometimes consider using Stairs** | 82.2% [3036] | 81.8% [3027] | 76.0% [2799] |
| | **Median Stair Travel (floors)** | 5.1 | 5.3 | 4.0 |
| | **Mean Stair Travel (floors)** | 6.7 | 7.0 | 5.3 |
| | **Total Frequency** | 3694 | 3700 | 3684 |

## *Evacuation Usage*

The evacuation section of the survey was intended to investigate whether participants would consider using a lift to evacuate if they were informed that it was acceptable to do so during an emergency, and if so, identify and quantify influencing factors that would cause them to redirect to use the stairs. For the evacuation base scenario the following additional information was provided to the participants:

- You are travelling alone.

- You have been instructed that it is acceptable to use either a lift or stairs to evacuate from your building in emergency situations. During an evacuation you are free to choose to use a lift or stairs.

Of the participants who answered whether they would consider using a lift to evacuate (N=467), approximately a third (33.0% (154))) said that they would consider using a lift. Thus, two thirds of the participants would not consider using a lift to evacuate, even though they knew it was acceptable to do so.

Of the participants who would consider using a lift and answered whether or not they would always use a lift (152), a small proportion (7.2% (11)) said that they would always use a lift. Of the 154 participants who would consider using a lift, 78.6% (121) replied that the height of the floor they were on would influence their decision. These participants were then asked to specify a maximum/minimum number of floors above/below which they would not consider using the lift. Of the participants who specified a maximum number of floors they would be prepared to travel by lift (120), 46.7% (56) answered that there was no maximum number of floors, 22.5% (27) answered 100+ floors, and the remaining 30.8% (37) specified a varying number of floors with an average maximum of 21.9 floors. Of the participants who specified a minimum number of floors they would be prepared to travel by lift (121), 9.9% (12) answered that there was no minimum number of floors, 0.83% (1) answered 100+, and the remaining 89.3% (108) specified a varying number of floors with an average minimum of 8.4 floors. When asked if the height of the building would influence their decision to use a lift (N=136), almost two thirds (65.4% (89)) said that the height of the building would influence their decision. Of this group (N=86), 80.2% (69) said that the higher the building the more likely they would be to use a lift.

For the remaining evacuation related questions, the following additional scenario information was provided:

- You are instructed to evacuate from a multi-storey building.
- It is not a drill but you are not in immediate danger.
- You have a choice to use one of the 4 lifts servicing your floor or the stairs.
- Each lift has a capacity of 10 people.
- The lift waiting area on your floor is crowded with people.

Participants were then asked, given that they were located on progressively higher floor ranges in the building, would they consider using a lift to evacuate and if so, after arriving in the lift waiting area, what level of crowd size/density already waiting for the lift would cause them to redirect to use the stairs. To quantify the crowd density, six different crowd densities (ranging from A to F) were presented to the participants based on graphics generated by the vrEXODUS software, three of which are presented in Fig. 1. Participants where then asked for each floor range to specify the crowd density that would deter them from waiting for a lift and to estimate, providing the crowd density was below that stated level, how long they would be prepared to wait in the crowd for a lift before they decided to use the stairs.

A: 10 people, $0.13 p/m^2$ | C: 80 people, $1.0 p/m^2$ | F: 200 people, $2.5 p/m^2$

**Fig. 1. Three of the six Crowd Levels in the lift waiting area**

As with the circulation based questions, answers to the various questions were further categorised according to building familiarity. While there were some differences due to building familiarity, due to space limitations the responses have been collapsed into a single category and the average results are presented here (see Fig. 2 and Table 3).

**Table 3. Frequency of participant responses that would consider using a lift as a function of crowd density (familiar and unfamiliar combined)**

| Floor Range Location | Proportion of participants that would consider waiting to use lift on a given floor range | | Of participants that would initially choose to use a lift, the crowd density in the lift waiting area that would cause a proportion of those participants to redirect to use the stairs. | | | | | | | | |
|---|---|---|---|---|---|---|---|---|---|---|---|
| | Yes | No | # | Doesn't Matter | A 0.13 $p/m^2$ | B 0.5 $p/m^2$ | C 1.0 $p/m^2$ | D 1.5 $p/m^2$ | E 2.0 $p/m^2$ | F 2.5 $p/m^2$ | F+ 2.5 $p/m^2$+ |
| 2-10 | 11.3% [39] | 88.7% [306] | 38 | 15.8% [6] | 18.4% [7] | 44.7% [17] | 73.7% [28] | 78.9% [30] | 78.9% [30] | 84.2% [32] | 84.2% [32] |
| 11-20 | 33.3% [114] | 66.7% [228] | 114 | 10.5% [12] | 14.0% [16] | 31.6% [36] | 60.5% [69] | 84.2% [96] | 86.8% [99] | 89.5% [102] | 89.5% [102] |
| 21-30 | 63.5% [216] | 36.5% [124] | 214 | 7.0% [15] | 5.6% [12] | 25.7% [55] | 61.7% [132] | 82.2% [176] | 89.7% [192] | 93.0% [199] | 93.0% [199] |
| 31-40 | 77.8% [256] | 22.2% [73] | 252 | 9.9% [25] | 2.8% [7] | 19.0% [48] | 49.2% [124] | 77.8% [196] | 86.9% [219] | 90.1% [227] | 90.1% [227] |
| 41-50 | 79.0% [260] | 21.0% [69] | 254 | 9.1% [23] | 2.8% [7] | 14.6% [37] | 39.0% [99] | 64.2% [163] | 83.5% [212] | 90.6% [230] | 90.9% [231] |
| 51-60 | 80.5% [265] | 19.5% [64] | 257 | 10.5% [27] | 3.1% [8] | 13.2% [34] | 33.5% [86] | 56.4% [145] | 72.8% [187] | 84.8% [218] | 89.5% [230] |

Presented in Table 3 is the overall proportion of participants that would consider using a lift/stair for each floor range. As the floor height increases the proportion of participants that would consider using the lift also increases. We note that approximately 10% of the population would use a lift even if located on the lowest floors i.e. 2-10. The proportion of the population that would use the lift increases to approximately 80% at floor range 31-40 and remains at this level for the higher floor ranges. This suggests that when located on or above floors 21-30, the majority of people on each floor would elect to use the lift compared to the stairs. Above floor 30, approximately 20% of the population are not prepared to use the lifts to evacuate irrespective of floor height.

In addition, presented within Table 3 is the cumulative proportion of those participants that would choose to redirect to use the stairs based on crowd density within the lift waiting area. We note that of those prepared to wait to use the lift given a crowd in the lift waiting area, an average 10.5% of the population would be prepared to wait for the lift, regardless of floor height or crowd density. Furthermore, the average crowd density that participants would be prepared to tolerate before redirecting to the stairs increases as the floor height increases. For a floor height of 2-10, 70% of the population waiting for the lift would redirect to the stairs when the average congestion levels are between B and C (0.5 p/m$^2$ and 1.0 p/m$^2$), while for a floor height of 21-30, this increases to between C and D (1.0 p/m$^2$ and 1.5 p/m$^2$) and for a floor height of 51-60, this increases to between D and E (1.5 p/m$^2$ and 2.0 p/m$^2$). Participants who would consider using a lift for a given floor range were asked, providing the crowd level did not reach or exceed the density which would cause them to redirect, what was the maximum time they would be prepared to wait to use a lift. For each floor range a small number of participants (0%-7%) stated that they would wait for a lift for "as long as it takes" with an average proportion of 5.8% for all floor ranges. In addition, a small number of participants (6.1% (14)) said that they would not be prepared to wait for a lift, regardless of floor height.

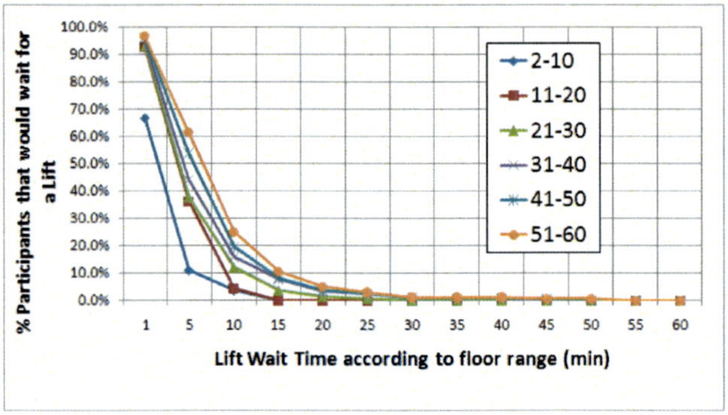


**Fig. 2. Cumulative proportion of participants that would wait for a lift (grouped into 5 min intervals) for each floor range**

In Fig. 2 the normalised cumulative frequency distribution of the remaining participants who specified the maximum time they would be prepared to wait for a lift on each floor range can be seen. As the floor height increases the proportion of participants that would wait a longer amount of time in the lift waiting area also approximately increases. This reflects participants increased tolerance to waiting a longer amount of time for a lift on progressively higher floors in light of the added travel time and energy expenditure they would need to expend travelling on the stairs. Fig. 2 also suggests that the majority of participants who initially chose to use a lift in floor range 2-40 would only be prepared to wait between 1-5 minutes for a lift before redirecting to the stairs. For the floor range 40-60 the majority of participants would be prepared to wait between 5-10 minutes. For all floor ranges, approximately less than 10% of participants would be prepared to wait more than 15 minutes for a lift before redirecting to the stairs; highlighting participants intolerance to wait long periods of time for a lift during an evacuation.

## Conclusion

This paper has presented an analysis of data collected from participant responses to an online survey in order to gain an understanding of human factors associated with lift/stair selection in both circulation and evacuation scenarios. In normal circulation conditions, between 90%-85% of the survey population would be prepared to use the stairs to travel down/up. On average participants were prepared to walk 6.7/4.2 floors in the down/up direction respectively. Results suggest that a queue in the lift waiting area does not influence these numbers greatly however, travelling in groups does. When travelling is a small group (up to four people), the percentage of the survey population prepared to use the stairs to travel down/up decreases to 80%/76% and the distance they are prepared to walk down/up decreases to 4.8/3.2 floors.

In evacuation conditions, despite being informed that the lifts were a safe and acceptable option, two thirds of the sample (308) said they would not consider using a lift to evacuate. This suggests that if buildings are being designed on the assumption that occupants will utilise lifts for evacuation, an extensive training campaign will be essential. This poses difficulties for buildings that are largely frequented by casual visitors. Of the participants whom would consider using a lift (152), less than 10% said that they would always use a lift, while over 75% (121) said that the height of the floor they were on would influence their decision to use a lift. The height of the building was also a significant factor in determining whether or not they would use the lift. Of the participants who specified a maximum number of floors they would be prepared to travel by lift (120), almost 70% (83) effectively indicated that there was no maximum number of floors while of those specifying a minimum number of floors, almost 90% (108) specified a varying minimum number of floors with an average minimum of 8.4 floors. As the floor height increases the proportion of participants that would consider using the lift increases. Approximately 10% of the population would use a lift even if located below the $10^{th}$ floor. The proportion of the population that would use the lift increases to approximately 80% up to floor 40 and remains at this level even for higher floors. This suggests that approximately 20% of the population will not use a lift to evacuate irrespective of floor height. A very small proportion of participants stated that they would wait in a lift waiting area regardless of crowd density and/or would wait for "as long as it takes" for a lift to service their floor. However, the majority of participants indicated there was a critical level of crowd density in the lift waiting area which, if reached or exceeded, they would redirect to the stairs. Furthermore, this critical density appears to increase as the floor height increases; reflecting the decreased attractiveness of using the stairs on progressively higher floors. The majority of participants also specified a finite time they would be prepared to wait for a lift; while this was dependent on floor height (the higher the floor, the longer the acceptable wait time), less than 10% of participants were prepared to wait more than 15 minutes regardless of floor height.

These results clearly show that in evacuation situations, building occupants are prepared to utilise lifts for evacuation but that this is strongly dependent on floor height, crowd density and expected lift wait time. Participants in the study clearly exhibit anticipatory behaviour and would expect a given level of service from an lift system during an evacuation. Further analysis of the survey data is currently underway examining the impact of pedestrian characteristics such as age, gender, country, building familiarity etc on both circulation and evacuation behaviours. The data is being used to enhance the agent based model associated with lift usage within the evacuation modelling software buildingEXODUS.

## References


1. Sekizawa, A., Ebihara, M., Notake, H., Kubota, K., Nakano, M., Ohmiya, Y., Kaneko, H.: Occupants' behaviour in response to the high rise apartment fire in Hiroshima City. Fire and Materials (1999)
2. Kinsey, M.J., Galea, E.R., Lawrence, P.J.: Investigating the use of lifts for high-rise building evacuation through computer simulation. Proc 4th Int Symp on Human Behaviour in Fire Conference, pp. 85-96 (2009)
3. Galea, E.R., Shields, J., Canter, D., Boyce, K., Day, R., Hulse, L., Siddiqui, A., Summerfield, L., Marselle, M., Greenall, P.: Methodologies employed in the Collection, Retrieval and Storage of Human Factors Information. Derived from First Hand Accounts of Survivors of the WTC Disaster of 11 September 2001, Journal of Applied Fire Science. Vol. 15, Number 4, 253-276 (2006)
4. Klote, J.H., Alvord, D.M., Levin, B.M., Groner, N.E.: "Feasibility and Design Considerations of Evacuation by Lifts." NISTIR 4870 (1992)
5. Groner, N.E., Levin, B.M.: Human Factors Considerations in the Potential for Using Lifts in Building Emergency Evacuation Plans. NIST 94-656 (1992)
6. Heyes, E.: "Human Behaviour Considerations." Proc 4th Int Symp on Human Behaviour in Fire Conference, pp. 73-84 (2009)

# Elevator Evacuation Algorithms

**M-L. Siikonen and J. S. Sorsa**

KONE Corporation
P.O. Box 7, 02151 Espoo, Finland
Corresponding author: Marja-Liisa.Siikonen@kone.com

**Abstract** In emergency situations, the practice has been to return elevators to the exit discharge level, and then shut down. After that, the elevators are not available for the building occupants until the emergency is over. In this paper, we study the use of the elevators in an emergency evacuation. We first introduce the theoretical egress time and the handling capacity calculation for an elevator group, which is based on floor-by-floor evacuation. We also describe a specialized elevator evacuation algorithm, which automatically, with or without landing call information, dispatches elevators to the occupied floors and shuttles passengers to the rescue level. The algorithm serves floors in a priority order and detects floor occupancy automatically. We compare this algorithm to a normal algorithm and to two staircases in evacuation. For that purpose, we run simulations of test buildings with realistic transport arrangements and obtain performance measures such as crowding levels of the lobbies, passenger service times and total evacuation time. On this basis, we propose the best algorithm for different types of emergencies.

## Introduction

It is estimated that over 800 buildings exceeding 200 meters will be built in 2012 [1]. In the beginning of 2010, the 828-meter tall Burj Khalifa with 160 floors was opened. It exceeds the height of the previous record holder, Taipei 101, by over 300 meters, and Taipei 101 lost its five-year leadership as the tallest building in the world. In tall buildings such as these two examples, evacuation plans are tailored specifically for the building. People travel to the safety area using either protected elevators all the way [2, 3] or, first stairways to refuge floors and then protected elevators to the safety area [4, 5]. In an emergency, elevators are driven either by attendants or an automatic evacuation algorithm. In fire evacuation, in addition to enhanced or protected elevators, fire-, smoke- and water-proof elevator lobbies have to be designed [6, 7].

In tall buildings, typically two stairways are reserved for evacuation. The capacity of a stairway depends only on its width. Consequently, stairway capacity is constant but egress time increases since the total population increases by the number of floors [8]. Elevators are planned according to the total population, which makes egress time when using elevators independent of the number of floors. Ac-

R.D. Peacock et al. (eds.), *Pedestrian and Evacuation Dynamics*,
DOI 10.1007/978-1-4419-9725-8_57, © Springer Science+Business Media, LLC 2011

cordingly, the total evacuation is faster by stairways than by elevators only up to a certain population density [9]. For high population densities, phased evacuation by stairways is used to avoid congestion. In another study, it was found that theoretically elevator evacuation becomes faster than two stairways in buildings having more than 2 500–3 000 occupants [10]. In residential buildings, the population rarely exceeds 2 000 persons; hence, in most such cases, stairways are the fastest way out.

Our challenge here is to determine the best methods for self-evacuation in buildings of more than 200 meters in height. The best evacuation methods for different emergency scenarios can be found by using simulation [11], and then selected as an integral part of the building security plan. A default evacuation mode is automatically switched on by the alarm system. Evacuation continues up to one to two hours depending on the fire-resistance time and emergency power resources.

In this paper, we study elevator evacuation algorithms in emergency scenarios. Throughout the paper we assume that the elevators can be used in total evacuation: either the emergency is a non-fire situation, or the elevators are designed as safe to use during a fire. Since fundamental elevator performance is set by the elevator cycle time from start to start, we review the parameters that define it. By using these parameters, we formulate equations for evaluating elevator group performance both in up-peak and in floor-by-floor evacuation. Our results show how evacuation time relates to up-peak performance, according to which elevators are currently planned. After the theoretical part, we describe elevator evacuation algorithms, simulate complete building evacuation with them, and then compare the obtained results to a benchmark evacuation with two staircases.

## Elevator Cycle Time

The cycle time, $T_C$, is the time between two elevator starts, which consists of the flight time, $t_f$, and the stop time, $t_s$, and is further divided into the components shown schematically in Figure 1 below.

The stop time consists of door operation delays, start delay and passenger transfer time. Start delay, $t_{st}$, is spent closing the safety circuit before the elevator starts to move. When the elevator approaches the floor level, the doors can be opened in advance some 15–30 mm before reaching the floor. This advance opening time, $t_{a\text{-}do}$, reduces the stop time. The door opening time, $t_{do}$, is defined as the time until the doors are 800 mm open, which is assumed to be enough for passengers to move in or out of the elevator. The doors stay open for a dwell time but in the calculations it is assumed that photocell beam is always broken by the passengers. After the beam is restored, it still takes the time equal to the photocell delay, $t_{ph}$, until the doors start to close. During the stop, $M$ passengers transfer to/from the elevator, each taking the transfer time $t_m$. Finally, the door closing time, $t_{dc}$, is defined as the time until the doors are closed and locked. Typical values of the delays included in the stop time are shown in Table 1 below.

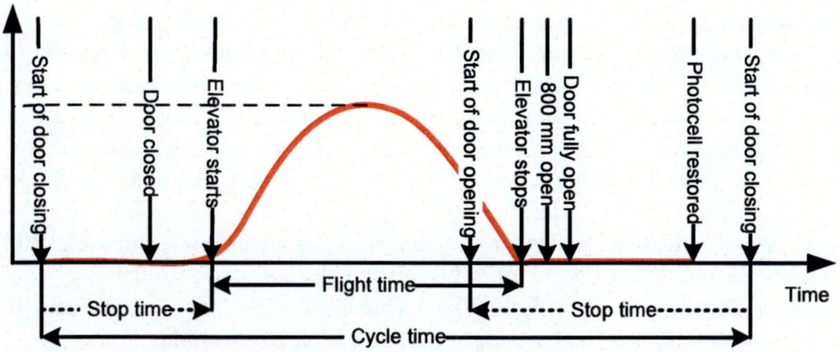


**Fig. 1. Cycle time components**

**Table 1. Typical values of the cycle time components**

| Start delay (s) | Door opening time (s) | Door closing time (s) | Advance door opening time (s) | Photocell delay (s) | Transfer time (s) |
|---|---|---|---|---|---|
| 0.7 | 1.2 | 3.0 | 0.7 | 0.9 | 1.0 |

The flight time depends on the travel distance, $s$, rated speed, acceleration, and jerk of the elevator, $v$, $a$, and $k$, respectively [12]. The minimum distance needed by the elevator to reach rated speed is

$$s_{\min} = v^2/a + va/k \, . \tag{1}$$

The flight time, depending on whether the rated speed is reached or not, is

$$t_v = \begin{cases} s/v + v/a + a/k, & \text{if } s \geq s_{\min}, \\ \sqrt{4s/a + (a/k)^2} + a/k, & \text{if } s < s_{\min}. \end{cases} \tag{2}$$

Finally, the cycle time is

$$T_C = t_v - t_{ado} + t_s = t_v - t_{ado} + t_{dc} + t_{st} + t_{do} + Mt_m + t_{ph} \, . \tag{3}$$

## Elevator Planning and Egress

The core of tall buildings, including elevator shafts and stairwells, is optimized to occupy as little of the floor area as possible. Usually elevator groups are zoned so that one elevator group serves particular floors of the building. If a building is higher than 20–25 floors, it is divided, for instance, between two elevator groups serving low- and high-rise zones, both starting from the main lobby. In this manner, the building can be divided even into ten zones. In mega-high-rise buildings, however, this kind of banking results in a large core area and the rentable area becomes too small. To overcome this problem, mega-high-rise buildings are most often designed with sky lobbies. In the sky lobby arrangement, shuttle elevators transport passengers from the main lobby to the sky lobby. From the sky lobby, the passengers travel to their destinations with local elevators. This reduces the building core area so that the building is still economically feasible.

Elevator groups are usually planned for the morning up-peak traffic since it is the most demanding situation for the elevators. Up-peak handling capacity is calculated from the roundtrip time, $T_{RT}$, during which the elevator loads $M$ persons in the main lobby, serves an expected number of car call stops, $S$, and expresses back to the main lobby from the expected reversal floor $H$. It is customary to set $M$ equal to 80% of the rated capacity of the elevator. The up-peak roundtrip time with an average floor-to-floor distance $s$ becomes [13]

$$T_{RT} = 2Hs/v + (S+1)t_s + 2Mt_m \,. \tag{4}$$

The handling capacity of an elevator group with $L$ elevators is

$$HC = 300ML/T_{RT} \,. \tag{5}$$

The handling capacity, given in persons per five minutes, is usually scaled relative to the building population and given as a percentage. This relative handling capacity is used as a design criterion. Typically, a handling capacity of 12–16% of the population in five minutes is required in offices, but only 5–7.5% in residential buildings. This reflects the fact that traffic demand in offices is much higher than in residential buildings. In addition, the building filling time can be calculated from the relative handling capacity. The criteria given above correspond to filling times of 31–42 minutes in offices and 68–100 minutes in residential buildings.

Evacuation is not generally considered in elevator planning since elevators are currently shut down during an emergency situation. Buildings with heavy down-peaks, such as hotels in Islamic countries with specific praying times, make an exception to the rule. With a full collective control system, elevator-group handling capacity in down-peak is 1.5–1.8 times the up-peak handling capacity without congestion. In down-peak, the elevator dispatching algorithm has more degrees of freedom in serving the calls, which results in shorter building emptying times.

The emptying, or egress time of a building with $L$ elevators can be calculated from the roundtrips, where the elevators transport population $P_i$ from floor $i$ to the level of discharge [14]. In each roundtrip, $M_{i,j}$ persons are transported. As in the up-peak case, we assume that $M_{i,j}$ is at most 80% of the rated elevator capacity. The number of roundtrips needed to empty floor $i$ is

$$J_i = \lceil P_i / M \rceil, \tag{6}$$

and the total egress time, $T_E$, to evacuate $N$ floors to the level of discharge is

$$T_E = \sum_{i=1}^{N} \sum_{j=1}^{J_i} T_{RT,i,j} / L. \tag{7}$$

By counting the total number of roundtrips, an average egress roundtrip time can be calculated. If this roundtrip time is substituted into equation (5) then we get the elevator egress handling capacity. The egress times vary in office buildings from 20 to 28 minutes for the above-mentioned up-peak handling capacity criterion, and from 44 to 67 minutes in residential buildings. Table 2 shows the calculated egress times and handling capacities in conjunction with the corresponding up-peak figures. According to the results, the theoretical egress handling capacity is 2.5 times the up-peak handling capacity. The reason is the number of car call stops in up-peak, in evacuation situation elevators shuttle between two floors.

**Table 2. Calculated elevator group egress and up-peak performance**

| Egress Time (min) | Evacuation HC (%/5 min) | Filling Time (min) | Up-Peak HC (%/ 5 min) |
|---|---|---|---|
| 10 | 64 | 25 | 20 |
| 20 | 32 | 50 | 10 |
| 26.8 | 24 | 66.7 | 7.5 |
| 40 | 16 | 100 | 5 |
| 60 | 12 | 133.3 | 3.75 |
| 80 | 8 | 200 | 2.5 |

## Elevator Evacuation Algorithms

Three evacuation types have been recognized: fractional, staged, and total evacuation [6]. An elevator evacuation algorithm based on one of these evacuation types can automatically switch on when an alarm occurs. During evacuation it is critical to communicate correct and up to date information to the evacuees. In the case where the elevators are used in evacuation, this becomes even more important. People then have to choose whether to wait for the elevator, or to start de-

scending the stairs. Elevator displays and voice announcements can be used to inform the status of the elevator group, for example: "Elevator evacuation in progress", "Next elevator arrives in 5 minutes", or "2 elevators arriving, room for 40 persons". If such information is available, the occupants are able to decide their means of egress.. Later, authorized personnel can change the mode [15]. Automatic elevator evacuation is interrupted if smoke, heat or water is detected in the lobbies, the elevator shafts or the machine rooms. After the emergency, an authorized people such as firemen, return the elevators back to their normal operation.
In *fractional evacuation*, a focused group of building occupants such as disabled people is rescued by the firemen using attendant drive. In *staged evacuation*, first the emergency floor and a couple of floors above and below it are rescued in a priority order, while the rest of the building occupants may remain at their floors waiting for further instructions. After the first stage of the evacuation has completed, the rest of the building can be emptied in a *total evacuation*.

Complete evacuation with self evacuation mode is more efficient compared to attendant-driven manual modes since the dispatching algorithm has information of all the served floors simultaneously. In a fire situation, automatic dispatching algorithm can operate with or without call buttons depending on whether the elevators are protected or not. In a non-fire emergency situation, it is possible to serve landing calls as in normal operation. During evacuation, elevators accept destination calls only to the level of exit discharge.

The *normal operation mode* of the existing elevator dispatching algorithms can also be used for the evacuation. In the normal mode, however, the dispatching algorithm gives its own priorities in serving the floors. *Interconnected Full Collective (IFC)* control has already been used by the relay control systems. In IFC, an empty elevator is always sent to the topmost landing call. If a car is already serving that call, the next vacant elevator is sent to the second highest call. Then landing calls are served one by one on the way down so that only one car stops to each landing call at a time. In *modern full collective control systems*, mathematical methods, such as forecasting or *Genetic Algorithm (GA)* [16], are used in the optimization. Different optimization objectives can be set to minimize, for example, average passenger waiting or journey time. The optimization results in about equal service at all floors. If all the people waiting do not fit in a car, they give a new landing call after the fully loaded car has left. If the car is not full, the elevator serves the next landing call on the way towards the rescue floor. *Destination Control (DC)* is an elevator system where passengers give the destination call using a numeric keypad already at the lobby [17]. In evacuation, a floor warden can use the keypad to enter the number of evacuees at the lobby. Then, the dispatching algorithm can to send automatically as many cars as are needed to empty each floor.

In the *floor-by-floor algorithm* mode the floors are evacuated one by one starting from the floor with the highest priority, e.g. from the emergency floor or the topmost floor. If landing calls exist, elevators are dispatched to the corresponding floors in priority order. The algorithm takes into account slow reaction times since elevators are dispatched to occupied floors when requested. Then, elevator may stop to several landing calls during a trip depending on a set load limit (Fig. 2)

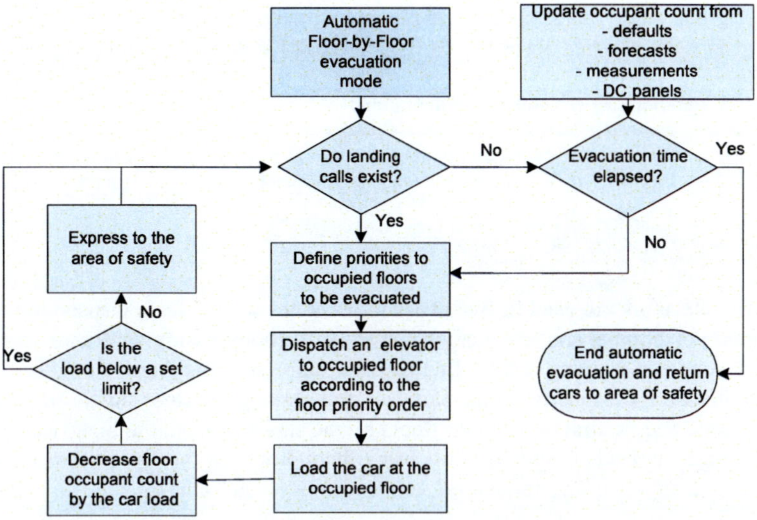


**Fig. 2. Flow chart for total evacuation floor-by-floor**

In case there are no landing calls, floors are served according to the occupant count, which can be based either on default values, measured people, forecasts or external inputs. External input can be, for example, processed image data, or exact figure dialed by the floor warden using the destination control keypad in the lobby. According to the number of occupants, the dispatching algorithm sends as many cars as needed to rescue the waiting occupants from the floor. The algorithm keeps track of the number of occupants on each floor by decreasing the occupant count by the car load each time the elevator leaves the floor. For safety reasons, the algorithm dispatches elevators to the floors in priority order until the evacuation mode is turned off, or automatically after a defined time, e.g. two hours.

## Simulation Study using Evacuation Algorithms

In the simulation benchmark study, egress times and passenger service times are compared in three building zones for a total evacuation scenario. The Building Traffic Simulator, BTS, [18, 19] is used to simulate the scenarios. In the simulation, occupant arrival time at the exits is adjustable, here passengers arrive at elevator or stairway lobbies within 30 seconds. The egress time is defined as a time from the arrival time until the moment when all people have been rescued to the safety area. Passenger journey time consists of passengers' waiting times and egress times in staircases and with elevators. The building data as well as the elevator group sizes, loads, speeds and other parameters are shown in Table 3. Elevator groups are planned according to 15% up-peak handling capacity criterion.

**Table 3. Building and elevator group definitions**

| Zone | Served Floors | Travel (m) | Population (persons) | Group size (elevators) | Car Size (persons) | Speed (m/s) | Acceleration ($m/s^2$) | Jerk ($m/s^3$) |
|---|---|---|---|---|---|---|---|---|
| Low-rise | 0, 1-20 | 66 | 1440 | 7 | 24 | 3.5 | 1.0 | 1.6 |
| High-rise | 0, 41-60 | 198 | 1440 | 8 | 24 | 8.0 | 1.2 | 2.0 |

The elevators are used in five evacuation modes with diverse dispatching algorithms. For comparison, evacuation with two 1 200mm-wide staircases is simulated. A person queuing at a landing enters a staircase if his journey time is longer than the one's descending in the staircase and passing the floor. In zoned evacuation mode people first descend 20 floors by the stairway from the high-rise area to the refuge floor where the elevators bring them down; in the low-rise area people use only stairways. The simulation results are shown in Table 4.

**Table 4. Egress and passenger service times with different evacuation algorithms**

| Zone | Parameter (min) | GA normal | GA Evac mode | Fl-by-Fl algorithm | Fl-by-Fl Eq.7 | Refuge Floor | Stairs |
|---|---|---|---|---|---|---|---|
| | Egress time | 13.3 | 14.1 | 14.5 | 13.1 | 11.4 | 11.4 |
| Low-rise | Waiting time | 5.5 | 6.3 | 6.7 | 6.7 | 0.3 | 0.3 |
| | Journey time | 6.8 | 7.4 | 8.1 | 7.4 | 6.5 | 6.5 |
| | Egress time | 15.2 | 16.8 | 16.9 | 14.7 | 14.7 | 23.5 |
| High-rise | Waiting time | 6.3 | 6.9 | 7.4 | 7.6 | 1.7 | 0.4 |
| | Journey time | 7.7 | 8.3 | 8.7 | 8.6 | 7.7 | 17.5 |

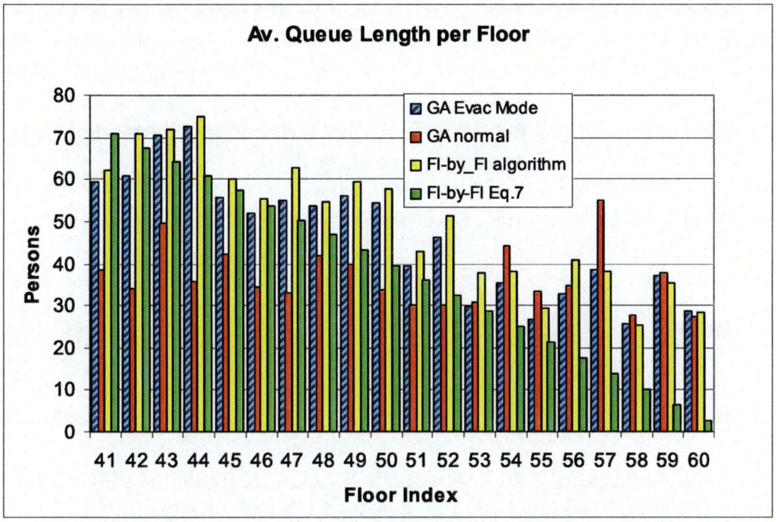


**Fig. 3. Average passenger queues per floor in the high-rise area**

The combined use of stairs and elevators with refuge floor is the fastest in all rises. Stairways are the fastest in the low-rise area, but in the high-rise elevators are clearly faster. Modern normal GA algorithm gives the shortest passenger waiting and journey times. Also average queue lengths over time are the most balanced with normal GA algorithm, as shown in Figure 3.

## Conclusion

In tall buildings, the number of emergency elevators, and the dispatching algorithm have a direct effect on how many people can be rescued in an emergency situation. The one-hour Required Safe Egress Time (RSET) has been proposed [2]. This target could be reached in most existing buildings if all the passenger elevators were used during evacuation. In 200+ meters high buildings people may have to use several transportation devices on the way down and a one-hour criterion can be too tight. According to this paper elevator group egress time varies from 10 to 60 minutes depending on the designed up-peak handling capacity, but more precise results are obtained by simulation.

As a conclusion of this study, in complete evacuation the egress times are the shortest with staircases up to 20-40 floors depending on the population density per floor. A combined usage of stairways and elevators gives the shortest egress times, which is in line with earlier studies [10, 20]. The normal modern dispatching algorithms provide passengers the fastest egress and the best service level.

For self-evacuation in a fire situation, floor-by-floor evacuation algorithm is needed to set priority order to the floors to be evacuated. The floor-by-floor mode is more efficient, if it also serves landing calls instead of knowing just the number of people waiting at the floor. The algorithm utilizes all the available information to evacuate the building as fast as possible but is not dependent on any specific piece of information. With proper signalization devices the control system can communicate of the status of the elevator evacuation to the occupants since the dispatching algorithm is capable of estimating the future position of the elevators.

## References

1. Tallest Buildings Completed in 2009. Council on Tall Buildings and Urban Habitat. http://ctbuh.org/HighRiseInfo/TallestDatabase/TallestBuildingsCompleted/2009BuildingsCompleted/tabid/1353/language/en-US/Default.aspx (2010)
2. Bukowski, R.W.: Emergency Egress Strategies for Buildings, Interflam '07. International Interflam Conference, 11th Proceedings, September 3-5, 159-168, London, England (2007)
3. Fortune, J.: Emergency Building Evacuations via Elevators. Proceedings of the CIB-CTBUH International Conference on Tall Buildings, 3-5 February, India (2010)
4. Ariff, A.: Review of Evacuation Procedures of Petronas Twin Towers. Proceedings of the CIB-CTBUH International Conference on Tall Buildings, 20-23 October, CIB Publication No: 290, Malaysia (2003)
5. Hsiung, K.H., Chien, S.W., Huang, P.T.: A study on high rise building fire evacuation strategies for Taipei 101 Financial Centre. 4th International Symposium on Human Behaviour in Fire, Cambridge, UK, 51- 60 (2009)
6. Council on Tall Buildings and Urban Habitat: Emergency evacuation – Elevator Systems Guideline. Chicago: CTBUH, ISBN 0-939493-21-7 (2004)
7. Klote, J.H., Levin, B.M., Groner, N.E.: Emergency Elevator Evacuation Systems. Fire and Accessibility, 2nd Symposium, Proceedings, April 19 & 21, New York, NY, pp. 131-150, American Society of Mechanical Engineers (1995)
8. Pauls, J.: Elevators and Stairs for Evacuation: Comparisons and Combinations. ASME Workshop on the Use of Elevators in Fires and Other Emergencies. Atlanta, Georgia (2004)
9. Galea, E.R., Sharp, G., Lawrence, P.J., Holden, R.: Approximating the Evacuation of the World Trade Center North Tower using Computer Simulation. Journal of Fire Protection Engineering. **18**, 85-115 (2008)
10. Siikonen, M.L., Hakonen, H.: Efficient Evacuation Methods in Tall Buildings. Elevator World. **8**, 117-123 (2003)
11. Bärlund, K., Mäkelä, M., Kattainen, A., Siikonen, M.L.: Evacuation Mode for Total Building Evacuation. Building Safety Journal. **6**, 36-39 (2006).

12.Roschier, N.R., Kaakinen, M.J.: New formulae for elevator round trip calculation. Supplement to Elevator World for ACIST members. **19**, 189-197 (1979)

13.CIBSE D: Transportation systems in buildings. The Chartered Institution of Building Services Engineers, London (2005)

14.Klote, J.H.: Routine for Analysis of the People Movement Time for Elevator Evacuation. NISTIR 4730 (1991)

15.Bukowski, R.W.: Emergency egress from ultra tall buildings. CTBUH 8th World Congress (2008)

16.Tyni, T., Ylinen, J.: Genetic algorithms in elevator car routing problem. In: Spector, L. et al (eds.), Proceedings of the Genetic and Evolutionary Computation Conference, pp. 1413-1422. Morgan Kaufman Publishers (2001)

17.Sorsa, J., Hakonen, H., Siikonen, M.L.: Elevator selection with destination control system, Elevator World. **54**, 148-155 (2006)

# Experiments for the Feasibility Study of the Evacuation by Moving Escalator in Public Space

**E. Okamoto, Y. Hasemi, S. Moriyama and N. Okada**

Department of Architecture, Waseda University, Tokyo, 169-8555, Japan
try58red3753@toki.waseda.jp

**Abstract** This paper reports experiments on the reaction of pedestrians to the sudden stop and restarting of an escalator to explore the applicability of escalators moving toward safer side. Deceleration rate at the stop and acceleration at the restarting were controlled so that optimum operation to keep the safety of pedestrians were sought. 32 experiments were conducted on each of 10 subjects(6 females and 4 males) by changing the moving/still mode of the escalator, moving direction of escalator(upward/downward), deceleration and acceleration rates, walking/still standing modes of the subjects, and burden conditions. The experiments indicate restarting generally safer than stopping, downward operation generally safer than upward one, and significance of the influence of the burden condition to the safety of pedestrians at emergency stopping and restarting. The difference whether the pedestrian walks or stands still did not cause significant difference in the pedestrians' safety. The experiments also indicate significant effectiveness of the reduction of the deceleration rate at emergency stopping from the current 0.61 $m/s^2$ to 0.43 $m/s^2$ and the acceleration rate at emergency restarting from the current 0.17 $m/s^2$ to 0.10 $m/s^2$ for the improvement of the pedestrians' safety.

## Introduction

Escalators are becoming a common means of vertical pedestrian traffic in public facilities, and we must anticipate accidental or intended use of escalators for evacuation at the event of fire or any other emergency. However, there are still number of questions about the feasibility of smooth evacuation with escalators, including how escalators have to be operated and how the performance of escalators should be designed. Our previous experiments with long and short escalators actually in service [1] quantified walking velocity, flow coefficient and other basic parameters for the assessment of the capability of escalators for upward group evacuation, and revealed general effectiveness of escalators moving toward safer direction for

R.D. Peacock et al. (eds.), *Pedestrian and Evacuation Dynamics*,
DOI 10.1007/978-1-4419-9725-8_58, 

the group evacuation especially from underground facilities. The benefit of evacuating with moving escalator is particularly pronounced for elderly and others who have basic difficulty in walking upward on stairway.

On the other hand, for such application of escalator for evacuation, an escalator originally running toward hazard site, at least needs to be stopped first and then restarted toward the safer side. Keeping pedestrians' safety during these operations is worth studying because few research and development has been conducted on the active use of escalators for evacuation and safe stopping of escalator is an important issue to prevent accidents in an emergency even if the evacuation with escalators is limited to the stopped mode of escalator. Also it is important to note that when we consider evacuation through escalator, we must anticipate that pedestrians carrying bags do not abandon them even at emergency partly because leaving bags on escalator steps would block the evacuation of following pedestrians and partly because bags on the steps might damage the escalator operation if the bags are jammed into the gaps of the steps. The burdens may restrict the pedestrians' reaction and may further affect the safety of pedestrians at the stopping and the restarting.

The present study aims at seeking conditions for safe stopping and restarting of escalator for its emergency use. For this purpose, a series of experiments are conducted with subjects bearing bags or baby-simulating dolls using experimental escalators with velocity controller. Times to full-stopping and to full-running are varied and the safety is assessed according to the physical reaction and the mental perception of the subjects.

## EXPERIMENTAL DESIGN AND PROCEDURE

### *Experimental Facility*

The experiments were conducted using one-story high experimental inverter controlled escalators in a technical center of an escalator manufacturer. Two escalators of the same dimension are used (Figure 1), and both upward and downward running modes are examined. Both escalators run at the current standard velocity, 0.5 m/s, but one stops at the current standard rate of deceleration, 0.61 m/s2( 0.86 s from 0.0 m/s to 0.5 m/s), and restarts at the current standard rate of acceleration, 0.17 m/s2 (3.04 s from 0.5 m/s to 0.0 m/s), and the another is controlled to stop at reduced rate of deceleration, 0.43 m/s2( 1.17 s from 0.5 m/s to 0.0 m/s), and to restart at reduced rate of acceleration, 0.10 m/s2( 4.45 s from 0.5 m/s to 0.0 m/s).

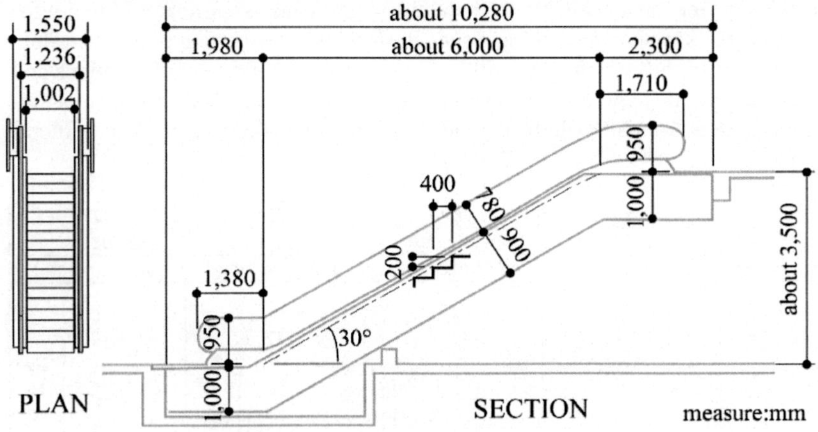


**Fig. 1. Dimensions of the test escalator [2]**

## *Conditions of the Subjects*

**1) Selection of the Subjects:** 10 subjects, 6 male and 4 female, all university students from 20 years old to 30 years old were hired. Subjects had not been announced of the full experimental information to ensure their unpreparedness. Subjects lined turning on back not to see other subjects on the way to the experiments.
**2) Burden and Use of Handrail:** Since escalators are common in railway stations, public underground buildings, and other open public facilities, the passengers are essentially on the travel from facility to other facility for business, shopping or amusement. Many of them must have bags and some might be accompanied by their family. We should try to reflect these conditions in the experiments.

Burden to a pedestrian such as bags and carried baby may prevent keeping balance during an unexpected change of the moving velocity of an escalator, whereas handrails are available for assisting the pedestrians on an escalator to keep their balance. Burden and availability of handrail are important elements that may affect the safety of pedestrians during an unexpected stopping or restarting at the event of an emergency. According to the recent investigation on the escalators in subway stations, 71.9 % of all the pedestrians have bags [3]; 58.6 % carried the bag(s) with one hand, and another 13.3 % had bags on both hands. It is natural that they would not abandon bags at the event of such emergency as fire not directly visible from the escalator. Also although there is no organized investigation, it is common that babies or small kids are carried by hand of the parents at railway stations; carrying a baby can restrict both hands and must be still harder for keeping balance than carrying bags. These pedestrians' conditions with significant burden and without free hand available for handrails could cause significant difficulty for pedestrians to keep balance during a sudden stopping or restarting of the escalator. We reproduced the combinations of the bag- or baby- carrying configurations and the dependence on handrail as shown in Figure 2. The "baby doll" is 6 kg to simulate an average 6-months old baby, and each bag weighs 3 kg according to literature [3].

| | A | B | C | D |
|---|---|---|---|---|
| handrail | × not use | × not use | × not use | O use |
| doll/bag | doll | bags both hands | bag one hand | bag one hand |
| subject | 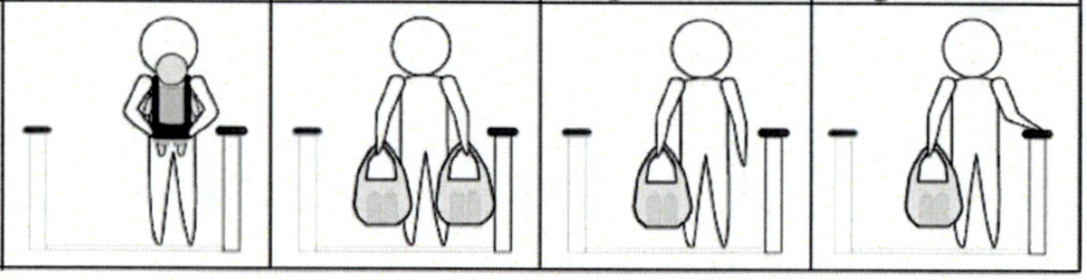 | | | |

**Fig. 2. The combinations of handrail and doll or bag**

**3) Use of "baby band":** When the subjects carried the baby-doll, female subjects held the doll with "baby band" (Fig. 3), and males held the doll without it, because women with baby in railway station are most likely to go shopping and want to keep their hands free for carrying goods. But we scarcely see a man carrying baby with baby band.

**Fig. 3. Female subjects holding baby-doll with "baby band"**

## *Experimental Procedure and Conditions*

Each subject waited alone in front of the entrance of the escalator until the signal for start turned on, and then rode on the moving escalator. While the subject was on the escalator, the escalator was suddenly stopped without warning, and after a while the escalator was restarted toward the other direction. While sound signal may be given to escalator pedestrians at an emergency, such warning was not given at these experiments partly because the current sound guidance system does not seem to be enough loud to deliver emergency guidance in crowded terminal station and partly because it is anticipated that pedestrians wearing earbuds are not aware of any sound guidance. Behavior of the subject throughout his/her travel from the entrance to the exit of the escalator was monitored with 4 video cameras from different directions. 32 tests in total were conducted (Table 2).

**Table 1: Test conditions**

| | Case | Acceleration/deceleration | Trend | Subject condition | Handrail burden | |
|---|---|---|---|---|---|---|
| | | | | | Handrail | Burden |
| 1 | 1-A | Current rate | up | walking | × not use | doll |
| | 1-B | Deceleration: 0.61m/s$^2$ | | | × not use | bag both hands |
| | 1-C | Acceleration: 0.17 m/s$^2$ | | | × not use | bag one hand |
| | 1-D | | | | use | bag one hand |
| 2 | 2-A | | down | still | × not use | doll |
| | 2-B | | | | × not use | bag both hands |
| | 2-C | | | | × not use | bag one hand |
| | 2-D | | | | use | bag one hand |
| 3 | 3-A | | up | walking | × not use | doll |
| | 3-B | | | | × not use | bag both hands |
| | 3-C | | | | × not use | bag one hand |
| | 3-D | | | | use | bag one hand |
| 4 | 4-A | | down | still | × not use | doll |
| | 4-B | | | | × not use | bag both hands |
| | 4-C | | | | × not use | bag one hand |
| | 4-D | | | | use | bag one hand |
| 5 | 5-A | Reduced rate | up | walking | × not use | doll |
| | 5-B | Deceleration: 0.43 m/s$^2$ | | | × not use | bag both hands |
| | 5-C | Acceleration: 0.10 m/s$^2$ | | | × not use | bag one hand |
| | 5-D | | | | use | bag one hand |
| 6 | 6-A | | down | still | × not use | doll |
| | 6-B | | | | × not use | bag both hands |
| | 6-C | | | | × not use | bag one hand |
| | 6-D | | | | use | bag one hand |
| 7 | 7-A | | up | walking | × not use | doll |
| | 7-B | | | | × not use | bag both hands |
| | 7-C | | | | × not use | bag one hand |
| | 7-D | | | | use | bag one hand |
| 8 | 8-A | | down | still | × not use | doll |
| | 8-B | | | | × not use | bag both hands |
| | 8-C | | | | × not use | bag one hand |
| | 8-D | | | | use | bag one hand |

### *Measures for the Safety of Pedestrians by Stopping and Restarting of Escalator*

We tried to visualize the safety of pedestrians at the stopping and the restarting of escalator by the physical and mental impacts to the subjects. These impacts are captured by the visual observation (video) of the behavior of each subject and by the post-test vote on the danger-perception. If the escalator is suddenly stopped or started without warning, a pedestrian on an escalator may take any "defense action" against potential hazard. We collected various defense actions from the videos of the experiments, and classified them into 6 levels as shown in Table 3. Also the subjects were asked to deliver danger perception after each experiment by questionnaire; the danger perception was classified into the 5 levels as shown in Table 4. Suitability of the emergency operations for various pedestrian conditions is discussed on the basis of the comparison of the number of actions and votes for each level.

**Table 2. Classification of typical Defense Actions**

| | Defense Action |
|---|---|
| 1 | No movement |
| 2 | Pause (subjects are walking at start) |
| 3 | Grasp handrail (subjects do not use handrail at start) |
| 4 | Lose balace |
| 5 | Take a step forward |
| 6 | Take both legs forward |

**Table 3. Danger Perception level classification**

| | Danger-Perception |
|---|---|
| 1 | Do not feel tension |
| 2 | Hardly feel tension |
| 3 | Feel tension weakly |
| 4 | Feel tension |
| 5 | Feel tension too much |

## Experimental Results and Discussion

Let us summarize the experimental results, hazard-avoiding actions and danger perceptions, first by dividing the experiments into the stopping mode and the restarting ones.

## *Action and Perception at the Stopping of Escalator*

Figure 3 and Figure 4 are a summary of the levels of the actions and the votes of perception at the escalator stopping experiments.

**(1) Behavioral Impacts and Defense Actions against Sudden Stopping:** The left and right columns of Figure 4 are a summary of the actions of the subjects on the upward and downward escalators stopping at the current standard and reduced rates of deceleration respectively. These indicate that, when an escalator is stopped at the current standard deceleration, considerable portion of the subjects, either if they were walking or still, had to take any defense action; it is particularly pronounced for the upward running mode. Effectiveness of handrails is found to be limited for the still pedestrians with single bag; the weak effectiveness of the handrails for walking pedestrians is attributed to the height of the handrails notably higher than those of normal staircases, probably for safety reasons, causing general difficulty for walking pedestrians to apply their hand on the handrail.

Occurrence of defense actions are significantly decreased if the deceleration rate is reduced to $0.10 m/s^2$; no subjects except for one carrying the baby-doll or bags on both hands take any defensive actions. Since "unexpected stopping" of a crowded escalator can be inevitable at the event of any emergency including fires, falling of a pedestrian and earthquake, this fact indicates general importance of

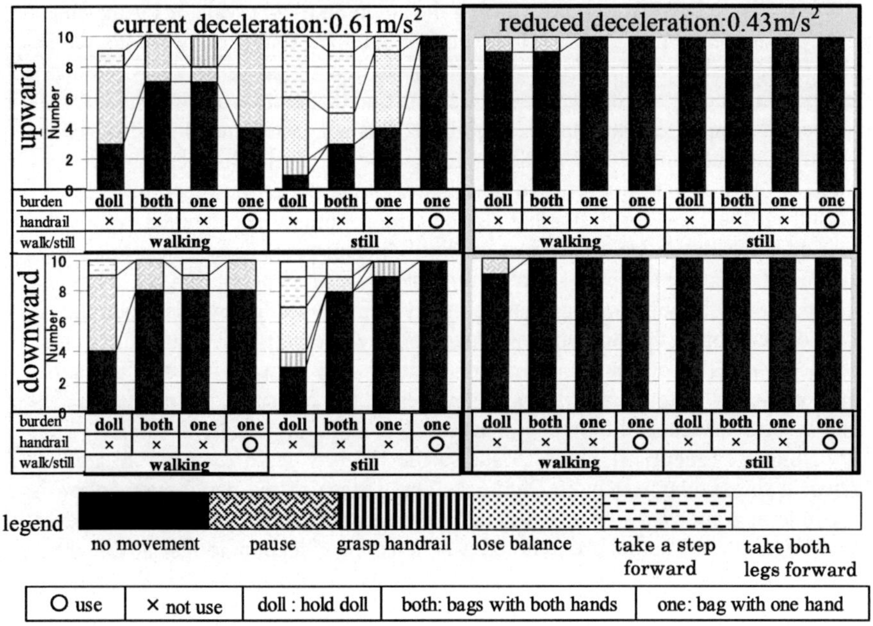


Figure 4. action at the stopping of escalator

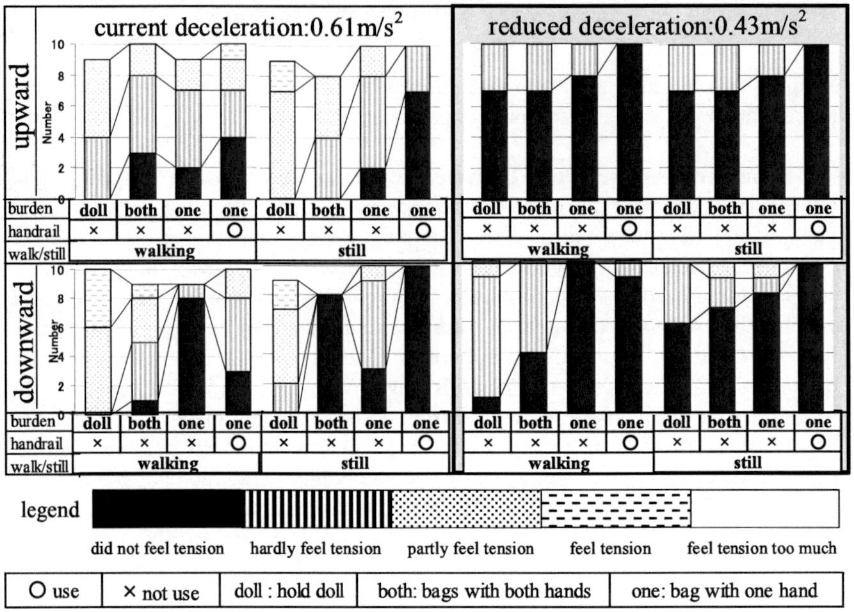


Figure 5. danger perception at the stopping of escalator

the reduction of deceleration rate during the stopping of an escalator for the emergency management in general. The subjects with baby-doll or two bags, either walking or still, had to take some defense action if the escalator is suddenly stopped at the current standard deceleration: this difficulty was significantly improved by simply reducing the deceleration. This suggests general need of reducing the deceleration of any escalator for public use from disaster management point of view. It is also worthy to note that walking pedestrians tend to pause especially if the escalator is stopped at the current standard deceleration. While the present experiments were executed on single independent pedestrians, this result suggests that sudden stopping of a "foregoing" pedestrian of a group could cause a bump of the following ones and successively dominoes of the whole group unless the deceleration at the stopping is reduced or any alternative to reduce the impact for pedestrians is introduced. Similarly, taking a step forward is essentially difficult on a crowded escalator; such action seen in the present experiments may result in the occurrence of bumping on a crowded escalator actually in service.

**(2) Danger-Perception:** Figure 5 shows the levels votes of perception summarized in the similar way with Figure 4. The pattern of the danger-perception is found to be somewhat similar with that of the defense actions, but it indicates even those who do not yet take any defense actions could feel some fear at the unexpected stopping of the escalator even if the deceleration of the escalator is reduced. This effect is particularly pronounced for baby-carrying pedestrians.

## *Action and Perception at the Restarting of Escalator*

Figure 6 and Figure 7 are a summary of the levels of the actions and the votes of perception at the escalator restarting experiments.

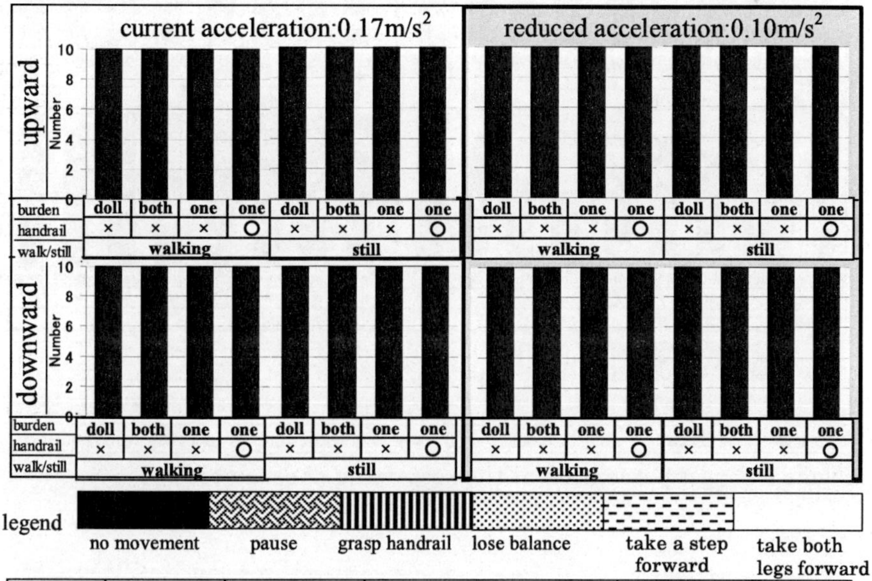


| | | | | |
|---|---|---|---|---|
| O use | × not use | doll : hold doll | both: bags with both hands | one: bag with one hand |

Figure 6. action at the restarting of escalator

**(1) Behavioral Impacts and Defense Actions against Sudden Stopping:** There was no subject taking significant defense action irrespective of the moving direction, the rate of deceleration or the walking/still mode of pedestrians. It indicates the unexpected restarting of an escalator generally is safer than is stopping.

**(2) Danger-Perception:** Considerable portion of the subjects with baby-doll or two bags felt some fear when the escalator was restarted at the current standard acceleration even if they did not take any defense actions. This reaction was significantly depressed by the reduction of the acceleration.

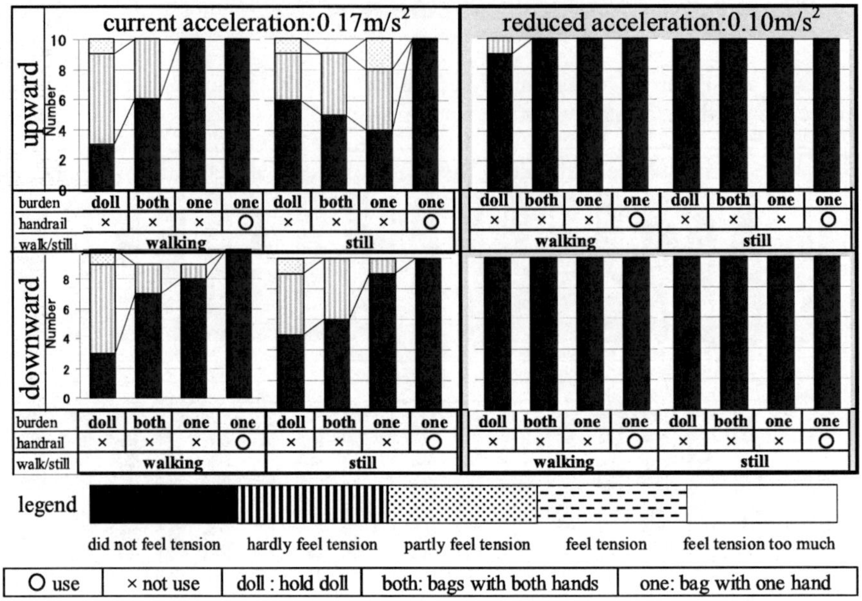

Figure 7. danger-perception at the restarting of escalator

## Conclusions

From the present experiments, following conclusions can be drawn.

The present experiments indicate that use of the escalator with the current deceleration acceleration for the stopping and the restarting at the event of an emergency could cause notable fear and reactions of pedestrians, and could cause subsequent accidents/incidents. The physical and mental impact of sudden operation is particularly pronounced at the stopping of escalator. Prohibition of walking on escalator does not seem to make good improvement of the situation for those carrying baby or two bags. The behavioral and mental impact to pedestrians do not seem to be resolved significantly. This risk during an emergency operation of escalator can be thoroughly resolved if the deceleration at the stopping and the acceleration at the restarting, at the event of an emergency, are reduced (deceleration 0.61 $m/s^2$→: 0.43 $m/s^2$ ,acceleration: 0.17 $m/s^2$→: 0.10 $m/s^2$).

Some of the defense actions such as pausing seen during the experiments are believed to cause bumping or other impact to surrounding pedestrians if the escalator is crowded, Existence of surrounding pedestrians may also influence the reaction to emergency, which should need further study to establish escalator as a safe means for emergency. We have recently run another series of escalator experiments with small groups of subjects to verify such effect. From mechanical performance point view, it is important to develop technical guidelines for the escalators available for emergency evacuation as escalator might bear large crowd when it is used for evacuation.

**Acknowledgments** The authors would like to acknowledge cooperation of the escalator manufacturer who offered the test facility though they declined our offer for the disclosure of their name in the publication. We also would like to acknowledge the assistance of the members of Hasemi Lab., Waseda University in the experiments. This study is supported by the JSPS Science Research Fund.

## References

1 Okada, N., Hasemi, Y., Moriyama, S., Hirakawa, K., Takemori, K., Hebiishi, T., Lu, Y.: Feasibility of Upward Evacuation by Escalator-an Experimental Study. Proceedings of 4th International Symposium on HUMAN BEHAVIOUR IN FIRE Conference, Cambridge, UK, pp.161-172 (2009)

2 Toshiba Escalator Kindmover (TE series) design construction observation. TOSHIBA Elevator Corporation, pp.3 (in Japanese)

3 Tokyo Fire Department Life Safety Division, Accident Prevention Measure Related with Escalator (2005) (in Japanese)

# Real Events

# Assessing Crowd Dynamics and Spectator Safety in Seated Area at a Football Stadium

**S.Y.Z. Au[1], J. Gilroy and R.A. Haslam**

[1] Human Interactions Limited
13 Meadow Lane, Fearnhead, Warrington, Cheshire, WA2 0PR UK
Corresponding author: zac.au@human-interactions.co.uk

**Abstract** This paper presents a study at the Manchester United Football Stadium at Old Trafford, UK. The purpose of the study was to assess the safety risks associated with spectators standing in an all-seated stadium. Information gathering was a key part of the study; this paper looks at the information gathering exercises undertaken and how they contributed to the risk assessment. It also presents some of the key study findings. It may not be immediately apparent that standing in a stadium that is not overcrowded could be an issue. But there are in fact a variety of movements and behavioral factors involved when people stand. It is these movements and behaviors and their interactions with the physical environment that was our focus of investigation. Overall, the study has found that the risk of standing varies depending on the situation and the physical design of the seated area. There are a number of risk factors involved; we believe that many of them can be controlled through stadium design and crowd management.

## Introduction

Football (or soccer) are the most popular games in the UK. Top level matches are often attended by several tens of thousands of spectators. There have been a number of crowd related accidents and disasters in the history of football. In particular, the Hillsborough Stadium Disaster in 1989, which resulted in the deaths of 96 people, is considered one of the worst football disasters. Following this disaster and the subsequent inquiry by Lord Justice Taylor [1], it has become a legal requirement in the UK to provide all-seated accommodation in all Premier League and First Division football stadia. In terms of safety, the rationale is that with one person per seat, people are not subjected to crowd pressure and those managing crowd numbers will know exactly how many there are in each area without having to count them in or assess crowd density by visual impression.

With this requirement, the intention is that spectators should be seated for the duration of the match. However, in reality many spectators stand up in seated areas, sometimes persistently for long periods of time. The national and local licensing and enforcement authorities were concerned about the safety implications of this behavior. Their perception was that standing in areas designed for sitting is

R.D. Peacock et al. (eds.), *Pedestrian and Evacuation Dynamics*,
DOI 10.1007/978-1-4419-9725-8_59, © Springer Science+Business Media, LLC 2011

"less safe". But it was not entirely clear how and to what extent it is "less safe". This paper presents a study commissioned by one of the local authorities to examine this issue at the Manchester United Football Stadium at Old Trafford, UK. Due to length restriction, it is not intention of this paper to detail the methods used for each part of the study or all its findings. Instead, it focuses on our approach to tackle the main challenge of this study, namely information gathering, and highlights some of the key study findings.

## Objectives of the Study

The overall objectives of the study were as follow:

1. To assess the safety risks associated with spectator standing in areas designed for seating only at the said stadium.
2. To assess the adequacy of the existing provisions in place at the time.
3. To propose additional safety measures if appropriate so that risks are kept as low as reasonably practicable.

Three types of standing were assessed: persistent standing for prolonged periods; standing at moments of excitement; and "standing" on entering or leaving seated areas.

## The Challenge

The risk assessment aspect of the study was relatively straightforward. It was based on a methodology developed for the UK Health and Safety Executive specifically for assessing crowd safety risks [2]. In line with general risk assessment principles, it involved the following: (a) identifying the safety hazards that could arise from standing; (b) establishing their causes and consequences; (c) identifying existing mitigating factors and precautions already in place;, and (d) evaluating the risks involved by assessing the likelihood of occurrence for each hazard and the severity of its consequences.

Information and data were required to facilitate each part of the assessment but they were scarce. For example, Pauls [3] is one of the few researchers to have looked at the risk of falling in sports grounds but the focus of his work tends to be on steps and stairs. Au et al [4] and Smith [5] recognize falling as a risk but give little details on this problem. The main UK guide on safety at sports grounds [6] provides guidance on the physical design of stadia including spectator seating, but it had not given adequate consideration to standing in seated areas. This could be because the law requires them to sit; to provide guidance that makes allowance for standing could be seen as condoning it. The only relevant information from published literature was the research on trips/slips/falls, which provided an insight into the factors that could contribute to the risk of falling. Using the behavioral classification derived by Hill et al [7], they include: the design of the environment; individual interactions with the environment (e.g. carrying items); behavior that modifies the environment (e.g. littering, spillage of drink); and behavior affecting the individuals (e.g. standing, jumping up, etc.). These factors were taken into consideration in the study, but much more information was still required. Thus a key challenge was to understand the issues and the hazards involved and to establish a basis for risk evaluation through information gathering and investigation.

## Information Gathering

In theory, risk data could be generated through extensive controlled experiments on different seated area designs with different types of standing, etc. However, this would be a huge undertaking which was simply not possible within the scope, timescale and resources of this study. Therefore, the information needed for the assessment had to come from existing sources. We identified the following sources where relevant information could be obtained: historical records and statistics; existing knowledge of the stakeholders; the experience of those who frequently attended matches; and examination of the seated area designs and crowd behaviors. None of these sources on their own would provide a clear answer to how "risky" standing is. The requirement was therefore to build up a pool of information, in all shapes and forms, from all available sources. To this end, an extensive information gathering program was undertaken. It consisted of:

- Site survey
- Review of internal literature and post match reports
- Consultation with stakeholders
- Observations of crowd behaviors and activities
- Questionnaire survey of front line staff and spectators
- Ergonomics assessment of the seated areas

### *Site Survey*

The purpose of this survey was to collect background information about the stadium to facilitate the rest of the study. Site visits were carried out when the stadium was not in use to survey the physical design and obtain key dimensions of the seated areas. This stadium is one of the largest in the UK with a capacity of 68,000 at the time of the study. It has been renovated and built up in sections over the years; therefore the physical designs in different sections vary. As it was not possible to assess the entire stadium within the available time and resources, four areas were identified for the study to focus on. They were the East Stand lower tier, West Stand tier 2, North Stand tier 3, and the Away Supporters Section. This selection was made on the basis that: (a) they were representative of the design characteristics of the rest of the stadium; and (b) they were areas where standing was more widespread or deemed more of a concern (e.g. due to their height and gradient).

## *Review of Internal Literature and Post Match Reports*

Internal papers and reports produced by the stakeholders presented their existing knowledge on standing. They provided valuable background information about the problem and their views and analyses of the issues involved. Care was taken to recognize that sometimes such information could be rather subjective.
Following every match at the stadium, Manchester United Football Club (MUFC) also produces a post match report on the key events that occurred before, during and after the match. It also contains qualitative information and statistical data such as match type and category, attendance figures, number of ejections and arrests, injury figures, information relating to persistent standing, information on any crowd behavior issues, the circumstances surrounding the match and any events could have affected crowd behavior. Collation and analysis of this data enabled us to look at matches over a period of time to identify any trends or relationships between standing and other possible risk factors such as match type and category, anti-social behavior, locations in the stadium, whether alcohol was on sale, etc. Initially, it was also hoped that the injury data would give an indication on the risk of injury associated with standing. However, this was not possible as only the nature of the injuries was recorded. This made it hard to determine whether an injury was standing related or due to other causes.

## *Consultation*

This exercise allowed us to further probe the stakeholders' knowledge, perceptions and anecdotal details about spectator standing; but this time through detailed discussions. This information is often venue-specific and highly relevant to the subject matter concerned. But it can also be subjective and does not necessarily provide a complete picture. Hence it is important that people with different viewpoints and backgrounds are consulted. In this study, consultation meetings were held with MUFC, MUFC supporters associations, the local authority, the national Football Licensing Authority and the local emergency services. The aims were to:

(a) obtain their views on standing; (b) investigate the reasons for standing; (c) discuss the perceived dangers and safety issues; and (d) identify potential solutions and practices elsewhere that could address the issues. Stakeholders were engaged in open discussions around each of these topics. These discussions provided valuable and varied information about the potential hazards, the background and reasoning of standing, the existing precautions in place and possible solutions to reduce the risk. At the time of the study, standing in all seated stadia was a hotly debated issue attracting much public and media attention in the UK. In this climate, it was not only technically important but also politically necessary to consult different views.

## *Observations*

Observations allowed us to study in detail crowd behaviors and movements, and their interactions with the physical design of the seated areas in different circumstances during a match. This enabled identification of potentially dangerous behaviors and the manners in which they could cause harm and injuries. Therefore, observations are very important in understanding behavior related problems and identifying the hazards. Experience shows that they often reveal information that may be unexpected or would otherwise not be found by other means.

Post match reports by MUFC and match inspection reports by the local authority indicated that the amount of standing could vary between match types. Hence several different match types were selected for observations. They included European and domestic matches as well as big/important and less high profile matches. Observers were deployed to the four areas previously identified. They were each briefed by being given guidance about their tasks and guidance to ensure that they conducted observations safely and without hindering MUFC's operations. The success of such observations depends to a large extent on the quality of the observers. As such, briefing was rigorous and only observers with human factors, health & safety and/or crowd safety backgrounds were employed.

## *Questionnaire Survey*

This survey aimed to collect quantifiable information about the risk of standing from the front line staff and the spectators (MUFC supporters). These people attended matches frequently and had detailed first-hand knowledge on the events and occurrences on the ground. Thus they are a valuable source of information. The data collected provided an important basis for risk evaluation in the risk assessment. In the questionnaire, respondents were asked from a list of 13 hazard scenarios to indicate how frequently they had experienced or witnessed each scenario (i.e. never, once or twice a season, several times a season or nearly every match). The scenario list was devised from hazard information collected from previous exercises. Respondents were also asked to identify the areas in which they normally sit or work so that data could be collated in relation to the relevant part of the stadium.

The questionnaire also invited the respondents to enter any additional comments. The comments received helped to identify some of the underlying causes of standing and gain further insight into the risk factors that cause people to fall. Whilst it is acknowledged that there are weaknesses to using questionnaires, but within the time and resources available, the use of questionnaires was the best achievable way to capture a large amount of information quickly.

## *Ergonomics Assessment*

Spectator standing (and associated movements) generally occurs in the confines of a seated area and therefore the physical design of that area can be an important risk factor. For this reason, an Ergonomics assessment was conducted to determine whether and how risk could be affected by design features such as seat dimensions, clearways, rear seat heights, flooring, and front barriers. Dimensions of seated areas, information on spectator behaviors and movements and anthropometrical data were used for the assessment. The results provided an analysis of how physical design attributes could give rise to safety hazards and affect the risks. They also provided recommendations on how the physical design of seated areas could be improved to reduce the risks.

## *Facilitating Risk Assessment*

Figure 1 illustrates how these exercises were relevant to the final risk assessment. In general, each exercise generated different forms of information from different sources that input to various parts of the risk assessment. Within the risk assessment, the identification of safety hazards was primarily based on the observation findings as well as information from internal papers and reports, consultation and the Ergonomics assessment. The underlying and immediate causes of the hazards were established from a wide range of sources, including consultation, observations, post match reports, open comments from the questionnaire survey, and the Ergonomics assessment. Information on existing precautions was obtained from the stakeholders through internal literature and consultation. Finally in risk evaluation, an established rating system was used [2]. The likelihood of a hazard occurring was rated based on a list of "rationale" drawn up from the findings from all exercises, although prime consideration was given to the results of the questionnaire survey and the Ergonomics assessment. The severity of the consequences was rated based on the physical environment in which the hazard may take place, such as the height and gradient of the stand and any fixtures and fittings that may exacerbate or reduce the severity of the hazards. The level of risk was then determined from the likelihood and severity ratings.

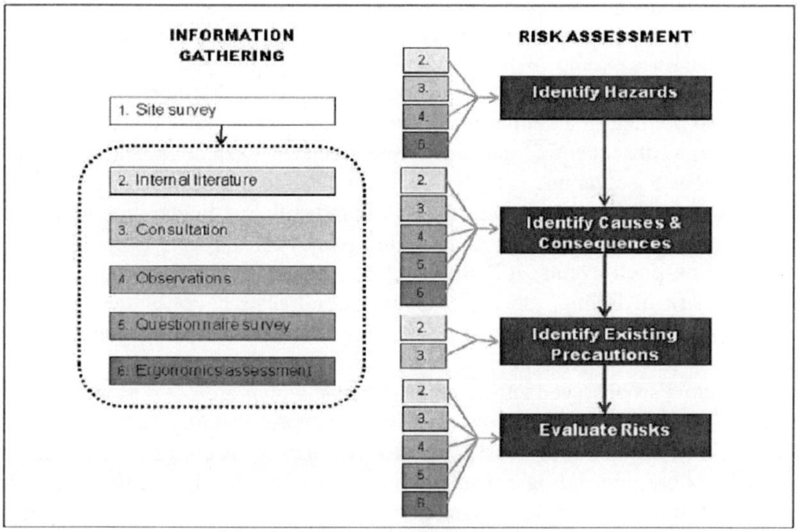


**Figure 1: Input to risk assessment**

## Study Findings

This study set out to establish the safety risk of standing in the seated areas of a stadium. But it found that the risk cannot be generalized, instead it can vary from time to time and from places to places. Three types of standing were investigated and they were found to pose different hazards and risk levels. This is largely because different types of standing have associated with them different movements and behaviors; which are one of the key factors affecting risk. The physical environment in which standing occurs is another key risk factor. In general, the study found that standing in the upper tiers (i.e. tier 2 and tier 3) of the stadium tended to pose a higher risk. This is not just because they were higher up per se. Clearly the consequence of falling from height is very severe. Perhaps more importantly, it was the constrictive dimensions of the seated areas and how well they catered for spectator movements that were the main risk contributory factor.

## *Types of Standing and Associated Behaviors*

Trips/slips/falls are the main hazards associated with standing. However, the likelihood of them occurring depends on the nature and the amount of movements involved. Of the three types of standing considered, persistent standing per se posed a low risk when people stood fairly still. As shown in Table 1 below, over 96% of respondents in the questionnaire survey indicated that they had either never seen or experienced a trip, slip or fall incident or saw it only once or twice a season when people were just standing. Our observations and Ergonomics assessment findings also pointed to a similar conclusion. However, persistent standing could lead to other "consequential" hazards. Some related hazardous behaviors were noted from our observations. (1) People stood close to the front edge of the row. This is not normally a problem when people stand still, but they are more vulnerable to tripping and falling when more vigorous movements were adopted, e.g. in response to on-pitch events. (2) Standing on seats is clearly less stable and increases the risk of falling, especially given that all seats have folding seat pans. Those who stood on seats were mainly children as this was the only way that they could see the pitch when people in front of them stood up. (3) Both adults and children were also observed sitting on seat backs, which are not designed to provide support for sitting. This suggested that of those standing, some spectators were unable or unwilling to stand for prolonged period. A concern was raised in internal literature and during consultation that people who do not want to stand were forced to stand up due to persistent standing. In fact, there was a general consent amongst those consulted that most people prefer sitting. However, when people in front stand, others will also have to stand in order to see the pitch. Our findings show that this is not just a customer care issue but also one that can have safety implications. (4) Intoxication is another risk factor. Despite the measures in place to turn away those who were excessively drunk at the turnstiles, both past records and our observations showed that some would get through into the stadium. When intoxicated, it is more likely for people to trip or fall when standing then when sitting.

Because of these "consequential" hazards, we concluded that persistent standing in seated areas is less safe than sitting although the risk is still relatively low.

Standing at moments of excitement and when entering/leaving seated areas involve more movements and hence the risk of trip, slip or fall increases. In particular, standing at moments of excitement was found to pose the highest level of risk. Moments of excitement refer to events on the pitch that cause high emotion amongst spectators (e.g. a goal). In response to such events, spectator movements can become very vigorous indeed. The movements observed included jumping upward and forwards at the same time, jumping up and down with both feet off the ground, vigorous upper body movements whilst jumping, jumping and turning at the same time, and jumping and swaying as a group. These are common occurrences. Furthermore, the focus of the spectators was on reacting to the on-pitch event and they could forget where they are. The Ergonomics assessment concluded that the likelihood of falling would increase as a result of the limited space available in the seated areas to cater for such movements or subsequently to allow people to recover balance. This is in line with the results of the questionnaire survey (Table 1), which indicates that trips/slips/falls occurred much more frequently at moments of excitement. Our observations also identified other hazards concerning moments of excitement. The one that gave rise to the most concern was people leaning on/bend double over the front barrier of the upper tiers during these moments. This behavior was not widespread but the consequence of falling can be very severe. Understanding spectator movements and behavior is important in helping us to specify design criterion for future designs.

People entering and leaving seated areas occurs not just before and after a match and at half time, it also occurs throughout the match. This involves passing other people, whether they are standing or sitting. Using anthropometrical data (of the British adult population), our Ergonomics assessment showed that there was little room in the seated areas for people to pass each other. The only way to make this happen was if the individuals corporate by standing and leaning backwards or forwards; which increases the likelihood of falling. In addition, the study had also identified the following behavioral factors that could affect the risk: people carrying large items; people with hot drinks (sometimes with several hot drinks); people looking elsewhere (e.g. at the pitch) rather than where they are going; and people moving at speed or climbing over seats onto the adjacent row. The questionnaire survey results in Table 1 suggested that trips, slips and falls when entering/leaving seated areas can be nearly as likely to occur as at moments of excitement.

**Table 1. Questionnaire survey findings – percentage of respondents who witnessed or experienced slips/trips/falls**

| | Never | Once or twice/season | Several times/season | Almost every match |
|---|---|---|---|---|
| Persistence standing | 83.8% | 12.8% | 2.7% | 0.7% |
| Moments of high excitement | 44.4% | 34.3% | 15.3% | 6% |
| Entering/leaving seated areas | 55.2% | 30.6% | 12.5% | 1.7% |

## *Physical Design of Seated Areas*

The risk of standing was found to vary between different parts of the stadium due, to a large extent, to the fact that they were built at different times and therefore have different physical designs. Taking into account the spectator movements and behaviors identified above, the three main design attributes that could have the most impact on safety risk are believed to be the clearway and seat size, the flooring, and the front barrier.

Clearway refers to the distance between the foremost projection of a seat when it is in the up position and the back of the seat in front. As previously discussed, the likelihood (and the risk) of falling increases when there is only limited clearway for spectator movements at moments of excitement, people passing when entering/leaving the seated area, and people moving to recover balance. With a typical seating row depth in the stadium being merely 40mm above the $50^{th}$ percentile buttock to front of knee dimension for British males, the study found that space was very restricted not just for movements but even when people are seated. It is worth noting that in the questionnaire survey, many respondents commented that the seats were uncomfortable to sit in over time, which caused them to want to stand. Some also commented that the seats were generally too close together and there was not enough room to sit. Others reported that the lack of space and the pressure against their legs from the seat in front caused their legs and feet to "go numb". Whilst these may affect only some spectators, it is worth noting that once someone in front stands up, all of the spectators behind will also have to stand in order to see the pitch. Hence, it will only take a minority of spectators to stand up to make standing a widespread behavior.

In terms of flooring, its coefficient of friction was not measured in this study. But it is worth noting that we found in our observations that even within the seated areas, floor surface was often affected by spillage of liquids or the present of contaminants such as food or ice-cream. Consequently, the smooth concrete surface found in all of the four areas examined could become slippery.

The height of barriers in front of the seated areas on the upper tiers can be a difficult issue to address. On one hand the barriers need to be sufficiently high to protect against falling. On the other hand they have to be low enough not to obstruct spectators' views. At the time of the study, barrier heights in all four areas examined well exceeded the minimum height of 800mm recommended in the relevant guide [6]. But this minimum height assumes spectators are seated and it is less than the estimated centre of gravity of some standing adults. Therefore, front barriers designed to this standard only provided a limited protection against falling; especially when bearing in mind the observed behavior of people leaning over front barriers at moments of excitement.

Finally, it should be noted that seated area design should not only cater for spectator movements but also for the staff who works there to manage spectator safety. To this end, the lack of suitable space for crowd monitoring was a concern because this meant that front line staff members were often unable to see spectators clearly without blocking someone's view to the pitch. The lack of space between seating rows could also create problems (and cause delays) for security and medical staff to access the seated areas, some of whom could be carrying equipment.

### *Underlying Reasons for Standing*

The significance of understanding the underlying reasons is so that risk mitigations can be devised to address not just the symptoms but also the causes of standing. The restricted space in the seated areas and uncomfortable seating reported above is a prominent factor. But in the questionnaire survey, the most common reason cited for standing or wanting to stand was that it improved the atmosphere of the game for the supporters. It appears that the type of match being attended is also a strong factor influencing persistent standing. Questionnaire survey and stakeholder comments both suggested that it tended to occur more during tense matches and important matches where the match results are important to the club.

The amount of persistent standing also tended to vary in different sections of the stadium. The away supporter section was reported to be a main problem area. There could be several reasons for this. One reason could be that the away supporters, being the minority in the stadium and surrounded by the rival MUFC supporters, feel the need to be more vocal and more active to support and encourage their team. In the case of European away supporters, this could also be because they are used to standing in their own countries. Consequently, the adjacent section was also affected as spectators there became more likely to stand up in response. In some cases, historical factors could also play a part. For example, it was reported that in an area where there tended to be more people standing, many spectators there were from the part of stadium that used to be a standing area. It is clear from these findings that some of the reasons for standing are down to cultural or historical factors, which could take time and further efforts to change. Others such as uncomfortable seating will have to be addressed through stadium design.

## Conclusion

This study was a pioneering attempt to examine the issue of standing in seated areas of a stadium. It found that the safety risk of standing can vary from scenario to scenario and from location to location. There are many factors involved but this paper focuses on two main risk factors; namely spectator movements and behaviors, and stadium design. Taking these factors into consideration, the study concluded that persistent standing per se is not necessarily risky, however when the associated consequential hazards are taken into consideration, standing is perhaps less safe than many people may think. The risk of trips/slips/falls becomes much higher whenever movements are involved, particularly at moments of excitement and when people entering or leaving seated areas. Understanding spectator movements and behavior is important in helping us to specify design guidance and criterion for future designs.

Many of the risk factors were not immediately apparent at the outset of the study, as there was a lack of previous research. To identify these factors and generate enough information to support and facilitate the risk assessment was a major challenge. As such, information gathering was a crucial aspect of this study. This paper outlines the types of information collected in the study, the methods used to gather them and how they were used to help us to understand the hazards, the causes and the risks of standing. We believe that a similar methodology could also be useful in comparable contexts.

The findings of this study have provided better understanding about the safety implication of standing in seated areas. We were able to identify the potential hazards and understand the causes and the risk factors. However, how best to address the risks and produce practical solutions to some of the problems identified will require further research. Furthermore, although venue-specific, this study identified a number of issues that could have much wider implications for stadium design in general. Hence we believe that there is scope for expanding the study into a more generic context with a view to improving existing design guidance and criterion.

## References


1. Lord Justice Taylor: The Hillsborough Stadium Disaster Inquiry Final Report. HMSO, London (1990)
2. Au, S.Y.Z.: Research to Develop a Methodology for the Assessment of Risks to Crowd Safety in Public Venues. Health and Safety Executive Contract Research Report CRR 204. HMSO, London (1998)
3. Pauls, J.: Safety standards, requirements and litigation in relation to building use and safety, especially safety from falls involving stairs. Safety Science. 14, 125-154 (1991)
4. Au, S.Y.Z, Ryan, M.C., Carey, M.S., Whalley, S.P.: Managing Crowd Safety in Public Venues: A study to generate guidance for venue owners and enforcing authority inspectors. Health and Safety Executive Contract Research Report CRR 53. HMSO, London (1993)
5. Smith, R.A.: Density, velocity and flow relationship for closely packed crowds. Safety Science. 18, 321-327 (1995)
6. Scottish Office and Department of National Heritage: Guide to Safety at Sports Grounds (4$^{th}$ edition). HMSO, London (1997)
7. Hill, L.D., Haslam, R.A., Howarth, P.A., Brooke-Wavell, K., Sloane, J.E.: Safety of Older People on Stairs: Behavioural Factors. DTI ref 00/788. Department of Trade and Industry, London (2000)

# Stay or Go? Human Behavior in Major Evacuations

**P. F. Johnson[1], C. E. Johnson[2] and C. Sutherland[3].**

1. Arup, 1 Nicholson Street, Melbourne, VIC, 3147 Australia
2. Latrobe University, School of Psychological Sciences, Bundoora, VIC, Australia
3. Monash University, Department of Business Law and Taxation, Caulfield, VIC, Australia

Corresponding author: peter.johnson@arup.com.au

**Abstract** Human behaviour in building evacuation and use of elevators has been studied more extensively in recent years. There is also a move to explore concepts of "protect-in-place", thus avoiding evacuation in some circumstances. This raises the matter of decision making as to whether to "stay or go". In the fields of natural hazards, such as bushfires and floods, this decision to "stay or go" becomes equally important for safety. A comparison of literature sources on human behaviour and decision making in buildings with recent research on bushfires and floods highlights some common factors critical to decision making as to evacuation or not. These factors include emergency preparedness, situation awareness and trusted information systems. This paper suggests that greater understanding of group behaviour and socio-cultural differences is required if more effective emergency management is to be achieved.

## Introduction

There is significant research occurring internationally into the performance of occupant evacuation of buildings during fire and other major emergencies. Recent research evidence by Averill et al [1], Gershon [2] and Galea et al [3], based on human behaviour in events such as WTC 9/11, has brought a new understanding on evacuation, including the use of elevators.

At the same time, there is strong interest in building safety on the concept of "defend in place", sometimes called "protect in place". The proposition is whether, at least in some circumstances, buildings can be designed and constructed so that occupants can safely remain within the building during fire and other emergencies. This research has been summarized recently by Barber [4] for the Fire Protection Association, Australia.

Another whole field of emergency management in relation to human behaviour, decision making and people movement occurs in the area of natural hazards. This includes the domains of bushfires (wildfires), hurricanes and floods. In such

R.D. Peacock et al. (eds.), *Pedestrian and Evacuation Dynamics*,
DOI 10.1007/978-1-4419-9725-8_60, 

events, people may often choose to either "stay and defend" their property and themselves, or evacuate. These decisions can have significant life safety consequences.

The loss of 173 lives on "Black Saturday", 7 February 2009, in the Victorian bushfires [5] has brought the question of "stay or go" into sharp focus as public policy. Equally, the study of survivors of the 2005 Hurricane Katrina in the US has highlighted the difficulties in decision making, and the resistance of many to evacuate, despite official policies mandating evacuation in the event of major community wide threats to life [6].

This paper looks at some of the recent research on evacuation and "defend in place" for buildings, and attempts to draw parallels with some of the similar decision making and human behaviour required under "stay or go" regimes in the field of natural hazards. The implications for public policy for people management in emergencies are highlighted as well as the differences in human behaviour between building safety and natural hazard safety.

## Building Evacuation

The latest research in human behaviour and use of elevators for evacuation was extensively presented at the 4th International Symposium on Human Behaviour in Fire held in Cambridge in July 2009 [7]. The research presented on the WTC 9/11 disaster was summarized in a report published in the Fire Australia journal [8]. From that summary, a number of key concepts emerged in relation to building occupant decision making and evacuation behaviour. They included:

- Many people had pre-existing medical conditions, obesity or other disabilities which affected their decision making or movement speeds
- The majority of people had never evacuated the building using the stairs before, did not know all the stair locations, or where they would lead
- Numbers of people demonstrated fatigue behaviour
- Strong "group' influence meant that many waited for others to make decisions to leave or stay before acting
- There was a poor "safety climate" or "safety culture" amongst the building populations
- The buildings lacked damage hardened communication systems and their failure contributed to lengthy pre-movement times
- Overall, there was a lack of "situation awareness" and emergency preparedness.

As examples, Gershon's WTC research [2], based on semi-structured interviews with over 1000 survivors, showed data including 26% of people had a pre-existing disability or other medical conditions that they felt created problems for them to use the stairs for evacuation, 94% had never evacuated those buildings by stairs, and 89% did not know where the stairs would lead.

If this is typical of many buildings, then emergency evacuation preparedness policies and procedures are clearly not resulting in adequate knowledge or action by building occupants. And it raises the question of whether we should try to evacuate buildings at all? Or under what circumstances is evacuation an appropriate practice??

Galea et al [3] summed up a series of recommendations in relation to building design and evacuation as follows:

- There is a need to harden means of egress and communication systems
- There is a need to design high rise buildings for full evacuation, including for non-fire emergencies
- Response times and effective decision making can be improved by better emergency procedures, more effective training, and more informative and reliable communication systems

This last point on communication systems was highlighted in the panel discussion at Cambridge on "situation awareness". It was widely agreed that people need reliable, trusted information sources and clear instructions if they are to make and act on good decisions in a fire or other emergency.

The use of elevators was extensively debated in Cambridge [7, 9, 10]. Through the simultaneous use of stairs and elevators for evacuation, simulations and actual evacuation studies suggest a reduction in total evacuation time of 40 – 50%. Again, some key points to emerge were:

- The need for well protected stairs and elevator systems
- The importance of emergency preparedness, with training in the use of elevators and stairs for all building occupants
- The critical provision of protected communication systems and trusted information sources to enable prompt decision making, i.e., creation of good situation awareness.

Research by Heyes [11] through surveys and interviews confirmed the view that, for commercial buildings, occupants will be inclined to use elevators if given sufficient training and information about elevators and their protection.

## Protect-in-Place for Buildings

Recent research by Barber [4] in high-rise buildings has suggested an alternative to evacuation. This alternative is to remain in place in a protected area of a building rather than evacuate, ie. stay rather than go.

In a commercial building, there are generally well established procedures, systems and a management team to direct evacuations. While building codes such as the Building Code of Australia [12] may set requirements for emergency warning systems and evacuation plans for residential buildings, in practice many people will fail to evacuate. And in some countries such as the UK and Hong Kong, evacuation is not encouraged in residential buildings, although protected stairs and other fire safety measures are provided. They effectively practise "protect-in-place".

In a survey of occupants of three high rise residential buildings in Australia, Barber found that only 23% evacuated the last time the fire alarm sounded in their building. From Barber's research into decision making of Australian residents, he found the decision to evacuate was dependent upon:

- Additional cues such as the presence of flame or smoke
- The arrival of fire brigades on the ground or in the building
- People's perception of the effectiveness of the fire protection features within the building

Barber also noted a number of other factors which influenced a resident's decision to stay or evacuate, including:

- Previous experience of false alarms
- Floor height of residence
- Lack of understanding of alarms, fire plans
- Behaviour of neighbours, and
- Age, family and cultural factors

Importantly, Barber's research showed that, for those who eventually chose to evacuate, they generally did so at around 3-7 minutes after the alarm. This was potentially the worst possible period to evacuate because at that time, if there had been a fire, it may have grown significantly and to a dangerous level, but was not yet to a point where it was fully controlled by sprinklers or the fire brigade.

This problem of failure to evacuate or lateness of evacuation prompted Barber to question whether it is really necessary to evacuate residents for a fire emergency in well protected high rise buildings at all, unless directed by the fire brigade. "Well protected" implies the building has strong structural protection and compartmentation, automatic sprinklers, and other measures typical of new high rise residential buildings.

Barber believes that the principle of "protect in place" would be successful in Australia residential buildings with their high levels of protection, and be a better alternative to largely unsuccessful evacuation policies and procedures currently promoted. This approach would also make for much simpler decision making for most people, provided training and communication systems are effective.

## Victorian Bushfires 2009

In the bushfire (wildfire) situation, decision making to "stay or go" is also critically important, although the timeframes compared to building fires may be different. The Victorian bushfires of "Black Saturday" on 7 February 2009 were devastating. The terrible outcomes highlighted the critical safety importance of these decisions of whether to evacuate or not, and whether these matters should be made mandatory for whole communities or left to individuals.

In Victoria the motivation to evacuate or not under bushfire threat is embodied within a public policy which has been abbreviated to "stay or go". This simplification has reduced the nuances of the more detailed official advice: rural residents have been encouraged to either stay and defend their properties and themselves, if they considered their properties defendable, or to go early and evacuate well before any fire arrives on days of extreme fire danger.

The underlying emphasis on individuals making informed choices to stay and defend or evacuate early has been consistent with the Victorian legislation in which property owners and tenants have the legal right to stay and defend their property, even if threatened by fire or other natural hazards [13]. While wider powers are available in some other states of Australia to order mandatory evacuations, the evidence seems to be that these powers have been rarely used in Australia [14].

The massive loss of lives and property led to the creation of a Royal Commission or major public enquiry, headed by Mr. Justice Teague. Based on the Commission's Interim Report [5], the evidence seems to indicate that:

- Despite widespread public advice that catastrophic fire conditions were expected, many people were not prepared for the conditions on the day
- Many assumed they could defend their property or use their buildings to protect themselves, but failed to understand the severity of the conditions
- With the many fires moving rapidly, the public warning and information systems were overloaded or failed, and many people received insufficient warning of fire approaching
- Seeing flames seemed to be the trigger for many to evacuate, which is considered the worst possible period for evacuation
- With public utilities such as power, water and communication systems failing, many had no resilient fall back position or plan

In their Interim Report, the Royal Commissioners concluded:

- People need to better understand the considerable effort needed to prepare a property to make it defendable
- The safest course is to evacuate or relocate well away from any fire area, but this must be done early
- People need to be fit and mentally robust to stay and defend, and it is not appropriate for children, the elderly or the infirm

- Advice, training and education need considerable improvement.
- People need to have options to "stay or go"; including the availability of local areas of refuge hardened against fire
- More accurate and timely warnings and information is required for good decision making. People near to hear, understand and act.

The work of the Royal Commission has been strengthened by the input of research work conducted by the Bushfire Co-operative Research Centre (Bushfire CRC). The Bushfire CRC were able to gather time-critical data in more than 600 semi-structured interviews with survivors. The interviews focused on fire behaviour, human behaviour and building performance prior to and during the fires.

In June 2009, the Bushfire CRC Interim Report [15], preliminary findings showed:

- Half the households reported at least one household member who intended to stay and defend
- A significant number of residents intended to stay and see what the bushfires were like before deciding whether to stay or go
- More than half (55%) reported a household member left because of the fire conditions, often late
- Some of those who stayed to defend their property reported a range of factors which influenced their ability to defend their property and survive themselves, including heat exhaustion, dehydration, and pre-existing medical conditions such as asthma and arthritis.

The Bushfire CRC Interim Report identified a number of recommendations concerning human behaviour and decision making in this bushfire emergency, including:

- The importance of planning and preparedness for evacuation, or to stay and defend
- The critical nature of warnings and other information, and the timeliness and quality of that information
- The perception of threat and situational awareness which motivates decision making
- The need for good education and training on the potential fire threats and appropriate responses in bushfires for all in the community

It appears that a failure to understand the fire threat and conditions likely to be experienced in a large bushfire, the physical and psychological effort required to defend and survive in such conditions, the timing required for good decision making, as well as the lack of adequate warnings and appropriate information all contributed to the many deaths and injuries on Black Saturday.

## Hurricane Katrina

Another relevant piece of research that provides insight into the decision making required for people to evacuate or protect themselves in place is that undertaken by Stephens et al [6], with interviews of survivors as well as relief workers from Hurricane Katrina in the USA in 2005.

In response to rising flood waters and threats to life from Hurricane Katrina, officials in New Orleans instituted a mandatory evacuation. Despite the official call for evacuation of all residents, the survivor interviews showed:

- More than 50% chose not to evacuate
- Those who chose not to evacuate were generally from lower socio-economic groups
- Among the non-leavers, only 50% owned a vehicle, few knew people outside the city, and many had limited financial or other resources
- Many who did not leave exhibited strong reliance on group decision making, lacked independent thinking, and felt they could not take individual control of the situation.
- Those who did evacuate were more typically middle class (90%) and owned a vehicle (100%)

There were some strong views among relief workers and other observers about those who chose to stay or leave. These views included:

- Those who chose to stay were often described as passive (lazy and dependent), irresponsible (careless, negligent) and inflexible (stubborn)
- On the other hand, those who evacuated were described as independent (self reliant, in control), responsible (conscientious), and action oriented (prepared, planning)
- Leavers were described as "influencing agents", and stayers as lacking "agency" and less sensible in their actions.

This research has shown that US emergency management was based on the misconception that all survivors were "free agents" who were unconstrained by their socio-cultural environment. For many in the lower socio-economic groups, they exhibited strong interdependence and relied upon others for decision making, as well as having limited options due to their lack of resources and positions in the socio-economic structure.

This research may hold part of the answer to why so many people stay rather than evacuate in events such as building fires, the Victorian bushfires and other emergencies, even where policies, procedures and training appear to have been provided.

## Discussion

The review of disasters such as WTC (9/11), the 2009 Victorian bushfires, Hurricane Katrina and other research has highlighted that considerable improvement is needed in terms of understanding people safety in emergencies. Human behaviour and better decision making is at the heart of the issue.

Public policies, codes and standards suggest we have the procedures and knowledge on how to prepare for evacuation or how to advise people on how to protect them from fires and similar disasters. However, the reality seems to be quite different, and human behaviour research perhaps offers the psychological and sociological understanding and tools to significantly improve outcomes. The research reviewed in this paper suggests some key areas for attention, and in most cases, areas for further research. An integrated approach to safety is required, which includes:

Well Protected Buildings – it is only realistic to encourage people to remain in place and not evacuate if the buildings are well protected, and people and the building are "defendable". For urban buildings, this means structurally strong, sprinkler protected, well protected stairs (and elevators if used for egress), hardened communication systems, and resilient power, water and lighting systems. In bushfire prone areas, understanding topography, adjacent bushland fire loads, alternative escape paths, possibly safe refuges, fire brigade access and other measures are all important.

Emergency Preparedness – whether people are encouraged to stay or go, they need to understand the implications of both options through solid emergency planning, procedures, training and awareness of the threat through education. People must comprehend the experience of staying: what it might be like to be in a building that has a fire, or what the heat, sound, visibility and other factors will be like if they stay.

Trusted Communications and Warning Systems – people must believe they can trust the alarm signals and the communication systems and make decisions accordingly. They must also have an understanding of the meanings of signals and alarms. Without the appropriate cues and knowledge of what they mean for safety, people may wait and see, or evacuate inappropriately. Situation awareness is critical to safety success.

Personal Expectations – it is clear that people and communities must understand that many may have to make early choices to stay in a protected building, if residential high rise, or to evacuate early if in a bushfire prone area. This applies to young families, the obese, the elderly, and those with a disability. Planning must account for a sizeable percentage of the population in these categories

Group Behaviour – most importantly, the Hurricane Katrina research shows that, for many, group behaviour is a very strong influence, and many lack individual decision making capacity and control. This must be recognized in planning and emergency response, not as "poor behaviour", but as a reality that must be addressed as part of all management strategies. In emergency management, "one size can not fit all."

This issue of socio-cultural differences between people may be at the root of some of the problems in past disasters. It may be that many professionals write codes and standards, prepare emergency plans and procedures, and train and educate communities based on their own values and norms that place major emphasis on people being able to make decisions for themselves in isolation from others in the communities. Inadvertently, this may disregard and dis-enfranchise many in the community. A better understanding of these aspects of human behaviour may be one key to better outcomes in emergencies where the decision is to evacuate or not.

## Conclusions

This paper has examined some of the research undertaken on the underlying concepts of evacuation or protect in place in a range of emergency situations.

While we appear to have seemingly well established policies and procedures for evacuation or to protect people in place, the reality is that in a number of major disaster events, there appears to have been a failure of those systems.

Recent research has highlighted the importance of well protected buildings for those who choose to stay. In addition, the effectiveness of decisions is influenced by the considerable consequences of: emergency preparedness, situation awareness through trusted communication and information systems, and personal education and training.

We need a far better psychological understanding of group behaviour and the impacts of socio-cultural differences on decision making if we are to prepare emergency management plans and actions that fully serve all members of our communities.

## References

1. Averill, J., Peacock, R., Kuligowski, E., Reneke, R., Mileti, D., Grove, N., Nelson, H., Proulx, G.: Federal Investigation of the Evacuation of the World Trade Center on September 11, 2001. 4th International Symposium on Human Behaviour in Fire – Conference Proceedings, 13-15 July (2009)
2. Gershon, R.: The World Trade Center Evacuation Center: Factors Associated with Evacuation Time and Injury. 4th International Symposium on Human Behaviour in Fire – Conference Proceedings, 13-15 July (2009)

3. Galea, E., Hulse, L., Day, R., Siddiqui, A., Sharp,G.: The UK WTYC 9/11 Evacuation Study: An Overview of the Methodologies Employed and Some Analysis Relating to Fatigue, Stair Travel Speeds, and Occupant Response. 4th International Symposium on Human Behaviour in Fire – Conference Proceedings, 13-15 July. Cambridge (2009)
4. Barber, D.J.: 'Strategies for Occupant Response to Fire in High Rise Residential Buildings.' Fire Protection Association, Australia (April 2009)
5. Victorian Bushfires Royal Commission, Interim Report (August 2009)
6. Stephens, N., Hamedani, M.Y., Markus, H.R., Bergsieker, H.B., Eloul, L.: 'Why Did They "Choose" To Stay – Perspective of Hurricane Katrina Observers and Survivors.' Psychological Science. Vol III, No II (2009)
7. 4th International Symposium on Human Behaviour in Fire – Conference Proceedings, 13-15 July. Interscience Communications, Cambridge (2009)
8. Human Behaviour in Fire – A Summary of the Cambridge Conference. Fire Australia Journal (Autumn 2009)
9. International Building Code, International Code Council (2009)
10. Kinsey, M., Galea, E., Lawrence, P.: Investigating the Use of Elevators for High-Rise Building Evacuation through Computer Simulation. 4th International Symposium on Human Behaviour in Fire – Conference Proceedings, 13-15 July. Cambridge (2009)
11. Heyes, E., Spearpoint, M.: Lifts for Evacuation – Human Behaviour Considerations. 4th International Symposium on Human Behaviour in Fire – Conference Proceedings, 13 – 15 July. Cambridge (2009)
12. Building Code of Australia. Australian Buildings Codes Board (2009)
13. Loh, E.: 'Evacuation Powers of Emergency Workers and Emergency Service Organizations in Australia.' The Australian Journal of Emergency Management. Vol.22, No.4 (November 2007)
14. Rush, J., Doyle, R., Richards, M., Nichols, L.: Submissions of Counsel Assisting the 2009 Victorian Bushfires Royal Commission (June 2009)
15. Bushfire CRC: 'Victorian 2009 Bushfire – Research Response.' Interim Report. Bushfire Co-operative Research Centre, Melbourne (June 2009)

# Analysis of Occupant Behavior During a High-rise Office Building Fire

**E. D. Kuligowski and B. L. Hoskins**

National Institute of Standards and Technology, USA
Contact email: erica.kuligowski@nist.gov

**Abstract** Survey responses from occupants involved in a 32-story high-rise building evacuation during a fire were collected and analyzed to study the pre-evacuation period. Multiple regression models were used to test whether specific occupant, building, and environmental factors predicted pre-evacuation times. This study found that the actions taken by occupants during this period, i.e., waiting, helping, and preparation actions, was a main factor that significantly increased pre-evacuation time. In addition, although the action of seeking information did not significantly influence pre-evacuation time, interaction effects were found among certain groups of occupants who sought information. Of those who sought additional information, older adults took less time seeking information (per action) than younger adults, occupants with disabilities took more time seeking information (per action) than occupants without disabilities, and occupants who perceived risk/danger took less time seeking information (per action) than those who did not perceive risk. This study also found that the initial location of the occupant (floor) also significantly influenced pre-evacuation time, likely due to the information that occupants received on these floors. The effects of these factors on pre-evacuation times are quantified by stating how much pre-evacuation time was required for each factor (i.e., action taken or floor location).

## Introduction

The level of safety that a building affords occupants during a fire is determined in one of two ways: 1) the use of prescriptive codes and 2) the use of performance-based codes or design. In the first case, an engineer consults a set of building and/or fire codes to ensure that the building has the required number and width of exits, appropriate number and spacing of suppression and detection systems, etc. However, very few of these regulations are based on a scientific understanding of how occupants will use the building and take action during an emergency (i.e., occupant behavior). In the second case, an engineer determines whether the time when untenable conditions develop in the building (ASET or available safe egress time) exceeds the time needed for occupants to evacuate the building or a portion of the building (RSET or required safe egress time). Engineers use tools ranging in sophistication from back-of-the-envelope evacuation calculations to computer-

R.D. Peacock et al. (eds.), *Pedestrian and Evacuation Dynamics*,
DOI 10.1007/978-1-4419-9725-8_61, 

based evacuation simulation models. However, due to a lack of data on occupant behavior during fire emergencies, these tools often simplify each evacuation scenario such that occupant behavior is mostly ignored and only the time required for occupant movement is accounted for.

In order to achieve a more accurate assessment of the life safety provided by a building (and its systems) during a fire, a better understanding of *occupant behavior during evacuation* is required. The purpose of this paper is to identify the factors that influenced pre-evacuation (or delay) time of occupants during an actual fire evacuation and to quantify the factors by stating how much pre-evacuation time was gained or lost due to each factor. Further analysis, similar to Kuligowski and Mileti [1], will validate the larger evacuation theory from this event. This preliminary analysis is the first step in a larger project.

## 2008 Office Building Fire [2]

In 2008, a fire occurred within the building envelope, outside of the 1$^{st}$ floor mezzanine walls, of a 32-story office building in the United States. This evacuation took place in the winter months in the late afternoon. Some occupants had already left for the day or were on their way out while others were still working. On a typical day, the building houses a population of approximately 4400 people in over 25 different companies. Some floors contained occupants from only one company while other floors contained occupants from multiple companies.

The fire alarm system is designed as a selective evacuation system and provides different emergency messages to different floors of the building. Shortly after the fire was detected, occupants below the 5$^{th}$ floor received the automatic, prerecorded voice alarm message to evacuate from the emergency voice/alarm communication system as per the original system design. Initially, occupants on the 5$^{th}$ through 32$^{nd}$ floors received the "safe zone" message that informed them that they were in a safe location and to wait for further instructions. Shortly after that initial message (less than 5 minutes), the alarm system received a second alarm initiated from a 6$^{th}$ floor manual pull station alarm which resulted in the 5$^{th}$ through 7$^{th}$ floor occupants consequently receiving a second automatic pre-recorded message. This message was an alarm message which informed the occupants to evacuate down three floors and wait on that floor. In addition, occupants on the 8$^{th}$ floor and above also received a second automatic pre-recorded message that they were in a safe location and to wait for further instructions. Approximately 15 minutes later; a “live” voice announcement initiated from the fire command center, informed all the occupants to evacuate the building. Some occupants had previously evacuated the building, while others remained (and likely heard this message). In addition to the instructions on what to do, occupants received other environmental cues from the incident, including seeing smoke (especially below the 8$^{th}$ floor).

## Pre-evacuation Theory from Buildings

The pre-evacuation time period for an occupant, sometimes called pre-movement time or pre-response time, is the time beginning when the occupant is alerted that something may be wrong and ending when the occupant begins purposive movement within the stair or exit*. For some emergencies, occupants are instructed to travel to another place within the building (e.g., three floors below), and the pre-evacuation time period is the time spent before purposive movement to a place of safety begins. For most occupants, depending on the building, the pre-evacuation period is spent on their building floor and evacuation time is spent in the stairs or exit path.

Whereas some RSET calculations assume a pre-evacuation time of zero for a building population, human behavior research of past emergencies shows that occupants spend time, sometimes a significant amount of time, engaging in actions during the pre-evacuation period [3]. Studies have collected data on pre-evacuation time for occupants evacuating buildings during emergencies and evacuation drills [4]. Resulting from these studies are distributions of pre-evacuation time data for use in RSET calculations. These data are frequently categorized by the type of building from which they were collected, including university buildings [5], hospitals [6], retail stores [7], apartment or residential buildings [8], office buildings [9,10,11], and hotel buildings [12], for example.

After further analysis of the pre-evacuation period, research shows that occupants perform a variety of activities [13,14]. These actions can include investigating the event, warning others, searching for others, getting personal items, and preparing to leave. All of the actions mentioned take time to complete; however, there is very little, if any, data available on the time necessary to complete each type of pre-evacuation action. For this reason, much of the RSET calculations still rely on overall time distribution data to describe the entire pre-evacuation period.

Additionally, research has been performed that identify the factors that are likely to influence (i.e., increase or decrease) overall pre-evacuation time [3,15]. These factors can originate from the actions that occupants perform during this period, the environment (including the building and the fire), and occupant characteristics. First, theory shows that occupants who take certain actions during the pre-evacuation period have increased overall pre-evacuation time [1]. Pre-evacuation actions and pre-evacuation delays have been linked in the fire field for decades [14]. Research has determined that actions performed during this period increase an occupants' delay time [16], and certain actions in particular, i.e., searching for information and confirming information about an event, have been identified as ones that increase pre-evacuation delays [20]. Therefore, it is hypothesized that each action type performed by occupants increases their overall pre-evacuation time.

---

* This definition of pre-evacuation time is for this analysis only. Other research on pre-evacuation time may define the boundaries of this time period differently.

An occupant's pre-evacuation time can also be influenced by environmental factors, specifically the information that people receive about the event [17] and the source of that information [7]. Theory shows that occupants who receive information on what to do during an incident [18], especially when this information is given via an emergency voice/alarm communication system rather than simply hearing sound tones from an audible notification appliance (e.g., horn, bell, etc. [19], are more likely to follow this information. Therefore, it is hypothesized that occupants in the 32-story building were more likely to report a longer overall pre-evacuation time when given instructions via the emergency voice/alarm communication system to wait on their floor.

Also, theory has shown that occupant characteristics influence pre-evacuation actions, which can influence pre-evacuation times [3]. Pre-event occupant factors (i.e., factors that exist prior to the event occurring), such as demographics (i.e., older occupants and gender) [13], occupant disabilities [5], and previous experience/training in emergencies [20], can influence an occupant's engagement in pre-evacuation actions, which can then influence his/her overall pre-evacuation time. Also, some research has found that an occupant's perception of risk, a factor that occurs in relation to the event (known as an event factor), decreases an occupant's overall pre-evacuation time [11], while other research has not found a strong influence of risk on pre-evacuation timing [1]. Therefore, it is hypothesized that older occupants and occupants with disabilities will have longer pre-evacuation times; whereas occupants with previous experience/training and those with a higher perception of risk/threat will have lower overall pre-evacuation times. The data will be examined to see if there was any difference between pre-evacuation timing of men and women, since the previous research shows differences in pre-evacuation actions only.

## Methods of Data Collection and Analysis

Building management of a 32-story high-rise office building sent out anonymous, self-administered questionnaires via email to the occupants of this building who were evacuated due to a fire incident. The purpose of these questionnaires was to obtain information on 1) the background of the occupant (occupant demographics, previous training and education in fire safety, and previous experience in fire evacuations), 2) actions and decisions made by the occupant on his/her floor during the building evacuation, and 3) actions and decisions made by the occupant during the building evacuation via the stairs and/or elevators [2]. These data were collected via paper questionnaires and out of 4400 occupants in the building, over 670 survey responses were received (a response rate of 15 %)†.

The sample for this study is a convenience sample, containing data from only those evacuees who decided to participate. Of the 670 surveys received, 375 occupants were included in the analyzed sample. Participants were removed from the sample due to the following reasons: 1) they were not present in the building during the event or were already leaving at the time of the incident, and therefore did not participate in the pre-evacuation period of the event, 2) they did not provide answers to any or all of the survey questions included in the final regression model, or 3) they identified themselves as evacuation coordinators. The original sample contained too few evacuation coordinators to include in this study's sample. Also, additional participants were removed from the sample due to response errors on the dependent variable [2].

*Measurement*

A total of 10 independent variables were included in the model to predict pre-evacuation time. These variables fell into four categories: pre-evacuation actions, environmental cues, event-based occupant factors, and pre-event occupant factors. Descriptive statistics are provided for each model variable in Table 1.

The dependent variable, pre-evacuation time, was measured by asking respondents how much time passed from the moment that they were alerted to the incident until they entered the exit (i.e., the stair, elevator, or exit pathway to the door). Answers to this question were coded as time in minutes.

---

† This response rate is underestimated due to the fact that an unknown number of the 4400 occupants had already left work for the day and were not in the building when the fire occurred.

**Table 1: Descriptive statistics for model variables**

| Variable name | Mean | Standard Deviation | Measure | Range |
|---|---|---|---|---|
| | **Dependent Variable** | | | |
| Pre-evacuation time | 10.42 | 7.63 | Self-reported time (minutes) | 1 to 45 |
| | **Independent Variables** | | | |
| | **Pre-evacuation Actions** | | | |
| Wait/Cont. Action | 0.93 | 0.79 | Number of actions | 0 to 2 |
| Seek Information | 0.66 | 0.83 | Number of actions | 0 to 4 |
| Help | 0.25 | 0.51 | Number of actions | 0 to 2 |
| Prepare | 1.81 | 1.01 | Number of actions | 0 to 5 |
| | **Environmental Factors** | | | |
| Floor ($8^{th}$ floor and above) | 0.65 | --- | Yes/No | 0 or 1 |
| | **Event Occupant Factors** | | | |
| Perceived Risk or Threat | 0.53 | --- | Yes/No | 0 or 1 |
| | **Pre-event Occupant Factors** | | | |
| Gender | 0.60 | --- | Woman = 1 | 0 or 1 |
| Age < 36 | 0.24 | --- | Yes/No | 0 or 1 |
| Age 36-45 | 0.21 | --- | Yes/No | 0 or 1 |
| Age 46-55 | 0.33 | --- | Yes/No | 0 or 1 |
| Age > 55 | 0.22 | --- | Yes/No | 0 or 1 |
| Disability | 0.21 | --- | Yes/No | 0 or 1 |
| Training | 0.68 | --- | Yes/No | 0 or 1 |

For the pre-evacuation actions category, respondents were asked to identify any and all of the pre-evacuation actions that they performed before moving into the exit. These actions were categorized as Waiting, Seeking information, Helping, and Preparing. Waiting was measured by asking respondents whether they engaged in any of the following two actions: 1) continuing the activity in which they were engaged prior to alarm and 2) waiting (e.g., for instructions). Answers to these questions were each coded as dummy variables (0 or 1) and then added to form a scale ranging from 0 to 2. Seeking information was measured by asking respondents whether they engaged in any of the following actions, including looking around, investigating the source of the alert, seeking more information about the alarm or event, and discussing with others about the event. Answers to these questions were each coded as dummy variables (0 or 1) and then added to form a scale ranging from 0 to 4. Helping was measured by asking respondents whether they engaged in any of the following actions, including giving instructions to others on what to do and looking for others in the building. Answers to these questions were each coded as dummy variables (0 or 1) and then added to form a scale ranging from 0 to 2. Preparing was measured by asking respondents whether they engaged any of the following actions, including gathering coat/shoes (i.e., getting dressed) or gathering valuables, gathering emergency coordinator supplies, saving files or turning off the computer, securing files or documents, and securing the office/room/space (e.g., shutting the door, turning off the lights, etc.). Answers to these questions were each coded as dummy variables (0 or 1) and then added to form a scale ranging from 0 to 5.

For the environmental factor, the floor on which occupants were located at the initial stages of the evacuation was measured. Floor was measured by asking respondents which floor they were located when they first became aware of a fire in the building from the initial alarm (ranging from 1 to 32). Then, since different information was provided to the occupants on the 1) $8^{th}$ floor and above (i.e., to wait on their floor) and 2) below the $8^{th}$ floor (e.g., to evacuate), a dummy variable was created to represent occupants' location on the 8th floor and above (given a value of 1) and occupants located below the $8^{th}$ floor (given a value of 0).

Occupant factors were also measured in the questionnaire. Perceived risk was measured by asking occupants if they felt at risk at any time before entering the stair and answers were coded as yes or no. Also, occupants were asked a series of demographic questions: 1) their gender, 2) their age (using the following categories: 18 to 25, 26 to 35, 36 to 45, 46 to 55, 56 to 65, and 66+ years old), 3) if they had any medical or physical conditions that made evacuation more difficult and answers were coded as yes or no, and 4) if they had any evacuation training in the building, including evacuation or practice emergency drills (yes or no).

## Limitations

There are four main limitations of this study. First, this analysis used a convenience sampling technique. Therefore, it is inappropriate to generalize these findings to the larger population in the building as well as other office populations. Second, pre-evacuation time was reported by the respondents as an open-ended question in the questionnaire. As with any post-fire evacuation study, occupants' memories of the event may be imperfect. Thus, there may be some error between their actual pre-evacuation time and the time they reported. Third, the questionnaire asked only *if* actions/factors took place, not *when* they took place. Any variable that was subject to change with time, e.g., perception of risk, might appear insignificant in the model, but could have altered the pre-evacuation time. Last, this questionnaire did not directly ask about the occupant's management role in the building or whether fire cues (e.g., fire or smoke) caused occupants to become aware of the incident. The questionnaire will be modified for future events to account for additional influential variables.

## Findings

SPSS Version 12.0.1‡ was used to estimate the linear regression model gauging the net effects of actions, environmental factors, and occupant characteristics on reported overall pre-evacuation times (Table 2). The models are organized in hierarchical fashion. Model 1 includes the main independent variables, including floor and actions. In progression, Model 2 includes all event and pre-event occupant characteristics. This tiered model framework allows for the introduction of the main independent variables and then subsequent sets of variables to determine if the initial relationships still holds true. Table 2 includes the unstandardized and standardized coefficients for each variable as well as the standard error, in parenthesis. In each model, zero-order correlation matrices were examined for evidence of multicollinearity (when two or more independent variables are highly correlated), which was not found to be a source of bias. Also, plots of the standardized residuals against standardized predicted values as well as plots to test the normality of the residuals were examined to find that the assumptions of linearity and homoscedasticity (i.e., the error term does not vary with the values of the independent variable) were met.

Of the ten variables in the regression model, four were significant at the 0.10 level or better. The coefficients of the model can be interpreted as the increase in pre-evacuation time (in minutes) for each integer increase in the independent variable. The final regression model (Model 2 in Table 2) explained 29 % of the pre-evacuation times included in this study.

‡ Certain commercial entities, equipment, or materials may be identified in this document in order to describe an experimental procedure or concept adequately. Such identification is not intended to imply recommendation or endorsement by the National Institute of Standards and Technology, nor is it intended to imply that the entities, materials, or equipment are necessarily the best available for the purpose.

**Table 2: Dependent variable - Pre-evacuation Time**

| | Model 1 | Model 2 |
|---|---|---|
| **Pre-evacuation actions** | | |
| Wait/Cont. Action | 1.58/0.17*** | 1.56/0.16*** |
| | (0.46) | (0.47) |
| Seek Info | 0.35/0.04 | 0.24/0.03 |
| | (0.44) | (0.45) |
| Help | 1.16/0.08$^+$ | 1.16/0.08$^+$ |
| | (0.67) | (0.68) |
| Prepare | 0.51/0.07 | 0.59/0.08$^+$ |
| | (0.33) | (0.34) |
| **Environmental factors** | | |
| Floor | 7.37/0.46*** | 7.31/0.46*** |
| | (0.72) | (0.74) |
| **Event occupant factors** | | |
| Perception of Risk | | -0.01/-0.00 |
| | | (0.71) |
| **Pre-event occupant factors** | | |
| Gender (Woman = 1) | | -0.81/-0.05 |
| | | (0.72) |
| Age (36 to 45) | | 0.60/0.03 |
| | | (1.00) |
| Age (46 to 55) | | 0.45/0.03 |
| | | (0.92) |
| Age (>55) | | 0.05/0.00 |
| | | (1.05) |
| Disability | | 0.73/0.04 |
| | | (0.87) |
| Training | | -1.41/-0.09$^+$ |
| | | (0.74) |
| Intercept | 2.76** | 3.75** |
| | (0.86) | (1.26) |
| Adjusted $R^2$ | 0.29 | 0.29 |

Note: Cell entries represent unstandardized / standardized slope coefficients with standard errors in parentheses. For each model, n=375. + p<0.10, * p<0.05; ** p<0.01; *** p<0.001

Model 2 shows that the types of actions performed by the occupant and the floor (or information received by the occupants) influenced their pre-evacuation time, more so than occupant factors. Waiting (e.g., continuing to work or waiting) significantly increased pre-evacuation times by 1.56 min per waiting action. These individuals waited to start their pre-evacuation activities and thus their total pre-evacuation time was greater than individuals who did not wait. Also, each action to seek information added 0.24 minutes to the pre-evacuation time, however, this result was insignificant. Occupants who engaged in helping actions had increased pre-evacuation times of 1.16 minutes per reported helping action, significant to the 0.10 level. Last, each preparation action that the respondents reported performing increased their pre-evacuation time by 0.59 minutes, significant at the 0.10 level.

The environmental factor (floor) influenced occupant pre-evacuation times in Model 2. Occupants on floors that received an initial recorded message to wait on their floor and no other messages from the emergency voice/alarm communication system until the live voice message to evacuate to wait on their floor required 7.31 more minutes of pre-evacuation time than those on other floors (that did not receive this set of messages). Finally, all of the occupant characteristics, except past experience were insignificant in Model 2, meaning that, for the average occupant, most occupant factors alone did not directly influence the occupant's pre-evacuation time. Occupants with experience in training/drills spent 1.41 minutes less of pre-evacuation time than those without training, also significant to the 0.10 level.

Whereas the waiting variable was found to significantly increase pre-evacuation times to the 0.001 level, the relationships between pre-evacuation times and the other three action variables, seeking information, helping, and preparing, were either barely significant or not significant at all. And since theory has shown that performing actions increases pre-evacuation time, it became important to further understand this result. Steps were taken to investigate whether the combination of individual factors with these action variables influenced pre-evacuation time, known as interaction effects.

Interaction terms were used to identify the actual factors that influenced pre-evacuation time in this event. Results show that the age, disabilities, and risk perception factors influenced pre-evacuation time through interactions with the following action variables: Seeking information and Helping. Only those results that are significant to at least the 0.10 level are listed here.

*Seeking information:* Of the occupants who sought information, a specific age group, reported disability, and perceived risk were identified as factors that significantly influenced pre-evacuation times. First, occupants 46 to 55 years old who sought additional information spent approximately 2.50 less minutes of pre-evacuation time per action of seeking additional information when compared with those younger than 45 ($p<0.05$). Also, individuals with a reported disability took 2.06 more minutes of pre-evacuation time per action of seeking information compared with occupants who did not report a disability ($p<0.10$). And, individuals who reported perceiving risk during the pre-evacuation period took 2.04 less minutes of pre-evacuation time per action of seeking information compared with those who did not perceive risk ($p<0.05$).

*Helping:* Of the occupants who helped others, a specific age group was identified as a significant influence of pre-evacuation time. Occupants above 55 years old spent 3.2 more minutes of pre-evacuation time per helping action than those younger than 45 ($p<0.10$).

Overall, pre-evacuation actions take time to complete. How much time, depends on the action itself and for some actions, on characteristics of the occupants performing the actions. Performing waiting actions, helping, and preparing for evacuation, regardless of occupant type, significantly increased pre-evacuation times. Also, the location of occupants (i.e., on the 8$^{th}$ floor and above) also significantly increased pre-evacuation times, regardless of occupant type, most likely because they were told initially that they were in a safe location.

On the other hand, certain occupant groups that engaged in seeking information took more or less pre-evacuation time (per action) than others outside of these groups. For this reason, the relationship between pre-evacuation time and seeking information appears insignificant in Models 1 and 2 (see Table 2). For the age variable, occupants over 45 years took less time to seek additional information (per action) as well as more time to help (per action) when compared with younger occupants performing the same actions. It is possible that occupants over 45 years old had a managerial role or seniority in the organization, or had a different motivation, which required them to allot different amounts of time to specific pre-evacuation actions when compared with younger occupants. Also, occupants reporting a disability spent more time seeking information (per action) than those without a reported disability. It is possible that occupants with a disability required more time confirming the event before moving to an exit compared with nondisabled occupants. Last, occupants who reported a level of perceived risk during the pre-evacuation period spent less time seeking information (per action) than those who did not perceive risk. In this case, occupants with higher levels of perceived risk may not have required as much time to gather information, but rather relied on their risk perceptions to begin movement to the exit.

## Conclusion

This study found that the main influential factors of pre-evacuation times were actions taken during the pre-evacuation period and initial floor location (likely due to the information that occupants received on these floors). For some actions, specifically seeking information and helping, occupant factors (before or during the event) combined with performing these actions showed significant differences in pre-evacuation times. This study also quantified the relationships between factors and pre-evacuation time, showing how much pre-evacuation time was required for each factor or combination of factors. Validation of this study and additional work to identify the factors that influence occupants to perform certain pre-evacuation actions is needed so that a predictive model of the pre-evacuation period can eventually be developed. With improvements in our understanding of human behavior in building fires, building codes, evacuation models, and individuals who use these models will be better equipped to ensure the safety of occupants in buildings around the world.

**Acknowledgments** The authors would like to thank the U. S. General Services Administration (GSA) for the support of this work through the following project: "Research on Enhanced Performance Requirements for the Use of Elevators and Stairwells for Occupant Egress and Fire Service Access."

## References

1 Kuligowski, E.D., Mileti, D.S.: "Modeling pre-evacuation delay by occupants in World Trade Center Towers1 and 2on September 11, 2001." *Fire Safety Journal.* 44:487-496 (2009)

2 Kuligowski, E.D., Hoskins, B.L.: "Occupant Behavior in a High-rise Office Building Fire." NIST Technical Note (forthcoming). National Institute of Standards and Technology, Gaithersburg, MD (2010)

3 Proulx, G.: "Evacuation Time." In: *The SFPE Handbook of Fire Protection Engineering Third Edition*, edited by DiNenno, P.J., National Fire Protection Association, pp. 3-355 – 3-372. Quincy (2008)

4 Purser, D.A., Bensilum, M.: "Quantification of Behavior for Engineering Design Standards and Escape Time Calculations." *Safety Science* 38:157-182 (2001)

5 Gwynne, S.M.V.: "Optimizing Fire Alarm Notification for High Risk Groups: Notification Effectiveness for Large Groups." Prepared for The Fire Protection Research Foundation. Quincy (June 2007)

6 Gwynne, S., Galea, E.R., Parke, J., Hickson, J.: "Collection of Pre-evacuation Times from Evacuation Trials involving a Hospital Outpatient Area and a University Library Facility." In: *Fire Safety Science – Proceedings of the Seventh International Symposium*, Interscience Communications Ltd., London, England, pp 877-888 (2003)

7 Shields, T.J., Boyce, K.E.: "A Study of evacuation from large retails stores." *Fire Safety Journal.* 35:25-49 (2000)

8 Proulx, G., Latour, J.C, McLaurin, J.W.: *Housing Evacuation of Mixed Abilities Occupants in High-rise Buildings*. Internal Report No. 706. National Research Council of Canada, Ottawa, Ontario (1995)

9 Proulx, G., Reid, I.M.A., Cavan, N.R.: *Human Behavior Study, Cook County Administration Building Fire, October 17, 2003 – Chicago, IL.* National Research Council of Canada, Ottawa, Canada (2004)

10 Averill, J.D., Mileti, D.S., Peacock, R.D., Kuligowski, E.D., Groner, N., Proulx, G., Reneke, P.A., Nelson, H.E.: "Federal Building and Fire Safety Investi-

gation of the World Trade Center Disaster: Occupant Behavior, Egress, and Emergency Communications." *Report NCSTAR 1-7*. National Institute of Standards and Technology, Gaithersburg (2005)

11 Brennan, P: "Timing Response in Real Fires." In: *Fire Safety Science – Proceedings of the Fifth International Symposium,* Interscience Communications Ltd., London, England, pp. 807-818 (1997)

12 Bryan, J.L.: *An Examination and Analysis of the Dynamics of the Human Behavior in the MGM Grand Hotel Fire*. National Fire Protection Association, Quincy (1983)

13 Bryan, J.L.: "Behavioral Response to Fire and Smoke." In: DiNenno, P.J., *The SFPE Handbook of Fire Protection Engineering Third Edition*, pp. 3-320 – 3-354. National Fire Protection Association, Quincy (2008)

14 Canter, D., Breaux, J., Sime, J.: "Domestic, Multiple Occupancy and Hospital Fires." In: Canter, D. (ed.), *Fires and Human Behaviour*, pp. 117-136. John Wiley & Sons, New York (1980)

15 Kuligowski, E.D.: "The Process of Human Behavior in Fires." NIST Technical Note 1632. National Institute of Standards and Technology, Gaithersburg (2009)

16 Proulx, G.: Evacuation time and movement in apartment buildings. *Fire Safety Journal*. 24:229-246 (2005)

17 Mileti, D.S., Sorensen, J.H.: "Communication of Emergency Public Warnings: A Social Science Perspective and State-of-the-Art Assessment." Oak Ridge National Laboratory, U.S. Department of Energy, Oak Ridge (1990)

18 Mileti, D.S., Bandy R., Bourque, L.B., Johnson, A., Kano, M., Peek, L., Sutton, J., Wood, M.: "Annotated Bibliography for Public Risk Communication on Warnings for Public Protective Actions Response and Public Education." http://www.colorado.edu/hazards/publications/informer/infrmr2/pubhazbibann.pdf . (2006)

19 Proulx, G.: "How to initiate evacuation movement in public buildings." *Facilities*. 17:331-335, (1999)

20 Fahy, R.F., Proulx, G.: "Human Behavior in the World Trade Center Evacuation." In: Hasemi, Y. (ed.), *Fire Safety Science -- Proceedings of the Fifth International Symposium* Interscience Communications Ltd., London, England (1997)

# Regulation / Engineering Guidance

# Accessibility and Evacuation Planning – Similarities and Differences

**J. Bendel, and H. Klüpfel**

Access Unlimited, 18 Arlozorov St. Jerusalem 92181, Israel
TraffGo HT GmbH, Bismarckstraße 142a, 47057 Duisburg, Germany

**Abstract** Twenty–twenty four percent (20-24 %) of the world population are people with disabilities. This population has special access needs for performing activities of daily living. This is especially important in case of an emergency. The evacuation concept for a building must take people with disabilities into consideration by means of including specific parameters as an integral aspect in the evacuation simulation model.

This paper presents the accessibility parameters to be considered in evacuation simulations. Examples of tools used in the two disciplines (evacuation and accessibility planning) are provided for illustration: The PedGo evacuation simulation package [Kluepfel, 2003], a model which does only implicitly take into account disabilities; and a decision support system for evaluating accessibility of facilities [Bendel, 2006] illustrates the accessibility approach to facility planning.

Recommendations for integration of accessibility and evacuation analysis summarize the paper.

## Introduction

Twenty–twenty four percent (20-24 %) of the world population are people with disabilities (U.S. Census Bureau, 2005; Australian Bureau of Statistics, 2004; EU commission, 2003). This population has special access needs for performing activities of daily living. This is especially true and important in case of an emergency. The evacuation concept for a building must take this into account. Planning should not assume that individuals with obvious physical disabilities are the only ones who will have difficulties evacuating. Individuals with a range of medical conditions — hidden disabilities including asthma or heart conditions — or any occupant who is recuperating from an injury or surgery might have difficulty using steps.

Despite of its importance, most current evacuation simulation models do not take into consideration many aspects of evacuation specific to people with disabilities and the elderly [Pelechano, 2008]. And they do not adequately address individuals with disabilities in their simulated populations Christiansen [2008].

R.D. Peacock et al. (eds.), *Pedestrian and Evacuation Dynamics*,
DOI 10.1007/978-1-4419-9725-8_62, 

## *The "Time" Factor in Accessibility and Evacuation*

Evacuation is not the same as accessibility, as the 'time'-factor plays a more restrictive role in the case of evacuation. There is an available safe egress time ASET, which defines the upper limit for the required safe egress time RSET (RSET < ASET) Nevertheless, evacuation and accessibility have much in common and they complement each other. For example, the spatial configuration and occupancy of rooms, obstacles, route marking as well as other factors related to various types of disabilities need to be explicitly taken into account in both fields.

This paper presents the accessibility parameters to be considered in evacuation simulations. The PedGo evacuation simulation package [Kluepfel, 2003] is used as an example for a model which does only implicitly take into account disabilities. In comparison to the simulation, a decision support system for evaluating accessibility of facilities [Bendel, 2006] will show the similarities and differences between evacuation and accessibility.

## Anthropometrics, Ergonomics, and Pedestrian Movement

The physical, sensory and mental fitness of the populations in western society changes, as well as the age distribution. Age, as well as other factors like obesity, medical conditions etc. have an impact on the walking speed. These factors also affect accessibility needs as well as evacuation performance. Therefore, the special needs as well as the predicted demographic changes have to be taken into account for pedestrian planning and design.

There is a scarcity on data for the movement of persons with disabilities. Concerning wheelchair users, Buchmüller (2007) gives a value of 0.5m/s for the movement speed. The SFPE Handbook of Fire Protection Engineering provides additional information on the movement speed of persons with disabilities. General information on the evacuation of wheelchair users is contained in the "Emergency Evacuation Planning Guide for people with disabilities" (NFPA, 2009).

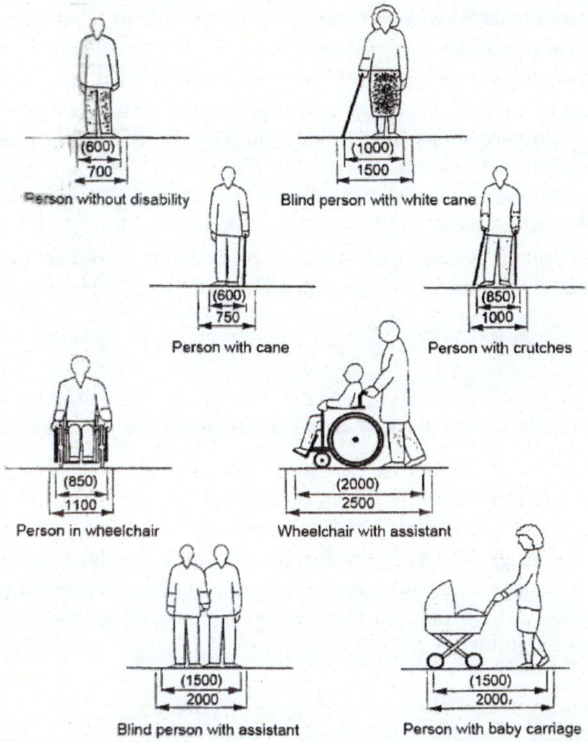


**Figure 1: Spatial Needs for persons with special needs [Buchmüller, 2007].**

## Accessibility and Evacuability

Accessibility and Evacuability correspond to and complement each other. Nevertheless, the two fields have remained quite separate in the past.

**Accessibility** is a general term used to describe the degree to which a product (e.g., device, service, or environment) is accessible by as many people as possible. Accessibility is strongly related to universal design when the approach involves "direct access." This is about making the environment accessible to all people (whether they have a disability or not).

However, accessibility is often used to focus on people with disabilities (including people with mobility sensory and mental impairments) and their right of access to entities on an equal basis with the rest of the general population. In order to achieve barrier free environment, accessibility standards and regulations were created. The standards focus on minimum requirements to create an architectural and communication barrier free environment

**Emergency evacuation** on the other hand is the immediate and rapid movement of people away from the threat or actual occurrence of a hazard.

Laws and regulations regarding evacuation focus on life safety codes including fire protection. However, individuals with disabilities are disproportionately affected by conditions in the built environment especially during an evacuation event (US Fire Administration, 1999). An accessible facility by definition should provide for improved evacuation routes and procedures, therefore it is essential to address as many as possible accessibility elements in the evacuation evaluation models.

## *Accessibility elements which have an impact on evacuation:*

In many countries evacuation regulations are included to some extent in the accessibility standards or guidelines. There is much similarity among the guidelines in the different countries. The main differences are in the details required for specific elements for example door width in one country must be minimum 80 cm, while in another country the minimum required is 75 cm of clear width. In this paper we'll use as an example the ADA Accessibility Guidelines (ADAAG, 2002) in the USA.

The chapter on *Accessible Means of Egress* (ADAAG 4.1.3(9), 4.3.10) includes minimum number of egress routes required to be accessible (based on life safety code requirements). Means of egress is defined as:

> "A continuous and unobstructed way of exit travel from any point in a building or facility to a public way. A means of egress comprises vertical and horizontal travel and may include intervening room spaces, doorways, hallways, corridors, passageways, balconies, ramps, stairs, enclosures, lobbies, horizontal exit, courts and yards."

The egress routes must comply with the criteria for accessible routes. An accessible route does not include stairs, steps, or escalators

### Accessibility elements to be considered:

**Accessible routes**: i.e.; Length of route (Travel distance), width, Passing Space, Head Room, Surface Textures (such as carpet, slippery surface, uneven surface etc), Slopes, doors (i.e. not revolving doors, including elements such as clear width, maneuvering clearances at doors, two doors in series, thresholds at doorways, door hardware: handles, pulls, latches, locks, door closers, door opening force, automatic doors and power-assisted doors and the treatment of elevation changes: a curb ramp, ramp, elevator, or platform lift respectively. Elevators, the standard means of access between floors are typically taken out of service in emergencies for safety purposes. ADAAG addresses this situation through requirements for areas of rescue assistance with two-way communication devices (voice and visible signal requirement such as a button that lights) or horizontal exits. (ADAAG 4.1.3(9), 4.3.11).

**Evacuation elevators**, which are recognized by the model building codes although are not included in the current ADAAG, offer an additional solution. These are elevators that are specially designed to remain functional in emergencies. Possibly also emergency personnel may be able to operate standard elevators in certain emergencies. Meaning the route to the elevator must be considered as well as means of egress, as well as the specifications for accessible elevators (i.e., automatic operation, call buttons, hall lanterns (visible and audible signals), door protective and reopening device, door and signal timing for hall calls, door delay for car calls, and the floor plan of elevator cars.

**Stairways** although steps and stairs are not considered accessible means of egress, they can be used for evacuation by various types of people with disabilities for example visually or hearing impaired and some mobility impaired who are not wheelchair users. Therefore specifications for stairs mentioned in the accessibility guidelines must also be considered: i.e. width, number of steps, treads and risers, nosings, handrails, detectable warnings at stairs.

**Alarms** ADAAG (4.1.3(14), 4.28) provides specifications for emergency alarms so that they are accessible to persons with disabilities, including those with sensory impairments. i.e. audible and visual features which address intensity, flash rate, mounting location in all general usage and common use areas including meeting and conference rooms, classrooms, cafeterias, employee break rooms, dressing rooms, examination rooms and similar spaces

**Signage** (ADAAG 4.1.3(16), 4.30): i.e. character proportion, character height, raised and brailed characters and pictorial symbol signs (pictograms), finish and contrast, mounting location and height, symbols of accessibility, illumination levels, especially of information and directional signs.

Most of these elements may have an impact on the evacuation and the movement speed of evacuating a facility.

Many of these specifications are part of the building design, and therefore can be addressed more easily in the design based evacuation simulation model. However, elements which mostly do not appear in the building design such as surface textures (carpet, slippery surface), door hardware, signage, alarms etc. should be also included.

## *Evaluating accessibility*

All accessibility elements in the built environment must be analyzed in order to determine the level of their compliance with the accessibility standards. The standards are the minimum requirements to enable people with different types of disabilities to use the facilities in general and in case of an emergency and need for evacuation in particular. A thorough accessibility survey not only will identify barriers to access but will highlight emergency evacuation problem as well. Accessibility evaluation is needed during the planning stage of a new building as well as for existing facilities to identify required modifications. Defining accessibility in the built environment is complicated. Many details are needed to meet the needs of people with a wide range of disabilities. Each detail does not stand alone; therefore, the sum of the non-compliant items is not always an indication of accessibility. For example, the general width of a corridor may meet only the minimum requirements, the door width and the opening direction of the doors (in direction of the corridor) comply with the guidelines. Yet the corridor may still not be accessible due to the combination of the two complying elements (door and corridor). In this case when the doors are open they may cause an obstacle, as well as decrease the width of the corridor, especially when there are many doors involved. The decision support system for evaluating accessibility of facilities [Bendel, 2006] presents a computerized audit tool and a decision support system model for evaluating accessibility of existing public facilities, and grading them. The system focuses on the interaction between the individual and the environment, as well as possible combinations between the different elements of design. This system highlights items in the facility requiring adaptation or upgrading to better meet the user's needs. The decision support and grading system operates on formulae and equations for data analysis to define accessibility grades. The equations consist of Boolean operators ("and", "or", "not") combining the accessibility components.

There are two sets of formulae. The first set is formulae defining the accessibility of each specific element or space in the building separately (i.e. parking, entrance, offices, staircase etc.), with different equations each one relating to the four major types of disability groups: wheelchair users, users of crutches or walkers, individuals with visual impairments and individuals with auditory impairments. In other words, the elevator's accessibility is graded specifically for the visually impaired, for the wheelchair users etc. Therefore the grade for the visually impaired, for example will be a combination of all elements of the elevator which are relevant to this group, i.e. specifications of accessible signage, availability of tactile elements in front of the entrance door and on control buttons audible signage, etc. The accessibility of the elevator for wheelchair users will integrate elements such as door opening, size of compartment, height of control buttons etc. The second set of formulae defines weighted grades of the whole facility according to the type of the building/site and to the type of disability. As behavior of people in public buildings is related to the type of service given in the building, the usability of the building is analyzed accordingly. Therefore, the formulae for a bank, for example, will differ from those of a sports facility. The relative importance of each element and the connection between elements are considered in the definition of the formulae. For example the toilets are an element with high relative importance for people who visit a sports facility, but of low relative importance for banking. The overall weighted grade of the building's accessibility for each type of disability is computed based on all the services available and the building environment, using the grades of each element as described above.

## Evacuation Model for Persons with Disabilities

The BUMPEE model Christiansen [2008] is an evacuation model taking persons with disabilities explicitly into account. This is a person-based simulation classifying the built environment according to environmental characteristics and simulating a heterogeneous population according to variation in individual criteria representing the diversity and prevalence of disability in the population.

However, the impact of the building design as a whole and integration of the different elements of the building i.e. location of elements, direction signs, direction of doors etc. as well as behavior under pressure, behavior that have an impact on others are not used in the model.

### *Evacuation Simulation*

Fig. 2 shows the layout of the building used for the sample evacuation simulation. There are altogether 1019 persons in the building, five of them being wheelchair users.

Each person possesses individual parameters. The following parameters are used in this example: Velocity ($V_{max}$): The maximum free walking speed; Patience: The maximum time a person stands still (e.g. in a congestion) before changing her route and attempting to find another escape route; Sway: The accuracy, with which a person follows the gradient of the potential; Reaction: The duration a person needs to respond to the evacuation signal, e.g. start moving (= premovement time); Dawdle: The probability, for a person to reduce his walking speed, e.g. to stand still for the rest of a sub time step; Inertia: The force with which persons try to continue on their walking direction, i.e. resist change of current walking direction.

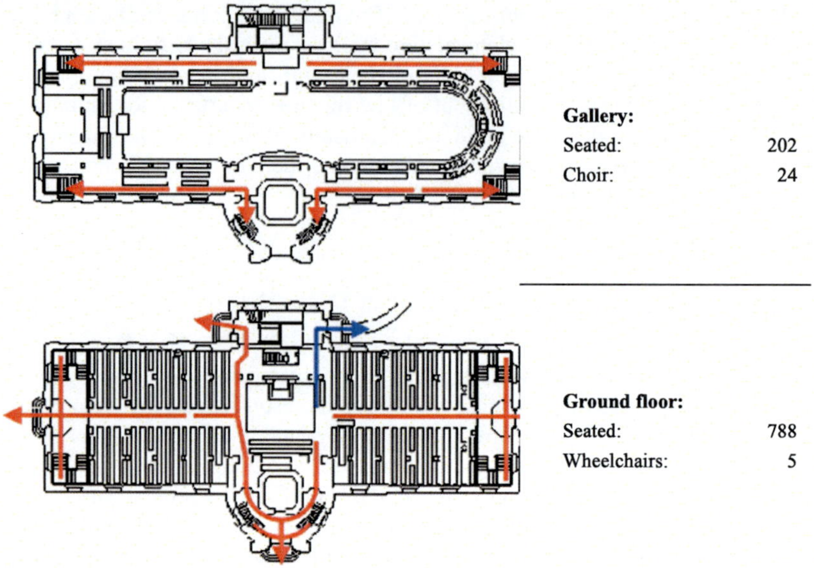


**Fig. 2. Layout of the two-storey building. The evacuation route for the persons with disabilities (wheelchair users) is shown in blue (via the ramp).**

The parameters of the standard population are shown in the following table.

**Table 1: Population parameters (Standard Population)**

| | Min | Max | Mean | Std.-Dev. | Unit |
|---|---|---|---|---|---|
| Velocity | 0.8 | 2.0 | 1.2 | 1 | m/s |
| Patience | 5000 | 5000 | - | - | s |
| Sway | 1 | 5 | 3 | 2 | |
| Reaction | 0 | 60 | 30 | 300 | s |
| Dawdle | 0 | 30 | 15 | 5 | % |
| Inertia | 1 | 5 | 3 | 2 | % |

The next table only provides the parameters that are different for the persons with special needs. All other parameters are as in **Table 1**.

**Table 2: Population parameters for persons with disabilities (wheelchair users).**

| | | | | | |
|---|---|---|---|---|---|
| Velocity | 0.8 | 0.8 | - | - | m/s |
| Sway | 1 | 1 | - | - | |
| Dawdle. | 0 | 50 | 25 | 250 | % |

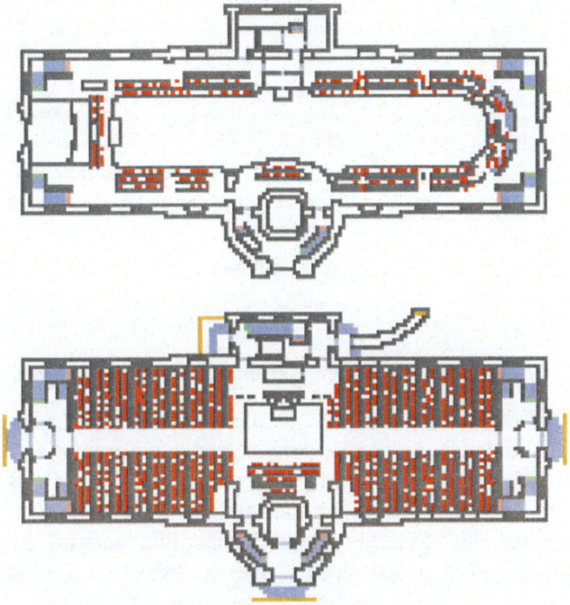

**Fig. 3. Initial distribution of persons.**

The wheelchair users had been assigned a maximum free velocity of 0.8 m/s, whereas the abled population has a normally (Gaussian) distributed range from 0.8 m/s to 2.0 m/s. The initial distribution of persons (red dots) is shown in Fig. 3.

For this project, several scenarios had been defined with different availability of exits and routes. For the sake of this paper, we focus on scenario 1, where all exits are available (this is the reference scenario, cf. Fig. 2).

Based on this distribution and the parameters stated above, the evacuation of the building is simulated. As can be seen from the evacuation curve (number of persons evacuated versus time, cf.), the evacuation of the wheelchair users which are seated close to the ramp in the ground floor (cf. Fig. 2) is faster than for the general population. This is mainly due to the fact that there are only five wheelchair users and they are seated close to the exit which is used specifically for them.

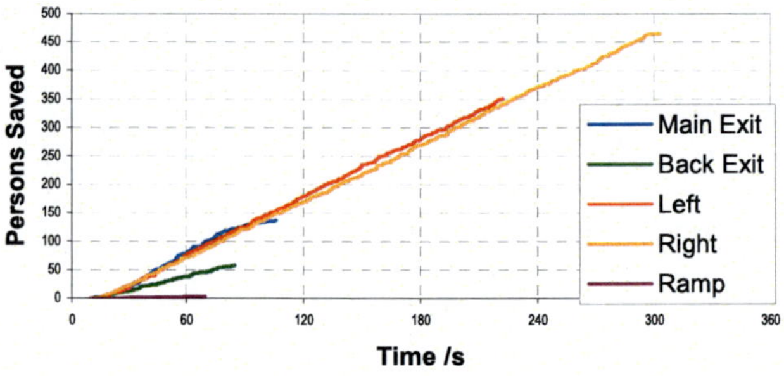


**Fig. 4. Results for the evacuation of standard and wheelchair population.**

Based on this distribution and the parameters stated above, the evacuation of the building is simulated. As can be seen from the evacuation curve (number of persons evacuated versus time, cf.), the evacuation of the wheelchair users which are seated close to the ramp in the ground floor (cf. Fig. 2) is faster than for the general population. This is mainly due to the fact that there are only five wheelchair users and they are seated close to the exit which is used specifically for them.

## Summary, Conclusion, and Recommendations

In this paper we presented a summary of the similarities and differences between accessibility and evacuation. In general, a universal building designed will also perform well in an evacuation scenario. The main difference is the RSET < ASET requirement, i.e. the available safe egress time determined by the hazard (e.g., a fire) must be longer than the required safe egress time. This criterion is usually not required in the case of access to a building, whereas the simulation model is not based on the interaction and integration of the building elements which view the building as a whole unit.

Recommendations for integration of accessibility and evacuation analysis:

- Information items should be added to the CAD plan, for example location of signage to reduce trial and error behavior during evacuation.
- The accessibility grades – the results of the accessibility analysis should be incorporated in the simulation model. This would allow taking into account accessibility parameters in a straightforward manner.
- Accessible facilities therefore should also have an impact on parameters used in the PedGo evacuation simulation model presented above

The demographic factors mentioned in the introduction will inevitably change the composition of societies of many if not most countries in the world. A prominent example is the need for rest during the evacuation of high-rise buildings. Furthermore, many mobility impaired persons are not able to evacuate a building. One consequence will probably be the use of elevators for evacuation. The field of accessibility or barrier free planning and design has a lot to offer for the field of evacuation planning. Therefore, practices developed in the former field

## References

1. ADA Accessibility guidelines for buildings and facilities (ADAAG) as amended through September 2002. www.access-board.gov/adaag/html/adaag.htm
2. Australian Bureau of Statistics: Disability Ageing and Carers Summary of Findings, 2003 (Cat No. 4430.0). From http://www.abs.gov.au/Ausstats/ (2004)
3. Bendel, J.: "Decision Support System for Evaluating Accessibility of Facilities." The Israeli Journal of Occupational Therapy, Vol.15, No.3 (2006)
4. Buchmüller, S., Weidmann, U.: Parameters of pedestrians, pedestrian facilities, and walking facilities. Technical report 132. IVT, ETH Zurich (2007)
5. Christensen, K., Sasaki Y.: Agent-Based Emergency Evacuation Simulation for Individuals with Disabilities in the Population. Journal of Artificial Societies and Social Simulation. Vol. 11, No. 3.9 (June 2008)
6. DiNenno, P. (ed.): SFPE Handbook of Fire Protection Engineering (2nd & 3rd Edition). National Fire Protection Association (1995, 2002)
7. Fruin, J.: Pedestrian Planning and Design. Metropolitan Association of Urban Designers and Environmental Planners, New York (1971)
8. International Maritime Organization: Guidelines for Evacuation Analyses for New and Existing Passenger Ships. MSC.1/Circ.1238 (2007)
9. Klüpfel, H.: A Cellular Automaton Model for Crowd Movement and Egress Simulation. Dissertation, Universität Duisburg (2003)
10. National Disability Authority: Promoting Safe Egress and Evacuation for People with Disabilities. Dublin. ISBN: 978-1-870499-18-7. http://evacmod.net/?q=node/1914 (2008)
11. NFPA Evacuation Guide http://www.nfpa.org/assets/files/PDF/Forms/EvacuationGuide.pdf (2009)
12. Miyzaki, K., Katuhra, M., Hiroshi, M., Koichi, H.: Evacuation Simulation for Disabled People in Passenger Ship. National Maritime Research Institute, Japan
13. Pelechano, N., Malkaw, A.: Evacuation simulation models: Challenges in modeling high rise building evacuation with cellular automata approaches. Automation in Construction. Volume 17, Issue 4, Pages 377-385 (May 2008)

14. Stein, J.: Evacuation: The ADA Link. Maintenance Solutions. (November 2003)
15. U.S. Census Bureau: Survey of Income and Program Participation. 2004 Panel, Wave 5 (June–September 2005)
16. U.S. FIRE ADMINISTRATION: Removing the Barriers; A Fire Safety Fact sheet for People with Disabilities and their Caregivers. Office of Fire Management Programs, Emmetsburg, MD (September 1999)
17. 2010: A Europe Accessible For All. Report from the Group of Experts set up by the European Commission, October 2003

# Micro-Simulation Modeling of Persons with Reduced Mobility: Is the London Framework Applicable in North America and Does it Affect Modeling Output

**D. A. Fisher and A. Jenkins**

Arup Consulting Inc
2 Bloor Street East, Suite 2400
Toronto, Ontario Canada, M4W, 1A8
daniel.fisher@arup.com, andrew.jenkins@arup.com

**Abstract** Transit authorities globally are developing their networks to make them more accessible for Persons with Reduced Mobility (PRM's) whilst demographic and lifestyle trends suggest that the proportion of PRM's within the population who are likely to make use of these newly accessible networks will increase over time. In the UK, London Underground has developed a modeling framework that incorporates PRM movement into pedestrian micro-simulation models. Network Rail, the UK mainline station operator has also been undertaking PRM surveys to determine the characteristics of PRM demand at mainline stations and this data validates the approach adopted by LUL which requires a location specific understanding of PRM populations, incorporates spatial and temporal variation, and is coupled to location specific characteristics (for example, Network Rail long distance passengers are less familiar with their station environment than commuters and require more information, signage and management. They are more likely to travel in groups and be encumbered).

Anecdotal evidence from studies completed in London suggests modeling output is sensitive to the volume of PRM's modeled. To date, no detailed reviews have been undertaken to assess this sensitivity. This paper reviews the PRM modeling methodology used in London and assesses its applicability in the America's and elsewhere. The relationship between maximum flow rates and the average speed and footprint of entities is tested through the use of Legion micro-simulation models and described through the application of multi-variable linear regression to the output. Impacts of modeling PRM's are described and implications for Space Planning Performance Metrics and design recommendations are reviewed.

R.D. Peacock et al. (eds.), *Pedestrian and Evacuation Dynamics*,
DOI 10.1007/978-1-4419-9725-8_63, 

## Introduction

Developing inclusive public infrastructure that fully meets the diverse needs of existing and future users is of great importance. In recent years and on both sides of the Atlantic, disability discrimination legislation such as the Americans with Disabilities Act of 1990 and the UK Disability Discrimination Act of 1995 has been the spur to progress in this area. Nowhere is this more evident than in the field of public transportation, where significant progress has been made towards creating barrier free access for passengers (for example, all of London's 8,000+ buses operated on TfL contracts are accessible, as are the 6,000+ buses in the in the MTA fleet, New York). In Canada, similar federal legislation does not exist and the Council of Canadians with Disabilities has published a study [1] that is heavily critical of the Canadian approach to the provision of accessible transport. Commitments have been made by Prime Minister Harper [2] to create federal legislation similar to the ADA but none exists at the current time. Whilst progress towards barrier free accessibility within the transport system is less advanced in Canada, it is none the less happening.

The creation of fully accessible transit networks is anticipated to alter the characteristics of the passenger population. This in turn has implications for capacity analysis and therefore the design and planning of transit. A simple example is provided through examination of the passenger capacity of low floor buses. In the UK, licensing regulations require bus manufactures to determine the safe maximum passenger load for each configuration of bus they manufacture. This information must be displayed within the bus. Historically, the overall maximum capacity comprised the maximum number of seated passengers + the maximum number of standees. With the advent of low floor vehicles, it was recognized that dual purpose wheelchair/standee space, together with the use of tip up seats within this space, would render a single maximum capacity figure inappropriate as more passengers could be safely carried if the wheelchair space and tip up seats were not occupied. The solution was to require manufacturers to determine maximum safe capacities in both the "with" and "without" wheelchair passenger scenario. This alters maximum capacity by up to 10%; for example a single decker MAN 18.220 with Alexander ALX300 bodywork in the fleet of Stagecoach UK is licensed to carry a maximum of either 39 seated passengers + 1 Wheelchair passenger + 28 standees or alternatively 37 seated passengers + 38 standees, a difference of 7 passengers. By implication this variability in maximum capacity affects the "planning capacity" used by transit planners to determine peak headways. "Planning capacity" is often defined as a percentage of maximum capacity, for example within TfL London Buses service planning methodology [3] or may be based on seating capacity, for example the "Level of Service" bands referenced in the American Transit Capacity and Quality of Service Manual [4]. To add a further level of complexity when deciding how best to plan service capacity to reflect this change in the characteristics of the passenger population, use of transit by wheelchair users and others with oversize space requirements is unlikely to be spread evenly across all routes in a network and across all periods of operation.

Bus services are relatively flexible and responsive to local needs and bus drivers and passengers alike provide immediate and vocal feedback to management when capacity problems start to occur. Bus transit service providers may be proactive in assessing the capacity implications of developing accessible networks, but in practice can afford to be re-active.

The same flexibility to alter built infrastructure does not exist and therefore those responsible for planning for new or revitalized transit station infrastructure should proactively plan for the implications of barrier free access. In the transit terminal context, the relationship between passenger demand and design capacity is significantly more complex and requires forward planning. Conflict always exists between the competing pressures of minimizing the spatial requirements (and therefore cost) and ensuring adequate spatial provision for the needs of passenger today and tomorrow. Optimum design will reflect an appropriate balance between these pressures; it occupies a relatively narrow band and is difficult to identify. Just as many of today's passengers in the London suburbs enjoy the built legacy of Charles Holden but suffer in the central area from the inadequate spatial planning of his Victorian predecessors, so future generations of transit passengers will live with the legacy of what we build today. As a result, significant time and effort is spent in pursuit of the correct cost/size balance through the iterative process of design and capacity analysis.

Capacity analysis may best be summarized as the combination of passenger demand and spatial performance modeling, with the latter task typically completed (at least for larger projects) using pedestrian micro-simulation modeling software such as Legion or Mass Motion. As micro-simulation technology becomes ever more sophisticated, and in parallel, there is ever greater pressure for designers to understand future year spatial performance in greater detail, efforts have been made to incorporate a more detailed understanding of pedestrian movement in micro-simulation models. London Underground completed a study [5] to develop a modeling framework that enables the characteristics of Passengers with Reduced Mobility (PRM's) to be more accurately modeled using Legion software.

## The London Framework: Defining a PRM Population

The standard approach when building micro-simulation models is to define speed and entity size profiles for the entire modeled population. Individual entities within the modeled population are randomly assigned speed and size characteristics in accordance with the population distribution. Pearce, Powell, Duff, Anigbogu and Kerr [5] suggest a methodology that seeks to allow modelers to generate a more realistic population profile. A 3 stage approach is suggested which involves 1) the definition of subsets within the population, 2) description of their movement characteristics in terms of speed distribution and footprint size and 3) estimations of their quantities within the population with reference to spatial and temporal variance. With this data, pedestrian flow modelers are able to construct models to a finer level of detail which should generate greater confidence in modeling output. For example, dis-aggregation of the population allows modelers to incorporate routing or behavioral differences between groups within models (eg, wheelchair users would always be assigned to step free routes, passengers with young children are likely to pause before and/or after ticket barriers to retrieve and then secure tickets, etc). It is believed that altering the speed and size profiles of model populations may alter modeling output and new work has been completed to test this assumption. Before detailing that work, a summary of the London Framework is provided, alongside a commentary on its applicability to similar studies outside London.

## *PRM Categories and Movement Characteristics*

**Table 1. Passenger size movement characteristics [5]**

| Category | Mean Speed | Mean Size (sqm) | Routing Characteristics |
|---|---|---|---|
| Non-PRM | 1.53 m/s | 0.087sqm | No restrictions |
| Wheelchair User | 0.58 m/s | 0.683sqm | Always use accessible routes |
| Passengers with permanent or temporary Physical Impairments | 0.80 m/s | 0.102sqm | Preference for accessible routes |
| Non-disabled Passengers with medium sized Luggage | 1.53 m/s | 0.393sqm | Preference for accessible routes |
| Non-disabled Passengers with Large Luggage | 1.32 m/s | 0.562sqm | Preference for accessible routes |
| Adults with Young Children (including Pushchairs) | 1.37 m/s | 0.682sqm | Preference for accessible routes |

Table 1 describes the PRM categories proposed by Pearce, Powell, Duff, Anigbogu and Kerr [5] and currently used by London Underground and the wider TfL transport planning community in London when performing pedestrian modeling studies.

## PRM Categories: Applicability in North America

Where available, comparable data in North America demonstrates a reasonable if not precise fit with the data in table 1, for example the passenger space requirements described in the TCRP Transit Capacity and Quality of Service Manual [4]. Additional categories could be added to provide a more comprehensive list and/or better reflect local conditions. In particular, passengers travelling with bicycles are a familiar sight on systems which permit their carriage. The grouping of permanent and temporarily physically impaired passengers is a cover-all grouping that comprises a very wide range of mobility impairment and it may be desirable to disaggregate some of the sub-groups, in particular, those with significant sight impairment. The additional benefits in terms of accuracy of model output to be gained from further dis-aggregation of the population will be a function of local circumstance and in all cases, speed and size profiles should be agreed locally. None the less, the PRM categorization suggested, subject to appropriate local modification, is universal and could be applied in any station modeling situation globally.

## PRM Population Size

The framework and data proposed by Pearce, Powell, Duff, Anigbogu and Kerr [5] has been adopted by London Underground and coupled with additional survey data to better define the number of PRM's for station modeling purposes [6]. The survey data was aggregated into 6 different station typologies (City, Inner Suburb, Outer Suburb, Shopping, Terminus and Tourist). Differentiation by time period (am peak, midday, pm peak and weekend) and by access or egress movement gives a total of 48 different potential PRM population profiles. Establishing the number of PRM passengers using the London system by station typology and time of day was a complex task requiring detailed local survey data. Variability in the volume of PRM movement within the system is extensive. Indeed, a PRM movement survey completed jointly by Crossrail, LUL and Network Rail at Paddington Station during the PM Peak of 5 August 2008 [5] indicated that wide differences in PRM volumes exist between different transit lines within the single Paddington Station site, further complicating the task.

An additional challenge is the need to assess changes in future year volume of PRM movement. These changes may occur in response to reduced barriers to travel within the transit system. They may also occur in response to reduced barriers to the activities that generate demand for travel. Socio-economic changes may also play a role in changing the mobility of the population. London Underground step free access programme was intended to bring step free access to 25% of the 270 stations on the network. The impacts of this programme were estimated to be the generation of 19 Million new passenger journeys per annum and benefits for a further 30 million existing passenger journeys per annum and this translated into a benefit cost ratio for the scheme of 2.7 to 1. [7] In the context of c.1 Billion LUL Trips annually, this represents growth of c. 1.9%

## *PRM Volumes: Applicability in North America*

In the case of the London guidance, PRM volumes have been derived from detailed analysis of available survey data that identifies the volumes of existing PRM movement by time of day and station location. Work to understand the future year implications of providing an accessible network has been completed at a network level and network forecasts are capable of disaggregation as required to inform local studies.

This offers a template for other transit authorities to use as a basis for similar studies. Many metropolitan transit authorities in North America face trends similar to those faced in London. It is reasonable therefore to assume that the quantity of PRM movement on North American systems will increase over time, and further, that the volume of additional PRM movement will be of a similar order of magnitude as that suggested by the London Underground modeling framework [6]. However, detailed forecasts of the type generated by London Underground can only be applicable in their local context.

## The Impact of Incorporating PRM Movement

Should transit planners expend time and effort to understand and then model PRM movement? Does it make any difference to the results?

Space planning guidance for transit systems in many North American cities, as in Europe, makes reference to the seminal and universal Fruin [8], specifically his description of pedestrian Levels of Service to define the acceptable parameters of spatial performance. These scales have the immense benefit of universal understanding, simplicity and have stood the test of time. In some cases, transit authorities reference maximum flow rates to generate key performance metrics, particularly in connection with acceptable platform clearance times. Fruin noted that wheelchair users and others with restricted mobility were relegated to the status of "disenfranchised citizens" [8] because of the lack of access to public space. He was ahead of his time in calling for improvements to the situation but as a result of the era in which he collected his data, minimal PRM movement would have been included in that data. This suggests that additional PRM movement may depress planning capacities. On the other hand, recent data published by MAIA Institute/Legion Limited [9] indicates that the average speed of movement amongst commuter populations in New York and London has increased since Fruin conducted his research (from an average of 1.35 m/s to 1.50 m/s) and this may counter-balance the effect of additional PRM movement.

## *Test Methodology*

A simple study methodology has been devised to assess the impact of variations in changes to speed and size profiles within Legion populations. Legion models have been used to assess the maximum flow rates achievable through stairway and walkway space.

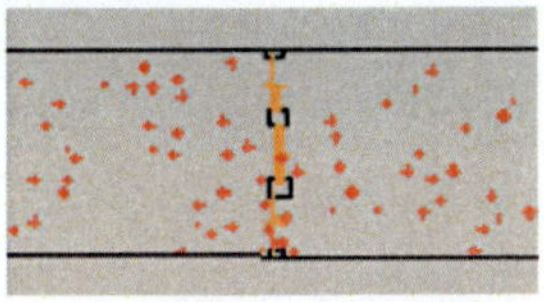

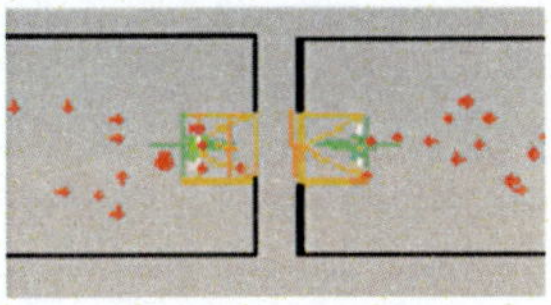

**Fig. 1. Walkway model with three 2 metre constrictions**
**Fig. 2. Stairway model with 2.4 metre stair**

For each geometry, 72 tests have been completed to test nine different speed distributions (with average speed ranging from 0.92 to 1.72 m/s), 4 different entity size (no luggage, small luggage, medium luggage and large luggage) settings, and one and two-way (50/50 distribution) flows. Stepped Arrival Profiles (each step of 3 minutes duration) have been used. Entity flows during each 3 minute time step are set to be consistent with flow rates expected at Fruin Level of Service thresholds, from A/B to E/F. Demand has then been set to exceed the theoretical maximum flow rates outlined by Fruin to ensure the tests generate the actual maximum modeling flow rate. Flow rates have been recorded throughout the simulations. Time Average density plots have also been recorded in the final minute of each 3 minute step to enable density to be reviewed during stabilized flow conditions. This data will allow the relationship between maximum flow rates and population characteristics to be explored. It is anticipated that the dependent variable (maximum flow rate) can be described by the explanatory variables (average population speed and average population size) through the application of multi-variable linear regression techniques to data from the tests. Further, if this examination indicates maximum flow rates are significantly influenced by PRM movement, a detailed review of the density plots can be completed to determine if there are implications for Level of Service output at lower density levels. It is noted that Legion functionality allows extensive flexibility for users to disaggregate populations and define speed profiles. However, one area of limited functionality is in the realm of specifying entity spatial characteristics. Legion offers users the ability to define entities with settings of either "small luggage", "medium luggage", "large luggage" or "no luggage". This means that the entity "footprint" described in table 1 can not be fully replicated.

## Results

For walkway tests, a visual review of the results appears to confirm that as speed increases, maximum flow increases, whilst as size of entity increases, maximum flow decreases. The data allows the relationship to be described in greater detail as shown by equation 1:

$$Y = 50.6 + 36.8X_1 - 150.8X_2$$

Where $Y$ = *Maximum Flow (per metre per minute)*
$X_1$ = *Average Population Speed (metres per second)*
$X_2$ = *Average Population size (square metres)*

This equation produces predicted values that exhibit a Pearson correlation coefficient of 0.84 with the output from the tests. For stairway tests, a visual review of the results appears to show a more variable pattern of results. However, using the same methodology, the relationship between the variables can be described with a similar degree of confidence as described by equation 2:

$$Y = 31.9 + 28.6X_1 - 119.6X_2$$

Where $Y$ = *Maximum Flow (per metre per minute)*
$X_1$ = *Average Population Speed (metres per second)*
$X_2$ = *Average Population size (square metres)*

This equation produces predicted values that exhibit a Pearson correlation coefficient of 0.80 with the output from the tests. The results from the fixed geometry tests completed for this paper allow descriptive relationships to be derived. These do not precisely describe the relationship between the variables, indeed it is acknowledged that the application of rigid relationships to describe pedestrian movement are generally best avoided in the field of pedestrian modeling as movement characteristics are almost always sensitive to local context. None the less, the equations produce predictions that are sufficiently well correlated with actual results to give confidence when using them to demonstrate order of magnitude flow rate impacts associated with changes in the speed and size distributions of passengers within a population.

## *Order of Magnitude Results in London Framework Distributions*

Equations 1 and 2 have been applied to the 48 London Underground PRM distributions. Of note, there is very minimal variability in average population speed amongst these distributions. The results indicate that, dependent on station typology, time of day and direction of flow, maximum modeled walkway flows will be in the range 83.5 to 92.7 pedestrians/metre/minute whilst stairway flows will occupy the range 57.1 to 64.4 pedestrian per metre per minute. The variability between flow rates is just over 10%. This suggests that there are implications for model output and therefore design recommendations flowing from such studies which incorporate PRM movement. Reporting metrics such as platform clearance times will be affected directly. Models with larger PRM populations will indicate failure conditions more quickly. A review of Level of Service plots at intermediate Fruin LoS threshold flow rates for the test case models indicate a marginal impact on density plots, although very few transit authorities enforce their own level of service standards rigidly so this is unlikely to have a significant impact. Of note, all the configurations detailed in Appendix B exhibit maximum flow rates greater than those suggested by Fruin [8]. It may be concluded that the increase in speed of the general population over recent decades [9] counter-acts the capacity reducing trends associated with greater numbers of passengers with reduced mobility. In so far as these tests can be used to judge (and limitations in methodology are discussed below), the Fruin scales remain conservative and applicable.

## *Study Limitations and Further Work*

As noted, there are limitations in the methodology employed. Most critically, real world data to enable validation of the models is not available. It is hoped that as further work is completed to understand the implications of PRM movement, these data gaps will gradually be filled. Further, Legion is not capable of accurately portraying PRM movement in 3D and the extent to which it is capable of accurate modeling of PRM movement in 2D is not clear. Large entities can only be represented as "with luggage" which does not enable sufficient flexibility for modelers to replicate desired entity sizes. In addition, Legion routing and behavior algorithms do not replicate PRM specific characteristics. Some of these are capable of input by modelers, others are not. A body of anecdotal evidence from practitioners in London suggests that blockages occur relatively quickly when large agents meet each other or when negotiating restricted space and that these conflicts do not self resolve although it is noted that large Legion entities should be capable of temporarily "fluxing" in tight spaces to squeeze past each other [10].

## Conclusions

The use of micro-simulation models during the design phase of infrastructure construction and reconstruction delivers tangible benefits and has become a mainstream activity during design development. As modeling software becomes ever more sophisticated, so greater demands are placed on modeling capability. Understanding PRM movement is high on the agenda for many Transit Authorities and their stakeholders. Effective studies require a combination of good local knowledge and data, the right tool for the job and good analytical skills amongst the modeling team. Features of behavior and movement unique to PRM's should be reflected in models, although in some instances, it is not possible to do this yet due to shortcomings in available technology. In many cases, such behaviors will be location specific and dependent on the configuration of the space in question. Some however are universal. A good example of the latter is the pause which most passengers with heavy luggage and strollers make to reconfigure themselves before passing through a ticket barrier. It is concluded from this study that the London modeling framework, with appropriate modification as described to suit local circumstances, is relevant to studies completed beyond London. It is concluded that modeling PRM movement does impact the results of modeling work and two equations are proposed to describe order of magnitude impacts on maximum flow rates for stairs and walkways. Further work is proposed to reduce known limitations of PRM modeling, both in terms of data collection and software development. Finally, it is concluded that the evergreen Fruin scales and maximum flow rates as applied by many transit authorities and others globally remain conservative and applicable.

**Acknowledgments** The authors wish to acknowledge the support provided by Arup to complete this work and extend their sincere thanks to Sandra Weddell, Rob Duff and Howard Wong (London Underground Ltd), Charles Harmer (Cross London Rail Links Ltd), Brett Little (TfL Street Management) and Alan Kerr (Beca, Auckland) for their assistance. The views expressed are those of the authors alone, as are the analytical interpretations.

## References

1. Baker, D.: Moving Backwards – Canada's State of Transportation Accessibility in an International Context. Council of Canadians with Disabilities (2005)
2. Harper, S.: Prime Minister Harper Outlines Agenda for a Stronger, Safer, Better Canada, www.pm.gc.ca. Ottawa, Ontario (6 February 2007)
3. Barry, J.: Bus Networks: Planning for Growth in London. Proceeding of the European Transport Conference/ Association of European Transport (2002)

4. TCRP Report 100: Transit Capacity and Quality of Service Manual (2nd Edition). Transportation Research Board (2003)
5. Pearce, L., Powell, P., Duff, R., Kerr, A. (LUL): Passengers with Reduced Mobility: Modeling Their Behavior in Underground Stations. Proceedings of the European Transport Conference (2008)
6. Wong, H.: Station Modeling With Legion – Best Practice Guide v2, LUL (2009)
7. Transport for London Business Plan 2006/7 – 2009/10. LU PJ21, p.156.
8. Fruin, J.J.: Pedestrian Planning and Design, (Revised Edition). Elevator World, Inc., Mobile, AL (1987)
9. Berrou, J.L., Beecham, J., Quaglia, P., Kagarlis, M.A., Gerodimos, A.: Calibration and Validation of the Legion Simulation Model Using Empirical Data. In: Waldau, N., Gattermann, P., Knoflacher, K, Schreckenberg, M. (eds.), Pedestrian and Evacuation Dynamics 2005. Springer (2007)
10. Legion Studio 2006 EP3: User Manual. Legion International Ltd (2007)

# Prospects for the Design of Cognitive Systems that Manage the Movement of Large Crowds

**N. E. Groner[1], E. Miller-Hooks[2], and L. Feng[2]**

[1] Department of Protection Management, John Jay College, City University of New York, New York, NY 10019, USA.
[2] Department of Civil and Environmental Engineering, University of Maryland, College Park, MD 20742, USA.

**Abstract** Environmental uncertainties and the complexities of human behavior create challenges in the design of adaptive crowd management systems. The authors suggest a computer-assisted management system based on principles of cognitive systems engineering, including distributed decision-making agents with variable goals, and feedback loops that facilitate adaptation. A computer-based tactical crowd advisor uses a behavioral model to suggest actions to the crowd manager based on the behavioral model's predictions of crowd movement. The crowd manager instructs security personnel and pedestrians, and observes the results, thus completing a feedback loop. An agent-based simulation is used to generate behavioral input so that the expected effects of physical and procedural features of alternative proposed systems. The basic principles of the cognitive systems approach adopted herein are illustrated using examples from an investigatory report of the Presidential Inauguration in Washington D.C. in January of 2009.

## Introduction

A cognitive systems approach can be used to guide the design of adaptive systems that manage pedestrian movement in complex situations with high levels of uncertainty. We use this approach to analyze and discuss how such adaptive systems could be designed and assembled for the management of crowds during a large scale public event. Cognitive systems are viewed as comprised of both physical artifacts (e.g., the built environment, computers, sensors, and communications hardware) and social artifacts (e.g., training methods, procedures and protocols). People *per se* are not components of cognitive systems; rather the cognitive systems are designed to support the goals of people [1,2]. These systems are assembled for the purpose of supporting the pursuit of the goals of various system users. In the context of crowd management, users are both the persons that comprise the crowd and the persons who manage the crowd. Users have various roles, and are geographically distributed. Feedback loops support tactical decisions that implement control strategies as situations evolve in unpredicted ways, for example, un-

R.D. Peacock et al. (eds.), *Pedestrian and Evacuation Dynamics*,
DOI 10.1007/978-1-4419-9725-8_64, 

foreseen medical emergencies, natural events such as thunderstorms, and intentional events such as acts of terrorism.

To illustrate the proposed approach, we examine relevant research and illustrate basic principles using examples from an investigatory report of the Presidential Inauguration in Washington D.C. in January of 2009 [3]. Some attendees with tickets failed to gain access, while others without tickets quickly gained access. No deaths resulted, but emergency responders would have had difficulties gaining access should there have been the need, and many attendees reported being frightened at the loss of control over their own movement.

A handful of modelers [4,5,6,7,8] have conceived architectures for on-line management and control of pedestrian and traffic movement. None of these works takes a cognitive systems approach, and, thus, they do not account for the possibility that the users will change the environment and their goals in response to situation awareness and instructions. Moreover, unlike in prior works, we discuss how individual behavioral modeling can be used to construct optimal instructions. We describe system architectures for both operational and tactical applications that can potentially provide adaptive controls to meet both pedestrian access and safety goals during events like the inauguration. We then describe an example of such a system, and how it can be used on-line to manage crowds and off-line to compare design alternatives.

## Attributes of a Cognitive System for Crowd Control

### *Cognitive systems support users' goals*

Conventional approaches for modeling crowds are based on a physical systems approach where pedestrians are viewed as physical objects, and pedestrian and traffic flow modeling is based on densities in static physical environments. Behaviors result from causal physical processes; people simply react to their environments in accordance to rules specified in the models. There is no need to consider the individual goals of people, although recent innovations in physical models base these rules on the presumed goals of individual agents. Moreover, such deterministic rules can be probabilistically applied, creating stochastic models.

In a cognitive systems approach, designs are based on supporting goals at both system and individual levels. As regards crowd management, the individuals in the crowd are users of the system, along with the persons who want to manage their behaviors. Systems are user-centered; the system is designed to support the overall goals of the people who use the system, although the goals of individuals are not optimized. Individuals in crowds pursue goals related to their reasons for being in the environment (e.g., moving towards a destination) and, when the need arises, people will also pursue goals related to self-preservation and the staying near and protecting valued persons and objects. Conflicting goals must sometimes be reconciled. For example, maximizing the pedestrian flow can conflict with the system goal of providing rapid access for emergency services and with the individual goal of rendezvousing with friends.

In our example of the presidential inauguration, the design of the crowd management system sometimes failed to support important individual and system goals. At the systems level, some attendees never managed to reach locations where they could view the inauguration, despite arriving early and following what few instructions were provided. Luckily, no fatalities or serious injuries were reported, despite reports that it would not have been possible to reach and remove casualties from many locations if they had occurred. At the individual level, there are numerous reports of persons who not only failed to view the inauguration, but were repeatedly frustrated and frightened by the extreme density and by a lack of information concerning how they might cope with the situation. Attendees reported that public safety personnel could or would not provide them with information relevant to their goals of finding restrooms, quicker access to the inauguration, escape from excessive crowding, or medical attention.

### *Cognitive systems support adaptation to unforeseeable scenarios*

Complex events, such as the presidential inauguration, inherently involve high levels of uncertainty. Unforeseen natural events (e.g., earthquakes, thunderstorms), technological events (e.g., vehicular accidents, chemical releases), and intentional acts of violence (e.g., fights, terrorist attacks) all contribute to inherent uncertainties. To improve efficiency and safety, designs that enable the adaptive management of pedestrians are needed where the events that affect the movement of people cannot be accurately predicted.

Problems with adapting to unforeseen situations were evident during the presidential inauguration. "Extreme overcrowding in the purple staging area...forced crowds [to use]...the northbound tube of the Third Street Tunnel..." preventing people in the tunnel from viewing the inauguration and impeding emergency access if it had been required [3, p. 6]. Similarly [3, p 6], a decision had been made that some silver ticketholders would be escorted to another entrance to relieve congestion, but "a miscommunication prevented this from occurring."

### *Cognitive systems facilitate shared understanding and feedback loops among distributed system users*

Coordinating and reconciling the pursuit of various goals is an important design challenge in managing crowds. Participants in crowded events have different roles and goals, and are distributed across a wide area. Adaptive decision-making requires an accurate and shared understanding of what is happening in such situations, and cognitive systems are designed to facilitate situation awareness by providing information to distributed decision-makers to maximize efficiency and effectiveness [9].

Crowd managers at the presidential inauguration had some situation awareness, but pedestrians in the crowd had little idea about what was happening. The infamous "purple ticket holders" have vociferously complained about a lack of information and the inability of public safety personnel to provide any useful information. A congressional report recommends that signage, public address systems, and personal communications from "'way finders,' volunteer staff and law enforcement officials equipped with loud speaker capability" provide much better information to crowd members at future inaugurations [3, p. 3].

Good situation awareness requires relevant accurate inputs that can come from a variety of sources, including embedded informants, such as security personnel. Video and sensor-based technologies, among other technologies, along with automated application of inference algorithms for crowd density and movement monitoring and traffic flow property estimation, can provide information about pedestrian flows. In the future, there may be opportunities to use cellular phones to calculate movement even where other technologies, e.g., CCTV cameras, are unavailable. Force sensors can be used to alert authorities where excessive dynamic pressure is exerted against crowd control barriers.

Computer programs can support centralized control functions by interpreting raw data gathered by sensors and informants, and presenting that information in understandable ways. This information can be used to make decisions about changing the availability of routes, moving barriers and gates, and changing message signs that suggest alternative paths that can improve pedestrian's travel times. The choice of routes for emergency vehicles can be dynamically altered according to distance and crowd densities.

## Designing a System Architecture for Adaptive Control of Large Crowds

We suggest the use of a cognitive systems approach to design a computer-based adaptive control system to facilitate crowd movement during large-scale events. The system will exploit sensor, video and other widely available technologies, along with inference algorithms for data interpretation, for improved situation awareness with respect to both the physical infrastructure and crowd and personnel movement. Existing available communication technologies will provide a conduit for the provision of continuously updated information and guidance to pedestrians (i.e., people in the crowd) and security personnel. While designed for on-line application, the proposed system will have applicability in off-line assessment of potential crowd control strategies during pre-event planning stages.

## *System architecture and framework*

The capabilities of the proposed system stem from two core functions: (1) real-time monitoring and prediction of the state of the physical infrastructure and growth and movement of the crowd, conditions which evolve dynamically in response to external forces, and (2) development of updated behavior-based optimal instructions created with the use of system state information and on-line predictions. Information flow required to support these functions and associated feedback mechanisms are depicted in Figure 1.

Sensors, video, devices that facilitate asset tracking, and other technologies can provide real-time data related to crowd movement and growth or contraction, permit the monitoring of the physical environment, and can be used to ascertain the location of security personnel and equipment. Additionally, security personnel and other embedded informants can provide input. As data is received, sensor and video inference algorithms (the Sensor-Driven Monitoring and Prediction Component) can be applied to quickly produce desired updated information and predictions. The continuously updated information on the dynamically evolving state of

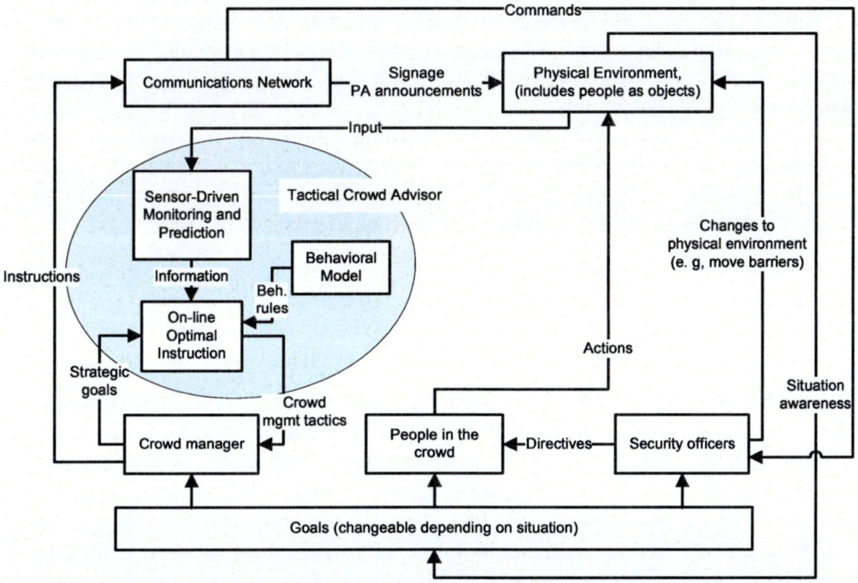


**Fig. 1. System architecture and information flow**

the physical environment, crowd location, density and size, and personnel activities, provide input to the On-line Optimal Instruction component. This component exploits this information and a network flow-based representation of the system in applying behavior-based optimization techniques to provide optimal control strategies (i.e., on-line instructions). Behavior rules associated with individual decision-making and group dynamics, such as, for example, propensity toward flocking, following, repulsion, attraction, and splitting, are embedded within the algorithmic steps of the On-line Optimal Instruction component. Crowd management suggestions developed by the On-line Optimal Instruction component are provided to the person or group in command (e.g., the Crowd Manager). The Crowd Manager can provide additional input to the On-line Optimal Instruction component in the form of strategic goals, completing a feedback loop. The Crowd Manager implements the control tactics through the Communication Network, communicating commands to security officers and directives to people in the crowd. This completes an information loop within the adaptive control framework.

The pedestrians respond to updated information from the command center, directives from security personnel, and changes to the physical environment that may arise naturally or as a result of actions taken to control crowd movement. Changes to the physical environment and crowd location and size, as well as asset locations, will be detected by the Sensor-Driven Monitoring and Prediction component beginning the next information loop.

The model shows pedestrians and security officers as separate agents that make decisions in response to changes in the environment. All agents, including the Crowd Manager, understand the environment according to their goals; that is, they have "situation awareness." Changes in the environment cause behavioral changes, as mediated by goals, which result in further changes in the environment, closing another feedback loop.

This system aids in alleviating over-crowding, improving flow rates, and regaining individual control of personal movement, thus, gaining greater efficiency from the existing physical capacity. Instructions that are provided are suited to individual priorities and goals, improving adherence to given directives.

## *Adaptations for off-line use*

A similar adaptive control framework can be employed in the off-line assessment of potential physical and procedural crowd control methods for use in pre-event planning. For this purpose, an agent-based simulation tool replicates the feedback gained through input from the pedestrian/crowd, physical environment, crowd manager and security officers. The agent-based simulation models individual responses to instructions, as mediated by personal goals and information within the individuals' visual fields. Agent-based simulation techniques have been proposed for use in off-line assessment of (i.e., pre-event planning for) crowd movement (e.g., [10,11,12,13,14,15,16]) and building evacuation (e.g., [17, 18]). Barros et al. [16] developed architecture to simulate crowds of virtual humans for on-line use, where an urban evacuation is called due to an accident involving petro-chemicals. In their approach, the scenario is played in the simulated world and actions are taken in the real-world based on predictions from the simulation. It is envisioned that the simulation component that would be used within the proposed cognitive and adaptive control framework would exploit agent-based modeling techniques from PetroSim.

Within the simulation component, every agent (representing a pedestrian, group of pedestrians, or security officer) is simulated as an autonomous entity with its own decision-making module. As in PetroSim, decisions are made based on status (e.g., agent profile, health situation, movement ability) and knowledge within a perception radius. Knowledge can be gleaned about the nearby environment and agents or may be pre-existing (e.g., knowledge of a nearest route). Unlike in PetroSim, feedback loops and optimal control strategies are incorporated. Instructions to pedestrians from the communication system or directly from security personnel can also be simulated. Each agent makes its own decisions as to how to respond to Knowledge, Status and Instructions that are received. Individual decision-making and behavior models (e.g., flocking, following, goal changing, and avoiding) can be embedded.

The framework for using the agent-based behavioral model is shown in Figure 2. The agent-based behavioral model generates feedback that would have been gathered by sensors and embedded informants in the on-line mode. By simulating externally induced changes to the physical environment, the Simulation component can also provide input that would have been received from the Sensor-Driven Monitoring and Prediction component; thus, providing input to the Online Optimal Instruction component.

The On-Line Optimal Instruction component exploits information from the behavioral model, and suggests instructions than can be relayed by Crowd Manager to guide pedestrian behaviors that improve system performance. Instructions to pedestrians could include moving barriers and gates, changing messages provided on variable message signs, and providing guidance via security officers. Strategies developed by the On-line Optimal Instructions component take into consideration simulated individual behavioral responses to the instructions. Agents in the simulation make decisions (e.g., to take a least time path to their destinations) that benefit only themselves and may not serve to optimize system goals.

## Conclusions

A cognitive systems approach to managing the movement of large crowds shows promise by its explicit recognition that decisions are distributed, not centralized, and may reflect competing goals. Adaptations are a function of understanding how unpredicted events can be expected to affect the behaviors of people in crowds, and that the effects of interventions need to be observed using feedback loops. The suggested system incorporates these features, and can be used both online to manage crowds and off-line to examine the expected effectiveness of various physical and procedural design features.

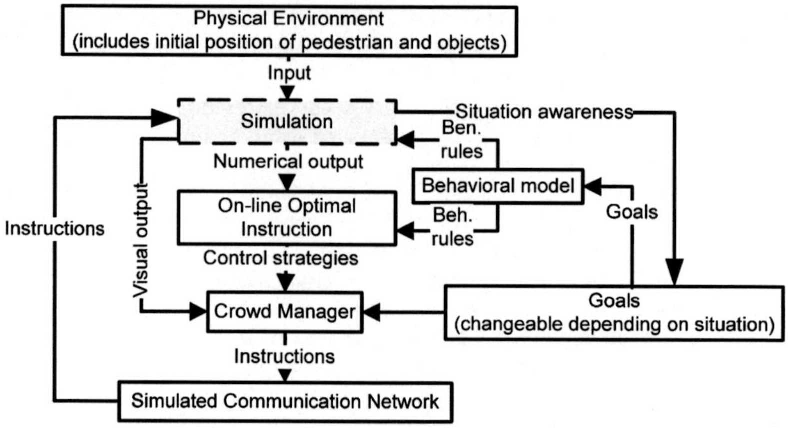


**Fig. 2. Adaptive control framework for off-line use**

## References


1. Groner, N.: Two Systems Safety Analyses Useful in the Design and Management of Security Systems. Journal of Applied Security Research. 3 (3/4), 351 – 372 (2008)
2. Hollnagle, E., Woods, D.D.: Joint Cognitive Systems: Foundations of Cognitive Systems Engineering. Taylor and Francis, Boca Raton, FL ( 2005)
3. Executive Summary of Findings: Multi-Agency Response to Concerns Raised by the Joint Congressional Committee on Inaugural Ceremonies for the 56th Presidential Inauguration (2009) Retrieved January, 24, 2010
4. Sugiman, T., Misumi, J.: Development of a New Evacuation Method for Emergencies: Control of Collective Behavior by Emergent Small Groups. Journal of Applied Psychology. **73**(1), 3-10 (1988)
5. Nakamiti, G., Freitas, R.: Adaptive, Real-Time Traffic Control Management. International Journal of Automotive Technology. **3**(3), 89-94 (2002)
6. Kamel, M.I., Fkry, M., Mashat, A. et al.: Monitoring, Surveillance and Control of the Crowds in the Holy Sites Using SCADA system. The Seventeenth National Computers Conference, Kingdom of Saudi Arabia (2004)
7. Liu, H.X., Ban, J.X., Ma, W. et al.: Model Reference Adaptive Control Framework for Real-Time Traffic Management under Emergency Evacuation. Journal of Urban Planning and Development. 133(1),43-50 (2007)

8. Toth, C., Grejner, D.A., Moafipoor, S.: Pedestrian tracking and navigation using neural networks and fuzzy logic. Proceeding of IEEE International Symposium on Intelligent Signal Processing (2007)
9. Endsley, M.R., Bolte, B., Jones, D.G.: Designing for situation awareness: An approach to user-centered design. Taylor & Francis, New York (2003)
10 Musse, S.R., Garat, F., Thalmann, D.: Guiding and Interacting with Virtual Crowds in Real Time. Proceedings of Workshop Eurographics Computer Animation and Simulation. Milan, Italy (1999)
11. Musse, S.R., Thalmann, D.: Hierarchical Model for Real Time Simulation of Virtual Human Crowds. IEEE Transactions on Visualization and Computer Graphics. **7**, 152-164 (2001)
12Ulicny, B., Thalmann, D.: Crowd Simulation for Interactive Virtual Environments and VR Training Systems. In: proceeds of Eurographics Workshop on Animation and Simulation, 163-170. Springer-Verlag (2001)
13. Ulicny, B., Thalmann, D.: Towards Interactive Real-time Crowd Behavior Simulation. Computer Graphics Forum. 21, 4, 767-775 (2002)
14. Cherif, F., Djedi, N.: A Framework to Simulate the Evacuation of A Crowd in Emergency Situations. Georgian Electronic Scientific Journal: Computer Science and Telecommunications. **1** (8), 17-27 (2006)
15. Cherif, F., Djedi, N.: An autonomous and guided crowd in panic situations. JCS&T. **7**(2), 134-140 (2007)
16. Barros, L.M., Da Silva, A.T. and Musse, S.R.: Petrosim: An architecture to manage virtual crowds in panic situations. Proceedings of Computer Animation and Social Agents, 111-120. ACM Press (2004)
17 MacDonald, M.: STEPS-Simulation of Transient Evacuation and Pedestrian Movements: User Manual. Croydon: STEPS-Simulation Group-Transportation. Mott MacDonald (2006)
18. Galea, E.R., Lawrence, P.J., Gwynne, S. et al: Building EXODUS v 4.06 user guide and technical manual. Fire Safety Eng. Group, University of Greenwich, London (2006)

# Risk Management at Major Events - Study of Behavioral Aspects and Implementation into the ASERI Microscopic Evacuation Model

**R. Könnecke and V. Schneider**

IST GmbH
60325 Frankfurt, Germany
Corresponding author: r.koennecke@ist-net.de

**Abstract** EVA is the acronym for an interdisciplinary research project addressing risk management of major events. The focus here is on the behavioral aspects of crowd movement at annual fairs, festivals, large parades or sport events. As a first step, video recordings at accentuated places are taken and analyzed. The respective findings are next included into the microscopic evacuation model ASERI developed by IST GmbH. The so expanded model is now in the process of validation against available empirical data and will be used to investigate crowd movement in normal and perturbed cases including emergency scenarios. Emphasis is put on social interactions and collective effects like group formation and persistence and the interaction of crowd flow with technical provisions and organizational measures. Ultimate goal of the EVA project is the development of a guideline on the evacuation of large scale event spaces in cooperation with German municipal fire brigades.

## Introduction

Assuring human safety is one of the major challenges in managing large scale events. The lack of information on limiting factors like number of visitors, available space and behavioral aspects makes it difficult to establish reliable safety concepts. All the more this is true in the case of an emergency that might occur during such an event. Hence a joint research project titled EVA, sponsored by the German Federal Ministry of Education and Research, was established in order to define and analyze parameter relevant for safety concepts of major events. EVA project members are the German Fire Protection Association (vfdb), University Paderborn, Fraunhofer-Institute ICT Pfitztal, IFR Institute of Fire and Emergency Technologies Dortmund, IST GmbH, Christian-Albrechts-University Kiel, VdS Schadenverhütung GmbH and Weller & Herden GmbH. The focus of this paper is on the data collected up to now from large scale annual fairs and festivals, the analysis of this data with respect to human behavior and the implementation of the respective aspects into the microscopic evacuation model ASERI.

R.D. Peacock et al. (eds.), *Pedestrian and Evacuation Dynamics*,
DOI 10.1007/978-1-4419-9725-8_65, 

## EVA project

Risk assessment of large scale events is the goal of the EVA project. Based on data collected from large annual fairs, festivals, parades or sport events methods are derived to process and evaluate these data in order to define the relevant parameter for the planning and organization of such events. Sub-models for collective effects in crowd movement and social behavior are formulated and implemented into the microscopic evacuation model ASERI. Questions of special importance are the formation and persistence of social groups, collective effects and dynamical crowd patterns associated with high occupant density, counterflow and the effect of interruptions and disturbances in the usual crowd flow (e.g. introduced by mobile barriers).

The data to be evaluated consists of videos, photographs and reports either made available by archived documents from the fire brigades or by material collected during the project. Existing video analysis technique has to be expanded and adapted in order to obtain results suitable for model calibration and validation. Furthermore the analysis has to comply with the requirements of data integrity.

## The ASERI evacuation model

ASERI is a computer model designed to simulate egress movement of people in geometrically complex spaces of arbitrary size on a microscopic basis. Occupant movement is modeled continuous in space, based on the present individual orientation, unrestricted walking speed and available egress route choice options. Exposure to asphyxiates, heat and smoke can be considered if required.

The geometrical scenario (buildings, facilities or urban units) is defined in a hierarchical way, based on interconnected levels. Each level is further subdivided using enclosures, rooms, corridors, stairs, ramps, ladders, tiers and safe areas as basic units. Obstacles, seat rows, recesses or signage and lighting can be specified inside units. It is therefore possible to model very large and geometrically complex buildings or spaces with tens of thousands of occupants (agents) moving simultaneously. Examples for the application of ASERI to large scale events can be found in [1-2].

Figure 1 presents an instantaneous view on the area of the “Cranger Kirmes” during a hypothetical evacuation process. The Cranger Kirmes is an annual festival with a core area (as depicted in figure 1) of about 22000 $m^2$ and a perimeter of 1.5 km. The density in the walking area (streets) is about 2 persons/$m^2$, the maximum number of persons in the sojourn zones is about 11000, yielding a maximum total of about 55000 persons included in the simulation.

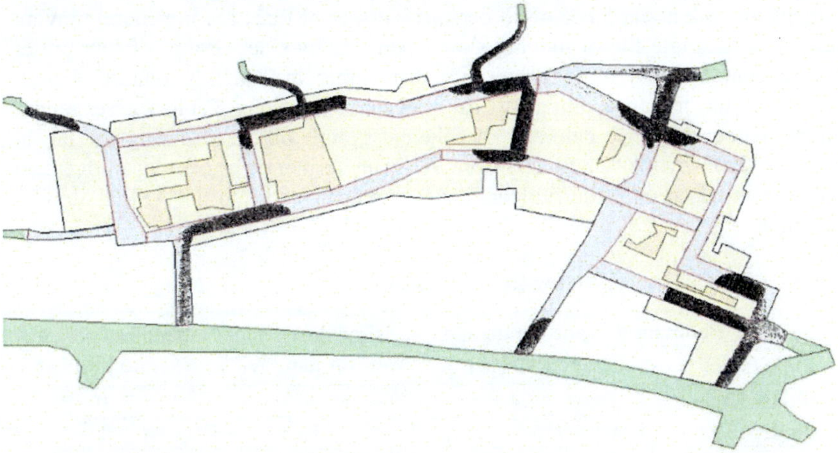

**Fig. 1. Snap-shot from an ASERI simulation of a large festival evacuation scenario.**

## Group behavior

Figure 2 summarizes observations made at the Cranger Kirmes 2009 by Dr. Oberhagemann (vfdb) within the EVA project. While in the density regime below 0.8 persons/m$^2$ the observed group speed is found to be in the region between 0.7 and 1.9 m/s, the speed at higher densities is substantially smaller (0.4 to 1.1 m/s).

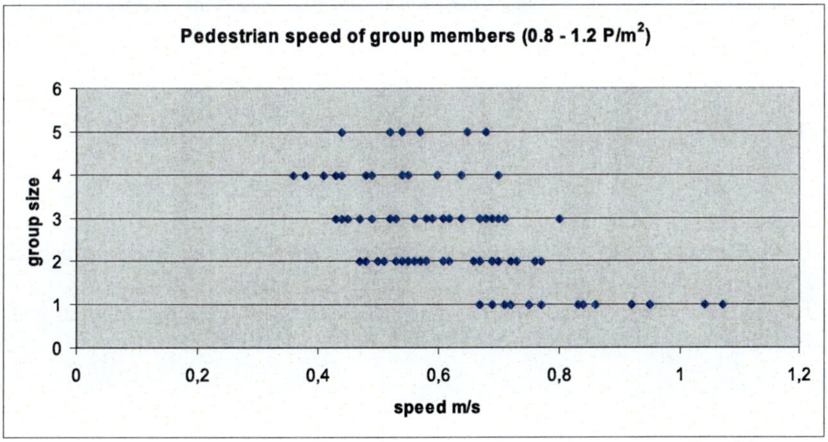


**Fig. 2. Group speed for occupant density between 0.8 and 1.2 persons/m$^2$.**

Groups of 2 are about 50 % of all persons. Groups of three are in general moving fast when in single file or moving slow when side by side. Groups of four prefer square contours that may temporarily dismember but quickly reunite. Larger groups are too few to obtain statistical relevant conclusions. Table 1 gives an example of the deference behavior of coherent groups observed during the fair at Soest in 2009 by Dr. Oberhagemann. A group formation and group movement mechanism based on these findings is now implemented into the ASERI simulation tool.

**Table 1. Group movement - deference vs. cohesion.**

| Behavior<br>Type (size) | accelerate and use gap [%] | slow down and wait for gap [%] | heedless penetration [%] |
|---|---|---|---|
| family (2) | 4 | 4 | 0 |
| family (3) | 4 | 15 | 1 |
| family (4) | 3 | 15 | 2 |
| male-male (2) | 4 | 14 | 2 |
| male-female (2) | 14 | 48 | 6 |
| female-female (2) | 10 | 10 | 7 |
| mixed (3) | 5 | 13 | 10 |
| mixed (4) | 3 | 21 | 8 |
| mixed (5 - 7) | 6 | 17 | 8 |
| total | 21 | 62 | 17 |

**Acknowledgments** This work has been supported by the German Federal Ministry of Education and Research BMBF. The authors thank Dr. Dirk Oberhagemann (vfdb) and all partners of the EVA project for their respective contributions and assistance.

## References

1. Schneider, V.: Simulating the Evacuation of Large Assembly Occupancies, Proceedings of the 2nd International Conference on Pedestrian and Evacuation Dynamics, pp. 319 - 330, University of Greenwich, London (2003)
2. Hosser, D., Forell, B. and Schneider, V.: Sicherheit in deutschen Stadien und Arenen - eine Nachlese, vfdb-Zeitschrift **1**, pp. 33 - 42 (2007)

# Posters

# Crowd Management Based on Scientific Research to Prevent Crowd Panic and Disasters

**S. Al Bosta**

Central Directorate of Development of Projects, MOMRA- Riyadh/ Kingdom of Saudi Arabia
Corresponding author: salimalbosta@web.de

**Abstract** This paper presents the developed concept of management and control to the pedestrian flow movements at Jamarat in the regular annual pilgrimage to Makkah (Hajj) season in Saudi Arabia. Every year, 3 to 4 millions of pilgrims perform their rituals in the course of extensively high restrictions, in the midst of a climax of limited /narrow space and time constraint. The Jamarat where pilgrims gather to perform a ritual stoning of pillars symbolizing the devil as part of the Hajj. The new Jamarat leveled building replaced the old ones. The project objective is to prevent crowd panic and to minimize the risk of crowd disasters. Management and control of pedestrian group movement to and/or inside Jamarat leveled building and area, using new experimental knowledge methodologies observed from the science of crowd dynamics, throughout anticipation and analysis of the pilgrim flow from low crowd density to extremely high crowd density, accompanied, attended, and escorted with an insider real-time-life video- scrutiny/ observation/analysis.

## Introduction

Hajj is the annual pilgrimage to Makkah in Saudi Arabia for each Muslim once a lifetime, whoever is able to find their way to it. Every year 3 to 4 millions of pilgrims are performing their rituals in the course of high restrictions with a climax of limited /narrow space and time constraint. The event has witnessed several catastrophic crowd tragic and disasters in the past. Jamarat is one of the focal points during Hajj, where millions of pilgrims must perform the "stoning of the devil" symbolic ritual on three successive days within 24 hours each day. About 3 Million people pass Jamarat area, with dimensions of about 100m width and 1000m length. The Ministry of Municipal and Rural Affairs in Saudi Arabia is currently constructing a new five- storey building for Al-Jamarat stoning rituals & practice

R.D. Peacock et al. (eds.), *Pedestrian and Evacuation Dynamics*,
DOI 10.1007/978-1-4419-9725-8_66, © Springer Science+Business Media, LLC 2011

[1]. The Jamarat project objective is to prevent crowd panic and to minimize the risk of crowd disasters (Fig 1).

**Fig. 1. Jamarat Building in Mina Makkah at Hajj 2009**

## *Concept of management and control to the pedestrian flow movements:*

### Planning and design of new Jamarat Building and infrastructures for pedestrians' roads

The aim of the new design of the Jamarat is to reduce critical overcrowding and prevent catastrophic stampedes. A multiple accesses points with 12 entrances and 14 exits in the design of the new Jamarat building should help to disseminate the human mass in acceptable flow paths leading to the different five levels in smoothly and safety way. The ground and first floor of Jamarat serve the pilgrims mass coming from the eastern side (Mina camps). The second floor is dedicated to cater for the western side (Makkah). The third floor should accommodate the pilgrims coming from the North (Moiesem). The fourth floor serves pilgrims from the southern side (Al Azizah) [1 & 2].The paths for pilgrim's streams are installed in one-way system. The pedestrians' roads with fences and clear sings direct the flow to there dedicated Jamarat levels and back routs (Fig. 2).

**Fig. 2. Pedestrians' flow in Jamarat plaza at Hajj 2009**

## Operation and management plan

Controlling movable and fixed cameras are installed for the monitoring and control of the pedestrian's flow inside the Jamarats building, surrounding areas, and in the persuading roads [4 &5]. Around 700 cameras provide real life–time photography about the situation on the main streets, avenues, and the Jamarat building for the operators' team for control, direction and redirection of the human streams. Together with, there are 90 fixed cameras for count up, measurement and calculation of pilgrims' numbers, density and flow speed. The cameras and software program are used to control the capacity of streets, plazas and Jamarat area. In defended points, the average number of people passing to Jamarat from all monitored directions per minute can be counted and analyzed [6]. The sum of mass flow in & out of Jamarat for all entrances, exits and for each Jamarat level can be added, by comparing the limit capacity. The measures based on the scientific research helped the Hajj run smoothly. Image-recognition software now tracks the flow of pilgrims and warns organizers to slow the arrival of pilgrims or to re-direct to the site when capacity limits approaches a critical value [5]. Scenario plans are prepared for the emergency case and exercised in advance.

Scheduling programs of registered pilgrims groups for Jamarat stoning have been implemented [7]. All 2 million registered pilgrims get a timetable and an assigned route in order to distribute them uniformly in the different Jamarat building levels in a timely manner. More-than-a-million unregistered participants irritate the scheduling plan. Awareness and guidance programs for pilgrims are entrenched. The last 4 years the pedestrian's stream ran comfortably all the way.

**Analysis of the real-time-life video**

During the Hajj in the last 4 years, the video-tracking software [6] was developed and used to measure the flow, and to total the number of pilgrims on the entrances/exits of Jamarat levels as well as on most roads leading to Jamarat. The video-analysis software has been improved, updated, and made increasingly accurate [8 -11].

Management and control of pedestrian pressure group movement, using new experimental knowledge methodologies observed from the science of crowd dynamics, throughout anticipation and analysis of the pilgrim flow from low crowd density to extremely high crowd density, accompanied, attended, and escorted with an insider real-time-life video- scrutiny/ observation/analysis. Through the evaluations of video recordings, improvement measurements for corridors, bottleneck areas, and intersections are achieved. The results from simulation models of pedestrian flows should be the avoiding of the long waiting times for people in the back, and shock waves due to impatience in cases of emergency evacuation.

## *Human Behavior in Crowd Pressures and Panic*

The computer simulations model of the real life-time video recordings [6] of the mob stampedes based on distances, sizes and velocities instead of emotional states but produces results similar to actual panics. When conditions of the crowd become critical, the flow drops significantly, this can cause stop-and-go waves and a further increase of the density until critical crowd conditions are reached. Then, "crowd turbulence" sets in, which may prompt shock waves and crowd disasters. The average individual speed does not go to zero even at local densities of 8-10 persons per square meter. The crowd can experience sudden changes in form of shock waves, crowd turbulence (crowd quakes) [4]. The pressure forces increases and add up, the bodies transmit the forces. In case of occurring barriers or obstacles in the climax of limited, narrow space and time constraint, additional dynamic reaction forces will be built. The stampede escalates quickly without any possibility to control or interferences.

With the scientific research of video analysis, it was possible to explain the physical behavior of the crowd critical movement from catastrophic stampedes in 2006 at Jamarat entrances where by then 364 people killed and 300 injured. The method of the "gas kinetic" simulations (crowd pressure) combines crowd density and the rate of change in the velocity of the flow. The critical overhang in crowd pressure correlates with the beginning of stop-and-go patterns and turbulence. The shock waves in the human mass can produce pressure strong enough to crush a person, break a brick wall or bend steel [9]. The behavior of the flow and mass movement cannot be compared with the fluid movement. The fluid cannot acknowledge the fear or pain. Nor could it chose a motion, and cannot trip up, collapse or fall. The fluid streams naturally fill up the 3-dimension space equally at the same time, opposite of irrational human mass.

In panic situations, people in high congestion stampedes cannot think about the next reaction or measurement. People create friction and violate personal space to save their lives. People's desire to leave faster, result in a contradictory outcome they leave slower instead hence the tragedy begins, the power and speed of each one swap into panic. Most Pilgrims at the begging of the mass belong to groups until the mass becomes with high density; they try to keep the group together once the turbulence and shock waves occur each individual try to save him self and concentrate on their lives, consequently victims collapse and are trampled creating further obstacles for others. The sudden use of blockades to stop the irrational movement of the moped mass can drive the people's hysteria.

The distinguished experience during pilgrimage within the flow of pedestrians to and from Jamarat, emphasizes and magnify one simple fact that the majority of pilgrims flow abide by the awareness of movement systems, road signs, and safety regulations regardless of their nationality, education, age, language and culture.

## References


1. Al-Abideen, H.Z.: The development of multistory Jamarat Bridge and surrounding area. 6th International Congress, 5-7 July, Global Construction, University of Dundee, Scotland (2005)
2. Serwill, D., Vollmer, R., Al Bosta, S., Tayara, A.: Design and Organization of the Jamarat Stoning Process at Hajj 1427-1430H. IVV (Ingenieurgruppe IVV), Aachen –Riyadh. 10 (2009)
3. Al-Abideen, H.Z., Al-Bosta, S.: Human crowd management, movement of pilgrims around Jamarat in Mina in Hajj 1426H. In: Workshop: Management and Control of Human Crowds, 5-8 Feb, Riyadh, Kingdom of Saudi Arabia (2006)
4. Al-Abideen, H.Z., Al-Bosta, S., Johansson, A., Helbing, D.: Scientific evaluation of videos showing hajjis into Jamarat area. Specialized Architectural Engineering Magazine. Ministry of Municipal and Rural Affairs, Kingdom of Saudi Arabia. 10 (December 2006)
5. Johansson, A., Helbing, D., Al-Abideen, H.Z., Al-Bosta, S: From crowd dynamics to crowd safety: A video-based analysis. Advances in Complex Systems. 11(4), 497(527) (2008)
6. Johansson, A.F.: Data-Driven Modeling of Pedestrian Crowds. Doctoral Thesis, Faculty of Traffic Sciences, Friedrich List "Technische Uni Dresden" (2009)
7. Helbing, D.: Scheduling, Management, and Safety of Human Crowds. In: Workshop Jeddah, Central Directorate for Development of Projects, Dec. 4-

11, Riyadh – Kingdom of Saudi Arabia. Ministry of Municipal and Rural Affairs, Jeddah (2005)
8. Helbing, D., Johansson, A., Al-Abideen, H.Z.: The dynamics of crowd disasters: An empirical study. Physical Review E 75. 046109 (2007)
9. Helbing, D., Johansson, A., Al-Abideen, H.Z.: Crowd turbulence: the physics of crowd disasters. In: The fifth International Conference on Nonlinear Mechanics (ICNM-V), pp. 967-969. Shanghai (2007)
10. Helbing, D., Johansson, A., Al-Abideen, H.Z., Al-Bosta, S.: Evaluations of video data of crowds around the Jamarat Bridge and safety implications for mass events. European Conference on Complex Systems. Dresden (2007) (on CD-ROM)
11. Al-Bosta, S.: Consequences of understanding of Jamarat stoning capacity. Workshop on the results of the program prepared for the scheduling and management of hajjis groups to Jamarat (2006)

# The Effect of Stair Width on Occupant Speed and Flow Rate for Egress of High Rise Buildings

**A. J. Blair and J. A. Milke**

University of Maryland
College Park, MD 20740 USA
Corresponding author: ablair2@umd.edu, milke@umd.edu

**Abstract** The current NFPA Life Safety Code and International Building Code have regulations of stairwell widths for high-rise buildings that are primarily based on a linear relationship of 7.6 mm of stair width per person based on the highest occupant load from a floor expected in the stairwell. This linear relationship between stairwell width and the speed and flow of occupants as they egress the building will be examined using data from four high-rise building evacuation drills that experience high density of occupants during egress and was collected by the National Institute of Standards and Technology.

## Introduction

Current U.S. model codes recognize a linear relationship between stair capacity and stair width, i.e. increasing the stair capacity by one person for every additional 7.6 mm. The International Building Code states that the minimum total width should not be less than the total occupant load serving that stair multiplied by 7.6 mm per occupant but specifies that each stair must be at least 1.12 m wide unless it serves less than fifty people [1]. NFPA101 states if the cumulative occupant load is less than 2,000 people the minimum stair width should be 1.12 m and if the occupant load is greater than or equal to 2,000 people the minimum clear width of the stairs should be 1.42 m. This code also stipulates that the width be calculated based on the linear relationship mentioned previously [2].

Prior to the 1988 edition of the Life Safety Code, a step function was used to address the relationship between effective width of stairways and flow down stairs. This approach used a lane model for evacuation flow, rather than the linear relationship in the current codes. The earlier codes specified a minimum width for stairways to be 1.12 m, which accounted for two .56 m lanes of people that would flow down the stairs at the same time [3].

A lack of scientific evidence for the units of exit prompted researchers such as Pauls, Fruin, Predtechenskii and Milinskii, and Templer, to study evacuations of people from buildings using stairs. As a result of their research, correlations were developed to express the relationships between flow, density, speed, and width (or effective width) inside a stairwell. As an example, Pauls studied evacuation from

R.D. Peacock et al. (eds.), *Pedestrian and Evacuation Dynamics*,
DOI 10.1007/978-1-4419-9725-8_67, 

high rise buildings as well as crowd movement and concluded that there was a linear relationship between flow and effective width of stairs. Pauls' findings indicated that for every increase in stair width, a corresponding increase in the flow rate was observed. Both Pauls and Fruin used their research of evacuation to estimate the minimum width for stairs. Pauls recommended a minimum nominal width 1.40 m while Fruin recommended a slightly larger minimum width of 1.52 m. Both of these estimates accounted for two lanes of people with 0.56 m shoulder width. Fruin added 0.1 m per person on each side to account for lateral body sway, while Pauls' recommendation was based on a 0.1 m lateral body sway by each person towards the outside of the stairwell and 0.1 m in between the two lanes. While these empirical recommendations resulted in changes being made to model building codes and the Life Safety Code, they are based on data that relates to observations and demographics from the 1970s [3].

This study will allow the evacuation community to gain insight into whether the current population of the U.S. behaves as people did thirty years ago and also gather quantitative data for high density evacuations with varying stair width to determine whether the current model code requirements accurately portray the evacuation needs of the occupants to safely and timely egress the building.

## High Rise Building Evacuation Data

The data that will be examined and analyzed for this research will be data pertaining to high-rise office building evacuations that were observed by the National Institute of Standards and Technology (NIST). This data was collected by video cameras located throughout the stairwells during fire drill evacuations of these office buildings. Using the data collected by each camera, NIST was able to measure the time each occupant exited the stairwell, the time each person passed a camera in the stairwell, and the floor of entry for each individual [4]. As a previous intern at NIST, Blair helped to collect some of this data from the video cameras. As each occupant exited the building, they were identified through a pixilated image by the color/type of clothing they were wearing. This occupant was then tracked and the times were recorded when they entered and exited the stairwell on every other floor above the exit floor. For instance if an occupant entered the stairwell on floor 9, they could be seen and the times relative to the fire alarm was noted on floors 7, 5, 3, and the exit floor 1.

As of April of 2009, NIST has collected data from fire evacuation drills of eight high-rise office buildings [5]. In each of the buildings, there was a range of the number of occupants who participated in the drill. Since this research will examine speed and flow with respect to the width of stairwells, only the buildings that experienced high densities of people during egress will be considered.

Of the eight buildings from which data was collected by NIST there were only four buildings in which density of people on the stairwells was considered to be high [5]. Considering these four buildings, NIST collected egress data from a total of ten stairwells. Two of these ten stairwells were from a 24 story office building and they were both 1.12 m wide. There was data gathered from two stairwells that were 1.27 m wide from a ten story office building. Data was collected from four stairwells of 1.12 m in width from an 18 story office building. Also, data was collected from two stairwells of 1.37 m width from a 31 story office building [5]. While all four buildings experience high densities during egress, they vary in both height as well as stairwell width.

## Data Analysis

Once the data from these four buildings are released to the public they will be analyzed. Since the data only contains descriptions of each occupant as well as the times they pass the cameras, the local speeds will need to be determined from the data. NIST also measured the stair riser height and tread depth, so the distance the occupant travels from floor to floor can be determined [4]. Since the time can also be calculated between these cameras, the local floor-to-floor speeds can be found for each occupant by dividing the distance that an occupant traveled by the time it took for them to travel this distance. The density (person/m$^2$) around each occupant will also be determined. Using this information, flows can be calculated for each occupant. These numbers can then be compared between the four different buildings to determine if stair width has an impact on either speed or flow during highly dense evacuations from high-rise buildings. This study will also examine and compare both floor-to-floor speeds as well as overall egress speeds of occupants descending the stairs of varying width.

Quantitatively this research will examine if Pauls' perception that flow rate is linearly proportional to stair width is an accurate approach. For example, this research will seek to explore if an increase of 15-25 cm in stairwell width results in an increase the flow rate of individuals evacuating or do individuals exude the same speeds and behavior? As the data has yet to be available from NIST, this research is ongoing, and thus more information will be provided about the results from this study on the poster during the PED Conference.

## References

1. International Building Code. International Code Council (2009)
2. NFPA 101 Life Safety Code Handbook, 11th Edition. Ron Cote (ed.), NFPA, Quincy (2009)
3. Pauls, J.L., Fruin, J.J., Zupan, J.M.: Minimum Stair Width for Evacuation, Overtaking Movement and Counterflow- Technical Bases and Suggestions for the Past, Present, Future. In: Waldau, N., Gattermann, P., Knoflacher, H.,

Schreckenberg, M. (eds.), Pedestrian and Evacuation Dynamics, pp. 57-69. Springer (2005)

4. Peacock, R.D., Averill, J.D., Kuligowski, E.D.: Stairwell Evacuation from Buildings: What We Know We Don't Know. In: 4th International Conference on Pedestrian Evacuation Dynamics, University of Wuppertal, Germany (2008)
5. Kuligowski, E.: Building Evacuation Data Collection in Tall Buildings. Presented at NIST Annual Fire Conference- People in Fire Workshop, Gaithersburg, MD (April 2004)

# Using Stereo Recordings to Extract Pedestrian Trajectories Automatically in Space

**M. Boltes[1], A. Seyfried[1], B. Steffen[1] and A. Schadschneider[2]**

[1] Forschungszentrum Jülich GmbH, JSC, 52425 Jülich, Germany
[2] Universität zu Köln, Institut für Theoretische Physik, 50937 Köln, Germany
Corresponding author: m.boltes@fz-juelich.de

**Abstract** For a proper understanding and modeling of pedestrian dynamics reliable empirical data are necessary for analysis and verification. Therefore we have performed a series of experiments with a large number of persons. For the time-efficient automatic extraction of accurate planar pedestrian trajectories we developed the program PeTrack. We now have extended the software to stereo recordings, which allows a direct height measurement without additional markers and to extract trajectories on stairs.

## Introduction

The available database for collective motion of pedestrians is small, inaccurate and highly contradictory [1]. To understand and to model pedestrian dynamics reliable empirical data is needed. We have performed an extensive series of well-defined experiments to study the movement of pedestrians in different situations. PeTrack has been developed to automatically extract trajectories from video recordings of movements on plane ground [2]. The program is able to handle lens distortion and high pedestrian density. To extract spatial trajectories in environments such as stairways we have extended the software to stereo recordings. This also allows the measurement of peoples' height directly without special marker, which was necessary to segregate the perspective distortion. The exact trajectories allow a detailed analysis of movement and verification of microscopic models in space and time [3]. The set of trajectories of all pedestrians provide data like velocity, flow, density and individual distances at any time and position, thus lane formation and local densities can now be analyzed [4].

## Experiments

Some of our new experiments have been designed to analyze the movement inside a stadium. Figure 1 (left) shows 300 spectators leaving a grandstand in a stadium

R.D. Peacock et al. (eds.), *Pedestrian and Evacuation Dynamics*,
DOI 10.1007/978-1-4419-9725-8_68, 

through one port along steep stairs. The colored shirts indicate the placement inside the block. The long exposure time strikingly shows the movement.

Stereo cameras were mounted on dollies overhead. One view of a stereo recording can be seen in figure 2 (left). With the two views of a stereo camera a height field can be calculated. Figure 2 (right) is the grey scaled one of the left view. To detect the complete routes of the persons we used multiple overlapping stereo cameras, synchronized to give continuous trajectories.

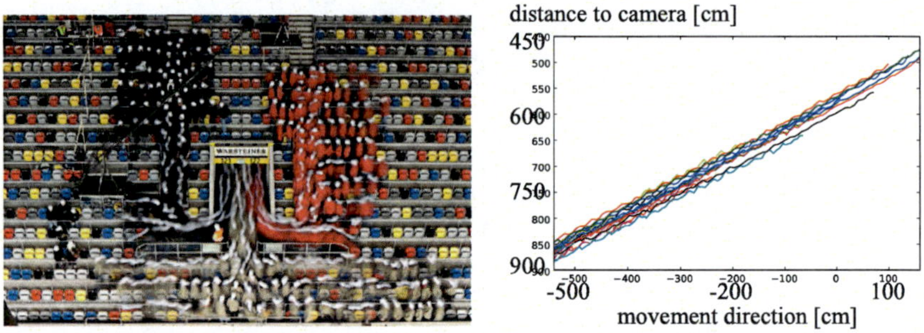


**Fig. 1. Left: 300 pedestrians leave a grandstand in a stadium through one port. Right: Trajectories from side view of 30 people moving downstairs.**

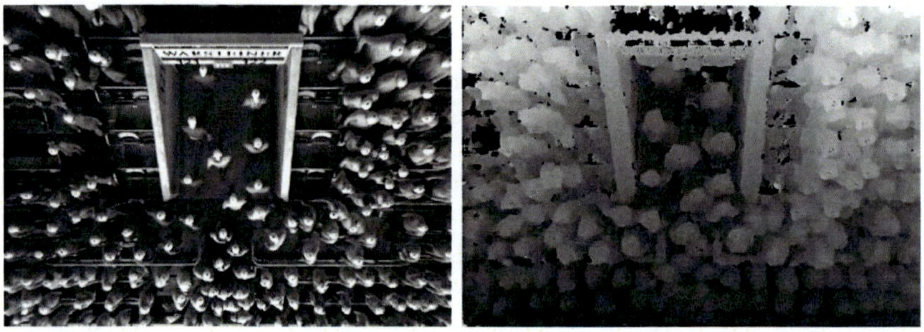

**Fig. 2. Left: One view of a stereo camera mounted overhead. Right: Grey scaled height field of the left view. For the black pixel a height value could not be calculated.**

## Results

The disparity map of the two views of a stereo camera with offset $b$ allows the calculation of the distance $h$ perpendicular to the camera of every pixel seen by both views: $h(d) = b \cdot f \cdot w / d$, where $f$ is the normalized focal length, $w$ the number of pixel in offset direction and $d$ the disparity between both views [5]. Figure 2 (right) gives the grey scaled perspective height field of the left view. The brighter the pixel is, the nearer the object is to the camera at that position. For the black pixel a height value could not be calculated.

To increase robustness, the detection of a person is done by markers. The position in space is calculated from the perspective height field. With normal video recordings we had to take into account the perspective view to get exact trajectories of pedestrians with different heights. Stereo recordings now directly provide the real position of the head. Figure 1 (right) shows the trajectories of 30 people moving downstairs from side view. The number of trajectories accumulates downstairs, because the experiment starts with the people standing on the stairs already.

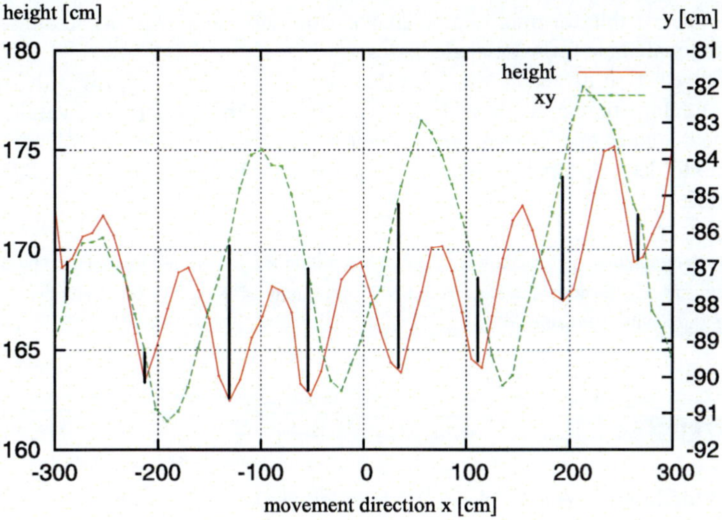


**Fig. 3. Trajectory of one person from top and side view showing the correlation between step sequence and height variation.**

The microscopic quality of the height measurement is precise enough to see the step sequence in the height variation. Figure 3 shows the oscillating movement of one person on a plane ground and the height variation along the same moving direction. The bars between the curves show the correlation between both values. The height is minimal at the position where body shifting moves from one to the other leg. At this position the height of a person with legs of length $l$ and step size $s$ is about $l - \sqrt{l^2 - s^2/4}$ lower than in straight position (here: ca. 5 cm). The frequency of the height variation is twice that of the step sequence.

Figure 3 also shows that the measured height is sensitive to incorrect rectification, because the cameras were mounted high in order to cover a large area. E.g. in 6.5 m distance like in the corresponding experiment a disparity of one pixel results in a difference of 20 cm in the measured height. In figure 3 the height is more than

5 cm greater at the cameras' boundary than directly underneath. Thus we only use the average of measured heights for plane experiments, which distribution matches exactly the distribution of the height of the pedestrians gained by the evaluation of questionnaires handed out to the test persons.

## Outlook

To get more accurate distance measurements also to the border the stereo views may have to move apart e.g. in form of separated cameras. Errors bigger than 5 cm are not acceptable for microscopic analysis, because the error will be the same in the motional plane for wide angle lenses.

The detection of people has been done by markers. This improves the robustness of automatic extraction of trajectories, but markerless detection would facilitate field studies and increase the amount of video data. Stereo recordings should be able to detect people by searching for forms of head and shoulders inside the height field.

**Acknowledgements** This study was performed within the project Hermes funded by the Federal Ministry of Education and Research (BMBF) Program on "Research for Civil Security - Protecting and Saving Human Life".

## References

1. Schadschneider, A., Klingsch, W., Klüpfel, H., Kretz, T., Rogsch, C., Seyfried, A.: Evacuation dynamics: Empirical results, modeling and application. Encyclopedia of Complexity and System Science, p. 3142. Springer (2009)
2. Boltes, M., Seyfried, A., Steffen, B., Schadschneider, A.: Automatic Extraction of Pedestrian Trajectories from Video Recordings. Proc. Pedestrian and Evacuation Dynamics 2008, p. 43-54. Springer (2010)
3. Steffen, B., Boltes, M., Seyfried, A.: Improved Methods for Checking Forces Based Models of Pedestrian Dynamics.
4. Liddle, J., Seyfried, A., Steffen, B.: Analysis of bottleneck motion using Voronoi diagrams.
5. Scharstein, D., Szeliski, R.: A taxonomy and evaluation of dense two-frame stereo correspondence algorithms, International journal of computer vision. Vol. 47, p. 7-42. Springer (2002)

# A Methodology to Calibrate Pedestrian Walker Models Using Multiple-Objectives

**M. C. Campanella, S. P. Hoogendoorn, and W. Daamen**

Delft University of Technology, Transport & Planning Department
Stevinweg 1, 2628CN Delft, the Netherlands
Corresponding author: m.c.campanella@tudelft.nl

**Abstract** In this paper we present a generic methodology for calibrating walking models and we discuss some results of its application. This methodology aims to describe all aspects of the parameter estimation process of walker models, while making it applicable to the calibration of any walker model. The base of the methodology is the simultaneous calibration of several aspects of pedestrian traffic to improve the robustness and generic application of the model. Applying the methodology to the Nomad model showed that combining different flow configurations indeed improved substantially the quality and significance of the parameters.

## Introduction

Each pedestrian has its own, unique, walking behaviour requiring different modelling outcomes to reflect this inter-pedestrian heterogeneity. One way to account for heterogeneity is to use a single walker model for the total population, but to assign different parameter sets for different pedestrians. Most models would allow for this approach and some authors estimated parameter distributions [1, 2] rather than average parameter values [3, 4]. A problem with calibrated models is the risk of specialisation. A model that is calibrated using a specific flow configuration may not perform well in other flow configurations. Ideally, walker models should perform well in all situations, but such a generic model is not available yet. However, current models may deal with parameter sets estimated for different flow configurations or a combination of those. Literature on calibration showed that no generally accepted methodology and guidelines for calibration of traffic models and in particular pedestrian models exists. The main contribution of this paper is a methodology that focuses on (but is not limited to) estimating parameters reflecting the inter-pedestrian heterogeneity. Furthermore, it also incorporates the ability of using simultaneously different flow configurations and different aspects of the pedestrian traffic in calibration. Another added value of this methodology is that it enables to compare the quality of different parameter sets. This then allows making the appropriate choice of parameter sets before validation procedures.

R.D. Peacock et al. (eds.), *Pedestrian and Evacuation Dynamics*,
DOI 10.1007/978-1-4419-9725-8_69, © Springer Science+Business Media, LLC 2011

## Generalised Calibration Methodology

Figure 1 shows the calibration methodology scheme where different calibration scenarios are used to estimate the model parameters. Each scenario receives the same parameter set $\theta$ and calculates the error $\varepsilon_n$. These scenarios represent those aspects of the traffic in which we are interested and the error represents the difference between the predicted and revealed aspects. For instance, the position of one simulated pedestrian can be directly compared with the *correct* position obtained from trajectories in the reference data. The error $\varepsilon$ would then be a measure of all the errors at certain moments along the trajectory. Another possible aspect would be the difference between estimated and revealed fundamental diagram relations. Any aspect that represents meaningful pedestrian behaviour can be used to build a scenario: instantaneous accelerations, distribution of headways, or bottleneck capacities. The various errors are combined in a multiple-objective function $\psi$. The combined error is then submitted to an optimisation algorithm that will use stopping criteria to assess the state of the optimisation. While the criteria are not met, the optimisation algorithm generates a new parameter set. This parameter set is applied to the model and new simulations are run within each scenario, repeating the cycle. Otherwise, the optimisation is finished. The optimised parameter set $\hat{\theta}$ is then submitted to a parameter analysis and if shown to be significant the parameter set $\theta^*$is considered good to be used or validated.

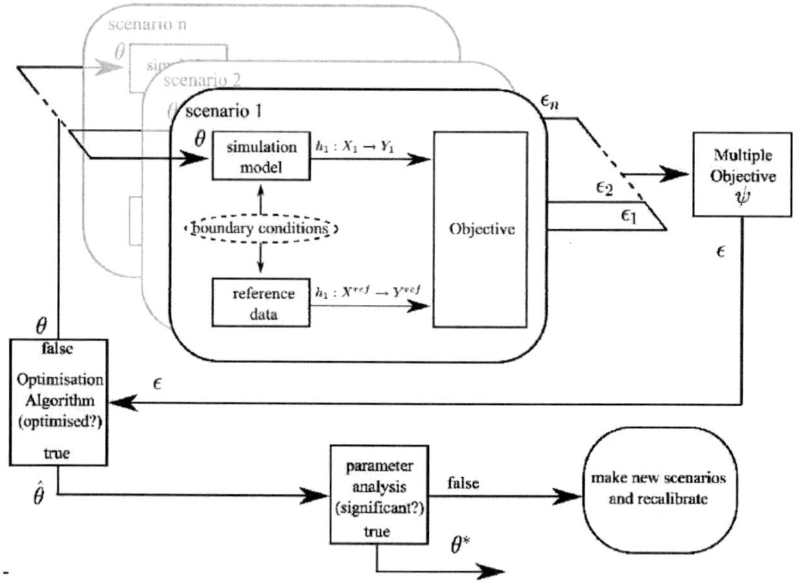


**Fig. 1. Calibration methodology scheme**

To obtain a model for generic application, several scenarios with different trajectories can be used simultaneously. The resulting parameter set will be the best compromise between the different flow configurations. A parameter that influences the outcome of the predictions very little is not significant or is irrelevant. The parameter is not significant if the reference trajectories do not contain enough information to estimate it properly. For example, suppose that the trajectories were collected in conditions of very low pedestrian densities. These trajectories will present a low probability of situations in which interaction between pedestrians occurs. Consequently, the model will be indifferent to the values of the parameters affecting the interaction behaviours. This situation is also referred to as *poorness* of data. A parameter is irrelevant if it does not influence the outcomes of any trajectories and consequently it should not be present in the model (not in figure1).

## Some Applications of the Methodology

In this section we investigated what is the influence of flow configurations for estimating correct parameters of the Nomad model. To answer this question we applied the methodology using synthetic trajectories in which parameters are known. We created a scenario that measured the errors of pedestrian accelerations simulated at regular intervals developed by [1]. Four parameters of the set $\theta$ were estimated: A0 and R0 are the parameters responsible for the interaction behaviour, T the parameter responsible for the pedestrians to stay along their intended path and AW is the parameter responsible for interactions with obstacles. Three different flow configurations were compared: a bidirectional corridor, a unidirectional flow with a narrow bottleneck and a 90° crossing flow. For each flow configuration we calibrated the model 25 times with one single trajectory (one pedestrian) chosen randomly. This resulted in a distribution of 25 optimised sets of parameters $\hat{\theta}$.

**Table 1. The results per flow configuration of four parameters estimated with the methodology and the mean of the minimum errors for the 25 calibrations**

| Parameters $\Theta$ | | | | | | | | | Error ε |
|---|---|---|---|---|---|---|---|---|---|
| | Mean | | | | Deviation | | | | |
| | **A0** | **R0** | **T** | **AW** | **A0** | **R0** | **T** | **AW** | **Mean** |
| **Correct values** | **10** | **0.16** | **0.25** | **20** | **0.7** | **0.02** | **0.04** | **0.0** | |
| Bidirectional | 8.5 | 0.17 | 0.27 | 6.1 | 1.5 | 0.03 | 0.07 | 7.5 | 434 |
| Crossing | 8.4 | 0.16 | 0.29 | 2.4 | 1.9 | 0.02 | 0.07 | 2.5 | 387 |
| Bottleneck | 7.2 | 0.15 | 0.40 | 16 | 2.0 | 0.03 | 0.20 | 10 | 1204 |
| Multiple-flows | 8.5 | 0.16 | 0.32 | 16 | 1.4 | 0.03 | 0.08 | 4.9 | 2442 |

Table 1 shows that the mean and deviations of the parameters A0 and T were better estimated using the bidirectional corridor and the crossing flows than the unidirectional flow. Furthermore, table 1 shows that the mean of the optimal error for these flows was significantly smaller in comparison with the bottleneck flow indicating that the trajectories were better estimated. However, the unidirectional flow permitted a better estimation of AW (some outliers caused the variance to be very large). For the bidirectional corridor and crossing flows this parameter was tested and shown not significant. This happened because the flow configurations presented insufficient situations in which pedestrians were close to obstacles. To improve this situation 25 new calibrations were performed with a multiple-objective function that simulated all three configurations simultaneously and added the errors. In this calibration three parameters were well estimated (except T was not good), which indicated that the richness of the individual flows was captured by the multiple-objective estimation (table 1).

## Conclusions and Future Work

In this paper we presented a generalised calibration methodology composed of calibration scenarios that represent aspects of pedestrian traffic. By defining calibration components these can easily be investigated and modified when parameters are not significant. An application of the methodology on Nomad showed that unidirectional flows produce less optimal parameters. Furthermore, non significant parameters can be estimated by combining the results of several flow configurations in multiple-objective estimations. The presented methodology greatly improves the confidence in the results of pedestrian walker models. Models calibrated according to it can be made more precise, general and robust. In the future this methodology can be applied on real trajectories allowing the study of heterogeneity and variations of behaviours in different flow configurations.

## References


1. Hoogendoorn, S., Daamen, W.: A Novel Calibration Approach of Microscopic Pedestrian Models. In: Timmermans, H. (ed.), Pedestrian Behaviour. Emerald Group, pp. 195-214 (2010)
2. Johansson, A., Helbing, D., Shukla, P.: Specification of the social force pedestrian model by evolutionary adjustment to video tracking data. Advances in Complex Systems. **10**(2): p. 271-288 (2007)
3. Campanella, M., Hoogendoorn, S., Daamen, W.: Improving the Nomad microscopic walker model. In: 12th IFAC Symposium on Control in Transportation Systems (CTS'09) (2009)

4. Steiner, A., Philipp, M., Schmid, A.: Parameter Estimation for Model. In: 7th Swiss Transport Research Conference (STRC) (2007)

# Influence of Doorway Width on Emergency Door Capacity

**W. Daamen and S. P. Hoogendoorn**

Delft University of Technology
Faculty of Civil Engineering and Geosciences
Department of Transport & Planning
PO Box 5048, 2600 GA Delft, the Netherlands
Corresponding author: w.daamen@tudelft.nl

**Abstract** New large-scale laboratory experiments have been performed to investigate the capacity of emergency doors during evacuation conditions. Varying doorway widths showed that only the experiment with the widest doorway (275 cm) resulted in a capacity lower than the capacity found in previous experiments (2.25 P/m/s). However, this was not due to the wide door, but to a population composition with less children.

## Introduction

In order to determine the capacity of emergency doors experimental research has been performed [1]. In these experiments, the relation between capacity and four independent variables doorway width, population composition, light intensity, the presence of an open door, and stress level have been investigated. This paper focuses on the relation between doorway width and capacity. In [1], the set up of the experiments is described in more detail, as well as the methodology to calculate the capacities and the relations between capacity and population composition and stress level.

## Doorway Width

During the experiments the doorway width has been varied between 50 cm and 275 cm. All these experiments have been performed with an average population (25% children, 55% adults and 20% elderly of 65 years and older), a normal light intensity (200 lux) and without the presence of an open door. Fig. 1 shows the results of these experiments.

For each repetition of the experiment, the observed capacity is shown in the figure. The type of marker indicates the stress level (no stress, slow-whoop signal or a combination of slow-whoop signal and stroboscope light), while the green

R.D. Peacock et al. (eds.), *Pedestrian and Evacuation Dynamics*,
DOI 10.1007/978-1-4419-9725-8_70, © Springer Science+Business Media, LLC 2011

star represents the average capacity per experiment over all stress levels. In addition, the current threshold capacity from the Dutch design guidelines has been indicated (C = 2.25 P/m/s = 135 P/min).

The figure shows that only the experiment with the widest doorway results in an average capacity lower than the threshold value from the design guidelines. Furthermore, the high capacity of the doorway of 220 cm is remarkable, as well as the large difference between the repetitions in this experiment.

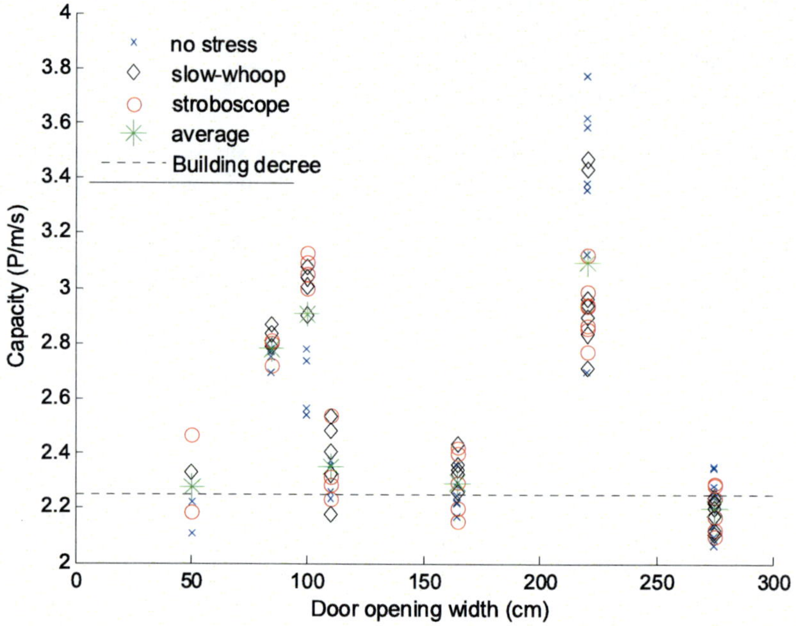


**Fig. 1. Capacity as function of doorway width under similar conditions**

During the day, we found that the capacities were higher than we expected while designing the experiment. So, when performing the experiment with the widest doorway, we increased the number of adults and elderly. Since the non-participating children were involved in an education program, they could not join the experiment. This led to a population composition with smaller share of children. As shown in [1], the presence of children increases the capacity, which implies that the changes in population composition led to the lower capacity. The high capacity found for the 220 cm wide doorway could be explained by looking at the time of day the experiments were performed, see Fig. 2.

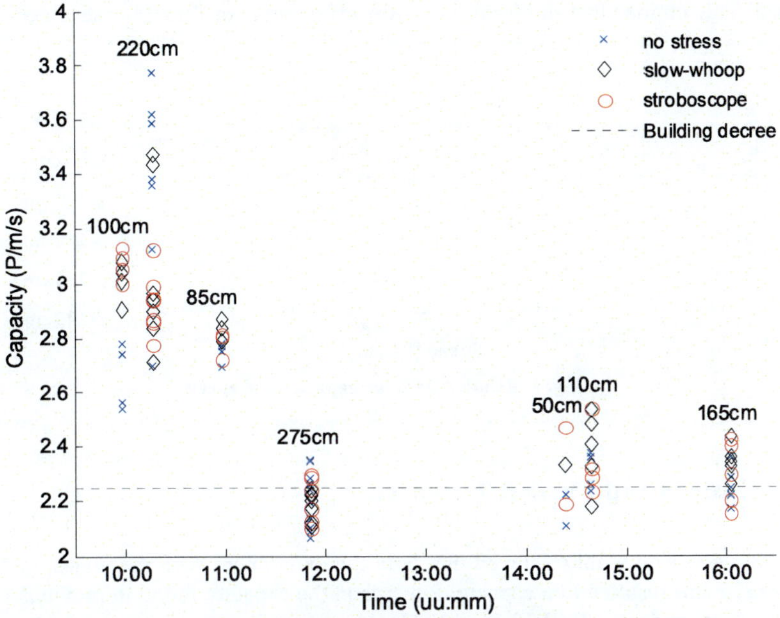


**Fig. 2. Capacity as function of time of day**

Fig. 2 shows that both experiments with the largest variance in capacity have taken place at the beginning of the morning. Moreover, the figure shows that the stress level has an opposite effect for both experiments: for the first experiment the capacity is lowest for the lowest stress level, while in the second experiment the highest capacities occur at the lowest stress level. The other experiments do not show such a clear effect of the various stress level. This leads to the conclusion that the difference is not structural and can be attributed to the conditions (enthusiasm) during the first experiments. Although the number and the distribution of the participants over the three groups (children, adults, elderly) are equal for all experiments, other persons participated. While the participants of the first experiment did not know what to deal with, the participants of the second experiment could wait and see what was happening. In the first repetitions (without stress and with slow-whoop respectively) this lead to pushy behavior, which was clearly visible in the video images. One of the important observations was the high speed of the participants. Fig. 3 gives an overview of the relation between capacity and speed for the different doorway widths (each marker indicates a single repetition). First of all, we see that the repetitions for each doorway width are clustered, implying that the conditions were similar for the different repetitions. Furthermore, the figure shows that when the speed is increasing, also the capacity increases. One of the explanations of the higher capacity is therefore the higher possible speed. This implies that the faster-is-slower effect was not found in these experiments.

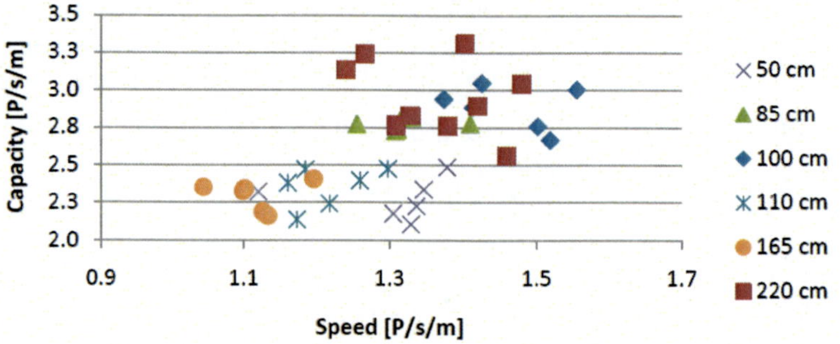


**Fig. 3. Relation between capacity and speed**

## Conclusions and Further Research

Varying doorway widths showed that only the experiment with the widest doorway (275 cm) resulted in a capacity lower than the capacity found in previous research with students [2] (2.25 P/m/s). The high capacity of a doorway of 220 cm (on average 3.09 P/m/s) is remarkable as well as the high variance in the repetitions of this experiment. This is most likely caused by the difference in behavior of the participants during the day. In the first experiments participants have a much stronger drive to pass the door than in the experiments later that day. This leads to more pushing and higher speeds. Further investigations showed that higher capacities occurred when the speeds were higher as well. This implies that in these experiments the 'faster-is-slower' effect did not occur.

In future research, we will focus on the individual behavior of the participants using the trajectory data derived from the video images. One of the aspects to focus on is the detailed behavior of persons passing the doorway (e.g. number of people passing simultaneously) and the density and speed distributions just upstream of the doorway. In addition, we will look at other parts of the total evacuation process (pre-evacuation, route choice, walking towards the exit), since this has a direct influence on the arrival pattern of pedestrians at the emergency door, and thus whether or not capacity of the door will be reached.

## References

1. Daamen, W., Hoogendoorn, S.P.: Emergency door capacity: influence of population composition and stress level. In: Proceedings of the PED2010 Conference, to appear (2010)
2. Peschl, I.A.S.Z.: Evacuation Capacity of Door Openings in Panic Situations. Bouw, **26**, 62–67 (1971) (In Dutch)

# An Example of Complex Pedestrian Route Choice

**F. Gräßle and T. Kretz**

PTV Planung Transport Verkehr AG
Stumpfstr. 1, 76131 Karlsruhe
Corresponding author: Tobias.Kretz@ptv.de

**Abstract** Pedestrian route choice is a complex, situation- and population-dependent issue. In this contribution an example is presented, where pedestrians can choose among two seemingly very similar alternatives. The choice ratio is not even close to being balanced, but almost all pedestrians choose the same alternative. A number of possible causes for this are given.

## Introduction and Motivation

Route choice of pedestrians in general is complex for a number of reasons. Pedestrians can stick to the links and nodes of the street network or they can ignore it and plan their path in 2d. A second issue is the set of determinant factors for route choice. While vehicle drivers strongly prefer the quickest route, for pedestrians it bears some attractiveness as well, but sometimes they may find a more scenic or safer route or one with paved tracks more attractive. After all there is one simplification for walking as mode of traffic compared to vehicle drivers: the quickest path most of the time only slightly deviates from the shortest, as pedestrian densities in public space normally are fairly small. The situation changes for events, where large crowds gather intentionally or in buildings [1, 2]. The first thing that comes to mind, why understanding pedestrian route choice would be necessary, is the planning of large crowd events. But understanding what is considered to be element of an attractive route also helps to build cities that are considered to be attractive by the local population and thus create activity [3, 4]. And knowing which elements are helpful for orientation and navigation enables city planners to assist tourists to find their own way through the city [5, 6]. If it is justified to assume that the observed pedestrians know about the attractiveness and utility of the available route alternatives, the observed route choices give the actually preferred routes and thus the resulting data can be used to calculated route suggestions that meet the expectations of a user of a pedestrian navigation device [7]. To assess the

R.D. Peacock et al. (eds.), *Pedestrian and Evacuation Dynamics*,
DOI 10.1007/978-1-4419-9725-8_71, © Springer Science+Business Media, LLC 2011

value of a shop location in new retail trade infrastructure it is crucial to know about preferred routes to estimate the number of walk-in customers [8, 9].

## Geometry and Situation

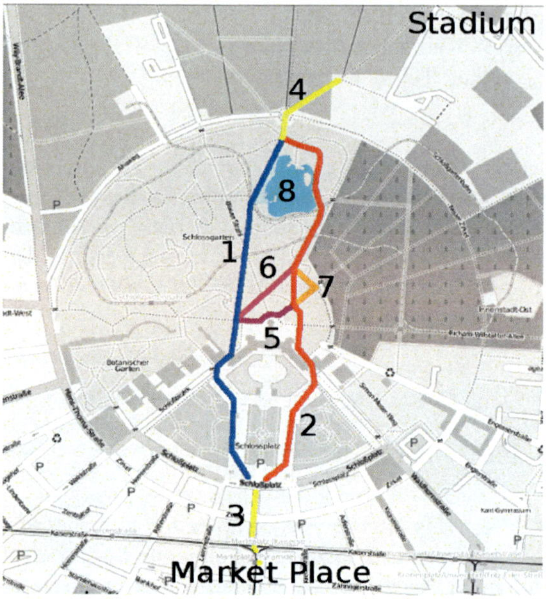


**Fig. 1. Location. [10, 11]**

Karlsruhe's main soccer stadium until today is neither connected directly to the tram system, nor does it have sufficient parking spaces. So, many fans decide to approach by tram and walk the remaining about 2 km to the stadium. One of the stations they alight from is“market place”. The market place is located almost exactly south of the central tower of Karlsruhe castle (at “5” in fig. 1), which itself is built symmetrically approximately 135 m each to east and west referred to the axis “castle tower to market place”. Therefore the castle poses an obstacle for pedestrians, who want to walk northbound from the market place to the stadium. Its function as an obstacle is weakened by one gate each in the castle's east as well as west wing about 77 m from the symmetry axis. These two gates are “fix points” for the soccer fans: only a few meters north of market place each fan has to decide, if he wants to pass the east or the west gate. Between the market place and the castle (i.e. south of the castle) lies the castle square. In the castle square there is no notable deviation from a perfect symmetry related to named symmetry axis.

This implies that the walking distance from market place to each of the two gates for practical purposes is identical. North of the castle the castle garden extends. There is a pond, which stretches a bit more to the east than the west ("8" in fig. 1). For this reason passing the pond on the west is the shorter and quicker route. East of the pond there is a paved walkway that leads northwards in a somewhat zig-zag-style. West of the pond there are no walkways, but a well trimmed meadow. At the north tip of the nearly circle shaped area of castle park and castle garden there is one single gate (the "north gate") exactly on the symmetry axis, which all fans have to pass. After this they have to turn north east to walk straightly to the stadium ("3" in fig. 1). From the point, where the fans have to decide, if they want to walk the eastern or the western route to the north gate, 895 m have been measured for the western route ("1" in fig. 1) and 926 m for the eastern route ("2" in fig. 1). The measurement has been done using a GPS device and simply walking both routes, trying to minimize the distance along each route wherever possible. The difference of 31 m is seemingly small. However, it is possible to correctly estimate the shorter route without tools - as a matter of fact realizing this difference was the initiation for this study and for measuring the distance with a tool.

## Observations and Discussion

Before the match almost 100% of the fans pass the pond on the east, i.e. on the longer route. There are even fans entering the castle garden through the western gate (some approaching from market place, some from western city quarters) and cross the symmetry axis south of the pond to continue their way on the eastern route. Some fans on the eastern route accept an additional detour to walk the whole route on a paved track ("7" in fig. 1). After the match, when the fans return a large share of fans passes the pond on the western side. But nearly 100% of them immediately after the western gate turn west to head for western city quarters. Those few fans walking from western gate to market place can be explained by people, who have walked with friends from the northern gate to western gate and then parted way, as their friends headed for western city quarters.

When asking, why almost no fan is choosing the shortest and quickest route, one has to bear in mind that a large share of fans walks this way frequently and knows the area quite well. On their way to the stadium time does not matter a lot and also not an extra 30 meters of distance. This may change after the match, when people are tired and maybe in a hurry to reach for home or a bar. Still, of those fans, who reach for market place, almost no one walks the shorter western route. This is even more astonishing as those fans, who pass the pond on the western side (as in the end they reach for western city quarters) show that there is a viable track. In addition on the way back the fans can orient themselves at the castle tower to walk the shortest track over the grass to the western gate [5]. Thus fans who head for market place either erroneously estimate the eastern route to be the shorter one, or still see utilities in the eastern route that more than balance the difference.

From just observing the fans one cannot tell the true reason. It's even doubtful that a reliable answer could be found with questionnaires. Such decisions are often done more or less unconsciously and when being asked, people often wonder about the reasons themselves. Therefore here possible reasons are listed, which in their diversity demonstrate the complexity of pedestrian route choice, even in this seemingly manageable almost "laboratory" example.

**False Estimation:** It might be that the zig-zag character of the eastern route makes it appear shorter than the western route, which is less structured.

**Utility and History:** On their way to the stadium one can observe fans on the eastern route, who accept an additional detour to avoid walking over grass at all ("7" in fig. 1), even at dry weather. And this on a very clear area, where both alternatives are entirely visible. These fans value a paved track a lot and accept large detours, if they can avoid leaving it. For other fans this might be less important, but still have some influence. As most fans walk to the stadium frequently, it might be that they behave on sunny, dry days, as they do on a rainy day, namely stick to the paved track.

**General Direction:** When the fans on the way to the stadium have to choose between eastern or western route, the stadium lies to the east. If - unconsciously - they ignore the fact that they have to pass the northern gate, the eastern path is more similar to the linear direction.

**Attachment instead of Equilibrium:** As the event does not begin with the match but also includes approaching the stadium, the fans tend to stick together. This induces a social attachment of the fans and would amplify any asymmetry in route choice.

## References

1. Kretz, T.: Pedestrian Traffic: on the Quickest Path. JSTAT, P03012 (2009)
2. Meschini, L., Gentile, G.: Simulating car-pedestrian interactions during mass events with DTA models: the case of Vancouver Winter Olympic Games. In: SIDT 2009 International Conference. Milan, Italy (2009)
3. Chang, D.: Spatial choice and preference in multilevel movement networks. Environment and Behavior. 34(5), p. 582. (2002)
4. Sarkar, S.: Qualitative evaluation of comfort needs in urban walkways in major activity centers. In: TRB annual meeting (2003)
5. Lynch, K.: The image of the city. MIT press (1960)

6. O'Connor, A., Zerger, A., Itami, B.: Geo-temporal tracking and analysis of tourist movement. Mathematics and Computers in Simulation. 69, p. 135 (2005)
7. Millonig, A., Gartner, G.: A Multi-Method Approach to the Interpretation of Pedestrian Spatio-Temporal Behaviour. In: Klingsch, W. et al. (eds), Pedestrian and Evacuation Dynamics 2009, p. 563. Springer-Verlag, Berlin, Heidelberg (2010)
8. Borgers, A., Timmermans, H.: A Model of Pedestrian Route Choice and Demand for Retail Facilities within Inner-City Shopping Areas. Geographical analysis. 18(2) (1986)
9. Haupt, T.: Verkehrs- und Raumplanung - Wirkungssimulation - Information: Praxis und Vision. In: Schrenk, M. (ed.),11th Conference on Urban Planning and Regional Development in the Information Society, p. 543. Vienna (2006)
10. Openstreetmap, This image is published under the cc-by-sa 2.0 license, independent of the license under which the work may be published in which this image and this contribution are embedded to http://openstreetmap.org/?lat=49.01362&lon=8.40454&zoom=16&layers=B000FTF
11. Graessle, F., Kretz, T.: An Example of Complex Pedestrian Route Choice. Color online at (extended) preprint version: http://arxiv.org/abs/1001.4047 (2010)

# Sensor-Assisted Support Tools for Live Evacuation

**S. Grindrod[1], S. Welch[1] and J. Fraser-Mitchell[2]**

[1]BRE Centre for Fire Safety Engineering, The University of Edinburgh, UK
[2]BRE Global, UK

**Abstract** This study looks at the possibility of exploiting sensor-linked modeling to provide information to occupants on the available routes within the building and help influence their final egress path to support efficient evacuation. In general, it is intended that the system will able to influence the occupants' egress route choices, even before they have decided on their final exit choice. The effectiveness and reliability of the system will be tested through a series of comparisons obtained by modeling evacuations using K-CRISP (Koo [5]) in conjunction with live data that will be gathered during project M*E*T*R*O (Nilsson [1]).

## Introduction

The effectiveness of an emergency evacuation following some sort of incident is often hindered by the lack of information that is available about the current conditions. It is important to provide occupants, who find themselves in an evacuation scenario, with all the required information needed to safely evacuate. However, with the increasing complexity of building layout and the great variety of activate and passive egress systems used, it is becoming more and more difficult to provide adequate information for safe egress without confusing the occupants.

The evacuation processes of cue validation and decision-making ("pre-movement behavior") are believed to be more decisive on survival than the actual movement speed. Deaths caused in situations which are sometimes labeled by the media as "Panic" scenarios are, in reality, generally caused by lack of information on egress routes, occupants being providing with insufficient or incorrect information. It is therefore important to provide the vital information, both accurately and in a format that is quick and easy to understand, during the pre-movement phase.

An underlying problem here is that many modern buildings do not conform to traditional architectural designs, on which the prescriptive codes are based. Thus, as the complexity of the architectural design increases there is potential for failure of egress designs derived from the prescriptive codes, leading to potential problems with emergency egress. The aim of the current work is to research and develop an intelligent system that can provide occupants with information on the

R.D. Peacock et al. (eds.), *Pedestrian and Evacuation Dynamics*,
DOI 10.1007/978-1-4419-9725-8_72, © Springer Science+Business Media, LLC 2011

most appropriate egress path, by exploiting sensor outputs obtained from within the building.

## Computer Models

Computer modeling programs are often used as a support tool for egress solutions based upon accepted building codes. One of the biggest issues with using simulation tools for egress is that due to the many uncertain inputs the user can often adjust model parameters to produce an answer that best suits their needs, i.e. the one giving them a building permit. This technique may provide results that do not practically resemble human behavior, as it is a very complex and difficult to represent computationally. Thus, raw model outputs should not be used on their own to justify designs but rather taken to be an indicative guide or a means of comparing alternative solutions.

The reasons for these inadequacies are because our initial assumptions are far too simple and often neglect important factors, like group mentality for example. Nevertheless, computer models can be useful in designing egress route widths and the number of exits that are needed within a building based on an assumption of an occupant mentality corresponding to "survival of the fittest".

Human behavior can sometimes be adequately modeled after an event has taken place if uncertain elements can be eliminated with the help of video recordings and other evidence taken from occupants who have escaped the fire. However, it is often near impossible to meaningfully predict how people will behave before an event occurs. A useful approach for prior analysis is Monte Carlo modeling, where each plausible scenario is considered and modeled to produce a series of outputs based on initial randomized inputs.

## Sensor-assisted Intelligent System

The current work seeks to build upon the research on sensor-linked modeling done within the FireGrid project, which showed the value of using sensor-linked fire model (Koo [5]) and its potential for interpreting human behavior (Fraser-Mitchell [4]). However, there are many challenges in representing human behavior aspects.

For example, an experimental study of three Ikea stores in Sweden showed the potential flaws in the way current codes represent human behavior. Emergency evacuations which demonstrated that people often walked past emergency exits without using them and chose to exit through the main entrance or main exit. This led to much longer evacuation times that those estimated via design codes.

The evacuation trial therefore failed because of a combination of human behavior and design error. Most of the exits were located at 90 degrees to the marked walking path in the store and therefore not clearly visible. Moreover, occupants within Ikea are accustomed to following a marked route that directs them around the store and through all of the different departments. This has become the normality and during evacuation occupants tended to follow the same path in the knowledge that it would eventually lead them to a familiar location.

If an "intelligent" egress system were available which used sensor-linked modeling occupants could be provided with accessible information on the most appropriate egress path, mitigating the above issue. The simulation tool would need to incorporate information using real time data on all possible egress routes, their capacities, and live information on areas impeded by smoke and fire hazards.

This system could be used in conjunction with wayfinding tools, e.g. lighting and audio recordings, to help guide the occupants towards exits. It would provide live information to the occupants while reducing confusion about alarms, therefore reducing the amount of time for egress movement to commence. If the system could reduce the reaction time of occupants and direct them to appropriate exits this would significantly increase the effectiveness of the fire safety design.

## Human Behavior Aspects

The ability to provide clear and concise information that is appropriate to the building can be a great challenge. It has been seen within experiments that evacuees do not make significant use of emergency signage during egress. They instead move towards familiar, but not necessarily safe, routes or they may follow someone who they think "looks like they know where they are going".

Death and serious injuries of occupants within a building typically arise not due to their direct fire exposure but by them remaining too long within the smoke. They struggle to find their way out quickly enough, and hence consuming excessive dosages of CO and other toxic gases. And it is not only smoke intoxication or thermal exposure that hinders egress, but rather the effects of poor visibility and irritation of the eyes, or a combination of all these factors. If the vision of the occupants is affected by smoke they are more likely to slow down or even stop, leading to a significant increase in the evacuation time.

Low density of smoke is a key factor in a successful evacuation. Smoke can easily block normal and emergency lights causing limited visibility, in some cases to less than 2 m, creating a nearly impossible situation for an unfamiliar occupant. Lighting systems themselves are prone to fail due to structural damage, power shutdowns and damage to circuitry, therefore, lighting systems are required not only to have good reliability, but also excellent visibility at close range.

## Wayfinding

The concept of wayfinding was developed in the late 1970s and concerns the use of fundamentally passive design tools that help and influence occupants to evacuate towards a safe location during an emergency scenario. A well designed wayfinding system should ideally be able to provided occupants with their current location and heading relative to known landmarks, their desired destination, descriptions of surrounding features and the general layout of the environment.

Wayfinding systems should cater for all ages and abilities to assimilate information and in particular should be easily understood by occupants who do not know a building's layout. Wayfinding designs should be components of efficient egress systems that are well articulated and aesthetically pleasing when not in use.

Novel wayfinding systems are often limited by the information required by the occupants during an evacuation (e.g. the type, amount and accuracy), the environment in which the system will operate and how the system integrates with current code-based fire strategies. A successful design will address these limitations while significantly increasing the success rate of an evacuation.

## Development and Validation

In the current work an intelligent egress system is being researched which can provide occupants with accessible information on the available routes within the building, and the most appropriate egress path based on live sensor readings. Development and validation will use data gathered from the range of intermediate and full-scale experiments to be conducted for Project M*E*T*R*O at Lund University, which involve using wayfinding systems within smoked-filled subway tunnels. These will look at the effects of smoke on visibility and whether a lighting wayfinding system can be effectively implemented within a tunnel situation.

Using an extension of the modeling program known as K-CRISP (developed by BRE) and the data obtained from the Project M*E*T*R*O the effects on the overall evacuation times of the station will be predicted, according to the nature of the information provided to occupants on the location of safe routes and exits at the early stages of an emergency. It is intended that the wayfinding system will be used to successfully manipulate the final exit choice of an occupant without causing them any stresses during the process. The results will provide an insight into the potential benefits of using sensor-linked tools to support emergency response, thereby contributing to the development of novel intelligent egress systems.

## References

1. Nilsson, D.: Exit choice in fire emergencies – Influencing choice of exit with flashing lights. The University of Lund (2009)
2. Hajibabai, L.: Spatial cognition and wayfinding strategy during building fire. In: Dynamics in Spatial Interactions. ICSC (2009)
3. Fraser-Mitchell, J., Koo, S.H., Welch, S.: Sensor-linked simulation for emergency response. In: Human Behavior in Fire Symposium 2009, pp. 561 – 566. BRE Global,University of Edinburgh, UK (2009)
4. Koo, S., Fraser-Mitchell, J. Upadhyay, R. Welch, S.: Sensor-linked fire simulation using a Monte-Carlo approach. Fire Safety Science 9, pp. 1389 – 1400. BRE Centre for Fire Safety Engineering, the University of Edinburgh (2009)
5. Sime, J.D.: Affiliative behavior during escape to building exits. J Environ Psychol (3rd ed.), pp. 21 – 41. Jonathan Sime Associates, Surrey, UK (1983)

# Mutual Information for the Detection of Crush

**P. J. Harding[1], M. Amos[1], and S. M. V. Gwynne[2]**

[1]Novel Computation Group, Manchester Metropolitan University, Manchester, U.K.

[2]Hughes Associates, Inc., London, UK

Corresponding Author: p.harding@mmu.ac.uk

**Abstract** This paper describes the application of Mutual Information to the detection of crush in a well-established model of pedestrian evacuation. We show that Mutual Information offers a computationally low-cost alternative to "expensive" physical force calculations for the detection of crush in evacuation simulations.

## Introduction

A number of software environments [1] exist for the simulation of large-scale egress situations, such as the evacuation of buildings, stadia and other enclosed spaces. These environments offer sophisticated tools for the analysis of human behaviour under evacuation conditions, and can recreate many of the social, environmental, structural, and psychological factors that may affect egress. Although such simulation environments can accurately model many aspects of crowd behaviour, they generally lack the capability to analyse the effects of the *physical forces* that build up within crowds. These forces can give rise to *crush conditions* (or simply *crush*), and are commonly cited as a major cause of injuries and fatalities during emergency evacuations [2-6]. The inclusion of crush analysis in simulations has traditionally been achieved by one of two methods;

1. **Implicit**: The implicit approach is the traditional method of qualifying the presence of crush within a simulation. It requires experienced engineers and technicians to analyse simulation output, such as population densities and environmental considerations, to ascertain the likelihood of crush becoming a danger during an evacuation. This is, however, a fundamental weakness of this technique - since it requires the knowledge of an expert analyst, the identification of crush is inherently subjective and difficult to automate.
2. **Explicit**: This approach requires the deployment of physical force calculations to quantify the level of crush that arises within a crowd. Most often based on traditional physical force equations, the explicit analysis of crush conditions offers a highly accurate measure of the force that exists within a simulation, but incurs a significant computational overhead.

R.D. Peacock et al. (eds.), *Pedestrian and Evacuation Dynamics*,
DOI 10.1007/978-1-4419-9725-8_73, © Springer Science+Business Media, LLC 2011

We propose the use of Mutual Information (MI) [7] as a new approach to the analysis of crush conditions within a simulation environment. MI is a probabilistic method of analysing order within variable sets, and offers the possibility of automated qualification of the presence of crush within a simulation, whilst requiring a fraction of the computational overhead required by physical force calculations. In this paper we first define the notion of crush, before discussing previous work on the analysis of crowd movements. We then give a formal definition of Mutual Information, before describing its application to an established model of pedestrian movement. We conclude with a discussion of possible future work.

## Crush

The danger presented by crush conditions has been recognised for some time as a major cause of injury and death during emergency situations [8,9]. The build-up of force within groups of people is known to be a major cause of *compressive asphyxia* (or *traumatic asphyxia*), which is the application of pressure on or about the chest or ribcage which leads to shortness of breath and, eventually, suffocation. These types of injury are characteristic of situations in which crush conditions are present.

There have been a number of situations where crush has caused a great number of injuries or deaths. Some of the most notable include the Hillsborough disaster [2], the Gothenburg dancehall fire [3], the E2 nightclub incident [4], the Station Nightclub fire [5], and the Mihong bridge disaster [6]. The precipitating factors for the formation of crush are many and varied, e.g. emergency evacuation due to fire (Gothenburg and the Station Nightclub) or poor event management (Hillsborough, Mihong bridge). The numerous causes of crush, and the dynamically changing nature of crowd behaviour, can therefore make it difficult to precisely define the parameters under which a situation may lead to crush conditions forming.

## Previous Work

After analysing video recordings of the Hajj pilgrimage in Saudi Arabia (2006), it was noted by Johansson *et al* [10] that the crowd exhibited a marked change in behaviour under certain conditions. This change in behaviour appeared to mark a *transition* between laminar ("smooth") flow of individuals, and a more *turbulent* flow. We therefore suggest that this latter state of crowd behaviour immediately precedes the formation of crush, and that its detection can therefore act as an indicator of imminent crush conditions. It has already been shown that Mutual Information may be used to identify phase transitions within a system of interacting, self-propelled particles [11]. This work focused on identifying kinetic phase transitions in the Scalar Noise Model [12] (SNM), a system of dimensionless particles that exhibits flocking behaviour under correct parameterisation. By measuring the MI of the system, it is possible to detect the point (the kinetic phase transition) at which the system moves from *chaotic* or stochastic behavioural characteristics to exhibiting signs of *order*.

We suggest that these transitions within evacuating crowds, from one state of collective behaviour to a qualitatively different behavioural state, may be considered analogous to the kinetic phase transition identified in the SNM. It is by treating the formation of crush as a phase transition (which can be identified within an evacuation) that we form the basis of applying the MI technique for crush analysis and detection.

## Mutual Information

Mutual Information (MI) is a statistical measure of the mutual dependence of two variables, and has been used extensively as an analytic technique [13,14]. Equation 1 expresses the mutual information (*I*) of two discrete signals (*A* and *B*).

$$I(A, B) = \sum_{i,j} P(a_i, b_j) \log_n \frac{P(a_i, b_j)}{P(a_i)P(b_j)}$$

$P(a_i)$ is the probability of *A* having the value $a_i$; $P(a_i, b_j)$ is the probability of *A* having the value $a_i$ *and B* having the value $b_j$. The base of the logarithm (*n*) defines the units in which the MI will be measured; this is commonly base 2, giving the MI in *bits*. In general terms, MI quantifies the measure of *interdependence* between two signals or variables; therefore, if *A* and *B* are entirely independent then $I(A,B) = 0$, but in all other cases MI is non-zero.

## Experimentation

The Social Forces Modem [15] (SFM) is a well-established framework for the simulation of pedestrian movement. A particle-based model, the SFM can accurately recreate many of the social, psychological and physical forces present within evacuating crowds. We use the Mutual Information technique to analyse a version of the SFM identical to that presented in [15], but with two important additions. Firstly, in the original model an injured agent forms an immovable obstacle, still able to exert force (both physical and *social*) on agents within their interaction radius. This behaviour causes problems during simulations, as it makes possible the creation of a *barricade* of injured agents between the evacuating mass and the only available exit. This causes the simulation to end with evacuees remaining in the structure. Simulations in which this occurs are declared void, the results unusable, and the experiments must be restarted. To counteract this issue, we add a rule dictating that when agents succumb to injury they are removed from the simulation after an arbitrary amount of time. This allows the increase in *force* that an injured agent may incur to be fully taken into account within the simulation, but prevents the *barricading* behaviour mentioned previously. Secondly, in order to obtain a baseline for the MI of the system (i.e. a null value), a period of "milling" is introduced. This takes the form of a 10 second "pre-evacuation" period inserted at the start of each experiment, during which agents have no clearly defined goal. This addition yields a baseline value for the MI in each simulation, i.e. the value of the MI for a random geo-spatial distribution of agents. Therefore all experiments show the start of the experiment, at $t = -10s$, with evacuation beginning at $t = 0s$ as per the standard model.

Tests are run using a combination of the spatial and directional data taken from the agents during multiple simulations of the Social Forces model. The MI is calculated as shown below:

$$I(X,\Theta) = \sum_{i,j} P(x_i,\theta_j) \log_2 \frac{P(x_i,\theta_j)}{P(x_i)P(\theta_j)}$$

$$I(Y,\Theta) = \sum_{i,j} P(y_i,\theta_j) \log_2 \frac{P(y_i,\theta_j)}{P(y_i)P(\theta_j)}$$

$$I = \frac{I(X,\Theta) + I(Y,\Theta)}{2}$$

With this approach the coordinate and directional data on each agent is analysed in such a way that the spatial clustering is abstracted from the analysis, i.e. the two positional variables are analysed separately. This counteracts problems associated with the measurement of the Euclidean distance between particles, in which the MI acts as a better metric of *clustering* than of the *alignment* of behavioural characteristics. This analysis relies solely on on the changing behaviours of the agents (more precisely, the changing velocity vectors), rather than their spatial clustering. The results are depicted in Figure 1.

We observe that the peak in MI is pronounced, with a large increase in the MI as the agent vectors become ordered (at $t > 0s$), displaying the characteristic rise in MI that would be expected as a system attains order. The more relevant characteristic of the mutual information - the severe drop in MI that identifies the deterioration of the system into a state of disorder ($1s < t < 3s$) - is also highly pronounced.

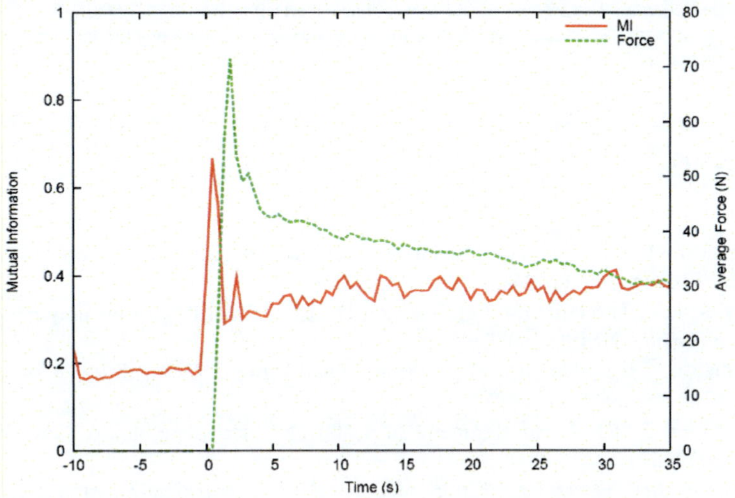


**Fig. 1 - MI and Average Force, as recorded against simulation time**

## Conclusions

Despite this work still being in at an early stage, our preliminary results show that by employing the Mutual Information technique it is possible to detect the point at which the formation of dangerously high levels of force become likely, without the need to calculate the *precise levels of force* present within a specific simulation. As this method of crush analysis negates the need for the actual calculation of physical forces acting between agents, it is thought that it provides a far less computationally expensive method of analysis, although the exact *cost saving* has yet to be calculated.

## Future Work

The investigation into the use of Mutual Information as a detector for crush is still in its infancy, but research into this area will continue, with a particular focus on the use of MI as an automated indicator of crush. Exact computational cost savings must also be calculated. These steps will make the case for the MI analysis to be regarded as a valid and effective alternative to traditional crush detection methods

**Acknowledgments** The authors would like to thank the Dalton Research Institute for supporting this work, and also Hughes Associates, Inc for allowing Dr Steve Gwynne the time to take part in this research.

## References

1. Kuligowski, E.: A comprehensive review of 28 evacuation models. http://fire.nist.gov/ (2004)
2. Taylor, L.J.: The Hillsborough Stadium Disaster. Final Report on Enquiry. UK Home Office, HMSO (1989)
3. Comaeau, E., Duval, R.: Dance hall fire. Gothenburg, Sweden. http://www.nfpa.org (1998)
4. Wilgoren, J.: 21 die in stampede of 1,500 at Chicago nightclub. New York Times (2003)
5. Grosshandler, W. et. al.: Report of the technical investigation of the station nightclub fire. http://fire.nist.gov/ (2007)
6. Zhena,W., Maoa, L., Yuanc, Z.: Analysis of trample disaster and a case study Mihong bridge fatality in China in 2004. Safety Science. 46(8):1255–1270 (2008)
7. Kraskov, A., Stoegbauer, H., Grassberger, P.: Estimating Mutual Information. Phys. Rev. E. **24** (2004)
8. Predtechenski, V., Milinski, A.: Planning for Foot Traffic Flow in Buildings. Amerino Publishing Co., New Delhi (1978)
9. Fruin, J.: Pedestrian Planning and Design, Metropolitan Association of Urban Designers and Environmental Planners. New York (1971)
10. Johansson, A., Helbing, D., Al-Abideen, H., Al-Bosta, S.: From crowd dynamics to crowd safety: A video-based analysis. Advances in Complex Systems (ACS). 11(04):497–527 (2008)
11. Wicks, R., Chapman, S., Dendy, R.: Mutual information as a tool for identifying phase transitions in dynamical complex systems with limited data. Physical Review E. 75(5) (2007)
12. Vicsek, T., Czirok, A., Ben-Jacob, E., Cohen, I., Schochet, O.: Novel type of phase transition within a system of self-driven particles. Phys. Rev. Lett. 75:1226–1229 (1995)
13. Jeong, J., Gore, J., Peterson, B.: Mutual information analysis of the EEG in patients with Alzheimer's disease. Neurophysiology. 112:827–835 (2001)
14. Battiti, R.: Using mutual information for selecting features in supervised neural net learning. IEEE Transactions on Neural Networks, 5 (1994)
15. Helbing, D., Farkas, I., Vicsek, T.: Simulating dynamical features of escape panic. Nature. 407:487–490 (2000)

# Evacuation Planning Tool (EPT) for Emergency, Event and Space Planning

**M. Harmon**[1] **and J. Joseph**[2]

[1]Manager, SIMLAB, US Secret Service, Washington, DC 20001 USA, Mark.Harmon@usss.dhs.gov
[2]VP Technology, Regal Decision Systems Inc. Secret Service, Nottingham, MD 21236 USA, jjoseph@regaldecision.com

**Abstract** The authors present an application of agent based simulation in evacuation planning and emergency response. The Evacuation Planning Tool (EPT) developed by Regal Decision Systems Inc for DHS is a PC based software application that consists of a 3D Editor, an agent based model, a 3D animation component and a reporting module. The EPT provides users with the ability to simulate and analyze a wide variety of evacuation scenarios. The EPT simulates movements of crowds and provides users with the ability to control security teams at a command level. Users can assess civilian responses, facility operations and security options related to a variety of disasters and threats in order to analyze and formulate evacuation strategies. Entities and groups of entities are created with decision branching intelligence and dynamic selection of destinations, routes, and behaviors. The behaviors are consistent with environmental influences such as threats, physical barriers, encumbrances, and commands.

## Introduction

The Secret Service, as the lead agency for securing National Special Security Events (NSSE), must accommodate evacuation plans as part of the overall security arrangements for an event or venue site. Evacuation Planning Tool (EPT) was developed to plan for event security and emergency conditions.

Because security measures may significantly impact location and movement of attendees, evaluating evacuation scenarios is necessary to examine potential hindrances to flow and to assess building features that affect expedited evacuation.

There are a number of parameters to consider in using agent-based simulation for modeling evacuation of public venues: population densities, security configurations, congestion, clutter and other factors. In addition, route selection, information propagation, group size and behavior, building design, and encumbrances affect overall evacuation time.

R.D. Peacock et al. (eds.), *Pedestrian and Evacuation Dynamics*,
DOI 10.1007/978-1-4419-9725-8_74, 

EPT was designed for ease of use on a laptop computer with 2D and 3D visualization for rapid assessment of evacuation strategies. EPT was developed by Regal Decision Systems with funding from the Dept. of Homeland Security and the Combating Terrorism Technical Support Office (DoD) and has been commercialized for public use.

Recent enhancements to EPT include modeling wheel chairs with attendants, tagging and tracking individuals, elevators and loading a building through checkpoints. Structures may be easily and simply modeled in 2D, or for greater visual impact, in 3D with iconic objects representing individuals. A large outdoor event with 75k entities, including vehicles, is being developed in EPT. The eventual goal is to simulate evacuation of a metropolitan city.

## EPT Capabilities

EPT can be used to simulate both evacuation and loading scenarios. The program provides the user with a range of parameters so that users can evaluate several "what-ifs". These parameters can be classified into the following groups:

1. Entity (person) and group parameters
2. Environment parameters
3. Scenario parameters

Entity (person) parameters include maximum speed of movement, health level, panic level, mobility, protection and awareness. Defaults are available for each of these parameters but advanced users can modify the data as necessary. Users can define "official personnel" entities who direct other entities away from regions with threats and also reduce their panic level. All entities have the capacity to transmit information (about threats and closed exits) to other entities, but "official personnel" can also reduce the panic level of entities as they pass. Users can define groups of entities with each group composed of a leader and a set of followers. The followers will stay close to the leader and take directions from the leader on which exits to choose.

Environment parameters include maximum allowable population, soft clutter and hard clutter, fences, threats and explosions. Soft clutter (movable chairs) reduces the capacity of an area while hard clutter (desks, tables) reduces capacity *and* hinders movement of entities. Fences can be used to cordon off certain areas if necessary. All these features of the environment can be placed in the EPT 3D editor using mouse click and drag. For threats and explosions, users enter data elements such as health and panic effects by entity type, timeframe for exposure, dissipation effects and location of effect.

To provide scenario parameter flexibility, EPT can be executed in three different modes: 1) Evacuation 2) Loading and 3) Combination. In the Evacuation mode, users designate exits for certain types of entities, determine if any particular entities have a "shelter in place" option and execute the simulation. The model uses A* to compute the shortest paths to intermediate destination taking into consideration final destinations (exits/shelter in place). In addition, each entity will continually recompute the path taken based on congestion, wait time, speed of movement, mobility, awareness, threats, explosions and group behavior.

EPT can be used in the loading mode to determine the number of security checkpoints necessary to allow patrons to enter the venue before an event (concert, speech, game). Alternatively, EPT can be used to determine the number of hours necessary to provide complete screening of all entities before an event given a set of resources.

EPT can simulate pedestrian entry and flow at airports, checkpoints, US consulates, etc. Users place processing stations (or facilities) and enter necessary parameters such as open/close schedules, process time distributions and available queuing area. EPT can also simulate escalators and elevators providing stakeholders with usage reports and bottle necks in pedestrian flow (Fig. 1).

**Fig. 1. EPT – Loading scenario**

In the combination mode, EPT can simulate a loading scenario in conjunction with an evacuation. For example, the user can set up a loading scenario where entities arrive at a stadium three hours before an event. The user can then trigger an explosion a few minutes or hours into the loading to evaluate the impact of people evacuating the building as others are trying to enter. Information propogation kicks in and entities that are trying to enter the venue also start evacuating.

## Case Study

Before EPT was used to conduct analyses, the underlying model was validated by comparing the simulation model results against video footage from an actual evacuation. The summary data obtained from the model was compared to the summary data from the video footage and the results were statistically equivalent.

EPT was used to evaluate several evacuation scenarios at the Xcel Energy Center in preparation for the Republican National Convention held in September of 2008. 16,405 entities in 5 different scenarios, listed below, were evaluated.

**Table 1. Scenarios evaluated in the case study**

| | Brief Scenario Description |
|---|---|
| Scenario A | Floor Stairs Closed, Loading Dock Closed |
| Scenario B | Floor Stairs Closed, Loading Dock Open |
| Scenario C | Floor Stairs Open, Loading Dock Closed |
| Scenario D | Floor Stairs Open, Loading Dock Open |
| Scenario E | Floor Stairs Open, Loading Dock closed, Floor Level Incident |

For each scenario, the overall evacuation time was recorded, bottle necks identified and mitigation steps recommended.

# Counterflow Model for FDS+Evac Simulations

**S. Heliövaara[1] and T. Korhonen[2]**

[1]Systems Analysis Laboratory, Aalto University
P.O. Box 11100, 00076 Aalto, Finland
[2]VTT Technical Research Centre of Finland
P.O. Box 1000, FI-02044 VTT, Finland
Corresponding author: simo.heliovaara@hut.fi

**Abstract** We present a new method for modeling counterflow situations in crowds. Agents, describing individual pedestrians, are set to avoid the moving directions where there is counterflow and prefer the directions with forward flow. In dense counterflow situations, people tend to move shoulder first to occupy less space in the moving direction. If the elliptical cross-section of a human body is considered in a crowd model, the rotational positions in which the agents move affect the counterflow. In our model, agents try to rotate their bodies in certain counterflow situations to move shoulder first. The model is implemented in the FDS+Evac simulation software. Test simulations show that it is able to create rather realistic simulations of counterflow.

## Introduction

Counterflow situations are common in moving human crowds. On sidewalks and in large public venues like railway stations, streams of people moving to different directions encounter repeatedly. In evacuation situations, most of the occupants tend to move to the same direction, but counterflow may occur, e.g., when firefighters try to enter the building.

The crowd dynamics model of Helbing *et al.* [1] is widely used and has been found to realistically simulate many phenomena occurring in real crowds. One of the downsides of the model is that agents moving in opposite directions are unable to dodge each other, and thus, unrealistic collisions occur in couterflow situations. Two recent articles [2, 3] present collision avoidance methods that can be added to Helbing's model. Both of these models give realistic appearing results in sparse crowds. However, in these methods, each agent is only able to dodge one other agent at a time, which may cause problems when crowd density increases.

We present a model for counterflow situations, where each agent observes its proximity and selects the moving direction with the smallest counterflow. Agents moving to the same direction create negative counterflow, and thus, the model also makes agents favor the directions with forward flow.

R.D. Peacock et al. (eds.), *Pedestrian and Evacuation Dynamics*,
DOI 10.1007/978-1-4419-9725-8_75, 

The cross-sectional shape of a human body is elliptical. Hence, the rotational positions of agents may affect counterflow, as agents moving shoulder first occupy less space in the moving direction. We consider this by describing the agents' body dimensions with three overlapping circles and by setting the agents to move shoulder first in certain counterflow situations.

The presented collision avoidance model is implemented in the FDS+Evac simulation software [4, 5, 6].

## Counterflow Model

The platform of our counterflow model is the crowd dynamics model of Helbing *et al.* [1], extended by the three-circle representation of agents [2, 4]. Nevertheless, similar approach could most likely also be applied to many other agent-based crowd models.

In the counter flow model, agents frequently update their desired moving directions. Each agent has three options on each update: to go straight ahead, to dodge to right, and to dodge to left. The agents make the decisions by observing the area in front of them and by selecting the direction with the least counterflow. This is done by dividing the area into three overlapping sectors and by giving each sector a score according to locations and moving directions of the agents within the sector. The agents moving to the same direction increase the score of the sector and the counterflow-agents decrease it. On each step, the direction with the highest score is selected and set to be the *desired moving direction* of the agent in the Helbing *et al* model. The range of the sectors varies between 1.5 m and 3 m, according to the velocity of the agent. The features of the model are illustrated in Figure 1.

In situations of strong counterflow, the *desired body angles* of the agents are changed to make them try to move shoulder first. Also the *motive forces* are increased to make the agents more determined in their movement.

## Simulation Results

The performance of the counterflow model was tested with FDS+Evac-simulations in the IMO test geometry 8 [6], where a 2 m wide corridor connects two rooms. In the initial situation, 40 agents were located in both rooms and their goal was to pass the corridor to the other room.

Snapshots of the test simulations are presented in Figure 2. When the counterflow model is not applied, the two streams create an impassable jam in the corridor. Using the counterflow model, the flow in the corridor is rather smooth and all agents have passed the corridor in about 50 seconds.

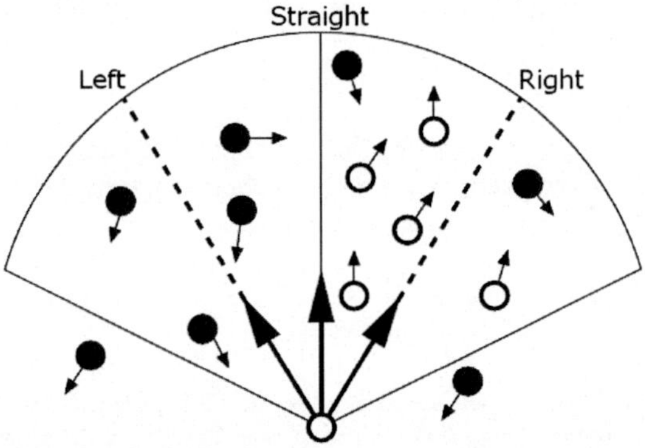


**Fig. 1. Illustration of the counterflow model. The agent making the decision is in the origin of the sectors. The large arrows denote the three options of moving direction for the agent. The sectors related to the moving directions are overlapping, as the dashed lines denote the edges of the middle sector and the solid lines the edges of the left and right sectors. The smaller arrows denote the moving directions of the other agents, and thus, the white agents increase the scores of the sectors they are in and the black agents decrease them.**

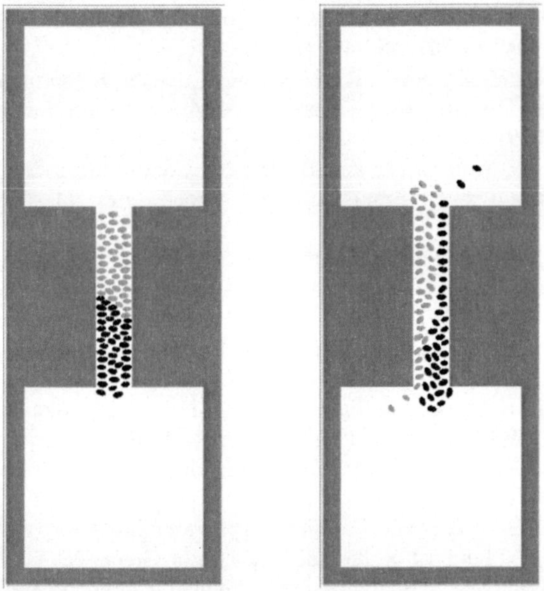

**Fig. 2. Simulation snapshots. The counterflow model is used in the right figure. The left figure uses the original model of Helbing *et al.* The gray agents are heading to the bottom room and the black agents to the top room.**

## Conclusions

In order to realistically model counterflow with the crowd dynamics model of Helbing *et al.,* a method to describe the interaction between agents moving to opposite directions is necessary. We present a short-range model, where agents adjust their walking directions and rotate their bodies to avoid collisions with the oncoming agents.

Test simulations show that the presented model is able to eliminate the unrealistic jams occurring with the original model.

**Acknowledgments** This work has been funded by the VTT Technical Research Centre of Finland, the Finnish Funding Agency for Technology and Innovation, the Finnish Fire Protection Fund, the Ministry of the Environment, and the Academy of Finland.

## References

1. Helbing, D., Farkas, I., and Vicsek, T.: Simulating Dynamical Features of Escape Panic, Nature, **407**, 487-490 (2000)
2. Smith, A., James, C., Jones, R., Langston, P., Lester, E., and Drury, J.: Modelling Contra-Flow in Crowd Dynamics DEM Simulation, Safety Science, **47**, 395-404 (2009)
3. Pelechano, N., Allbeck, J., and Badler, N.: Controlling Individual Agents in High-Density Crowd Simulation. In: Proceedings of the 2007 ACM SIGGRAPH/Eurographics symposium on computer animation, 99-108 (2007)
4. Korhonen, T., Hostikka, S., Heliövaara, S., Ehtamo, H., and Matikainen, K.: FDS+Evac: Evacuation Module for Fire Dynamics Simulator. In: Proceedings of the Interflam2007: 11th International Conference on Fire Science and Engineering, Interscience Communications Limited, London, UK, 1443-1448 (2007)
5. Hostikka, S., Korhonen, T., Paloposki, T., Rinne, T., Matikainen, K., and Heliövaara, S.: Development and Validation of FDS+Evac for Evacuation Simulations, Project Summary Report, VTT Research Notes 2421, VTT Technical Research Centre of Finland (2007)
6. Korhonen, T., Hostikka, S.: Fire Dynamics Simulator with Evacuation: FDS+Evac – Technical Reference and User's Guide, VTT Working Papers 119, VTT Technical Research Centre of Finland, Espoo, Finland (2009)
7. Guidelines for Evacuation Analyses for new and Existing Passenger Ships, MSC/Circ. 1238, International Maritime Organization, London, UK (2007)

# A Study on Evacuation Simulation after Earthquake in Consumer Facilities

**A. Hokugo, T. Kaneko, A. Sekizawa, S. Kakegawa and H. Notake**

Research Center for Urban Safety and Security, Kobe Univeristy
1-1 Rokkodai, Nada, Kobe, 657-8501 Japan
Corresponding author: hokugo@kobe-u.ac.jp
University of Toky
Institute of Technology, Shimizu Corporation

**Abstract** This study attempted to re-create the impact of evacuation obstacles such as moved or tipped furniture and scattered goods on people's evacuation behavior in a building at the time of an earthquake in an evacuation simulation. We conducted an earthquake evacuation simulation of the time required to complete evacuation and evacuation behavior with simulation conditions based on the information obtained through the interviews with representatives of four consumer facilities which actually suffered damage from an earthquake, focusing on people's behavior both when the building was shaking and after the earthquake, the reaction of staff, damage inside the building and evacuation behavior in order to seek the impact of damage inside the building, assuming the tipped or moved furniture and scattered goods would partially block an evacuation route and thus cause a reduction in walking speed. As a result, a phenomenon that people are crowded into sales spaces and passageways causing a traffic jam was re-created on the assumption that evacuation obstacles such as furniture and goods would slow down walking speed.

## Introduction

Studies on the action of objects inside a building, as a matter of course, focus not only on the changes in the building's interior environment but also on the relationship between changes in the interior and human casualty or the interior damage and people. There is, however, no study found which links the building's interior change to people's evacuation behavior. On the other hand, there are many studies on evacuation behavior taking the impact of earthquake fire into consideration, which is probably because the fire breakout can cause human fatality. However, changes to the interior environment of buildings such as moving and tipping of fixtures and furniture can also narrow or block evacuation routes. When the action of objects inside buildings during an earthquake is examined, it is also necessary to consider people's evacuation behavior after the earthquake as well as damage to human beings.

R.D. Peacock et al. (eds.), *Pedestrian and Evacuation Dynamics*,
DOI 10.1007/978-1-4419-9725-8_76, © Springer Science+Business Media, LLC 2011

The authors conducted a computer simulation[1] of an earthquake evacuation considering the floorplan, pre-earthquake furniture layout, interior damage such as furniture tipping and moving during a quake and people who are attempting to evacuate.

## Setup of Calculation Conditions

### *Space Model and Situation of Damage Caused by Earthquake*

Floorplan was set as Figure 1 based on some actual consumer facilities[2] as reference. The reference facilities that had experienced damage from an earthquake kindly provided us with photos of the damage at that time. The simulated damage from an earthquake such as furniture moving or tipping and goods' scattering was based on the photos. Figure 2 to Figure 5 are examples of damage

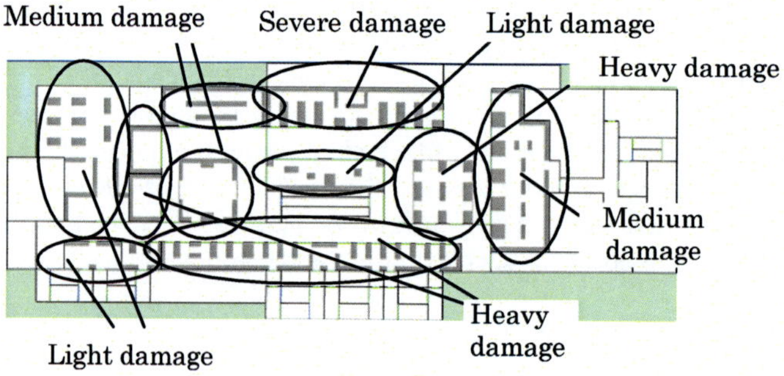


**Fig.1. Floor plan and damage situation at the time of an earthquake**

### *Number of Evacuees*

The earthquake the reference facilities experienced occurred at around 11.00 am and the population in the buildings was not very high, but it was not possible to ascertain the exact number of people in the buildings. The total number of the evacuees on the floor was, therefore, set as 630 based on the Japanese Verification Methods for Determining Safe Evacuation of a Floor and Building (in selling areas: 0.5 person/m², in passages: 0.25 person/m²).

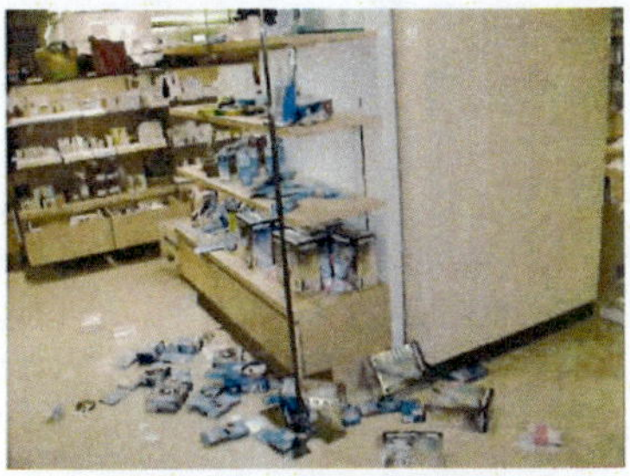

**Fig. 2. Light damage**

**Fig. 3. Medium damage**

**Fig. 4. Heavy damage**

**Fig. 5. Severe damage**

## *Walking Speed*

The damage situation was categorized as no damage, light damage, medium damage, heavy damage and severe damage, then the walking speed was set according to the degree of damage as no damage (1.0m/s), light damage (0.55m/s), medium damage (0.38m/s), heavy damage (0.32m/s) and severe damage (0.27m/s) using the walking experiment data as shown in Fig. 6.

**Fig. 6. Walking experiment in goods' scattering passageway**

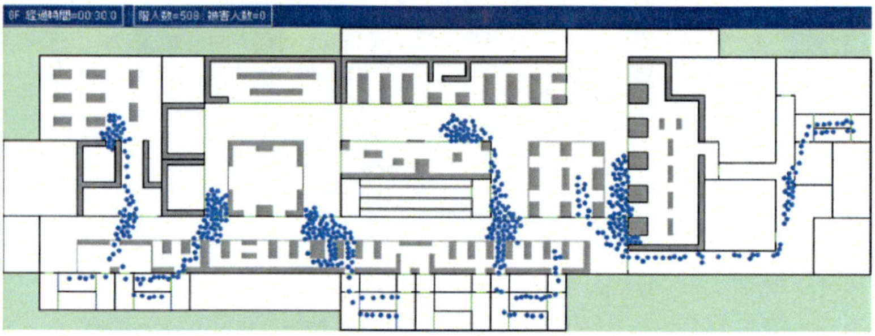


**Fig. 7. Snapshot of floor evacuation**
**(30 seconds has passed since evacuation started)**

## Conclusion

In this study, the impact of evacuation obstacles on evacuation situation was examined. Evacuation completion time was longer in the case where furniture is taken into consideration than in the case where furniture is not considered, since the evacuation route is limited in the former case. Also evacuation completion time was much longer in the case where the post-earthquake evacuation obstacles such as tipped or moved furniture and scattered goods were considered. Evacuation calculation should be made with possible post-earthquake evacuation obstacles being taken into consideration if people have to go through a selling space where evacuation obstacles would likely occur after an earthquake to get to escape stairs, like those in the floorplan used in the evacuation simulation in this study.
The accumulation of queuing evacuees during the earthquake evacuation caused by the presence of furniture is also pointed out in this study. In today's consumer facilities, however, furniture and fixtures are laid out closely together, for instance in an exhibition space, where a large number of people gather on a daily basis. The furniture layout in a shop is created mainly from the management viewpoint. When the possibility of an earthquake is considered, however, viewpoint of disaster reduction should be regarded as important.

## References

1. Kakegawa, S., Notake, H., Sekizawa, A., Ebihara, M.: Evacuation and Smoke Movement Interactive Simulation Model. In: Proceedings of 7th Asia-Oceania Symposium on Fire Science and Technology (2007)
2. Harada, Y., Hokugo, A., Sekizawa, A., Kakegawa, S.: A Study on Earthquake Evacuation from Buildings Considering Furniture Tipping. In: Proceedings of the 2008 JAFSE annual symposium, pp.46-49 (2008)

# Introducing Emotion Modelling to Agent-Based Pedestrian Circulation Simulation

**C. Hollmann, P. J. Lawrence and E. R. Galea**

Fire Safety Engineering Group, University of Greenwich
London, SE10 9LS, UK
Corresponding author: C.Hollmann@gre.ac.uk

**Abstract** In agent-based pedestrian circulation models, the simulation of the pedestrian-environment interaction is mostly achieved by imposing on each agent a predefined list of goal locations which the agent visits in turn. However, in reality human behaviour in complex environments is highly dynamic and fixed plans are often changed and adjusted according to emergent conditions and the person's individual interpretation of these events, in particular the amount of time available to achieve all the desired tasks. In this paper we present a prototype emotion model implemented within the buildingEXODUS evacuation and pedestrian dynamics software which enables simulated agents to react to perceived time pressures by modifying their behaviour. The model is demonstrated using a circulation scenario within a rail terminal.

## Introduction

Pedestrian circulation modelling tools have been developed with the intention to simulate human behaviour in various environments under normal usage conditions. Most of these models consider the physical movement of agents through a complex building layout and impose a set of goals on each agent which requires the agents to visit certain features within the structure, e.g. retail outlets, catering outlets, ticket offices, etc. The agents visit each location in turn, perhaps spending an arbitrary amount of time 'engaged' in activities at each location, often not taking into account external factors which may impact the viability or desirability of undertaking the prescribed activities. However, external influences such as congestion or time limitations may have an impact on the agents' short and long-term planning with regard to achieving their goals, and therefore need to be addressed within the simulation.

It is well known that emotions, motivations and feelings have a direct impact on human decision making as individuals attempt to cope with emergent events [1]. As a result the modelling of emotions has been introduced into areas such as computer graphics [2] and agenda scheduling [3]. There have also been approaches to introduce emotions into simulated human decision making associated with emergency evacuation [4] and pedestrian circulation [5]. Emotion modelling is

R.D. Peacock et al. (eds.), *Pedestrian and Evacuation Dynamics*,
DOI 10.1007/978-1-4419-9725-8_77, 

used to augment the agent's context awareness and hence their capability to evaluate the current situation, come to a conclusion and take appropriate actions. In this paper, we present a basic emotion model with the aim to enhance the simulation of pedestrian behaviour in multi-purpose environments. The model has been incorporated into the existing agent-based evacuation and pedestrian dynamics simulation software, buildingEXODUS [6]. Here we provide a brief overview of our prototype emotion model and demonstrate its capabilities in a pedestrian circulation scenario within a large rail terminal.

## The Emotion Model within buildingEXODUS

The new prototype emotion model monitors the time available for each agent to complete a set of assigned tasks and the estimated time required to complete those tasks. As the available time decreases the agent might have to decide to increase their movement speed and if necessary drop some or all of the assigned tasks while attempting to complete the critical assigned tasks. Within buildingEXODUS, pedestrian circulation movement is primarily determined by a list of tasks assigned to each agent (the *Itinerary*) which the agent completes as part of the simulation. Within the new prototype, these tasks are assigned to different categories: ***Compulsory Tasks***, these are tasks that must be completed e.g. 'purchase a train ticket'; ***Time Critical Tasks,*** these are tasks which must be completed at a particular time e.g. 'board the train before 17.45' and ***Elective Tasks***, these are tasks which are completely voluntary and while desirable to complete are not essential e.g. 'purchase a newspaper'. The agent's behaviour is therefore mostly governed by those tasks which are compulsory or time critical.

**Table 1. Behavioural responses triggered by the agent's individual time evaluation.**

| Time Assessment | Urgency State | Physical Behaviour | Itinerary Adjustment |
|---|---|---|---|
| $ERT_1 \leq AT$ | 0 | Continue as normal | - |
| $ERT_2 \leq AT < ERT_1$ | 1 | Escalate speed + drive, Reduce patience | - |
| $ERT_3 \leq AT < ERT_2$ | 1 | Escalate speed + drive, Reduce patience | Drop least important elective task and reassess situation with new itinerary |
| $AT < ERT_3$ | 1 | Escalate speed + drive, Reduce patience | Drop all remaining unessential tasks and reassess situation with new itinerary |

In the prototype version of the emotion model, a new personal parameter has been introduced which reflects the agents' attempt at fulfilling their assigned itinerary within the given constraints: the ***Urgency***. If the agent estimates that they have sufficient time to complete all of their assigned tasks, their Urgency is 0. As soon as they perceive that they have insufficient time to complete all of their remaining tasks their Urgency is increased to 1. The agent parameters drive, patience and walk speed have been modified to allow agents to vary these parameters according to their state of urgency. In addition, the tasks on the agents' itinerary have been prioritised, reflecting not only the compulsory, time critical and elective nature of the tasks, but also a ranking within each category. The task ranking determines the order in which the agents can drop their elective tasks if required. At dynamic time intervals, the agent calculates the amount of time remaining until their next critical time task i.e. the ***Available Time (AT)***, as well as estimates the amount of time that is required to complete all the remaining tasks i.e. the ***Estimated Required Time (ERT).*** Three threshold ERT's are determined ($ERT_i$) which are based on the individual agent's walk speed (slow or fast walk), perceived congestion considerations and appropriate safety or comfort factors. The agent then classifies their AT with respect to the ERT thresholds and adjusts their state of urgency and behavioural response as shown in Table 1.

## The Emotion Model applied in a Station Scenario

A comprehensive rail station geometry has been developed in buildingEXODUS to demonstrate the functionality of the emotion model in a complex multipurpose environment. The demonstration case geometry comprises two floors and thirteen platforms as well as a variety of retail and catering outlets which can be visited by the agents to accomplish elective tasks prior to boarding the required train. In this example we examine an off-peak station scenario involving about 7,750 agents and a total simulated time period of approximately two hours. Here we use the term ***Foot Passengers*** to represent those agents who arrive at the station via an external access point and leave the station on their assigned trains and ***Train Passengers*** to represent those agents who arrive at the station by train. A foot passenger's itinerary comprises a compulsory route i.e. an access point, by which the agent enters the geometry; a ticket office, where the agent purchases a ticket and a platform with an assigned train departure time. In addition to these compulsory and time critical tasks, foot passengers are assigned a random number of up to two elective tasks which they may want to undertake while at the station.

**Table 2. Average percentage of foot passengers who have skipped none, one or two of their initially assigned number of elective tasks.**

| Number of elective tasks initially assigned | Skipped no elective tasks (%) | Skipped one elective task (%) | Skipped two elective tasks (%) |
|---|---|---|---|
| One task | 78 | 22 | - |
| Two tasks | 50 | 8 | 42 |

To evaluate the impact of the prototype emotion model upon the agents' behaviour, we examine for each individual agent the proportion of the agent's elapsed time which the agent has spent in an urgent state i.e. the ***Urgency Ratio***. The average Urgency Ratio for a foot passenger was 14%, ranging from 0 to 65%. The large range clearly indicates that the individual agents experienced highly variable conditions. Furthermore, while on average 45% of the foot passengers did not enter an elevated urgency state, approximately 25% of the foot passengers spent more than 25% of their elapsed time in the station in an urgent state. Of those foot passengers assigned one and two elective tasks, 78% and 50% dropped none of their tasks respectively (see Table 2). To further assess the impact of the prototype emotion model the simulation was repeated with the emotion model disabled. We find that the agent's ability to alter their assigned itinerary has a strong influence on the foot passengers' punctuality, with no agents missing their assigned train when the emotion model is active and an average of approximately 22% of foot passengers missing their initially assigned train when the emotion model is deactivated. These results also demonstrate that only 22% of the foot passengers in fact needed to become urgent in order to catch their assigned train. The fact that 55% of the foot passengers entered an elevated urgency state indicates that a large number of agents were "rushing" needlessly – as is often the case in reality!

## Conclusions

This work has introduced a prototype emotion model into agent based circulation simulation and demonstrated its impact on time critical behaviour which on the surface appears to produce realistic behaviours. However, further work is required to; assess the model, introduce refinements to the urgency escalation parameter, introduce fuzziness into the ERT, provide more representative safety factor distributions (optimistic as well as pessimistic safety factors) and to the modelling of short and long term planning and decision making behaviour.

## References


1. Dolan, R.J.: Emotion, Cognition, and Behavior. Science. 298(5596):1191–1194 (2002)
2. Shao, W., Terzopoulos, D.: Autonomous pedestrians. Graphical Models. 69(5-6):246–274 (2007)
3. Gärling, T., Gillholm, R., Montgomery, W.: The role of anticipated time pressure in activity scheduling. Transportation. 26:173–191 (1999)
4. Pan, X., Han, C.S., Dauber, K., Law, K.H.: A multi-agent based framework for the simulation of human and social behaviors during emergency evacuations. AI & Society. 22(2):113–132 (2007)
5. Dijkstra, J., Jessurun, J., de Vries, B., Timmermans, H.: Agent Architecture for Simulating Pedestrians in the Built Environment. In: Fourth International Workshop on Agents in Traffic and Transportation at AAMAS-06. Hakodate Japan: ACM, pp. 8–16 (2006)
6. Galea, E.R, Sharp, G., Lawrence, P.J., Holden, R.: Approximating the Evacuation of the World Trade Center North Tower using Computer Simulation. Journal of Fire Protection Engineering. Vol. 18 (2), 85-115 (2008)

# A Study of Density of the Person in a Classroom for Building Evacuation Safety Regulations in Korea

**H. Youn[1], Y. Hwang[1], and Y. Kwon[2]**

[1]Korea Institute of Construction Technology, Goyang-si, Gyeonggi-do, Republic of Korea
[2]Hoseo Universty*
Corresponding author: yhj@kict.re.kr

**Abstract** Density of the person in a classroom is one of the important factors to evaluate building evacuation safety regulation if being fire a building. It provides the number of persons in the classroom for the unit area in according to building purposes on Code or Standard in the evacuation overseas countries of USA or UK and in Japan also, institute of Fire organizes a research committee and investigates the number of persons in the classroom for every service under cooperation of Tokyo Fire Department and Construction Companies, and prescribes to consider them for design a building.

However, it is approved only a specified criterion as a law without a criterion in according to building purposes in Korea, and then it has limitation to evaluate whether the person in a classroom has evacuation stability in multiple buildings and some of part might not be suitable with the condition in Korea, even to fit the density criterion of the person in a classroom of overseas simply, because it might be difference a structural characteristic for using and composing for every country even though buildings have the same purposes.

For that, this study presented density of the person in a classroom after investigating survey research of buildings that has high risk to make injuries to persons if being fire in a large shopping mall, multiple buildings and a large multiple theater for setting up the density criterion of the person in a classroom that is suitable to the condition in Korea..

R.D. Peacock et al. (eds.), *Pedestrian and Evacuation Dynamics*,
DOI 10.1007/978-1-4419-9725-8_78, 

## A Notion and Research Method on Occupant Density in Internal and External

### *Definition on a Notion of Occupant Density in Internal and External*

It is the occupant density that calculates the number of expected occupancies who will be inside of building as the most important factor if we assume the evacuation behavior and assess the safety performance on fire. The value of occupant density has to be fixed according with the building use so that we can calculate qualitatively the evacuation capacity of the stairs and a hallway over the evacuation routes, can design reasonably to ensure the life safety of an occupant on fire.

There are different ways to calculate the occupant density by every country, and it is not fixed the way to calculate the occupant density of a building in the case of internal. There is only the grid method in which calculates the number of persons at the place where a great number of people crowds in the beaches, but this method has difficulty to calculate the precisely number of persons because it is to suppose the whole of persons to count the average number of persons for every sector after dividing the sectors.

Therefore, first, this study established the methodology through considering the literature, such as USA', UK' and Japan', which conducted the advanced research for calculating the occupant density that reflected the own nature in internal, and used the research method in which executed the field survey and the sensor survey and compared them. So then this study analyzed the number of occupancy.

### *Research on Calculating the Occupant Density*

This study primarily selected the building(wholesale mart, the complex building, the multi-complex cinema) as the subject that is possible to make large loss of life because the frequency of use and the risk factors are higher on fire for calculating occupant density in internal.

This research was conducted for the period when the customers were most concentrated among 1years. It used the method that based on survey methodology in which conducted the field survey and the sensor parallel.

**Table 1. Comparison on definition of the available people & calculation method in Internal and external**

| . | USA | Korea | Japan |
|---|---|---|---|
| Provisons | NFPA Life Safety Code NFPA | Law for setting and maintaining the fire facilities & safety control | Enforcement regulations of Fire Service Act |
| Characteristic | Safe Level presented by Fire protection association (voluntary coverage) | Mandate by Law | Mandate by Law |
| Definition of A.P | Occupants load, There is no difference between the occupant density and the definition of the number of persons to be admitted and it presents the number of persons to be admitted as an area hold by every person. | That the number of persons to be admitted on the Fire Services Act generally means minimum for setting Fire service system. | That the number of persons to be admitted on the Fire Services Act generally means minimum for setting Fire service system. |
| Calculation method of A.P | Assume that there is at least one person for every unit area given by density parameter.<br>It is divided into the numerical value marked by total area and by Net area. | Calculate the number of persons to be admitted with the result that the areas for every facility divided by every constant value. | Calculate the number of persons to be admitted with the result that the number of employees for every facility divided by constant value.<br>Except, evacuation safe Test-test which is performance evacuation safe design doesn't use the number of persons to be admitted but the value of occupant density. |

- Field survey: for measuring the occupancy persons, it used the measurement methods in which the research persons were placed in every entrance and moving walk who calculates the number of coming and going, and measured the occupancy persons in whole building and in every floor, excepted the children and infants who is impossible to walk by themselves.
- Sensor survey: The value of sensor survey was measured the number of customer who via the checkout for purchasing the products, which predicted the occupancy persons to apply correction value of three times to the value of sensor survey.

**Table 2. Research method and research period on occupant density [persons/m²]**

| Section | Research Facility | Research method & period |
|---|---|---|
| Wholesale Mart | Facility A | Field Survey : 2007. 09. 15~16<br>Sensor Survey: 2007. 09. 01~30 |
| | Facility B | Sensor Survey: 2007. 09. 01~30 |
| | Facility C | Field Survey : 2008. 01. 27<br>Sensor Survey : 2007. 09. 01~30, 2008. 01. 01~31 |
| | Facility D | Sensor Survey : 2007. 09. 01~30, 2008. 01. 01~31 |
| Complex Building | Facility E | Field Survey : 2008. 03. 29, 2008. 10. 25 |
| | Facility F | Field Survey :2008. 05. 31 |
| Multi-Complex Cinema | Facility G | Field Survey : 2008.09.27 |
| | Facility H | Field Survey: 2008.07.12 |

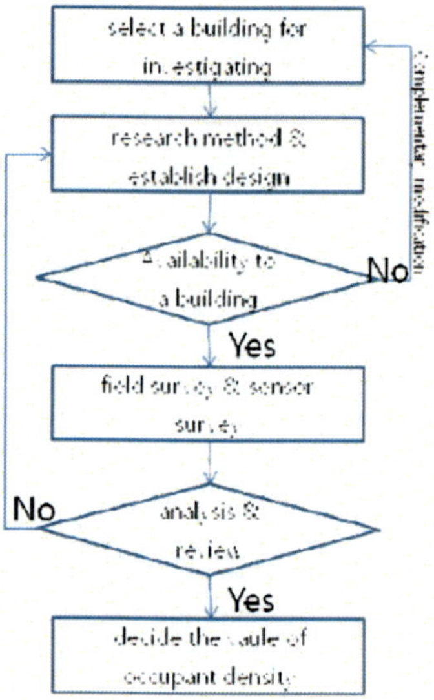


**Fig. 1. Research method**

## The Result of Research and the Analysis on Occupant Density

This researcher conducted the survey of occupant density on the wholesale mart, the complex building and the multi-complex cinema.

This study presented the result that the average occupant density is 0.41[persons/㎡], and the maximum occupant density is 0.46[persons/㎡], which was on the facility B in the Metropolitan area.

In the case of wholesale mart, the occupant density was presented 100:80 in the result that compared the field measurement survey with the sensor survey, the value of sensor survey was low, and it was presented the similar appearance if applying the correction value, 1.25folds.

There are many differences with 0.43[persons/㎡] and 1.00[persons/㎡] in the result of two researches.

It was changed delicately according with the research environment(the research time, the climate, the location) because it has a limit to ensure adequacy of the survey due to the nature of complex building in internal.

When the researcher compared the survey result of occupant density in internal with one in external, this study presented a insignificant difference, but such as difference, is difference with the survey environment and the research method only, it represented the similar result.

**Table 3. The result of occupant density in Internal & comparing with external**

| | Korea | USA | Japan | UK |
|---|---|---|---|---|
| Wholesale Mart | Facility A 0.36<br>Facility B 0.46<br>Facility C 0.42<br>Facility E 0.38<br>Average 0.41 | shop 0.36.<br>a)ground floor 0.36.<br>b)basement 0.27.<br>c)multiple under-ground 0.18.<br>d)the other floor 0.20~<br>mall 0.35.<br>store 0.036.<br>kitchen 0.11. | shop (including a passage) 0.50<br>continuous stores<br>a) shop 0.50<br>b) passage 0.25 | shop (including a department store, supermarket)0.50<br>shop(furniture, cycle, etc.) 0.14<br>mall 01.33<br>store 033<br>kitchen 0.1 |
| Complex Building | Facility E 0.43<br>Faciltity F 1.00<br>Average 0.72 | safety floor sale area 2.8<br>more than 2nd floor sale area 3.7<br>basement sale area 2.8 | shop 0.5<br>restaurant 0.7<br>shop 0.5<br>passage 0.25<br>meeting hall 1.5 | space for shopping center 2.0<br>department store 7.0 |
| Multi complex cinema | Facility G 0.43<br>Facility H 0.51<br>Average 0.47 | number of fixed seats | seats 1.5 | number of fixed seats |

However, in the future, for making the own occupant density standard like overseas', the study has to be progressed to find the way to solve the nature of the domestic buildings which are delicate according with the survey environment, such as the survey date, the climate and the location, and to study the segmentation of the domestic buildings and the establish of occupant density research.

## Conclusion and Moving Foward

Calculation of the occupant density in a building is the most important factor during the evacuation plan, the value of occupant density has to be fixed according with applications of a building so that we can calculate reasonably the evacuation capacity as well minimize the injuries to persons due to on fire.

Nevertheless, it never have been executed such the study on the calculation of occupant density in internal, only it has utilized the evacuation density standard from the advanced countries. Therefore, for calculating the occupant density reflected by the own nature in internal, first, this study established the research methodology over considering the literatures of which is in USA, UK and Japan, ect where conducted the advanced research and performed the field survey and the sensor survey on a building in internal, so that assumed the occupancy persons with comparing mutual.

Yet, this study presented that to calculate the occupant density to apply the statical methods within short-term has a limitation because the occupant density is very different according with the survey date, the survey environment, the climate and the location of a building.

So then, for making the standard guidance with researching on the actual condition of the occupant density in domestic buildings from now on, it will have to be made systematic and concrete the program of evacuation safety design as follows:

- Need to establish the research scope system of occupant density in internal on the location, service, time and environment to the research objects.
- Need to survey the concrete and delicate occupant density for every service and contracture.
- Need to suggest a program that is possible to apply to the building in which the use and the occupant density is severe changed.
- Need to discuss and review for the result of investigating and analyzing with a public hearing hold by the different facilities.

## References

1. Building Code of Australia. AUBRCC. 12 (1988)
2. Securite Contre L'Incendie-dons les etablissement recevant du public. Journal Officiel De La Republique Francaise. No.1477-1 (1984) (English language translation)

3. Securite Contre L'Incendie Tome II, Journal Officiel De La Republique Francaise, No.1011-II (1982) (English language translation)
4. The Building Regulations (1991-Approved Document B). HMSO (1991)
5. British Standard 5585 - Fire Precautions in the design, construction and use of building (Parts 1, 2, 3, 6 & 10). BSI (1990,1985, 1983,1991,1991)
6. NFPA 101 Life Safety Code. NFPA (1994)
7. Uniform Building Code. ICBO (1994)
8. Building Code, State of Wisconsin. Industrial Commission of Wisconsin (1918)
9. Report of Commitee on Safety to Life. In: Proceedings of 23th Annual Meeting. NFPA (1919)
10. Report of Commitee on Safety to Life. In: Proceedings of 38th Annual Meeting. NFPA (1934)
11. Report on Means of Escape from Fire. Building Industries National Council (1935)
12. Report on Means of Escape in case of Fire. Building Industries National Council (1945)
13. A Study on the Architectural Evacuation Guideline and the Improvement of Design Technology. KICT. 5 (2008)

# Experiments on Egress of Persons with Mobile Disability in Train Car

**K. Jong-Hoon[1], K. Woon-Hyung[1], R. Sam-Kew[2], L. Duck-Hee[3], J. Woo-Sung[3], and H. Kyung[4]**

[1]Kyungmin College, Ganeung 3, Uijeoungbusi, Gyeonggido, Korea
[2]Kwangwoon University
[3]Korea Railroad Research Institute
[4]Korea Institute of Construction Technology
Corresponding author:aina47@hanmail.net

**Abstract** Korean railroad company is continuously improving the service for the disabled. They also install and reform many facilities for them. However, It is not considered the importance of egress when fire occurred in a cabin Now, Fire hazard analysis and risk assessment are introduced into domestic regulations for improving fire safety performance of train car and railroad facilities. Considering the fact that fire safety evaluation process doesn't fully include disabled person, fundamental egress data for them is very important for life safety aspect. In this paper, Egress experiments and analysis were conducted to measure the movement speed of the disabled with Mugunghwa and Saemaul.

## Introduction

After Daegu subway fire in 2003, Many efforts for railway fire safety are conducting in Korea. One of these works is to construct the guideline for fire hazard assessment for rail passenger car using fire modeling. The main purpose of fire hazard assessment is to find the design point to improve the fire safety of rail passenger car. Especially, it is most important to prevent the loss of life. The evaluation of life safety requires a balanced comparison of predicted fire condition and egress prediction. For that reason, train car designer should show that the time required to escape (RSET) will be less than the time available to escape (ASET).1 Egress model can be the most effective tool for estimating egress time on train car fire. If reliable outcome is expected from egress model, reliable data is needed. However, the data that can be used in Korea has problems. One is most data measured in buildings, the other is that the data is taken from western people or Japanese, not Korean. For this reason, this experiment was conducted and analyzed to obtain reliable data for persons with mobile disability in Korea.

R.D. Peacock et al. (eds.), *Pedestrian and Evacuation Dynamics*,
DOI 10.1007/978-1-4419-9725-8_79, 

## Survey of the Disabled Egress on Korean Train Car

The survey of Korean trains and subway cars, Mugunghwa, Saemaul, and KTX was conducted in order to inquire the egress of the disabled. The way of survey is to collect and analyze train car plans. Wheel chair user is able to get on and off all train cars. Wheel chair user can move through car to car and corridor on subway car and Mugunghwa. However, it is impossible to move through corridor in Saemaul and KTX. In the move from car to platform, height of car floor level and platform level are the same. Passenger can use the stair to get on and off the train as Mugunghwa, Saemaul, and KTX. If the train stop at somewhere between the stations, people obviously jump down to track side to get off the train car. Height between car floor and track side is about 1.2~1.3m. Wheelchair user is not able to get off the train by itself if train stop at the track.

## Experimental Method

The experiment was conducted in Korail Seoul branch suseak maintenance facility in Seoul, Korea. Train cars was stopped for the maintenance in that facility. A total of 25 persons took part in the experiment as an evacuee. Participants aged from 20 to 40 years. two severely handicapped persons are joined as a experiment participants. Egress time was measured by experiment assistances using CC-TV camera and stop watch.

## Experiments on Mugunghwa Car

The movement distance between start line and finish line was 54m. Wheelchair user can move through corridor, door, and gangway on Mugunghwa car. Especially, door and gangway has enough space to pass it. However, wheelchair user with assistant spent much time to pass through door and gangway on the experiment. Wheelchair user should be carried in the assistant's arm if it get off the train. The time of carrying was measured.

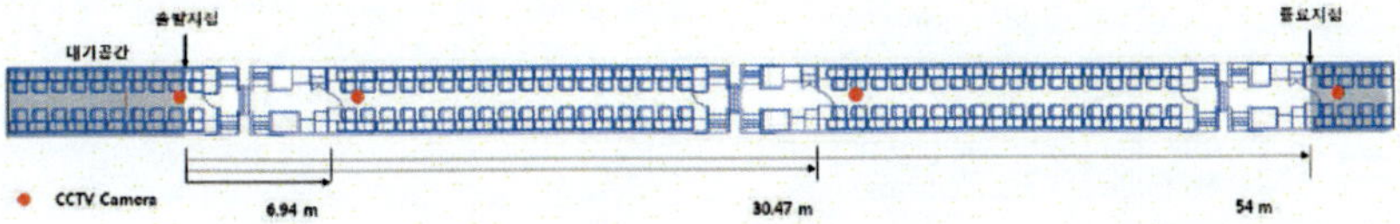


**Fig. 1. Plan of Mugunghwa Car**

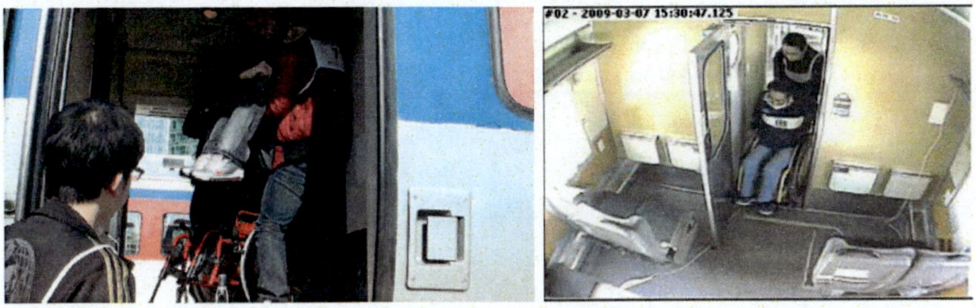


**Fig. 2. Experiment scenes**

**Table 1. Movement time and speed on Mugunghwa Car**

| Subject Group | Average time(s) | Speed(m/s) |
|---|---|---|
| Waking (Male 20's) | 38.18 | 1.41 |
| Waking (Female 20's) | 38.12 | 1.42 |
| Waking (Male 30-40's) | 38.81 | 1.39 |
| Crutches User (male)* | 102 | 0.53 |
| Crutches User (Female)* | 83.16 | 0.65 |
| Crutches User and assistant | 64.58 | 0.84 |
| wheelchair User with assistant (severely handicapped person) | 124 | 0.44 |
| wheelchair User with assistant | 136.69 | 0.40 |

**Table 2. Wheel chair user and helper group movement times to get out of the car**

| Subject Group | Time (s) |
|---|---|
| Carrying male handicapped person in male assistant's arms | 28.07 |
| Carrying female handicapped person in male assistant's arms (1st) | 22.56 |
| Carrying female handicapped person in male assistant's arms (2nd) | 10.97 |
| Carrying female handicapped person in male assistant's arms (3rd) | 11.00 |

## Experiments on Saemaul Car

The movement distance between start line and finish line was 35.4m. Wheelchair user is not able to move through corridor on Saemaul car. The travel time and speed of Person without disability and Crutches users were measured.

**Table 3. Movement time and speed on Saemaul Car**

| Subject Group | Average time(s) | Speed(m/s) |
|---|---|---|
| Walking (Male 20's) | 25.82 | 1.37 |
| Walking (Female 20's) | 22.66 | 1.56 |
| Walking (Male 30-40's) | 26.75 | 1.32 |
| Crutches User (Male) | 56.94 | 0.62 |
| Crutches User (Female) | 60.94 | 0.58 |

**Table 4. Comparison walking speed data**

| disability | Walking Speed (m/s) | | |
|---|---|---|---|
| | Western[1] | Japan [3] | Experiment Result |
| without disability | 1.25 | 1.0 | 1.32~1.56 |
| crutches | 0.94 | - | 0.53~0.65 |
| Electric Wheelchair | 0.89 | - | - |
| Manual Wheelchair | 0.69 | 0.5 | - |
| Assisted manual wheel chair | 1.30 | - | 0.4~0.44 |

## Conclusion

In this experiment, wheelchair user was not able to pass door and gangway on the train car. wheelchair user with assistant spent much time to pass through it. The reason is the floor between car to car is not flat and the space is not enough wide to pass. The result of comparison of experiment data with data from Fire SERT and Japanese show the difference. Especially, Korean participants as assisted manual wheel chair were slower than western people.

## References

1. Proulx, G.: Movement of People: The Evacuation Timing. In: SFPE Handbook of Fire Protection Engineering, pp. 3-360 (2002)
2. Boyce, K.E., Shields, T.J., Silcock, G.W.H.: Toward the Characterization of building occupancies for fire safety engineering prevalence, type, and mobility of disabled people. Fire Technology. 35, pp. 42 (1999)
3. SFPE Japan Chapter: Fire Safety Design of Apartment Building Japanese Case Study. In: 7th International Conference on Performance-based Codes and Fire Safety Design Methods, pp. 29 (2008)

# On Time Scaling and Validation of a Stochastic CA Pedestrian Dynamics Model

**E. Kirik[1], T. Yurgel'yan[2], and D. Krouglov[3]**

[1]Institute of Computational Modeling SB RAS
Akademgorodok, Krasnoyarsk, Russia, 660036
[2]Siberian Federal University, Institute of Mathematics
[3]V.N. Sukachev Institute of forest SB RAS
Corresponding author: kirik@icm.krasn.ru

**Abstract** This paper deals with a problem of time scaling and validation of a mathematical model of a pedestrian flow. We focus on stochastic cellular automata approach. What kind of tests should be applied to say that model ``works"? In this paper some our tests and time scaling observations are presented.

## Introduction

Modeling of pedestrian dynamics is an actual problem today. Different approaches from the social force model based on differential equations to stochastic CA models are developed [1]. They reproduce many collective properties and are an important basis for pedestrian modeling. But there are still things to be done.

There is still no common set of test to verify models from uniform positions. Validation of models with fundamental diagrams doesn't solve this problem completely. For example, diffusion of the flow can't be reproduced by the model and can't be checked consequently.

Here we present our attempt to investigate dynamics of our model and some observations about time scaling. Note that very simple case studies are used; they don't allow for all aspects of model dynamics to be pronounced. So it is only starting (but obligatory) point for complex investigation of the model.

## Model

The model is stochastic discrete CA model and supposes short-term decisions made by the pedestrians [3]. From the comprehensive theory of pedestrian dynam-

R.D. Peacock et al. (eds.), *Pedestrian and Evacuation Dynamics*,
DOI 10.1007/978-1-4419-9725-8_80, © Springer Science+Business Media, LLC 2011

ics [1] such model may be refereed to tactical level. (The model was presented at the PED2008.)

The space (plane) is known and sampled into cells $40cm \times 40cm$ which can either be empty or occupied by one pedestrian (particle) only [2]. Cells may be occupied by walls and other nonmovable obstacles.

The model imports idea of a map (static floor field $S$) from the floor field (FF) CA model [2] that provides pedestrians with information about ways to exits. Our field S increases radially from exit cells. It doesn't evolve with time and isn't changed by the presence of the particles.

A target point for each pedestrian is the nearest exit. Each particle can move to one of four its next-neighbor cells or to stay in present cell (the von Neumann neighborhood) at each discrete time step $t \to t+1$; i.e., $v_{\max} = 1[step]$.

A typical scheme for stochastic CA models is used in our model. There is step of some preliminary calculations (field $S$ is computed). Then at each time step transition probabilities are calculated, and directions are chosen. If there are more then one candidates to one cell a conflict resolution procedure is applied, and then a simultaneous transition of all particles is made.

Because of a restricted volume we omit here update rules and probability formulas. We only note here that normal (not emergent) directed evacuation was investigated; i.e., pedestrian sees, knows, and wants to go to the exit very much. This supposes strong influence of the static floor field $S$, therefore $k_S = 4$ ($k_S$ is a sensitivity parameter of the field $S$).

## *Case study*

The following case studies were used, see Fig.1. We considered long rooms $2m \times 50m$, $2m \times 100m$, and $2m \times 150m$. Set of initial numbers of particles $N$ was considered, see Table 1. For each room for each initial density $\rho_0$ 100-500 runs were made (initial positions of particles were fixed).

Observables obtained during experiments are the following: total evacuation time (in steps) distribution (as an estimate of the total evacuation time ($T_{tot}^{50}$, $T_{tot}^{100}$, $T_{tot}^{150}$) we use mode of each time distribution); direction frequencies distribution over series of experiment for each $\rho_0$ for each room; $\widetilde{T}_{tot}^{50} = T_{tot}^{50}[st]$, $\widetilde{T}_{tot}^{100} = T_{tot}^{100} - 125[st]$, $\widetilde{T}_{tot}^{150} = T_{tot}^{150} - 250[st]$ (see Table 1). We calculated the flows $J^{50} = N/\widetilde{T}^{50}/2$, $J^{100} = N/\widetilde{T}^{100}/2$, $J^{150} = N/\widetilde{T}^{150}/2$.

**Table 1.**

| $\rho_0$ [1/m²] | $\rho_0$ | $N$ | $\tilde{T}_{tot}^{50}$ | $\tilde{T}_{tot}^{100}$ | $\tilde{T}_{tot}^{150}$ |
|---|---|---|---|---|---|
| 0,25 | 0,04 | 25 | 130 | 133 | 141 |
| 1 | 0,16 | 100 | 132 | 138 | 141 |
| 2 | 0,32 | 200 | 140 | 150 | 157 |
| 3 | 0,48 | 300 | 154 | 170 | 184 |
| 3,5 | 0,56 | 350 | 166 | 193 | 206 |
| 4 | 0,64 | 400 | 190 | 213 | 229 |
| 4,5 | 0,72 | 450 | 211 | 234 | 247 |
| 5 | 0,80 | 500 | 232 | 261 | 275 |
| 5,75 | 0,92 | 575 | 265 | 291 | 311 |

For the 50 m room we have $\min \tilde{T}_{tot}^{50}(\rho_0^{\min}) = 125$, $\min \tilde{T}_{tot}^{50}(\rho_0^{\max}) = 250$ (due to exclusion principle). Shifts that the model gives (see Table 1, column $\tilde{T}_{tot}^{50}$) are coursed by not strictly one-dimensional motion and stochastic nature of the model. Nevertheless Figure 2 shows strong prevalence of the right side movements with gradually increasing ``No" leaving present position under increasing density. Left direction (opposite to exit) has a miserable value and decreases with increasing initial density. Such behavior of virtual people may be refereed to realistic.

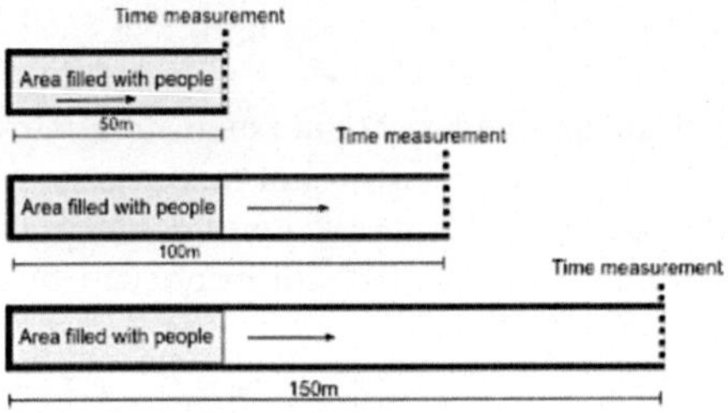


**Fig. 1. Case study**

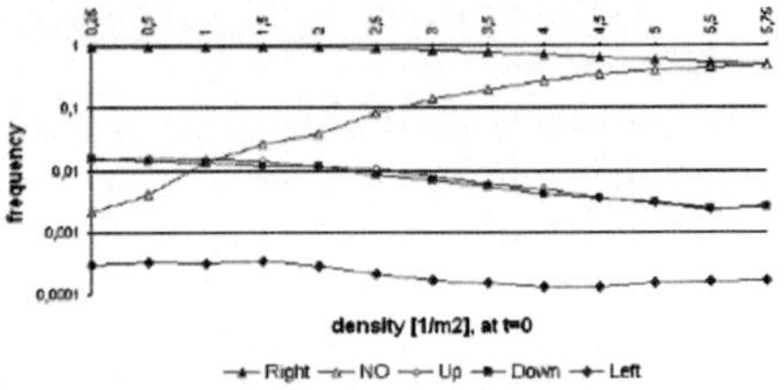


**Fig. 2. Direction frequency distribution, 50 m.**

Experiments with 100m and 150m rooms show diffusion of the flow (decompression). In such environment a real people flow has diffusion and it's more pronounced with increasing initial density. Divergence of data from columns $\tilde{T}_{tot}^{100}$, $\tilde{T}_{tot}^{150}$ with corresponding values from column $\tilde{T}_{tot}^{50}$ says that the model realizes the diffusion. Diffusion doesn't present directly in the model. This effect is simulated indirectly, and the main contribution is given by the stochasticity of the model. Figure 3 shows that flows $J^{100}$, $J^{150}$ are lower then $J^{50}$ and it becomes more pronounced under higher $\rho_0$; non of the flows reach maximal possible value 1.252.

Figure 4 gives some observations on time scaling problem in CA models. For the 50 m room line 3 is the flow $J = N/T/2 \approx v\rho$ after the data of Predtechenskii and Milinskii, which may be considered as an upper bound of the flow in this case study. Line 1 is the flow $J^{50} = N/\widetilde{T}^{50}/2$, which may be considered as a low bound of the flow. Line 4 gives flow $J^{50} = N/(\widetilde{T}^{50}_{tot} \cdot 0.3)/2$, which is unrealistically higher than the estimated upper bound. Therefore timescale $\Delta t = 0.3[s] = 0.4[m]/1.3[m/s]$ is not proper starting with $\rho_0 > 1[per/m^2]$. Line 2 corresponds to the flow $J^{50} = N/(\widetilde{T}^{50}_{tot} \cdot \Delta t(\rho_0))/2$, where $\Delta t(\rho_0) = \begin{cases} 0.4/v(\rho_0), & \rho_0 \le 3; \\ 0.4/v(\rho = 3), & \rho_0 > 3. \end{cases}$

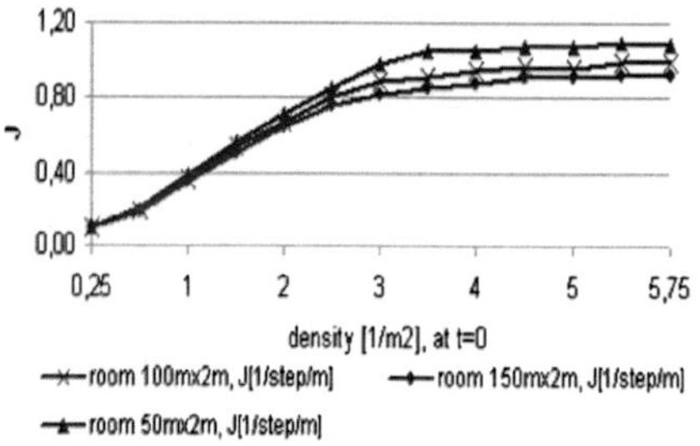


**Fig. 3.**

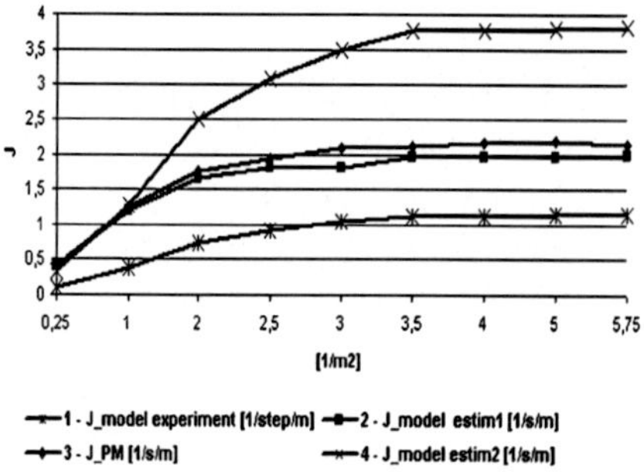


**Fig. 4.**

Thus our experiments (tests) show that model (under certain parameters) provides directed movement of the particles, diffusion of the flow. For this case study one better method of time scaling is proposed. But these results don't describe all the dynamical properties of the model. Turning flows are not still mentioned, and influence of turnings to model dynamics is not investigated.

## References


1. Schadschneider, A., Klingsch, W., Kluepfel, H., Kretz, T., Rogsch, C., Seyfried A.: Evacuation Dynamics: Empirical Results, Modeling and Applications. Encyclopedia of Complexity and System Science. Springer (2009
2. Schadschneider, A., Seyfried, A.: Validation of CA models of pedestrian dynamics with fundamental diagrams. Cybernetics and Systems. **40** (5), 367-389 (2009)
3. Kirik, E., Yurgel'yan, T., Krouglov, D.: The Shortest Time and/or the Shortest Path Strategies in a CA FF Pedestrian Dynamics Model/ Journal of Sib. Fed. University, Mathematics and Physics Series. **2** (3), 271–278 (2009)

# FDS+Evac: Modelling Pedestrian Movement in Crowds

**T. Korhonen and S. Heliövaara**

VTT Technical Research Centre of Finland
P.O. Box 1000, FI-02044 VTT, Finland
Corresponding author: timo.korhonen@vtt.fi
Systems Analysis Laboratory, Aalto University,
P.O. Box 11100, FI-00076 Aalto, Finland

**Abstract** The recent developments of the evacuation programme FDS+Evac are presented including a counterflow model for continuum crowd modelling. The major new additions are described and some verification and validation results are presented including an analysis of specific flows through doors and corridors. The presented additions are implemented in the latest version of FDS+Evac. This version is embedded in the computational fluid dynamics based fire simulation programme Fire Dynamics Simulator, which is freely obtainable from the FDS web page http://fire.nist.gov/fds/.

## Introduction

Performance based fire codes allow the use of numerical simulation of fire and evacuation processes to be used to improve fire safety in buildings. However, the usability of some current evacuation models is limited because they do not take into account the individual properties and decision making processes of humans, the dynamics of large crowds, and the interaction between fire and people. One specific property is the ability of the model to simulate the counterflow of humans in dense crowds.

In this article, recent developments of the agent based evacuation modelling programme FDS+Evac [1] are presented. The programme models the evacuation process of a building floor as a 2D system, where autonomous agents simulating the escaping humans are moving according to equations of motion and decision making processes. The different floors of the building can be connected through stairs. FDS+Evac follows the trajectories of the agents in continuous space and time. The agents choose their target exits using an algorithm, where the familiarity and visibility of the exits are considered together with the queues at the exits. FDS+Evac is coupled with the computational fluid dynamics based fire simulation programme Fire Dynamics Simulator (FDS) [2] and the smoke and toxic gas information is used to reduce the movement speeds, calculate intoxication effects, and modify the exit choice of the agents.

R.D. Peacock et al. (eds.), *Pedestrian and Evacuation Dynamics*,
DOI 10.1007/978-1-4419-9725-8_81, 

## FDS+Evac Model

The method of Helbing et al. [3-6] is used as the starting point for the pedestrian movement method used in FDS+Evac. The method introduces so-called "social force", which is used to keep reasonable distances between pedestrians and between pedestrians and walls. For a description of the method, see the papers by Helbing et al. and references therein. For the modification of a one-circle representation of the elliptical cross sectional shape of a human body to a three-circle one, where one large circle describes the torso and two smaller ones the shoulders, see the papers by Langston et al. [7] and Korhonen and Hostikka [1].

The original agent movement model introduced by Helbing et al. is not well suited for situations, where agents are going to different directions and their paths are crossing or opposite to each other. To overcome this deficiency of the model, a counterflow model was implemented in FDS+Evac [8].

Also a model for an entire staircase was added to the programme. Now the agents are treated in stairs similarly to the horizontal floors, i.e., using the equations of motion. This enables the modelling of merging flows in staircases easily and it also enables counterflow situations, i.e., the agents can have different target floors inside the staircase. Thus, it can be used to model the effect of fire fighters ascending the stairs while the evacuees are descending the stairs. Counterflow movement can also arise in maritime applications, where the personnel may have different duties during the evacuation process and the passengers may have different target embarkation stations at different levels of the ship.

## Model Validation

In Figure 1 the results for specific flows through doors and corridors calculated by FDS+Evac are compared to some other simulation programmes and hand calculation formulas. The simulation results of the programmes MASSEgress and Simulex are extracted from Pan's thesis [9] and the hand calculation formula for corridor flow is from the SFPE Handbook [10]. The FDS+Evac simulations were performed with two different sets of parameters, the default ones and a set that produces a more rapid flow. It is seen that FDS+Evac is able to produce reasonable flows in corridors and through doors.

The counterflow model was tested using the counterflow test case included in the IMO test cases for evacuation programmes for maritime applications [11]. There two 10 m × 10 m rooms are connected by one 2 m wide and 10 m long corridor. The other room is populated with 100 agents and the other with 0, 10, 50, and 100 agents. The results of FDS+Evac simulations including the counterflow model were 44 s, 69 s, 128 s, and 205 s for the cases, respectively. The results of FDS+Evac without the counterflow model were 41 s and 305 s for the cases, where there were 0 and 10 agents in the other room. For the other cases jams were formed, which produced a really slow movement or no movement at all in the corridor [1].

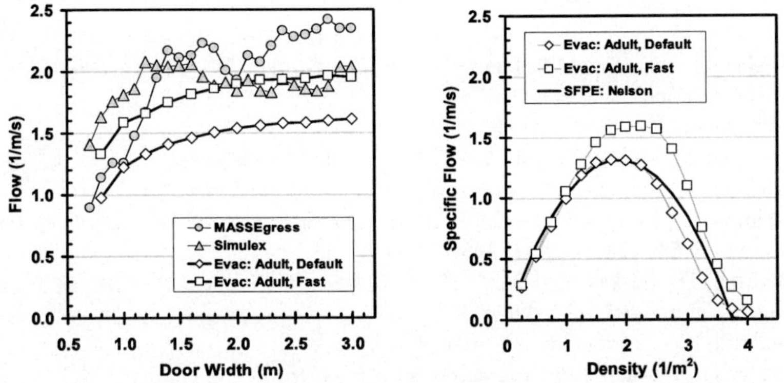


**Fig. 1. The specific flows through doors (on left) and corridors (on right) calculated using FDS+Evac compared to other simulation programmes and SFPE Handbook values.**

## Conclusions

FDS+Evac model is and has been able to predict the specific flows through doors and corridors. Qualitative analysis shows that for the situations, where there is counterflow or the trajectories of the agents are crossing and the density is reasonably high, the newly added counterflow model produces much more realistic behaviour than FDS+Evac without the counterflow model.

The presented models are implemented in the latest version of computational tool for evacuation modelling, FDS+Evac, which is embedded inside FDS (version 5.4.3). The executable and the source code are obtainable from the FDS web page at http://fire.nist.gov/fds/. The documentation of FDS+Evac is found at the web page http://www.vtt.fi/fdsevac/.

**Acknowledgments** The development work of FDS+Evac has been funded by VTT Technical Research Centre of Finland, the Finnish Funding Agency for Technology and Innovation, the Finnish Fire Protection Fund, the Ministry of the Environment, and the Academy of Finland.

## References


1. Korhonen, T., Hostikka, S.: Fire Dynamics Simulator with Evacuation: FDS+Evac - Technical Reference and User's Guide. VTT Working Papers 119, VTT Technical Research Centre of Finland. Espoo, Finland (2009)
2. McGrattan, K., Klein, B., Hostikka, S., Floyd, J.: Fire Dynamics Simulator (version 5) User's Guide. NIST Special Publication 1019-5. Gaithersburg (2008)
3. Helbing, D., Molnár, P.: Social Force Model for Pedestrian Dynamics. Physical Review E. **51**, 4282–4286 (1995)
4. Helbing, D., Farkas, I., Vicsek, T.: Simulating Dynamical Features of Escape Panic. Nature. **407**, 487–490 (2000)
5. Helbing, D., Farkas, I., Molnár, P., Vicsek, T.: Simulating of Pedestrian Crowds in Normal and Evacuation Situations. In: Schreckenberg, M., Sharma, S. D. (Eds.), Pedestrian and Evacuation Dynamics, pp. 21–58. Springer, Berlin (2002)
6. Werner, T., Helbing, D.: The Social Force Pedestrian Model Applied to Real Life Scenarios. In: Pedestrian and Evacuation Dynamics – Proceedings of the Second International Conference, pp. 17–26. University of Greenwich, London (2003)
7. Langston, P.A., Masling, R., Asmar, B. N.: Crowd Dynamics Discrete Element Multi-Circle Model. Safety Science. **44**, 395–417 (2006)
8. Heliövaara, S., Korhonen, T.: Counterflow Model for FDS+Evac Simulations, *ibid.*
9. Pan, X.: Computational Modeling of Human and Social Behaviors for Emergency Egress Analysis. Dissertation, Stanford University. Palo Alto (2006)
10. Gwynne, S.M.V., Rosenbaum, E.R.: Employing the Hydraulic Model in Assessing Emergency Movement. In: DiNenno, P. J., Drysdale, D., Beyler, C.L., Walton, W.D., Custer, R.L.P., Hall, J.R., Jr., Watts, J.M., Jr. (Eds.), The SFPE Handbook of Fire Protection Engineering (Fourth ed.), pp. 3-373-3-396. Society of Fire Protection Engineers, Bethesda (2008)
11. Guidelines for Evacuation Analyses for New and Existing Passenger Ships, MSC/Circ. 1238. International Maritime Organization, London (2007)

# Behaviour and Perception-based Pedestrian Evacuation Simulation

**T. Kretz[1], G. Mayer[1], and A. Mühlberger[2]**

[1]PTV Planung Transport Verkehr AG, Kriegerstraße 15, D-70191 Stuttgart
[2]Psychologie I, Universität Würzburg, Marcusstraße 9-11, D-97070 Würzburg
Corresponding author: georg.mayer@ptv.de

**Abstract** This contribution reports on the research project SKRIBT and some of its results. An evacuation simulation based on VISSIM's pedestrian dynamics simulation was developed, that - with high time resolution - integrates results from studies on behavior in stress and crisis situations, results from CFD models for e.g. fire dynamics simulations, and considers visibility of signage and — adding a psychological model — its cognition. A crucial issue is the cognition of smoke or fire by the occupant and his / her resulting spontaneous or deliberate reaction to this episode.

## Introduction and Motivation

To estimate numbers of affected occupants in cases of emergency evacuations, simulation models have been developed that already for a number of years not only perform microscopic simulations of pedestrian dynamics, but also calculate effects from physical hazards on the occupants — mostly by means of fractional effective dose models. But to our knowledge there is no model that considers perception and behavior in such special environments more than to a rudimentary degree.

As part of the research project SKRIBT [1] an evacuation simulation based on VISSIM's pedestrian dynamics simulation [2] was developed, that - with high time resolution - integrates results from studies on behavior in stress and crisis situations, results from CFD models for e.g. fire dynamics simulations, and considers visibility of signage and — adding a psychological model — its cognition. A crucial issue is the cognition of smoke or fire by the occupant and his / her resulting spontaneous or deliberate reaction to this episode.

R.D. Peacock et al. (eds.), *Pedestrian and Evacuation Dynamics*,
DOI 10.1007/978-1-4419-9725-8_82, 

## Simulation of Pedestrians

For the calculation of damage degree and countermeasure efficacy one needs models with high resolution in time and space to determine the influences on the occupants. This was done using numerical calculation schemes (CFD models).

The simulation of pedestrian dynamics was done with VISSIM, using the Social Force Model [3]. The latest version of the model [4] was extended, so it can be applied for the simulation of conflict areas (interaction of pedestrians and vehicles), multi-storey buildings, pedestrians as passengers of public transport, and waiting queues.

Route choice and pre-movement time depend for one on a calculation of visibility of exit signage (and the routes that are linked with a sign) and after by the principal availability of information that has been calculated by a behavioral model. The successful reaction on external influences requires both the perception and the cognition of the situation and information. Apart from individual parameters, this depends on the local conditions.

The calculation of visibility considers the orientation of an occupant in space, his body height, obstacles like trucks standing still, smoke concentration along the line of sight, and a sign's luminosity. Not considered are occlusion by moving objects like other occupants or vehicles. This calculation is rather straight forward, as it is based on long-known and well-defined physical and geometrical laws and states. The only biological and therefore individual constant entering is the threshold at which a sign can be distinguished from its background.

The behavioral and decision model contrary to that has been developed newly in cooperation with the University of Würzburg[1]. It is based on a factor concept that adds and weighs different factors influencing a decision to start evacuation or the ability to become aware of a sign that is visible to a specific occupant. Additionally walking speed is influenced by visibility conditions [5].

The cognition of smoke (resp. fire) by the occupants mostly takes place before any instructions by the tunnel operators or rescue services are given. The model calculates the amount of psychical stress the occupant suffers from, depending on internal (individuals′ state, fearfulness, behavioral routines, knowledge, …) and external parameters (intensity of danger, social influence, …), by means of a "potential for reaction". It triggers a decision progress to flee, that leads to either an impulsive or reflective action. Both types are distinguished by the model, and the resulting paths of action considered.

---

1 Paul Pauli, Andreas Mühlberger, Silke Eder, Johanna Brütting, University of Würzburg, Department of Psychology, Biological Psychology, Clinical Psychology and Psychotherapy, Marcusstr. 9-11, D-97070 Würzburg, Germany

Finally, to determine the damage amount on individual occupants the Fractional Effective Dose (FED) model by Purser is applied [**Error! Reference source not found.**].

## Application and Results

The system has been tested extensively and, as a SKRIBT project, been applied to a tunnel incident scenario. There routing is simple, as there is only one fully developed dimension. Occupants need to decide when to start moving and can either head towards the tunnel portal or recognize and use emergency exits.

In this example, two 1,200 m long tunnels have been modeled. The first one has a constant gradient of 3 %, the second one a de- and an ascending leg. The incident takes place in the middle of the tunnel. Emergency exits are 300 m apart.

So far, the scenarios "tunnel fire" - due to the discharge and ignition of fuel or propane - and "toxic gas" - modeling the discharge of chlorine or ammonia - have been analyzed. In all cases, both a spontaneous discharge with major consequences as well as a continuous discharge with minor effects have been applied.

Figure 1 shows a tunnel fire (burning fuel of a road tanker). It is apparent how people recognize the incident either by direct view of the flames or the cognition of smoke, which spreads with a velocity of approx. 7 m/s, due to the stack effect. People start to flee, triggering other people still in their vehicles to do the same. They are hindered by diminished visibility and breathing problems. It is assumed that detection and general alarm, takes place 60 s after the incident. It is assumed that then all people flee. After about 60 s, people begin to queue up at the emergency exits. Fatal casualties and unconscious persons are depicted in red.

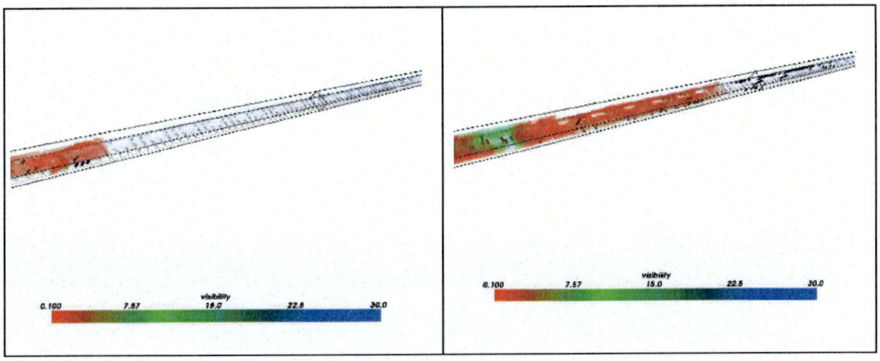


**Fig. 1. Tunnel fire (fuel) after 5 s (left) and 60 s (right)**

Figure 2 shows the spontaneous discharge of 21 t chlorine. Chlorine appears in high concentrations as yellow-green fumes. In tunnel lighting conditions, these are not easily discernible. Therefore, it is assumed that only people nearest to the incident – those who are able to see that a vehicle transporting hazardous goods leaking some kind of substance is involved – start to flee. These trigger others to follow, but the process takes more time than in the first example. The effects of inhaling chlorine are more severe than for smoke. On the other hand, visibility – affecting orientation as well as perceptibility of emergency signs and exits – is not impaired as it would be by smoke, and the distribution of the gas is slower. As no heat is involved, no incident detection takes place in this scenario. In the end, the resulting number of casualties is about the same as in the first example.

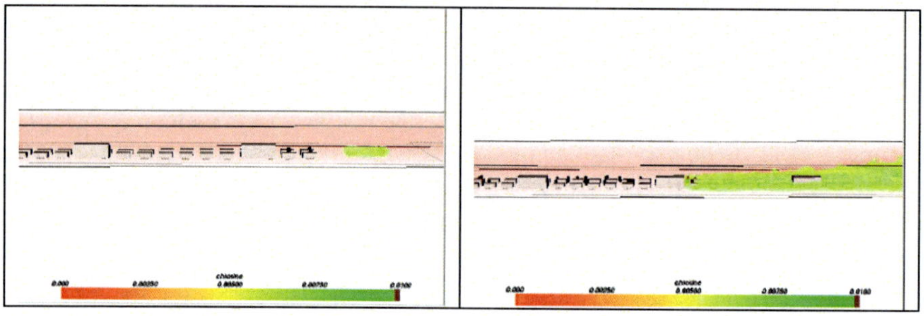

**Fig. 2. Spontaneous discharge of 21 t chlorine after 5 s (l.) and 60 s (r.)**

## Summary

The model allows the realistic reproduction of episodes and respective escape situations. Its main applications will be the determination of extents of loss within risk analysis and the development of escape and evacuation concepts.

**Acknowledgements** SKRIBT is funded by the German Ministry of Education and Science (BMBF). The authors thank Dmitri Danewitz and Peter Ehrhardt for SKRIBT-related project work and software development.

## References

1.SKRIBT – Schutz kritischer Brücken und Tunnel im Zuge von Straßen. http://www.skribt.org/
2.PTV Planung Transport Verkehr AG – VISSIM 5.20 User Manual (2009)
3.Helbing, D., Farkas, I., Vicsek, T.: Nature. 407, 487-490 (2000)
4.Johansson, A., Helbing, D., Shukla, P.K.: Advances in Complex Systems. 10(4):271-288 (2009)
5.Eder, S., Brütting, J, Mühlberger, A., Pauli, P.: Human Behavior in Tunnel Accidents. In: Proceedings of the Human Behavior in Fire Symposium (2009)
6.Purser, D.: SFPE Handbook of Fire Protection Engineering, 4th edition (2008)

# Analysis of Bottleneck Motion Using Voronoi Diagrams

**J. Liddle, A. Seyfried, and B. Steffen**

Forschungszentrum Jülich GmbH, JSC
52425 Jülich, Germany
Corresponding author: j.liddle@fz-juelich.de

**Abstract** Standard definitions of the density exhibit large fluctuations when the size of the measurement area is comparable with the size of a pedestrian. An alternative measurement method exists where a personal space, calculated through the Voronoi diagram, is assigned to each pedestrian. In this contribution this method is applied to an experiment studying motion through a bottleneck and the reduced fluctuations demonstrated. The integrated density also permits examination on much smaller spatial scales than the standard definition, the insights into the pedestrian motion this provides are discussed.

## Introduction

The standard definition of the density in a measurement area $A$, of area $|A|$, containing $N$ pedestrians,

$$\rho_S = \frac{N}{|A|},$$

exhibits large fluctuations when the size of the measurement area is comparable with the size of a pedestrian. These fluctuations are typically of the same order as the density measurement itself. By averaging over time these fluctuations can be removed at the expense of reducing the time resolution of the measurements.

An alternative definition exists, the integrated density [1], where a density distribution (step function) is assigned to each pedestrian, this density distribution is calculated through the Voronoi diagram [2]. At a given time, ti, we have at set of positions for each pedestrian $\{\vec{x}_1(t_i), \vec{x}_2(t_i), \cdots, \vec{x}_M(t_i)\}$. We compute the Voronoi diagram for these points obtaining a set of cells, Ai, for each pedestrian i. These cells can be thought of as the personal space belonging to each pedestrian. With these cells a density distribution can be defined for each pedestrian.

R.D. Peacock et al. (eds.), *Pedestrian and Evacuation Dynamics*,
DOI 10.1007/978-1-4419-9725-8_83, 

$$\rho_i(\vec{x}) = \begin{cases} \frac{1}{|A|} & \text{if } \vec{x} \in A_i \\ 0 & \text{otherwise} \end{cases}$$

and the density inside a measurement area, $|A|$, is defined as

$$\rho_V = \frac{\int p(\vec{x})\,d\vec{x}}{|A|} \quad \text{where } p(\vec{x}) = \sum \rho_i(\vec{x}).$$

This method provides several advantages. The reduced fluctuations mean an instantaneous estimate of the density is possible and the presence of non-stationary states can be unambiguously detected, which is not possible with the standard method, see figure 1. The integrated density can also provide microscopic information about the density, which will inform the development of microscopic models.

Our experiment was performed in 2006 in the wardroom of the 'Bergische Kaserne Düsseldorf', with a test group of soldiers [3]. The experimental setup allowed the influence of the bottleneck width and length to be probed. In this proceeding the experiment with varying width (between 90 to 250 cm), b, is considered. Wider bottlenecks with more test persons were studied than in previous attempts [4-6].

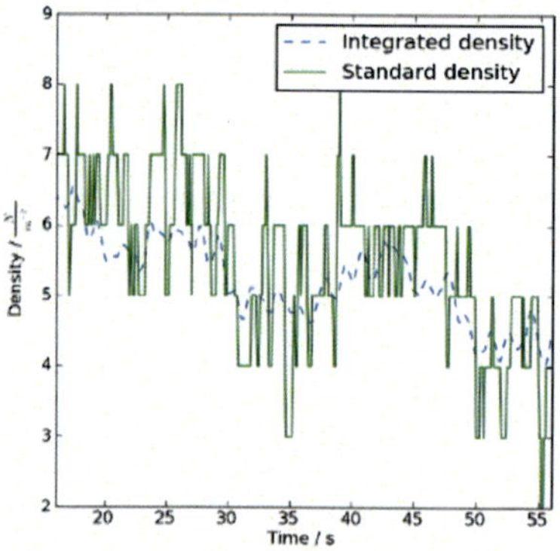


b = 120 cm

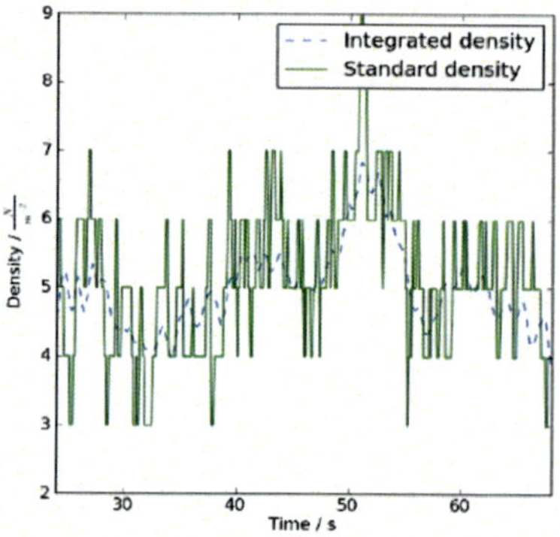


b = 140 cm

**Fig. 1. Density computed using standard and integral definitions. It can be clearly seen that the integrated definition fluctuates less than the standard definition. In the left figure the downward trend in the density is clearly visible when using the integrated density.**

In some of our experiments no steady state in the density was observed. The existence of the non-steady states can probably be attributed to the presence of highly motivated persons positioned at the front of the pack. The problems arising from trends in the density have not been discussed before, most likely due to insufficient resolution of the measurement method.

## Application

In figure 1 the density time series for two experiments are shown. It is immediately obvious that fluctuations are greatly reduced, and that an accurate estimate of the density can be obtained from a single frame.

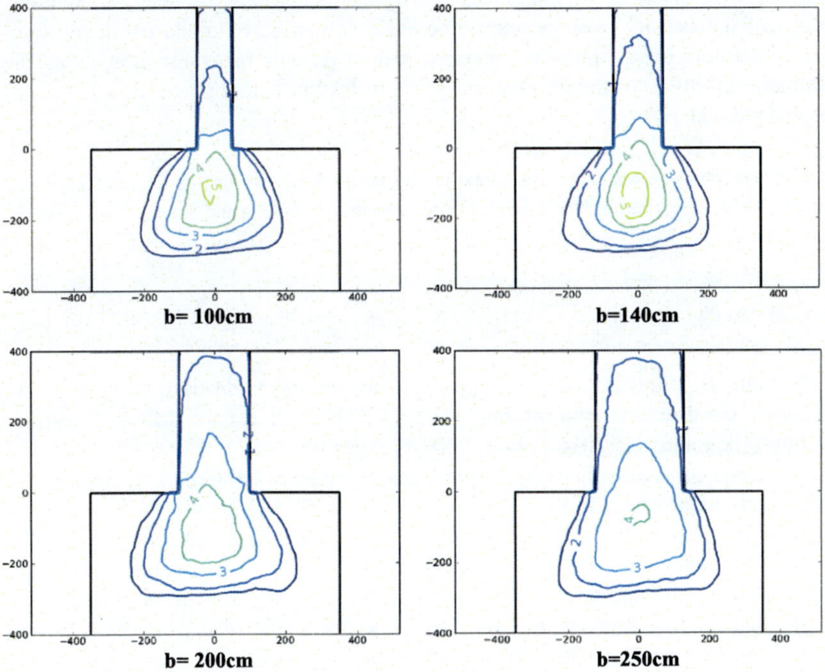

b= 100cm b=140cm

b= 200cm b=250cm

**Fig. 2. Density maps for bottlenecks of varying width. Average values during steady state.**

In addition to reduced fluctuations we can also obtain density measurements on small spatial scales, not obtainable with the standard definition. By calculating the integrated density over small (10 cm) regions and averaging over all configurations (during the steady state phase), maps of the density over the experimental area can be obtained. In figure 2 maps of the density are shown for a variety of widths, spanning the range of bottleneck widths studied. These maps reveal details of the density distribution like shape of congested areas. Firstly we note that the peak of the density lies around 125 cm from the bottleneck entrance, for all widths tested. As would be expected the peak of the density is sharpest in front of the narrowest bottlenecks. The shape of the high density region is also revealed, for the narrowest bottlenecks the high density region remains outside the bottleneck and for widest bottlenecks this region extends to the interior of the bottleneck.

## Conclusions

It has been shown that the integrated density exhibits greatly reduced fluctuations. Maps of the density over the experimental area can be obtained using the integrated method; these cannot be obtained, from available measurements, using the standard definition of the density and serve to further demonstrate the advantages of this method.

**Acknowledgments:** This study was supported by the Federal Ministry of Education and Research (BMBF), program on "Research for Civil Security - Protecting and Saving Human Life".

## References

1. Steffen, B., Seyfried, A.: Methods for measuring pedestrian density, flow, speed and direction with minimal scatter. Physica A, accepted for publication. http://dx.doi.org/10.1016/j.physa.2009.12.015 (2010)
2. Voronoi, G.: Nouvelles applications des paramètres continus à la théorie des formes quadratiques. *J. reine angew. Math.* **133**, 97-178 (1907)
3. Seyfried, A., Boltes, M., Kähler, J., Klingsch, W., Portz, A., Rupprecht, T., Schadschneider, A., Steffen, B., Winkens, A.: Enhanced empirical data for the fundamental diagram and the flow through bottlenecks. In: Klingsch, W., Rogsch, C., Schadschneider, A., Schreckenberg, M. (eds), Pedestrian and Evacuation Dynamics 2008, p. 145-156. Springer (2010)
4. Hoogendorn, S., Daamen, W.: Pedestrian Behavior at Bottlenecks. Transportation Science. **39/5,** 147-159 (2005)
5. Kretz, T., Grünebohm, A., Schreckenberg, M. J.: Experimental study of pedestrian flow through a bottleneck. J. Stat. Mech. P10014 (2006)
6. Seyfried, A., Rupprecht, T., Passon, O., Steffen, B., Klingsch, W., Boltes, M.: New Insights into Pedestrian Flow Through Bottlenecks. Transportation Science. **43/5,** 395-406 (2009)

# Human Guiding, Turning Theory into Practice

**J. Lindström, L. Barna, P. Stevn, and R. Eggleton**

MariMils Oy
Ohtolankatu 3 FI-01510 Vantaa, Finland
Corresponding author: juha.lindstrom@marimils.fi

**Abstract** We introduce an intelligent lighting system for dynamic pedestrian guiding. Its benefits in emergency situations are demonstrated through a simulated evacuation scenario. A pilot system for passenger flow optimization is presented as an example of a non-emergency application. The system is useful for implementing solutions for pedestrian guiding found through modeling and simulations.

## Introduction

The last few decades have seen a rising interest in pedestrian behavioral modeling. Significant progress has been made on the theoretical side resulting in increasingly more accurate models [1]. A particularly important scenario is emergency evacuation where accurate models and careful planning can save lives.

Traditional emergency guidance systems have focused on one emergency type only, fire. However, modern emergency planning should also take other kinds of hazards into consideration such as black outs, terrorist threats, flooding, chemical leakage, hurricanes and earthquakes. Depending on the situation, different kinds of threats might require different kinds of actions.

Modern evacuation models, theories and simulators provide the designers and engineers with tools to find optimal evacuation plans for different emergency scenarios. However, up until now, the tools to actually implement these scenarios have been severely limited. The traditional exit signs and photo luminescent stripes can only point to the nearest exit, which sometimes may not be the safest one. In the worst case scenario, the threat is actually located within the planned evacuation path, and the static signage could guide people in the wrong direction.

We have developed an intelligent lighting system that avoids these problems. The system is called MILS® and it is based on dynamically controlled LEDs embedded into thin, durable and fire resistant thermoplastic polyurethanes (TPU) stripes. The MILS® stripe can display static, flashing and moving patterns of light and it comes in different colors and intensities. As the human visual perception is very sensitive to motion, the moving light patterns are an efficient way to notify and guide people in the right direction [2]. Figure 1 illustrates the stripe in action.

R.D. Peacock et al. (eds.), *Pedestrian and Evacuation Dynamics*,
DOI 10.1007/978-1-4419-9725-8_84, © Springer Science+Business Media, LLC 2011

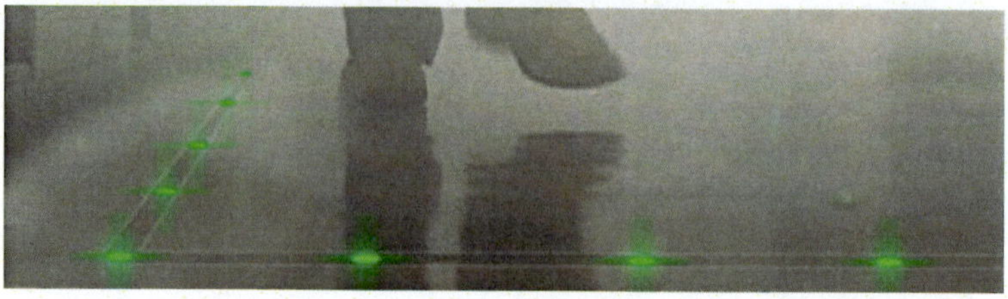

**Fig. 1. Dynamic MILS® stripes mounted on the floor for increased visibility during fire**

An important non-emergency use case for a dynamic pedestrian guiding system is flow optimization. When there are multiple bottlenecks limiting the flow, such as gates or desks, the guiding system can try to distribute the pedestrians evenly across them. If there is only one bottleneck, the system should try control the flow to preventing jamming, because a jam can further decrease the flow [3].

## System Overview

The architecture of a basic MILS® system is illustrated in figure 2. The system is controlled by the MILS® System Control Unit (SCU), which is an industrial PC running the control software. The SCU is connected to external systems like fire alarm systems that provide input to the MILS® system. Based on this input the control software chooses the right guidance scenario from a predefined list.

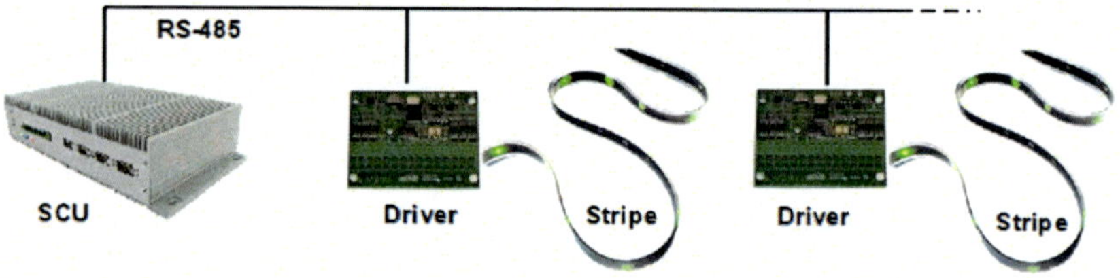


**Fig. 2. MILS® system architecture**

The SCU controls the lighting drivers over an RS-485 bus that can contain tens of drivers and span up to one kilometer in length. Depending on the model, each driver can control 1-4 LED stripes or up to 12 light signs or panels.

## Evacuation Simulation

In order to demonstrate the importance of a dynamic emergency guidance system, we used the FDS+Evac fire dynamics and evacuation simulator [4] to simulate two evacuation scenarios: one with a dynamic guidance system and one without it. Both scenarios start with 25 people randomly located in the four rooms in the middle of the area. The people are expected to have entered the area through the lower left door. The fire is located in the upper left corridor and is modeled as having heat release rate per unit area (HRRPUA) of 1000 kW/m² and the basic stoichiometry of polyurethane.

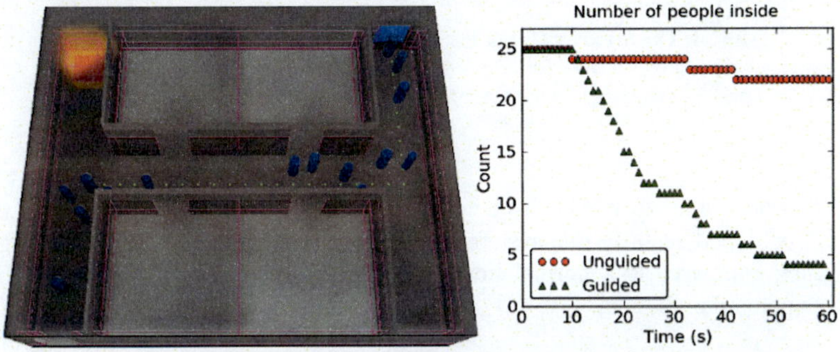


**Fig. 3. Simulated evacuation scenario and the results of the simulation**

In the first scenario, the people are expected to know only the door through which they entered with probability 1. In the second scenario the effect of the guidance system is modeled as follows. First, as the light pattern in the floor moves towards the upper right exit, all the people know this exit with probability 1. Second, to make the people prefer the guided exit instead of the one they entered, the probability of knowing the lower left exit is decreased to 0.5. Figure 3 illustrates the second scenario at time 21s and the results of the simulation. The benefits of the guidance system are evident.

## Security Check Passenger Flow Optimization

A pilot MILS® system has been successfully tested in real-life environment at the main Departures Terminal at Helsinki-Vantaa International Airport, Finland. The system (Figure 4) included two sections of ELSI™ motion sensors and MILS® stripes embedded in two rubber mats and four MILS® stripes, placed vertically on the security gate and two side poles placed in front of the gate. Access to the gate was controlled automatically in response to input from the sensors, or manually with remote controls, adjusting the color and motion of the stripes.

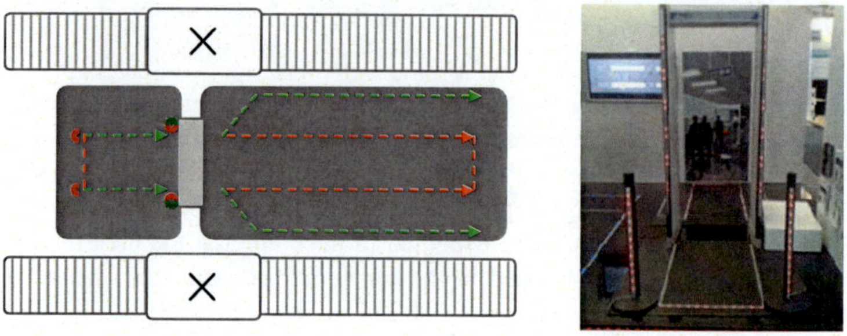

System Design Realization

**Fig. 4. The Security Check Passenger Flow Optimization System**

Following a one week test period concentrated on the terminal's rush hours, the MILS® equipped security line system showed 15% average increase in passenger flow rate, compared to a normal line with similar dimensions and security gate model operated in parallel.

## Conclusions

We have created a modular intelligent lighting system for dynamic guiding of pedestrians. The system can be used in emergency as well as non-emergency situations to optimize pedestrian traffic. MILS® provides pedestrian traffic designers and engineers with efficient tools useful to implement optimal guiding in every situation, turning theory into practice.

## References

1. Kuligowski, E.D., Peacock, R.D.: Review of Building Evacuation Models. Technical Note 1471. Natl. Inst. Stand. Technol., Gaithersburg (2005)

2. Akita, J. et al.: Dynamic Lighting Sign System for Way-Finding by People with Low Vision. In: Computers Helping People with Special Needs: 9th International Conference (ICCHP) (2004)
3. Schadschneider, A., Klingsch, W., Kluepfel, H., Kretz, T., Rogsch, C., Seyfried, A.: Evacuation dynamics: empirical results, modeling and applications. arXiv:0802.1620v1 (2008)
4. Korhonen, T., Hostikka, S.: Fire Dynamics Simulator with Evacuation: FDS+Evac Technical Reference and User's Guide. VTT Technical Research Centre of Finland (2009)

# Modeling Evacuation in Selected Types of Buildings and an Analysis of the Achieved Results

**M. Lopušniak**

TU Kosice, faculty of civil engineering
Vysokoskolska 4, 042 00 Kosice, Slovakia
Corresponding author: martin.lopusniak@tuke.sk

**Abstract** Requirements for ever larger, architecturally complex, structurally challenging and economically efficient buildings, but with a constantly growing number of users for buildings creates a situation that in the event of an emergency scenarios (fire) can cause significant complications for rescuing people. Ensuring the safe evacuation of persons from buildings is a fundamental requirement in the design of fire safety works. Current standardization requirements are difficult to enforce in the design of existing multifunctional buildings. The paper presents the results of people-evacuation modeling in two buildings in Slovakia, which were only recently completed. The paper presents a comparison of calculation results with standardization values. Part of the contribution is also a discussion of the results and their interpretation. The conclusion allows us to use the results of the evacuation modeling after changing the actual requirements.

## Introduction

The present state of the building design uses an essential normative approach to engineering question solutions in Slovakia. This approach is built on the experience and information from projects realized in the past [1]. Other data is obtained from the buildings operation.

The application of modern access to engineering question solutions has limits in historical contexts connected with unsatisfactory technical equipment and expert knowledge. Those barriers remain till present in the form of inaccessibility to new knowledge and technologies. It is not possible to say that it is of no interest in new and present situations, only that their interpretation is slow and is met with opposition.

Requirements' regarding fire safety improves with foreign capital entrance in a different way than that which is usual. It is influenced by the optimization of investment costs connected with construction. Because any testing is problem in the fire safety area, the better way is to use various computing tools. An evacuation model represents powerful engineering instruments which, together with another

R.D. Peacock et al. (eds.), *Pedestrian and Evacuation Dynamics*,
DOI 10.1007/978-1-4419-9725-8_85, 

30 accessible evacuation models [2], allow the solution of important parts of building design.

## Present state

At present the computing model used in Slovakia is based on evacuation time according to the following formula:

$$t_u = \frac{0{,}75.l_u}{v_u} + \frac{E.s}{K_u.u}$$

| | | | |
|---|---|---|---|
| $v_u$ | - | Walking speed (20, 25, 30*) | (m/min) |
| E | - | Number of evacuated persons | (-) |
| s | - | Coefficient of evacuation condition | (-) |
| u | - | Number of escape units, one unit 0,55 m | (-) |
| $l_u$ | - | Length of escape route | (m) |
| Ku | - | Capacity of escape route | (1/m) |

* walking speed: upstairs (20), downstairs (25), direct (30); the worst value is selected

Calculated evacuation time must be shorter than evacuation time determined in the standard. Allowed evacuation time depends on the level of the evacuation route safety and people that are located on this route. The allowed evacuation time can be from 2 to 30 minutes.

The determination of the number of evacuating people also involves a collision point [3]. Their quantity doesn't come from real numbers but from a designed number of people multiplied by a coefficient of 1.5. This access is not correct and leads to the over-dimensionality of people number.

## Calculation

Two buildings were chosen for comparison. Building A was selected as an administrative building. This building was chosen for its simplicity of evacuation conditions and as an intensive demonstration possibility of evacuation model. Building B was chosen as a 12-story hotel. The enclosure of buildings is shown on the following pictures.

The above mentioned method and evacuation model bEXODUS [4] were used for mutual comparison. Default values were used for the model calculation [5].

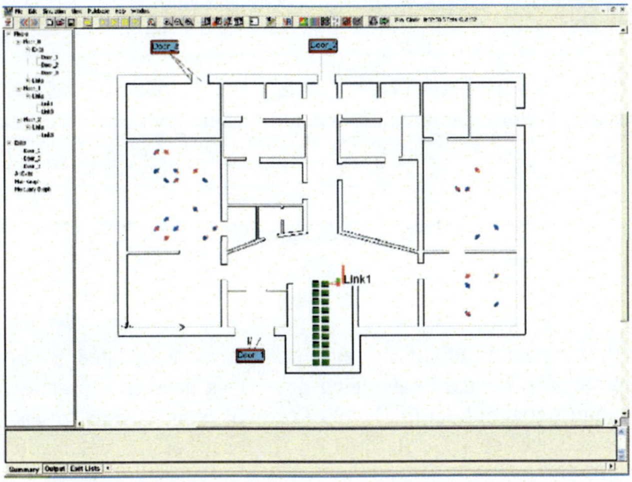


**Fig. 1. Floor plan of the building A from bEXODUS**

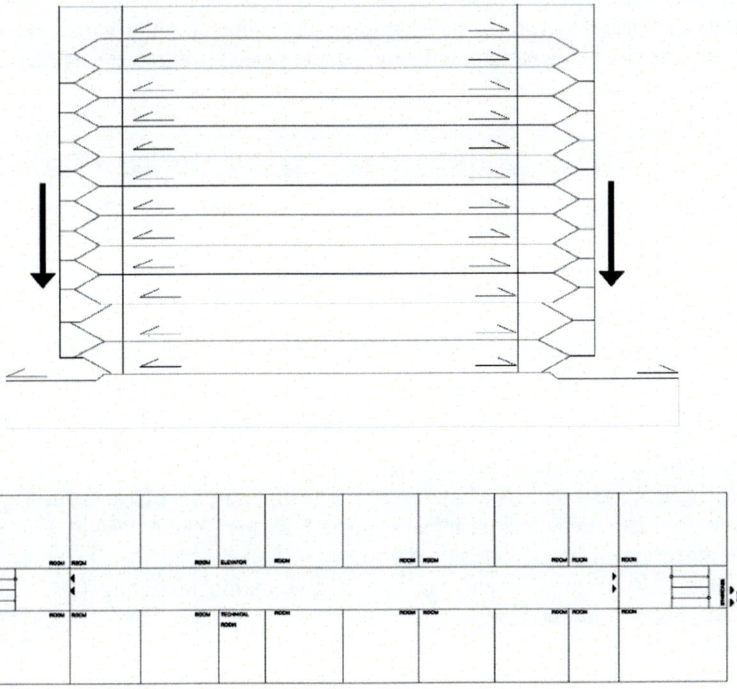

**Fig. 2. Section view and floor plan of the building B**

## Discussion

The calculation results are shown on the table 1. As the results show, the calculation model usage is approaching values that are reached by standard calculation.

**Table 1. Comparison of the time for evacuation**

| Building | Standard calculation | Allowed tyme | bEXODUS |
|---|---|---|---|
| Building A | 2,1 min | 4 min | 2,3 min |
| Building B | 11,97 min | 10 min | 12,45 min |

The substantial answer is not if the evacuation time calculated by bEXODUS model is approaching standard calculated time. This answer is practically unimportant. It is interpreted only at the level of present requirements for escape route design. These requirements are defined on the basis of the calculation method used in Slovakia.

The calculation model application with definitions of various entrance data (pre time for evacuation) will provide a lot of information and results. Those results can't be comparable with present requirements, because requirements don't come from calculation methodology.

It is necessary to take in new criterion that allow for the wider transfer of present data from evacuation modeling, people behavior and technical evacuation equipment to the practice [6].

## References

1. Gwynne, S., Galea, E.R., Owen, M., Lawrence, P.J., Filippidis, L.: A Review of the Methodologies Used in Evacuation Modelling Fire and Materials. Vol. 23, 6, 383-389 (1999)
2. Kuligowski, E.D., Peacock, R.D.: Review of Building Evacuation Models. Technical Note 1471. Natl. Inst. Stand. Technol., Gaithersburg (2005)
3. Lopušniak, M.: Modelling of escape routes according to occupancy, economy and level of safety in Slovak republic. In: Pedestrian and Evacuation Dynamics, pp. 733-737. Springer, Berlin (2008)
4. Galea, E.R., Gwynne, S., Lawrence, P.J., Filippidis, I., Blackshields D., Cooney, D.: User guide and technical manual, buildingEXODUS V4.0. Fire Safety Engineering Group, University of Greenwich, London (2008)
5. Galea, E.R.: Principles and Practice of Evacuation Modeling (10$^{th}$ Edition). CMS Press, London (2009)
6. Lopušniak, M.: Computer Simulation Methods in the Evacuation. In: FIRECO 2009, pp. 1-5. Fire Safety and Expertise Institute MV SR (2009)

# Experimental Investigation on the Pair Interaction Between Pedestrians

**J. Ma[1,2], W. G. Song[1], S. M. Lo[2], G. X. Liao[1] and K. K. Yuen[2]**

[1] State Key Laboratory of Fire Science,
University of Science and Technology of China, Hefei 230026, P. R. China
[2] Department of Building and Construction,
City University of Hong Kong, Tat Chee Avenue, Kowloon Tong, Hong Kong

**Abstract** Both experimental and numerical studies indicate local interaction among pedestrians plays an important role in determining the self-organization of pedestrian traffic. However, little is know to the form of the interaction. We explore the fundamental interaction between pedestrians by performing controlled experiments in a long narrow channel where pedestrians were asked to first walking freely in the channel, then to evade a standing still pedestrian in the middle of the channel. Trajectories of these pedestrians extracted from the video-recordings of the experiments were then used to analyze detailed microscopic moving features. Averaged change of speed as well as the change of direction of these pedestrians was then calculated. An angular and metric distance based interaction map reviewing the laws ruling the behavioral changes when a pedestrian interacts with others was detailed. This study provides further insights to the fundamental individual level interaction between pedestrians for understanding the pedestrian dynamic and is expected to be helpful in evacuation analysis.

## Introduction

Self-organized pedestrian crowd behaviors such as lane formation are believed to be the result of local interaction among them. As a result, different models have been constructed by assuming a metric-distance related force form, as shown in the 'social-force' model [1], to represent the interaction. However, the knowledge about the nature of the interaction is still rare, neither to the force nor to the form itself, especially when facing field observations with almost distinct conclusions [2].

With the development of digital image processing, it is now possible for us to study detailed pedestrian moving characteristics. Antonini studied decision making process when walking in crowds [3]. Helbing et al. investigated the disaster of crowds during Hajj pilgrimage to Mecca [4] and pointed out when the pedestrian density come across a critical value it might trigger trampling. Johansson et al. specified the social force model by analyzing video recording of pedestrian mov-

R.D. Peacock et al. (eds.), *Pedestrian and Evacuation Dynamics*,
DOI 10.1007/978-1-4419-9725-8_86, © Springer Science+Business Media, LLC 2011

ing trajectories [5]. Liu et al. also extracted parameters of moving pedestrians for evacuation based on digital image processing [6].

In this paper, we follow the experimental procedure proposed by Moussaïd et al. [7] to investigate the fundamental interaction law in pedestrian flows by controlled experiments and digital image processing.

## Experimental Procedure

For the convenience of comparison, the experimental procedure is the same as that of the Ref. [7]. A corridor with a length of 8 m, a height of 1.65 m, and a width of 1.75 m, has been built to perform the experiments. The scheme of the experimental setup could be found in Fig.1 (a).

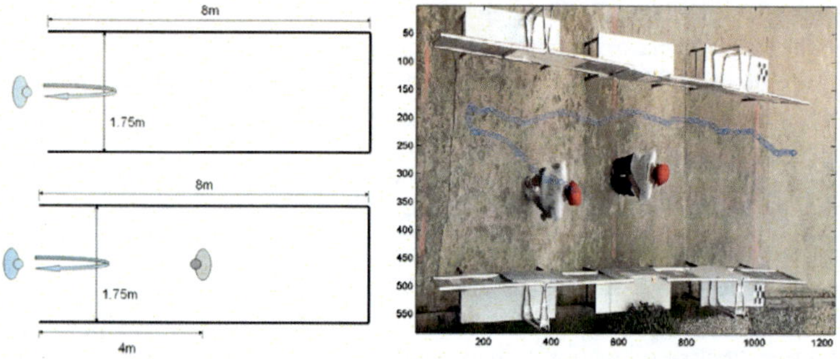


**Fig. 1(a). Scheme of the experimental procedure and (b): trajectory of a typical participant**

**Table 1. Pedestrians' information**

| Number of participants | | Mean height/cm | | Standard Deviation | |
|---|---|---|---|---|---|
| Female | 5 | 160 | 169.85 | 8.15 | 8.62 |
| Male | 15 | 173 | | 5.96 | |

Pedestrians participated in the experiments were randomly selected in the Campus of USTC. Detailed information can be found in Table 1. They were asked to take part in the following two experiments. In Condition I, each pedestrian walk alone in the corridor from the entrance to the end then back again. In Condition II, we first let a pedestrian stand still in the middle of the corridor, i.e., at the location of the yellow cross as shown in Fig.1 (b). We then ask another pedestrian to repeat what they do in Condition I.

## Results and Analysis

The trajectories of the pedestrians, as shown in Fig.1 (b), were extracted following the method in Ref [6]. We first study the acceleration process of a single pedestrian. We show in Fig.2 (a) the averaged instantaneous speed of a typical participant. It is easy for us to find that the acceleration process follows an exponential increase, which means the acceleration of pedestrian can be described by two important parameters, i.e., the desired velocity and relaxation time.

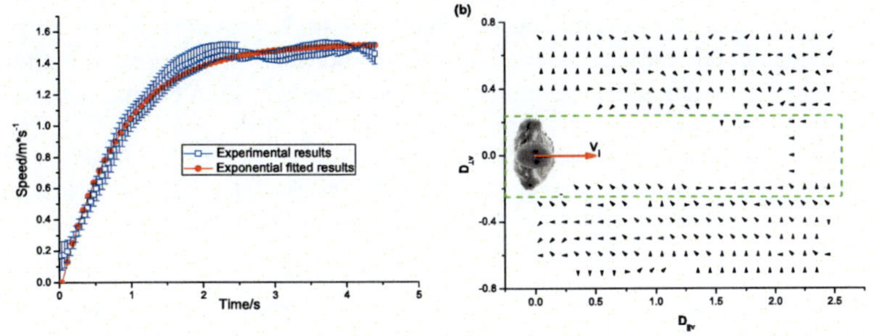


**Fig. 2. (a): Instantaneous speed of a typical participant and (b): Interaction map between pair pedestrians**

We then quantify interaction among pair pedestrians by explore the effects of the presence of pedestrian *j* in front of the other moving pedestrian *i*. As described in social force model, moving characteristics of a pedestrian can be described by the following equation,

$$\vec{f}_{ij} = \frac{d\vec{v}_i}{dt} - \vec{f}_i^0 - \vec{f}_i^{wall} \tag{1}$$

Where $\vec{f}_{ij}$ means the force on pedestrian *i* as a result of pedestrian *j*, $\vec{f}_i^0$ represents the self-driven behavior which can be described by the above acceleration process, and $\vec{f}_i^{wall}$ means the forces on pedestrian *i* as a result of the corridor wall. Johansson et al. specified this force in Ref [5]. We calculated the forces $\vec{f}_{ij}$ along all the trajectories and at last get an interaction map among pair pedestrians, which is shown in Fig.2 (b). As we can find in this figure, the effect of the force from pedestrian *j* on pedestrian *i* makes pedestrian *i* changes velocity. In specific, when the pedestrian *j* present on the left-forward direction of pedestrian *i*, the force from pedestrian *j* makes pedestrian *i* to move to the right hand side, and when the pedestrian *j* present on the right-forward direction, the force makes pedestrian $i$ moves to the left hand side. The presence of pedestrian $j$ in front of pedestrian $i$ makes pedestrian $i$ decelerate to avoid potential collision.

## Summary

Acceleration of single pedestrian as well as velocity change as a result of the interaction between pair pedestrians were analyzed. An angular and metric distance based interaction map reviewing the laws ruling the behavioral changes when a pedestrian interacts with others was detailed. This study provides further insights to the fundamental individual level interaction between pedestrians for understanding the pedestrian dynamic and is expected to be helpful in evacuation analysis.

**Acknowledgments** The study is supported by NNSF (No.50678164), PNCET (NCET-08-0518), NS TPP (No.2006BAK06B00) and the RGC, Hong Kong (No. CityU118708).

## References

1. Helbing, D., Farkas, I., Vicsek, T.: Simulating dynamical features of escape panic. Nature. 407(6803):487-490 (2000)
2. Ballerini, M., Calbibbo, N., Candeleir, R., et al..: Interaction ruling animal collective behavior depends on topological rather than metric distance: Evidence from a field study. Proceedings of the National Academy of Sciences of the United States of America, 105(4):1232-1237 (2008)
3. Antonini, G., Bierlaire, M., Weber, M.: Discrete choice models of pedestrian walking behavior. Transportation Research Part B: Methodological. 40(8):667-687 (2006)
4. Helbing, D., Johansson, A., Al-Abideen H.Z.: Dynamics of crowd disasters: An empirical study. Physical Review E, 75(4) (2007)
5. Johansson, A., Helbing, D., Shukla, P.: Specification of the social force pedestrian model by evolutionary adjustment to video tracking data. Advances in Complex Systems. 10(Supplementary Issue 2):271-288 (2008)
6. Liu, X., Song, W., Zhang, J.: Extraction and quantitative analysis of microscopic evacuation characteristics based on digital image processing. Physica A: Statistical Mechanics and its Applications. 388(13):2717-2726 (2009)
7. Moussaïd, M., Helbing, D., Garnier, S., Johansson, A., Combe, M., Theraulaz G.: Experimental study of the behavioural mechanisms underlying self-organization in human crowds. Proceedings of the Royal Society B: Biological Sciences. 276(1668):2755-2762 (2009)

# Emergency Situations in Nightclubs: A Discussion on How to Improve the Fire Safety Strategies Through The Use of Evacuation Modeling Analysis

**R. Machado**[1] **and M. Anderson**[2]

[1]CAPTIA SYMONDS, 90-98 Goswell Rd., London EC1V 7DF UK
[2]AECOM, 1 Tanfield, Edinburgh, EH3 5DA, Scotland
Corresponding author: rodrigotmj@gmail.com

**Abstract** In the past few years, the world has witnessed severe fires in nightclubs, which have caused many human losses. This paper promotes a methodology which uses evacuation modeling analysis for improving the fire safety strategies (FSS) in nightclubs. The methodology is presented through a study case which represents a nightclub scenario. The results have shown that the use of evacuation modeling analysis can help immensely fire protection engineers to develop their FSS reports, incorporating the human behavior factor; rather than only considering the standard safety factors, such as travel distance. The same methodology can be also applied for fire safety management, which is another key-issue to be properly addressed for fire safety purposes in nightclubs. It is expected that this paper can bring some additional light to the way emergency evacuations are planned and addressed is in nightclubs.

## Introduction

In the past few years, the world has witnessed severe fires in nightclubs. For instance, the Gothenburg nightclub fire in Sweden (1998), the Station nightclub fire in the USA (2003) and the most recent one, the Bangkok nightclub fire in Thailand during the New Year's celebration of 2009. All these fires have caused many human losses. For this reason, fire safety in nightclubs should be addressed distinctly, given their specific environment. In reality, there are many factors which make fire safety in nightclubs challenging to be achieved. These factors are substantially different from ordinary buildings, including also assemblies; and they vary from the physical building structure to the occupants' profile. For instance, the architectural features of nightclubs are very often composed by complex geo-

R.D. Peacock et al. (eds.), *Pedestrian and Evacuation Dynamics*,
DOI 10.1007/978-1-4419-9725-8_87, 

metries (which quite often look like real labyrinths). In addition to that, frequently, the occupants are not familiar with the actual environment. Besides that, the use of alcohol changes completely the occupants' responses as well as their physical mobility. There are other relevant factors that can be also mentioned, such as the inappropriate illumination; the noisy environment and the high population density. Furthermore, the fire safety strategies for nightclubs, which will define their final designs, must be developed considering these factors.

In the other hand, the fire safety strategies (FSS) should comply with current legislation requirements which normally cover main aspects, such as: means of warning and escape; internal fire spread; external fire spread and access and facilities for the fire service. In fact, all of these requirements will be dependent on the type of occupancy which the building is classified. And for the particular case of nightclubs, most fire safety codes, including the principal building regulations document for fire safety in Wales and England, the Approved Document B (AD B) as well as the NFPA 101 Life Safety Code for the USA (which is probably the fire safety code most used globally speaking) classify nightclubs as "assembly and recreation" types of occupancy. This category is the same used to define other enclosures such as clinics and libraries. This clearly does not seem to be appropriate, because nightclubs premises are completely different from clinics and libraries in many aspects (for not saying all).

Therefore, this paper promotes a methodology which uses evacuation modeling analysis for improving the fire safety strategies for nightclubs. One of the parameters used in the proposed methodology is the use of the evacuation times as the core criterion for evaluating the occupants' safety rather than prescriptive requirements such as the maximum travel distances. The methodology is presented through a study case which represents a nightclub scenario. The results have shown that the use of evacuation modeling analysis can help fire protection engineers and designers to develop their FSS reports satisfactorily, i.e., not underestimating the occupants' behavior, which plays a core-rule within the whole evacuation efficiency. (*The occupants' behavior, not only in terms of physical aspects, but also in terms of psychological aspects should be considered when developing fire safety strategies, for nightclubs and enclosures in general*). The same methodology can be also applied for fire safety management, which is another very important issue to be properly addressed in nightclubs.

## Fire Safety in Nightclubs X Fire Safety Codes

The fire design of nightclubs, and any type of enclosure, should comply with current legislation and this includes, amongst others: means of warning and escape; internal fire spread (linings); internal fire spread (structure); external fire spread and access and facilities for the fire service.

In the last few decades, alternative solutions to the current ones suggested by the requirements within prescriptive codes have been started to be adopted. These alternative solutions are commonly called as fire engineering solutions (or as performance-based solutions). Nevertheless, when defining the FSS, engineers do use the prescriptive codes as the main guidance for basic and initial planning. For example, such codes are largely applied when defining the purpose for which the building or compartment of a building is intended to be used.

Within this context, the question which can be asked is: how do the fire safety codes address fire safety in night clubs?

In other words, since that the fire safety of an enclosure starts to be planned based on the information from the FSS reports, it is important to understand how the fire safety codes "understand" nightclubs premises. This is mentioned, because the FSS reports rely mainly on the information contained within the fire safety codes for defining the type of occupancy which a building would be classified, as mentioned previously. In fact, the type of occupancy is a crucial parameter for the whole FSS; because important issues such as the wall and floor compartmentation, space separation between buildings, the requirement of sprinklers systems, amongst other issues, will be dependent on the type of occupancy.

As mentioned before, in the UK (restricting to Wales and England), the AD B classifies nightclubs, in its Table D1 of Appendix D, as "assembly and recreation" types of occupancy. This category is the same used to define other enclosures such as clinics and libraries. This clearly does not seem to be appropriate, since these mentioned occupancies are completely different in many aspects. These aspects vary from the physical building structure to the occupants' profile as mentioned in the previous sections. In reality, the AD B is the main source of information when defining the Fire Safety Strategies (FSS) for buildings in Wales and England. This consideration appears to be the same for some other fire safety codes, such as the NFPA 101 Life Safety Code for the USA; which is probably the fire safety code most used globally speaking.

In order to investigate this issue, a nightclub scenario was built up using a robust and validated evacuation model: STEPS. The basis of this well-known evacuation model, STEPS, has frequently described in many other publications; and therefore it will not be described here in this paper.

The study case is explained further in the next section.

## Study Case

The study case was based on a geometry in which the population density is $2p/m^2$. This means that the occupants will potentially have only 50% of freedom for their escape movement. Two exits of equal sizes, 1m in width, were placed on the perimeter of the enclosure. Two scenarios were considered: Scenario 1 (based on the AD B requirements): The angle between the two escape routes leading to the two exits is more than 45° and the maximum travel distance is less or equal to 45m; and Scenario 2 (based on the evacuation modelling analysis): The angle between the two escape routes leading to the two exits is less than 45° and the maximum travel distance is greater than 45m.

In total, 100 simulations were performed: 50 simulations for scenario 1 and 50 simulations for scenario 2. For each simulation, the occupants' locations within the enclosure were randomized. This procedure intended to reduce bias in the data. In terms of psychological aspects, the occupants were modelled in a way that they would avoid congestion; they were aware of the exits' locations; their pre-movement times were 0 to 30 seconds and they were allowed to overtake those slower occupants.

## Evacuation Modeling Simulation Results and Discussion

**Table 1. Evacuation modeling simulation results**

| Scenario | Average Time Spent during the Movement (TSM) (in seconds) | Average Time spent during Queuing towards the Exits (TQE) (in seconds) |
|---|---|---|
| Scenario 1 | 25.5 | 30.0 |
| Scenario 2 | 29.0 | 14.7 |

The results for both scenarios are presented in Table 1. Based on Table 1, the average evacuation time for scenario 1 was 55.5 seconds; while for scenario 2 was 43.7 second. It is important to see that the TQE played a major rule on the evacuation efficiency rather than the TSM. This is an important issue to be pointed, since the fire safety codes define the travel distances as one of the main safety factors when developing FSS. Based on that, the methodology proposed in this paper is that evacuation modeling analysis should be considered to be performed for challenging enclosures, such as nightclubs when defining the FSS reports, see Figure 1.

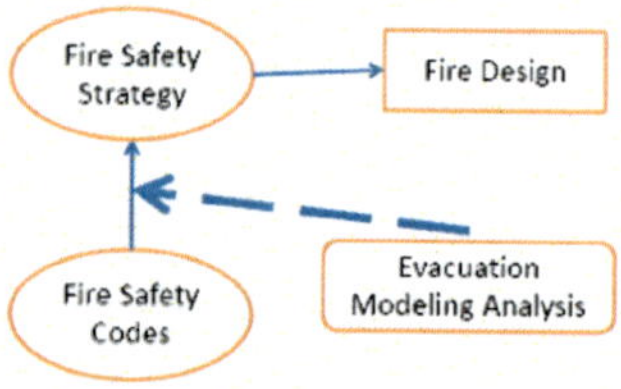


**Fig. 1. Proposed methodology for developing fire safety strategy for nightclubs**

# Efficiently Using Micro-Simulation to Inform Facility Design – A Case Study in Managing Complexity

**E. Morrow**

Arup Canada Inc.,
2 Bloor St. West
Toronto, ON M4W 1A8 Canada
Corresponding author: erin.morrow@arup.com

**Abstract** At major transit terminals large, volumes of people, intricate operational procedures, and complex built environments present significant challenges to effective pedestrian facility design. Transbay Terminal in San Francisco is a prime example of layered complexity. It is a multi-modal transit terminal designed to serve commuter rail, commuter bus, local bus, and eventually high speed rail passengers in downtown San Francisco. With the existing facility already near capacity and new transit modes being planned it is essential that the terminal facility is upgraded. This paper will present a new crowd simulation technology called MassMotion and describe how this toolset was applied to inform the design of a new Transbay terminal.

## Introduction

Pedestrian micro-simulation enables understanding of complex design problems and confidence in design solutions. The MassMotion pedestrian simulation system has been designed from the ground up to provide planners and designers with the tools to predict the performance of their designs. This is accomplished through a holistic consideration of:

- Dynamic crowd interactions
- Network assignment and predicted loads
- Capacity planning for predicted loads

MassMotion is an autonomous agent based crowd system that operates within three dimensional virtual environments. The following sections will describe the basic architecture of the MassMotion crowd simulation system and then explore

R.D. Peacock et al. (eds.), *Pedestrian and Evacuation Dynamics*,
DOI 10.1007/978-1-4419-9725-8_88, © Springer Science+Business Media, LLC 2011

how this tool can be used to enhance the planning and design process using the Transbay Terminal project as a case study.

## Architecture

MassMotion separates the calculation of crowd activity into two distinct processes. The first component is referred to as "reflexive" and governs the individual agents' basic movements and responses to the environment. This reflexive component navigates the agents through open space while avoiding obstacles and other agents. The second component is referred to as "contemplative" and governs the agents' network path planning between origins and destinations. This component analyzes distance, congestion, and terrain to develop costs for all available routes to the agent goal and to select an appropriate route based on these costs.

### *Reflexive agent motion*

The reflexive component of MassMotion agent movement is broken down into spatial analysis and movement toward areas of high utility. As described by Kuffner1 each individual agent is made aware of their environment through bit map representations of free and obstructed space on all walk-able surfaces of the 3D simulation environment. This approach uses a 2D projection of all static obstacles within a defined volume to map obstructed areas and then uses a modified version of Dijkstra's2 algorithm to describe all complete paths between a starting and goal location within the map. Each agent is also aware of other agents within their immediate neighbourhood using a global space partitioning structure to improve the efficiency of neighbor discovery within a specified range.

Using a combination of information from the path/obstruction map and the positions and velocities of the neighbouring agents, each agent determines their best available target location for the next frame of the simulation and adjusts their velocity and orientation to achieve that position. This calculation is executed at five frames per second of simulated time which is frequent enough to allow agents to adjust to dynamically changing conditions within the environment without encroaching on locations occupied by obstructions or other agents.

While the computational methods used are well documented, the practice of pedestrian planning requires that the results of the simulation of agent motion conform to industry standards. The Level of Service (LOS) standard developed by Fruin3 for pedestrian planning and design defines performance thresholds A through F. These thresholds describe the expected motion and interactions between people in a crowd for densities ranging from completely unimpeded in free space (LOS A) to packed shoulder to shoulder and chest to chest (LOS F). The reflexive motion of MassMotion agents has been calibrated to the Fruin walkway and bulk queuing LOS standards for level passages, restricted passages, stairs, and escalators. While the reflexive agent motion of the MassMotion system does not in and of itself help to manage complexity, it is the foundation of reliable reporting on crowd conditions within a complex network.

## *Network Construction & Assignment*

The contemplative component of the MassMotion system is based on a sparse network description of the overall simulation environment.

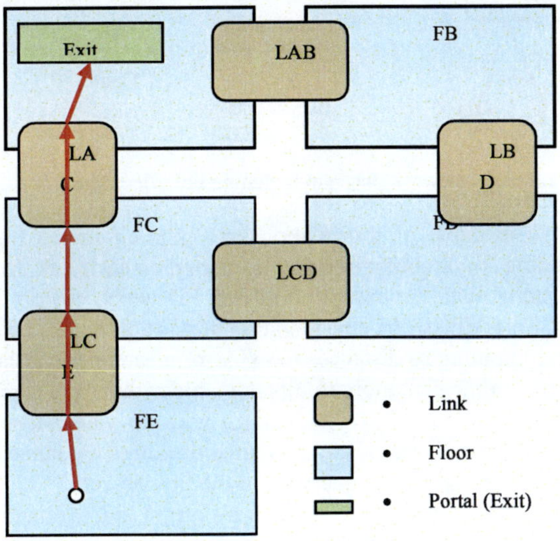


**Fig. 1. Example of MassMotion environment comprised of floors and links**

As shown in Figure 1 the environment is broken up into a series of floors and links with floors typically representing rooms, plazas, corridors, etc. and links representing doorways, thresholds, stairs, escalators, etc. The floors and links of the environment model constitute the nodes in the sparse network.

The first way in which MassMotion helps to manage the complexity of simulating a pedestrian environment is in the automatic association of floors and adjacent links. Simple geometric tests are carried out to establish which two floors are connected by a specific link. For example, in Figure 1 floors FE and FC are implicitly connected by link LCE which touches both floors. In this way the modeler is relieved of the task of defining the organization of the network beyond the geometric definition of the environment and specifying the type of each element/node.

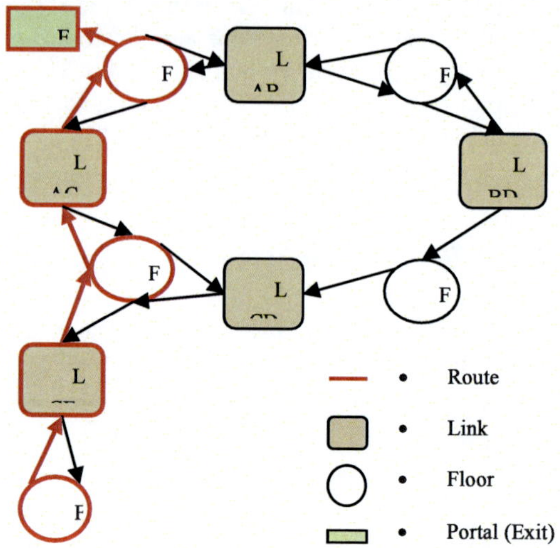


**Fig. 2. Sparse network diagram including link directions**

Figure 2 shows an example of the relationships implicit in the floors-links sparse network including the direction of travel permitted on each link. In this case all links are bi-directional with the exception of link FD which could represent a turnstile or escalator. To facilitate the consideration of possible routes by agents in the simulation a pre-computing of distances to exits from each node is carried out (again using a modified implementation of Dijkstra's algorithm) and stored on the link nodes. Directionality is considered and routes with no viable forward path (e.g. LCD to LBD ) are ignored. These pre-computed distances are made available to the agents at run time through direct querying of the link objects.

In MassMotion all agent journeys are defined by origin and destination pairs. Each agent is given autonomy over the route it will take between its origin and destination points. This is the second significant way that MassMotion helps manage complexity. Because the software will manage the network assignment of agents on an individual basis there is no need for the modeler to specify assignments at junctions. In Figures 1 and 2 for example, the shortest distance path from floor FE to the exit on floor FA is via floor FC. This is particularly advantageous when simulating complex interconnected environments, where the effort required to manually assign volume splits at junctions per destination would result in an unmanageable number of permutations to be defined.

The route self-assignment process at any junction will be based on the perceived cost of all available routes. Available is defined as a route that leads to the agent's ultimate goal without using a previously traversed node. Cost perception is randomized per agent through the use of randomized weights for the cost components of routes. The simplified algorithm for total route cost is as follows:

$$cost = W_D * \left(\frac{D_G}{V}\right) + W_Q * Q + W_L * L$$

Where,
$W_D$ = Distance Weight (random agent property)
$D_G$ = Total distance from agent position to ultimate goal
$V$ = Agent's desired velocity (random agent property)
$W_Q$ = Queue Weight (random agent property)
$Q$ = Expected time in queue before reaching link entrance
$W_L$ = Link Traversal Weight (random agent property)
$L$ = Link Type Cost (level, ramp, stair, etc.)

Based on the results of the costing of all available routes, the agent will generate a probability list with the best cost route having the greatest chance of being selected and the worst cost route having the least chance of being selected. The agent then uses a randomizing function to select their route. This process, including the randomized selection of a rote based on probability has been shown4 to result in statistically similar network activity as surveyed at complex, high volume transit facilities. A significant advantage of the MassMotion system, that results from the automatic organization and costing of routes through a pedestrian network, is that design alternatives may be explored by simply replacing or modifying environment geometry. The sparse node network will update itself based on the new geometric relationships, while the availability and cost of routes within the network will likewise adjust to new structure. As pedestrian planning and design work is fundamentally concerned with the testing and refinement of design ideas this ability to rapidly adjust design models and run simulations is exceptionally valuable both in terms of effort saved and confidence gained through extensive testing.

## Transbay Terminal Case Study

The design for a new Transbay Terminal in San Francisco needed to demonstrate effectiveness in three key areas of pedestrian activity:

- Transit boardings and alightings
- Interchange between transit modes
- Neighbourhood impacts

The new terminal will be a multi-modal transit hub designed to serve commuter rail, commuter bus, local bus, and eventually high speed rail passengers in downtown San Francisco. The variety of modes, degree of interchange between modes, and density of the surrounding urban fabric required an analysis approach that would consider the interaction of people with disparate destinations and patterns of movement within a complex environment. MassMotion models were constructed to assist the design team in analyzing the proposed layout of the station. The intended use of the model was to predict:

- Capacity of platforms and vertical circulation
- Demand on internal circulation routes
- Neighbourhood dispersion patterns

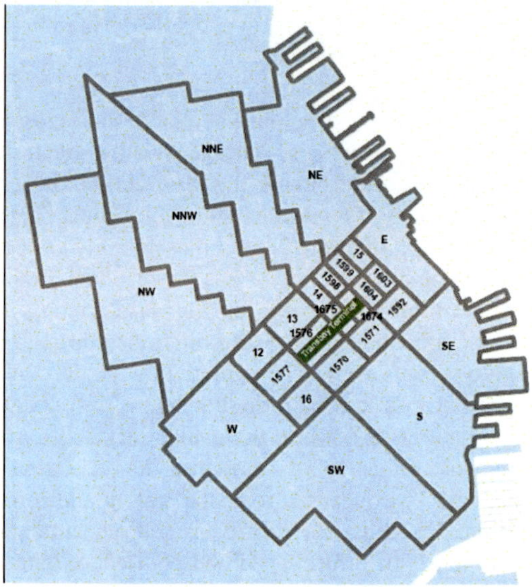


**Fig. 3. Transbay neighbourhood TAZs showing aggregations outside of immediate site**

## *Input Data*

Projected ridership and transit schedules were provided to the design team for inclusion in the simulations. Much of this data was derived from regional transit models that provided distributions between various transit modes and the surrounding Traffic Analysis Zones (TAZs). The TAZs are based on census data including residential and workplace densities which inform the regional transit models in terms of mode assignment and volume of traffic.

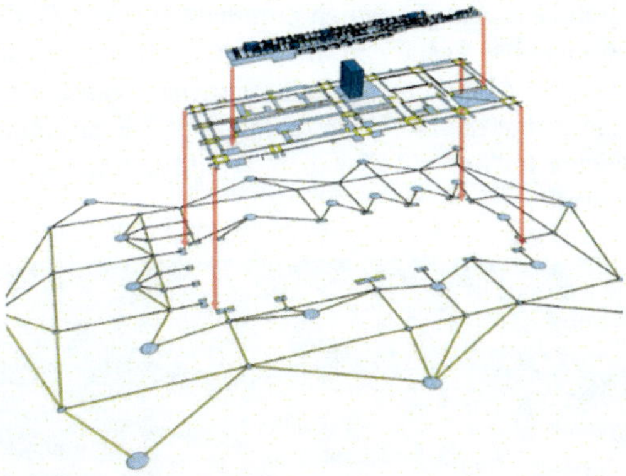

**Fig. 4. Exploded network diagram of Transbay MassMotion model showing terminal building, immediate site, and surrounding neighbouhood layers**

A matrix of origin and destination (O/D) pairs was developed that described the relationships between the various transit modes at the terminal building and the block in the surrounding neighbourhood.

## *3D Model of the Terminal and Neighbourhood*

With the origin and destinations entered into the MassMotion model based on transit activity (and including schedule timings for train and bus arrivals and departures) a 3D model of the terminal and neighborhood was developed. As with the O/D pairs, the construction of the 3D model was based on incremental reductions in level of detail as distance from the terminal building increased. As shown in Figure 4, all public circulation spaces and elements were modeled within the terminal building while sidewalks and crosswalks were modeled in the immediate site area. A simplified 3D representation of the aggregated TAZs from the outer neighbourhood (as defined in the O/D matrices) was developed that would provide reasonable approximations of overall distances without incurring a significant burden of modeling effort.

## *Results*

The simulation results indicated that early designs of the terminal building contained problem areas from a pedestrian circulation point of view due to insufficient channel widths in what were predicted to be high volume routes. Subsequent design iterations contained changes which eliminated areas of congestion according to the simulation results. At the end of schematic design the simulations were predicting that there would be no significant concerns regarding the boarding and alighting of passengers, that the internal circulation of the terminal building would accommodate projected traffic, and that there would not be significant impact to neighbourhood sidewalks.

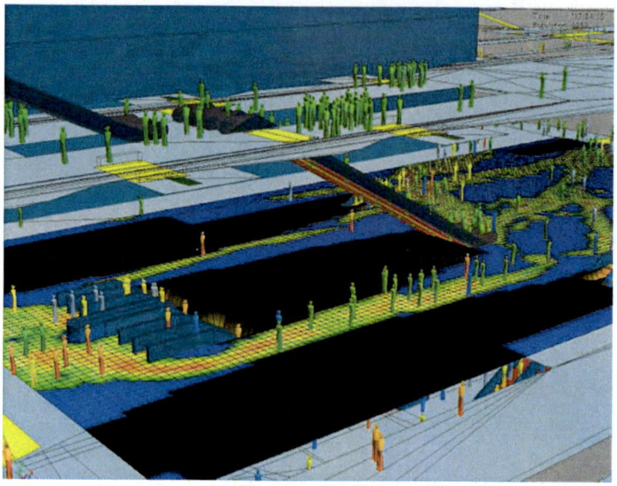

**Fig. 5. Screen capture of running simulation including average crowd density mapping**

It is clear from the Transbay case study that a simulation tool that minimizes the amount of modeler effort and provides a predictive view of design effectiveness is exceptionally valuable to the planning and design process. It enables the design team to devote less time to modeling more time to analysis and alternatives exploration which in turn increases confidence in the effectiveness of the design.

In addition to providing the design team with analysis of particular conditions and comparisons with desired outcomes it turns out that there are significant communications advantages to the MassMotion system. During the design process it became standard practice to bring the simulation model to design meeting for on the spot querying of particular issues and to gain insight into the overall functioning of the pedestrian network. The 3D models and animated motion of the agents provided a clear depiction of projected conditions and in a number of cases eliminated extra design effort and construction expenses that might have been required without such a tool.

## Next steps

While MassMotion has proven to be a very valuable design and capacity planning tool on Transbay and on similar projects around the world there are performance issues that need to be addressed. The modeling of thousands of individual agents in a complex virtual environment is computationally intense and there are hardware and software opportunities such as multi-core processors and 64-bit addressing that should be explored. In addition there are specific implementation approaches which could be done in alternative and perhaps more efficient ways. The work done by Helbing et al.5, Lakoba et al.6 and others on Social Forces looks very promising as an efficient means of describing the reflexive component of agent motion. Work has already begun on the next generation of MassMotion to address some of these possibilities.

**Acknowledgments** Much gratitude is extended to Wendy Tao and Andrew McCulloch for their work on developing the input data and simulation models for Transbay terminal designs and to Micah Zarnke for his ongoing contribution to the development and documentation of the MassMotion suite of simulation and analysis tools. The development of MassMotion has been made possible by ongoing research funding provided by the Design and Technical Executive of Arup and Oasys.

## References

1. Kuffner Jr., J.J.: Goal-Directed Navigation for Animated Characters Using Real-Time Path Planning and Control. In: Proceedings of CAPTECH '98 (November 1998)
2. Fruin, J.J.: Community Walkability. In: Pedestrian Planning & Design , Revised Edition. Elevator World (1987)
3. Dijkstra, E.W.: "A note on two problems in connexion with graphs." Numerische Mathematik, 1: 269–271 (1959)
4. Morrow, E.: Mass Motion: Simulating Human Behavior. In: Proceedings of the Design for Human Performance Conference. Royal Society, London (January 2009)
5. Helbing, D., Molnár, P.: Social force model for pedestrian dynamics II. Institute of Theoretical Physics, University of Stuttgart, Germany (January 1995)
6. Taras, I., Lakoba, D., Kaup, J., Finkelstein, N.M.: "Modifications of the Helbing-Molnár-Farkas-Vicsek Social Force Model for Pedestrian Evolution." SIMULATION, 81, 339- 352 (2005)

# Next Generation Paradigms in Pedestrian Modeling

**D.W. Peterson**

Ove Arup and Partners
New York, NY 10013 USA
Corresponding author: daniel.peterson@arup.com

**Abstract** This paper discusses the development of object oriented, agent based facility simulation tools geared to examining not just physical movement performance, but operational interactions from overarching facility command and control to individual human reactions as experienced during both normative and extreme scenarios.

The paper examines how the next generation of facility modeling will make use of agents moving within 3-dimensional space physically interacting with each other using onboard stochastic rules that define their actions and movement geometries. Human agents will have both a physical kinematic presence as well as visual and auditory capabilities to apprehend their environment.

The suite of agents involved in the "facility" model will expand from simply representing human actors and simple mechanisms such as escalators and elevators to include the "equipment" pedestrians and facility operators use, both large and small: from cell phones, to wheelie luggage, to signage, to kiosks and stores and their contents, to trains, buses and cars, to their controlling signaling and their centralized command and control centers, to the air and smoke they breathe.

Expansion of the physical components that can independently interact will facilitate much more realistic next generation analyses of facility operations under exceptional conditions including the disorienting effects of fire, smoke, noise, poor lighting, miscommunication, equipment failure, etc. Such "deep" models will produce new understandings of the interaction of humans in the physical environment as distorted under extreme event conditions.

The basis of this next generation modeling lies in the creation of autonomous agents – human, non-human and systematic, stochastically parameterized to respond and interact with the other agents in their environment on the basis of behavioral rule sets informed by their perceptual capabilities of "seeing," "hearing," and otherwise sampling their environment.

An object oriented agent based approach to pedestrian modeling will provide a platform to develop through research deeper understandings of the underlying behaviors that direct human performance and response under stressful situations and can drive facility design and operational protocols to assure competence and capability in untoward situations.

R.D. Peacock et al. (eds.), *Pedestrian and Evacuation Dynamics*,
DOI 10.1007/978-1-4419-9725-8_89, © Springer Science+Business Media, LLC 2011

## Agent-based Modeling

It is suggested that sufficient computing power is now generally available to support the modeling of observation and decision making activities by numbers of autonomous agents similar to the instantaneous volumes (5,000 – 50,000) found in typical large urban pedestrian oriented facilities (terminals, airports, street environments, arenas). These prime human representative agents can be possessed with direct observational and interpretive faculties similar to the binocular apprehension of the environment by the human vision system, auditory system, and olfactory system. Sampling rates on the order of 2-15 Hz. combined with narrow-wide perception discrimination and binocular integration can lead to robust spatial location within static peripheral environment surroundings while moving, as well as characterization and trajectory identification of other active entities moving within the environment space.

Agents themselves can exhibit adequate degrees of freedom to grossly replicate human kinematic movements across the range of gaits available. These can be invoked to provide normative progression in response to purpose based behavioral objectives, as well as collision avoidance actions (side-step, twist, etc.) through more direct "sub-conscious" apprehension of agents within the field far, medial and immediately ahead and the observed angular rates of change of incoming observed agents and elements.

Human agents themselves can be expressed and observed within a three-dimensional model space full formed of muscular-skeletal directed surfaces covered within a diverse wardrobe that correlates with stochastic behavioral "personality" types. Human agents can posses a bi-level set of controllable physical attributes, namely gross physiology and visage. Control is expressed of gross kinematics (head, torso and limbs), and fine kinematics – facial muscles yielding expressions, hand and finger motions (as required), and eye movement.

Utilizing object co-recognition within the binocular observations, human agents are capable of observing the other agents and entities within the simulated three-dimensional model and developing likely trajectories for proximate and further afield agents. Movement cluing is supplemented with interruption of gross observation to afford brief (0.5s to 1s) focused observation of approaching near space faces to read facial element patterning (eye orientations, brows, lips, jowl positions, etc. to ascertain intended incipient near space movement, as well as correlation with specific known identities, potential for engagement – either positive or threatening, or other behavioral states interpreted from the observed facial element orientations (expressions).

Agents likewise can make use of environmental noise and the phenomenon of echolocation to confirm spatial location within the physical environment and to identify event states. Agents can utilize speech recognition to identify (with varying degrees of imperfection) spoken information and use syntactical rules to identify meaning and appropriate responsive actions.

Agents thus can exist as "real" entities, can see and be "seen" by other agents, can hear the world around them, and posses a suite of behavioral directed correlates including clothing, hair, facial expression and gait.

The directed-ness of movement can likewise be behaviorally influenced including: modifying the directional consistency to far field wayfinding derived target acquisition, engaging in circumambulation to avoid observed overconcentration of other agents and elements or to advantage (by short lateral movements for instance) more proximate access to oversaturated limited access facilities (escalators, doors, stairs, etc), and exhibiting altruistic behavior – pausing for others, giving way, interleaving (zipper) constrained access, "ladies first", and other cultural, gender and normative rules.

Under extreme stress situations, the behavioral response can manifest human correlates such as tunnel vision, diminished auditory processing, loss of wayfinding identification in spite of observation, overreliance on known routes, extreme altruism or loss of behavioral restraints (every man for himself).

Agents, of course, are not limited to human constructs. They can include vehicles (automobiles, trains, bicycles, etc.), which exhibit a limited set of geometric and behavioral capabilities to move through space intrinsic to the travelways they move upon and provide for the seating or standing of human agents as well as control by endowed human agents (drivers, train engineer, taxi driver, cyclist, etc.). The travelways, which themselves contain laterally limiting and forward progression controls (curbing, stripes, rails, signals, signage, etc), define a movement space for the vehicular agents.

Agents today can also include various supplemental devices enabled by human agents. These include active objects such as cell phones, iPods, watches, Blackberrys, hand or shoulder bags, wheelies, and luggage. Passive objects such as newspapers, books, etc. can be included in the simulation, to affect observation capabilities and elicit distractions from directly accessing intended objectives. These agents, co-attached to and variant among the human agents can enable or impede their hosts. Specific behavioral responses are elicited by activation of these co-attached agents, which in turn, affect goal achievement.

Similar active and passive agents can be fixed within the physical environment. These include various forms of signage/signaling including direct wayfinding signage, textures, surfaces and styles that are indicative of location within a particular complex, intercoms and public address systems, bells, sirens, whistles and other audio generating elements that possess coded meaning. Lighting or the lack thereof can temper the acquisition of visual information, resulting in missed information, or misinterpretation, which can lead to poor or erroneous judgments or inappropriate behavioral responses. Various facility and street furniture can stimulate agent behaviors, whether avoidance or use (sit and read the paper then throw it away).

The system of human agents within a facility can include both the users (commuters, pedestrians, fans, attendees), as well as the controlling agents (facility managers, staff, dispatchers, police, security, ticket vendors and takers, ushers) The latter can be supplemented with first responders (fire, police, EMS, OEM, DHS), operational staff (drivers, conductors, train engineers, brakemen, pilots, crew, ground crew, etc) and facility staff who may have specific duties (vending, collecting garbage, cleaning, etc.).

These supervisory agents can have at their disposal the simulated suite of observational enhancement tools (cameras, audio pickup, facial recognition, counters, sensors, etc) as well as internal and external communication tools (public address, variable message signage, train/flight information displays, emergency bells & klaxons, etc.). Endowed with behaviors developed from expert knowledge and operational protocols, these controlling agents can initiate cascading directives through their chains of command to effect transformative perceptions and consequent behaviors amongst the controlled publics.

Two other realms of agent simulation are also germane to the simulation of the public or quasi-public realm. The first is the panalopy of experiential elements typical of personal commerce and engagement while enjoying or transiting within public facilities. This includes the consuming experience, notably shopping and eating. Highly behaviorally oriented, the myriad elements of a typical newsstand, kiosk, bookstore, convenience store, pharmacy, or sandwich shop, restaurant, pizza joint, bar, etc. are readily simulate-able with their content map-able to a variety of personality types that can be correlated with other behavioral traits. The recognition, selection, acquisition and consumption of goods can be a useful and valuable avenue of exploration for the marketing and merchandising communities in the design of their facilities. The effectiveness of the circulation patterns can be refined to maximize particular consumptive behaviors within target populations. In considering the variety of experiences typical of controlled environments, it is noted that restroom actions are a prevalent activity of patrons, particularly captive patrons (stadia, airports, etc).

The second realm of agent simulation is rather nebulous and concerns the facility atmosphere and the possible agents it might carry. These include the normal conditions of clean air at comfortable temperatures, in which case it is not necessarily a simulated item. On the other hand, for simulations that are concerned with smoke, fire, terroristic attacks, or with disease transmission and airborne pathogens, the inclusion of a lung function with a binocular olfactory system in human agents with consequent odor detection and smoke or other element incapacitation can render realistic simulations wherein a causative agent has been introduced (e.g., fire, smoke, radiation, chemical agent, etc). These causative agents can behave according to classical rules of development and propagation, consuming materials based upon their content and producing appropriate types of smoke and other toxins. The air handling systems of the facility and the natural air movements that may occur in the facility, tempered by heat can operate within the simulation to reduce visual acuity, obfuscate signage or exits, incapacitate or kill human agents or distort behaviors according to established precepts (loss of cognitive thinking, reversion to basal drives, inappropriate decision making, crowd following, stampeding, etc.)

It is in this last area, the transformation of normative behavior to baser primitives (echoing Le Bon), that this author finds particularly intriguing. A simulation has some value if everything goes right. In fact much of the effort in crafting simulations lies in producing an acceptable normative example. Perhaps the larger value in the simulation exercise is identifying what happens when things go wrong: the fire door is left open, the crowd follows its leader down a set of stairs with inward opening doors, the escalator delivers an unending stream into an undersized platform while the train disgorges a thousand revelers to a parade, those in command are buried in a Clausewitzian fog, the lights go out and the emergency back-ups weren't maintained, bad decisions get made, posts are abandoned, radios are on the wrong frequency, someone pulls the emergency brake. It goes on and on. Often, small events can precipitate extreme reactions putting thousands of lives at risk. Somebody stumbles, Bethnal Green.

A highly interactive agent based simulation platform with real world correlates to human interactions has the capability to explore the what-ifs in facilities where operators and authorities have line responsibility for providing safe and stable environments. The work program suggested here is sizable, however, the development and deployment of human agents continues apace both within the gaming as well as the facility management, military, entertainment and academic communities. That we can see this synergy presently developing suggests that perhaps the paradigms of modeling described herein are within reach of those of this generation, and that all that is lacking is sufficient collaborative depth and coordinated purpose to enable them to come to broader fruition.

**Acknowledgments** The author thanks NIST and the organizers of the Ped2010 conference for the opportunity to contribute the foregoing speculations.

# Next Steps for Agent-Based Simulations of Mass Egress

**D. A. Samuelson**

InfoLogix, Inc.
Annandale, VA 22003 USA
Corresponding author: samuelsondoug@yahoo.com

**Abstract** We consider how to address some of the most significant challenges at present in agent-based simulations of mass egress and evacuation. These include more detailed depiction of agents' movements and decisions; more acurate representations of agents' behavior, including group movements and the consequent delays; and more attention to long-term effects of injury and exposure, such as respiratory, digestive and immune system disorders and mood disorders, which may have a physical component cause.

## Introduction

Modeling pedestrian dynamics and evacuations has progressed dramatically in the past decade. Most of the presentations at PED2001 were mathematical or anecdotal; agent-based simulations have become increasingly prominent, to the point of dominating the field. The rapidly improving computer simulation capabilities offer great opportunity, but they also impose some restraints, as it is natural to do what is easier and neglect the topics that are still difficult. It is timely, therefore, to review some of the most important current challenges and to suggest some approaches to them.

## Challenges in Modeling Methods

With respect to techniques in agent-based simulation modeling, some major challenges and proposed approaches are:

- *Improve the trade-off between level of detail and execution speed in models of pedestrian behavior and related events.* Using a limited set of paths for each person-agent, identified at the beginning of the simulation, offers substantial improvements in speed with modest loss of detail. This made it possible for the Homeland Security Institute (HSI) / Redfish stadium model, [3, 5] to run with

R.D. Peacock et al. (eds.), *Pedestrian and Evacuation Dynamics*,
DOI 10.1007/978-1-4419-9725-8_90, 

up to 70,000 person-agents in real time. Additional improvements in visualization and depiction, along with general running-time progress, have made better, faster and more detailed models feasible since then. The need will continue, however, for well-chosen trade-offs in model design, to preserve real-time performance while adding needed details.

- *Incorporate lessons learned from recent real events, such as the 2008 U. S. Presidential nominating conventions, the 2009 U.S. Presidential Inauguration and the 2008 Hajj.* As the analysis of the HSI stadium model suggested, better guidance to available exits, presented in a way that reassures people that they will get out safely, substantially alters their behavior and lowers the risk of secondary (crush and trampling) casualties. Including this in the model requires adding "level of urgency" parameters to person-agents, which in turn modify the agent's perceived path distance when paths are partially obstructed: less urgency means more willingness to go around obstacle, some of which may be other people. The models of the conventions developed by Regal Decision Systems for the U.S. Secret Service [1] highlighted the importance of which exits are blocked, such as by TV equipment, and how well alternate exit routes are marked or announced.
- *Integrate modeling catastrophic evacuation and research on normal movement.* The HSI / Redfish subway model used the recently developed Continuum Crowds algorithm, which more accurately represents non-emergency adaptations to counter-flow. This model, however, exhibited some behaviors that suggest Continuum Crowds does not capture some of the behavioral changes that occur under urgency. The extent to which people will avoid bumping into each other seems to change more in real life than it does in this algorithm. Again, the inclusion and calibration of perceived urgency in person-agents is indicated.
- *Incorporate and evaluate automatic control methods; connect simulation modeling to classic methods such as queuing theory (for example, how to handle a combined queue).* Control methods considered in most models to date are relatively simple and unchanging during the event, mostly focused on changing person-agents' perceptions of path lengths and obstructions. More sophisticated route selection and redirection can be built in with little change in the remainder of the model, simply by embedding more complex control code in the updating of route selection and indications of changing path lengths.
- *Include group movements in person-agent behavior.* Family groups and groups of people moving a disabled person tend to remain together, moving at the speed of the slowest member of the group and occupying a clump of space that inhibits movement of other people. Person-agents, in addition to being programmed to exhibit more individual variation in speed and endurance, can be programmed to identify themselves with a group and resist separation. This is one of the more difficult enhancements to current models, and data on real-event behaviors are sparse.

## Challenges in Scope and Focus of Modeling

*Assess what pre-event information and training are valuable (for example, the public awareness programs California uses for earthquakes and various Midwestern states use for tornadoes.)* This can be modeled by including in person-agents a parameter varying the time from receipt of new information until the expected behavioral change occurs. Pre-event preparation reduces this time interval. Information from actual events, such as California earthquakes (where many people know where to seek shelter) and evacuations after forced airplane landings, also indicates that choice of behavior, not just speed of choice, depends critically on prior information provided to the public.

*Include such effects as toxic plumes and falling debris.* This is a straightforward addition of such effects to the model, distributed appropriately among patches, with calculations of resulting casualties, as in the shrapnel effects calculations in the HSI / Redfish subway model. The University of Greenwich stadium area evacuation model incorporates some of these effects, [6,7] and there are models available from the U. S. Naval Research Laboratory that offer a good start in depicting plume effects. [1]

*Take account of injuries and stress (respiratory distress, toxic substance ingestion, traumatic scenes) which may induce long-term adverse effects.* [4] Length of exposure to toxins, amount of time and exertion required to escape, and number of casualties in the person-agent's vicinity can be computed for person-agents and incorporated into the output measures of the model, along with such basic information as number of casualties and average time to leave the area.

*Include treatment options (e.g., for smoke inhalation, the U.S. practice of sending people to hospital versus the German practice of anti-inflammatory treatment on the spot) in the output measures of effects.* Pulmonologist Dr. Hubert Lollgen discussed these findings in his plenary address at PED2008. [2] This is more an interpretive than a pure modeling issue, but, again, adding more output measures to the models will facilitate better analysis of overall effects of events.

*Use the same modeling framework to guide development, especially the display options, of real-time information systems used in incident management.* Emergency managers increasingly demand good real-time visual displays of problem areas and remaining resources. Current technology does not support integrating such information systems and behavioral modeling in the same computer program. However, to the extent that analytical / training models look similar to intended real-time information systems, model developers can aid incident managers in relating what they learn in training exercises to what they would see in real-time incident management.

## Challenges in Evaluating Models

*Explore validation / calibration issues, especially for rare catastrophic situations.* Agent-based simulations generally present unusual difficulties in validation, as the complexity of the simulations approaches that of the real situations being modeled. Also, available data for real rare, catastrophic events are likely be inadequate to assess many options in the model, and opportunities to experiment with real events are limited. Therefore, such modeling requires an increasing emphasis on validation of parts, such as the crowd movement algorithm. Another useful validation activity is presenting stories of how an event might unfold, based on movie representations generated by the model, to subject matter experts and having these experts judge whether important elements seem to be misstated or missing.

## References

1. Harmon, M., Joseph, J.: Evacuation Planning Tool (EPT) for Emergency, Event and Space Planning. Presentation scheduled for PED2010 (2010)
2. Loellgen, H.: Injury of Lung and Heart after Inhalation of Toxic Substances. In: Klingsch et al. (ed.), Pedestrian Dynamics and Evacuation 2008. Springer (2010)
3. Samuelson, D., Parker, M., Miller, L., Zimmerman, A., Blacksten, H.R., Pommerenck, S.: Agent-Based Simulation of Mass Egress After an Improvised Explosive Device Attack. Homeland Security Institute Final Report to the Science and Technology Directorate, HSI Document Number RP06-IOA-31-03. U.S. Department of Homeland Security (February 2007)
4. Samuelson, D.: Mass Egress and Post-Disaster Responses. OR/MS Today. **34**, 5: 20-24 (October 2007)
5. Samuelson, D., Parker, M., Zimmerman, A., Guerin, S., Thorp, J., Densmore, O.: Agent-Based Simulations of Mass Egress After an Improvised Explosive Device Attack. In: Proceedings of the European Conference on Complex Systens. Dresden, Germany (October 2007)
6. Veeraswamy, A., Galea, E.R., Lawrence, P.: Implementation of Cognitive Mapping, Spatial Representation and Wayfinding Behaviors of People within Evacuation Modeling Tools. In: Proceedings of the 4th International Symposium on Human Behavior in Fire, Robinson College, pp. 501-512. Cambridge, UK (July 2009)
7. Xie, H., Filippidis, L., Galea, E., Blackshields, D., Lawrence, P.J.: Experimental Study of the Effectiveness of Emergency Signage. In: Proceedings of the 4th International Symposium on Human Behavior in Fire, Robinson College, pp. 289-300. Cambridge, UK (July 2009)

# Experimental Study of Pedestrian Flow in the Channel through Bottleneck

**W. Song[1], J. Zhang[1, 2] and A. Seyfried[3]**

[1]University of Science and Technology of China, Hefei, 230026, P.R.China
[2]University of Wuppertal, Institute for Building Material Technology and Fire Safety Science, Pauluskirchstr. 7, D-42285 Wuppertal, Germany
[3]Forschungszentrum Jülich GmbH (JSC) D-52425 Jülich, Germany
Corresponding author: wgsong@ustc.edu.cn

**Abstract** In this study, experiments of pedestrian flow through a bottleneck are performed by changing the width of the bottleneck, *b*, from 0.5 *m* to 1.4 *m*. The time dependence of the pedestrian density in front of and inside the bottleneck is studied. We calculated the velocity of each pedestrian passing through the measurement section inside the bottleneck and the corresponding mean density, and the fundamental diagram is obtained. The relation between the flow rate *J* and the width *b* is studied and compared with the results of other researchers.

## Introduction

Experimental results are used to calibrate and validate current models. For architects and planners as well as for the management of evacuations it is more important to obtain quantitative results than qualitative ones. For this reason the accuracy and the reliability of the models needs to be verified before making quantitative predictions, like the total evacuation time etc. However, the experimental data are seriously insufficient and contradictory [1]. A number of experiments and egress drills have been conducted. One example is bottleneck experiments [2-5]. Most typical phenomena, such as "zipper effect" and the shape of the jam in front of the bottleneck are studied. In this paper, we present results of a series of pedestrian experiments through a bottleneck. The time dependence of density, the fundamental diagram and other data are obtained.

## Experiment Setting and Results

The experimental scenario, as shown in Fig. 1, is similar to Ref. [3] to enable an easy comparison. The only difference is a small ground gradient of 2.7°. The experimental setup is built up with tables and publicity boards to prevent pedestrians putting their body out of the corridor. There are 62 participants, 13 females and 49

R.D. Peacock et al. (eds.), *Pedestrian and Evacuation Dynamics*,
DOI 10.1007/978-1-4419-9725-8_91, 

males aging from 17 to 35, participated in the 10 groups experiment. The width of the bottleneck was increased from the minimal value of $b = 0.5\ m$ in steps of 0.1 *m* to a maximal value of $b = 1.4\ m$. At the beginning of the experiment, all the participants were uniformly distributed in a 3 *m*×3 *m* area 3 meters away from the bottleneck; and were asked to move towards and through the bottleneck at normal speed upon hearing a starting command. Each run was recorded by two video cameras situated vertically above the bottleneck.

**Fig. 1. Snapshot of the experimental setup**

Firstly, we calculated the time dependence of the density in front of and inside the bottleneck. Here the density is defined as the number of the pedestrians in the measurement section divided by the area of the section. In order to study the density evolution in front of the bottleneck, we choose a *1 m* x *1 m* area close to the bottleneck as the measurement area, while we chose an area *2* meters in length inside the bottleneck to study the density. Both of these sections have been marked before the experiment and are easy to distinguish from the video.

We calculated the number of pedestrians in the measurement section per second and then obtained the density evolution (as shown in Fig. 2). The mean values for the stationary state over time agree for different width b within the variance. The mean value over b is 4.8 $1/m^2$. This is in good agreement with the results of the experiment performed in Germany [3]. But maximum values are higher for smaller widths, see e.g. b = 0.5 m in Fig. 2 (left), and reach 8 $1/m^2$. Inside the bottleneck the density retains a constant value all the time for b = 0.5 m and 0.6 m, with relatively large fluctuations when b > 0.7 m. That may be due to layer formation inside the bottleneck. For b < 0.7 m, only a stationary single file inside the bottleneck exists. Inside the bottleneck the mean value over b is 2.4 $1/m^2$. This is significantly higher than the densities reported in [3].

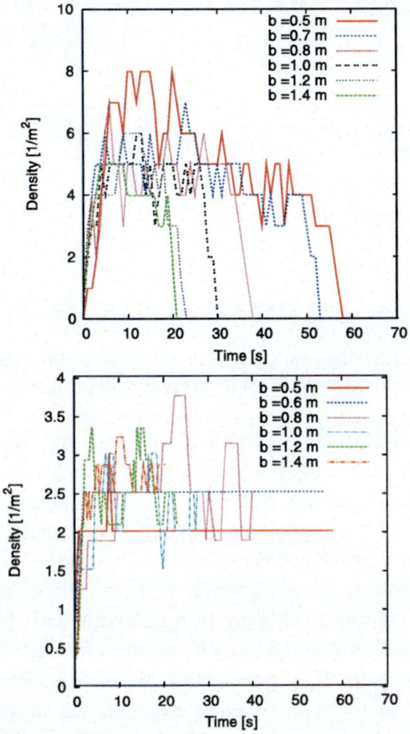


**Fig. 2. Time dependence of density in front of (left) and inside (right) the bottleneck**

To determine the fundamental diagram, we calculate the density and velocity for all experiments inside the bottleneck. The density is determined at the time a test person enters the measurement section of $l = 2$ *m*. The corresponding velocity is calculated by the entrance and exit times $v = l/(t_{out}-t_{in})$. It is observed that the velocity presents a decline with increasing density and the maximum of the velocity is larger than 2 *m/s* due to the first people entering the bottleneck without restriction. With increasing density, the change of the velocity slows and retains a value. The mean speed for densities larger than 2.5 $1/m^2$ is 0.88 *m/s* and smaller than the stationary state velocities presented in [3].

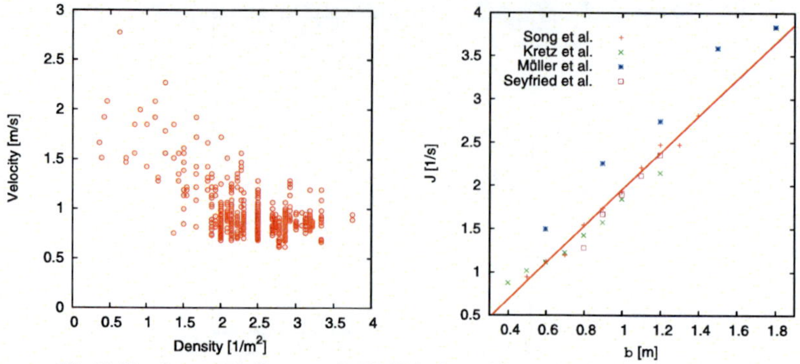

**Fig. 3. Fundamental diagram (left). Relation of flow rate against b in comparison with experimental results presented in [3-5] (right).**

We compare the relations between flow rate and the width of the bottleneck with others experiments (see Fig. 3 right). It is shown that the flow rate increases monotonously with the width for $b > 0.8\ m$. In the majority of analysis, it was concluded that details of the bottleneck geometry and position only play a minor role whereas the initial density in front of the bottleneck has a major impact [3]. It can be seen that the flow is neither a linear function of the width, $b$, of a bottleneck nor grows in a stepwise manner. In agreement with [4] the slope of $J(b)$ changes between $b = 0.7$ m and 0.8 m. We assume that lane formation inside the bottleneck is responsible for this, because the number of lanes inside the bottleneck changes from one to two. When $b$ exceeds 1.3 $m$ the number of lanes changes from two to three. However, the transition is not as clear as it was previously because the space is larger and pedestrians can move freer. In this case, the effect of lane formation on the flow becomes weaker. Hence $b = 0.8\ m$ can be thought as a critical width. The flow at this point exhibits differences between this study and [2]. In contrast to this study, we observed that the number of lanes is two. However, it becomes one at later stage with the decrease of the density. Even if the width of 0.8 $m$ allows two lanes to exist, it is not the most comfortable state for pedestrians and they still prefer to move in one single file if possible. There are several possible explanations for these differences.

## Conclusion

We presented results from an experiment to study pedestrian flow through a bottleneck under normal conditions. Characteristics such as density, velocity and flow were analyzed. The comparison of the characteristics with experimental results gained in Germany [3] shows that densities in front of the bottleneck are equal. Inside the bottleneck the density is higher while the velocity is smaller. The fundamental diagram is analyzed inside the bottleneck, which shows that the congestion induced by the bottleneck can lead to a stationary state inside. After comparing the flow rate with other experiments, it can be concluded that the flow is a linear function for b > 0.8 m but shows an edge between b = 0.7 m and 0.8 m. The formation of lanes inside the bottleneck has some effects on its increasing manner, and these effects will become weaker with the increasing width of the bottleneck. For b = 0.8 m we observed two lanes inside the bottleneck. This disagrees with the results in [3].

**Acknowledgement:** The study is supported by China National Natural Science Foundation (No. 50678164), Program for New Century Excellent Talents in University (NCET-08-0518), National Science and Technology Pillar Program (No.2006BAK06B00).

## References

1. Schadschneider, A., Klingsch, W., Klüpfel, H., Kretz, T., Rogsch, C., Seyfried. A.: Evacuation dynamics: Empirical results, modeling and applications. In: Meyers, R.A. (Ed), Encyclopedia of Complexity and System Science, pp. 3142. Springer (2009)
2. Hoogendoorn, S.P., Daamen, W.: Pedestrian Behavior at Bottlenecks. Transportation Science. **39/2**, p. 147-159 (2005)
3. Seyfried, A., Steffen, B., Winkens, T., Boltes, M., Klingsch, W.: Empirical data for pedestrian flow through bottlenecks In: Appert-Rolland, C. et. al (eds.) Traffic and Granular Flow '07, p. 189-199. Springer (2009)
4. Kretz, T., Grünebohm, A., Schreckenberg, M.: Experimental study of pedestrian flow through a bottleneck. J. Stat. Mech., **10,** P10014 (2006)
5. Müller, K.: Zur Gestaltung und Bemessung von Fluchtwegen für die Evakuierung von Personen aus Bauwerken auf der Grundlage von Modellversuchen. PhD Thesis, Technische Hochschule Magdeburg, Germany (1981)

# Occupant Wayfinding in Multi-storey Buildings

**J. S. Sorsa and M-L. Siikonen**

KONE Corporation
P.O. Box 7, 02151 Espoo, Finland
Corresponding author: Janne.Sorsa@kone.com

**Abstract** Earlier, wayfinding has been studied only in the horizontal context, e.g., in urban navigation. In this paper, we extend route selection preferences to vertical travel arising in multi-storey buildings. We describe our vertical wayfinding model using multi-attribute utility theory. The model is implemented in a simulation tool that is used in designing building transports as well as for evacuation studies in tall buildings. We show how the path selection determines not only the individual decisions but also the end result of the simulation.

## Introduction

In multi-storey buildings, occupants use various means of vertical transportation such as elevators, escalators or stairs, on their path from the origin to the destination. The path consists of vertical trips and horizontal movement in between. The occupants select the path from multiple alternatives and either follow it to their destinations or change it at choice points; we call this vertical wayfinding. Wayfinding is a well-known problem in the horizontal context: Golledge studied path selection criteria in urban navigation [1]; Veeraswamy et al. [2] implemented wayfinding in an evacuation simulation of one-storey buildings. Vertical wayfinding has not been studied earlier, except in [3-5] although it was not identified as wayfinding. In this paper, we relate the path selection criteria of [1] to the vertical wayfinding and describe the path selection problem using multi-attribute utility theory [6].

A wayfinding model is an integral part of any simulation tool for multi-storey buildings, whether targeted at evacuation or circulation. A simple selection of the nearest exit to the next floor is not sufficient owing to trip-chaining, which requires knowledge of the follow-up connections as well as their type. As described in [1], people use, in addition to the typical shortest path, also other criteria in path selection. If, in the vertical context, only the shortest path criterion is used, the occupants always choose the fastest route. Eventually, this leads to the situation where some transports become overcrowded while others are not used at all. Everyday experience suggests that people do not behave like this but are tempted to look for alternative paths. Thus, the wayfinding model in a simulation also has to consider other criteria so that the virtual agents choose realistic routes. Using si-

R.D. Peacock et al. (eds.), *Pedestrian and Evacuation Dynamics*,
DOI 10.1007/978-1-4419-9725-8_92, © Springer Science+Business Media, LLC 2011

mulations, we show the effect of the wayfinding model on individual path selection decisions and on the aggregate simulation results.

## Vertical Wayfinding in Building Traffic Simulator

Here, we describe the vertical wayfinding model implemented in the Building Traffic Simulator [7]. Virtual agents need to find their way from their origins to their destinations located on different floors. First, each agent selects an *initial path*. They follow this path until reaching a *choice point*, where they observe dynamically changing information such as queue lengths of the exits leading to other floors. Then, the agents reconsider the earlier path selection decision. If the current path seems unfavourable compared to other alternatives, the agent changes it. The path choice is re-evaluated also if an agent gets stuck in a queue for a time we call *frustration time*. Finally, the agents reach their destinations; the actual routes they followed are defined by the path selection decisions.

The path selection is a multi-attribute utility problem. The decision criteria of the vertical wayfinding are listed in Table 1 below, where we also relate them to the criteria of [1]. The last three of our criteria are unique for the vertical wayfinding. The travel distance criterion concerns short trips of up to five floors. For example, escalators are the most preferred transport for one-floor travel but for travels of five or more floors elevators become the most preferred transport [8].

**Table 1. Decision criteria of the vertical wayfinding model**

| Criteria | Criteria in [1] | Rationale |
|---|---|---|
| Travel time | Fastest route | The fastest route from current location to the destination |
| Walking time | Fastest route | The shortest walking distance between transports |
| Waiting time | Avoiding congestion | The shortest wait in front of the exits near current location |
| Transfer | Fewest turns | The least number of transports |
| Direction | First noticed | Preference for paths towards the destination. |
| History | N/A | Preference for non-visited locations |
| Transport type | N/A | Preference for transport types |
| Travel distance | N/A | Transport type preference for short travel distances |

Spatial knowledge of the building is encoded in graph $G = (N, A)$, where nodes $i \in N$ represent locations and arcs $(i, j) \in A$ represent agents' either horizontal or vertical travel. Let node $i \in N$ denote an agent's current location, where he contemplates how to reach his destination. Assume now that there are $J$ nodes adjacent to $i$. From each of these nodes, the shortest path $P_j$ is evaluated to the destination. A utility function $u_b(f_b(P_j)) = (1 + f_b - \min(f_b))^{-1}$ determines the utility of path $P_j$ with respect to the decision criterion $b$, $0 < u_b(\cdot) \leq 1$. The functions $f_b(P_j) \geq 0$ measure the value of the path $P_j$. The weighted sum of $u_b(f_b(P_j))$ over all decision criteria $b$

is the multi-attribute utility $u(P_j)$ of path $P_j$. The path selection problem is to find path $P$ that maximizes the multi-attribute utility $u(P)$ over the $J$ shortest paths.

## Numerical Experiments

In this section, we study the effect of route cost and queuing criteria on the path selection. We adopt the simulation experiment of an underground station [9]. We assume that 400 passengers leave the train within one minute and head for the concourse level 4.2 m above the platform. The levels are connected by one escalator and two stairways, all having a width of 1.1 m. The speed of the escalator is 0.5 m/s. We also add an elevator with a speed of 1 m/s and a capacity of 13 persons.

We simulate totally four scenarios with varying transports: the escalator and the stairs in Scenario 1 and 2; all the transports in Scenario 3 and 4. In addition, we vary the route preferences. In Scenario 1 and 3, the utility weights are 1 and 0 for the travel and the waiting time, respectively, but 0.5 and 0.5 in Scenario 3 and 4.

In Table 2 below, we show the actual values of the decision criteria for an individual decision of an agent: travel time (*TT*) from the origin to the destination and waiting time (*WT*) in front of the exits on the platform level. The agent chooses the route giving the maximum multi-attribute utility *u*. The results show clearly the effect of blindly preferring the fastest route in scenarios 1 and 3: the agent has the greatest utility of choosing the escalator and joins the escalator queue no matter how long it is. In contrast, in scenarios 2 and 4, where both travel and waiting time have positive weights, the expected waiting times are almost the same for all the transports; the agent chooses the shortest wait in the expense of a bit longer travel time resulting in the best overall utility.

**Table 2. Criteria measures of an individual path selection decision**

| Route | Scenario 1 | | | Scenario 2 | | | Scenario 3 | | | Scenario 4 | | |
|---|---|---|---|---|---|---|---|---|---|---|---|---|
| | *TT* (s) | *WT* (s) | *u* | *TT* (s) | *WT* (s) | *u* | *TT* (s) | *WT* (s) | *u* | *TT* (s) | *WT* (s) | *u* |
| Escalator | 21.2 | 101.5 | **1.00** | 21.2 | 37.5 | 0.59 | 21.2 | 101.5 | **1.00** | 21.2 | 31.4 | 0.58 |
| Stair 1 | 25.1 | 0 | 0.20 | 25.1 | 34.7 | 0.28 | 25.1 | 0 | 0.20 | 25.1 | 27.4 | 0.30 |
| Stair 2 | 25.1 | 0 | 0.20 | 25.1 | 32.9 | **0.60** | 25.1 | 0 | 0.20 | 25.1 | 27.4 | 0.30 |
| Elevator | - | - | - | - | - | - | 26.1 | 0 | 0.17 | 26.1 | 25.9 | **0.59** |

Table 3 below shows the simulation results: number of agents that use each transport, *N*, the average waiting time, *AWT*, and, in the bottom of the table, also the total time elapsed until all the agents reach the concourse level. The total time is the shortest in scenarios 2 and 4, about half of the time of Scenario 1 and 3. In addition, the transports are used about equally in scenarios 2 and 4. Hence, the path selection preferences have a great impact on the simulation results.

**Table 3. Simulation results: transport usage, average waiting time and total time**

| Transport | Scenario 1 | | Scenario 2 | | Scenario 3 | | Scenario 4 | |
|---|---|---|---|---|---|---|---|---|
| | *N* (pers.) | *AWT* (s) | *N* (pers.) | *AWT* (s) | *N* (pers.) | *AWT* (s) | *N* (pers.) | *AWT* (s) |
| Escalator | 400 | 74.5 | 213 | 25.3 | 400 | 74.5 | 196 | 20.5 |
| Stair 1 | 0 | 0 | 93 | 15.5 | 0 | 0 | 88 | 12.6 |
| Stair 2 | 0 | 0 | 94 | 15.7 | 0 | 0 | 88 | 12.6 |
| Elevator | - | - | - | - | 0 | 0 | 28 | 12.8 |
| Total time | 4 min 23 s | | 2 min 41 s | | 4 min 23 s | | 2 min 34 s | |

Although we have analyzed only a rather simple example of vertical wayfinding, the model is feasible for buildings of any height and complexity. Furthermore, it can be extended to include landmarks on the horizontal plane such as rooms, corridors, doors, and turnstiles. Our results show that the path selection determines the end result, which alleviates the importance of wayfinding in simulation.

## References

1. Golledge, R.G.: Path Selection and Route Preference in Human Navigation: A Progress Report. In: Frank, A.U., Kuhn, W. (eds.), Lecture Notes in Computer Science 988, pp. 207-222. Springer (1995)
2. Veeraswamy, A., Lawrence, P.J., Galea, E.R.: Implementation of Cognitive Mapping, Spatial Representation and Wayfinding Behaviours of People within Evacuation Modelling Tools. In: Human Behaviour in Fire 2009, pp. 501-512. Interscience Communications Limited (2009)
3. Susi, T., Sorsa, J., Siikonen, M.L.: Passenger Behaviour in Elevator Simulation. Elevatori. **34**, 28-37 (2005)
4. Sorsa, J., Hakonen, H., Susi, T., Siikonen, M.L.: Study of human factors in vertical evacuation. In: Proceedings of International Symposium on Elevator Evacuation during High-rise Fires. Shanghai Fire Research Institute (2007)
5. Siikonen, M.L., Sorsa, J.S.: Evacuation Simulation and Human Behaviour Models in Tall Buildings. In: Klingsch, W.W.F. et al (eds.), Pedestrian and Evacuation Dynamics 2008, pp. 653-658. Springer (2010)
6. Keeney, R.L., Raiffa, H.: Decisions with Multiple Objectives: Preferences and Value Tradeoffs. Cambridge University Press (1993)
7. Siikonen, M.L., Susi, T., Hakonen, J.: Passenger Traffic Flow Simulation in Tall Buildings. Elevator World. **49**, 117-123 (2001)
8. Barney, G.C.: Elevator Traffic Handbook. Spon Press (2003)
9. Kinsey, M.J., Galea, E.R., Lawrence, P.J.: Extended Model of Pedestrian Escalator Behaviour Based on Data Collected within a Chinese Underground Station. In: Human Behaviour in Fire 2009, pp. 501-512. Interscience Communications Limited (2009)

# Improved Methods for Checking Forces Based Models of Pedestrian Dynamics

**B. Steffen, M. Boltes and A. Seyfried**

Forschungszentrum Jülich GmbH, JSC
52425 Jülich, Germany
Corresponding author: b.steffen@fz-juelich.de

**Abstract** The force based models of pedestrian dynamics like the social force model or the centrifugal force model can demonstrate and sometimes explain many features of the collective behaviour of pedestrian crowds as self organisation effects based on fairly simple microscopic rules. However, experiments that have been done either gave only collective data like averages of flux and density, or treated only very simple situations like two person interactions. The progress of digital cameras and of image processing during recent years now allows the measurement of pedestrian movements on a ``microscopic" scale with low costs. Comparing forces derived from measured trajectories with calculated ones for the same situation allows a microscopic analysis of models and shows what parameters are important in special situations. Applying these methods to bottleneck experiments shows the importance of removing the effects of head movement and of stepping from the trajectories before calculating forces.

## Introduction

For the estimation of required evacuation time from buildings, macroscopic methods [1, 2] have a long history and good reliability for simple building geometries. For complicated buildings, defining the critical bottlenecks and performing the calculations may be a formidable task, so the need for computer calculations became obvious as buildings got larger and more complicated. For about 20 years now microscopic modeling of pedestrian flows has been done. These models may be based on cellular automata [3, 4], rules of movement [5, 6] or forces [7, 8]. Of these, the force based models present the most ambitious approach, as they try to give realistic individual trajectories for arbitrary situations. All three approaches rely on self organization of a large number of pedestrians in an environment through individual reactions on local situations. If the microscopic movement follows the correct statistics, the macroscopic quantities will be described correctly, too. Therefore these approaches have uses not only in evacuation situations, but also in the estimation of levels of service for normal operation. A big problem is the verification of the models. Even if macroscopic quantities like evacuation time or flow through a facility come out correct - and measurements are expensive,

R.D. Peacock et al. (eds.), *Pedestrian and Evacuation Dynamics*,
DOI 10.1007/978-1-4419-9725-8_93, © Springer Science+Business Media, LLC 2011

scarce and often contradictory [9,10] - this does not prove that the individual trajectories are anywhere near realistic. Proper verification works at the level of detail the model provides. In the force based models this means that the forces have to be correct. Of course, this can be only in a statistical sense, as individual features of persons cannot be predicted.

During the last few years, the video techniques have advanced so that individual trajectories can be experimentally determined even for fairly dense groups of people [11]. This allows to compare calculated and measured forces in a large number of simple situations, and thus try finding out weather the force model captures the influences on movement acting in these situations.

## Calculating forces from overhead video data

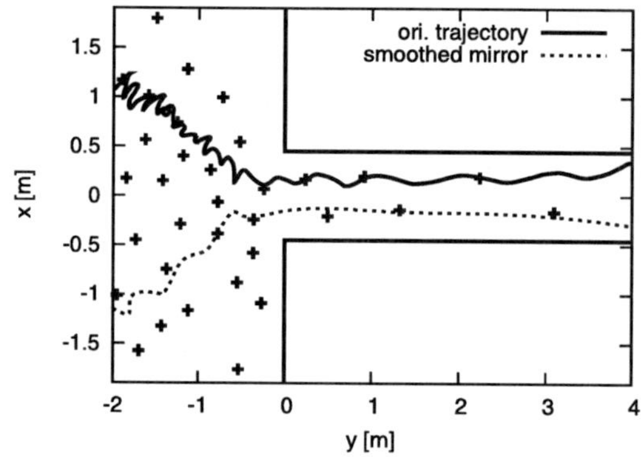


**Fig. 1. Positions of persons at a bottleneck (+), one individual trajectory and the mirror image of the corresponding smoothed trajectory.**

For a number of scenarios (walking through bottlenecks, around corners, bidirectional movement in corridors and others) we have performed experiments with overhead video recording [13]. Some of these recordings were done using stereo cameras, others with persons wearing markers indicating their size, so that the perspective distortion could be corrected and very accurate (errors < 5 cm) trajectories were obtained. E.g. for the bottleneck experiment, a typical situation is given in Fig. 1. From the trajectories, the forces are obtained by differentiating twice. Unfortunately, differentiating amplifies the noise of the trajectory data, and the result (Fig. 2) is certainly not what any model wants to predict.

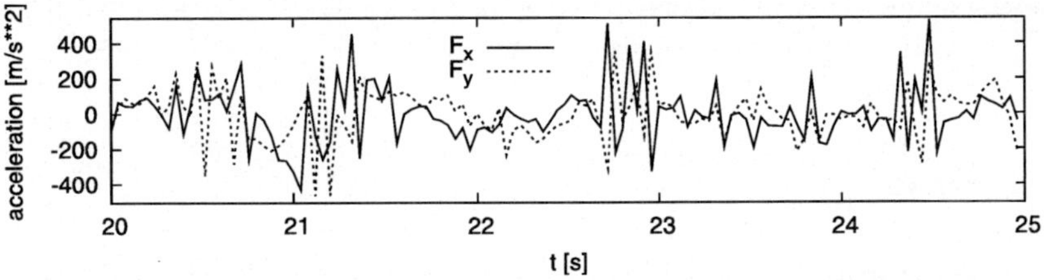


**Fig. 2. Forces in x and y- direction from trajectory (see Fig. 1) of a person for 5 s.**

The forces could be smoothed by a moving average, but that can introduce a small drift error in the recalculated trajectory. A trajectory x(t) = t-sin(t) gives a force x"(t) = sin(t), averaged over length $2\pi$ this gives simply x"(t)=x(t)=0. So the smoothing should be done with the positions, not the forces. One problem here is that force based simulations attempt to model a center of mass (principal) movement, while the trajectories contain at least three different scales: the principal movement, the stepping movement, and a head movement. Filtering out the latter is no problem, but for the stepping movement, step detection is required unless an averaging over times/distances of many steps is done. This is described in [13].

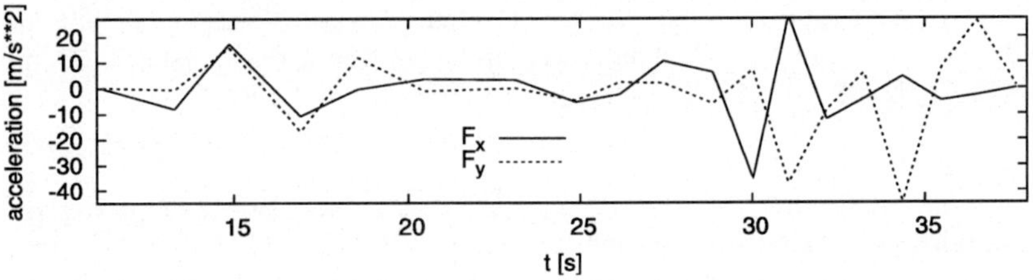


**Fig. 3. Forces for same trajectory as in Figs. 1 and 2 after double step based smoothing.**

The method there is good for microscopic measurements of density and velocity, for smoothing of the force we take double steps. Now the stepping movement is still visible but no longer prominent in the forces, and they are suitable for comparison with calculations from models. We tried fitting the trajectory with forces where the vector components are parallel and orthogonal to the direction of movement and multiplication is component-wise.

$$\vec{F} = \vec{a}\left(\vec{v}_d - \vec{v}\right) + \vec{b}\sum \frac{\Delta\vec{x}}{\left\|\Delta\vec{x}\right\|^2} + \vec{c}\sum \frac{\Delta\vec{v}}{\left\|\Delta\vec{v}\right\|^2} + \vec{d}\,\frac{\Delta\vec{w}}{\left\|\Delta\vec{w}\right\|^2}$$

In this equation $\vec{a}, \vec{b}, \vec{c}, \vec{d}$ are free parameter, while $\vec{x}$ is the position, $\vec{v}$ the velocity, $\vec{v}_d$ the intended velocity and $\vec{w}$ points on the walls. The first two sums run over all immediate neighbors in forward direction, the last one over points on the walls in forward direction. The results show that this simple model explains at best about half of the forces. Factor analysis may give hints for improving the model, and doing this for a large variety of experiments may finally give a general force based model of pedestrian movement, but this is a task for years of research.

## References

1. Predtechenskii, V.M., Milinskii A.I.: Planning for Foot Traffic Flow in Buildings. Amerind Publishing, New Dehli (1978) (Translated from Russian)
2. Nelson, H.E., Mowrer, F.W.: Emergency Movement. In: DiNenno, P.J. (ed.), The SFPE Handbook of Fire Protection Engineering, pp. 3-367-3-380. Society of Fire Protection Engineers, Bethesda (2002)
3. Kirchner, A., Schadschneider, A.: Simulation of evacuation processes using a bionics-inspired cellular automaton model for pedestrian dynamics. Physica A. **312**, p. 260-276 (2002)
4. Blue, V., Adler, J.: Cellular automata microsimulation for modeling bidirectional pedestrian walkways. Transp. Res. B, **35**, p. 293-312 (2001)
5. Antonini, G., Bierlaire, M., Weber, M.: Discrete choice models of pedestrian walking behaviour, Transp. Res. B, **40**, 667-687 (2006)
6. Crooks, A.; Castle, C., Batty, M.: Key challenges in agent-based modelling for geo-spatial simulation Computers, Environment and Urban Systems. **32**, p. 417-430 (2008)
7. Helbing, D., Molnàr, P.: Social force model for pedestrian dynamics. Phys. Rev. E **51**, 4282-4286 (1995)
8. Chraibi, M., Seyfried, A., Schadschneider, A., Mackens, W.: Quantitative Description of Pedestrian Dynamics with a Force-based Model. In: 2009 IEEE/WIC/ACM International Joint Conference on Web Intelligence and Intelligent Agent Technology. 3, 583-586 (2009)
9. Seyfried, A., Steffen, B., Winkens, A., Boltes, M., Klingsch, W.: Empirical data for pedestrian flow through bottlenecks. In: Proc. TGF 2007, p. 189-199. Springer (2009)
10. Rogsch, C., Klingsch, W., Seyfried, A.: How reliable are commercial software-tools for evacuation simulation? In: Proc. Interflam 2007, p. 235-246. Intersience (2007)
11. Boltes, M., Seyfried, A., Steffen B. and Schadschneider A.: Automatic Extraction of Pedestrian Trajectories from Video Recordings. In: Proc. PED 2008, p. 43-54. Springer (2010)
12. Steffen, B., Seyfried, A.: Methods for measuring pedestrian density, flow, speed and direction with minimal scatter. Physica A (to appear 2010). DOI: 10.1016/j.physa.2009.12.015
13. Seyfried, A., Boltes, M., Kähler, J., Klingsch, W., Portz, A., Rupprecht, T., Schadschneider, A., Steffen, B., Winkens, A.: Enhanced Empirical Data for the Fundamental Diagram and the Flow Through Bottlenecks. In: Proc. PED 2008, p. 145-156. Springer (2010)

# Pathfinder: An Agent-Based Egress Simulator

**C. Thornton, R. O'Konski, B. Hardeman, and D. Swenson**

Thunderhead Engineering Consultants, Inc.
403 Poyntz Ave, STE B
Manhattan, KS 66502 USA
Corresponding author: thornton@thunderheadeng.com

**Abstract** Pathfinder is an agent-based egress simulator. It uses a combination of steering behaviors and physical constraints to simulate large-scale occupant behavior and evacuation times based on the movement of individual occupants or agents. In this poster, we present algorithms used in Pathfinder, validation and verification results, and a review of Pathfinder's graphical user interface and output visualization features.

## Introduction

Pathfinder is an agent-based egress simulator that is designed to meet the practical needs of fire protection engineers who work with increasingly complex building models. Pathfinder's simulation model takes advantage of advances in agent-based approaches to movement modeling that make it possible to capture more complex behavior and interactions between occupants. In addition, Pathfinder provides tools that make it easy for modelers to create simulation input from existing data and view results using high-quality visualization techniques.

This paper presents a review of the major components of this modeling system. A brief overview of the verification and validation approach is also presented as well as a look forward to new features users can expect in coming versions.

## Movement Model

Pathfinder uses agent-based artificial intelligence. This approach attempts to model the behavior of large groups by simulating the behaviors and interactions of individual occupants. Each occupant has individual traits, goals, and perceptions and can take unique actions based on that data. Such systems allow realistic behavior to emerge as occupants move and self-organize.

Pathfinder moves occupants in continuous 3D space using a triangulated mesh. This movement mesh represents areas where occupants can walk, and the triangu-

R.D. Peacock et al. (eds.), *Pedestrian and Evacuation Dynamics*,
DOI 10.1007/978-1-4419-9725-8_94, 

lated geometry allows it to accommodate arbitrary obstructions and curved paths with great accuracy.

At each time step, Pathfinder agents evaluate the effectiveness of moving in several sample directions using a technique called inverse steering [1]. Agents select the minimal-cost direction based on a combination of several values, including the proposed movement direction compared to shortest path angle and the chance of colliding with other agents.

Pathfinder's default simulator mode ("steering" mode) provides the necessary elements to allow complex human behavior to emerge as a byproduct of the interactions of all the occupants' individual steering behaviors. Pathfinder includes an alternate mode ("SFPE" mode) designed to produce results very similar to the static flow calculations presented in the SFPE Handbook [2]. This mode can be used to automate comparisons of the steering mode to a well-known reference point.

## User Interface

Pathfinder provides several tools for creating the mesh needed by the movement system. It is possible to create the mesh directly using drawing tools within the software. Alternately, users can import geometry from DXF files, PyroSim models, and FDS input files. An automatic floor extraction tool makes it possible to extract the mesh from imported geometry using a flood fill technique. In cases where the imported geometry contains 3D solid objects (e.g. FDS input files) or background images, that data can be displayed along with occupant movement visualization when viewing animated results.

To make it easier to manage occupant parameters, Pathfinder includes a profile system that controls settings for speed, initial delay, size, and appearance for groups of occupants. Each occupant refers to its profile to establish default values for these settings. All settings can also be overridden for specific occupants. Occupant characteristics can be constant or generated using statistical methods. Occupants can also be assigned specific exits to help simulate varying levels of familiarity with a building's exit system.

Pathfinder provides real-time output visualization for 3D results (see Figure 1). It operates much like a video player in that it allows users to play, pause, and stop playback as well as providing additional playback features. This output visualization is allows users to navigate through models and change view settings in ways that make analysis of complex structures more convenient. A live recording feature makes it possible for users to create video files based on what they see during results analysis.

The results visualization system was designed specifically to accommodate large models with tens of thousands of occupants. Data streaming is used to allow the visualization system to load only a small portion of the results files. This makes it possible to view long-running simulations that may have many gigabytes of animation data.

**Fig. 1. Occupants walk down a stairway during an evacuation scenario. The geometry for this simulation was imported from an FDS model.**

## Verification and Validation

Pathfinder is subject to an ongoing validation process based on current and emerging movement research. To verify that individual elements of the simulator are functioning properly, simulation results are routinely compared with hand calculations. To validate the overall behavior of the simulator, real evacuation scenarios are recreated in Pathfinder and the results are compared with data gathered by independent researchers. The results of this verification and validation work are available in Pathfinder's Verification and Validation document [3].

For simple, well-defined scenarios, it is possible to pre-calculate occupants' velocities and exit times based on the equations given in the Pathfinder's Technical Reference Manual [4]. These calculations are then compared to the corresponding simulation results to ensure Pathfinder is behaving according to specification. For some elements, such as doorway flow rates, verification tests are only appropriate for the SFPE modes since the Steering mode does not specifically model doorway flow rate - rather this value emerges from the interaction of other simulation elements and is better checked with a validation test.

An example of validation work is a test based on the research performed by Seyfried et al. [5]. This test consists of multiple scenarios designed to measure the relationship between flow rate and corridor width. Using Pathfinder, it is possible to reconstruct an identical set of scenarios and compare the results directly. Results of this and other validation tests are presented with experimental results in Pathfinder's Verification and Validation document along with analysis to demonstrate the appropriateness of the simulation model in similar scenarios.

## Future Work

The next major step in Pathfinder development is integration with Fire Dynamics Simulator (FDS) results. Users will be able to provide an FDS results set as input for the simulator and Pathfinder will accumulate fractional effective dose values for each occupant. In addition, behaviors will be added to allow occupants to change movement speeds and goals based on conditions indicated by fire data.

In addition to fire model integration, support will be added to Pathfinder for dynamic geometry. Examples include doors that open and close as well as moving walkways and elevators.

## References

1. Amor, H.B., Murray, J., Obst, O.: Fast, Neat, and Under Control: Arbitrating Between Steering Behaviors. In: Rabin, S. (ed.), AI Game Programming Wisdom 3, pp. 221-232 (2006)
2. Nelson, H. E., Mowrer, F.W.: Emergency Movement. In: DiNenno, P.J., Walton, W.D. (eds.), The SFPE Handbook of Fire Protection Engineering (Third ed.), pp. 3-367-3-380. Society of Fire Protection Engineers, Bethesda (2002)
3. Thunderhead Engineering. Verification and Validation: Pathfinder 2009.1.0417 (May 2009 ed.). Thunderhead Engineering Consultants, Inc., Manhattan (2009)
4. Thunderhead Engineering. Pathfinder Technical Reference (May 2009 ed.).Thunderhead Engineering Consultants, Inc., Manhattan (2009)
5. Seyfried, A. et al.: Capacity Estimation for Emergency Exits and Bottlenecks. In: Proceedings of the Interflam, September 2007, London, England. Interscience Communications, London (2007)

# Sonic Speed on Pedestrian Dynamics: Relation between Sonic Speed and Density

**A. Tomoeda[1,2], D. Yanagisawa[3,4] and K. Nishinari[2,5]**

[1] Meiji Institute for Advanced Study of Mathematical Sciences, Meiji University, Higashi Mita, Tama-ku, Kawasaki, Kanagawa, 214-8571, JAPAN
[2] Research Center for Advanced Science and Technology, The University of Tokyo, Komaba 4-6-1, Meguro-ku, Tokyo, 153-8904, JAPAN
[3] Department of Aeronautics and Astronautics, School of Engineering, The University of Tokyo, Komaba 4-6-1, Meguro-ku, Tokyo, 153-8904, JAPAN
[4] The Japan Society for the Promotion of Science, Kojimachi, Chiyoda-ku, Tokyo, 102-8471, JAPAN
[5] PRESTO, Japan Science and Technology Corporation, Komaba 4-6-1, Meguro-ku, Tokyo, 153-8904, JAPAN
Corresponding author: atom@isc.meiji.ac.jp

**Abstract** This paper discusses the sonic speed on pedestrian dynamics, in which the flow of pedestrians at bottlenecks shows some similarity to a supersonic airflow in aerodynamics at narrow passage. In order to consider the analogy between these two dynamics, we have investigated the propagation speed of starting wave of pedestrians by performing walking experiments along a line. The propagation speed of successive reaction of pedestrians is regarded as the sonic speed in pedestrian dynamics in terms of propagations of a perturbation. The experimental result shows that the sonic speed in pedestrian dynamics is inversely proportional to the power of the density of pedestrians like in aerodynamics, if the pressure in pedestrian dynamics is assumed as a constant. Moreover, we have found that the power index obtained from the numerical simulations based on the stochastic cellular automaton model is the same with the power index in experiments.

## Introduction

Various kinds of researches on the dynamics of self-driven particles (SDP) such as vehicles and pedestrians have been progressing during the last few decades [1, 2]. One of the essential features of SDP is that they do not necessarily obey the Newton's law of motion, mainly because of their psychological or social origin. Therefore, the behaviors of SDP are not well described by the usual theory based on the Newtonian mechanics. Nevertheless, the behaviors in SDP dynamics sometimes have similarities with those in Newtonian particles, such as wave motion and domain walls [1, 2]. Especially, when pedestrians walk through the bottleneck, the flow decreases as the width of an exit decreases, as well as the supersonic airflow

R.D. Peacock et al. (eds.), *Pedestrian and Evacuation Dynamics*,
DOI 10.1007/978-1-4419-9725-8_95, 

in aerodynamics [3]. In order to open a new paradigm in fluid dynamics, it becomes significant to consider the analogy between these two dynamics. In this contribution, the sonic speed of pedestrian dynamics is investigated by both experiments and numerical simulations.

## Experiments

In our experiments, we have measured the propagation speed of starting wave under several densities which is decided by the initial number of pedestrians along a line [4, 5]. Here we would like to point out that this starting wave plays a significant role for also queuing systems [6]. At first, we make a long straight passage (30 meters) and put marks with a distance of 0.5 meters between them. As an initial condition of pedestrians, all pedestrians stand in line with same headway distance. After that, the leader of queue starts to walk by the cue and then we measure the elapsed time until the last pedestrian starts to walk. Thus, we have obtained the density and the propagation speed of successive reaction which is derived from the length of initial queue divided by the elapsed time. In order to investigate the analogy between aerodynamics and pedestrian dynamics, we assume the relation between the sonic speed and the density of pedestrians as the form

$$a(\rho) = \alpha\rho^{-\beta}, \tag{1}$$

where $\rho, a(\rho)$ are the density and the sonic speed and $\alpha, \beta$ indicate positive parameters. As is well known, the parameter $\beta$ is 0.5 in the aerodynamics. Note that, this parameter value and the form (1) is satisfied in the case of an isentropic flow, so-called *barotropic flow*. In this contribution, since the stating wave is considered as an expansion wave, we assume the flow of pedestrians is an isentropic flow. Fitting our experimental data as (1), we have obtained the parameter values $\alpha \simeq 3.01, \beta \simeq 1.35$ . This fitting function and the experimental data are shown in Fig. 1, where each plot corresponds to one experimental data. Comparing the fitting function and the data, we have found that the fitting function based on the form (1) is quite suitable for describing the relation between the density of pedestrians and sonic speed in pedestrian dynamics. If the power index $\beta$ equals to 1, the sonic speed is linearly-proportional to the headway (reciprocal of the density). However, the index of our result do not equals to 1. It would appear that pedestrians move forward more smoothly as the density decreases, since they have a lot of perspectives.

## Mathematical Model and Discussions

In this section, we explain our mathematical model based on the cellular automaton approach in detail. Let us imagine that the passage is partitioned into $L$ identical cells that each cell can accommodate at most one pedestrian at a time. The length of each cell corresponds to 0.5 meters by considering the reasonable volume exclusion effect of pedestrians. Unlike the usual stochastic cellular automaton model such as ASEP [7], in our model, only if the next cell is empty and predecessor had already moved, following pedestrian can move forward with probability $p$ which depends on their headway (h). This hopping probability of pedestrians is given by Optimal Velocity (OV) function, which is often introduced into the mathematical model for vehicular traffic as a desired velocity depending on headway distance [8]. In this study, we have derived the OV function from the data of pedestrians' walk on a circular passage way (Fig. 2) [9]. As shown in Fig. 3, this OV function is scaled so that the free hopping probability satisfies p=1, if h=5, since the maximum headway in our experiment is 5 cells.

As a significant result of simulations, we have obtained the similar plot to the experiment as shown in Fig. 4. Each plot corresponds to the average velocity after 100 iterations of numerical simulations. In a similar way to fit the simulation data as the form (1), we have obtained the parameter values $\alpha \simeq 1.71, \beta \simeq 1.34$. The power index $\beta$ from numerical simulations is a good agreement with the one from experiments. However, the coefficient $\alpha$ is slightly different to the experimental data. In our experiments, the following pedestrian behaves sensitively for the predecessor, since pedestrians are ready to walk. Whereas, the data from a circuit experiments includes the delay to start to walk, since pedestrians are not so sensitive. Therefore, the coefficient $\alpha$ becomes a low value.

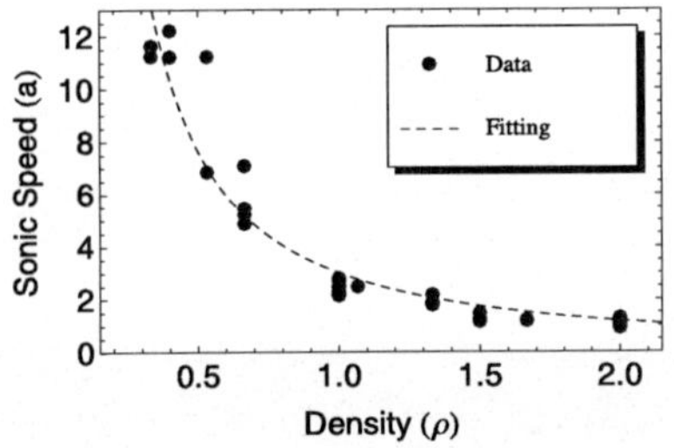


**Fig. 1. Experimental results of sonic speed**

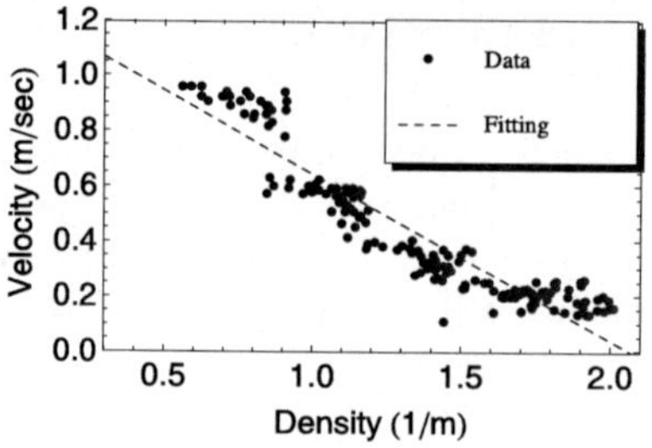


**Fig. 2. Data in ref. [9] and its fitting line**

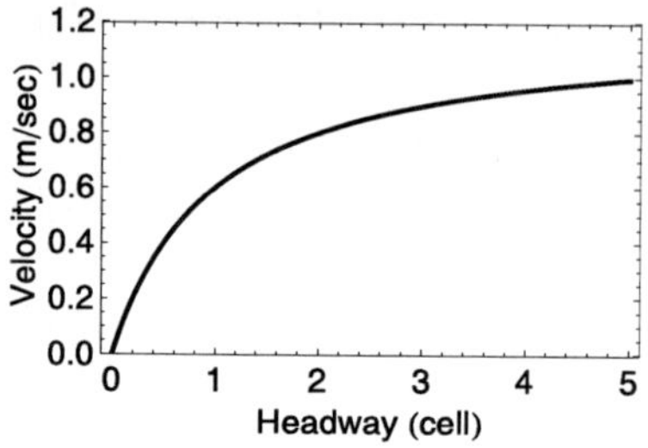


**Fig. 3. Optimal Velocity (OV) function in pedestrian dynamics (after scaling)**

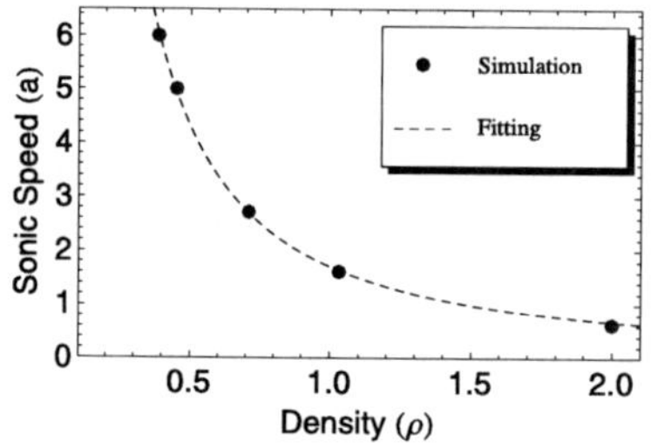


**Fig. 4. Simulation results of sonic speed**

## Conclusions

In this contribution, we have investigated the sonic speed in pedestrian dynamics by both the walking experiments and numerical simulations based on the stochastic cellular automaton. If the flow of pedestrians is assumed as an isentropic flow, the sonic speed is considered by the power function depending on the density of pedestrians in terms of the analogy of the sonic wave in aerodynamics. We have found the power index is about 1.35 and have verified the value by the numerical simulations.

**Acknowledgments** We thank Kozo Keikaku Engineering Inc. in Japan for the assistance of the experiments. This work is financially supported by Grant-in-Aid for Young Scientists from Meiji University in Japan, the Japan Society for the Promotion of Science and the Japan Science and Technology Agency.

## References

1. Chowdhury, D. et al.: Statistical physics of vehicular traffic and some related systems. Phys. Rep. **329**, 199 (2000)
2. Helbing, D.: Traffic and related self-driven many-particle systems, Rev. Mod. Phys., **73**, 1067 (2001)
3. Anderson, J.D.: Modern Compressible Flow: With Historical Perspective. Mcgraw-Hill Series in Aeronautical and Aerospace Engineering (2004)
4. Rogsch, C.: Start Waves and Pedestrian Movement – An Experimental Study, Pedestrian and Evacuation Dynamics 2008. Springer (2010)
5. Tomoeda, A. et al.: Analysis on the Propagation Speed of Pedestrian Reaction: the Velocity of Starting wave and Stopping wave. Pedestrian and Evacuation Dynamics 2008. Springer (2010)
6. Yanagisawa, D. et al.: Methods for Improving Efficiency of Queuing Systems (2010)
7. Derrida, B.: An exactly soluble non-equilibrium system: The asymmetric simple exclusion process. Phys. Rep. **301**, 65 (1998)
8. Bando, M. et al.: Dynamical model of traffic congestion and numerical simulation, Phys. Rev. E, **51**, 1035 (1995)
9. Seyfried, A. et al.: The fundamental diagram of pedestrian movement revisited. J. Stat. Mech., 10002 (2005)

# Sensitivity Visualization of Circulation under Congestion and Blockage

**G. Wurzer , M. Ausserer, H. Hinneberg, C. Illera and A. Rosic**

Vienna University of Technology, Vienna, Austria
Corresponding author: wurzer@iemar.tuwien.ac.at

**Abstract** In the context of circulation design for large buildings (e.g. hospitals, airports), the question of sensitivity of the path network against congestions and blockages naturally arises. To date, the answer to this question would require planners to use a simulation package, which is, however, almost never done in the early stages of building design. We therefore propose a novel visual planning tool that enables architects to estimate the impact of disturbances on the building circulation without having to use a simulation package. Our approach is integrated into a common CAD system and visualizes changes in the path-time relationship of adjacent functional areas under the effects of impeded accessibility.

## Introduction

Our planning tool is a graph which visualizes functional areas as nodes and connections between functional areas as edges (refer to Fig. 1). A functional area is a collection of pedestrian sources and thresholds (either leading to the next functional area or being exits). Each pedestrian source models the usage of a space in a time span t0 to t1. Usage data is of high interest in complex buildings which are designed around pre-determined processes (e.g. hospitals, airports). Our goal is to try to visualize flows along the circulation (both in and between functional areas), the target audience being architects who want to get "insight, not data" (as once proposed by Hamming).

Our program is integrated with a CAD system. As starting point, we take a basic CAD drawing that depicts walls and thresholds as input. This drawing is imported as raster image with a resolution of 0.5m, and must then be attributed with functional areas, pedestrian sources and threshold points in a pre-step (Fig. 1, left). A cell-space simulation then computes flows from the sources to the exits in the set time span, giving us:

- the number of people crossing each circulation between two functional areas
- the duration of travel along this circulation
- This information forms the basis for our visualization, which will now be described

R.D. Peacock et al. (eds.), *Pedestrian and Evacuation Dynamics*,
DOI 10.1007/978-1-4419-9725-8_96, 

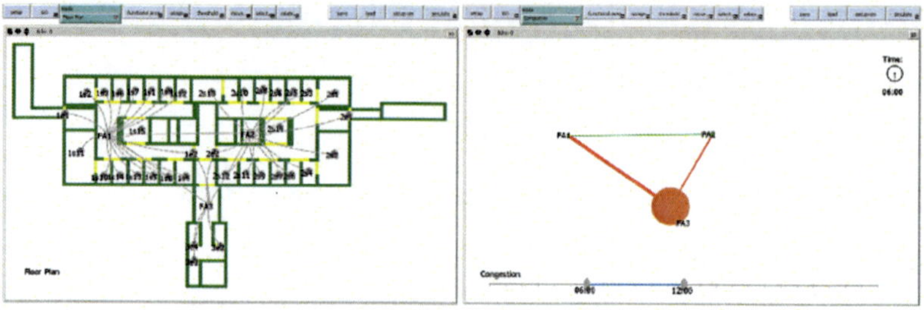

**Fig. 1. Based on an attributed floor plan (left), we perform a visualization of sensitivity against congestion and blockages (right). Nodes are used for functional areas and edges for circulation. The colour of each edge gives the sensitivity.**

## *Visualization of sensitivity against congestion*

The thickness of each edge (see Fig. 1, right) depicts the throughput of a circulation, i.e. the number of people crossing it in the set time-span. The same goes for the size of each functional area node, which tells the amount of people that have entered it. We have opted use the absolute value of the throughput, scaled by a user-defined constant, which gives us good results.

The duration of travel along the circulation is depicted as edge color. However, we do not use this value directly, but first look at the whole time-span that is set. It is the difference between the maximum travel times that is taken gives us a measure of congestion in both functional areas (nodes) and the circulation in between them (edges), which we map to a red-green color gradient. In this sense, attaining "thick green lines" (high throughput, little sensitivity against congestion) is the preferable goal. The mapping of differences in travel times onto colors is obtained by taking ranges and mapping these to discrete colors (classification of absolute values).

## *Visualization of sensitivity against blockages*

In a pre-step, the simulation runs through the set time span and records, for each functional area node and circulation edge, the maximum travel time. Then, the algorithm subsequently removes an edge from the circulation graph and performs the simulation again. The difference between the recorded and the current maximum duration time gives a measure of the sensitivity against blockages in the chosen time span, which is again mapped to a red-green color scale in the same fashion. The throughput is visualized in the same way as before, as thickness of circulation edges and functional area nodes.

## Underlying simulation

Our simulation uses a raster image that was exported from the CAD drawing as input into a cell-space simulation algorithm, which performs the actual work of generating the data that is to be visualized. Furthermore, we require that, for each pedestrian source, there is usage data (in the form of a spreadsheet table). This usage data tells the simulation engine how many agents to generate at a given time in which pedestrian source area (see Fig. 2). The agents are then simulated using the model of Blue and Adler [1] which is extended with a higher-level exit route choice function, in order to support choosing exits on the bases of functional areas. Furthermore, the extension is also responsible to disallow access to functional areas that are currently marked as "blocked".

Each performed simulation run measures when and how many agents cross area borders. These resulting values are recorded for later use in the visualization. When crossing thresholds, agents must choose their next target using our extension algorithm that considers adjacent functional areas. If there is such an adjacent functional area, the agent resets his internal clock and crosses the functional area to find an exit. In due course, the number of agents crossing the circulation is incremented by one. Upon reaching the exit, the agent records the total time of travel for the circulation he has just crossed, and the choosing of the next functional area continue. If there is no further functional area to go to, the agent is taken out of the simulation (the exit can then be considered as safe area).

## Previous Work

Our work employs usage data of functional areas to aid the planning process. Previous work in this context recorded and simulated building user's activities [2] in order to assess the building design's performance. Extending CAD systems with user activities has also previously been researched in [3][4][5]. Furthermore, our work focuses on providing meaningful visualizations of simulation data, which have been considered in [6][7]. From the view of pedestrian dynamics community, architectural considerations have been previously brought forward by [8][9].

## Conclusions

We have brought forward the idea of a novel diagram that lets an architect assess the effects of congestions and blockages on the planned circulation. Our concept is integrated into a regular CAD system and can be used during the early stages of building design. We are confident that, using our approach, architects can further improve the circulative system (i.e. add redundant paths, or increase capacity) in order to design safer.

We are aware of several points that our approach does not address and that require a justification. First of all, our underlying simulation is rather simple (for example, we have not considered taking any response times into account). In the context of this paper, simulation is merely a basis for gathering data, which is to be fed into a higher-level visualization. It is clear that more elaborate forms of simulation algorithms that have been in existence for a long time could be added. As our prototype will be published under an open-source license, we encourage interested researchers to do so.

## References


1. Dijkstra, J., Timmermans, H., Vries, B.: Virtual Reality Based Simulation of User Behavior within the Built Environment to Support the Early Stages of Building Design. CIB Conference (2001)
2. Ekholm, A., Fridqvist, S.: Concepts of Space in Computer Based Product Modelling and Design. eCAADe Conference (1997)
3. Wurzer, G.: Systems – Constraining Functions through Processes (and Vice Versa). eCAADe Conference (2009)
4. Wurzer, G., Loffreda, G., Fioravanti, A., Trento, A.: Function versus Action. Submitted to eCAADe Conference (2010)
5. Wurzer, G.: Visualizing the Human Form for Simulation and Planning. Pedestrian and Evacuation Dynamics Conference (2008)
6. VDI 3633 Part 11: Guideline on Simulation and visualization, Verein Deutscher Ingenieure (2005)
7. Illera, C., Fink, M., Hinneberg, H., Kath, K., Waldau, N., Rosic, A., Wurzer, G.: NO_PANIC – Escape and Panic in Buildings. Pedestrian and Evacuation Dynamics Conference (2008)
8. Wurzer, G., Hinneberg, H., Illera, C., Swoboda, S., Jonas, A.: Architectural Patterns Enabling Reconfigurable Exit Routes for Complex Buildings. Traffic and Granular Flow (2009)

# Integration of Human Evacuation Route Optimization Model and Fire Prevention and Control System

**P. Zhang[1], P. Gang[2], G. Li[1], and B. Chen[1]**

[1]School of Resources and Civil Engineering, Northeastern University, Shenyang, Liaoning 110003, P R China
[2]Shenyang Fire Research Institute of Public Security Ministry, Shenyang, Liaoning 110034, PR China
Corresponding author：bczhangph@sina.com.cn; zhangpeihong@mail.neu.edu.cn

**Abstract** Traditional evacuation induction system is unidirectional and unchangeable, directing to the nearer egress without any change considering the dynamic spread of fire scene. Intelligent evacuation induction system aims at dynamically directing to the safest and most efficient evacuate route far from fire sources, which improves the traditional and passive evacuation idea to active idea. Some intelligent evacuation induction systems have been developed in-and-abroad. However, they can simply realize the function of two-way directional evacuation on the level of local evacuation safety. To achieve real intelligent evacuation on the level of global evacuation safety, a linkage control is realized based on the integration of Human Evacuation Route Optimization Model (HEROM) and the Fire Prevention and Joint Control System (FPJCS). A new intelligent evacuation induction system is established and a case study is illustrated in this poster. It is established on the basis of information distribution between fire spreading scenario and the action status of fire prevention and control facilities in building. Based on dynamic relationship database technology, data transfer is realized between AACA based HEROM and FPJCS. The optimized evacuation route database of building is updated dynamically and timely with consideration of the fire spread scenario. For example, once a fire shutter blocked an evacuation corridor, the ever safe egress became inaccessible, the optimized evacuation route is updated and saved in an order file, the accessibility information of relevant node is recorded and transferred to change the induction direction of the evacuation signs.

## Introduction

Traditional evacuation induction system is unidirectional and unchangeable, directing to nearer egress without any change considering the dynamic spread of fire scenarios. This always results in huge crowd congestion and serious death hazards

R.D. Peacock et al. (eds.), *Pedestrian and Evacuation Dynamics*,
DOI 10.1007/978-1-4419-9725-8_97, 

near egress, which has already become dangerous and inaccessible in building fire. Intelligent evacuation induction system aims at finding the safest and most efficient evacuate route far from fire sources, which improves the traditional passive evacuation idea to active idea. Some intelligent evacuation induction systems have been developed in-and-abroad. However, most of them mainly focused on sensor control system and the information technology issues related to its implementation[1-3], they can simply realize the function of two-way directional evacuation on the level of local evacuation safety.

To realize real intelligent evacuation on the level of global evacuation optimization and safety, AACA based Human Evacuation Route Optimization Model (HEROM)[4,5] is integrated with the Fire Prevention –Joint Control System. HEROM based intelligent evacuation induction system can help a lot to realize performance-based fire safety evacuation with the collaboration of fire prevention and control system.

## The integration step of AACA based HEROM and Fire Prevention -Joint Control System (FPJCS)

The whole integration step is illustrated as Fig. 1. Firstly, according to the fire alarm action status in building, fire prevention and joint control system (FPJCS) updates the data in a file named as FireSensor.dat. In the data, the first column illustrates the action status of fire sensor, the second column illustrates the machine number, the third column illustrates the circuit number, and the fourth column illustrates the number of the fire sensor. For example, data (1,01,01,026) in FireSensor.dat illustrates that No.026 fire sensor of No.01 machine in the No.01 circuit has detected fire. While, if No.030 fire sensor of No.02 machine in the No.01 circuit does not detect fire, it is illustrated as data (0,02,01,030) in FireSensor.dat.

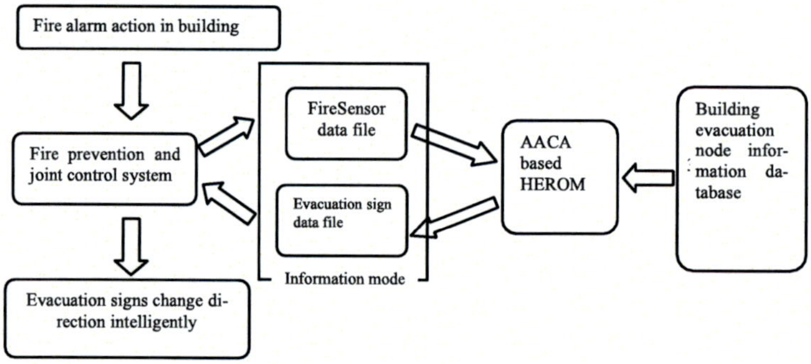


**Fig. 1. The integration step of HEROM and FPJCS**

Once FireSensor.data is updated, building node information database, which is named as *.mdb, is updated accordingly based on the relationship database technology. Datas in FireSensor.dat and *.mdb are transferred to AACA based HEROM, the optimal evacuation route can be obtained after certain cycles of interation.

The optimatization result is transferred to Eva_Marker.dat. For example, data (1,01,01,001) in Eva_Marker.dat illustrates that No.001 evacuation sign of No.01 machine in No.01 circuit should direct to left after route optimization searching process. data (2,01,01,002) in Eva_Marker.dat illustrates that No.002 evacuation sign of No.01 machine in No.01 circuit should direct to right after route optimization searching process.

The data in file Eva_Marker.dat is transferred to FPJCS, which intelligently change the direction of evacuation signs. In the running process of algorithm, the ID number of fire detector and alarmer is used to establish the relationship and data transfer among FireSensor.dat, *.mdb, as well as Fire_Marker.dat.

## Case study of the integrated intelligent evacuation sign system

The integrated intelligent evacuation sign system is applied in a hotel with $3400m^2$ of building area. In normal status when there is no fire, the evacuation sign direction is illustrated as Fig. 2 after route optimization process by AACA based HEROM.

Assuming fire broken out initially in a node next to a stair at the left down corner of the hotel and spreaded to corridor immediately. With the development of fire, four nodes are not accessible, and the revevent nodes are shown as red ones. FireSensor.dat updated with time going and transferred data to hotel.mdb. The intelligent evacuation sign direction responses based on the integration of AACA based HEROM and FJPCS. The simulation result was shown in Fig .3.

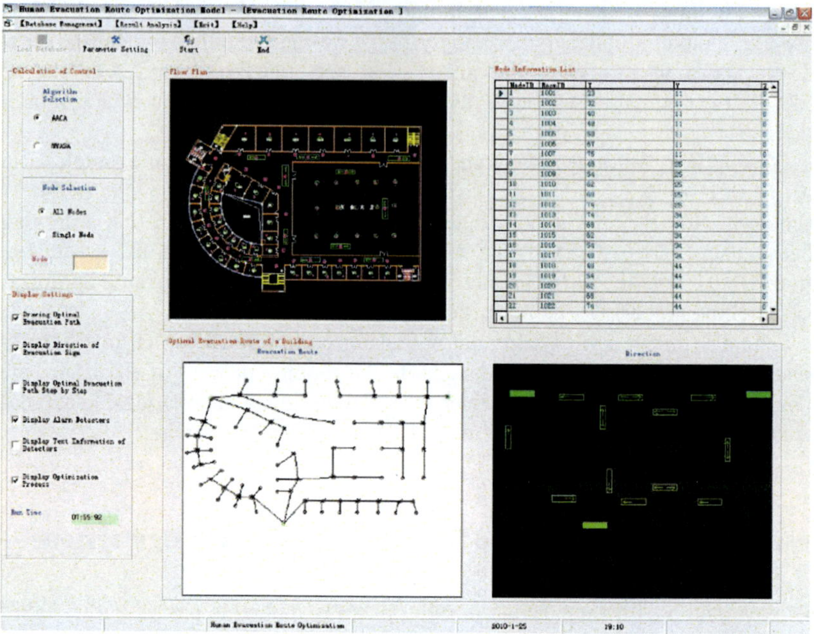

**Fig. 2.** **Optimization evacuation sign direction in no-fire status**

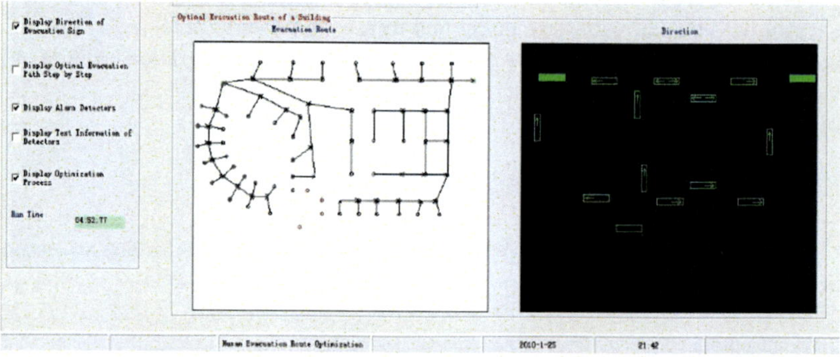

**Fig. 3.** **Evacuation sign direction in late fire**

## Conclusions

Based on the data transfer between fire alarm system of FPJCS and AACA based HEROM, the initial status database of building realize real-time updates dynamically with the spread scenario of fire. After the process of route optimization, the optimized evacuation route is updated and saved in an order file, the order file is then transferred to FPJCS, a new intelligent evacuation induction system is established as a result. Further research will be focused on evacuation drill and system test to perfect the optimization model and the integration technology.

## References

1. Maymí, F., Rodríguez-Martínez, M., Qian, Y., Manz, P.C.: Ancile: Pervasively Shared. Situational Awareness. IEEE Internet Computing (January / February 2008)
2. Kachroo, P., Wadoo, S.A., Al-nasur, S.J., Shende, A.R., Singh, M.P., Ozbay, K.: Information technology requirements for intelligent evacuation systems [J]. World Review of Intermodal Transportation Research. 2009, 2(2/3): 127 - 144.
3. Gosalia, M., Lin, K., Redfern, A., et al.: Smoke: Mote Powered Fire Evacuation. CS199: Sensor Network Systems. http://cents.cs.berkeley.edu/retreats/winter_2005/nest-1-05-fire.ppt (Fall 2004)
4. Zhang, P., Zhang, Y., Mei, Z., et al.: Adaptive Ant Colony Algorithm Based Evacuation Route Optimization Model and Application in Large Public Building Fire[J]. Journal of Shenyang Jianzhu University (Natural Science). 24(6): 1055-1059 (2008)
5. Mei, Z., Dong, W., Pan, G., Zhang, P., Zhang, Y:. Adaptive Ant Colony Algorithm Based on Evacuation Wayfinding Optimization Model in Building Fire[J]. Journal of Shenyang Jianzhu University (Natural Science). 24(4): 671-674 (2008)